*Student Study Guide and Solutions Manual*
for Atkins and Jones's

# CHEMICAL PRINCIPLES

*The Quest for Insight*

THIRD EDITION

## John Krenos and Joseph Potenza
RUTGERS, THE STATE UNIVERSITY OF NEW JERSEY

## Lynn Koplitz and Thomas Spence
LOYOLA UNIVERSITY

 W. H. Freeman and Company
New York

ISBN: 0-7167-0740-3

EAN: 9780716707400

Printed in the United States of America

Third printing

W. H. Freeman and Company
41 Madison Avenue
New York, NY 10010
Houndmills, Basingstoke RG2I 6XS England

www.whfreeman.com

# CONTENTS

Preface ............................................................................................ v
Acknowledgements ........................................................................ vii

## STUDY GUIDE                                                                    SG-

1   Atoms: The Quantum World ........................................................ 1
2   Chemical Bonds .......................................................................... 15
3   Molecular Shape and Structure .................................................. 25
4   The Properties of Gases .............................................................. 35
5   Liquids and Solids ...................................................................... 45
6   Thermodynamics: The First Law ................................................ 57
7   Thermodynamics: The Second and Third Laws .......................... 71
8   Physical Equilibria ..................................................................... 85
9   Chemical Equilibria .................................................................... 101
10  Acids and Bases ......................................................................... 111
11  Aqueous Equilibria ..................................................................... 125
12  Electrochemistry ......................................................................... 135
13  Chemical Kinetics ...................................................................... 149
14  The Elements: The First Four Main Groups .............................. 165
15  The Elements: The Last Four Main Groups ............................... 181
16  The Elements: The $d$ Block ...................................................... 195
17  Nuclear Chemistry ..................................................................... 209
18  Organic Chemistry I: The Hyrdocarbons .................................. 221
19  Organic Chemistry II: Polymers and Biological Compounds ..... 231

## SOLUTIONS MANUAL                                                              SM-

    Fundamentals ............................................................................. 3
1   Atoms: The Quantum World ...................................................... 46
2   Chemical Bonds .......................................................................... 63
3   Molecular Shape and Structure .................................................. 82
4   The Properties of Gases .............................................................. 100
5   Liquids and Solids ...................................................................... 127
6   Thermodynamics: The First Law ................................................ 149
7   Thermodynamics: The Second and Third Laws .......................... 175
8   Physical Equilibria ..................................................................... 202
9   Chemical Equilibria .................................................................... 240
10  Acids and Bases ......................................................................... 278
11  Aqueous Equilibria ..................................................................... 318
12  Electrochemistry ......................................................................... 359
13  Chemical Kinetics ...................................................................... 397
14  The Elements: The First Four Main Groups .............................. 424
15  The Elements: The Last Four Main Groups ............................... 443
16  The Elements: The $d$ Block ...................................................... 460
17  Nuclear Chemistry ..................................................................... 478
18  Organic Chemistry I: The Hyrdocarbons .................................. 500
19  Organic Chemistry II: Polymers and Biological Compounds ..... 519

# PREFACE

This Study Guide and Solutions Manual accompanies the textbook *Chemical Prin...*
*Quest for Insight,* Third Edition, by Peter Atkins and Loretta Jones — an authoritativ...
thorough introduction to chemistry for students anticipating careers in science or engine...
disciplines. We have followed the order of topics in the textbook chapter for chapter. The para...
between the symbols, concepts, and style of this supplement and the textbook enables the reader to
easily move back and forth between the two.

Much of the Study Guide is presented in outline style with material highlighted by bullets,
arrows, and tables. In general, bullets offset major items of importance for each section and
arrows provide explanatory material, descriptive material, or some examples to reinforce a
concept. This telegraphic style should help the student obtain a broad perspective of large blocks
of material in a relatively short period of time, and may prove particularly useful in preparing for
examinations. We believe that the Study Guide will be most useful following a careful reading of
the text. Chapter **Sections** follow those in the text and contain descriptive material (bullets and
arrows) as well as numerous examples. Many examples are worked out in detail. While we have
covered most of the material in the text, we have not attempted to be encyclopedic. To help obtain
a broad overview of the material, important equations are highlighted in boxes. Students may find
this aspect of the guide particularly useful before exams. Where appropriate, we have introduced
supplementary tables either to amplify material in the text, to clarify it further, or to summarize a
body of material. Sprinkled throughout the guide are **Notes;** in the main, these are designed to
point out common pitfalls.

Some important material in the previous edition of the separate Study Guide supplement
has been placed on the textbook web site in the form of downloadable Adobe Acrobat© files. In
particular, a **Brief Review of Symbols, Units, and Number Conventions,** sections with
**Overviews,** lists of **Key Concepts,** and numbered **Examples** are updated and improved on the
web site. The review of symbols, units and number conventions is written in summary/tabular
form and is intended to be a resource to the student when reading the text or solving problems.
Most chapters begin with an **Overview,** which provides a summary of the main ideas and concepts
to be introduced and, in some instances, the **Goals** of the chapter. Each major topic in a chapter
begins with a list of **Key Concepts.** After reading the text and the Study Guide, students should
understand and be able to define these concepts, most of which are defined in the guide or in the
glossary at the back of the text. Many examples are worked out in detail. In some instances, to
reinforce a given concept, a given worked-out example is followed by a similar one for which only
the answer is shown. **Examples** are numbered by section rather than in numerical order so that if
students need to brush up on a particular topic, it should be relatively easy to do so.

The Solutions Manual section contains solutions and answers to the odd-numbered
problems in the textbook including the sections on Fundamentals. The rules of significant figures
have been adhered to in reporting the numerical answers for all exercises. In exercises with
multiple parts, the properly rounded values were used for subsequent calculations.

# ACKNOWLEDGMENTS

The Study Guide authors, John Krenos and Joseph Potenza, are indebted to many individuals who have made significant contributions to it.

First and foremost, we wish to thank Beth Van Assen for a thorough and incisive reading of the drafts. Her critical suggestions led to improved readability and scientific accuracy of the Study Guide. Many examples and important sections of descriptive text were clarified and expanded with her help. Her encouragement, patience, and persistence were essential to the completion of this project.

We also thank two of our colleagues at Rutgers for critically reading three chapters. Harvey Schugar carefully reviewed Chapter 16. Our knowledge and appreciation of inorganic chemistry were broadened greatly by his comments and analysis. Spencer Knapp critiqued Chapters 18 and 19 and gave us a quick tutorial on modern organic chemistry. Many of the chemical structures were prepared with his help.

David Becker, author of the Study Guide for the textbook *Chemistry: Molecules, Matter, and Change* by Peter Atkins and Loretta Jones, provided inspiration for the layout and format of our worked-out examples (now available on the textbook web site). We were also guided by his **Pitfalls** sections that alert students to commonly encountered misperceptions and traps. We attempted to incorporate some of his insights into our comments (**Notes**) for students.

The copy editors, Jodi Simpson and Alice Allen, did an excellent job in making the material clear, complete, and in harmony with the textbook. In particular, we thank Jodi Simpson for help in developing a workable, consistent format for the telegraphic style.

We also wish to acknowledge the help and encouragement of the staff at W. H. Freeman and Company. We especially thank Jessica Fiorillo (chemistry editor) for asking us to author this guide, Jodi Isman (project editor) for guiding us through the laborious process of developing a manuscript from scratch and also for helping with the editing, and Shawn Churchman (supplements editor) for encouraging us to keep to a tight schedule.

Finally, we extol the efforts of the authors of the textbook, Peter Atkins and Loretta Jones, in creating a new approach to teaching general chemistry. Beginning an introductory chemistry textbook with quantum theory is a logical, but challenging approach. The authors succeed by treating the fundamentals of chemistry (an "as needed" review) in a separate section and by minimizing the coverage of a great deal of material (mostly historical) presented in other books. In our opinion, this approach leads to the development of all the major topics in chemistry enlightened and enlivened in the first instance by the molecular viewpoint.

# Chapter 1   Atoms:  The Quantum World
## Observing Atoms  (Sections 1.1–1.6)

### 1.1   The Characteristics of Electromagnetic Radiation

- **Oscillating amplitude of electric and magnetic field**   $\rightarrow$   Wave characterized by *wavelength* and *frequency*

**Distance behavior  (fixed time *t*)**        **Time behavior  (fixed position *x*)**

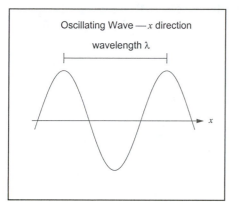  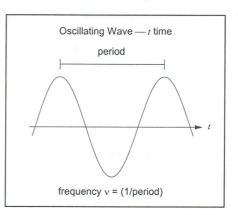

Speed of light (distance/time) = wavelength / period = wavelength × frequency

$$\boxed{c = \lambda \nu}$$        speed of light = wavelength × frequency

SI units:        $(\text{m·s}^{-1})$        $(\text{m})$        $(\text{s}^{-1})$        $[1 \text{ Hz (hertz)} = 1 \text{ s}^{-1}]$

**Note:**   The speed of light c ($c_0$ in vacuum $\approx 3.00 \times 10^8$ m·s$^{-1}$) depends on the medium it travels in. Medium effects on wavelength in the visible region are small (beyond 3 significant figures).

### 1.2   Radiation, Quanta, and Photons

- **Black body**
    - $\rightarrow$ Perfect absorber and emitter of radiation
    - $\rightarrow$ Intensity of radiation for a series of temperatures
    - $\rightarrow$ Stefan–Boltzmann law:   $$\boxed{\frac{\text{Power emitted (watts)}}{\text{Surface area (meter}^2)} = \text{constant} \times T^4}$$
    - $\rightarrow$ Wavelength corresponding to maximum intensity = $\lambda_{max}$
    - $\rightarrow$ Wien's law:   $\boxed{T\lambda_{max} = \frac{1}{5}c_2}$   where $c_2 = 1.44 \times 10^{-2}$ K·m   (second radiation constant)

        At higher temperature, maximum intensity of radiation shifts to lower wavelength.

- **Energy of a quantum (packet) of light (generally called a photon)**
    - $\rightarrow$ Postulated by Max Planck to explain black body radiation
    - $\rightarrow$ Resolved the "ultraviolet catastrophe" of classical physics, which predicted intense ultraviolet radiation for all heated objects ($T > 0$)

→ Quantization of electromagnetic radiation

$$E = h\nu$$

photon energy = Planck constant × photon frequency

SI units:    (J)    ($h = 6.6261 \times 10^{-34}$ J·s)    ($\text{s}^{-1}$)

- **Photoelectric effect**
  - → Ejection of electrons from a metal surface exposed to photons of sufficient energy
  - → Indicates that light behaves as a particle

$$E_K = h\nu - \Phi$$

$E_K$ = kinetic energy of the ejected electron, $\Phi$ = threshold energy (work function) required for electron ejection from the metal surface, and $h\nu$ = photon energy

  - → $h\nu \geq \Phi$ required for electron ejection

- **Wave behavior of light**    → Diffraction and interference effects of superimposed waves (*constructive* and *destructive*)

## 1.3  Wave–Particle Duality of Matter

- **Matter has wave properties**
  - → Proposed by Louis de Broglie
  - → Consider matter with mass $m$ and velocity $v$
  - → Such matter behaves as a wave with a characteristic wavelength

$$\lambda = \frac{h}{mv}$$

de Broglie wavelength for a particle with linear momentum $p = mv$

## 1.4  Heisenberg Uncertainty Principle

- **Complementarity of location ($x$) and momentum ($p$)**
  - → Uncertainty in $x$ is $\Delta x$; uncertainty in $p$ is $\Delta p$
  - → Limitation of knowledge

$$\Delta p \Delta x \geq \hbar/2$$    Heisenberg uncertainty principle, where $\hbar = h/2\pi$

  - → $\hbar$ is called "h bar"    $\hbar = 1.0546 \times 10^{-34}$ J·s

  - → Refutes classical physics on the atomic scale

## 1.5  Wavefunctions and Energy Levels

- **Classical trajectories** → Precisely defined paths
- **Wavefunction** $\psi$    → Gives *probable* position of particle with mass $m$
- **Born interpretation** → Probability of finding particle in a region is proportional to $\psi^2$
- **Schrödinger equation**
  - → Allows calculation of $\psi$ by solving a differential equation
  - → $H\Psi = E\Psi$; $H$ is called the Hamiltonian

- **Particle in a box**
  - → Mass $m$ confined between two rigid walls a distance $L$ apart
  - → $\psi = 0$ outside the box and at the walls (boundary condition)

$$\Psi_n(x) = \left(\frac{2}{L}\right)^{1/2} \sin\left(\frac{n\pi x}{L}\right) \qquad n = 1, 2, \ldots$$

$\Psi_n(x)$ = wavefunction that satisfies the Schrödinger equation between the box limits. $n$ is a *quantum number*.

**Note:** A node is a point in the box where $\psi = 0$ and $\psi$ *changes* sign. $\boxed{\psi^2 \geq 0,\ \text{always}}$

$$E_n = \frac{n^2 h^2}{8mL^2}$$

$E_n$ = allowed energy values of a particle in a box
**Note:** $n = 1$ gives the zero-point energy $E_1$
$E_1 \neq 0$ implies residual motion

$$\Delta E = E_{n+1} - E_n = \frac{(2n+1)h^2}{8mL^2}$$

Energy difference between two neighboring levels

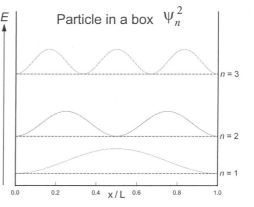

- **Probability as a function of position in the box**
  - → Plot is shown for the first 3 levels
  - → $\psi^2$ as a function of the dimensionless variable $x/L$

## 1.6 Atomic Spectra and Energy Levels

- **Spectral lines** → Discharge lamp of hydrogen

  $H_2$ + electrical energy → $H + H^*$    [* ≡ asterisk denotes excited atom]

  $H^* \rightarrow H^{(*)} + h\nu$    [$^{(*)}$ ≡ denotes a less excited atom]

- **Lines form a discrete pattern** → Discrete energy levels

- **Bohr frequency condition**

$$h\nu = E_{\text{upper}} - E_{\text{lower}}$$

Relates photon energy to energy difference between two energy levels in an atom.

- **Hydrogen atom spectral lines**
  - → Johann Rydberg's general equation
  - → $n_2 = n_{\text{upper}}$ and $n_1 = n_{\text{lower}}$

$$\nu = \Re\left(\frac{1}{n_1^2} - \frac{1}{n_2^2}\right) \quad n_1 = 1, 2, \ldots \quad n_2 = n_1 + 1, n_1 + 2, \ldots$$

$\Re = 3.29 \times 10^{15}$ Hz = Rydberg constant

Rydberg expression reproduces pattern of lines in H atom emission spectrum. The value of $\Re$ is obtained empirically.

**Note:** Lines with a common $n_1$ can be grouped into a series and some have special names:

$n_1 = 1$ (Lyman), 2 (Balmer), 3 (Paschen), 4 (Brackett), 5 (Pfund)

## Models of Atoms   (Sections 1.7–1.10)

### 1.7  Principal Quantum Number

- **Charge on an electron** $\rightarrow q = -e = -1.602\,177\,33 \times 10^{-19}$ C

- **Vacuum permittivity** $\rightarrow \varepsilon_0 = 8.854\,187\,817 \times 10^{-12}$ $C^2 \cdot N^{-1} \cdot m^{-2}$

- **Coulomb potential energy**  $\boxed{V(r) = \dfrac{q_1 q_2}{4\pi\varepsilon_0 r}}$

$$V(r) = \frac{\text{product of charges on particles}}{4\pi \times \text{permittivity of free space} \times \text{distance between charges}} \qquad \text{(SI system)}$$

Units of $V(r)$:   $J = \dfrac{C^2}{(C^2 \cdot N^{-1} \cdot m^{-2})\,m} = N\cdot m = (kg \cdot m \cdot s^{-2})\,m = kg \cdot m^2 \cdot s^{-2}$

- **H atom and one-electron ion energy levels ($He^+$, $Li^{2+}$, $Be^{3+}$, *etc.*)**
  - $\rightarrow$ Solutions to the Schrödinger equation

$$E_n = -h\Re\left(\frac{Z^2}{n^2}\right) \qquad Z = 1 \ \& \ n = 1, 2, 3, \dots \qquad (n \text{ is dimensionless})$$

$$\Re = \frac{m_e e^4}{8h^3\varepsilon_0^2} = 3.289\,842 \times 10^{15} \text{ Hz} \qquad \text{Units:} \qquad \frac{kg \cdot C^4}{(J\cdot s)^3 (C^2 \cdot N^{-1} \cdot m^{-2})^2} = s^{-1} \equiv Hz$$

- $\rightarrow$ All quantum numbers are dimensionless.  With $\Re$ in units of frequency, the H atom energy-level equation has the same form as the Planck equation, $E = h\nu$.

- $\rightarrow$ $E_n$ = energy levels (states) of the H atom.  Note the *negative* sign.

- $\rightarrow$ $\Re$ = Rydberg constant, calculated exactly using Bohr theory *or* the Schrödinger equation

- $\rightarrow$ $Z$ = atomic number, equal to 1 for hydrogen

- $\rightarrow$ $n$ = principal quantum number

- $\rightarrow$ As $n$ increases, energy increases, the atom becomes less stable, and energy states become more closely spaced (more dense).

- $\rightarrow$ Integer $n$ varies from 1 (ground state) to higher integers (excited states) to $\infty$ (ionization).

- $\rightarrow$ Energies of H atom states vary from $-h\Re$ ($n = 1$) to 0 ($n = \infty$).  States with $E > 0$ are possible and correspond to an ionized atom in which the energy $> 0$ equals the kinetic energy of the electron.

## 1.8  Atomic Orbitals (AOs)

- **Definition of AO**

  → Wavefunction ($\Psi$, psi) describes an electron in an atom.

  → Orbital ($\Psi^2$) holds 0, 1, or 2 electrons.

  → Orbital can be viewed as a *cloud* within which the point density represents the *probability* of finding the electron at that point.

  → Orbital is specified by *three* quantum numbers ($n$, $\ell$, $m_\ell$; see below).

- **Wavefunction**

  → Fills all space

  → Depends on the *three* spherical coordinates: $r$, $\theta$, $\phi$

  → Written as a product of a radial [$R(r)$] and an angular [$Y(\theta,\phi)$] wavefunction; mathematically, $\Psi(r,\theta,\phi) = R(r)Y(\theta,\phi)$

- **Wavefunction for the H atom 2s orbital** → $n = 2$, $\ell = 0$, and $m_\ell = 0$

$$R(r) = \frac{\left(2 - \dfrac{r}{a_0}\right)e^{-r/2a_0}}{(2a_0)^{3/2}} \qquad Y(\theta,\phi) = (4\pi)^{-1/2} \qquad a_0 = \frac{4\pi\varepsilon_0\hbar^2}{m_e e^2} = \left\{ \begin{array}{c} 5.29177 \times 10^{-11}\text{ m} \\ \text{(Bohr radius)} \end{array} \right\}$$

$$\text{Units:}\quad R(r) = \text{m}^{-3/2} \qquad Y(\theta,\phi) = \text{none} \qquad a_0 = \frac{(\text{C}^2 \cdot \text{N}^{-1} \cdot \text{m}^{-2})\,(\text{J} \cdot \text{s})^2}{\text{kg} \cdot \text{C}^2} = \text{m}$$

## Three Quantum Numbers [$n$, $\ell$, $m_\ell$] Specify an Atomic Orbital

| Symbol | Name | Allowed Values | Constraints |
|--------|------|----------------|-------------|
| $n$ | Principal quantum number | $= 1, 2, 3, \ldots$ | Positive integer |
| $\ell$ | Orbital angular momentum quantum number | $= 0, 1, 2, \ldots, n-1$ | Each value of $n$ corresponds to $n$ allowed values of $\ell$. |
| $m_\ell$ | Magnetic quantum number | $= \ell, \ell-1, \ell-2, \ldots, -\ell$ <br> $= 0, \pm1, \pm2, \ldots, \pm\ell$ | Each value of $\ell$ corresponds to $(2\ell+1)$ allowed values of $m_\ell$. |

- **Terminology (nomenclature)**

  **shell:** AOs with the same $n$ value

  **subshell:** AOs with the same $n$ and $\ell$ values;

  $\ell = 0, 1, 2, 3$  equivalent to  $s$-, $p$-, $d$-, $f$-subshells, respectively, or

  | | | | |
  |---|---|---|---|
  | $s$-orbital | $\Rightarrow$ | $\ell = 0$ | $m_\ell = 0$ |
  | $p$-orbital | $\Rightarrow$ | $\ell = 1$ | $m_\ell = -1, 0,$ or $+1$ |
  | $d$-orbital | $\Rightarrow$ | $\ell = 2$ | $m_\ell = -2, -1, 0, +1,$ or $+2$ |
  | $f$-orbital | $\Rightarrow$ | $\ell = 3$ | $m_\ell = -3, -2, -1, 0, +1, +2,$ or $+3$ |

- **Physical significance of the wavefunction** $\psi(r,\theta,\phi)$

  For all atoms and molecules, $\psi^2(r,\theta,\phi)$ is proportional to the probability of finding the electron at a point $r, \theta, \phi$. We can also regard $\psi^2(r,\theta,\phi)$ as the electron density at point $r, \theta, \phi$.

- **Plot of electron density for the 2s-orbital of hydrogen**

Computer-generated electron density dot diagram for the hydrogen atom 2s-orbital. The nucleus is at the center of the square and the density of dots is proportional to the probability of finding the electron. Notice the location of the spherical node.

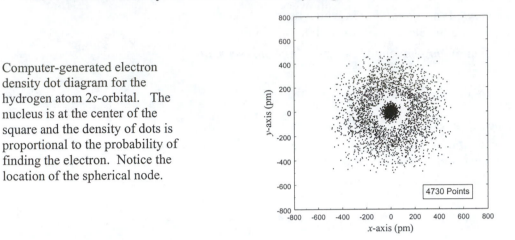

- **Concept of AOs** → Two interpretations are useful:

  1  Visualize AO as a *cloud of points,* with the density of points in a given volume proportional to probability of finding the electron in that volume.

  2  Visualize AO as a *surface* (boundary surface) within which there is a given probability of finding the electron. For example, the 95% boundary surface, within which the probability of finding an electron is 95%.

- **Boundary surfaces** → *Shapes* of atomic orbitals

  | | |
  |---|---|
  | *s*-orbital | *spherical* |
  | *p*-orbital | *dumbbell* or *peanut* |
  | *d*-orbital | *four-leaf clover* or *dumbbell* with equatorial *torus* (doughnut) |

- **Number and type of orbitals** → Follow from allowed values of quantum numbers

  ***s*-orbitals:** *s* means that $\ell = 0$; if $\ell = 0$, then $m_l = 0$. So, for each value of *n*, there is *one s*-orbital.

  > *one* for each $n$ : 1s, 2s, 3s, 4s, ...

  ***p*-orbitals** *p* means that $\ell = 1$; if $\ell = 1$, then $m_\ell = -1, 0, +1$, and $n > 1$. So, for each value of $n > 1$, there are *three p*-orbitals, $p_x, p_y, p_z$.

  > *three* for each $n > 1$ : $2p_x, 2p_y, 2p_z$; $3p_x, 3p_y, 3p_z$; ...

  The *np*-orbitals are referred to collectively as the 2p-orbitals, 3p-orbitals, and so on.

  ***d*-orbitals** *d* means that $\ell = 2$; if $\ell = 2$, then $m_\ell = -2, -1, 0, +1, +2$, and $n > 2$.

  So, for each value of $n > 2$, there are *five d*-orbitals, $d_{xy}, d_{xz}, d_{yz}, d_{x^2-y^2}$ and $d_{z^2}$.

  > *five* for each $n > 2$ :  $3d_{xy}, 3d_{xz}, 3d_{yz}, 3d_{x^2-y^2}$, and $3d_{z^2}$;
  >
  > $4d_{xy}, 4d_{xz}, 4d_{yz}, 4d_{x^2-y^2}$, and $4d_{z^2}$; ...

  The *nd*-orbitals are referred to collectively as the 3d-orbitals, 4d-orbitals, and so on.

## 1.9 Electron Spin

- **Spin states**
  - $\rightarrow$ An electron has *two* spin states, represented as $\uparrow$ and $\downarrow$ **or** $\alpha$ and $\beta$.
  - $\rightarrow$ Are described by the spin magnetic quantum number, $m_s$
  - $\rightarrow$ For an electron, only two values of $m_s$ are allowed: $+\frac{1}{2}$, $-\frac{1}{2}$.

## 1.10 The Electronic Structure of Hydrogen

- **Degeneracy of orbitals**

  **In the H atom**, orbitals in a given shell are *degenerate* (have the same energy).
  **For many-electron atoms**, this is not true and the energy of a given orbital depends on $n$ and $\ell$.

- **Ground and excited states of H**

  In the ground state, $n = 1$, and the electron is in the $1s$-orbital. The first excited state corresponds to $n = 2$, and the electron occupies one of the *four* possible orbitals with $n = 2$ ($2s$, $2p_x$, $2p_y$, or $2p_z$). Similar considerations hold for higher excited states.

- **Ionization of H**

  Absorption of a photon with energy $\geq h\Re$ ionizes the atom, creating a free electron and a free proton. Energy in excess of $h\Re$ is utilized as kinetic energy of the system.

- **Summary**

  The state of an electron in a H atom is defined by four quantum numbers $\{n, \ell, m_\ell, m_s\}$. As the value of $n$ increases, the size of the H atom increases.

---

# The Structures of Many-Electron Atoms
## (Sections 1.11–1.13)

## 1.11 Orbital Energies

- **Many-electron atoms**
  - $\rightarrow$ Atoms with more than one electron
  - $\rightarrow$ Coulomb potential energy equals the sum of *nucleus-electron* **attractions** and *electron-electron* **repulsions**.
  - $\rightarrow$ Schrödinger equation cannot be solved exactly.
  - $\rightarrow$ Accurate wavefunctions obtained numerically by using computers

- **Variation of energy of orbitals** $\rightarrow$ For orbitals in the same shell but in different subshells, a combination of *nucleus-electron* attraction and *electron-electron* repulsion influences the orbital energies.

- **Shielding and penetration** $\rightarrow$ Qualitative understanding of orbital energies in atoms

- **Shielding** $\rightarrow$ Each electron in an atom is *attracted* by the nucleus and *repelled* by all the other electrons. In effect, each electron feels a *reduced* nuclear charge ($Z_{eff}e$ = effective nuclear charge). The electron (orbital) is *shielded* to some extent from the nuclear charge and its energy is raised accordingly.

- **Penetration** → The *s-, p-, d-, ...* **orbitals** have different shapes and different electron density distributions. For a given *shell* (same value of the principal quantum number *n*), *s*-**electrons** tend to be closer to the nucleus than *p*-**electrons**, which are closer than *d*-**electrons**. We say that *s*-**electrons** are more *penetrating* than *p*-**electrons**, and *p*-**electrons** are more *penetrating* than *d*-**electrons**.

**Note:** Shielding and penetration can be understood qualitatively on the basis of the nucleus-electron potential energy term: $-(Ze)e/4\pi\varepsilon_0 r$, where $Ze$ is the nuclear charge and $r$ the nucleus-electron distance. *Shielded* electrons have the equivalent of a reduced $Z$ value ($Z_{eff} < Z$) and therefore *higher* energy; *penetrating* electrons have the equivalent of a reduced value of $r$ and therefore *lower* energy.

- **Review** → $Z$ has three equivalent meanings:
  Nuclear charge (actually, $Ze$)
  Atomic number
  Number of protons in the nucleus

- **Consequences of *shielding* and *penetration* in many-electron atoms**
  → Orbitals with the same *n* and different *l* values have *different* energies.
  → For a given *shell (n)*, subshell energies *increase* in the order: $ns < np < nd < nf$.
  **Example:** A 3*s*-electron is lower in energy than a 3*p*-electron, which is lower than a 3*d*-electron.
  → Orbitals within a given *subshell* have the *same* energy.
  **Example:** The five 3*d*-orbitals are degenerate for a given atom.
  → *Penetrating* orbitals of higher shells may be lower in energy than less *penetrating* orbitals of lower shells.
  **Example:** A penetrating 4*s*-electron may be lower in energy than a less penetrating 3*d*-electron.

## 1.12 The Building-Up Principle

- **Pauli exclusion principle**
  → No more than *two* electrons per orbital
  → *Two* electrons occupying a single orbital must have paired spins: $\boxed{\uparrow\downarrow}$
  → *Two* electrons in an atom may *not* have the same *four* quantum numbers.

- **Terminology**

  | | |
  |---|---|
  | **closed shell:** | Shell with maximum number of electrons allowed by the exclusion principle |
  | **valence electrons:** | Electrons in the outermost occupied shell of an atom; they occupy the shell with the largest value of *n* and are used to form chemical bonds. |
  | **electron configuration:** | List of all occupied *subshells* or *orbitals*, with the number of electrons in each indicated as a numerical superscript. Example: Li: $1s^2 2s^1$ |

- **Building-Up (*Aufbau*) Principle** → Order in which electrons are added to *subshells* and *orbitals* to yield the *electron configuration* of atoms

  **The ($n + \ell$) rule** → Order of filling subshells in *neutral atoms* is determined by filling those with the *lowest* values of ($n + \ell$) first. Subshells in a group with the same value of ($n + \ell$) are filled in the order of increasing *n*. (*topic not covered in the text*)

- **Usual filling order of subshells ( $n + \ell$ rule):** $1s < 2s < 2p < 3s < 3p < 4s < 3d < 4p < 5s < 4d < ...$

**Note:** Ionization, or subtle differences in shielding and penetration can change the order of the energy of subshells. *Thus, energy ordering of subshells is not fixed absolutely, but may vary from atom to atom, or from an atom to its ion.*

- **Orbitals within a subshell** → Hund's rule:

    Electrons add to *different* orbitals of a subshell
    with spins *parallel* until the subshell is half full.

- **Applicability** → The building-up principle in combination with the *Pauli exclusion principle* and *Hund's rule* accounts for the *ground-state* electron configurations of atoms. The principle is generally valid, **but there are exceptions.**

## 1.13  Electronic Structure and the Periodic Table

- **Order of filling subshells**
    - → Understanding the organization of the periodic table
    - → Straightforward determination of (most) electron configurations

- **Terminology** → Given in the following two tables

| | |
|---|---|
| **groups:** | Columns in the periodic table, labeled 1–18 horizontally |
| *s*-**block elements:** | Groups 1, 2;     *s*-subshell fills |
| *p*-**block elements:** | Groups 13–18; *p*-subshell fills |
| *d*-**block elements:** | Groups 3–12;   *d*-subshell fills |
| **transition elements:** | Groups 3–11;   *d*-subshell fills |
| **main-group elements:** | Groups 1, 2 and 13–18 |
| **lanthanides:** | 4*f*-subshell fills |
| **actinides:** | 5*f*-subshell fills |

**Note:** A transition element has a partially filled *d*-subshell either as the element or in any commonly occurring oxidation state. Thus Zn, Cd, Hg are not transition elements. The $(n-1)d$ electrons of the *d*-block elements are considered to be valence electrons. Groups 1, 2, and 13–18 are alternatively labeled with Roman numerals I–VIII which correspond to the number of valence electrons in the element.

| | |
|---|---|
| **valence shell:** | Outermost occupied shell (highest $n$) |
| **period:** | Row of the periodic table |
| **period number:** | Principal quantum number of valence shell |
| **Period 1:** | H, He;  1*s*-subshell fills |
| **Period 2:** | Li through Ne;  2*s*-, 2*p*-subshells fill |
| **Period 3:** | Na through Ar;  3*s*-, 3*p*-subshells fill |
| **Period 4:** (first long period) | K through Kr;  4*s*-, 3*d*-, 4*p*-subshells fill |

- **Periodic table** → Two forms displayed:
  One shows the elements, the other the *final* subshell filled.
  Look at the tables to understand the terminology.

## Periodicity of Elements in the Periodic Table

**Period** ↓          **Group (1–18)** →

| Period | 1 | 2 | 3 | 4 | 5 | 6 | 7 | 8 | 9 | 10 | 11 | 12 | 13 | 14 | 15 | 16 | 17 | 18 |
|---|---|---|---|---|---|---|---|---|---|---|---|---|---|---|---|---|---|---|
| 1 | H | 2 | | | | | | | | | | | 13 | 14 | 15 | 16 | 17 | He |
| 2 | Li | Be | | | | | | | | | | | B | C | N | O | F | Ne |
| 3 | Na | Mg | 3 | 4 | 5 | 6 | 7 | 8 | 9 | 10 | 11 | 12 | Al | Si | P | S | Cl | Ar |
| 4 | K | Ca | Sc | Ti | V | *Cr* | Mn | Fe | Co | Ni | *Cu* | Zn | Ga | Ge | As | Se | Br | Kr |
| 5 | Rb | Sr | Y | Zr | *Nb* | *Mo* | Tc | *Ru* | *Rh* | ***Pd*** | *Ag* | Cd | In | Sn | Sb | Te | I | Xe |
| 6 | Cs | Ba | Lu | Hf | Ta | W | Re | Os | Ir | *Pt* | *Au* | Hg | Tl | Pb | Bi | Po | At | Rn |
| 7 | Fr | Ra | Lr | Rf | Db | Sg | Bh | Hs | Mt | Ds | Uuu | Uub | | Uuq | | Uuh | | |

| | | | | | | | | | | | | | | |
|---|---|---|---|---|---|---|---|---|---|---|---|---|---|---|
| Lanthanides | *La* | *Ce* | Pr | Nd | Pm | Sm | Eu | *Gd* | Tb | Dy | Ho | Er | Tm | Yb |
| Actinides | *Ac* | ***Th*** | *Pa* | *U* | *Np* | Pu | Am | *Cm* | Bk | Cf | Es | Fm | Md | No |

- **Electron configurations of the elements** → See the Appendix section on *Ground-State Electron Configurations* in the text.

- **17 italicized elements**
  - → Exceptions to the $(n + \ell)$ rule
  - → Differ by the placement of *one* electron
    **Example:** Cr: $1s^2 2s^2 2p^6 3s^2 3p^6 4s^1 3d^5 = [\text{Ar}]\, 4s^1 3d^5$ (not...$4s^2 3d^4$ as might be expected)

- **2 bold, italicized elements** → Differ by the placement of *two* electrons (Pd, Th)
  **Pd**      We expect $[\text{Kr}]\,4d^8 5s^2$, but *actually* find $[\text{Kr}]\,4d^{10}$      *No 5s-subshell electrons*
  **Th**      We expect $[\text{Rn}]\,5f^2 7s^2$, but *actually* find $[\text{Rn}]\,6d^2 7s^2$      *No 5f-subshell electrons*

## Order of Filling Subshells in the Periodic Table

**Period** ↓          **Group (1–18)** →

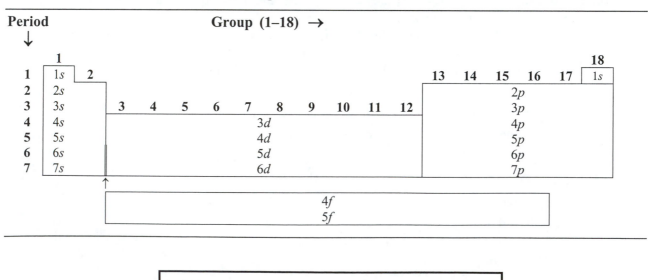

# The Periodicity of Atomic Properties
## (Sections 1.14–1.19)

### 1.14  Atomic Radius ($r$)

- **Definition of atomic radius**
  - $\rightarrow$  Half the distance between the centers of neighboring atoms (nuclei)
  - $\rightarrow$  For metallic elements, $r$ is determined for the solid.
    - **Example:** For solid Zn, 274 pm between nuclei, so $r$ (atomic radius) = 137 pm
  - $\rightarrow$  For nometallic elements, $r$ is determined for diatomic molecules (covalent bond).
    - **Example:** For $I_2$ molecules, 266 pm between nuclei, so $r$ (covalent radius) = 133 pm

- **General trends in radius with atomic number**
  - $\rightarrow$  $r$ decreases from *left* to *right* across a period  (*effective* nuclear charge increases)
  - $\rightarrow$  $r$ increases from *top* to *bottom* down a group  (change in valence electron principal quantum number and size of valence shell)

### 1.15  Ionic Radius

- **An ion's share of the distance between neighbors in an ionic solid (cation to anion)**
  - $\rightarrow$  Distance between the nuclei of a neighboring cation and anion is the sum of two ionic radii
  - $\rightarrow$  Radius of the oxide anion ($O^{2-}$) is 140 pm.
    - **Example:** Distance between Zn and O nuclei in zinc oxide is 223 pm; therefore the ionic radius of $Zn^{2+}$ is 83 pm.   $[r(Zn^{2+}) = 223 \text{ pm} - r(O^{2-})]$
  - $\rightarrow$  Cations are **smaller** than parent atoms; for example, Zn (133 pm) and $Zn^{2+}$ (83 pm)
  - $\rightarrow$  Anions are **larger** than parent atoms; for example, O (66 pm) and $O^{2-}$ (140 pm)

- **Isoelectronic atoms and ions**  $\rightarrow$  Atoms and ions with the same number of electrons
    - **Example:** $Cl^-$, Ar,  and $K^+$, and $Ca^{2+}$

### 1.16  Ionization Energy  ($I$)

- $I$ $\rightarrow$  Energy needed to remove an electron from a gas-phase atom in its lowest energy state

- **Symbol** $\rightarrow$  Number subscripts (*e.g.*, $I_1$, $I_2$) denote removal of successive electrons.

- **General periodic table trends in $I_1$ for the main-group elements**
  - $\rightarrow$  Increases from *left* to *right* across a period      ($Z_{\text{eff}}$ increases)
  - $\rightarrow$  Decreases from *top* to *bottom* down a group      (change in principal quantum number $n$ of valence electron)

- **Exceptions** $\rightarrow$  For Groups 2 and 15 in Periods 2–4, $I_1$ is *larger* than the neighboring *main group element* in Groups 13 and 16, respectively.  Repulsions between electrons in the same orbital and/or extra stability of completed and half completed subshells are responsible for these exceptions to the general trends.

- **Metals toward the lower left of the periodic table**
  - $\rightarrow$  Have low ionization energies
  - $\rightarrow$  Readily lose electrons to form cations

- **Nonmetals toward the upper right of the periodic table**
  - $\rightarrow$ Have high ionization energies
  - $\rightarrow$ Do not readily lose electrons

## 1.17 Electron Affinity ($E_{ea}$)

- **Energy *released* when an electron is *added* to a gas-phase atom**

- **Periodic table trends in $E_{ea}$ for the main-group elements**
  - $\rightarrow$ Increases from *left* to *right* across a period     ($Z_{eff}$ increases)
  - $\rightarrow$ Decreases from *top* to *bottom* down a group     (change in principal quantum number $n$ of valence electron)

  - $\rightarrow$ Generally, the same as for $I_1$ with the major exceptions displayed in Fig. 1.49

- **Summary of trends in $r$, $I_1$, and $E_{ea}$**
  - $\rightarrow$ All depend on $Z_{eff}$ and $n$ of outer subshell electrons.
  - $\rightarrow$ Recall that there are *many* exceptions to the general trends, as noted earlier.

## 1.18 Inert-Pair Effect

- **Tendency to form ions two units lower in charge than expected from the group number relation**

  - $\rightarrow$ Due in part to the different energies of the valence *p*- and *s*-electrons

  - $\rightarrow$ Important for the *lower* two members of Groups 13, 14, and 15, for example, Pb(IV) & Pb (II) or Sb(V) & Sb(III)

  - $\rightarrow$ Valence *s*-electrons are called a "lazy pair."

## 1.19 Diagonal Relationships

- **Diagonally related pairs of elements often show similar chemical properties.**
  - $\rightarrow$ Diagonal band of metalloids dividing metals from nonmetals
  - $\rightarrow$ Similarity of Li and Mg  (react directly with $N_2$ to form nitrides)
  - $\rightarrow$ Similarity of Be and Al  (both react with acids and bases)

# The Impact on Materials  (Sections 1.20–1.21)

## 1.20  The  Main-Group Elements  (Groups 1–2, 13–18)
- *s*-block elements (Groups 1 and 2)  $\rightarrow$  All are reactive metals (except H); they form *basic* oxides ($O^{2-}$),  peroxides ($O_2^{2-}$), or superoxides ($O_2^-$).

- **Compounds of *s*-block elements are ionic, except for beryllium.**

   **Notes:**  *Hydrogen* (a nonmetal) is placed by itself or *more usually* in Group 1, because its electronic configuration ($1s^1$) is similar to those of the alkali metals:  [noble gas] $ns^1$.

   *Helium*, with electronic configuration $1s^2$, is placed in Group 18, because its properties are similar to those of neon, argon, krypton, and xenon.

- ***p*-block elements  (Groups 13–18)**  →  Members are metals, metalloids, and nonmetals.

   **Metals:**     Group 13 (Al, Ga, In, Tl);  Group 14 (Sn, Pb);  Group 15 (Bi)

   **Note:**    These metals have relatively low $I_1$, but *larger* than those of the *s* block and *d* block.

   **Metalloids:**   Group 13 (B);  Group 14 (Si, Ge); Group 15 (As, Sb);  Group 16 (Te, Po)

   **Note:**    Metalloids have the *physical* appearance and properties of *metals*, but behave *chemically* as *nonmetals*.

   **Nonmetals:**   Group 14 (C); Group 15 (N, P); Group 16 (O, S, Se); Groups 17 & 18 (All)

   **Notes:**  High $E_{ea}$ in Groups 13–17; these atoms tend to *gain* electrons to complete their subshells. Group 18 noble-gas atoms are generally *nonreactive*, except for Kr and Xe, which form a few compounds.

## 1.21  The Transition Metals

- ***d*-block elements  (Groups 3–12)**  →  All are metals (*most* are transition metals), with  properties intermediate to those of *s*-block and *p*-block metals.

   **Note:**   *All* Group 12 cations retain the filled *d*-subshell.  For this reason, these elements are *not* classified as transition metals.  Recall that in this text *d*-orbital electrons are considered to be valence electrons for *all* the *d*-block elements.

- **Characteristically, many transition metals form compounds with a variety of oxidation states (oxidation numbers).**

- ***f*-block elements  (lanthanides and actinides)**

   → Rare on Earth

   → Lanthanides are incorporated in *superconducting* materials

   → Actinides are all *radioactive* elements

# Chapter 2   Chemical Bonds

## Ionic Bonds  (Sections 2.1–2.4)

### 2.1   Formation of Ionic Bonds

- **Ionic bond** → Electrostatic attraction (coulombic) of *oppositely* charged ions

- **Ionic model** →  Energy for the *formation* of ionic bonds is supplied *mainly* by coulombic attraction between *oppositely* charged ions.  This model provides a good description of bonding between metals (particularly *s*-block) and nonmetals.

- **Ionic solids**
    - → Are assemblies of cations and anions arranged in a regular array
    - → The ionic bond is *nondirectional*; each ion is bound to *all* its neighbors.
    - → Have *high* melting and boiling points
    - → Form electrolyte solutions if they dissolve in water

- **Crystalline solids** → Atoms, molecules, or ions arranged in a regular pattern

### 2.2   Interactions Between Ions

- **Coulomb potential energy, $E_P$**

$$E_{P,12} = \frac{(z_1 e) \times (z_2 e)}{4 \pi \varepsilon_0 r_{12}} = \frac{2.307\, z_1 z_2}{r_{12}} \times 10^{-19}\, \text{J·nm}$$

    $e$ is the elementary charge; $z_1$ and $z_2$ are the charges on the ions; $r_{12}$ is the distance in nanometers (nm) between the ions; $\varepsilon_0$ is the vacuum permittivity.

- **Lattice energy** → Potential energy difference: ions in solid *minus* ions infinitely far apart

- **Madelung constant ($A$)**

$$E_P = -A \times \frac{|z_1 z_2| N_A e^2}{4 \pi \varepsilon_0 d}$$

    $A$ is a positive integer; $d$ is the distance between the centers of nearest neighbors in the crystal.

    → Depends on the *arrangement* of ions in the solid

**Values of $A$** →  1.747 56 (NaCl structure)  [$d$ = 0.2798 nm   for NaCl  (actual value near 0 K)]
1.762 67 (CsCl structure)  [$d$ = 0.351 nm   for CsCl  (sum of ionic radii)]

- **Born-Meyer equation**

$$E_{P,min} = -\frac{N_A |z_A z_B| e^2}{4 \pi \varepsilon_0 d}\left(1 - \frac{d^*}{d}\right) A$$

    The constant $d^*$ is commonly taken to be 34.5 pm.  It derives from the repulsive effects of overlapping electron charge clouds.

- **Refractory material** → Substance that can withstand high temperatures;  MgO, for example

- **Coulomb interaction between ions in a solid** → *Large* when ions are **highly charged** and **small**

### 2.3   Electron Configurations of Ions

- **Cations** → Remove outermost electrons in the order  $np$, $ns$, $(n-1)d$
    - **Examples:** Iron (II), $Fe^{2+}$, [Ar] $3d^6$ and  iron (III), $Fe^{3+}$, [Ar]$3d^5$

- **Anions** → Add electrons until the next noble-gas configuration is reached.

  **Examples:** Carbide (methanide), $C^{4-}$, [He] $2s^2 2p^6$ or [Ne]　　　{octet}

  　　　　　　Hydride, $H^-$, $1s^2$ or [He]　　　　　　　　　{duplet}

- **Formulas of compounds composed of monatomic ions**
  - → Formulas are predicted by assuming that atoms forming cations lose all valence electrons and those forming anions gain electrons in the valence subshell(s) until each ion has an octet of electrons or a duplet in the case of H, He, and Be.
  - → For cations with variable valence, the Stock number, or oxidation number, is used (see Fundamentals section of text).
  - → Relative numbers of cations and anions are chosen to achieve electrical neutrality, using the smallest possible integers as subscripts.

## 2.4 Lewis Symbols (Atoms and Ions)

- **Valence electrons**
  - → Depicted as dots; a pair of dots represents two paired electrons
  - → Lewis symbols for neutral atoms in the first two periods of the Periodic Table:

    H·　　He:　　Li·　　Be:　　·B·　　·C·　　:N·　　:O·　　:F·　　:Ne:

    Note: Ground-state structures for B and C are :B· and :C· .

- **Variable valence**
  - → Ability of an element to form two or more ions with different oxidation numbers
  - → Displayed by many d-block and p-block elements

    **Examples:** Lead in lead(II) oxide, PbO, and in lead(IV) oxide, $PbO_2$

    　　　　　Iron in iron(II) oxide, FeO, and in iron(III) oxide, $Fe_2O_3$

---

# Covalent Bonds (Sections 2.5–2.9)

## 2.5 Nature of the Covalent Bond

- **Covalent bond**
  - → Pair of electrons *shared* between two atoms
  - → Located between two neighboring atoms and *binds* them together

    **Examples:** Nonmetallic elements such as $H_2$, $N_2$, $O_2$, $F_2$, $Cl_2$, $Br_2$, $I_2$, $P_4$, and $S_8$

## 2.6 Lewis Structures

- **Rules**
  - → Atoms attempt to complete duplets or octets by sharing pairs of valence electrons.
  - → Valence of an atom is the number of bonds it can form.
  - → A line (–) represents a shared pair of electrons.
  - → A lone pair of nonbonding electrons is represented by two dots (:).

**Example:** H–H, (single bond) duplet on each atom (valence of hydrogen = 1) and $:N\equiv N:$, (triple bond and lone pair) octet on each atom (valence of nitrogen = 3)

- **Lone pairs of electrons** $\rightarrow$ electron pairs not involved in bonding

    **Example:** The electrons indicated as dots in the Lewis structure of $N_2$ above

## 2.7 Lewis Structures for Polyatomic Species

- **Lewis structures**
    - $\rightarrow$ Show which atoms are bonded (atom connectivity) and which contain lone pairs of electrons
    - $\rightarrow$ Do not portray the *shape* of a molecule or ion

- **Rules**
    - $\rightarrow$ Count total number of valence electrons in the species.
    - $\rightarrow$ Arrange atoms next to bonded neighbors.
    - $\rightarrow$ Use minimum number of electrons to make all single bonds.
    - $\rightarrow$ Count the number of nonbonding electrons required to satisfy octets.
    - $\rightarrow$ Compare to the actual number of electrons left.
    - $\rightarrow$ If lacking a sufficient number of electrons to satisfy octets, make *one extra bond* for each deficit pair of electrons.
    - $\rightarrow$ Sharing pairs of electrons with a neighbor completes the octet or duplet.
    - $\rightarrow$ Each shared pair of electrons counts as one *covalent bond* (line).

        | One shared pair | **single bond** | (–) | Bond order = 1 |
        | Two shared pairs | **double bond** | (=) | Bond order = 2  (multiple bond) |
        | Three shared pairs | **triple bond** | ($\equiv$) | Bond order = 3  (multiple bond) |

- **Bond order** $\rightarrow$ Number of bonds that link a specific pair of atoms

- **Terminal atom** $\rightarrow$ Bonded to only one other atom

- **Central atom** $\rightarrow$ Bonded to at least two other atoms

- **Molecular ions** $\rightarrow$ Consist of *covalently* bonded atoms: $NH_4^+$, $Hg_2^{2+}$, $SO_4^{2-}$

- **Rules of thumb**
    - $\rightarrow$ Usually, the element with the lowest $I_1$ is a central atom, but electronegativity is a better indicator (see Section 2.13). For example, in HCN, carbon has the lowest $I_1$ and is the central atom. It is also less electronegative than nitrogen.
    - $\rightarrow$ Usually, there is a symmetrical arrangement about the central atom. For example, in $SO_2$, OSO is symmetrical, with S as the central atom and the two O atoms terminal.
    - $\rightarrow$ Oxoacids have H atoms bonded to O atoms;  $H_2SO_4$ is actually $(HO)_2SO_2$.

    **Examples:** Ethyne (acetylene), $C_2H_2$ (10 valence electrons):  $H-C\equiv C-H$

    Hydrogen cyanide, HCN (10 valence electrons):  $H-C\equiv N:$

    Ammonium ion, $NH_4^+$ (8 valence electrons):  $\left[\begin{array}{c} H \\ | \\ H-N-H \\ | \\ H \end{array}\right]^+$

## 2.8  Resonance

- **Multiple Lewis structures**
  - → Some molecules can be represented by different Lewis structures (*contributing structures*) in which the *locations* of the electrons, but not the nuclei, vary.
  - → Multiple Lewis structures used to represent a given species are called *resonance structures*.

- **Electron delocalization**  →  Species requiring multiple Lewis structures often involve *electron delocalization* with some electron pairs distributed over more than two atoms.

- **Blending of structures**  →  Double-headed arrows  ( ◄────► ) are used to relate contributing Lewis structures, indicating that a blend of the contributing structures is a better representation of the bonding than any one.

- **Resonance hybrid**  →  Blended structure of the contributing Lewis structures

    **Example:** $N_2O$, nitrous oxide, has $2(5) + 6 = 16$ valence electrons or eight pairs.

$$:\ddot{N}=N=\ddot{O}: \quad \longleftrightarrow \quad :N\equiv N-\ddot{\underset{\cdot\cdot}{O}}:$$

    Blending of two structures, both of which obey the octet rule.

    **Note:**  The central N atom is the least electronegative atom (Section 2.13).

    **Note:**  A triple bond to an O atom is found in species such as $BO^-$, CO, $NO^+$, and $O_2^{2+}$.

    **Example:**  $C_6H_6$, benzene, has $6(4) + 6(1) = 30$ valence electrons or 15 pairs.

            Kekulé structures                    Resonance hybrid

    **Note:**  Carbon atoms lie at the corners of the hexagon (the six C–H bonds radiating from each corner are not shown).  The last structure depicts six valence electrons *delocalized* around the ring.

## 2.9  Formal Charge

- **Formal charge**  →  An atom's number of valence electrons ($V$) minus the number of electrons assigned to it in a Lewis structure

- **Electron assignment in a Lewis structure**
  - → Atom possesses all of its lone pair electrons ($L$) and half of its bonding electrons ($S$) [$S$ means shared electrons].
  - → Formal charge $= V - (L + \frac{1}{2}S)$

- **Contribution of individual**  →  Structures with individual formal charges
  **Lewis  structures to a**       closest to **zero** usually have the lowest energy and
  **resonance hybrid**           are the major contributors.

**Example:** $N_2O$: $\ddot{\underset{-1}{N}}=\underset{+1}{N}=\underset{0}{\ddot{O}}\colon \longleftrightarrow \colon\underset{0}{N}\equiv\underset{+1}{N}-\underset{-1}{\ddot{O}}\colon \longleftrightarrow \colon\underset{-2}{\ddot{N}}-\underset{+1}{N}\equiv\underset{+1}{O}\colon$

(high formal charges)

- **Plausibility of isomers** → The isomer with lowest formal charges is *usually* preferred.

   **Examples:** $N_2O$, $\ddot{\underset{-1}{N}}=\underset{+1}{N}=\underset{0}{\ddot{O}}\colon$ and $\ddot{\underset{-1}{N}}=\underset{+2}{O}=\underset{-1}{\ddot{N}}\colon$

   (low formal charges)   (high formal charges)

   HCN, $\underset{0}{H}-\underset{0}{C}\equiv\underset{0}{N}\colon$ and $\underset{0}{H}-\underset{+1}{N}\equiv\underset{-1}{C}\colon$   (first structure with lower formal charges preferred)

## Summary

- **Formal charge**
  - → Indicates the extent to which atoms have gained or lost electrons in a Lewis structure (covalent bonding)
  - → Exaggerates the *covalent* character of bonds
  - → Structures with the lowest formal charges usually have the lowest energy (major contributors to the resonance hybrid).
  - → Isomers with lower formal charges are usually favored.

- **Oxidation number** → Exaggerates the *ionic* character of bonds

   **Example:** Carbon disulfide, $\colon\ddot{S}=C=\ddot{S}\colon$ Here, C has an oxidation number of +4 (all the electrons in the double bonds are assigned to the S atoms) and a formal charge of zero (the electrons in the double bonds are shared equally between the C and S atoms).

---

# Exceptions to the Octet Rule (Sections 2.10–2.12)

## 2.10 Radicals and Biradicals

- **Radicals**
  - → Species with an unpaired electron
  - → All species with an odd number of electrons are radicals.
  - → Are highly reactive; cause rancidity in foods, degradation of plastics in sunlight, and perhaps human aging

   **Examples:** $CH_3$ (methyl radical), OH (hydroxyl), OOH (hydrogenperoxyl), NO (nitric oxide), $NO_2$ (nitrogen dioxide), $O_3^-$ (ozonide ion)

   $\left[\colon\ddot{O}-\ddot{O}-\ddot{O}\colon\right]^- \longleftrightarrow \left[\colon\ddot{O}-\ddot{O}-\ddot{O}\colon\right]^- \longleftrightarrow \left[\colon\ddot{O}-\ddot{O}-\ddot{O}\colon\right]^-$

- **Biradicals** → Species containing *two* unpaired electrons
   **Examples:** O (oxygen atom) and $CH_2$ (methylene) [unpaired electrons on a single atom]
   $O_2$ (oxygen molecule) and larger organic molecules [on different atoms]

- **Antioxidant** → Species that reacts rapidly with radicals before they have a chance to do damage

## 2.11  Expanded Valence Shells

- **Expanded valence shells**
  - → More than eight electrons associated with an atom in a Lewis structure (expanded octet)
    A *hypervalent compound* contains an atom with more atoms attached to it than is permitted by the octet rule.  Empty *d*-orbitals are utilized to permit octet expansion.
  - → Electrons may be present as bonding pairs or lone pairs.
  - → Are characteristic of nonmetal atoms in *Period 3 or higher*
    *First and second period elements do not utilize expanded valence shells.*

- **Variable covalence**
  - → Ability to form different numbers of covalent bonds
  - → Elements showing variable covalence include:

**Valence Shell Occupancy**

| Elements | 8 electrons | 10 electrons | 12 electrons |
|----------|-------------|--------------|--------------|
| P | $PCl_3$ / $PCl_4^+$ | $PCl_5$ | $PCl_6^-$ |
| S | $SF_2$ | $SF_4$ | $SF_6$ |
| I | IF | $IF_3$ | $IF_5$ |

**Example:** Major Lewis structures for sulfuric acid, $H_2SO_4$ (32 valence electrons)

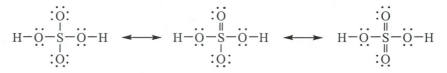

In the structure on the right, all atoms have zero formal charge and the S atom is surrounded by 12 electrons (expanded octet). *Because it has the lowest formal charges, this structure is expected to be the most favored one energetically and the one to make the greatest contribution to the resonance hybrid.*  An additional Lewis structure with one double is bond not shown.  Similar examples include $SO_4^{2-}$, $SO_2$, and $S_3O$ (S is the central atom).

## 2.12  The Unusual Structures of Some Group 13/III Compounds

- **Incomplete octet**
  - → Fewer than eight valence electrons on an atom in a Lewis structure

    **Example:** $BF_3$,

    The single-bonded structure with an incomplete octet makes the major contribution.
  - → Other compounds with incomplete octets are $BCl_3$ and $AlCl_3$, which is a vapor at very high temperature.

- **Coordinate covalent bond**  → One in which both electrons come from one atom

**Example:** Donation of a lone pair by one atom to create a bond that completes the octet of another atom with an incomplete octet. In the Lewis structure for the reaction of $BF_3$ with $NH_3$, both electrons in the B–N bond are derived from the N lone pair.

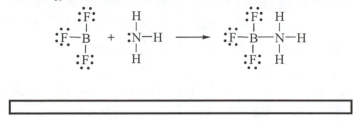

---

# Ionic versus Covalent Bonds (Sections 2.13–2.14)

## 2.13 Correcting the Covalent Model: Electronegativity

- **Bonds** → All molecules are resonance hybrids of pure covalent and ionic structures.

    **Example:** $H_2$ The structures are $H–H \longleftrightarrow H^+[H{:}]^- \longleftrightarrow [{:}H]^- H^+$

    Here, the two ionic structures make equal contributions to the resonance hybrid, but the single covalent structure is of major importance.

- **Partial charges**
    - → When the bonded atoms are different, the ionic structures are not energetically equivalent.
    - → Unequal sharing of electrons results in a **polar covalent bond**

    **Example:** HF The structures are $H–F \longleftrightarrow H^+[{:}\ddot{\underset{\cdot\cdot}{F}}{:}]^- \longleftrightarrow [{:}H]^-[\ddot{\underset{\cdot\cdot}{F}}{:}]^+$

- **Note:** The electron affinity of F is greater than that of H, resulting in a small negative charge on F and a corresponding small positive charge on H:

$$E_{ea}(F) > E_{ea}(H) \implies {}^{\delta+}H–F^{\delta-} \text{ (partial charges, } |\delta+| = |\delta-|)$$
$$E(H^+F^-) \ll E(H^-F^+) \implies H^-F^+ \text{ is a very } minor \text{ contributor}$$

- **Electric dipole** → A partial *positive* charge separated from an equal but *negative* partial charge

- **Electric dipole moment (μ)**
    - → Size of an electric dipole: partial charge times distance between charges
    - → Units: debye (D)
    - → $4.80\ D \equiv$ an electron (−) separated by 100 pm from a proton (+)
    - → $\boxed{\mu = (4.80\ D) \times \delta \times (\text{distance in pm}/100\ \text{pm})}$

- **Electronegativity (χ)**
    - → Electron-attracting power of an atom when it is part of a bond
    - → Mulliken scale: $\chi = \frac{1}{2}(I_1 + E_{ea})$
    - → Follows same periodic table trends as $I_1$ and $E_{ea}$
    - → *Increases from left to right and from bottom to top*
    - → Pauling numerical scale based on bond energies [$\chi(F) = 4.0$] is used in text (qualitatively similar to Mulliken's scale).
    - → See Section 2.13 in the text for the tabulation of values for various elements.

- **Rough rules of thumb**

$$(\chi_A - \chi_B) \geq 2 \qquad \text{bond is } \textit{essentially} \text{ ionic}$$
$$0.5 \leq (\chi_A - \chi_B) \leq 1.5 \qquad \text{bond is } \textit{polar} \text{ covalent}$$
$$(\chi_A - \chi_B) \leq 0.5 \qquad \text{bond is } \textit{essentially} \text{ covalent}$$

### 2.14 Correcting the Ionic Model: Polarizability

- **Ionic bonds** $\rightarrow$ All have *some* covalent character.
  A cation's positive charge attracts the electrons of an anion or atom in the direction of the cation (*distortion* of spherical electron cloud).

- **Highly *polarizable* atoms and ions** $\rightarrow$ Readily undergo a *large* distortion of their electron cloud
  Examples of polarizable species: large anions and atoms such as $I^-$, $Br^-$, $Cl^-$, I, Br, and Cl

- **Polarizing power**
  $\rightarrow$ Property of ions (and atoms) that cause large distortions of electron clouds
  $\rightarrow$ Increases with decreasing size and increasing charge of a cation
  Examples of species significant polarizing power: the small and/or highly charged cations $Li^+$, $Be^{2+}$, $Mg^{2+}$, and $Al^{3+}$

- **Significant covalent bonding character**
  $\rightarrow$ Bonds between highly polarizing cations and highly polarizable anions have significant covalent character.
  $\rightarrow$ The $Be^{2+}$ cation is highly polarizing and the Be–Cl bond is significantly covalent (even though there is an electronegativity difference of 1.6).
  $\rightarrow$ In the series AgCl to AgI, the bonds become more covalent as the polarizability (and size) of the anion increases ($Cl^- < Br^- < I^-$).

---

# The Strengths and Lengths of Covalent Bonds
## (Sections 2.15–2.17)

### 2.15 Bond Strengths

- **Dissociation energy (*D*)**
  $\rightarrow$ Energy required to separate bonded atoms in neutral molecules
  $\rightarrow$ Bond breaking is *homolytic,* which means that each atom retains half of the bonding electrons.
  $\rightarrow$ Determines the *strength* of a chemical bond
  The *greater* the dissociation energy, the *stronger* the bond.
  $\rightarrow$ *D* is defined exactly for diatomic (two-atom) molecules.
  For polyatomic (greater than two-atom) molecules, *D* also depends on the other bonds in the molecule. However, for many molecules, this dependence is slight.

- ***D* for several diatomic molecules**

$$\text{H–H(g)} \rightarrow \text{H(g)} + \text{H(g)} \qquad D = 432 \text{ kJ·mol}^{-1} \text{ (strong)}$$
$$\text{H–F(g)} \rightarrow \text{H(g)} + \text{F(g)} \qquad D = 562 \text{ kJ·mol}^{-1} \text{ (strong)}$$
$$\text{F–F(g)} \rightarrow \text{F(g)} + \text{F(g)} \qquad D = 155 \text{ kJ·mol}^{-1} \text{ (weak)}$$

$$I-I(g) \rightarrow I(g) + I(g) \qquad D = 149 \text{ kJ·mol}^{-1} \text{ (weak)}$$
$$O=O(g) \rightarrow O(g) + O(g) \qquad D = 494 \text{ kJ·mol}^{-1} \text{ (strong)}$$
$$N\equiv N(g) \rightarrow N(g) + N(g) \qquad D = 942 \text{ kJ·mol}^{-1} \text{ (very strong)}$$

## 2.16 Variation in Bond Strength

- **Bond strength** → For polyatomic molecules, bond strength is defined as the *average* dissociation energy for one type of bond found in different molecules.

  For example, the tabulated C–H single bond value is the *average* strength of such bonds in a selection of organic molecules, such as methane ($CH_4$), ethane ($C_2H_6$), and ethene ($C_2H_4$).

  **Note:** Values of average dissociation energies given in text Table 2.3 are actually those for the thermodynamic property of *bond enthalpy* introduced in Chapter 6. Thermodynamic values of *bond enthalpy* are measured at 298.15 K. The bond *dissociation energies* of diatomic molecules given in Section 2.15 apply at a temperature of 0 K (absolute zero).

- **Factors influencing bond strength**
  - → Bond multiplicity $\qquad$ (C≡C > C=C > C–C)
  - → Resonance $\qquad$ (C=C > C⋯C (benzene) > C–C)
  - → Lone pairs on neighboring atoms $\qquad$ (F–F < H–H)
  - → Atomic radii $\qquad$ (HF > HCl > HBr > HI)
  - (The smaller the radius, the stronger the bond.)

## 2.17 Bond Lengths

- **Bond length**
  - → Distance between the centers of two atoms linked by a covalent bond
  - → Helps determine the overall size and shape of a molecule
  - → Evaluated by using spectroscopic or x-ray diffraction (for solids) methods
  - → For bonds between the same elements, length is *inversely* proportional to strength.

- **Factors influencing bond length**
  - → Bond multiplicity $\qquad$ (C≡C < C=C < C–C)
  - → Resonance $\qquad$ (C=C < C⋯C (benzene) < C–C)
  - → Lone pairs on neighboring atoms $\qquad$ (F–F > H–H)
  - → Atomic radii $\qquad$ (HF < HCl < HBr < HI)
  - (the smaller the radius, the shorter the bond)

  **Note:** These trends are the *opposite* of the ones for bond strength.

- **Covalent radius**
  - → Contribution an atom makes to the length of a covalent bond
  - → *Half* the distance between the centers of neighboring atoms joined by a covalent bond (*for like atoms*)
  - → Covalent radii may be added together to estimate bond lengths in molecules.
  - → Tabulated values are *averages* of radii in polyatomic molecules.
  - → *Decreases* from left to right in the periodic table
  - → *Increases* in going down a group in the periodic table

$\rightarrow$ Decreases for a given atom with increasing multiple bond character

**Example:** Use the covalent radii given in text Fig. 2.21 to estimate the several bond lengths in the acetic acid molecule, $CH_3COOH$.

Lewis Structure:

$$\begin{array}{ccc} H & :\ddot{O} & \\ | & \| & \\ H-C-C-\ddot{O}-H \\ | & \\ H & \end{array}$$

| Bond type | Bond length estimate | Actual |
|-----------|----------------------|--------|
| C–H | 77 pm + 37 pm = 114 pm | ≈109 pm |
| C–C | 77 pm + 77 pm = 154 pm | ≈150 pm |
| C–O | 77 pm + 74 pm = 151 pm | 134 pm |
| C=O | 67 pm + 60 pm = 127 pm | 120 pm |
| O–H | 74 pm + 37 pm = 111 pm | 97 pm |

# Chapter 3   Molecular Shape and Structure

## The VSEPR Model  (Sections 3.1–3.3)

### 3.1  The Basic VSEPR (Valence-Shell Electron-Pair Repulsion) Model

- **Valence electrons about central atom(s)** → Controls shape of a molecule

- **Lewis structure** → Shows distribution of *valence* electrons in bonding pairs (bonds) and as lone pairs

- **Bonds, lone pairs** → Regions of high electron density that repel each other (Coulomb's law) by rotating about a central atom, thereby maximizing their separation

- **Bond angle(s)** → Angle(s) between bonds joining atom centers

- **Multiple bonds** → Treated as a *single* region of high electron concentration in VSEPR

- **Electron arrangement** → Ideal locations of bond pairs and lone pairs about a central atom (angles between electron pairs define the geometry)

### Typical Electron Arrangements in Molecules with *One* Central Atom

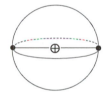   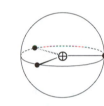   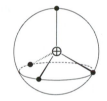

**Linear (180°)**          **Trigonal planar (120°)**          **Tetrahedral (109.5°)**

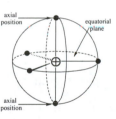          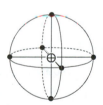

**Trigonal bipyramidal**          **Octahedral (90° and 180°)**
**(120° equatorial, 90° axial-equatorial)**

- **VSEPR formula, $AX_nE_m$**   (See Section 3.2 in the text)
  - → A = central atom
  - → $X_n$ = n atoms (*same* or *different*) bonded to central atom
  - → $E_m$ = m lone pairs on central atom
  - → Useful for generalizing types of structures

  **Note:** Molecular shape is defined by the location of the *atoms alone*.
  The bond angles are ∠ X–A–X.

## Typical Shapes of Molecules with *One* Central Atom*

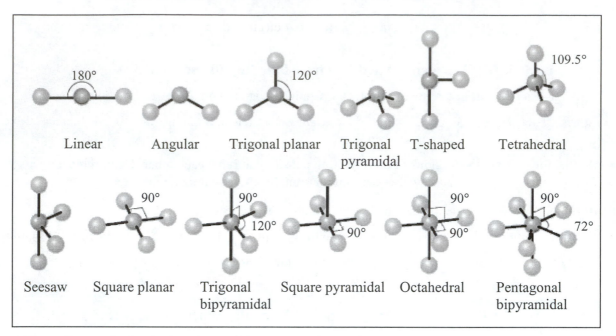

* These figures depict the locations of the atoms only, *not* lone pairs.

- **VSEPR (sometimes pronounced "vesper") method**  (See Toolbox 3.1 in the text)
  - → Write the Lewis structure(s).  If there are resonance structures, pick *any* one.
  - → Count the number of electron pairs (bonding and nonbonding) around the central atom(s). Treat a multiple bond as a *single* unit of high electron density.
  - → Identify the *electron arrangement*.  Place electron pairs as far apart as possible.
  - → Locate the *atoms* and classify the *shape* of the molecule.
  - → Optimize *bond angles* for molecules with *lone pairs* on the central atom(s) with the concept in mind that repulsions are in the order

    lone pair-lone pair > lone pair-bonding pair > bonding pair-bonding pair.

    **Example:**  Use the VSEPR model to predict the *shape* of tetrafluoroethene, $C_2F_4$.

    Lewis structure:

    There are three concentrations of electron density around each C atom, so the shape is *trigonal planar* about each C atom.  $C_2F_4$ is an $AX_3AX_3$ species.

### 3.2  Molecules with Lone Pairs on the Central Atom

- **VSEPR method for molecules with lone pairs ($m \neq 0$)**
  - → Lone pair electrons have preferred positions in certain electron arrangements.
  - → In the *trigonal bipyramidal* arrangement, lone pairs prefer the *equatorial* positions, in which electron repulsions are minimized.
  - → In the *octahedral* arrangement, all positions are *identical*.  The preferred electron arrangement for *two lone pairs* is the occupation of *opposite* corners of the octahedron.

**Example:** Use the VSEPR model to predict the shape of trichloroamine, $NCl_3$.

Lewis structure:
$$:\overset{\cdot\cdot}{Cl}-\overset{\cdot\cdot}{\underset{|}{N}}-\overset{\cdot\cdot}{Cl}:$$
$$:\overset{\cdot\cdot}{Cl}:$$

The 4 electron pairs around N yield a *tetrahedral* electron arrangement, while the shape is *trigonal pyramidal* (like a badminton birdie). Because lone pair-bonding pair repulsion is greater than bonding pair-bonding pair repulsion, the bond angles [$\angle$ Cl–N–Cl ] will be less than 109.5°.

## 3.3 Polar Molecules

- **Polar molecule** → Molecule with a *nonzero* dipole moment

- **Polar bond** → Bond with a *nonzero* dipole moment

- **Molecular dipole moment** → *Vector sum* of bond dipole moments

- **Representation of dipole** → Single arrow with head pointed toward the *positive* end of the dipole

    **Example:** Lewis structure: $H-\overset{\cdot\cdot}{\underset{\cdot\cdot}{Cl}}:$ and polar (dipole) representation: $H \leftarrow Cl$

- **Predicting molecular polarity**
    → Determine molecular *shape* using VSEPR theory.
    → Estimate electric dipole (bond) moments (text Section 2.13) or use text Fig. 3.7 to decide whether the molecular *symmetry* leads to a *cancellation* of bond moments (*nonpolar molecule*, no dipole moment) or not (*polar molecule*, nonzero dipole moment).

    **Example:** Is $XeF_2$ a polar or nonpolar molecule?

    Lewis structure: $:\overset{\cdot\cdot}{F}-\overset{\cdot\cdot}{\underset{\cdot\cdot}{Xe}}-\overset{\cdot\cdot}{F}:$ The electron arrangement is *trigonal bipyramidal;* the shape is *linear* (lone pairs prefer the equatorial positions). The bond angle is 180°, the bond dipoles cancel, and the molecule is *nonpolar.* $F \rightarrow Xe \leftarrow F$

---

# Valence-Bond (VB) Theory (Sections 3.4–3.8)

## 3.4 Sigma ($\sigma$) and Pi ($\pi$) Bonds

- **Two major types of *bonding* orbitals** → $\sigma$ and $\pi$

- **$\sigma$-orbital**
    → Has no *nodal surface* containing the interatomic (bond) axis
    → Is cylindrical or "sausage" shaped
    → Formed from overlap of two *s*-orbitals, an *s*-orbital and a *p*-orbital end-to-end, two *p*-orbitals end-to-end, a certain hybrid orbital and an *s*-orbital, a *p*-orbital end-to-end with a certain hybrid orbital, or two certain hybrid orbitals end-to-end

- **Example of σ-orbital formation** → Two 1*s*-orbitals of H atoms combine to form a σ-orbital of $H_2$

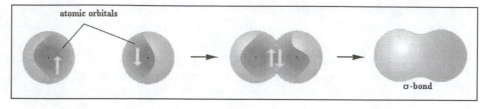

- **π-orbital**
  - → *Nodal plane* containing the interatomic (bond) axis
  - → Two cylindrical shapes (lobes), one above and the other below the nodal plane
  - → Formed from side-by-side overlap of two *p*-orbitals

- **Example of π-orbital formation** → Two $2p_x$-orbitals overlap side-by-side

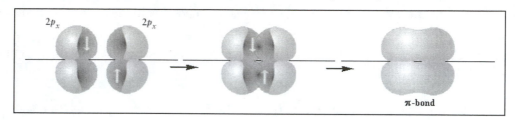

- **Electron occupancy**
  - → σ- and π-orbitals can hold 0, 1, or 2 electrons, corresponding to no bond, a half-strength bond, or a full covalent bond, respectively.
  - → In any orbital, spins of two electrons must be *paired* (opposite direction of arrows).

- **Types of bonds according to valence-bond theory**
  - → **Single bond** is a σ-bond.            [$H_2$, for example]
  - → **Double bond** is a σ-bond plus one π-bond.     [$O_2$, for example]
  - → **Triple bond** is a σ-bond plus two π-bonds.     [$N_2$, for example]

## 3.5 Hybridization of Orbitals *and* 3.6 Hybridization in More Complex Molecules

- **Hybrid orbitals**
  - → Produced by mixing (hybridizing) orbitals of a central atom
  - → A construct consistent with observed shapes and bonding in molecules

### Hybridization Schemes

| Electron Arrangement Around the Central Atom | Hybrid Orbitals (Number) | Angle(s) Between σ-Bonds | Example(s) [Underlined Atom] |
|---|---|---|---|
| linear | *sp* (two) | 180° | $\underline{Be}Cl_2$, $\underline{C}O_2$ |
| trigonal planar | $sp^2$ (three) | 120° | $\underline{B}F_3$, $\underline{C}H_3^+$ |
| tetrahedral | $sp^3$ (four) | 109.5° | $\underline{C}Cl_4$, $\underline{N}H_4^+$ |
| trigonal bipyramidal | $sp^3d$ (five) | 120°, 90°, and 180° | $\underline{P}Cl_5$ |
| octahedral | $sp^3d^2$ (six) | 90° and 180° | $\underline{S}F_6$ |

- **Determination of hybridization schemes**
  - → Write the Lewis structure(s). If there are resonance structures, pick *any* one.
  - → Determine the number of lone pairs and σ-bonds on the central atom.
  - → Determine the number of orbitals required for hybridization on the central atom to accommodate the lone pairs and σ-bonds.
  - → Determine the hybridization scheme from the preceding table.
  - → Use VSEPR rules to determine the molecular *shape* and *bond angle(s)*. (π-bonds are formed from the overlap of atomic orbitals [unhybridized] in a side-by-side arrangement).

    **Example:** Describe the bonding in $BF_3$ using VB theory and hybridization. The shape is trigonal planar (VSEPR) and the hybridization at B is $sp^2$ (see table above) and the three σ-bonds are formed from overlap of $B(sp^2)$ orbitals with F(2p) orbitals ($B(sp^2)$–F(2p) bonds).

- **Central atoms** → Each central atom can be represented using a hybridization scheme, as in the example below.

  **Example**: Use VB theory to describe the bonding and in and shape of acetaldehyde, a molecule with 2 central atoms.

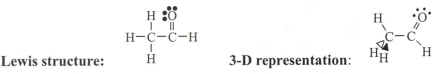

  **Lewis structure:**                                      **3-D representation**:

  **Bonds:** C–C σ-bonds from overlap of methyl C $sp^3$ with aldehyde C $sp^2$ hybrid orbital [$Csp^3$–$Csp^2$]. C=O σ-bond from overlap of aldehyde $sp^2$ hybrid orbital with methyl C $sp^3$ orbital. C=O π-bond from side-by-side overlap of C 2p and O 2p atomic orbitals. C–H single bonds: from overlap of H1s AOs and Csp³ hybrid orbitals (methyl C atom) and H1s AO and Csp² hybrid orbital.

  **Angles:** Three ∠ H–C–H ~ 109.5°, three ∠ H–C–C ~ 109.5°, one ∠ C–C=O ~ 120°, one ∠ O=C–H ~ 120°, and one ∠ C–C–H ~ 120°

  **Molecular shape:** Tetrahedral at methyl carbon atom, trigonal planar at aldehyde C atom

## 3.7 Bonding in Hydrocarbons

- **Alkanes ($C_nH_{2n+2}$, n = 1, 2, 3, ...)** → Characteristics: tetrahedral geometry and $sp^3$ hybridization at C atoms; all C–C and C–H single (σ) bonds; rotation allowed about C–C single bonds

  **Examples:** Methane, $CH_4$; and propane, $CH_3CH_2CH_3$

- **Alkenes ($C_nH_{2n}$, n = 2, 3, 4, ...)** → Characteristics: one C=C double bond (σ plus π); other C–C bonds and all C–H bonds single (σ); trigonal-planar geometry and $sp^2$ hybridization at double-bonded C atoms; tetrahedral geometry at other C ($sp^3$) atoms; rotation *not* allowed about double C=C bond; rotation allowed about C–C single bonds

  **Examples:** Ethene (ethylene), $H_2C=CH_2$; and propylene, $CH_3CH=CH_2$

- **Alkynes ($C_nH_{2n-2}$, n = 2, 3, 4, ...)** → Characteristics: one C≡C triple bond (σ plus two π); other C–C bonds and all C–H bonds (σ); linear geometry and *sp* hybridization at triple-bonded C atoms; tetrahedral geometry at other C ($sp^3$) atoms

  **Examples:** Ethyne (acetylene), HC≡CH; 2-butyne (dimethylacetylene), $H_3CC≡CCH_3$

- **Benzene ($C_6H_6$)**
  → **Characteristics**: *Planar* molecule with hexagonal C framework; trigonal-planar geometry and $sp^2$ hybridization at C atoms; σ framework has C–C and C–H single bonds; 2*p*-orbitals (one from each C atom) overlap side-by-side form *three π-bonds (six π-electrons) delocalized over the entire six C atom ring*
  → See Lewis resonance structures in Section 2.8.

### 3.8 Characteristics of Double Bonds

- **Double bonds**
  → Consist of *one* σ- and *one* π-bond; are always *stronger* than a single σ-bond:
    C=C is *weaker* than two single C–C σ-bonds
    N=N is *stronger* than two single N–N σ-bonds
    O=O is *stronger* than two single O–O σ-bonds

  → Formed readily by Period 2 elements (C, N, O)
  → Rarely found in Period 3 and higher period elements
  → Impart rigidity to molecules and influence molecular shape

  **Example:** Describe the structure of the carbonate anion in terms of hybrid orbitals, bond angles, and σ- and π-bonds.

  **Lewis resonance structures:**

  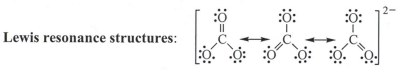

  Trigonal-planar geometry with $sp^2$ hybridization at C, 120° ∠ O–C–O angles, and $Csp^3$-$O2p$ single bonds. Remaining C 2*p*-orbital can overlap with O 2*p*-orbitals to form a π-bond in each of three ways corresponding to the resonance structures. Therefore, each carbon-oxygen bond can be viewed as one σ-bond and one-third of a π-bond.

# Molecular Orbital (MO) Theory (Sections 3.9–3.14)

### 3.9 Limitations of Lewis's Theory

- **Valence Bond (VB) deficiencies**
  → Cannot explain *paramagnetism* of $O_2$ (See text Box 3.2 for explanation of *paramagnetism* and *diamagnetism*)
  → Difficulty treating *electron-deficient* compounds such as diborane, $B_2H_6$
  → No simple explanation for *spectroscopic properties* of compounds such as color

- **Molecular Orbital (MO) advantages**
  - → Addresses *all* of the above shortcomings of VB theory
  - → Provides a *deeper* understanding of electron-pair bonds
  - → Accounts for the *structure and properties* of metals and semiconductors
  - → More facile for *computer calculations* than VB theory

## 3.10 Molecular Orbitals (MOs)

- **MO theory**
  - → In MO theory, electrons occupy MOs which are *delocalized* over the *entire* molecule.
  - → In VB theory, bonding electrons are *localized* between the two atoms.

- **Molecular Orbitals (MO)**
  - → Formed by superposition (**linear combination**) of atomic orbitals **(LCAO-MO)**
  - → **Bonding orbital** (constructive interference):
    *Increased* amplitude or electron density between atoms
  - → **Antibonding orbital** (destructive interference):
    *Decreased* amplitude or electron density between atoms

- **Types of Molecular Orbitals**
  - → **Bonding MO:** energy *lower* than that of the constituent AOs
  - → **Nonbonding MO:** energy *equal* to that of the constituent AOs
  - → **Antibonding MO:** energy *greater* than that of the constituent AOs

- **Rules for combining AOs to obtain MOs**
  - → $N$ atomic orbitals (AOs) yield $N$ molecular orbitals (MOs)
  - → Little overlap of inner shell AOs, and these MOs are usually *nonbonding.*

- **MO energy-level diagrams**
  - → Relative energies of original AOs and resulting MOs are shown schematically in energy-level diagrams.
  - → Arrows used to show electron spin and to indicate location of the electrons in the separated atoms and the molecule

    **Example:** Formation of MOs by LCAO method and MO energy-level diagram for the formation of $Li_2$ from Li atoms. Formation of bonding and antibonding orbitals:

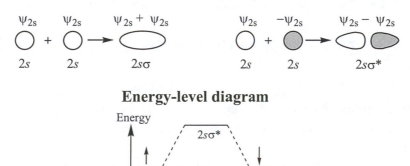

**Energy-level diagram**

### 3.11 Electron Configurations of Diatomic Molecules

- **Procedure for determining the electronic configuration of diatomic molecules**
  - → Construct all possible MOs from *valence-shell* AOs.
  - → Place *valence electrons* in the lowest energy, unoccupied MOs.
  - → Follow the **Pauli exclusion principle** and **Hund's rule** as for AOs: electrons have *spins paired* in a fully occupied molecular orbital and enter unoccupied *degenerate* orbitals with *parallel spins*.

- **Valence-shell MOs for Period 1 and Period 2 *homonuclear* diatomic molecules**

## Correlation Diagrams

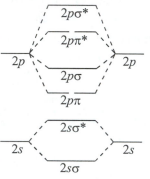
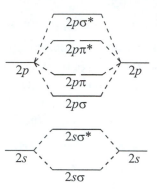

Schematic order of MO energy levels for $H_2$, $He_2$

Schematic order of MO energy levels for $Li_2$, $Be_2$, $B_2$, $C_2$, $N_2$

Schematic order of MO energy levels for $O_2$, $F_2$, $Ne_2$

- **Bond order (BO)** → $\boxed{\text{BO} = \frac{1}{2}(N - N^*)}$

  $N$ = total number of electrons in *bonding* MOs

  $N^*$ = total number of electrons in *antibonding* MOs

  **Examples:** BO in $Li_2 = 1$ (2 valence electrons in bonding $2s\sigma$ MO). BO in $H_2^- = \frac{1}{2}$ (2 electrons in $1s\sigma$ bonding MO, one electron in $1s\sigma^*$-antibonding MO). $O_2^+$ has 11 valence electrons, which are placed in the $2s\sigma$, $2s\sigma^*$, $2p\sigma$, two *degenerate* $2p\pi$, and one of the *degenerate* $2p\pi^*$ MOs. The valence electron configuration is $(2s\sigma)^2 (2s\sigma^*)^2 (2p\sigma)^2 (2p\pi)^4 (2p\pi^*)^1$. There is one unpaired electron. $N = 8$ and $N^* = 3$. Bond order $= \frac{1}{2}(8 - 3) = 5/2$.

### 3.12 Bonding in Heteronuclear Diatomic Molecules

- **Bond properties**
  - → Polar in nature
  - → Atomic orbital from the *more* electronegative atom is *lower* in energy.
  - → *Bonding* orbital has a *greater* contribution from the *more* electronegative atom.
  - → *Antibonding* orbital has a *greater* contribution from the *less* electronegative atom.
- **Result of bond properties is modified, unsymmetrical correlation diagrams.**
  - **Example:** Bonding in HCl using MO theory. The correlation diagram shows occupied valence orbitals for H and Cl atoms, resulting MOs from overlap of $1s$ H and $3p_z$ Cl orbitals, lone pair nonbonding orbitals $3p_x$ and $3p_y$ of Cl, and the nonbonding Cl $3s$ orbital.

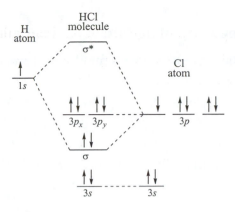

The electronic configuration is $(Cl\,3s)^2\,\sigma^2\,(Cl\,3p_x)^2\,(Cl\,3p_y)^2$.

## 3.13 Orbitals in Polyatomic Molecules

- **Bond properties**
  - → All atoms in the molecule are encompassed by the MOs.
  - → Electrons in MOs bond all the atoms in a molecule.
  - → Electrons are *delocalized* over all the atoms.
  - → Energies of the MOs are obtained by spectroscopy.

- **Effectively localized electrons** → While MO theory requires MOs to be *delocalized* over all atoms, electrons sometimes are *effectively localized* on an atom or group of atoms.

    **Example:** See the HCl example above with nonbonding $Cl\,3p_x$ and $Cl\,3p_y$ orbitals effectively localized on Cl.

- **Combining VB and MO theories** → VB and MO theory are sometimes jointly invoked to describe the bonding in complex molecules.

    **Example:** In benzene, *localized sp²* hybridization (hybrids of C 2s-, C 2p_x-, and C 2p_y-orbitals on each C) is used to describe the σ-bonding framework of the ring, and *delocalized* MOs (LCAOs of six C 2p_z-orbitals) to describe the π-bonding.

- **Triumphs of MO theory**
  - → Accounts for paramagnetism of $O_2$
  - → Accounts for bonding in electron-deficient molecules

    **Example:** In diborane ($B_2H_6$, with 12 valence electrons), *six* electron pairs bond *eight* atoms using *six* MOs. VB theory requires *eight* electron pairs and fails to describe the bonding.

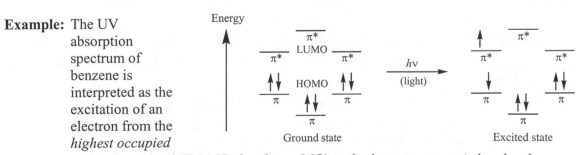

Diborane

  - → Provides a framework for interpreting visible and ultraviolet (UV) spectroscopy

    **Example:** The UV absorption spectrum of benzene is interpreted as the excitation of an electron from the *highest occupied*

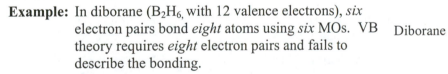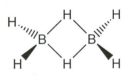

*molecular orbital* (**HOMO**, *bonding* π MO) to the *lowest unoccupied molecular orbital* (**LUMO**, *antibonding* π* MO*). The bond order of the π-system changes from 3 to 2.

### 3.14  Impact on Materials:  The Band Theory of Solids

- **MO theory** → Accounts for electrical properties of metals and semiconductors
- **Bands**
    - → Groups of MOs having closely spaced energy levels  (see text Fig. 3.43)
    - → Formed by the spatial overlap of a very large number of AOs
    - → Separate into two groups  (one mostly *bonding,* the other mostly *antibonding*)
- **Conduction band** →   Empty or partially filled band
    - **Example:** Formation of a conduction band in Na by the overlap of 3*s*-orbitals
- **Valence band** →   Completely filled band of MOs (insulators such as molecular solids)
- **Band gap** → An energy range between bands in which there are *no* orbitals
- **Electronic conductors** → Metals, semiconductors, and superconductors

| | |
|---|---|
| **Insulator:** | Does not conduct electricity:  substance with a full *valence band* **far** in energy from an empty *conduction band*  (large *band gap*) |
| **Metallic conductor:** | Current carried by *delocalized* electrons in *bands* <br> Conductivity *decreases* with *increasing* temperature <br> Substance with a partially filled *conduction band* (gap *not* relevant) |
| **Semiconductor:** | Current carried by *delocalized* electrons in *bands* <br> Conductivity *increases* with *increasing* temperature <br> Substance with a full *valence band* close in energy to an empty *conduction band* (small *band gap*):  excitation of some electrons from the *valence* to the *conduction band* yields conductivity. |
| **Superconductor:** | *Zero* resistance (infinite conductivity) to an electric current below a definite transition temperature |

- **Example:** The ionic solid $K_3(C_{60})$ below 19.3 K.  Ions in the solid are $K^+$ and $(C_{60})^{3-}$.  The molecule $C_{60}$ is buckminsterfullerene (buckyball).

- **Enhanced semiconductors** → Prepared by *doping,* that is, adding a small amount of impurity

- **n-type semiconductor** →  Adding some *valence electrons* to the *conduction band* (n = electrons added)
    - **Example:** A small amount of As (Group 15) added to Si  (Group 14):  extra electrons are transferred from As into the previously empty *conduction band* of Si.

- **p-type semiconductor** →  Removing some *valence electrons* from the *valence band*   (p = positive "holes" in the *valence band*)
    - **Example:** A small amount of In (Group 13) added to Si (Group 14):  valence electron deficit is created in the previously full *valence band* of Si, thereby becoming a *conduction band*.
- **p-n junctions**
    - → p-type semiconductor in contact with an n-type semiconductor
    - → Solid-state electronic devices
        - **Example:** One type of transistor is a p-n sandwich with two separate wires (*source* and *emitter*) connected to the n-side.  It acts as a *switch* in the following way:  when a positive charge is applied to a polysilicon contact (*base* wire) positioned between the *source* and *emitter* wires, a current flows from the *source* to the *emitter*.  Other examples include diodes and integrated circuits.

# Chapter 4   The Properties of Gases

## The Nature of Gases  (Sections 4.1–4.3)

### 4.1  Observing Gases

- **Gases** → Examples of *bulk matter*, forms of matter consisting of large numbers of molecules

- **Atmosphere** → Not uniform in composition, temperature ($T$), or density ($d$)

- **Physical properties** → Most gases behave similarly at low pressure

- **Compressibility**
    - → Ease with which a gas undergoes a volume ($V$) decrease
    - → Gas molecules are widely spaced

- **Expansivity**
    - → The ability of a gas to rapidly fill the space available to it
    - → Atomic and molecular gases move rapidly and respond quickly to changes in the available volume

### 4.2  Pressure

$$\text{Pressure} \equiv \frac{\text{force}}{\text{area}} \quad or \quad P = \frac{F}{A}$$

Opposing force from a confined gas

- **Barometer**
    - → A glass tube, sealed at one end, filled with liquid mercury, and inverted into a beaker also containing liquid mercury  (Torricelli)
    - → The height ($h$) of the liquid column is a measure of the atmospheric pressure ($P$) expressed as the length of the column for a particular liquid.
    Common pressure units:  mmHg,  cmHg,  $cmH_2O$

$$P = \frac{F}{A} = \frac{mg}{A} = \frac{(dV)g}{A} = \frac{(dhA)g}{A} = dhg$$

where $m$ = mass of liquid,  $g$ = acceleration of gravity
(9.806 65 m·s$^{-2}$ Earth's surface), and $d$ = density of liquid

- → The atmospheric pressure when the height of a column of mercury in a barometer is *exactly* 0.76 m at 0°C is 101 325 kg·m$^{-1}$·s$^{-2}$, a result which is obtained by using the value of the density of mercury = 13 595.1 kg·m$^{-3}$ and the gravitational constant at the surface of the earth.
    **Note:**  1 Pa (pascal) = 1 kg·m$^{-1}$·s$^{-2}$ = 1 N·m$^{-2}$   (SI unit of pressure)

- **Manometer  (see Fig. 4.5 in the text)**
    - → U-shaped tube filled with liquid connected to an experimental system
    - → "open-tube" pressure (system) = atmosphere when levels are equal
    - → "closed-tube" pressure (system) = difference in level heights

- **Gauge pressure** → Difference between the pressure inside a vessel containing a compressed gas (*e.g.*, a tire) and the atmospheric pressure

## 4.3  Alternative Units of Pressure

- **Atmosphere** → 1 atm = **760** Torr  (bold type = exact)
  
  1 atm = **101 325** Pa  = **101.325** kPa
  
  1 bar = $10^5$ Pa = **100** kPa
  
  1 Torr = 1 mmHg  (Not exact, but to better than $2 \times 10^{-7}$ Torr)

---

# The Gas Laws  (Sections 4.4–4.11)

## 4.4  Boyle's Law  (fixed amount of gas *n* at constant temperature *T*)

- **Gas behavior** → Boyle's law describes an isothermal system: $\Delta T = 0$ (constant *T*).

$$\text{Volume} \propto \frac{1}{\text{pressure}} \quad \text{or} \quad V \propto \frac{1}{P} \quad \text{or} \quad PV = \text{constant}$$

  The functional form is a *hyperbola*.

- **Changes between two conditions (1 and 2)**

$$P_1 V_1 = \text{constant} = P_2 V_2 \quad \text{or} \quad \frac{P_2}{P_1} = \frac{1/V_2}{1/V_1} = \frac{V_1}{V_2}$$

- **Deviations** → At low temperature

## 4.5  Charles's Law  (fixed amount of gas *n* at constant pressure *P*)

- **Gas behavior** → Charles's law describes an isobaric system: $\Delta P = 0$  (constant *P*).

$$\text{Volume} \propto \text{temperature} \quad \text{or} \quad V \propto T \quad \text{or} \quad V/T = \text{constant}$$

  The functional form is *linear*.

- **Kelvin temperature scale** → Two *fixed* points define the absolute temperature scale:

  $T \equiv 273.16$ K   The triple point of water, where ice, liquid, and vapor are in equilibrium (see Chapter 8).

  $T \equiv 0$ K   Defined from the limiting behavior of Charles's law. *V* approaches 0 at fixed *low* pressure (extrapolation).

- **Celsius temperature scale** → $t\,(°\text{C}) \equiv T(\text{K}) - 273.15$   (exactly)

  **Note:**  The freezing point of water at 1 atm external pressure is 0°C or 273.15 K. (The symbol *t* is also used for time.)

- **Changes between two conditions (1 and 2)**

$$V_1 / T_1 = \text{constant} = V_2 / T_2 \quad \text{or} \quad \frac{V_2}{V_1} = \frac{T_2}{T_1}$$

- **Deviations** $\rightarrow$ At high pressure

- **Another aspect of gas behavior (fixed amount of gas $n$ at constant volume $V$)**

  $\rightarrow$ $\boxed{\text{Pressure} \propto \text{temperature} \quad \text{or} \quad P \propto T \quad \text{or} \quad P/T = \text{constant}}$

  $\rightarrow$ This relation describes an *isochoric* system: $\Delta V = 0$ (constant $V$). The functional form is *linear*.

- **Changes between two conditions (1 and 2)**

  $$\boxed{P_1/T_1 = \text{constant} = P_2/T_2 \quad \text{or} \quad \frac{P_2}{P_1} = \frac{T_2}{T_1}}$$

- **Deviations** $\rightarrow$ For small volume

## 4.6 Avogadro's Principle (gas at constant pressure $P$ and temperature $T$)

- **Gas behavior** $\boxed{\text{Volume} \propto \text{amount} \quad \text{or} \quad V \propto n \quad \text{or} \quad V/n = \text{constant}}$

- **Avogadro's principle** $\rightarrow$ A given amount of gas molecules at a particular $P$ and $T$ occupies the same volume regardless of chemical identity.

- **Molar volume** $\boxed{V_{\mathrm{m}} = \dfrac{V}{n} = \text{constant} \qquad \text{for } \Delta P = 0 \text{ and } \Delta T = 0}$

- **Changes between two conditions (1 and 2)**

  $$\boxed{V_1/n_1 = \text{constant} = V_2/n_2 \quad \text{or} \quad \frac{V_2}{V_1} = \frac{n_2}{n_1}}$$

- **Deviations** $\rightarrow$ At high pressure and/or low temperature

- **Another aspect of gas behavior (gas at constant volume $V$ and temperature $T$)**

  $\boxed{\text{Pressure} \propto \text{amount} \quad \text{or} \quad P \propto n \quad \text{or} \quad P/n = \text{constant}}$

- **Changes between two conditions (1 and 2)**

  $$\boxed{P_1/n_1 = \text{constant} = P_2/n_2 \quad \text{or} \quad \frac{P_2}{P_1} = \frac{n_2}{n_1}}$$

- **Deviations** $\rightarrow$ For small volume and/or low temperature

## 4.7 The Ideal Gas Law

- **Gas behavior** $\rightarrow$ As $P$ approaches 0, all gases behave *ideally* (*limiting law*).

- **Derivation**

  $PV = \text{constant}_1 \qquad V = \text{constant}_2 \times T \qquad P = \text{constant}_3 \times n$

  $PV = (\text{constant}_3 \times n)(\text{constant}_2 \times T) = nRT \qquad$ ***equation of state***

- **Gas constant** $\rightarrow R = 8.205\,78 \times 10^{-2}\ \text{L·atm·K}^{-1}\text{·mol}^{-1}$

  $\rightarrow R = 8.314\,47\ \text{L·kPa·K}^{-1}\text{·mol}^{-1} = 8.314\,47\ \text{J·K}^{-1}\text{·mol}^{-1}$

## 4.8 Applications of the Ideal Gas Law

- **Changes between two conditions (1 and 2)**

$$\boxed{\frac{P_1 V_1}{n_1 T_1} = \frac{P_2 V_2}{n_2 T_2}}$$

Combined gas law

> **Note:** The more general form of the combined gas law includes the case where the amount of gas may change.

> **Note:** When using this relationship, remember that the actual units used in manipulating ratios of quantities are *not* important. Consistency in units is essential. *Absolute* temperature, however, *must* be used in all the equations.

- **Molar volume**   $\boxed{V_\mathrm{m} = \frac{RT}{P}}$

- **Standard ambient temperature and pressure (SATP)** $\rightarrow$ 298.15 K and 1 bar

- **Standard temperature and pressure (STP)** $\rightarrow$ 0°C and 1 atm

   [or 273.150 K and 1.013 25 bar]

- **Molar volume at STP**

$$\rightarrow V_\mathrm{m} = \frac{RT}{P} = \frac{(8.205\,78 \times 10^{-2}\ \mathrm{L \cdot atm \cdot K^{-1} \cdot mol^{-1}})(273.150\ \mathrm{K})}{(1\ \mathrm{atm})} = 22.4141\ \mathrm{L \cdot mol^{-1}}$$

## 4.9 Gas Density

- **Molar concentration**   $\boxed{\dfrac{n}{V} = \dfrac{1}{V_\mathrm{m}} = \dfrac{P}{RT}}$

- **Molar concentration at STP**

$$\rightarrow \frac{n}{V} = \frac{1}{V_\mathrm{m}} = \frac{1}{22.4141\ \mathrm{L \cdot mol^{-1}}} = 4.461\,48 \times 10^{-2}\ \mathrm{mol \cdot L^{-1}}$$

- **Density of a gas**   $\boxed{d = \dfrac{m}{V} = \dfrac{nM}{V} = \dfrac{MP}{RT}}$

- **Density of a gas at STP** $\rightarrow$ $d = \dfrac{m}{V} = \dfrac{nM}{V} = (4.461\,48 \times 10^{-2}\ \mathrm{mol \cdot L^{-1}}) \times M$

## 4.10 The Stoichiometry of Reacting Gases

- **Accounting for reaction volumes** $\rightarrow$ Use mole-to-mole calculations as described in Section L and convert to gas volume by using the ideal gas law.

- **Balanced chemical equation** → See Example 4.6 and the following self-test exercises in the text.

  **Example:** $NH_4NO_3 (s) \rightarrow N_2O (g) + 2 H_2O (g)$   A total of three moles of gas products is produced from one mole of solid.

## 4.11  Mixtures of Gases

- **Behavior** → Mixtures of nonreacting gases behave as if only a single component were present.

- **Partial pressure** → Pressure a gas would exert if it occupied the container alone

- **Law of partial pressures** → Total pressure of a gas mixture is the sum of the partial pressures of its components  [John Dalton]

  $P = P_A + P_B + \ldots$  for mixture containing A, B, …

- **Components at the same $T$ and $V$** → $n = n_A + n_B + \ldots$ = total amount of mixture

$$P = \frac{nRT}{V} \quad \text{and} \quad P_A = \frac{n_A RT}{V}$$

$$x_A = \frac{n_A}{n} = \frac{P_A}{P} = \text{mole fraction of A in the mixture}$$

Thus, $P_A = x_A P$

- **Humid gas** → For air, $P = P_{\text{dry air}} + P_{\text{water vapor}}$, and $P_{\text{water vapor}} = 47$ Torr at body temperature (37°C).
  **Note:** See Section 8.1

# Molecular Motion  (Section 4.12–4.14)

## 4.12  Diffusion and Effusion

- **Diffusion** → Gradual dispersal of one substance through another substance

- **Effusion**
  - → Escape of a gas through a small hole (or assembly of microscopic holes) into a vacuum
  - → Depends on molar mass and temperature of gas

- **Graham's law**
  - → Rate of effusion of a gas *at constant temperature* is *inversely* proportional to the square root of its molar mass $M$.

  For two gases A and B
  $$\frac{\text{Rate of effusion of A}}{\text{Rate of effusion of B}} = \sqrt{\frac{M_B}{M_A}}$$

  - → Rate of effusion is proportional to the average speed of the molecules in a gas *at constant temperature.*

  $$\frac{\text{Average speed of A molecules}}{\text{Average speed of B molecules}} = \sqrt{\frac{M_B}{M_A}}$$

→ Rate of effusion is *inversely* proportional to the time taken for given volumes or amounts of gas to effuse *at constant temperature*.

$$\frac{\text{Rate of effusion of A}}{\text{Rate of effusion of B}} = \frac{\text{time for B to effuse}}{\text{time for A to effuse}} = \sqrt{\frac{M_B}{M_A}}$$

- **Effusion at different temperatures** → Rate of effusion and average speed increase as the square root of the temperature.

$$\frac{\text{Rate of effusion at } T_2}{\text{Rate of effusion at } T_1} = \frac{\text{average speed of molecules at } T_2}{\text{average speed of molecules at } T_1} = \sqrt{\frac{T_2}{T_1}}$$

- **Combined relationship** → Average speed of molecules is *directly* proportional to the square root of the temperature and *inversely* proportional to the square root of the molar mass.

$$\text{Average speed of molecules in a gas} \propto \sqrt{\frac{T}{M}}$$

## 4.13 The Kinetic Model of Gases

- **Assumptions**

    1. Gas molecules are in continuous random motion.
    2. Gas molecules are infinitesimally small particles.
    3. Particles move in straight lines until they collide.
    4. Molecules do not influence one another except during collisions.

- **Wall collisions** → Consider molecules traveling only in one dimension $x$ with an average square velocity of $\langle v_x^2 \rangle$

- **Pressure on wall** → $P = \dfrac{Nm\langle v_x^2 \rangle}{V}$, where $N$ is the number of molecules, $m$ is the mass of each molecule, and $V$ is the volume of the container

- **Mean square speed** → $v_{\text{rms}}^2 = \langle v^2 \rangle = \langle v_x^2 + v_y^2 + v_z^2 \rangle = \langle v_x^2 \rangle + \langle v_y^2 \rangle + \langle v_z^2 \rangle = 3\langle v_x^2 \rangle$
    rms = root mean square

- **Pressure on wall** → $P = \dfrac{Nmv_{\text{rms}}^2}{3V}$ and $N = nN_A$, where $N_A$ is the Avogadro constant

    $P = \dfrac{nN_A mv_{\text{rms}}^2}{3V} = \dfrac{nMv_{\text{rms}}^2}{3V}$, where $M = mN_A$ is the molar mass

$$PV = \frac{nMv_{rms}^{2}}{3} = nRT$$   The term $nRT$ is from the ideal gas law.

- **Solve for $v_{rms}$**   $\quad v_{rms} = \sqrt{\frac{3RT}{M}} \quad$   $v_{rms}$ is the **root mean square** speed of the particle.

    **Note:** *Beware of units.* Use $R$ in units of $J \cdot K^{-1} \cdot mol^{-1}$ and convert $M$ to $kg \cdot mol^{-1}$. The resulting units of speed will then be $m \cdot s^{-1}$.

**Molar kinetic energy** $= \left(\frac{1}{2}mv_{rms}^{2}\right)N_A = \left(\frac{1}{2}m\right)\left(\frac{3RT}{M}\right)N_A = \frac{3}{2}RT$ , because $mN_A = M$

## 4.14 The Maxwell Distribution of Speeds

- **Symbols** → Let $v$ represent a particle's speed, $N$ the total number of particles, $f(v)$ the Maxwell distribution of speeds, $\Delta$ a *finite* change, and d an *infinitesimal* change (*not* density) in the following relationships.

- **For a *finite* range of speeds**

$$\Delta N = Nf(v)\Delta v, \text{ where } f(v) = 4\pi\left(\frac{M}{2\pi RT}\right)^{3/2} v^{2}e^{-Mv^{2}/2RT}$$

- **For an *infinitesimal* range of speeds**

$$\frac{dN}{N} = f(v)dv$$

    **Note:** Calculus can be used with the Maxwell distribution of speeds to obtain the following properties that are not mentioned in the text.

*Average* **speed** $\quad \langle v \rangle = \int_{0}^{\infty} vf(v)\,dv = \sqrt{\frac{8RT}{\pi M}}$

*Most probable* **speed** $\quad v_{mp} = \sqrt{\frac{2RT}{M}}$ , where $\frac{df(v)}{dv} = 0$   (maximum in distribution)

# Impact on Materials: Real Gases

## (Sections 4.15–4.17)

### 4.15 Deviations from Ideality

- **Definition** → Attractions and repulsions between atoms and molecules

- **Evidence** → Gases condense to liquids when cooled or compressed (attraction). Liquids are difficult to compress (repulsion).

- **Compression factor ($Z$)** → A measure of the effect of intermolecular forces

$$Z \equiv \frac{PV_m}{RT} = \frac{V_m}{V_m^{\text{ideal}}}$$

For an ideal gas, $Z = 1$
For $H_2$, $Z > 1$ at all $P$ (repulsions dominate)
For $NH_3$, $Z < 1$ at low $P$ (attractions dominate)

### 4.16 The Liquefaction of Gases

- **Joule–Thomson effect** → When attractive forces dominate, a real gas *cools* as it expands. (Compression and expansion through a small hole lowers temperature further.) Exceptions are He and $H_2$, for which repulsion dominates.

### 4.17 Equations of State of Real Gases

- **Virial equation** →

$$PV = nRT \left( 1 + \frac{B}{V_m} + \frac{C}{V_m^2} + \cdots \right)$$

$B$ = second virial coefficient
$C$ = third virial coefficient, *etc.*

Note: **Virial coefficients** depend on temperature and are found by fitting experimental data to the virial equation. This equation is more general than the van der Waals equation but is more difficult to use to make predictions.

- **van der Waals equation** →

$$\left( P + a\frac{n^2}{V^2} \right)(V - nb) = nRT$$

where $a$ and $b$ are the **van der Waals parameters** (determined experimentally).

- **Rearranged form** →

$$P = \frac{nRT}{V - nb} - a\frac{n^2}{V^2}$$

where $a$ accounts for the attractive effects and $b$ accounts for the repulsive effects.

- **Repulsion**
  - → Repulsive forces imply that molecules cannot overlap.
  - → Other molecules are *excluded* from the volume they occupy.
  - → The *effective* volume is then $V - nb$, not $V$.
  - → Thus, $b$ is a constant corresponding to the volume excluded per mole of molecules.
  - → The potential energy describing this situation is that of an *impenetrable* hard sphere.
  - → The units of $b$ are $L \cdot mol^{-1}$.

- **Attraction**
  - → Attractive forces lead to clustering of molecules in the gas phase.
  - → Clustering reduces the total number of gas phase species.
  - → The rate of collisions with the wall (pressure) is thereby reduced.
  - → Because this effect arises from attractions between *pairs* of molecules, it should be proportional to the *square* of the number of molecules per unit volume $(N/V)^2$ or equivalently, to the *square* of the molar concentration $(n/V)^2$.
  - → The pressure is predicted to be reduced by an amount $a(n/V)^2$, where $a$ is a positive constant that depends on the strength of the attractive forces.
  - → The potential energy describing this situation is that of long-range attraction with the form $-(1/r^6)$, where $r$ is the distance between a pair of gas phase molecules. This type of attraction primarily arises from *dispersion* or *London* forces (see Chapter 5).
  - → The units of $a$ are $atm \cdot L^2 \cdot mol^{-2}$.

# Chapter 5   Liquids and Solids

## Intermolecular Forces  (Sections 5.1–5.5)

### 5.1  Formation of Condensed Phases

- **Physical states** → Solid, liquid, and gas
- **Phase** → Form of matter uniform in both chemical composition and physical state
- **Condensed phase**
  - → Solid or liquid phase
    **Examples:**  Ag(s); Sn/Pb(s) alloy; $H_2O(s)$; $H_2O(l)$; 1% NaCl(s) in $H_2O(l)$
  - → Molecules (or atoms or ions) are close to each other all the time and intermolecular forces are of major importance.
- **Coulomb potential energy $E_P$**
  - → The interaction between two charges, $q_1$ and $q_2$, separated by distance $r$ is (see Chapter 1)  $\Rightarrow$  $\boxed{E_P \propto \dfrac{q_1 q_2}{r}}$
  - → Almost all intermolecular interactions can be traced back to this fundamental expression.

  **Note:**  The term *intermolecular* is used in a general way to include atoms and ions.

### 5.2  Ion-Dipole Forces

- **Hydration**
  - → Attachment of water molecules to ions (cations and anions)
  - → Water molecules are polar and have an electric dipole moment $\mu$.
  - → A small positive charge on each H atom attracts anions and a small negative charge on the O atom attracts cations.
- **Ion-dipole interaction**
  - → The potential energy (interaction) between an ion with charge $|z|$ and a polar molecule with dipole moment $\mu$ at a distance $r$ is  $\Rightarrow$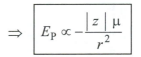
  - → For proper *alignment* of the ion and dipole, the interaction is  *attractive*: cations attract the *partial* negative charges and anions attract the *partial* positive charges on the polar molecule.
  - → Shorter range ($r^{-2}$) interaction than the Coulomb potential ($r^{-1}$)
  - → Polar molecule needs to be *almost* in contact with ion for substantial ion-dipole interaction.
- **Hydrated compounds**
  - → Ion-dipole interactions are much *weaker* than ion-ion interactions, but are relatively strong for *small, highly charged* cations
  - → Accounts for the formation of salt hydrates such as $CuSO_4 \cdot 5\,H_2O$ and $CrCl_3 \cdot 6\,H_2O$
- **Size effects**
  - → $Li^+$ and $Na^+$ (small) tend to form hydrated compounds.
  - → $K^+$, $Rb^+$, and $Cs^+$ (larger) tend not to form hydrates.
  - → $NH_4^+$ (143 pm) is similar in size to $Rb^+$ (149 pm) and forms *anhydrous* compounds.
- **Charge effects** → $Ba^{2+}$ and $K^+$ are similar in size, yet $Ba^{2+}$ (larger charge) forms *hydrates*.

## 5.3  Dipole-Dipole Forces

- **Dipole alignment**  → In solids, molecules with dipole moments tend to *align* with partial positive charge on one molecule near the partial negative charge on another.

- **Dipole-dipole interaction in solids**

  → The interaction between two polar molecules with dipole moments $\mu_1$ and $\mu_2$ *aligned* and separated by a distance $r$, is

  $$\Rightarrow \quad E_P \propto -\frac{\mu_1\,\mu_2}{r^3}$$

  → *Attractive* interaction in solids for head-to-tail *alignment* of dipoles
  → Shorter range ($r^{-3}$) interaction than ion-dipole ($r^{-2}$) or Coulomb potential ($r^{-1}$)
  → For significant interaction, polar molecules need to be *almost* in contact with each other.

- **Dipole-dipole interactions in gas-phase molecules**

  → Dipole-dipole interactions are much *weaker* in gases than in solids. The potential energy of interaction is

  $$\Rightarrow \quad E_P \propto -\frac{\mu_1^{\,2}\,\mu_2^{\,2}}{r^6}$$

  → Because gas molecules are in motion (rotating as well), they experience only a *weak* net attraction because of *occasional* alignment.

- **Dipole-dipole interactions in liquid-phase molecules**
  → Same potential energy relationship as in the gas phase, but the interaction is slightly *stronger* because the molecules are closer.
  → The liquid *boiling point* is a measure of the *strength* of the intermolecular forces in a liquid. The boiling point of isomers (see glossary for definition) is *often* related to the strength of their dipole-dipole interactions.
  → Typically, the *larger* the dipole moment, the *higher* the boiling point.

    **Example:** *Cis*-dibromoethene (dipole moment ≈ 2.4 D, bp 112.5°C) vs.
    *trans*-dibromoethene (zero dipole moment, bp 108°C)

## 5.4  London Forces

- **Nonpolar molecules** → Condensation of *nonpolar* molecules to form liquids implies the existence of a type of intermolecular interaction other than those described above.

- **London force**
  → Accounts for the attraction between any pair of ground-state molecules (*polar* or *nonpolar*)
  → Arises from *instantaneous* partial charges (*instantaneous* dipole moment) in one molecule *inducing* partial charges (dipole moment) in a neighboring one
  → Exists between atoms and *rotating* molecules as well
  → Strength depends on polarizability and shape of molecule.

    **Example:** Other things equal, rod-shaped molecules tend to have stronger London forces than spherical molecules because they can approach each other more closely.

- **Polarizability, α**
  → Of a molecule is related to the ease of deformation of its electron cloud
  → Is proportional to the total number of electrons in the molecule
  → Because the number of electrons correlates with molar mass (generally increases), polarizability does as well.
  → The potential energy between molecules (polar or nonpolar) with polarizability $\alpha_1$ and $\alpha_2$, separated by a distance $r$ is

  $$\Rightarrow \quad E_P \propto -\frac{\alpha_1\,\alpha_2}{r^6}$$

**Note:** Same ($r^{-6}$) interaction as dipole-dipole ($r^{-6}$) with *rotating* molecules, but the London interaction is *usually* of greater strength at normal temperatures.

- **Liquid boiling point**
  - → A measure of the strength of intermolecular forces in a liquid
  - → In comparing the interactions discussed in Sections 5.3 and 5.4, the *major* influence on boiling point in both *polar* and *nonpolar* molecules is the London force.

    **Example:** We can predict the relative boiling points of the nonpolar molecules: $F_2$, $Cl_2$, $Br_2$, and $I_2$. Boiling point correlates with polarizability, which depends on the number of electrons in the molecule. The boiling points are expected to increase in the order of $F_2$ (18 electrons), $Cl_2$ (34), $Br_2$ (70), and $I_2$ (106). The experimental boiling temperatures are: $F_2$ (−188°C), $Cl_2$ (−34°C), $Br_2$ (59°C), and $I_2$ (184°C).

- **Dipole–induced-dipole interaction**
  - → The potential energy between a polar molecule with dipole moment $\mu_1$ and a nonpolar molecule with polarizability $\alpha_2$ at a distance $r$ is $\Rightarrow$ 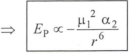 $E_P \propto -\dfrac{\mu_1^2 \, \alpha_2}{r^6}$
  - → Also applies to molecules that are both polar, each one can induce a dipole in the other
  - → Weaker interaction than dipole-dipole

    **Example:** Carbon dioxide ($\mu = 0$) dissolved in water ($\mu > 0$)

## 5.5 Hydrogen Bonding

- **Hydrogen bonds**
  - → Interaction *specific* to certain types of molecules (strong attractive forces)
  - → Account for unusually high boiling points in ammonia ($NH_3$, −33°C), water ($H_2O$, 100°C), and hydrogen fluoride (HF, 20°C)
  - → Arise from a H atom, *covalently* bonded to either an N, O, or F atom in *one* molecule, strongly attracted to a *lone pair* of electrons on an N, O, or F atom in *another* molecule
  - → Are *strongest* when the three atoms are in a *straight line*, and the distance between terminal atoms is within a particular range
  - → Common symbol for the H bond is three dots: $\cdots$
  - → *Strongest* intermolecular interaction between *neutral* molecules

    **Examples:** N–H$\cdots$:N hydrogen bonds between $NH_3$ molecules in pure $NH_3$
    O–H$\cdots$:O hydrogen bonds $H_2O$ molecules in pure $H_2O$

- **Hydrogen bonding in gas-phase molecules**
  - → Aggregation of some molecules persists in the vapor phase.
  - → In HF, fragments of *zigzag* chains and $(HF)_6$ rings are formed. In $CH_3COOH$ (acetic acid), *dimers* are formed. The abbreviated (no C–H bonds shown) Lewis structure for the dimer $(CH_3COOH)_2$ is $\Rightarrow$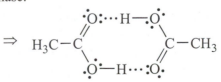

- **Importance of hydrogen bonding**
  - → Accounts for the open structure of solid water
  - → Maintains the shape of biological molecules
  - → Binds the two strands of DNA together

# Liquid Structure (Sections 5.6–5.7)

## 5.6 Order in Liquids

- **Liquid phase**
  - → Mobile molecules with restricted motion
  - → Between the extremes of gas and solid phases
- **Long-range order**
  - → Characteristic of a *crystalline solid*
  - → Atoms or molecules are arranged in an orderly pattern that is repeated over long distances.
- **Short-range order**
  - → Characteristic of the *liquid phase*
  - → Atoms or molecules are positioned in an orderly pattern at nearest-neighbor distances only.
  - → Local order is maintained by a continual process of forming and breaking nearest-neighbor interactions.

## 5.7 Viscosity and Surface Tension

- **Viscosity**
  - → Resistance of a substance to flow:  the *greater* the viscosity, the *slower* the flow.
  - → Viscous liquids include those with hydrogen bonding between molecules  [$H_3PO_4$ (phosphoric acid) and $C_3H_8O_3$ (glycerol): many H bonds], liquid phases of metals, and long chain molecules that can be entangled [hydrocarbon oil and greases ].

- **Viscosity and temperature**  →  *Usually* viscosity *decreases* with *increasing T*
  Exception: unusual behavior of sulfur (see text)

- **Surface tension (several definitions)**
  - → Tendency of surface molecules in a liquid to be pulled into the interior of the liquid by an imbalance in the intermolecular forces
  - → *Inward pull* that determines the resistance of a liquid to an increase in surface area
  - → A measure of the *force* that must be applied to *surface molecules* so that they experience the same *force* as molecules in the *interior* of the liquid.
  - → A measure of the tightness of the surface layer  [Symbol: $\gamma$ (gamma)   Units: $N \cdot m^{-1}$  or  $J \cdot m^{-2}$ ]

- **Capillary action**
  - → Rise of liquids up narrow tubes when the *adhesion* forces are greater than *cohesion* forces
  - → **adhesion:**  Forces that bind a *substance* to a *surface*
  - → **cohesion:**  Forces that bind *molecules* of a substance together to form a *bulk material*

- **Meniscus (curved surface that a liquid forms in a tube)**
  - → *Adhesive* forces greater than *cohesive* forces (forms a ∪ shape)
  - → *Cohesive* forces greater than *adhesive* forces (forms a ∩ shape)
  - → Glass surfaces have exposed O atoms and O–H groups to which hydrogen-bonded liquids like $H_2O$ can bind.  In this case, the *adhesive* forces are greater than the *cohesive* ones.  Water *wets* glass, forms a ∪ shape at the surface, and undergoes a capillary rise in a glass tube.
  - → Mercury liquid does not bind to glass surfaces.  The *cohesive* forces are greater than the *adhesive* ones.  Mercury does *not* wet glass, forms a ∩ shape at the surface, and undergoes a capillary *lowering* in a glass tube.

# Solid Structures (Sections 5.8–5.13)

## 5.8  Classification of Solids

- **Amorphous solid** $\rightarrow$ Atoms or molecules (neutral or charged) lie in random positions.

- **Crystalline solid**
  - $\rightarrow$ Atoms, ions, or molecules are associated with points in a **lattice,** which is an *orderly* array of equivalent points in three dimensions.
  - $\rightarrow$ Structure has long-range order.

- **Crystal faces** $\rightarrow$ Flat, well-defined planar surfaces with definite interplanar angles

- **Classification of crystalline solids**

| | | |
|---|---|---|
| **metallic solids:** | Cations in a sea of electrons  **Examples:** Fe(s), Li(s) | |
| **ionic solids:** | Mutual attractions of cations and anions  **Examples:** NaCl(s), $Ca(SO_4)_2$(s) | |
| **molecular solids:** | Discrete molecules held together by the *intermolecular forces* discussed earlier  **Examples:** sucrose, $C_{12}H_{22}O_{11}$(s); ice, $H_2O$(s) | |
| **network solids:** | Atoms bonded *covalently* to their neighbors throughout the entire solid  **Example:** diamond, C(d); graphite planes, C(gr) | |

## 5.9  Metallic Solids

- **Close-packed structures**
  - $\rightarrow$ Atoms occupy smallest total *volume* with the *least* amount of empty space.
  - $\rightarrow$ Metal atoms are treated as *spheres* with radii *r*.
  - $\rightarrow$ Type of metal determines whether a close-packed structure is assumed.

- **Close-packed layer**
  - $\rightarrow$ Atoms (*spheres*) in a planar arrangement with the *least* amount of empty space – contour of the layer has *dips* or *depressions*.
  - $\rightarrow$ Each atom in a layer has *six* nearest neighbors (hexagonal pattern).
  - $\rightarrow$ Layers stack such that bottoms of spheres in one layer fit in *dips* of the layer below.

- **Stacked layers** $\rightarrow$ In a close-packed structure, each atom has *three* nearest neighbors in the layer above, *six* in its original layer, and *three* in the layer below for a total of 12 nearest neighbors.

- **Coordination number** $\rightarrow$ Number of nearest neighbors of each atom in the solid

- **Two arrangements of stacking close-packed layers**

  | | | |
  |---|---|---|
  | **Pattern:** | ABABABAB… | hcp $\equiv$ *hexagonal close-packing* |
  | | ABCABCABC… | ccp $\equiv$ *cubic close-packing* |

- **Occupied space in hcp/ccp structure** $\rightarrow$ 74% of space is occupied by the spheres and 26% is *empty*.

- **Tetrahedral hole**
  - $\rightarrow$ Formed when a *dip* between *three* atoms in one layer is *covered* by *another* atom in an adjacent layer
  - $\rightarrow$ Two tetrahedral holes per atom in a close-packed structure (hcp or ccp)

- **Octahedral hole**
  - $\rightarrow$ Space between *six* atoms, *four* of which are at the corners of a square plane. The square plane is oriented 45° with respect to two close-packed layers. *Two* atoms in the square plane are in layer **A** and the other *two* in layer **B**. There is *one* additional atom in layer **A** that is *above* the

square plane and *another* in layer **B** that is *below* the square plane. The result is an octahedral arrangement of 6 atoms with a hole in the center.

→ One octahedral hole per atom in a close-packed structure (hcp or ccp)

*Figure of an octahedron and an octahedron rotated to show the relation to two close-packed layers*

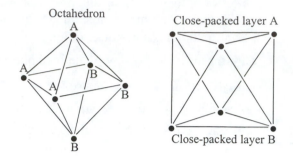

- **Body-centered cubic (bcc) structure**
  - → Coordination number of 8 (*not* close-packed)
  - → Spheres touch along the *body* diagonal of a cube.
  - → Can be converted into close-packed structures under high pressure

- **Primitive cubic (pc) structure**
  - → Coordination number of 6 (*not* close-packed)
  - → Spheres touch along the *edge* of the cube.
    Only one known example: Po (polonium)

| Crystal Structure | Coordination Number | Occupied Space | Examples |
|---|---|---|---|
| ccp (fcc) | 12 | 74% | Ca, Sr, Ni, Pd, Pt, Cu, Ag, Au, Al, Pb<br>Group 18: noble gases at low T, except He |
| hcp | 12 | 74% | Be, Mg, Ti, Co, Zn, Cd, Tl |
| bcc | 8 | 68% | Group 1: alkali metals, Ba, Cr, Mo, W, Fe |
| pc | 6 | 52% | Po (covalent character) |

## 5.10  Unit Cells

- **Lattice** → A regular array of equivalent points in three dimensions

- **Unit cell** → Smallest repeating unit that generates the full array of points

- **Crystal**
  - → Constructed by associating atoms, ions, or molecules with each lattice point
  - → **Metallic crystals:** One metal atom per lattice point  [ccp (fcc), bcc, pc]
    Two metal atoms per lattice point  [hcp]

- **Unit cells for metals**
  - → Cubic system:  pc, bcc, and fcc (ccp)
  - → Hexagonal system: *primitive unit cell* with *two* atoms per lattice point (hcp)

- **Unit cells in general**
  - → Edge lengths:  $a, b, c$
  - → Angles between two edges: $\alpha$ (between edges $b$ and $c$); $\beta$ (between $a$ and $c$); $\gamma$ (between $a$ and $b$)

- **Other crystal systems** →  Tetragonal, orthorhombic, rhombohedral, monoclinic, and triclinic

- **Bravais lattices**
  - → 14 basic patterns of arranging points in three dimensions
  - → Each pattern has a different unit cell (see figure following)

# The 14 Bravais Lattices

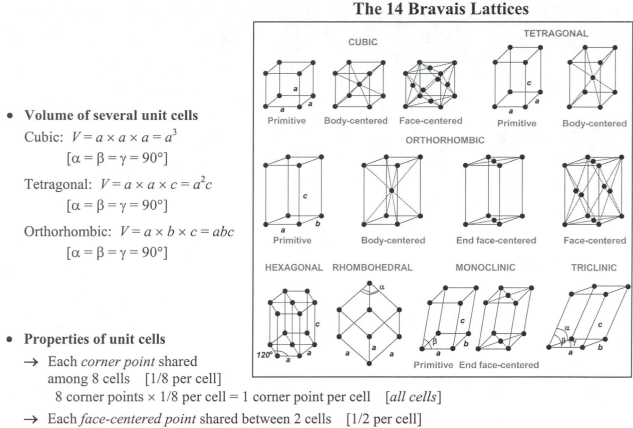

- **Volume of several unit cells**

  Cubic: $V = a \times a \times a = a^3$

  $[\alpha = \beta = \gamma = 90°]$

  Tetragonal: $V = a \times a \times c = a^2 c$

  $[\alpha = \beta = \gamma = 90°]$

  Orthorhombic: $V = a \times b \times c = abc$

  $[\alpha = \beta = \gamma = 90°]$

- **Properties of unit cells**

  → Each *corner point* shared
  among 8 cells    [1/8 per cell]

     8 corner points × 1/8 per cell = 1 corner point per cell    [*all cells*]

  → Each *face-centered point* shared between 2 cells    [1/2 per cell]

     6 face points × 1/2 per cell = 3 face points per cell    [*face-centered cells*]

     2 face points × 1/2 per cell = 1 face point per cell    [*end face-centered cells*]

  → Each *body-centered point* unshared    [1 per cell]

     1 body point × 1 per cell = 1 body point per cell    [*body-centered cells*]

- **Properties of *cubic system* unit cells** → *One* metal atom occupies each *lattice point*

  **Primitive**

     *One* metal atom (*sphere*) at each *corner point* in a unit cell

     [ *one* atom (radius *r* ) per cell (edge length *a* ) ]

     Spheres *touch* along an *edge:*   $a = 2r$

     Volume of the unit cell:   $V = a^3 = (2r)^3 = 8r^3$

     Density of a pc metal  (*M* = molar mass):

$$d = \frac{(1 \text{ atom per cell})(\text{mass of one atom})}{(\text{volume of one unit cell})} = \frac{M/N_A}{a^3} = \frac{M/N_A}{8r^3}$$

  **Body-centered**

     *One* metal atom (*sphere*) at each *corner point* in a unit cell

     *One* metal atom (*sphere*) at the *center* of the unit cell

     Total atoms in unit cell = 8 (1/8) + 1 (1) = 2

     [ *two* atoms (radius *r* ) per unit cell (edge length *a* ) ]

     Spheres *touch* along the *body diagonal:*   $\sqrt{3}a = 4r$   or   $a = \dfrac{4}{\sqrt{3}}r$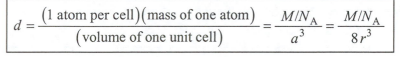

Volume of the unit cell: $V = a^3 = \dfrac{4^3}{3^{3/2}} r^3 \approx 12.3168 r^3$

Density of a bcc metal:

$$d = \frac{(2 \text{ atoms per cell})(\text{mass of one atom})}{(\text{volume of one unit cell})} = \frac{(2)(M/N_A)}{a^3} = \frac{(2)(M/N_A)}{(4^3/3^{3/2})r^3}$$

### Face-centered

*One* metal atom (*sphere*) at each *corner point* in a unit cell
*One* metal atom (*sphere*) at the *center* of each *face* in the unit cell
Total atoms in unit cell $= 8\,(1/8) + 6\,(1/2) = 4$
[*four* atoms (radius $r$) per unit cell (edge length $a$)]

Spheres *touch* along *face diagonal*: $\sqrt{2}\,a = 4r$ or $a = \dfrac{4}{\sqrt{2}} r = \sqrt{8}\,r$

Volume of the unit cell: $V = a^3 = 8^{3/2} r^3$
Density of a fcc (ccp) metal:

$$d = \frac{(4 \text{ atoms per cell})(\text{mass of one atom})}{(\text{volume of one unit cell})} = \frac{(4)(M/N_A)}{a^3} = \frac{(4)(M/N_A)}{(8^{3/2})r^3}$$

- **Calculations** → Determining the *unit cell* type from the *measured density* and *atomic radius* of metals in the *cubic* system; the *atomic radius* of a metal atom is determined from the crystal type and the edge length of a unit cell, which are both obtained by x-ray diffraction (see Major Technique 3 in text).

  **Example:** The *density* of lead is 11.34 g·cm$^{-3}$ and the *atomic radius* is 175 pm. Assuming the unit cell is *cubic,* we can determine the type of Bravais lattice. Using the equations given above, calculate the density of Pb assuming pc, bcc or fcc Bravais lattices. Results are: pc, 8.02 g·cm$^{-3}$; bcc, 10.4 g·cm$^{-3}$; fcc,11.3 g·cm$^{-3}$. The Bravais lattice is most likely fcc.

## 5.11 Ionic Structures

- **Model** [$r$(tetrahedral hole) $<$ $r$(octahedral hole) $<$ $r$(cubic hole)]
  - → Spheres of *different* radii and *opposite* charges representing cations and anions
  - → Larger spheres (*usually* anions) *customarily* occupy the unit cell lattice points. Smaller spheres (*usually* cations) fill *holes* in the unit cell.

- **Radius ratio** [use rule with caution, there are many exceptions]
  - → Ratio of radius of the *smaller* sphere to the radius of the *larger* sphere
  - → Defined for ions of the *same* charge number [$z_{cation} = |z_{anion}|$]

<table>
<tr><td colspan="2" align="center">Radius Ratio–Crystal Structure Correlations</td></tr>
<tr><td>radius ratio $\leq$ 0.414:</td><td>tetrahedral *holes* fill (fcc for *larger* spheres)<br>[*zinc-blende structure* (one form of ZnS)]</td></tr>
<tr><td>0.414 $<$ radius ratio $<$ 0.732:</td><td>octahedral *holes* fill (fcc for *larger* spheres)<br>[*rock-salt structure* (NaCl)]</td></tr>
<tr><td>radius ratio $\geq$ 0.732:</td><td>cubic *holes* fill (pc for *larger* spheres)<br>[*cesium-chloride structure* (CsCl)]</td></tr>
</table>

- **Coordination number** → Number of nearest-neighbor ions of *opposite* charge
- **Coordination of ionic solid** → Represented as (cation coordination number, anion coordination number)
- **Properties of the *ionic structure* of unit cells**

  **Zinc blende**
  → Radius ratio (ZnS) = $r(Zn^{2+})/r(S^{2-})$ = (74 pm)/(184 pm) = 0.40
  → $S^{2-}$ ions at the *corners* and *faces* of a fcc unit cell
     [*four* anions (atomic radius $r_{anion}$) per unit cell (edge length $a$)]
  → $Zn^{2+}$ ions in *four* of the *eight* tetrahedral holes in the unit cell
     [*four* cations (atomic radius $r_{cation}$) per unit cell (edge length $a$)]
  → Cation and anion spheres *touch* in the tetrahedral locations.
     Each cation has *four* nearest-neighbor anions, and each anion has *four* nearest-neighbor cations: (4,4)-coordination.

  **Rock salt**
  → Radius ratio (NaCl) = $r(Na^+)/r(Cl^-)$ = (102 pm)/(181 pm) = 0.564
  → $Cl^-$ ions at the *corners* and *faces* of a fcc unit cell
     [*four* anions (atomic radius $r_{anion}$) per cell (edge length $a$)]
  → $Na^+$ ions in all *four* octahedral holes in the unit cell
     [*four* cations (atomic radius $r_{cation}$) per cell (edge length $a$)]
  → Cation and anion spheres *touch* along the cell edges.
     Each cation has *six* nearest-neighbor anions, and each anion has *six* nearest-neighbor cations: (6,6)-coordination.
  → Spheres *touch* along the *edge:* $a = 2\,r_{cation} + 2\,r_{anion} = 566$ pm
  → Volume of the unit cell: $V = a^3 = 1.81 \times 10^8$ pm$^3$ = $1.81 \times 10^{-22}$ cm$^3$
  → Density of NaCl: $M$(58.44 g·mol$^{-1}$) and $d$(experimental) = 2.17 g·cm$^{-3}$

  $$d = \frac{(4\ \text{NaCl per cell})\,(M/N_A)}{a^3} = \frac{4\,(58.44/N_A)}{(1.81\times10^{-22})} = 2.14\,\text{g·cm}^{-3}$$

  **Cesium chloride**
  → Radius ratio (CsCl) = $r(Cs^+)/r(Cl^-)$ = (170 pm)/(181 pm) = 0.939
  → $Cl^-$ ions at the *corners* of a pc (*primitive cubic*) unit cell
     [*one* anion (radius $r_{anion}$) per cell (edge length $a$)]
  → $Cs^+$ ion at the *center* of the *cube* in the unit cell
     [*one* cation (radius $r_{cation}$) per cell (edge length $a$)]
  → Cation and anion spheres *touch* along the *body diagonal* of the cell.
     Each cation has *eight* nearest-neighbor anions and each anion has *eight* nearest-neighbor cations: (8,8)-coordination.
  → Spheres *touch* along the *body diagonal:* $\sqrt{3}\,a = 2\,r_{cation} + 2\,r_{anion} = 702$ pm   and   $a = 405$ pm
  → Volume of the unit cell: $V = a^3 = 6.64 \times 10^7$ pm$^3$ = $6.64 \times 10^{-23}$ cm$^3$
  → Density of CsCl: $M$(168.36 g·mol$^{-1}$) and $d$(experimental) = 3.99 g·cm$^{-3}$

  $$d = \frac{(1\ \text{CsCl per cell})\,(M/N_A)}{a^3} = \frac{1\,(168.36/N_A)}{(6.64\times10^{-23})} = 4.21\,\text{g·cm}^{-3}$$

- **Calculations**
  - → Determining the *ionic structure* in the *cubic* system from *ionic radii* using the *radius-ratio rule*
  - → Determining the *density* of the *ionic solid* from the *density* of the *unit cell*
    **Note:** Usually within 10% of the experimental value

### 5.12 Molecular Solids

- → Solid structures that reflect the nonspherical nature of their molecules and the relatively weak intermolecular forces that hold them together
- → Characterized by low melting temperatures and less hardness than ionic solids

### 5.13 Network Solids

- **Crystals**
  - → Atoms joined to neighbors by *strong* covalent bonds that form a network extending throughout the solid
  - → Characterized by high melting and boiling temperatures, and hard and rigid structures

- **Elemental network solids**
  - → Network solids formed from one element only, such as the allotropes graphite and diamond in which the carbon atoms are connected differently
  - → *Allotropes* are forms of an element with different solid structures.

- **Ceramics**
  - → Usually oxides with a network structure having great strength and stability, because covalent bonds must break to deform the crystal.
  - → Tend to shatter rather than bend under stress
    **Examples:** Quartz, silicates, and high-temperature superconductors

## The Impact on Materials  (Sections 5.14–5.16)

### 5.14 The Properties of Solids

- **Mobility of electrons**
  - → Responsible for characteristic luster and light reflectivity properties of a metal
  - → Accounts for *malleability, ductility,* and *electrical conductivity* of metals
  - → Produces a relatively strong metallic bond that leads to high melting points for most metals

- **Slip planes**
  - → Plane of atoms that, under stress, may slip or slide along an adjacent plane
  - → Characteristic of *close-packed* structures
  - → *Cubic close-packed* structures (ccp) have eight sets of slip planes in different directions.
  - → *Hexagonal close-packed* structures (hcp) have only one set of slip planes.
  - → Large number of slip planes accounts for *malleability* of ccp metallic crystals (may be bent, flattened, or pounded into different shapes). **Examples:** Coinage metals Cu, Ag, and Au
  - → Metallic crystals with the hcp structure tend to be brittle (only one slip plane). **Examples:** Zn, Cd

### 5.15 Alloys

- **Heterogeneous** → Mixture of crystalline phases; various samples have different compositions; for example, solders and mercury amalgams

- **Homogeneous** → Atoms of different elements are distributed uniformly; for example, brass, bronze, and coinage metals.

- **Substitutional**
  - → Atoms nearly the same size (< 15% difference) can substitute for each other more or less freely (mainly *d*-block elements)
  - → Lattice is distorted and electron flow is hindered.
  - → Alloy has lower thermal and electrical conductivity than pure element, but is harder and stronger.
  - → Homogeneous examples include the Cu-Zn alloy, brass (up to 40% Zn in Cu), and bronze (metal other than Zn or Ni in copper: casting bronze is 10% Sn and 5% Pb).

- **Interstitial**
  - → Small atoms (> 60% smaller than host atoms) can occupy holes or *interstices* in a lattice.
  - → Interstitial atoms interfere with electrical conductivity and the movement of atoms.
  - → Restricted motion makes the alloy harder and stronger than the host metal.
  - → Examples include low-carbon steel (C in Fe, which is soft enough to be stamped), high-carbon steel (hard and brittle unless subject to heat treatment), and stainless steel (a mixture of Fe with metals such as Cr and Ni, which aid in its resistance to corrosion).

## 5.16 Liquid Crystals

- → Substances that flow like viscous liquids, but molecules form a moderately ordered array similar to that in a crystal.
- → Examples of a *mesophase,* an intermediate state of matter with the fluid properties of a liquid and some molecular ordering similar to a crystal

- **Isotropic material** → Properties independent of the direction of measurement.
  Ordinary liquids are isotropic with viscosity values equal in every direction.

- **Anisotropic material** → Properties depend on the direction of measurement
  Certain rod-shaped molecules form liquid crystals, in which molecules are free to slide past one another along their axes but resist motion perpendicular to that direction.

- **Classes of liquid crystals** → Differ in the arrangement of the molecules
  See text Figs. 5.49 and 5.50.

  **nematic phase:** Molecules lie together in the same direction but are *staggered.*

  **smectic phase:** Molecules lie together in the same direction in *layers.*

  **cholesteric phase:** Molecules form *nematiclike layers*, but the molecules of neighboring layers are *rotated* with respect to each other. The resulting liquid crystal has a *helical* arrangement of molecules.

- **Thermotropic liquid crystals**
  - → Made by melting solid phase
  - → Exist over small temperature range between solid and liquid
    **Example:** *p*-azoxyanisole

- **Lyotropic liquid crystals**
  - → Layered structures produced by the action of a solvent on a solid or a liquid.
  - → Examples include cell membranes and aqueous solutions of detergents and lipids (fats).

# Chapter 6  Thermodynamics:
# The First Law

## Systems, States, and Energy  (Sections 6.1–6.8)

### Overview

- **Thermodynamics** → Branch of science concerned with the relationship between heat and other forms of energy

- **Laws of thermodynamics**
  - → Generalizations based on experience with *bulk* matter
  - → Not derivable, but understandable without recourse to knowledge of the behavior of atoms and molecules

- **First law of thermodynamics**
  - → Consequence of the **law of conservation of energy**
  - → Quantitative description of energy changes in physical (*e.g.*, phase changes) and chemical (*e.g.*, reactions) processes
  - → Energy changes in relation to heat and work:
    A paddlewheel stirrer immersed in a liquid produces heat from mechanical work (temperature rise in the liquid).
    A hot gas expanding in an insulated cylinder coupled to a flywheel produces mechanical work from heat (temperature drop in the gas).

- **Second law of thermodynamics**
  - → Criterion for spontaneity (see Chapter 7)
  - → Explains why some chemical reactions occur spontaneously and others do not

- **Statistical thermodynamics**
  - → Laws of thermodynamics reflected in the behavior of large numbers of atoms or molecules in a sample
  - → Link between the atomic level and the behavior of bulk matter

- **Chapter goals** → Gain insight into heat and work, and their relationship to energy changes in physical and chemical processes

### 6.1  Systems

- **The "universe" or "world"** → Composed of a system and its surroundings

- **System** → Portion of the universe of interest
  Examples include a beaker of ethanol, a frozen pond, 20 g of $CaCl_2(s)$, and the earth itself.

- **Surroundings**
  - → Remainder of the universe (everything that is *not* in the system)
  - → Where observations and measurements of the system are made
    For example, the observation of heat transferred to or from a system is made in the surroundings.

- **Boundary** → Dividing surface between the system and the surroundings
- **Types of systems**

  **Open system**
  - → Both *matter* and *energy* can be exchanged between the system and surroundings.
  - → An example is a half-liter of water in an open beaker. Matter can cross the boundary (water evaporates). Energy can cross the boundary (heat from the surroundings may enter the system).

  **Closed system**
  - → *Energy* can be exchanged between the system and surroundings, but *matter* cannot.
  - → An example is the refrigerant in an air-conditioning system. As the refrigerant expands, energy (heat) is withdrawn from the space to be cooled. As the refrigerant is compressed, energy (heat) is supplied to another space that is warmed. The mass of the refrigerant is unchanged.

  **Isolated system**
  - → Neither *matter* nor *energy* can be exchanged between the system and surroundings.
  - → An example of an *approximately* isolated system is a substance in a well-insulated container, such as ice in a Styrofoam ice-chest or coffee in a covered Styrofoam cup.

## 6.2 Work and Energy

- **Work, $w$**
  - → Fundamental thermodynamic property (see Table 6.1 in the text *Varieties of Work*)
  - → Movement against an opposing force
  - → Definition: work = force × distance
  - → Units: (force) *newton*, N; $1 \text{ N} = 1 \text{ kg·m·s}^{-2}$
    (work) *joule*, J; $1 \text{ J} = 1 \text{ N·m} = 1 \text{ kg·m}^2\text{·s}^{-2}$

- **Internal energy, $U$**
  - → Internal energy is the *total* capacity of a system to do work.
  - → *Only* a change in $U$ ($\Delta U = U_{final} - U_{initial}$) is measurable.
  - → An extensive property (see Section A in the text)
  - → For energy transferred to a system by doing work on the system, $\Delta U = w$

## 6.3 Expansion Work

- **Nonexpansion work** → Does not involve a change in volume
  Examples include electrical current and change of position.

- **Expansion work** → Change in volume of the system against an external pressure
  An example includes inflating a tire or balloon.

- **Calculating expansion work with a constant external pressure, $P_{ex}$**
  - → Consider a gas confined in a volume, $V$, by a piston with surface area, $A$.
  - → Work = force × distance = $F \times d = (P_{ex} \times A) \times d = P_{ex}\Delta V$ (see Section 4.2), where $d$ is the distance the piston is displaced (*not* the density) .
  - → $\boxed{w = -P_{ex}\Delta V}$  By convention, work done *on* the system is positive (+) in sign.
    Units:  $1 \text{ Pa·m}^3 = 1 \text{ kg·m}^{-1}\text{·s}^{-2} \times 1 \text{ m}^3 = 1 \text{ kg·m}^2\text{·s}^{-2} = 1 \text{ J}$
    $1 \text{ L·atm} = 10^{-3} \text{ m}^3 \times 101\,325 \text{ Pa} = 101.325 \text{ Pa·m}^3 = 101.325 \text{ J (exactly)}$

- **Free Expansion** $\rightarrow$ Expansion into a vacuum ($P_{ex} = 0$ and $w = 0$)

- **Reversible process**
  - $\rightarrow$ A process that can be reversed by an *infinitesimal* change in a variable
  - $\rightarrow$ For a change in pressure, the external pressure must always be infinitesimally different from the pressure of the system. A net change is effected by a series of infinitesimal changes in the external pressure followed by infinitesimal adjustments of the system.
  - $\rightarrow$ For a change in temperature, the temperature of the system must change by a series of infinitesimal amounts of either heat flow or of work done.
  - $\rightarrow$ Processes in which change occurs by *finite* amounts are *irreversible* in nature.

- **Reversible, isothermal expansion or compression of an ideal gas**
  - $\rightarrow$
  $$w = -nRT \ln \frac{V_{final}}{V_{initial}} = -nRT \ln \frac{P_{initial}}{P_{final}}$$
  - $\rightarrow$ The change in pressure is carried out in infinitesimal steps (text derivation in Section 6.3).

## 6.4 Heat

- **Heat, $q$**
  - $\rightarrow$ Fundamental thermodynamic property
  - $\rightarrow$ Energy transferred as a result of a temperature difference, $\Delta T = T_2 - T_1 > 0$
  - $\rightarrow$ Energy flows as *heat* from a region of high temperature, $T_2$, to one of low temperature, $T_1$.
  - $\rightarrow$ SI unit: *joule*, $1\ J = 1\ N{\cdot}m = 1\ kg{\cdot}m^2{\cdot}s^{-2}$
    Common unit:    The *calorie*, $1\ cal \equiv 4.184\ J$, is widely used in biochemistry, organic chemistry, and related fields. The *nutritional calorie*, Cal, is 1 kcal.

- **Sign of $q$ defined in terms of the system**
  - $\rightarrow$ If heat flows from the surroundings into the system, $q > 0$ (*endothermic* process).
  - $\rightarrow$ If heat flows from the system into the surroundings, $q < 0$ (*exothermic* process).
  - $\rightarrow$ If no heat flows between the system and surroundings, $q = 0$ (*adiabatic* process). For example, a thermally insulating (adiabatic) wall (Styrofoam or vacuum flask) prevents the passage of energy as heat.
  - $\rightarrow$ *Diathermic* (nonadiabatic) walls permit the transfer of energy as heat.

- **Internal energy, $U$**
  - $\rightarrow$ For energy transferred to a system by the flow of heat, $\Delta U = q$ ($w = 0$).
  - $\rightarrow$ If the system is initially at temperature $T_1$ and $\Delta T = T_2 - T_1 > 0$, then $q > 0$ and $\Delta U > 0$.

## 6.5 The Measurement of Heat

- **Heat capacity $C$ of a pure substance**
  - $\rightarrow$ Heat capacity of a pure substance is an *extensive* property.
  - $\rightarrow$ Amount of heat absorbed *by a sample of the substance* per degree Celsius rise in temperature

- **Specific heat capacity, $C_s$, of a pure substance**
  - $\rightarrow$ Specific heat capacity of a pure substance is an *intensive* property.
  - $\rightarrow$ Amount of heat absorbed *by one gram* of a substance per degree Celsius rise in temperature (values for common materials are given in Table 6.2 in the text)
  - $\rightarrow$ $C_s = C/m$   Units: $J{\cdot}(°C)^{-1}{\cdot}g^{-1}$ or $J{\cdot}K^{-1}{\cdot}g^{-1}$ (commonly, the *former*)

$\rightarrow$ $\boxed{q = C\Delta T = mC_\text{s}\Delta T}$

- **Molar heat capacity, $C_\text{m}$, of a pure substance**
  - $\rightarrow$ Molar heat capacity of a pure substance is an *intensive* property.
  - $\rightarrow$ Amount of heat absorbed *by one mole* of a substance per degree Celsius rise in temperature (values for some substances are given in Table 6.2 in the text)
  - $\rightarrow$ $C_\text{m} = C/n$    Units: $\text{J}\cdot(°\text{C})^{-1}\cdot\text{mol}^{-1}$ or $\text{J}\cdot\text{K}^{-1}\cdot\text{mol}^{-1}$ (commonly, the *latter*)
  - $\rightarrow$ $\boxed{q = C\Delta T = nC_\text{m}\Delta T}$
  - $\rightarrow$ Values of molar heat capacities increase with increasing molecular complexity. They depend on the temperature and the state of the substance. $C_\text{m}(\text{liquid}) > C_\text{m}(\text{solid})$

- **Calorimeter**
  - $\rightarrow$ Device used to monitor or measure heat transfer in a system by the temperature changes that take place in the surroundings
  - $\rightarrow$ In calorimetry, $q_\text{cal} = q_\text{surr}$
  - $\rightarrow$ For an exothermic process, $q_\text{cal} > 0$; the temperature of the calorimeter increases, $\Delta T > 0$.
  - $\rightarrow$ For an endothermic process, $q_\text{cal} < 0$; the temperature of the calorimeter decreases, $\Delta T < 0$.
  - $\rightarrow$ The heat capacity of a calorimeter, $C_\text{cal}$, is the amount of heat absorbed *by the calorimeter* per degree Celsius rise in temperature.
  - $\rightarrow$ $\boxed{q_\text{cal} = C_\text{cal}\Delta T}$
  - $\rightarrow$ A calorimeter has a heat capacity that is determined experimentally by the addition of a known amount of heat supplied electrically and the measurement of the resulting temperature rise in the calorimeter (see Example 6.3 in the text).
  
  Units of heat capacity: $\text{J}\cdot(°\text{C})^{-1}$ or $\text{J}\cdot\text{K}^{-1}$

## 6.6 The First Law

- **Combining heat and work**
  - $\rightarrow$ $\boxed{\Delta U = q + w}$
  - $\rightarrow$ Change in internal energy of a system is the *sum* of the heat *added* ($q > 0$) *to* or *removed* ($q < 0$) *from* the system *plus* the work done *on* ($w > 0$) or *by* ($w < 0$) the system.

- **First law of thermodynamics**
  - $\rightarrow$ The internal energy of an *isolated* system is constant.
  - $\rightarrow$ It is an extension of the law of conservation of energy.
  - $\rightarrow$ It is a generalization based on experience and cannot be proven.
    
    If the first law were false, one could construct a "perpetual motion machine" by starting with an isolated system, removing it from isolation, and letting it do work on the surroundings. It could then be placed in isolation again, and its internal energy allowed to return to the initial value. This feat has never been accomplished despite many attempts.

- **Constant volume process, $\Delta V = 0$**
  - $\rightarrow$ No expansion work is done
  - $\rightarrow$ $\Delta U = q$    (no work of any kind is done)
  - $\rightarrow$ Heat absorbed or released by a system equals the change in internal energy for a constant volume process if no other forms of work are done: $\Delta U = q$.

## 6.7  State Functions

- **State of a system** $\rightarrow$ Defined when all of its properties are fixed
- **State function**
    - $\rightarrow$ Property that depends only on the current state of the system
    - $\rightarrow$ Independent of the manner in which the state was prepared
      Examples of state functions include temperature, $T$; pressure, $P$; volume, $V$; and internal energy, $U$.  (Heat, $q$, and work, $w$, are *not* state functions.)
- **Path**
    - $\rightarrow$ Sequence of intermediate steps linking an initial and a final state
    - $\rightarrow$ Consider the transition from state 1 to state 2.  The changes in the state functions are independent of path.  The quantities $\Delta U$, $\Delta P$, $\Delta V$, and $\Delta T$ are uniquely determined.  The amount of heat transferred or the work done depends on the sequence of steps followed.

## 6.8  A Molecular Interlude: The Origin of Internal Energy

- **Internal energy, $U$** $\rightarrow$ Energy stored as potential and kinetic energy of molecules
- **Potential energy** $\rightarrow$ Energy an object has by virtue of its position
- **Kinetic energy**
    - $\rightarrow$ Energy an object has by virtue of its motion
    - $\rightarrow$ For atoms, kinetic energy is *translational* in nature.
    - $\rightarrow$ For molecules, kinetic energy has *translational, rotational,* and *vibrational* components.
      *Translation* is the motion of the molecule as a *whole*.
      (See Section 1.5 in the text for the particle-in-a-box translational energy levels.)
      *Rotation* and *vibration* are motions *within* the molecule, and do not change the center of gravity (mass) of the molecule.
      (See Box 2.2 in the text for the rigid-rotor rotational energy levels.)
      (See Major Technique 1 in the text for a description of vibrational motion.)

Translational, Rotational, and Vibrational Motion of $F_2$

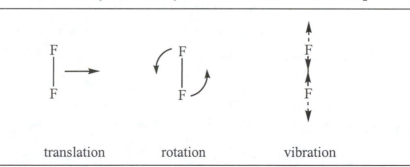

translation          rotation          vibration

- **Degrees of freedom**
    - $\rightarrow$ Modes of motion  (translational, rotational, and vibrational)
    - $\rightarrow$ *Mainly* translational and rotational degrees of freedom store internal energy at room temperature
- **Equipartition theorem**
    - $\rightarrow$ *Average* kinetic energy of each degree of freedom of a molecule in a sample at temperature $T$ is equal to $\frac{1}{2}kT$.  The Boltzmann constant $k = 1.380\,658 \times 10^{-23}$ J·K$^{-1}$

→ A molecule moving in *three* dimensions has *three* translational degrees of freedom.

$U(\text{translation}) = 3 \times \frac{1}{2}kT = \frac{3}{2}kT$

For 1 mol of molecules, $U_m(\text{translation}) = N_A(\frac{3}{2}kT) = \frac{3}{2}RT$   $(R = N_A k)$

→ A linear molecule has *two* rotational degrees of freedom.

$U(\text{rotation, linear}) = 2 \times \frac{1}{2}kT = kT$

For 1 mol of molecules, $U_m(\text{rotation, linear}) = N_A(kT) = RT$

→ A nonlinear molecule has *three* rotational degrees of freedom.

$U(\text{rotation, nonlinear}) = 3 \times \frac{1}{2}kT = \frac{3}{2}kT$

For 1 mol of molecules, $U_m(\text{rotation, nonlinear}) = N_A(\frac{3}{2}kT) = \frac{3}{2}RT$

→ A molecule of an *ideal gas* does *not* interact with its neighbors and the potential energy is 0. The internal energy is independent of volume and depends *only* on temperature, $U_m(T)$.

→ A molecule of a liquid, solid, or real gas does interact with its neighbors. The potential energy is an important component of the internal energy, which has a significant dependence on volume. The internal energy depends on both temperature and volume, $U_m(T,V)$.

---

# Enthalpy  (Sections 6.9–6.13)

## Overview

- **Heat transfer at constant volume, $q_V$ (notation not used in the text)**

  → $\boxed{\Delta U = q_V}$   for a system in which only expansion work is possible

  → Combustion reactions are studied in a bomb calorimeter of constant volume.

- **Heat transfer at constant pressure, $q_P$ (notation not used in the text)**

  → $\boxed{\Delta H = q_P}$   a new state function (*enthalpy*) for a system at constant pressure

  → Most chemical reactions occur at constant pressure largely because reaction vessels are usually open to the atmosphere.

## 6.9  Heat Transfers at Constant Pressure

- **Definition of enthalpy, $H$**

  → $\boxed{H = U + PV}$   a state function because $U$, $P$, and $V$ are state functions

  → Change in enthalpy of a system, $\Delta H = \Delta U + \Delta(PV)$
  At constant pressure, $\Delta H = \Delta U + P\Delta V = q_P + w + P\Delta V = q_P - P_{ex}\Delta V + P\Delta V = q_P$
  for a system open to the atmosphere, and for which only expansion work may occur

- **Heat change for a constant pressure process**

  → An *exothermic* process is one in which heat is released by the system into the surroundings, $\Delta H = q_P < 0$.

→ An *endothermic* process is one in which heat is absorbed by the system from the surroundings, $\Delta H = q_P > 0$.

## 6.10 Heat Capacities at Constant Volume and Constant Pressure

- **General definition of heat capacity** → $\boxed{C = \dfrac{q}{\Delta T}}$

  **At constant volume** → $\boxed{C_V = \dfrac{q_V}{\Delta T} = \dfrac{\Delta U}{\Delta T}}$　**At constant pressure** → $\boxed{C_P = \dfrac{q_P}{\Delta T} = \dfrac{\Delta H}{\Delta T}}$

- **Molar heat capacities**
  → $C_{V,m} = C_V/n$　and　$C_{P,m} = C_P/n$
  → $C_{V,m} \approx C_{P,m}$ *liquids and solids*, but $C_{P,m} > C_{V,m}$ *gases*

- **Molar heat capacity relationship for an ideal gas (see derivation in text)**
  → $C_P = C_V + nR$　*any amount of an ideal gas*
  → $\boxed{C_{P,m} = C_{V,m} + R}$　*one mole of an ideal gas*

## 6.11 A Molecular Interlude: The Origin of the Heat Capacities of Gases

- **Contributions to the molar heat capacity for a monatomic ideal gas**
  → Consider only translational motion.
  → $U_m = \dfrac{3}{2}RT$　and　$\Delta U_m = \dfrac{3}{2}R\Delta T$　(molar internal energy depends only on $T$)

  → $\boxed{C_{V,m} = \dfrac{\Delta U_m}{\Delta T} = \dfrac{\frac{3}{2}R\Delta T}{\Delta T} = \dfrac{3}{2}R}$ $= 12.471\,77$ J·K$^{-1}$·mol$^{-1}$　and　$\boxed{C_{P,m} = \dfrac{\Delta H_m}{\Delta T} = \dfrac{5}{2}R}$

- **Contributions to the molar heat capacity for a linear-molecule ideal gas**
  → Consider translational motion and 2 degrees of freedom in rotational motion.
  → $U_m = \dfrac{5}{2}RT$　and　$\Delta U_m = \dfrac{5}{2}R\Delta T$　(molar internal energy depends only on $T$)

  → $\boxed{C_{V,m} = \dfrac{\Delta U_m}{\Delta T} = \dfrac{\frac{5}{2}R\Delta T}{\Delta T} = \dfrac{5}{2}R}$ $= 20.786\,28$ J·K$^{-1}$·mol$^{-1}$　and　$\boxed{C_{P,m} = \dfrac{\Delta H_m}{\Delta T} = \dfrac{7}{2}R}$

- **Contributions to the molar heat capacity for a nonlinear-molecule ideal gas**
  → Consider translational motion and 3 degrees of freedom in rotational motion.
  → $U_m = 3RT$　and　$\Delta U_m = 3R\Delta T$　(molar internal energy depends only on $T$)

  → $\boxed{C_{V,m} = \dfrac{\Delta U_m}{\Delta T} = \dfrac{3R\Delta T}{\Delta T} = 3R}$ $= 24.943\,53$ J·K$^{-1}$·mol$^{-1}$　and　$\boxed{C_{P,m} = \dfrac{\Delta H_m}{\Delta T} = 4R}$

**Note:** The contribution from molecular rotation is for room temperature. At lower temperatures, this contribution diminishes, and the heat capacity approaches the value for a monatomic gas. At room temperature, a small contribution to the heat capacity from molecular vibrational motion leads to a value slightly larger than the ones given above.

## 6.12 The Enthalpy of Physical Change

- **Physical changes (phase transitions) at constant temperature and pressure**

  → Vaporization:    liquid → vapor    $\Delta H_{vap} = H_{vapor,m} - H_{liquid,m}$    $> 0$ *(endothermic)*
      Condensation:   vapor → liquid    $\Delta H_{cond} = H_{liquid,m} - H_{vapor,m}$    $< 0$ *(exothermic)*

  → Fusion (melting):   solid → liquid    $\Delta H_{fus} = H_{liquid,m} - H_{solid,m}$    $> 0$ *(endothermic)*
      Freezing:         liquid → solid    $\Delta H_{freez} = H_{solid,m} - H_{liquid,m}$    $< 0$ *(exothermic)*

  → Sublimation:    solid → vapor    $\Delta H_{sub} = H_{vapor,m} - H_{solid,m}$    $> 0$ *(endothermic)*
      Deposition:     vapor → solid    $\Delta H_{dep} = H_{solid,m} - H_{vapor,m}$    $< 0$ *(exothermic)*

  → $\Delta H_{vap} = -\Delta H_{cond}$      $\Delta H_{fus} = -\Delta H_{freez}$      $\Delta H_{sub} = -\Delta H_{dep}$

  → $\Delta H_{sub} = \Delta H_{fus} + \Delta H_{vap}$   (for the same $T$ and $P$)
  Values of $\Delta H_{fus}°$ and $\Delta H_{vap}°$ for selected substances are given in Table 6.3 in the text.
  Liquids may evaporate at any temperature, but they boil only at the temperature at which the vapor
  pressure of the liquid is equal to the external pressure of the atmosphere.

## 6.13 Heating Curves

- **Heating (cooling) curve**

  → The heating (cooling) curve is a graph showing the variation of the temperature, $T$, of a
  sample as heat is added (removed) at a constant rate.

  → Two types of behavior are seen in a heating curve. The temperature increase of a single phase to
  which heat is supplied has a positive slope that depends on the value of the heat capacity. The
  larger the heat capacity, the more gentle the slope. At the temperature of a phase transition, two
  phases are present and the slope is 0 (no temperature change). The *greater* the heat associated
  with the phase change, the *longer* the length of the flat line. As more heat is added to the
  substance, the relative amounts of the two phases change. The temperature does not rise again
  until only one phase remains.

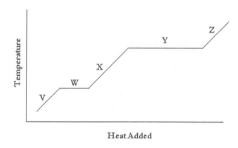

In this schematic heating curve, line V with positive slope
represents the heating of a solid phase. The flat line W shows
the phase transition between solid and liquid (fusion). Line X
represents the heating of the liquid after the solid has
completely melted. Line Y shows the phase transition between
liquid and vapor (boiling at constant external pressure).
Finally, line Z represents the heating of the vapor after the
liquid disappears. Because the enthalpy of vaporization is
greater than the enthalpy of fusion, the length of line Y is longer
than line W.

# The Enthalpy of Chemical Change (Sections 6.14–6.22)

## 6.14 Reaction Enthalpies

- **Thermochemical equation**

  → A chemical equation with a corresponding enthalpy change, $\Delta H$, expressed in kJ for the
  *stoichiometric* number of moles of each reactant and product

→ Applying the principles of thermodynamics to chemical change

- **Enthalpy of reaction, $\Delta H_r$**
  → The enthalpy change in a thermochemical reaction expressed in kJ·mol$^{-1}$, where "mol" refers to the number of moles of each substance as determined by the stoichiometric coefficient in the balanced equation
  → Combustion is reaction with oxygen, $O_2(g)$.

## 6.15 The Relation Between $\Delta H$ and $\Delta U$

- **Reactions with liquids and solids only**
  → Recall that $\Delta H = \Delta U + \Delta(PV)$
  → $\Delta H \approx \Delta U \implies q_P \approx q_V$
  → Only a small change in the volume of reactants compared to products at constant $P$

- **Reactions with ideal gases only**
  → $\Delta H = \Delta U + \Delta(PV) = \Delta U + \Delta(nRT) = \Delta U + (\Delta n)RT$     *constant temperature*
  → The equation is exact for ideal gases only.

- **Reactions with liquids, solids, and gases**
  → $\Delta H \approx \Delta U + (\Delta n_{gas})RT$     *constant temperature*
  → $\Delta n_{gas} = n_{final} - n_{initial}$ = change in the number of moles of gas between products and reactants
  → The equation is a very good approximation. The value of $(\Delta n_{gas})RT$ is usually much smaller than $\Delta U$ and $\Delta H$.

## 6.16 Standard Reaction Enthalpies

- **Standard conditions**
  → Values of reaction enthalpy, $\Delta H_r$, depend on pressure, temperature, and the physical state of each reactant and product.
  → A standard set of conditions is used to report reaction enthalpies.

- **Standard state**
  → Standard state of a substance is its pure form at a pressure of *exactly* 1 bar.
  → For a solute in a liquid solution, the standard state is a solute concentration of 1 mol·L$^{-1}$ at a pressure of *exactly* 1 bar.
  → Standard state of liquid water is pure water at 1 bar. Standard state of ice is pure ice at 1 bar. Standard state of water vapor is pure water vapor at 1 bar.

  **Note:** Compare the definition of standard state in thermodynamics to the definition of standard temperature and pressure, STP, for gases (Section 4.8). For gases, standard temperature is 0°C (273.15 K) and standard pressure is 1 atm (1.013 25 bar).

- **Standard reaction enthalpy, $\Delta H_r°$**
  → Reaction enthalpy when reactants in their standard states form products in their standard states
  → Degree symbol, °, is added to the reaction enthalpy.
  Reaction enthalpy has a *weak* pressure dependence, so it is a very good approximation to use standard enthalpy values even when the pressure is *not* exactly 1 bar.

- **Temperature convention**
  → Most thermochemical data are reported for 298.15 K (25°C).
  → Temperature convention is *not* part of the definition of the standard state.

### 6.17 Combining Reaction Enthalpies: Hess's Law

- **Hess's law**
  - → Overall reaction enthalpy is the sum of the reaction enthalpies of the steps into which a reaction can be divided.
  - → Consequence of the fact that the reaction enthalpy is path independent and depends only on the initial reactants and final products
  - → Used to calculate enthalpy changes for reactions that are difficult to carry out in the laboratory
  - → Used to calculate reaction enthalpies for unknown reactions

- **Using Hess's law**
  - → Write the thermochemical equation for the reaction whose reaction enthalpy is unknown.
  - → Write thermochemical equations for intermediate reaction steps whose reaction enthalpies are known. These steps must sum up to the overall reaction whose enthalpy is unknown.
  - → Sum the reaction enthalpies of the intermediate reaction steps to obtain the unknown reaction enthalpy (see Toolbox 6.1 in the text).

### 6.18 The Heat Output of Reactions

- **Heat output in a reaction**
  - → Heat is treated as a reactant or product in a stoichiometric relation.
  - → In an endothermic reaction, heat is treated as a reactant. In an exothermic reaction, heat is treated as a product.

- **Standard enthalpy of combustion, $\Delta H_c°$**
  - → Change in enthalpy per mole of a substance that is burned in a combustion reaction under standard conditions (see values in Table 6.4 in the text)
  - → Combustion of hydrocarbons is exothermic, and heat is treated as a product.

### 6.19 Standard Enthalpies of Formation

- **Treatment of chemical reactions**
  - → Possible thermochemical equations are nearly innumerable.
  - → Method to handle this problem is the use of a common reference for each substance. The reference is the *most stable* form of the elements.
  - → Standard enthalpies of formation of substances from their elements are tabulated and used to calculate standard reaction enthalpies for other types of reactions.

- **Standard enthalpy of formation, $\Delta H_f°$**
  - → Enthalpy change for the formation of one mole of a substance from the most stable form of its elements under standard conditions
  - → Elements (most stable form) → substance (one mole)      $\Delta H° = \Delta H_f°$
    For example,  $C(s) + 2 H_2(g) + \frac{1}{2}O_2(g) \rightarrow CH_3OH(l)$      $\Delta H_f° = -238.86 \text{ kJ.mol}^{-1}$
  - → Values of $\Delta H_f°$ at 298.15 K for many substances are listed in Appendix 2A.
  - → Most stable form of the elements at 1 bar (standard state) and 298.15 K (thermodynamic temperature convention):
    - Metals and semimetals are solids with atoms arranged in a lattice (see Chapter 5). The exception is Hg(l).
    - Nonmetal solids are B(s), C(s, graphite), P(s, white), S(s, rhombic), Se(s, black), and $I_2(s)$.

Monatomic gases are He(g), Ne(g), Ar(g), Kr(g), Xe(g), and Rn(g). *Group 18*
Diatomic gases are $H_2(g)$, $N_2(g)$, $O_2(g)$, $F_2(g)$, and $Cl_2(g)$.
Nonmetal liquid is $Br_2(l)$.

**Note:** P(s, red) is actually more stable than P(s, white), but the white form is chosen because it is easier to obtain pure. Also, some of the nonmetal solids have molecular forms, for example, $P_4$ and $S_8$.

- **Standard enthalpy of reaction, $\Delta H_r°$**
  - → Use the most stable form of the elements as a reference state to obtain enthalpy changes for any reaction, if $\Delta H_f°$ values are known for each reactant and product.

  - → $\boxed{\Delta H_r° = \sum n\Delta H_f°(\text{products}) - \sum n\Delta H_f°(\text{reactants})}$

  In this expression, values of $n$ are the stoichiometric coefficients and the symbol $\sum$ (sigma) means a summation. The procedure is to convert the reactants into the most stable form of their elements $[-\sum n\Delta H_f°(\text{reactants})]$ and then recombine the elements into products $[+\sum n\Delta H_f°(\text{products})]$.
  This is an application of Hess's law. Use $\Delta H_f°$ values in Appendix 2A to obtain $\Delta H_r°$.

## 6.20 The Born–Haber Cycle

- **Ionic solids**
  - → Cations and anions are arranged in a three-dimensional lattice (see Section 5.11) and held in place mainly by coulombic interactions (see Section 2.2).
  - → Enthalpy required to separate the ions in the solid into gas phase ions is the lattice enthalpy, $\Delta H_L = H_m(\text{ions, g}) - H_m(\text{ions, s}) > 0$. See values in Table 6.6 in the text.

- **Born–Haber cycle**
  - → Thermodynamic cycle constructed to evaluate the lattice enthalpy from other reactions, in which the enthalpy changes are known
  - → Steps in the Born–Haber cycle that start and end with the most stable form of the elements in amounts appropriate to form one mole of the ionic compound
  As an example, consider $CaF_2(s)$. Start with one mole of Ca(s) and one mole of $F_2(g)$.
  Two moles of F atoms are required.
  (1) *Atomize:* Atomization of the metal element that will form the cation and of the nonmetal element that will form the anion $\Delta H(\text{atomization}) > 0$

  $Ca(s) \rightarrow Ca(g)$ $\quad \Delta H_r° = \Delta H_f°[Ca(g)] = 178 \text{ kJ·mol}^{-1}$
  $F_2(g) \rightarrow 2 F(g)$ $\quad \Delta H_r° = 2\Delta H_f°[F(g)] = 2(79 \text{ kJ·mol}^{-1}) = 158 \text{ kJ·mol}^{-1}$
  $\Delta H(\text{atomization}) = (178 + 158) \text{ kJ·mol}^{-1} = +336 \text{ kJ·mol}^{-1}$

  (2) *Ionize (cation):* Ionize the gaseous metal atom to form the gaseous cation. Several ionization steps may be required. Because electrons created in the formation of cations in the ionization steps are consumed in the formation of anions in the electron-gain steps (3), energy and enthalpy are considered to be interchangeable when treating ionization and electron gain. $\Delta H(\text{ionization, cation}) > 0$

  $Ca(g) \rightarrow Ca^+(g) + e^-(g)$ $\quad \Delta H_r° = I_1[Ca] = 590 \text{ kJ·mol}^{-1}$
  $Ca^+(g) \rightarrow Ca^{2+}(g) + e^-(g)$ $\quad \Delta H_r° = I_2[Ca] = 1145 \text{ kJ·mol}^{-1}$
  $Ca(g) \rightarrow Ca^{2+}(g) + 2 e^-(g)$ $\quad \Delta H_r° = I_1 + I_2 = (590 + 1145) \text{ kJ·mol}^{-1} = 1735 \text{ kJ·mol}^{-1}$
  $\Delta H(\text{ionization, cation}) = +1735 \text{ kJ·mol}^{-1}$

(3) *Ionize (anion):* Attach electron(s) to the gaseous nonmetal atom to form the gaseous anion. Several electron affinity values may be required. Recall that for electron gain, $\Delta H = -E_{ea}$ for each electron added.

$\Delta H$ (ionization, anion) < or > 0

$$F(g) + e^-(g) \rightarrow F^-(g) \qquad \Delta H_r° = -E_{ea}[F] = -328 \text{ kJ·mol}^{-1}$$
$$2\,F(g) + 2\,e^-(g) \rightarrow 2\,F^-(g) \qquad \Delta H_r° = -2E_{ea} = -656 \text{ kJ·mol}^{-1}$$

$\Delta H$ (ionization, anion) = $-656$ kJ·mol$^{-1}$

(4) *Latticize:* Form the lattice of ions in the solid from the gaseous ions. The enthalpy of lattice formation is the negative of the lattice enthalpy, $\Delta H_L$.

$\Delta H$ (lattice formation) < 0

$$Ca^{2+}(g) + 2\,F^-(g) \rightarrow CaF_2(s) \qquad \Delta H_r° = -\Delta H_L \qquad value\ unknown$$

$\Delta H$ (lattice formation) = $-\Delta H_L$

(5) *Elementize:* Form the most stable form of the elements from the ionic solid.

$\Delta H$ (element formation) < 0

$$CaF_2(s) \rightarrow Ca(s) + F_2(g) \qquad \Delta H_r° = -\Delta H_f°[CaF_2(s)] = 1220 \text{ kJ·mol}^{-1}$$

$\Delta H$ (element formation) = $+1220$ kJ·mol$^{-1}$

(6) *Lattice enthalpy:* The sum of all the enthalpy changes for the complete cycle is 0.

$0 = \Delta H$ (atomization) $+ \Delta H$ (ionization, cation) $+ \Delta H$ (ionization, anion) $+ \Delta H$ (lattice formation) $+ \Delta H$ (element formation)

$$0 = 336 + 1735 + (-656) + (-\Delta H_L) + 1220 \text{ kJ·mol}^{-1} = 2635 \text{ kJ·mol}^{-1} - \Delta H_L$$

$\Delta H_L = +2635$ kJ·mol$^{-1}$ for $CaF_2(s) \rightarrow Ca^{2+}(g) + 2\,F^-(g)$

- **Summary** → Strength of interactions between ions in a solid is determined by the lattice enthalpy, which is obtained from a Born–Haber cycle.

## 6.21 Bond Enthalpies

- **Bond enthalpy, $\Delta H_B$**
  → Enthalpy change accompanying the breaking of a chemical bond in the gas phase
  → Difference between the standard molar enthalpy of the fragments of a molecule and the molecule itself in the gas phase
  → Reactants and products in their standard states (pure substance at 1 bar) at 298.15 K
  → Bond enthalpies for diatomic molecules are given in Table 6.7 in the text.

- **Mean (average) bond enthalpies**
  → In polyatomic molecules, the bond strength between a pair of atoms varies from molecule to molecule.
  → Variations are not very large, and the average values of bond enthalpies are a guide to the strength of a bond in any molecule containing the bond (see Table 6.8 in the text).

- **Using mean bond enthalpies**
  → Atoms in the gas phase are the reference states used to estimate enthalpy changes for any gaseous reaction. A value of $\Delta H_B$ is required for each bond in the reactant and product molecules.

→ $$\Delta H_r^\circ \approx \sum_{\text{reactants}} n\Delta H_B(\text{bonds broken}) - \sum_{\text{products}} n\Delta H_B(\text{bonds formed})$$

In this expression (*not in the text*), values of $n$ are the stoichiometric coefficients and the symbol $\sum$ (sigma) means a summation. The procedure is to convert the reactants into gaseous atoms $[+\sum n\Delta H_B(\text{bonds broken})]$ and then recombine the atoms into products $[-\sum n\Delta H_B(\text{bonds formed})]$.

This is an application of Hess's law.

→ Compare bond enthalpy to dissociation energy in Sections 2.15 and 2.16.

## 6.22 The Variation of Reaction Enthalpy with Temperature

- **Temperature dependence of reaction enthalpy**

  → $\Delta H_r^\circ$ needs to be measured at the temperature of interest.

  → Approximation method is possible from the value of $\Delta H_r^\circ$ measured at one temperature, if heat capacity data on the reactants and products are available.

- **Kirchhoff's law**

  → The difference in molar heat capacities of the products and reactants in a chemical reaction is

  $$\Delta C_P = \sum nC_{P,m}(\text{products}) - \sum nC_{P,m}(\text{reactants})$$

  → An estimation of the reaction enthalpy at a temperature of $T_2$, if the value at a temperature of $T_1$ and the difference in molar heat capacities are known, is

  $$\Delta H_{r,2}^\circ = \Delta H_{r,1}^\circ + \Delta C_P(T_2 - T_1)$$

  → A major assumption is that the molar heat capacities are independent of temperature.

  → The law in this form does not account for any phase changes. Separate steps must be added.

# Chapter 7 Thermodynamics:
# The Second and Third Laws

## Entropy (Sections 7.1–7.8)

### 7.1 Spontaneous Change

- **The big question:** What is the *cause* of *spontaneous change?*

- **Spontaneous (or *natural* ) change** → Occurs *without* an external influence, can be fast or slow

    **Examples:** Diamond converts to graphite (*infinitesimally slow rate*). Iron rusts in air (*very slow, but noticeable change*). $2H_2(g) + O_2(g) \rightarrow 2H_2O(g)$ (*very fast reaction or explosion*)

- **Nonspontaneous change** → Can be effected by using an external influence (using energy from the surroundings to do work on the system)

    **Examples:** Liquid water can be *electrolyzed* to form hydrogen and oxygen gases. Graphite can be converted to diamond under extremely high pressures.

### 7.2 Entropy and Disorder

- **Spontaneous changes** → Any spontaneous change is accompanied by an *increase* in the *disorder* of the universe (*system* plus the *surroundings*).

- **Entropy ($S$)** → A measure of disorder; *increase* in disorder leads to an *increase* in $S$

- **Second law**
    - → For a spontaneous change, entropy of an *isolated* system *increases*.
    - → For a spontaneous change, entropy of the *universe increases*.
    [The universe is considered to be a (somewhat large) isolated system.]

- **Macroscopic definition of entropy, infinitesimal change**
$$dS = \frac{dq_{rev}}{T}$$
(See Section 7.3)

- **Finite change, isothermal process**
$$\Delta S = \frac{q_{rev}}{T}, \text{ for constant } T$$

    - → Heat transfer processes are carried out *reversibly* (temperature of surroundings *equals* that of the system) to evaluate $\Delta S$.

- **Entropy is an extensive property** → Proportional to the amount of sample

- **Entropy is a state function** → Changing the path does not change $\Delta S$.

    **Note:** According to the definition of $\Delta S$, we must calculate $\Delta S$ by using a *reversible path*. The result then applies to *any* path, because $\Delta S$ is *independent of path*.

### 7.3 Changes in Entropy

- **Entropy increases**
    - → If a substance is heated (increase in thermal disorder)
    - → If the volume of a given amount of matter increases (increase in positional disorder)

- **Temperature dependence of entropy** (*any substance*)
  - → Let $n$ = number of moles of a substance, $C_P$ = heat capacity (J·K$^{-1}$) at constant $P$, $C_{P,m}$ = molar heat capacity (J·K$^{-1}$·mol$^{-1}$) at constant $P$, $C_V$ = heat capacity at constant $V$, $C_{V,m}$ = molar heat capacity at constant $V$, $T_1$ = initial temperature, and $T_2$ = final temperature.

  - → Isobaric heating of a substance (constant $P$)

  $$\Delta S = C_P \ln\frac{T_2}{T_1} = nC_{P,m} \ln\frac{T_2}{T_1}$$

  - → Isochoric heating of a substance (constant $V$)

  $$\Delta S = C_V \ln\frac{T_2}{T_1} = nC_{V,m} \ln\frac{T_2}{T_1}$$

  - → Above relationships assume that the heat capacity is *constant* over the range of $T_1$ to $T_2$.

    **Example:** If two moles of Fe are cooled from 300 K to 200 K at constant P, the entropy change is

    $$\Delta S = nC_{P,m} \ln\frac{T_2}{T_1} = (2 \text{ mol})(25.08 \text{ J·K}^{-1}\text{·mol}^{-1}) \ln\frac{200 \text{ K}}{300 \text{ K}} = -20.3 \text{ J·K}^{-1}; \ \Delta S < 0.$$

- **Volume dependence of entropy** (*ideal gas*)
  - → $n$ = number of moles, $R$ = gas constant (8.314 51 J·K$^{-1}$·mol$^{-1}$), $V_1$ = initial volume, and $V_2$ = final volume

  - → Isothermal volume change of an *ideal gas* (constant $T$)

  $$\Delta S = \frac{q_{\text{rev}}}{T} = -\frac{w_{\text{rev}}}{T} = nR \ln\frac{V_2}{V_1}$$

  - → Recall: $\Delta U = 0$ because the energy of an ideal gas depends only on its temperature (see Section 6.3).

- **Pressure dependence of entropy** (*ideal gas*)
  - → $n$ = number of moles, $P_1$ = initial pressure, and $P_2$ = final pressure

  - → Isothermal pressure change (*ideal gas*, constant $T$), $\Delta U = 0$, $\dfrac{V_2}{V_1} = \dfrac{P_1}{P_2}$ and

  $$\Delta S = nR \ln\frac{V_2}{V_1} = nR \ln\frac{P_1}{P_2}$$

    **Example:** Three moles of an ideal gas expand from 20 L to 80 L at constant $T$.

    $$\Delta S = nR \ln\frac{V_2}{V_1} = (3 \text{ mol})(8.314 51 \text{ J·K}^{-1}\text{·mol}^{-1}) \ln\frac{80 \text{ L}}{20 \text{ L}} = 34.6 \text{ J·K}^{-1}$$

    **Example:** Four moles of an ideal gas undergo a pressure increase from 0.500 atm to 1.75 atm at constant $T$.

    $$\Delta S = nR \ln\frac{P_1}{P_2} = (4 \text{ mol})(8.314 51 \text{ J·K}^{-1}\text{·mol}^{-1}) \ln\frac{0.500 \text{ atm}}{1.75 \text{ atm}} = -41.7 \text{ J·K}^{-1}$$

## 7.4 Entropy Changes Accompanying Changes of Physical State

- **Changes of physical state**
  - → fusion ≡ fus (solid to liquid); vaporization ≡ vap (liquid to vapor); sublimation ≡ sub (solid to vapor)
  - → solid-to-solid phase changes; for example, Sn(gray) → Sn(white)

- **At the transition temperature**
  - → Temperature of substance remains *constant* during change of physical state.
  - → Transfer of heat is *reversible*.
  - → Heat supplied is identified with *enthalpy change,* because pressure is constant ($q_{rev} = \Delta H$).

- **Entropy of vaporization**
  - → Normal boiling point, $T_b$: T at which liquid boils when P = 1 atm

  - → $$\boxed{\Delta S_{vap} = \frac{\Delta H_{vap}}{T_b}}$$ $\Delta S_{vap}$ is the *entropy of vaporization* (units: $J \cdot K^{-1} \cdot mol^{-1}$ or $kJ \cdot K^{-1} \cdot mol^{-1}$).

  - → $\Delta S_{vap}°$ = *standard entropy of vaporization* (liquid and vapor both pure and both at 1 bar)

- **Trouton's rule**
  - → For many liquids, $\Delta S_{vap}° \ (liq) \approx 85 \ J \cdot K^{-1} \cdot mol^{-1}$
  - → Rationale: approximately the same increase in *positional disorder* occurs when *any* liquid is vaporized (gas molecules are far apart and moving rapidly).
  - → Exceptions: *hydrogen-bonded* liquids, for example, water, $H_2O$ ($109 \ J \cdot K^{-1} \cdot mol^{-1}$), and methanol, $CH_3OH$ ($105 \ J \cdot K^{-1} \cdot mol^{-1}$)

- **Entropy of fusion (melting)**

  - → $$\boxed{\Delta S_{fus}° = \frac{\Delta H_{fus}°}{T_f}}$$ , where $T_f$ is the *melting point*

  - → $\Delta S_{fus}°$ is the *standard entropy of fusion.*
  - → $\Delta S_{fus}°$ is smaller than $\Delta S_{vap}°$, because a liquid is only slightly more *disordered* than its solid.

    **Example:** Methane melts *reversibly* at 1 bar and $-182.5°C : CH_4(s) \rightarrow CH_4(l)$. Under these conditions, the enthalpy of fusion is $+0.936 \ kJ \cdot mol^{-1}$. The standard (°) molar enthalpy of fusion is given by $\Delta S_{fus}° = \dfrac{\Delta H_{fus}°}{T_f} = \dfrac{936 \ J \cdot mol^{-1}}{90.65 \ K} = 10.3 \ J \cdot K^{-1} \cdot mol^{-1}$

    Because $\Delta S > 0$ for *melting*, the final state is more disordered than the initial state, as expected.

## 7.5 Molecular Interpretation of Entropy

- **Absolute value of the entropy** → If entropy $S$ is a measure of disorder, then a perfectly ordered state of matter (perfect crystal) *should* have *zero* entropy.

- **Third law of thermodynamics** → Entropies of perfect crystals are the same at T = 0 K.
  (Thermal motion almost ceases at $T = 0$ K, and by convention $S = 0$ for perfect crystals at 0 K.)

- **Boltzmann approach** → Entropy increases as the number of ways that molecules or atoms can be arranged in a sample (*at the same total energy*) increases.

- **Boltzmann formula** → $\boxed{S = k \ln W}$ ; $k$ = Boltzmann constant = $1.380\,658 \times 10^{-23} \ J \cdot K^{-1}$
  [ **Note:** $R = N_A k$ ]
  $W$ = number of *microstates* available to the system at a certain energy

- **Boltzmann entropy** → Also called statistical entropy

- **Microstate**
  - → *One* permissible arrangement of atoms or molecules in a sample with a given total energy.
  - → $W$ = total number of permissible arrangements corresponding to the same total energy, also called an **ensemble**
  - → If only one arrangement is possible, $W = 1$ and $S = 0$

- **Boltzmann interpretation** → Spontaneous change occurs toward more probable states.

- **Scaling $S$**
  - → For $N$ molecules or atoms, the number of *microstates* is related to the number of molecular *orientations* permissible: $W = (orientations)^N$ and

$$S = k \ln W = k \ln(orientations)^N = N\, k \ln(orientations)$$

  - → For $N_A$ molecules or atoms, $W = (orientations)^{N_A}$ and

$$S = k \ln(orientations)^{N_A} = N_A k \ln(orientations) = R \ln(orientations)$$

- **Residual entropy** → For some solids, $S > 0$ at $T = 0$ (These solids retain some *disorder*.)

  **Note:** Units of $S$ are usually $J \cdot K^{-1} \cdot mol^{-1}$.

  **Example:** The residual entropy of CO(s) at 0 K is 4.6 $J \cdot K^{-1} \cdot mol^{-1}$. The number of *orientations* of CO molecules in the crystal is calculated as

  $$S = R \ln(orientations); \quad (orientations) = e^{S/R} = e^{(4.6/8.314\,51)} = e^{(0.553\,25)} = 1.7$$

  This is less than 2, the number expected for a random orientation (*disorder*) of CO molecules at 0 K (see text Fig. 7.5), suggesting some ordering at 0 K, possibly caused by alignment of the small permanent dipoles of neighboring CO molecules.

## 7.6  Equivalence of Statistical and Thermodynamic Entropies

- **Boltzmann's *molecular interpretation* compared with the *thermodynamic approach***

- **Qualitative comparison**

  **Volume increase in an ideal gas:**

  **thermodynamic approach** → $\Delta S = nR \ln \dfrac{V_2}{V_1}$;  if $V_2 > V_1$, then $\Delta S > 0$

  **molecular approach** → Consider a gas container to be a box. From particle-in-a-box theory, as $V$ increases, energy levels accessible to gas molecules pack more closely together and $W$ increases (see text Fig. 7.7).

  So, $\Delta S = S_{V_2} - S_{V_1} = k \ln \dfrac{W_2}{W_1} > 0$

  **Temperature increase in an ideal gas:**

  **thermodynamic approach** → $\Delta S = nC_{(V \text{ or } P,m)} \ln \dfrac{T_2}{T_1}$;  if $T_2 > T_1$, then $\Delta S > 0$

**molecular approach** → From particle-in-a-box theory, at low $T$, gas molecules occupy a small number of energy levels ($W$ = small); at higher $T$, more energy levels are available ($W$ = larger) (see text Fig. 7.8).

$$\text{So, } \Delta S = S_{T_2} - S_{T_1} = k \ln\frac{W_2}{W_1} > 0$$

**Note:** For liquids and solids, similar reasoning for the $T$ dependence of $S$ applies.

- **Quantitative comparison**
  - → Assume that the number of microstates available to any molecule is proportional to the volume available to it: $W = \text{constant} \times V$.
  - → For N molecules, $W = (\text{constant} \times V)^N$
  - → For an expansion of $N_A$ molecules from $V_1$ to $V_2$,

  $$\boxed{\Delta S = N_A k \ln\left(\frac{\text{constant} \times V_2}{\text{constant} \times V_1}\right) = R \ln\left(\frac{V_2}{V_1}\right)}$$

  which is *identical* to the thermodynamic expression.

  **Summary** → Macroscopic and microscopic approaches give the same predictions for $\Delta S$. The Boltzmann approach provides deep insight into entropy on the molecular level in terms of the energy states available to a system.

## 7.7 Standard Molar Entropies, $S_m°(T)$

- **Use of different formalisms**
  - → Use Boltzmann formalism to *calculate* entropy (sometimes difficult to do).
  - → Use thermodynamic formalism to *measure* entropy.

- **Standard ($P = 1$ bar) molar entropies of *pure* substances, $S_m°(T)$**
  - → Determined experimentally from *heat capacity* data ($C_{P,m}$), *enthalpy* data ($\Delta H°$) for phase changes, and the third law of thermodynamics
  - → For a gas at final temperature $T$ (assuming only one solid phase)

$$S_m°(T) = S_m°(0) + \int_0^{T_f} \frac{C_{P,m}(\text{solid})dT}{T} + \frac{\Delta H_{fus}}{T_f} + \int_{T_f}^{T_b} \frac{C_{P,m}(\text{liquid})dT}{T} + \frac{\Delta H_{vap}}{T_b} + \int_{T_b}^{T} \frac{C_{P,m}(\text{gas})dT}{T}$$

  - → According to the third law, $S_m°(0) = 0$ and

$$\boxed{S_m°(T) = \int_0^{T_f} \frac{C_{P,m}(\text{solid})dT}{T} + \frac{\Delta H_{fus}}{T_f} + \int_{T_f}^{T_b} \frac{C_{P,m}(\text{liquid})dT}{T} + \frac{\Delta H_{vap}}{T_b} + \int_{T_b}^{T} \frac{C_{P,m}(\text{gas})dT}{T}}$$

  **Note:** For solids, only the first term in the equation is used if *only* one solid phase exists. For liquids, the first three terms are used. For gases, all five terms are required.

  **Example:** Plots of $C_{P,m}$ as a function of $T$ and $C_{P,m}/T$ as a function of $T$ for solid copper Cu are shown on the following page. In the second graph, the area under the curve ($T = 0$ to $T = 298.15$ K) yields the experimental value of the entropy for Cu: [$S_m°(298.15$ K$) = 33.15$ J·K$^{-1}$·mol$^{-1}$].

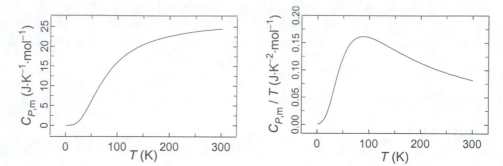

- **Appendix 2A**
  - → Lists of values of *experimental* standard molar entropies at 25°C for many substances (*standard* ≡ *pure* substance at 1 bar)

  - → Entropies of gases tend to be larger than those of solids or liquids ($S_{gas} > S_{liquid}$ *or* $S_{solid}$), as expected from the association of entropy with *randomness* and *disorder*.

  - → Other things equal, molar entropy increases with molar mass. Heavier species have more vibrational energy levels available to them than lighter ones, so $W$ and $S$ are larger. (See text Fig. 7.11 for use of particle-in-a-box model to understand this.)

## 7.8 Standard Reaction Entropies, $\Delta S_r^\circ$

- **Standard reaction entropies, $\Delta S_r^\circ$** → Determine similarly to calculation of standard enthalpy of reaction, $\Delta H_r^\circ$, for a chemical reaction.

- **For any chemical reaction** →
$$\Delta S_r^\circ = \sum n S_m^\circ (\text{products}) - \sum n S_m^\circ (\text{reactants})$$

  **Example:** Let us calculate $\Delta S_r^\circ$ for the reaction $2\,\text{Li(s)} + \text{Cl}_2(g) \rightarrow 2\,\text{LiCl(s)}$. Using $S_m^\circ$ values from Appendix 2A, we obtain

  $$\Delta S_r^\circ = \sum n S_m^\circ (\text{products}) - \sum n S_m^\circ (\text{reactants}) = 2\,S_m^\circ (\text{LiCl}) - [2\,S_m^\circ (\text{Li}) + S_m^\circ (\text{Cl}_2)]$$
  $$= 2(59.33) - [2(29.12) + 222.96] = -162.54 \ \text{J·K}^{-1}\text{·mol}^{-1}$$

  $\Delta S_r^\circ < 0$ as expected for net decrease in moles of gas when reactants form products

- **Generalizations**
  - → Entropy of gases dominates change in reaction entropy.
  - → $\Delta S_r^\circ$ is *positive* if there is a net *production* of gas in a reaction.
  - → $\Delta S_r^\circ$ is *negative* if there is a net *consumption* of gas in a reaction.

- **Properties of $S$ and $\Delta S$** → $S$ is an *extensive* property of state, like enthalpy.

- **For any process or change in state**
  - → *Reverse* the process, *change* the sign of $\Delta S$.
  - → *Change* the amounts of all materials in a process; make a *proportional change* in the value of $\Delta S$.
  - → Add *two* reactions together to get a *third* one; *add* the $\Delta S$ values for the first *two* reactions to get $\Delta S$ for the *third* reaction.

# Global Changes in Entropy  (Sections 7.9–7.11)

## 7.9  The Surroundings

- **Second law** → For *spontaneous* change, the entropy of an *isolated* system *increases*.

- *Any* **spontaneous change**
  - → Entropy of system *plus* surroundings *increases*.
  - → Criterion for spontaneity *includes* the *surroundings;*  $\Delta S_{tot} = \Delta S + \Delta S_{surr}$

- **Criterion**
  - → A process is *spontaneous* as written if $\Delta S_{tot} > 0$.
  - → A process is *nonspontaneous* as written if $\Delta S_{tot} < 0$.  The *reverse* process is spontaneous.
  - → For a system at *equilibrium,* $\Delta S_{tot} = 0$  (see Section 7.11)

- **Calculating $\Delta S_{surr}$ for a process at constant $T$ and $P$** →  $\boxed{\Delta S_{surr} = \dfrac{q_{surr}}{T} = -\dfrac{\Delta H}{T}}$

  **Note:**  $\Delta H$ is the enthalpy change for the *system*.  At constant pressure, the *heat* associated with the process is transferred *reversibly* to the surroundings.  The heat capacity of the surroundings is vast, so its temperature remains constant.  The process itself may be *reversible* or *irreversible*.

  **Example:** The entropy change of the surroundings when 1 mol of $H_2O(l)$ vaporizes at 25°C  is

  $$\Delta S_{surr} = \frac{q_{surr}}{T} = -\frac{\Delta H_{vap}}{T} = -\frac{40700 \text{ J} \cdot \text{mol}^{-1}}{(273.15 + 25)\,\text{K}} = -137 \text{ J} \cdot \text{mol}^{-1} \cdot \text{K}^{-1}$$

  The entropy of the surroundings decreases while that of the system increases.

## 7.10  Overall Change in Entropy

- **Spontaneity depends on $\Delta S$ and $\Delta S_{surr}$** → Four **cases** arise because *both* $\Delta S$ and $\Delta S_{surr}$ can be either positive (+) or negative (−).

| Case | [ $\Delta S_{tot}$ | = | $\Delta S$ | + | $\Delta S_{surr}$ ] | Spontaneity |
|------|------|---|------|---|------|-------------|
| 1 | + | | + | | + | *always* spontaneous |
| 2 | ? | | − | | + | spontaneous, if $\lvert \Delta S_{surr} \rvert > \lvert \Delta S \rvert$ |
| 3 | ? | | + | | − | spontaneous, if $\lvert \Delta S \rvert > \lvert \Delta S_{surr} \rvert$ |
| 4 | − | | − | | − | *never* spontaneous |

- **Application to chemical reactions**
  - → *Exothermic* reactions ($\Delta S_{surr} > 0$) correspond to Cases 1 and 2, that is, *spontaneous* if $\Delta S > 0$ or if $\lvert \Delta S_{surr} \rvert > \lvert \Delta S \rvert$ when $\Delta S < 0$.
  - → *Endothermic* reactions ($\Delta S_{surr} < 0$) correspond to Cases 3 and 4, that is, *spontaneous* only if $\Delta S > 0$ *and* $\lvert \Delta S \rvert > \lvert \Delta S_{surr} \rvert$.

- **For a given change in state (same $\Delta S$), the *path* can affect $\Delta S_{surr}$, $\Delta S_{tot}$, and *spontaneity***
  - → Comparison of a *reversible* and an *irreversible* isothermal expansion of an ideal gas for the same initial and final states (see text Example 7.12):

(1) $\Delta S$ (*irreversible*) = $\Delta S$ (*reversible*)  (entropy is a state function)

(2) $\Delta U = q + w = 0$ and $\Delta H = 0$, because energy and enthalpy of an ideal gas depend only on $T$, and $\Delta T = 0$ for this process.

(3) $q$ *and* $w$ are different for the two processes ($q$ and $w$ are *path* functions).  Heat given off to the surroundings ($-q = q_{surr}$) is different for the two processes.

(4) A *reversible* process does the *maximum* work $w$ (see Chapter 6).  Because $q + w = 0$, a *reversible* process delivers *less* heat to the surroundings.

(5) Because $\Delta S_{surr} = q_{surr} / T$,  $\Delta S_{surr}$ (*irreversible*) > $\Delta S_{surr}$ (*reversible*)

(6) Thus, $\Delta S_{tot}$ (*irreversible*) > $\Delta S_{tot}$ (*reversible*) = 0

(7) The *reversible* process corresponds to one in which the system is only *infinitesimally* removed from equilibrium as the process proceeds, whereas the *irreversible* process [$\Delta S_{tot}$(*irreversible*) > 0] is *spontaneous*.

## 7.11 Equilibrium

- **System at equilibrium** $\rightarrow$ No tendency to change in forward or reverse direction without an external influence

- **Types of equilibrium**

    **thermal:**  No tendency for heat to flow in or out of the system
    An example is an aluminum rod at room temperature.

    **mechanical:**  No tendency for any part of a system to move
    An example is an undeformed spring with no tendency to stretch or compress.

    **physical:**  Two phases of a substance at the transition temperature with no tendency for either phase to increase in mass
    An example is steam and water at the normal boiling point (100°C).

    **chemical:**  A mixture of reactants and products with no *net* tendency for either to form
    An example is a saturated solution of sucrose in water in contact with solid sucrose.

- **Universal thermodynamic criterion for equilibrium**

    $\rightarrow$ $\Delta S_{tot} = 0$ for any system at equilibrium.  If this were not true, $\Delta S_{tot}$ would be greater than 0 in either the forward or the reverse direction and *spontaneous change* would occur until $\Delta S_{tot} = 0$.

    $\rightarrow$ Total entropy change may be calculated to determine whether a system is at equilibrium.

# Free Energy  (Sections 7.12–7.16)

## 7.12 Focusing on the System

- **$\Delta G$, an alternative criterion for *spontaneity***

    (1) $\Delta S_{tot} = \Delta S + \Delta S_{surr}$   (always true)
    (2) If $P$ and $T$ are constant, $\Delta S_{surr} = -\Delta H / T$
    (3) If $P$ and $T$ are constant, $\Delta S_{tot} = \Delta S - (\Delta H / T)$  (function of *system only*)
    (4) Definition of Gibbs free energy, $G \equiv H - TS$
    (5) If $T$ is constant, $\Delta G = \Delta H - \Delta(TS) = \Delta H - T\Delta S$
    (6) According to (5) and (3),  $\Delta G = -T\Delta S_{tot}$

- **Summary**
  - → If a process at constant $T$ and $P$ is *spontaneous* ($\Delta S_{tot} > 0$), then $\Delta G < 0$.
  - → If a process at constant $T$ and $P$ is *nonspontaneous* ($\Delta S_{tot} < 0$), then $\Delta G > 0$.
  - → If a process at constant $T$ and $P$ is at *equilibrium* ($\Delta S_{tot} = 0$), then $\Delta G = 0$ as well.

- **Dependence of spontaneity on $\Delta S$ and $\Delta H$ of a system at constant $T$ and $P$**
  - → Four cases arise because *both* $\Delta S$ and $\Delta H$ can be *either* positive (+) *or* negative (−).

| Case | [ $\Delta G$ | $=$ | $\Delta H$ | $-$ | $T\Delta S$ ] | Spontaneity |
|------|--------------|-----|------------|-----|---------------|-------------|
| 1 | − | | − | | + | *always* spontaneous |
| 2 | ? | | − | | − | spontaneous, if $|\Delta H| > T|\Delta S|$ |
| 3 | ? | | + | | + | spontaneous, if $T|\Delta S| > |\Delta H|$ |
| 4 | − | | + | | − | *never* spontaneous |

**Case 1**
- → *Exothermic* process or reaction ($\Delta H < 0$)
- → *Enthalpy* and *entropy* favor *spontaneity*.
  *Enthalpy-* and *entropy*-driven process or reaction
- → *Temperature* dependence: if $\Delta H$ and $\Delta S$ are $T$ independent, the reaction is *spontaneous* at all $T$.

**Case 2**
- → *Exothermic* process or reaction ($\Delta H < 0$)
- → *Only enthalpy* favors *spontaneity*.
  *Enthalpy*-driven process or reaction
- → *Temperature* dependence: if $\Delta H$ and $\Delta S$ are $T$ independent, the reaction is *spontaneous* at low $T$ (*enthalpy* "wins") but *nonspontaneous* at high $T$. A crossover temperature exists at which $\Delta H = T\Delta S$ and the system is at equilibrium.

*Case 3*
- → *Endothermic* process or reaction ($\Delta H > 0$)
- → *Only entropy* favors *spontaneity*.
  *Entropy*-driven process or reaction
- → *Temperature* dependence: if $\Delta H$ and $\Delta S$ are $T$ independent, the reaction is *nonspontaneous* at low $T$ and *spontaneous* at high $T$ (*entropy* "wins"). A crossover temperature exists at which $\Delta H = T\Delta S$ and the system is at equilibrium.

**Case 4**
- → *Endothermic* process or reaction ($\Delta H > 0$)
- → *Nonspontaneous* process or reaction ($\Delta G > 0$)
- → *Temperature* dependence: if $\Delta H$ and $\Delta S$ are $T$ independent, then the reaction is *nonspontaneous* at all $T$.

- **Temperature dependence of $G$ for a *pure substance***
  - → $G = H - TS$
  - → Free energy of a substance decreases with increasing $T$.
  - → $S_{m,solid} < S_{m,liq} < S_{m,gas}$

→ So $G_{m,solid}$ decreases more slowly than $G_{m,liq}$ which decreases more slowly than $G_{m,gas}$.
→ Basis for thermodynamic understanding of melting and vaporization (see text Fig. 7.24)

## 7.13 Reaction Free Energy, $\Delta G_r$

- **Chemical reactions** → $\Delta G_r$ and $\Delta G_r°$ are defined in a manner similar to the reaction enthalpy, $\Delta H_r$, and standard reaction enthalpy, $\Delta H_r°$.

- **Definitions (n = stoiciometric coefficient)**
  → If $G_m$ is the molar free energy of a reactant or product, then the reaction free energy is

  $$\Delta G_r = \sum n G_m(\text{products}) - \sum n G_m(\text{reactants})$$

  → If $G_m°$ is the standard molar free energy, the standard reaction free energy is

  $$\Delta G_r° = \sum n G_m°(\text{products}) - \sum n G_m°(\text{reactants})$$

  → Values of $G_m$ or $G_m°$ cannot be determined directly. So, the equations given above *cannot* be used to determine $\Delta G_r$ and $\Delta G_r°$.
  → Values of $\Delta G_r$ and $\Delta G_r°$ are determined from the free energies of *formation*, $\Delta G_f$ and $\Delta G_f°$.

- **Standard free energy of formation, $\Delta G_f°$**
  → Standard free energy of *formation* of a compound or element
  → Free energy change for the formation of one mole of a compound from the most stable form of its elements under standard conditions (1 bar)
  → $\Delta G_f° \equiv 0$ for *all elements* in their most stable form (same convention as enthalpy)

- **Calculating $\Delta G_f°$ for a given compound**
  1. *Write and balance the formation reaction.*
     One mole of compound on the product side and the most stable form of the elements with appropriate coefficients on the reactant side

  2. *Calculate $\Delta H_f°$ and $\Delta S_f°$ for the reaction by using the data in Appendix 2A.*
     The value of $\Delta S_f°$ is obtained from the standard molar entropy values as follows:
     $\Delta S_f° = S_m°(\text{compound}) - \sum n S_m°(\text{reactants})$.

  3. *Solve for the formation of one mole of compound, using the expression*

  $$\Delta G_f° = \Delta H_f° - T\Delta S_f°$$
  $$= \Delta H_f°(\text{compound}) - T\left[S_m°(\text{compound}) - \sum n S_m°(\text{reactants})\right]$$

- **Thermodynamically *stable* compound**
  → *Negative* standard free energy of formation ($\Delta G_f° < 0$)
  → Thermodynamic tendency to *form* from its elements
  → Examples are ores such as $Al_2O_3(s)$, $Fe_2O_3(s)$.

- **Thermodynamically *unstable* compound**
  → *Positive* standard free energy of formation ($\Delta G_f° > 0$)
  → Thermodynamic tendency to *decompose* into its elements
  → Examples are some hydrocarbons such as acetylene ($C_2H_2$) and benzene ($C_6H_6$).

- **Properties of thermodynamically *unstable* compounds**
  - → *Many* decompose into their elements over a *long* time span.
  - → *Thermodynamically unstable* substances (like liquid octane and diamond) that decompose into their elements *slowly* are said to be *kinetically stable*.

- **Classification of thermodynamically *unstable* compounds**

  **labile:**      Decompose *or* react readily, for example, TNT and NO

  **nonlabile:**   Decompose *or* react slowly, for example, liquid octane

  **inert:**       Show virtually no reactivity *or* decomposition, for example, diamond

- **Calculating the standard reaction free energy $\Delta G_r°$ for a given chemical reaction**
  1. *Write the balanced chemical equation for the reaction.*
  2. *Calculate $\Delta H_r°$ and $\Delta S_r°$ for the reaction by using the data in Appendix 2A.*
  3. *Solve for $\Delta G_r°$, using the expression*

$$\Delta G_r° = \Delta H_r° - T\Delta S_r°$$
$$= \left[\sum n\Delta H_f°(\text{products}) - \sum n\Delta H_f°(\text{reactants})\right] - T\left[\sum nS_m°(\text{products}) - \sum nS_m°(\text{reactants})\right]$$

  4. *Alternatively, calculate $\Delta G_r°$ from the $\Delta G_f°$ values listed in Appendix 2A.*

$$\Delta G_r° = \sum n\Delta G_f°(\text{products}) - \sum n\Delta G_f°(\text{reactants})$$

  This equation yields a faster result, but does *not* allow for an estimation of the *temperature* dependence of $\Delta G_r°$.

  **Example:** Calculate $\Delta G_r°$ at 25°C for the reaction
  $$2\,Al_2O_3(s) + 3\,C(\text{graphite}) \rightarrow 3\,CO_2(g) + 4\,Al(s).$$

  $\Delta G_r° = \sum n\Delta G_f°(\text{products}) - \sum n\Delta G_f°(\text{reactants})$

  $\quad = 3\Delta G_f°[CO_2(g)] + 4\Delta G_f°[Al(s)] - \{2\Delta G_f°[Al_2O_3(s)] + 3\Delta G_f°[C(\text{graphite})]\}$

  $\quad = 3(-394.36\ \text{kJ·mol}^{-1}) - 2(-1582.3\ \text{kJ·mol}^{-1}) = 1981.5\ \text{kJ·mol}^{-1}$

  Since $\Delta G_r°$ is positive, the reaction is not *spontaneous* under standard conditions.

## 7.14 Free Energy and Nonexpansion Work

- **Processes at constant $T$ and $P$**

  **Expansion work**
  - → $dw = -P\,dV$  (*infinitesimal* change in volume, see Chapter 6)
  - → $w = -P\Delta V$  (*finite* change in volume, constant opposing $P$)
  - → Expansion work is performed against an opposing pressure.
  - → The sign of $w$ is defined in terms of the *system*:

    If work is done *on* the system *by* the surroundings, $w > 0$.

    If work is done *by* the system *on* the surroundings, $w < 0$.

  **Nonexpansion work**
  - → $dw_e$  (*infinitesimal* change, subscript "e" stands for extra)
  - → $w_e$  (*finite* change)

→ Any other type of work, including electrical work, mechanical work, work of muscular contraction, work involved in neuronal signaling, and that of chemical synthesis (making chemical bonds)

- **Relationship between nonexpansion work and (finite) free energy changes**
  → For a reversible process, $w_e = \Delta G$ (constant $P$ and $T$)
  → $w_e$ = *maximum* nonexpansion work obtainable from a constant P, T process
  → *Reversible* process: maximum amount of work is done by system on the surroundings (negative sign of $w$).
  → All real systems are *irreversible* and the maximum nonexpansion work is never obtained.

  **Example:** The maximum electrical work obtainable at 1 bar and 25°C by burning one mole of propane in a fuel cell with *excess* oxygen according to the equation
  $$C_3H_8(g) + 5\,O_2(g) \rightarrow 3\,CO_2(g) + 4\,H_2O(l) \text{ is given by}$$
  $$w_e = \Delta G_r^\circ = \sum n\Delta G_f^\circ(\text{products}) - \sum n\Delta G_f^\circ(\text{reactants})$$
  $$= 3\Delta G_f^\circ[CO_2(g)] + 4\Delta G_f^\circ[H_2O(l)] - \{\Delta G_f^\circ[C_3H_8(g)] + 5\Delta G_f^\circ[O_2(g)]\}$$
  $$= 3(-394.36 \text{ kJ·mol}^{-1}) + 4(-237.13 \text{ kJ·mol}^{-1}) - (-23.49 \text{ kJ·mol}^{-1})$$
  $$= -2108.11 \text{ kJ for 1 mol of } C_3H_8(g)$$
  The negative value for $w_e$ indicates work done on the surroundings.

## 7.15 The Effect of Temperature

- **For any chemical reaction at constant $P$ and $T$**

$$\boxed{\Delta G_r^\circ = \Delta H_r^\circ - T\Delta S_r^\circ = \left[\sum n\Delta H_f^\circ(\text{prod}) - \sum n\Delta H_f^\circ(\text{react})\right] - T\left[\sum nS_m^\circ(\text{prod}) - \sum nS_m^\circ(\text{react})\right]}$$

  → Cases 1–4 from the table in Section 7.12 of this *Study Guide* apply to any chemical reaction, assuming that $\Delta H_r^\circ$ and $\Delta S_r^\circ$ are independent of temperature.
  → In this case, the temperature dependence of $\Delta G_r^\circ$ arises from the $(-T\Delta S_r^\circ)$ term.

  **Example:** To calculate the temperature range over which the reaction $H_2(g) + \frac{1}{2}O_2(g) \rightarrow H_2O(g)$ is spontaneous, use $\Delta G_r^\circ = \Delta H_r^\circ - T\Delta S_r^\circ$. Set $\Delta G_r^\circ = 0$ to determine the crossover temperature.

  $$T_{crossover} = \frac{\Delta H_r^\circ}{\Delta S_r^\circ}$$

  $$= \frac{\Delta H_f^\circ[H_2O(g)]}{S_m^\circ[H_2O(g)] - \{S_m^\circ[H_2(g)] + (1/2)S_m^\circ[O_2(g)]\}}$$

  $$= \frac{-241.82 \text{ kJ·mol}^{-1}}{0.188\,83 \text{ kJ·K}^{-1}\cdot\text{mol}^{-1} - \{0.130\,68 + (1/2)(0.205\,14)\}\text{kJ·K}^{-1}\cdot\text{mol}^{-1}}$$

  $$= \frac{-241.82 \text{ kJ·mol}^{-1}}{-0.044\,42 \text{ kJ·K}^{-1}\cdot\text{mol}^{-1}} = 5444 \text{ K}$$

  In this case, enthalpy favors *spontaneity*; entropy does not. At 298.15 K, the reaction is spontaneous [$\Delta G_r^\circ = \Delta H_r^\circ - T\Delta S_r^\circ = -241.82 - (298.15)(-0.044\,42) = -228.58$ kJ mol$^{-1}$]. It is spontaneous until $T = 5444$ K. Above 5444 K, $H_2O(g)$ should decompose *spontaneously* into its elements.

## 7.16 Free Energy Changes in Biological Systems

- **Metabolism** → In metabolic processes, reaction steps may have a *positive* (unfavorable) reaction free energy. They can be coupled to spontaneous reactions to make a net reaction spontaneous.

- **Concept of coupled chemical reactions**
  → A way to make *nonspontaneous* reactions occur without changing temperature or pressure
  → A prominent process in biological systems, but also found in nonbiological ones
  → If reaction (1) has a *positive* reaction free energy $[\Delta G_r^\circ (1) > 0]$ (*nonspontaneous*), it can be coupled to reaction (2) with a *more negative* reaction free energy $[\Delta G_r^\circ (2) < 0]$, such that reaction (3), the sum of the two reactions, has a *negative* reaction free energy $[\Delta G_r^\circ (3) < 0]$:
  $\Delta G_r^\circ (3) = \Delta G_r^\circ (1) + \Delta G_r^\circ (2) < 0$   (overall reaction is *spontaneous*).

  **Example:** The sugar glucose (a food) is converted to pyruvate in a series of steps. The overall process may be represented as:

  glucose + other reactants → 2 pyruvate + other products; $\Delta G_r^\circ = -80.6$ kJ·mol$^{-1}$.

  The relatively large *negative* value of $\Delta G_r^\circ$ indicates that the overall reaction occurs spontaneously under *biochemical* standard conditions. The *biochemical standard state* corresponds more nearly to typical conditions in a cellular environment. The first step in the metabolic process is the conversion of glucose into glucose-6-phosphate:

  glucose + phosphate → glucose-6-phosphate + H$_2$O;   (l)  $\Delta G_r^\circ (1) = +14.3$ kJ·mol$^{-1}$.

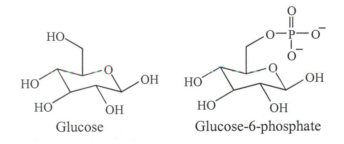

Glucose                Glucose-6-phosphate

  The *positive* standard free energy implies *nonspontaneity* for reaction (1). To generate glucose-6-phosphate, reaction (1) is coupled with (2), the hydrolysis of adenosine triphosphate (ATP) to yield adenosine diphosphate (ADP). Reaction (2) has a favorable $\Delta G_r^\circ < 0$ value (*spontaneous*):

  ATP + H$_2$O → ADP + inorganic phosphate;  (2)  $\Delta G_r^\circ (2) = -31.0$ kJ·mol$^{-1}$.

  Adding (coupling) reactions (1) and (2) gives a net spontaneous reaction (3):
  glucose + ATP → glucose-6-phosphate + ADP  (3)
  $\Delta G_r^\circ (3) = \Delta G_r^\circ (1) + \Delta G_r^\circ (2) = -16.7$ kJ·mol$^{-1}$

**Note:** Reactions in metabolic pathways are *catalyzed* by *enzymes*. Step (1) is catalyzed by the enzyme hexokinase. Biological catalysts are discussed in Chapter 13 and sugars in Chapter 19.

# Chapter 8  Physical Equilibria

## Phases and Phase Transitions  (Sections 8.1–8.7)

### 8.1  Vapor Pressure

- **Liquids** $\rightarrow$  Evaporate to form a gas or vapor

  Puddle of rainwater evaporates.  The reaction $H_2O(l) \rightarrow H_2O(g)$ goes to completion and the puddle disappears.  Phase equilibrium is *not* attained here.

- **Gases** $\rightarrow$  Condense to form a liquid or solid at sufficiently low temperature

  Water vapor condenses to form rain.  Steam condenses on cold surfaces.

- **Equilibrium** $\rightarrow$  In a closed system, both liquid and gas phases exist in *equilibrium*.

  Represent equilibrium system using double-headed arrow: $H_2O(l) \rightleftharpoons H_2O(g)$

- **Liquid-gas phase equilibrium** $\rightarrow$  *Dynamic equilibrium*

  Rate of evaporation equals rate of condensation.

  **Example:** In a *covered* jar half-filled with water at room temperature, the liquid and gas phases are in equilibrium, $H_2O(l) \rightleftharpoons H_2O(g)$.  Water vapor exhibits a characteristic equilibrium pressure, its *vapor pressure*.

- **Vapor pressure, *P***
  - $\rightarrow$ Characteristic pressure of a vapor above a *confined* liquid or solid when they are in dynamic equilibrium (*closed system*)
  - $\rightarrow$ Depends on temperature, increasing rapidly with increasing temperature
  - $\rightarrow$ Different vapor pressure for different liquids and solids

    **Examples:** Several solids with measurable vapor pressures are camphor, naphthalene, paradichlorobenzene (mothballs), and dry ice [$CO_2(s)$].

    Virtually all liquids have measurable vapor pressures, but some [*e.g.*, $Hg(l)$] may be very small at room temperature.

- **Sublimation** $\rightarrow$  Evaporation of solid to form a gas   [$CO_2(s) \rightarrow CO_2(g)$]

- **Volatility**
  - $\rightarrow$ Related to the ability to evaporate
  - $\rightarrow$ Liquids with high vapor pressures at a given temperature are *volatile* substances.

- **Characteristics of equilibrium between phases**
  - $\rightarrow$ *Dynamic equilibrium* occurs when molecules enter and leave the individual phases at the same rate, and the total amount in each phase remains unchanged.
  - $\rightarrow$ Molar free energies of the individual phases are equal.

    substance (phase 1) $\rightleftharpoons$ substance (phase 2),  $\Delta G_m = 0$

### 8.2  Volatility and Intermolecular Forces

- *Increasing* **the strength of intermolecular forces in a liquid**
  - $\rightarrow$ *Decreases* the volatility
  - $\rightarrow$ *Decreases* the vapor pressure at a given temperature
  - $\rightarrow$ *Increases* the normal boiling point, $T_b$ (defined in Section 8.4)

- **London forces**
  - → More massive molecules are less volatile.
  - → More electrons in a molecule produce stronger forces.
- **Dipolar forces**
  - → For molecules with the same number of electrons, dipolar molecules are associated with less volatile substances than are molecules with only London forces.
  - → A *greater* dipole moment yields a *lower* volatility.
- **Hydrogen bonding** → Molecules that form H-bonds may produce even less volatile substances than dipolar molecules.
- **Ionic forces** → Salts are essentially nonvolatile.

  **Note:** As with many rules in chemistry, the above qualitative guidelines must be applied cautiously.
- **Quantitative approach to vapor pressure** (see derivation in text)
  - → $\ln P = \dfrac{-\Delta G_{vap}°}{RT}$ *and* $\Delta G_{vap}° = \Delta H_{vap}° - T\Delta S_{vap}°$ *lead to*

$$\ln P = \frac{-\Delta H_{vap}°}{RT} + \frac{\Delta S_{vap}°}{R}$$

- **Insight from the above equation**
  - → Because $\Delta S_{vap}°$ is about the same for *all* liquids (Trouton's rule), the vapor pressure of a liquid depends mainly on $\Delta H_{vap}°$, which is always a positive quantity.
  - → *Stronger* intermolecular forces result in a *larger* $\Delta H_{vap}°$ and a *decrease* in vapor pressure.
  - → For a given liquid, an *increase* in temperature results in an *increase* in vapor pressure.

## 8.3 The Variation of Vapor Pressure with Temperature

- **Solve the vapor pressure equation for two temperatures** ($P_2$ at $T_2$ and $P_1$ at $T_1$)

  - → $$\ln \frac{P_2}{P_1} = \frac{\Delta H_{vap}°}{R}\left(\frac{1}{T_1} - \frac{1}{T_2}\right)$$   Clausius–Clapeyron equation

- **Uses of the Clausius-Clapeyron equation**
  - → Measure the vapor pressure of a liquid at two temperatures ($P_2$ at $T_2$ and $P_1$ at $T_1$) to *estimate* the standard enthalpy of vaporization $\Delta H_{vap}°$.

$$\Delta H_{vap}° \approx \left(\frac{R}{\left(T_1^{-1} - T_2^{-1}\right)}\right)\ln\frac{P_2}{P_1}$$

  - → Measure $\Delta H_{vap}°$ and the vapor pressure of a liquid at one temperature ($P_1$ at $T_1$) to *estimate* the vapor pressure at any different temperature ($P_2$ at $T_2$).
  - → Measure $\Delta H_{vap}°$ and the vapor pressure at one temperature ($P_1$ at $T_1$). *Estimate* the normal boiling point of the liquid ($T_b = T_2$, when $P_2 = 1$ atm) or the standard boiling point of the liquid ($T_2$, when $P_2 = 1$ bar).

→ Determine $\Delta H_{vap}°$ and the vapor pressure at the normal boiling temperature ($P_1$ at $T_b$). *Estimate* the vapor pressure at a different temperature ($P_2$ at $T_2$).

**Note:** In all cases, the term *estimate* is used because the assumption that $\Delta H_{vap}°$ is independent of temperature is not strictly true; it is an approximation.

## 8.4 Boiling

- **Liquid in an open container** → Rate of vaporization is greater than the rate of condensation, so the liquid evaporates.

- **Boiling** → Vapor pressure of liquid equals atmospheric pressure. Rapid vaporization occurs throughout the entire liquid.

- **Boiling point** → Temperature at which the liquid begins to boil

- **Normal boiling point** → Boiling point at 1 atm, $T_b$
  Within experimental error, $T_b = 99.974°C \approx 100°C$ for water

- **Standard boiling point** → Boiling point at 1 bar

- **Effect of pressure on the boiling point**
  → An *increase* of pressure on a liquid leads to an *increase* in the boiling point.
  The pressure cooker makes use of this fact.

  → A *decrease* of pressure on a liquid leads to a *decrease* in the boiling point.
  This helps explain why water boils at a lower temperature on a mountaintop. It also accounts for vacuum distillation at low temperatures to purify liquids that decompose at higher temperatures.

## 8.5 Freezing and Melting

- **Freezing and melting** → Two common phase transitions, one the reverse of the other

- **Freezing (melting)** → Solidification of a liquid (liquefaction of a solid)

- **Freezing (melting) point** → Temperature at which liquid freezes (solid melts)
  For a given substance, freezing and melting points are identical.

- **Normal freezing point, $T_f$** → Temperature at which solid begins to freeze at 1 atm
  Within experimental error, $T_f = 0°C$ and 1 atm for water

- **Standard freezing point** → Temperature at which solid begins to freeze at 1 bar
  Difference between $T_f$ and the standard freezing point is very small.

- **Supercooled liquid**
  → A pure liquid may exist below its freezing point if it is cooled extremely slowly.
  → Thermodynamically unstable with respect to formation of solid
  → If heat is withdrawn slowly from pure water, the temperature may drop below 0°C, indicating supercooling. At some point, a tiny crystal (nucleus) of ice forms, and the entire sample crystallizes rapidly to form ice whose temperature rises from the heat released by sudden freezing.

- **Pressure dependence of the freezing point:**
  **Most substances** → Solid phase is more dense than the liquid phase (smaller molar volume), so the solid *sinks* as it forms. Increasing pressure favors the phase with

the smallest density; thus, the freezing point *increases* with *increasing* pressure.

**Few substances** → Solid phase is less dense than the liquid phase (larger molar volume) and the solid *floats* as it forms. An increase in pressure favors the phase with the greatest density; thus, the freezing point *decreases* with *increasing* pressure. Examples include water and bismuth.

Note: The *anomalous* behavior of water is critical for the survival of certain species. If ice sank, bodies of water would freeze solid in winter, killing aquatic life.

## 8.6  Phase Diagrams

- **Component**
  - → A single substance, for example, aluminum, octane, water, or sodium chloride
  - → A chemically independent species

- **Single-component phase diagram** → Map showing the most stable phase of a substance at different pressures and temperatures

- **Phase boundary**
  - → Lines separating regions on a phase diagram
  - → Represents a set of *P* and *T* values for which *two* phases coexist in dynamic equilibrium

- **Triple point**
  - → Point where *three* phase boundaries intersect
  - → Corresponds to a single value of *P* and *T* for which *three* phases coexist in dynamic equilibrium

- **Critical point** → High-temperature terminus of the liquid-vapor phase boundary

- **Critical temperature, $T_c$** → The temperature above which a gas cannot condense into a liquid
  Only one phase is observed above $T_c$.

Note: Critical point and critical temperature are introduced in Section 8.7 in the text. The definitions given above are repeated later.

### Phase Diagram of Water

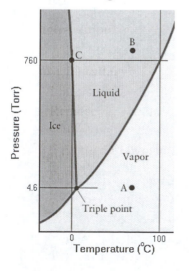

Note the regions (bounded areas) labeled ice (solid), liquid, and vapor (gas). In each area, only a single phase (ice, water, or water vapor) is stable. Within each region, pressure and/or temperature may be varied independently with no accompanying phase transition. Two independent variables, or *two* degrees of freedom, characterize such a region. The three lines that separate the regions define the phase boundaries of solid-liquid, liquid-vapor, and solid-vapor. At a boundary, only *P* or *T* may be varied if the two phases are to remain in equilibrium. Only one independent variable, or *one* degree of freedom, exists for water in states corresponding to the boundary. All three phases coexist at the *triple point* with its specific *P* and *T,* or *zero* degrees of freedom.

- **Slope of the solid-liquid phase boundary line**
  - → Positive for most substances:
    Solid sinks because it is more dense than liquid.
    At constant temperature, a pressure increase yields no phase change for the solid.
    At constant temperature, a pressure increase may cause the liquid to solidify.
  - → Negative for $H_2O$:
    Because the solid is less dense than the liquid, ice floats.
    At constant temperature, a pressure increase may cause ice to melt.
    At constant temperature, a pressure increase yields no phase change for the liquid.

- **Increasing the pressure on graphite to make diamond and liquid carbon**
  - → What occurs when the pressure on graphite at 3000 K is increased from 1 bar to 500 bar?

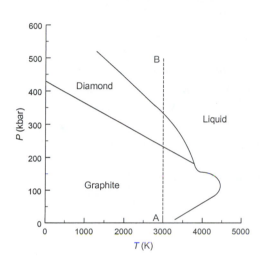

In the initial state of graphite (point A), one phase exists. Temperature is held constant (3000 K), and the pressure is allowed to vary freely. The C(graphite)-C(diamond) phase boundary is reached at approximately 230 kbar. At this point, C(diamond) forms and exists in equilibrium with C(graphite). All the C(graphite) is eventually converted to C(diamond). (This process may be quite slow). In the diamond region, one phase exists. Additional increases in the pressure on C(diamond) result in the equilibrium of C(diamond) and liquid carbon at about 400 kbar. When all the diamond has melted, the pressure on C(l) is free to increase to attain the final state (point B).

**Note:** There are three allotropes of carbon: graphite, diamond, and buckminsterfullerene ($C_{60}$); the latter, discovered in 1985, is composed of soccer-ball-shaped molecules. The thermodynamic stability of buckminsterfullerene has not yet been determined. The validity of its inclusion on the C phase diagram is, therefore, uncertain. (Metastable phases, such as supercooled water, do not appear on phase diagrams.) The crystal structure is face-centered cubic with $C_{60}$ molecules at the corners and faces of a cubic unit cell. The unit cell is shown below.

## 8.7 Critical Properties

- **Critical point** → Terminus of the liquid-gas phase boundary at high temperature

- **Critical temperature** → The temperature above which a vapor (gas) cannot condense into a liquid. Only one phase is observed above $T_c$.

- **Supercritical fluid** → Substance above its critical temperature, $T_c$

### Phase Diagram of Water

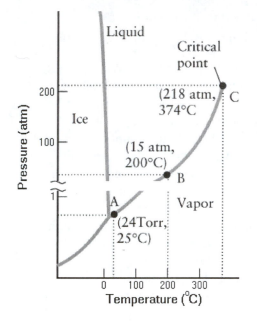

This discontinuous diagram shows both low and high pressure regions. Point A represents typical "room temperature" conditions. Liquid water has a vapor pressure of 23.76 Torr (0.031 26 atm) at 25°C (298.15 K). At 200°C (473.15 K), the vapor pressure is 15 atm (point B). As the temperature is increased further, the densities of the liquid and vapor in equilibrium approach one another and become nearly equal at the critical point (point C). Above the critical temperature, only one phase with the properties of a very dense vapor remains. The *critical pressure*, $P_c$, is the vapor pressure measured at the critical temperature. The temperature, pressure, and density values for water at the critical point are $T_c = 374.1$°C (647 K), $P_c = 218.3$ atm, and $d_c = 0.32$ g·cm$^{-3}$, respectively.

- **Intermolecular forces** → Both $T_c$ and $T_b$ have a tendency to *increase* with the *increasing* strength of intermolecular forces.

| Substance | $T_c$ (K) | $T_b$ (K) |
|---|---|---|
| He (helium) | 5.2 | 4.3 |
| Ar (argon) | 150 | 88 |
| Xe (xenon) | 290 | 166 |
| NH$_3$ (ammonia) | 405 | 240 |
| H$_2$O (water) | 647 | 373 |

**Note:** The strength of the London forces *increases* with increasing molar mass (number of electrons) of a noble gas atom. With *stronger* intermolecular forces, *higher* critical and normal boiling temperatures are expected. Ammonia and water form strong hydrogen bonds. Water forms more hydrogen bonds per molecule than does ammonia. Therefore, water has critical and normal boiling temperatures *higher* than those of ammonia. In making these comparisons, care must be taken.

# Solubility  (Sections 8.8–8.13)

## 8.8  The Molecular Nature of Dissolving

- **Two component solution** $\rightarrow$ Contains one solvent and one solute species

- **Interactions** $\rightarrow$ Solvent-solvent, solute-solute, and solute-solvent

- **Unsaturated solution**
  - $\rightarrow$ All solute added to solvent dissolves.
  - $\rightarrow$ Amount of solute dissolved in the solvent is *less* than the equilibrium amount.

- **Saturated solution**
  - $\rightarrow$ Solubility limit of solute has been reached, with any additional solute present as a precipitate.
  - $\rightarrow$ The *equilibrium* amount of solute has been dissolved in solvent.
  - $\rightarrow$ Dissolved and undissolved solute molecules are in *dynamic* equilibrium; in other words, the rate of dissolution equal the rate of precipitation.

- **Supersaturated solution**
  - $\rightarrow$ Under certain conditions, an amount of solute greater than the equilibrium amount (solubility limit) can be dissolved in a solvent.
  - $\rightarrow$ Because *more* than the equilibrium amount is dissolved, the system is thermodynamically unstable.
    A slight disturbance (seed crystal) causes precipitation and rapid return to equilibrium.

- **Solubility limit** $\rightarrow$ Depends on the nature of both the solute and the solvent

- **Molar solubility** $\rightarrow$ Molar concentration of a saturated solution of a substance
  Units:  (moles of solute)/(liter of solution)  or  $mol \cdot L^{-1} \equiv M$
  For example, a saturated aqueous solution of AgCl has a molar concentration of $1.33 \times 10^{-5}$ $mol \cdot L^{-1}$ or $1.33 \times 10^{-5}$ M at 25°C.

- **Gram solubility** $\rightarrow$ Mass concentration of a saturated solution of a substance
  Units:  (grams of solute)/(liter of solution)  or  $g \cdot L^{-1}$
  For example, a saturated aqueous solution of AgCl has a mass concentration of $1.91 \times 10^{-3}$ $g \cdot L^{-1}$ at 25°C.

- **Molal solubility** $\rightarrow$ Molality of a saturated solution of a substance
  Units:  (moles of solute)/(kg of solvent) or $mol \cdot kg^{-1}$
  For example, a saturated aqueous solution of AgCl has a molality of $1.33 \times 10^{-5}$ $mol \cdot kg^{-1}$ at 25°C.

> **Note:** For dilute solutions, the molar and molal solubilities will be essentially equal as in the example above.  This is because 1 L of water has a mass very nearly equal to 1 kg at 25°C and the mass of the solute is very small compared with that of the solvent.  In concentrated solutions, the mass of the solute becomes an appreciable fraction of the mass of the solution, and a liter of solution contains substantially less than 1 kg of water. In such a solution, molality and molarity can differ considerably.

## 8.9 The Like-Dissolves-Like Rule

- **Rule** → If solute-solute and solvent-solvent intermolecular forces (London, dipole, hydrogen-bonding, or ionic) are similar, larger solubilities are predicted. *Lesser* solubilities are expected if these forces are dissimilar.

- **Use** → A *qualitative* guide to predict and understand the solubility of various solute species in different solvents

   **Example:** Oil is composed of long chain hydrocarbons. That oil and water do not mix is an observation of everyday life. Oil molecules are bound together by London forces, whereas water associates primarily by hydrogen bonds. The like-dissolves-like rule suggests a solvent with only cohesive London forces is needed to dissolve oil. Gasoline, a mixture of shorter chain hydrocarbons such as heptane and octane, is a possible solvent, as is benzene, an unsaturated cyclic hydrocarbon.

   **Example:** Glucose, $C_6H_{12}O_6$, has five –OH groups capable of forming hydrogen bonds. It is expected to be soluble in hydrogen-bonding solvents such as water. Conversely, it should be insoluble in nonpolar solvents such as hexane.

   **Example:** Potassium iodide, KI, is an ionic compound. Because both water and ammonia are highly polar molecules, they are expected to be effective in solvating both $K^+$ and $I^-$ ions. Ethyl alcohol molecules are less polar than water, so KI is expected to be less soluble in ethanol than in water. A similar argument can be used to explain the *slight* solubility of KI in acetone.

- **Hydrophilic** → Water-attracting

- **Hydrophobic** → Water-repelling

- **Soaps** → Long chain molecules, with hydrophobic and hydrophilic ends, which are soluble in both polar and nonpolar solvents

- **Surfactant** → Surface-active molecule with a hydrophilic head and a hydrophobic tail

- **Micelle** → Spherical aggregation of surfactant molecules with hydrophobic ends in the interior and hydrophilic ends on the surface

   **Note:** The term hydrophobic is somewhat of a misnomer. In actuality, an *attraction* between solute and solvent molecules *always* exists. In the case of water, the solute-solvent interaction disrupts the local structure of water, which is dominated by hydrogen bonding. The stronger the solute-$H_2O$ interaction, the greater the likelihood that the local solvent structure will be disrupted by hydrated solute molecules, increasing the solubility of the solute.

## 8.10 Pressure and Gas Solubility: Henry's Law

- **Pressure dependence of gas solubility:**
  **Qualitative features**
   → For a gas and liquid in a container, an increase in gas pressure leads to an increase in its solubility in the liquid.
   → Gas molecules strike the liquid surface and some dissolve. An increase in gas pressure leads to an increase in the number of impacts per unit time, thereby increasing solubility.

→ In a gas mixture, the solubility of each component depends on its partial pressure because molecules strike the surface independently of one another.

**Quantitative features**

→ Henry's law $\boxed{s = k_{\mathrm{H}} P}$

→ The solubility, $s$, of a gas in a liquid is directly proportional to the partial pressure, $P$, of the gas above the liquid.

→ $s \equiv$ molar solubility (see Overview on page 162)

→ $k_{\mathrm{H}} \equiv$ Henry's law constant, a function of $T$ (see Section 8.11), the gas, and the solvent

→ Units: $s$ (mol·L$^{-1}$), $k_{\mathrm{H}}$ (mol·L$^{-1}$·atm$^{-1}$), and $P$ (atm)

> **Example:** Estimate the molar solubility and gram solubility of $O_2$ in dry air dissolved in a liter of water open to the atmosphere at 20°C. Assume that air is 20.95% $O_2$ by volume. The pressure of the atmosphere is 1 atm, thus the partial pressure of $O_2$ is 0.2095 atm.
>
> $s\,(O_2) = (1.3 \times 10^{-3}\ \mathrm{mol·L^{-1}·atm^{-1}})(0.2095\ \mathrm{atm}) = 2.7 \times 10^{-4}\ \mathrm{mol·L^{-1}}$
>
> $s \times M = (2.7 \times 10^{-4}\ \mathrm{mol·L^{-1}})(32.0\ \mathrm{g·mol^{-1}}) = 8.6 \times 10^{-3}\ \mathrm{g·L^{-1}}$

## 8.11 Temperature and Solubility

- **Dependence of molar solubility on temperature**

  → For solid and liquid solutes, solubility *usually* increases with increasing temperature, but for some salts, for example $Li_2CO_3$, solubility decreases with increasing temperature (see text Fig. 8.21).

  → For gaseous solutes, solubility *usually* decreases with increasing temperature.

  → Despite these trends, solubility behavior can sometimes appear complex, as with sodium sulfate, $Na_2SO_4$, whose solubility in water increases then decreases (text Fig. 8.21). In this instance, complex solubility behavior arises because different hydrated forms of sodium sulfate precipitate at different temperatures.

  **Examples:**

  1. The solute that precipitates from a saturated solution may be different from that of the solid initially dissolved. At room temperature, anhydrous potassium hydroxide, KOH(s), readily dissolves in water. The solid that precipitates from a saturated solution of KOH is the dihydrate, KOH·2 H$_2$O. Therefore, the chemical equilibrium in the saturated solution is

     KOH·2 H$_2$O(s) $\rightleftharpoons$ K$^+$(aq, saturated) + OH$^-$(aq, saturated).

  2. From the diagram on the next page, notice that between −50 and 160°C, the solubility of LiCl displays three discontinuities, corresponding to temperatures at which one solid hydrated form is transformed into another. The four regions between the three discontinuities correspond, from low to high temperature, to precipitation of LiCl·3 H$_2$O, LiCl·2 H$_2$O, LiCl·H$_2$O, and LiCl, respectively. Note also that saturated aqueous solutions freeze at lower temperatures and boil at higher temperatures than pure water (see Section 8.16), accounting for the large temperature range of these solutions.

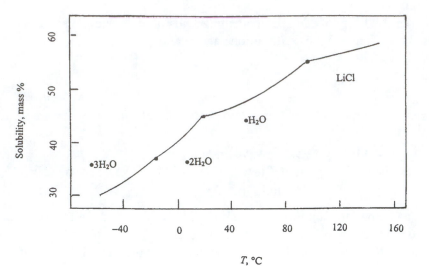

## 8.12 The Enthalpy of Solution

- **Solutions of ionic substances** $\rightarrow$ $A_mB_n(s) \rightarrow mA^{n+}(aq) + nB^{m-}(aq)$

- **Enthalpy of solution, $\Delta H_{sol}$**
  - $\rightarrow$ Enthalpy change per mole of substance dissolved
  - $\rightarrow$ Depends on *concentration* of solute

- **Limiting enthalpy of solution**
  - $\rightarrow$ Refers to the formation of a very dilute solution
    Values are given in Table 8.6 of the text.
  - $\rightarrow$ Use of the limiting enthalpy of solution avoids complications arising from interionic interactions that occur in more concentrated solutions, because ions are far apart in very dilute solutions.

  **Note:** All topics that follow refer to the limiting enthalpy condition.

- **Nature of the formation of solutions of ionic substances**
  - $\rightarrow$ Conceptualized as a two step process: sublimation of an ionic solid to form gaseous ions followed by the solvation of the gaseous ions to form an ionic solution
  - $\rightarrow$ The enthalpy change for the sublimation step is designated the lattice enthalpy, $\Delta H_L$ (see Table 6.6 in the text).
  - $\rightarrow$ The enthalpy change for the second step (formation of hydrated ions from the gas phase ions) is designated as the enthalpy of hydration, $\Delta H_{hyd}$ (see Table 8.7 in the text).
  - $\rightarrow$ $\Delta H_L$ *always* has a positive value, whereas $\Delta H_{hyd}$ *always* has a negative value.
  - $\rightarrow$ $\Delta H_{sol} = \Delta H_L + \Delta H_{hyd}$ is, then, the difference of two numbers, both of which are typically quite large.

- **Endothermic process:** $\Delta H_L > |\Delta H_{hyd}|$

- **Exothermic process:** $\Delta H_L < |\Delta H_{hyd}|$

  **Note:** For small, highly charged ions, both $\Delta H_L$ and $|\Delta H_{hyd}|$ have large values.

## 8.13 The Free Energy of Solution

- **Solutions of ionic substances** $\rightarrow$ $A_mB_n(s) \rightarrow mA^{n+}(aq) + nB^{m-}(aq)$

- **Free energy of solution, $\Delta G_{sol}$**
  - $\rightarrow$ Free energy change per mole of substance dissolved
  - $\rightarrow$ Depends on *concentration* of solute
  - $\rightarrow$ $\Delta G_{sol} = \Delta H_{sol} - T\Delta S_{sol}$ at constant $T$
- **Nature of solubility**
  - $\rightarrow$ A substance will dissolve if $\Delta G_{sol} < 0$ and will continue to dissolve until a saturated solution, for which $\Delta G_{sol} = 0$, is obtained.
  - $\rightarrow$ Both $\Delta H_{sol}$ and $\Delta S_{sol}$ change with increasing concentration to make $\Delta G_{sol}$ more positive.
  - $\rightarrow$ If the solute is consumed before $\Delta G_{sol}$ reaches 0, an unsaturated solution results, with the potential to dissolve additional solute.
  - $\rightarrow$ With excess solute present, $\Delta G_{sol}$ reaches 0 and a saturated solution results.
  - $\rightarrow$ For endothermic enthalpies of solution, the increase in entropy of solution drives the solubility process.
  - $\rightarrow$ A substance with a *large* endothermic enthalpy of solution is *usually* insoluble.

---

# Colligative Properties (Sections 8.14–8.17)

## 8.14 Molality

- **Molality, molarity, mole fraction** $\rightarrow$ Different concentration units used for quantitative treatment of colligative properties

- **Molality, $m$**
  - $\rightarrow$ Moles of solute per kilogram of solvent
  - $\rightarrow$ Independent of temperature (solute and solvent given as masses or mass equivalent)
  - $\rightarrow$ Units of moles per kilogram ($mol \cdot kg^{-1}$)
  - $\rightarrow$ Used when relative number of molecules of components is to be emphasized

  $$\text{For a binary solution,} \quad m_{solute} = \frac{n_{solute}}{kilograms_{solvent}}$$

  **Note:** The letter $m$ is used to designate mass and molality. Care should be taken to avoid confusing the two.

- **Mole fraction, $x$**
  - $\rightarrow$ Ratio of moles of solute to the total number of moles of all species in a mixture
  - $\rightarrow$ Independent of temperature
  - $\rightarrow$ Dimensionless quantity
  - $\rightarrow$ Used when relative number of molecules of components is to be emphasized

  $$\text{For a binary solution,} \quad x_{solute} = \frac{n_{solute}}{n_{solute} + n_{solvent}}$$

- **Molarity**
  - $\rightarrow$ Moles of solute divided by total volume of *solution*

→ Temperature dependent (volume of solution changes with temperature)

→ Units of moles per liter ($mol \cdot L^{-1}$)

$$\text{For a binary solution,} \quad \text{Molarity}_{\text{solute}} = \frac{n_{\text{solute}}}{V_{\text{solution}}}$$

- **Converting from one concentration unit to another**
  → The molar mass of one component is required to convert *molality* to *mole fraction* or *mole fraction* to *molality*.
  → The density of the solution is required to convert *molarity* to *molality* or *vice versa*.

  **Note:** When converting from one concentration unit to another, it is convenient to assume one of the following: (1) the solution has a volume of one liter (*molarity* ⇔ *molality*), (2) the total amount of solvent and solute is one mole (*mole fraction* ⇔ *molality*), or (3) the mass of the solvent in the solution is one kilogram (*molality* ⇔ *mole fraction*).

## 8.15 Vapor-Pressure Lowering

- **Qualitative features**
  → The vapor pressure of a solvent in equilibrium with a solution containing a *nonvolatile solute* (for example, sucrose or sodium chloride in water) is lower than that of the pure solvent.
  → For an *ideal solution* or a sufficiently dilute *real* solution, the vapor pressure of any volatile component is proportional to its mole fraction in solution.

- **Quantitative features**
  → Raoult's law $\boxed{P = x_{\text{solvent}} P_{\text{pure}}}$

  → The vapor pressure $P$ of a solvent is equal to the product of its mole fraction in solution, $x_{\text{solvent}}$, and its vapor pressure when pure, $P_{\text{pure}}$. The quantities $P$ and $P_{\text{pure}}$ are measured at the same temperature.

- **Ideal solution**
  → An *ideal* solution is a hypothetical solution that obeys Raoult's law exactly for all concentrations of solute.
  → Solute-solvent interactions are the same as solvent-solvent interactions; therefore, the enthalpy of solution, $\Delta H_{\text{sol}}$, is zero.
  → Entropy of solution, $\Delta S_{\text{sol}} > 0$
  → Free energy of solution, $\Delta G_{\text{sol}} < 0$, leading to a lowering of vapor pressure of the solvent (see Section 8.2)
  → The process of solution formation is entropy driven.
  → Solutions approximating ideality are typically formed by similar solute and solvent species, such as hexane and heptane.
  → Real solutions do not obey Raoult's law at all concentrations, but they do follow Raoult's law in the limit of low solute concentration (dilute solution).

- **Nonideal (real) solution**
  → Solution that does not obey Raoult's law at a certain concentration

→ Solute-solvent interactions differ from solvent-solvent interactions.

→ $\Delta H_{sol}$ is not equal to zero.

→ Real solutions approximate ideal behavior at concentrations below $10^{-1}$ mol·kg$^{-1}$ for *nonelectrolyte* solutions and below $10^{-2}$ mol·kg$^{-1}$ for *electrolyte* solutions.

→ Real solutions tend to behave ideally as the solute concentration approaches zero.

### 8.16 Boiling-Point Elevation and Freezing-Point Depression

- **Boiling-point elevation**

  → A nonvolatile solute lowers the vapor pressure of the solvent. As a result, a solution containing a nonvolatile solute will not boil at the normal boiling point of the pure solvent. The temperature must be increased above that value to bring the vapor pressure of the solution to atmospheric pressure. Therefore, boiling is achieved at a higher temperature (the boiling point is elevated).

  → Arises from the influence of the solute on the entropy of the solvent

  → Quantitatively, $\boxed{\text{boiling-point elevation} = k_b \times \text{molality}}$

  → The boiling-point constant, $k_b$, depends on the solvent and has units of K·kg·mol$^{-1}$ (Table 8.8).

  → The boiling-point elevation equation holds for nonvolatile solutes in dilute solutions that are approximately ideal.

- **Freezing-point depression**

  → Vapor-pressure lowering of a solution decreases the triple-point temperature, the intersection of the liquid-vapor (vapor pressure) and solid-vapor phase boundaries. The solid-liquid phase boundary originating at the triple point is moved slightly to the left on the phase diagram (the solid-vapor boundary is unchanged). The freezing temperature of the solution is thereby lowered (the freezing-point is depressed).

  → Arises from the influence of the solute on the entropy of the solvent.

  → Proportional to the molality of the solute

  → Quantitatively, $\boxed{\text{freezing-point depression} = k_f \times \text{molality}}$

  → The freezing-point constant, $k_f$, depends on the solvent and has units of K·kg·mol$^{-1}$ (Table 8.8).

  → The freezing-point depression equation holds for nonvolatile solutes in dilute solutions that are approximately ideal.

- **Freezing-point depression corrected for ionization or aggregation of the solute**

  → $\boxed{\text{freezing-point depression} = i\, k_f \times \text{molality}}$

  → The van't Hoff factor, $i$, determined experimentally, is an adjustment used to treat *electrolytes* (*e.g., ionic solids,* strong *acids* and *bases*) that *dissociate* and molecular solutes that *aggregate* (*e.g.,* acetic acid dimers).

  → For electrolytes, $i$ is the number of moles of ions formed by each mole of solute dissolved in 1 kg of solvent if all ions behave independently. In aqueous solutions, this only occurs in very dilute solutions; with increasing concentration, interionic effects cause $i$ to be smaller than the value expected for complete ionization.

  → For molecular aggregates, $i$ may be used to estimate the average size of the aggregates.

### 8.17 Osmosis

- **Qualitative features**
  - → *Osmosis:* The tendency of a solvent to flow through a membrane into a more concentrated solution. It is used to determine an unknown molar mass, particularly for large molecules such as polymers and proteins.
  - → *Osmometry* is a technique used to determine the molar mass of a solute, if the mass concentration is measured.
  - → A *semipermeable membrane* allows only certain types of molecules to pass through. Typical membranes allow passage of water and small molecules, but not large molecules or ions.
  - → In osmosis, the free energy of the solution is lower than that of the solvent; hence, dilution is a *spontaneous process*. The free energy of the solution can be increased by applying pressure on the solution. The increased pressure at equilibrium is called the *osmotic pressure, $\Pi$*.
  - → In a static apparatus open to the atmosphere, pressure is applied by the increased height of the raised column of solution caused by the flow of solvent through the membrane into the solution. In a dynamic apparatus in a closed system, pressure is applied by the increased force of a piston confining the solution (preventing solvent flow).
  - → If the external pressure $P < \Pi$, osmosis (dilution) is *spontaneous*.
    If $P = \Pi$, the system is at equilibrium (no net flow).
    If $P > \Pi$, *reverse osmosis* occurs (flow of solvent in solution to the pure solvent).
    Reverse osmosis is used to purify seawater.

- **Quantitative features**
  - → van't Hoff equation $\boxed{\Pi = i\,RT\mathrm{M}}$

  - → $\mathrm{M}$ = molarity of the solution $= \dfrac{\text{moles of solute}}{\text{liters of solution}}$

    $i$ = van't Hoff factor, $R$ = gas constant, $T$ = temperature in kelvin
    Use units of $\Pi$ in atm and of $R$ in $\mathrm{L\cdot atm\cdot K^{-1}\cdot mol^{-1}}$; $i$ is dimensionless.

---

# Binary Liquid Mixtures (Sections 8.18–8.20)

### 8.18 The Vapor Pressure of a Binary Liquid Mixture

- **Ideal solution with two volatile components A and B; liquid and vapor phases in equilibrium**
  - → Let $P_{A,pure}$ = vapor pressure of pure A and $P_{B,pure}$ = vapor pressure of pure B
  - → In an ideal solution, each component obeys Raoult's law at all concentrations.
  - → $P_A = x_{A,liquid} P_{A,pure}$, where $x_{A,liquid}$ = mole fraction of A in the liquid mixture
  - → $P_B = x_{B,liquid} P_{B,pure}$, where $x_{B,liquid}$ = mole fraction of B in the liquid mixture
  - → For a binary mixture, $x_{A,liquid} + x_{B,liquid} = 1$
  - → $P_A$ and $P_B$ are the partial pressures of A and B, respectively, in the vapor above the solution. $P_{total}$ is the total pressure above the solution. Dalton's law of partial pressures for ideal gases is $P_{total} = P_A + P_B$ (see Section 4.11).

$\rightarrow$ Combination of Raoult's law and Dalton's law produces the expression

$$P_{\text{total}} = P_A + P_B = x_{A,\text{liquid}}P_{A,\text{pure}} + x_{B,\text{liquid}}P_{B,\text{pure}}$$

$\rightarrow$ Nearly ideal solutions with two volatile components are formed when the components are very similar (hexane/octane and benzene/toluene), and both components are completely soluble in each other (miscible). The composition may range from $0 < x < 1$; the labels "solute" and "solvent" are not usually used.

- **Composition of the vapor above a binary ideal solution**
  - $\rightarrow$ Each vapor component obeys Dalton's law.
  - $\rightarrow$ $P_A = x_{A,\text{vapor}}P$, where $x_{A,\text{vapor}}$ = mole fraction of A in the vapor mixture and $P = P_{\text{total}}$
  - $\rightarrow$ $P_B = x_{B,\text{vapor}}P$, where $x_{B,\text{vapor}}$ = mole fraction of B in the vapor mixture
  - $\rightarrow$ For a binary mixture, $x_{A,\text{vapor}} + x_{B,\text{vapor}} = 1$
  - $\rightarrow$ Combination of Raoult's law and Dalton's law yields an expression for $x_{A,\text{vapor}}$

$$x_{A,\text{vapor}} = \frac{P_A}{P} = \frac{P_A}{P_A + P_B} = \frac{x_{A,\text{liquid}}P_{A,\text{pure}}}{x_{A,\text{liquid}}P_{A,\text{pure}} + x_{B,\text{liquid}}P_{B,\text{pure}}}$$

  - $\rightarrow$ If $P_{A,\text{pure}} \neq P_{B,\text{pure}}$, then $x_{A,\text{vapor}} \neq x_{A,\text{liquid}}$ and $x_{B,\text{vapor}} \neq x_{B,\text{liquid}}$
    The liquid and vapor compositions differ and the vapor is always *richer* in the more volatile component.

### 8.19 Distillation

- **Temperature-composition phase diagram (ideal solution)**
  - $\rightarrow$ Plot of the temperature of an equilibrium mixture *vs.* composition (mole fraction of one component)
  - $\rightarrow$ Two curves are plotted on the same diagram, $T$ *vs.* $x_{\text{liquid}}$ and $T$ *vs.* $x_{\text{vapor}}$.
  - $\rightarrow$ At a given temperature, the composition of each phase at equilibrium is given.
  - $\rightarrow$ See Fig. 8.37 in the text for a graphical illustration of a temperature-composition diagram.

- **Distillation** $\rightarrow$ Purification of a liquid by evaporation and condensation

- **Distillate** $\rightarrow$ Vapor produced during distillation that is condensed and collected in the final stage

- **Fractional distillation**
  - $\rightarrow$ Continuous separation (purification) of two or more liquids by repeated evaporation and condensation steps on a vertical column
  - $\rightarrow$ Each step takes place on a fractionating column in small increments.
  - $\rightarrow$ Liquid and vapor are in equilibrium at each point in the column, but their compositions vary with height as does the temperature.
  - $\rightarrow$ At the top of the column, condensed vapor (distillate) is collected in a series of samples or *fractions*; the most volatile fraction is collected first.
  - $\rightarrow$ Progress may be displayed on a temperature-composition phase diagram.

## 8.20 Azeotropes

- **Nonideal solutions with volatile components**
  - → Most mixtures of liquids are not ideal.
  - → For *nonideal solutions* with *volatile* components: Raoult's law is not obeyed, the enthalpy of mixing, $\Delta H_{mix} \neq 0$, and solute-solvent interactions are different from solvent-solvent interactions.

- **Azeotrope**
  - → A solution that, like a pure liquid, distills at a constant temperature without a change in composition.
  - → At the azeotrope temperature, $x_{liquid} = x_{vapor}$ for each component

- **Positive deviation from Raoult's law**
  - → Vapor pressure of the mixture is *greater* than the value predicted by Raoult's law.
  - → Enthalpy of mixing is endothermic, $\Delta H_{mix} > 0$.
  - → Solute-solvent interactions are *weaker* than solute-solute interactions.
  - → See Fig. 8.40a for an illustration of the vapor-pressure behavior of a mixture of ethanol and benzene. To make an ethanol-benzene solution, strong hydrogen bonds in ethanol are broken and replaced by weaker ethanol-benzene London interactions. Weaker interactions *increase* the vapor pressure and *decrease* the boiling temperature.

- **Minimum-boiling azeotrope**
  - → Forms if the minimum boiling temperature is less than that of each pure liquid
  - → Distillation yields azeotrope as the distillate (see Fig. 8.41 in text).

- **Negative deviation from Raoult's law**
  - → Vapor pressure of the mixture is *smaller* than the value predicted by Raoult's law.
  - → Enthalpy of mixing is exothermic, $\Delta H_{mix} < 0$.
  - → Solute-solvent interactions are *stronger* than solute-solute interactions
  - → See Fig. 8.40b for an illustration of the vapor-pressure behavior of a mixture of acetone and chloroform. In the acetone-chloroform solution, strong interactions, similar to hydrogen bonding, occur between a lone pair of electrons on the O atom of acetone and the H atom of chloroform. Stronger solute-solvent interactions *decrease* the vapor pressure and *increase* the boiling temperature.

- **Maximum-boiling azeotrope**
  - → Forms if the maximum boiling temperature is greater than each of the pure liquids
  - → Distillation yields one pure component as the distillate (see Fig. 8.42 in text).

# Chapter 9   Chemical Equilibria

## Reactions at Equilibrium  (Sections 9.1–9.3)

### 9.1  The Reversibility of Reactions

- **Chemical reaction**
  - → Reaction mixture approaches a state of dynamic equilibrium.
  - → At equilibrium, rates of the forward and reverse reactions are equal.

    **Example:** Forward reaction:   $H_2(g) + I_2(g) \rightarrow 2\,HI(g)$

    Reverse reaction:   $2\,HI(g) \rightarrow H_2(g) + I_2(g)$

    Dynamic equilibrium:   $H_2(g) + I_2(g) \rightleftharpoons 2\,HI(g)$  *or*  $2\,HI(g) \rightleftharpoons H_2(g) + I_2(g)$

    At chemical equilibrium, the reaction may be represented either way.

### 9.2  Equilibrium and the Law of Mass Action

- **Law of mass action**
  - → The composition of a reaction mixture can be expressed in terms of an *equilibrium constant K* that is unitless.
  - → The equilibrium constant $K$ has the form:

    $$K = \left\{ \frac{\text{activities of products}}{\text{activities of reactants}} \right\}_{\text{equilibrium}}$$

- *Activity* **of substance J,** $a_J$
  - → Partial pressure or concentration of a substance relative to its standard value
  - → Pure number that is unitless
  - → Idealized systems

    $$a_J = \frac{P_J}{P_J{}^\circ} \quad \text{for an ideal gas}$$

    partial pressure of the gas (bar) divided by the pressure of the gas in its standard state (1 bar)

    $$a_J = \frac{[J]}{[J]^\circ} \quad \text{for a solute in a dilute solution}$$

    molarity of substance J divided its molarity in the standard state (1 mol·L$^{-1}$)

    $$a_J = 1 \quad \text{for a pure solid or liquid}$$

  - → Because the standard state values of activity are unity, they are omitted in the expression for activity and numerically.

    $$a_J = \frac{P_J}{P_J{}^\circ} = \frac{P_J}{1} = P_J \quad \text{or} \quad \boxed{a_J = P_J \text{ for an ideal gas}}$$

    Similarly, $\boxed{a_J = [J] \text{ for a solute in a dilute solution}}$  and

    $\boxed{a_J = 1 \text{ for a pure solid or liquid}}$

- *Activity coefficient,* $\gamma_J$
  - → Pure number that is unitless

→ Equal to 1 for ideal gases and solutes in dilute solutions

→ Accounts for intermolecular interactions

→ Used for concentrated solutions and gases that do not behave ideally

$a_J = \gamma_J P_J$  for a real gas

$a_J = \gamma_J[J]$  for any concentration of solute in a solution

$a_J = 1$  for a pure solid or liquid

- **Form of the equilibrium constant**

  → For a general reaction: $a\,A + b\,B \rightarrow c\,C + d\,D$

$$K = \left(\frac{a_C^c a_D^d}{a_A^a a_B^b}\right)_{equilibrium}$$  law of mass action, $K$ is unitless

  → $K$ is the ratio of the products raised to powers equal to their stoichiometric coefficient to the corresponding expression for the reactants.

  → Use activities given in the boxed equations on the previous page to write expressions for $K$.

  → Distinguish between *homogeneous* and *heterogeneous* equilibria.

- **Homogeneous equilibria** → Reactants and products all in the same phase

- **Heterogeneous equilibria** → Reactants and products with different phases
  Activities of pure (or nearly pure) solids and liquids are set equal to 1.

## 9.3  The Thermodynamics Origin of Equilibrium Constants

- **$\Delta G_r$ as a function of the composition of the reaction mixture**

  → $G_m(J)$ is the *molar free energy* of the reactant or product J.

  → $G_m°(J)$ is the *molar free energy* of the reactant or product J *in the standard state*.

  → $\boxed{G_m(J) = G_m°(J) + RT \ln a_J}$  for any substance in any state

  → $\boxed{G_m(J) = G_m°(J) + RT \ln P_J}$  for ideal gases from Section 8.2

  → $\boxed{G_m(J) = G_m°(J) + RT \ln[J]}$  for solutes in dilute solution

  → $\boxed{G_m(J) = G_m°(J)}$  for pure solids and liquids ($a = 1$)

- **$\Delta G_m$ for changes in pressure or concentration in idealized systems**

  → Standard state of a substance is the pure form at 1 bar.

  → Standard state of a solute is the concentration of $1\ mol \cdot L^{-1}$.

- **$\Delta G_r$ for chemical reactions (see Section 7.13)**

  → Generalized chemical reaction: $a\,A + b\,B \rightarrow c\,C + d\,D$

  → $\boxed{\Delta G_r° = \sum n G_m°(\text{products}) - \sum n G_m°(\text{reactants})}$

$\rightarrow$ $\Delta G_r^\circ$, the *standard reaction free energy*, is the difference in molar free energies of the products and reactants in their standard states. The stoichiometric coefficients, $n$, are used as pure numbers.

$\rightarrow$ Evaluated by $\boxed{\Delta G_r^\circ = \sum n \Delta G_f^\circ (\text{products}) - \sum n \Delta G_f^\circ (\text{reactants})}$

$\rightarrow$ Free energies of formation, $\Delta G_f^\circ$, for a variety of substances are given in Appendix 2A.

$\rightarrow$ $\boxed{\Delta G_r = \sum n G_m (\text{products}) - \sum n G_m (\text{reactants})}$

$\rightarrow$ $\Delta G_r$ is the reaction free energy at any definite, fixed composition of the reaction mixture.

$\rightarrow$ Evaluated by $\boxed{\Delta G_r = \Delta G_r^\circ + RT \ln Q}$

$\rightarrow$ $Q$ is called the **reaction quotient.**

> Real systems: $\boxed{Q = \dfrac{a_C^c a_D^d}{a_A^a a_B^b}}$

$\rightarrow$ For ideal gases or dilute solutions, $Q$ reduces to:

> Ideal gases: $\boxed{Q = \dfrac{P_C^c P_D^d}{P_A^a P_B^b}}$     $Q$ is unitless.

> Dilute solutions: $\boxed{Q = \dfrac{[C]^c [D]^d}{[A]^a [B]^b}}$     $Q$ is unitless.

- **Chemical Equilibrium**
  $\rightarrow$ $\Delta G_r = 0$ and $Q = K$ (equilibrium constant)
  $\rightarrow$ $K$ has the same form as $Q$, but it is determined uniquely by the equilibrium composition of the reaction system.
  $\rightarrow$ If $Q < K$, reactants are in excess and reaction proceeds towards products ($\rightarrow$).
  $\rightarrow$ If $Q > K$, products are in excess and reaction proceeds towards reactants ($\leftarrow$).
  $\rightarrow$ If $Q = K$, reactants and products are at chemical equilibrium ($\rightleftharpoons$).

  $\rightarrow$ Real systems: $\boxed{K = \left( \dfrac{a_C^c a_D^d}{a_A^a a_B^b} \right)_{\text{equilibrium}}}$     law of mass action, $K$ is unitless.

  $\rightarrow$ Activities can be expressed in concentration units multiplied by an activity coefficient.

  $\rightarrow$ For a gas, $a_J = \dfrac{\gamma_J P_J}{P^\circ} = \gamma_J P_J$ numerically. For a solute, $a_J = \dfrac{\gamma_J [J]}{[J]^\circ} = [J]$ numerically.

- **Evaluation of $K$** $\rightarrow$ $\boxed{\Delta G_r^\circ = -RT \ln K}$

- **Thermodynamically,** $K = e^{-\Delta H_r^\circ / RT} e^{\Delta S_r^\circ / R}$

- **In a chemical reaction, products are favored if**
  $\rightarrow$ $K$ is large

→ The reaction is exothermic in the standard state (negative $\Delta H_r^\circ$, the more negative the better)

→ The reaction has a positive standard entropy of reaction ($\Delta S_r^\circ > 0$, final state more random)

---

# Equilibrium Constants  (Sections 9.4–9.8)

## 9.4  The Equilibrium Constant in Terms of Molar Concentrations of Gases

- **Gas-phase equilibria  (ideal gas reaction mixture)**

  → $a\,\mathrm{A\,(g)} + b\,\mathrm{B\,(g)} \rightleftharpoons c\,\mathrm{C\,(g)} + d\,\mathrm{D\,(g)}$   (general gas phase reaction at equilibrium)

  → $\boxed{K = \left(\dfrac{P_C^c P_D^d}{P_A^a P_B^b}\right)_{\text{equilibrium}}}$   ideal gases (activity = partial pressure)

  → $\boxed{K_c = \left(\dfrac{[\mathrm{C}]^c [\mathrm{D}]^d}{[\mathrm{A}]^a [\mathrm{B}]^b}\right)_{\text{equilibrium}}}$   ideal gases (activity = molar concentration)

  → Both $K$ and $K_c$ are unitless.

  → In general, $P_J = \dfrac{n_J RT}{V} = RT \dfrac{n_J}{V} = RT\,[\mathrm{J}]$

  → Relationship between $K$ and $K_c$ as derived in the text: $\boxed{K = (RT)^{\Delta n}\,K_c}$

  → $\Delta n = (c + d) - (a + b) = \Sigma(\text{coefficients of products}) - \Sigma(\text{coefficients of reactants})$

## 9.5  Alternative Forms of the Equilibrium Constant

- **Dependence of $K$ on the direction of reaction and the balancing coefficients**

  → $a\,\mathrm{A\,(g)} + b\,\mathrm{B\,(g)} \rightleftharpoons c\,\mathrm{C\,(g)} + d\,\mathrm{D\,(g)}$     $\Rightarrow$   $K_1 = \left(\dfrac{a_C^c a_D^d}{a_A^a a_B^b}\right)_{\text{equilibrium}}$

  → $c\,\mathrm{C\,(g)} + d\,\mathrm{D\,(g)} \rightleftharpoons a\,\mathrm{A\,(g)} + b\,\mathrm{B\,(g)}$     $\Rightarrow$   $K_2 = \left(\dfrac{a_A^a a_B^b}{a_C^c a_D^d}\right)_{\text{equilibrium}} = \dfrac{1}{K_1} = K_1^{-1}$

  → $na\,\mathrm{A\,(g)} + nb\,\mathrm{B\,(g)} \rightleftharpoons nc\,\mathrm{C\,(g)} + nd\,\mathrm{D\,(g)}$ $\Rightarrow$ $K_3 = \left(\dfrac{a_C^{nc} a_D^{nd}}{a_A^{na} a_B^{nb}}\right)_{\text{equilibrium}} = K_1^{n}$

- **Combining chemical reactions**
  → Adding reaction 1 with equilibrium constant $K_1$ to reaction 2 with equilibrium constant $K_2$ yields a combined reaction with equilibrium constant $K$ that is the *product* of $K_1$ and $K_2$ ($\Delta G_1^\circ + \Delta G_2^\circ$) leads to $K = K_1 \times K_2$

→ Subtracting reaction 2 with individual equilibrium constant $K_2$ from reaction 1 with equilibrium constant $K_1$ yields a combined reaction with equilibrium constant $K$ that is the *quotient* of $K_1$ and $K_2$.

$(\Delta G_1^\circ - \Delta G_2^\circ)$ leads to $K = K_1 \div K_2 = K_1 \times (K_2)^{-1}$

## 9.6 The Extent of Reaction

- **Guidelines for attaining equilibrium**
  - → If $K > 10^3$, products are favored.
  - → If $10^{-3} < K < 10^3$, neither reactants nor products are strongly favored.
  - → If $K < 10^{-3}$, reactants are favored.

    **Example:** At a certain temperature, suppose that the equilibrium concentrations of nitrogen and ammonia in the reaction

    $$N_2\,(g) + 3\,H_2\,(s) \rightleftharpoons 2\,NH_3\,(g)$$

    were found to be 0.11 M and 1.5 M, respectively. What is the equilibrium concentration of $H_2$ if $K_c = 4.4 \times 10^4$?

    The value of $K_c$ is in the range that strongly favors products. Solving the equilibrium constant expression requires the evaluation of the cube root of the concentration of $H_2$. Use the $y^x$ key on your calculator.

    $$K_c = \left(\frac{[NH_3]^2}{[N_2][H_2]^3}\right)_{eq} \quad \text{and} \quad [H_2]^3 = \frac{[NH_3]^2}{K_c[N_2]}$$

    $$[H_2] = \sqrt[3]{\frac{[NH_3]^2}{K_c[N_2]}} = \left(\frac{(1.5)^2}{(4.4\times10^4)(0.11)}\right)^{1/3} = \left(4.65\times10^{-4}\right)^{1/3} = 0.077 \text{ M}$$

## 9.7 The Direction of Reaction

- **For the general reaction** $a\,A\,(g) + b\,B\,(g) \rightleftharpoons c\,C\,(g) + d\,D\,(g)$

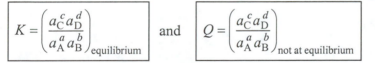

$$K = \left(\frac{a_C^c\, a_D^d}{a_A^a\, a_B^b}\right)_{\text{equilibrium}} \quad \text{and} \quad Q = \left(\frac{a_C^c\, a_D^d}{a_A^a\, a_B^b}\right)_{\text{not at equilibrium}}$$

- **Guidelines for approach to equilibrium**
  - → $Q$ versus $K$ (see Example 9.3d in this *Study Guide*)
  - → If $Q > K$, the concentration of products is too high or the concentration of reactants is too low. Reaction proceeds in the reverse direction, toward reactants.
  - → If $Q < K$, the concentration of reactants is too high or the concentration of products is too low. Reaction proceeds to form products in the forward direction.
  - → If $Q = K$, the mixture has its equilibrium composition and no change occurs.

## 9.8 Using Equilibrium Constants

- **Equilibrium table** → Very helpful in solving all types of equilibrium problems
- **Format and use of the equilibrium table:**
  **Step 1** Reactant and product species taking part in the reaction are identified.

**Step 2** The initial composition of the system is listed.

**Step 3** Changes needed to reach equilibrium are listed in terms of *one unknown quantity.*

**Step 4** The equilibrium composition in terms of one unknown is entered on the last line.

**Step 5** The equilibrium composition is entered into the equilibrium constant expression and the unknown quantity (equilibrium constant or reaction species concentration) is determined.

**Step 6** Approximations may sometimes be used to simplify the calculations.

- **Calculating equilibrium constants from pressures or concentrations**

    **Example:** A sample of $HI\,(g)$ is placed in a flask at a pressure of 0.100 bar. After equilibrium is attained, the partial pressure of $HI\,(g)$ is 0.050 bar. Evaluate $K$ for the reaction

    $$2\,HI\,(g) \rightleftharpoons H_2\,(g) + I_2\,(g)$$

    Set up an equilibrium table. In the initial system, line 1, there are no products so some must form to reach equilibrium. On line 2, let $x$ be the change in pressure of $I_2$ required to reach equilibrium. From the stoichiometric coefficients, the change in the pressure of $H_2$ is $x$ while the change in pressure of HI is $-2x$ (HI *disappears twice as fast as iodine appears*). Lines 1 and 2 of the table are added to give the equilibrium pressures of the reactants and products in terms of $x$. The value of $x$ is determined from the change in pressure of HI $(-2x)$ from the initial to the equilibrium value. Using the calculated value of $x$, the equilibrium pressures from line 3 of the equilibrium table are calculated and entered into the equilibrium constant expression to obtain $K$.

    | | Species | | |
    |---|---|---|---|
    | | HI | $H_2$ | $I_2$ |
    | 1. Initial pressure | 0.100 | 0 | 0 |
    | 2. Change in pressure | $-2x$ | $+x$ | $+x$ |
    | 3. Equilibrium pressure | $0.100 - 2x$ | $x$ | $x$ |

    $P_{HI} = 0.100 - 2x = 0.050$   and   $x = 0.025 = P_{H_2} = P_{I_2}$

    $$K = \frac{P_{H_2} P_{I_2}}{P_{HI}^2} = \frac{(0.025)^2}{(0.050)^2} = 0.25$$

- **Calculating the equilibrium composition**

    **Example:** A sample of ethane, $C_2H_6\,(g)$, is placed in a flask at a pressure of 2.00 bar at 900 K. The equilibrium constant $K$ for the gas phase dehydrogenation of ethane to form ethylene

    $$C_2H_6\,(g) \rightleftharpoons C_2H_4\,(g) + H_2\,(g)$$

    is 0.050 at 900 K. Determine the composition of the equilibrium mixture and the percent decomposition of $C_2H_6$.

    Set up an equilibrium table and let $x$ be the unknown equilibrium pressure of each product, $C_2H_4$ and $H_2$. The equilibrium pressure of $C_2H_6$ is then $2.00 - x$. Solve for $x$ by rearranging the equilibrium constant expression. The new form is a quadratic equation (see Example 9.7 in the text), which is solved exactly by the quadratic formula. The equilibrium pressures can then be obtained.

The percent decomposition is the loss of reactant divided by the initial pressure times 100% or $x/2.00 \times 100\%$.

| | Species | | |
| --- | --- | --- | --- |
| | $C_2H_6$ | $C_2H_4$ | $H_2$ |
| 1. Initial pressure | 2.00 | 0 | 0 |
| 2. Change in pressure | $-x$ | $+x$ | $+x$ |
| 3. Equilibrium pressure | $2.00 - x$ | $x$ | $x$ |

$$K = \frac{P_{C_2H_4}P_{H_2}}{P_{C_2H_6}} = \frac{(x)^2}{(2.00 - x)} = 0.050$$

Rearranging yields a quadratic equation: $x^2 + (0.050)x - 0.10 = 0$

Recall: $ax^2 + bx + c = 0 \Rightarrow x = \dfrac{-b \pm \sqrt{b^2 - 4ac}}{2a}$ (exact solution)

Then, $a = 1$, $b = 0.050$, and $c = -0.10$.

$$x = \frac{-(0.050) \pm \sqrt{(0.050)^2 - 4(1)(-0.10)}}{2(1)} = \frac{-(0.050) \pm \sqrt{0.4025}}{2}$$

$$= \frac{-(0.050) \pm (0.6344)}{2} = \frac{-(0.050) + (0.6344)}{2} = \frac{0.5844}{2} = 0.29$$

The negative root is rejected because it is unphysical.

$P_{C_2H_4} = P_{H_2} = x = 0.29$ bar   and   $P_{C_2H_6} = 2.00 - x = 1.71$ bar

$$\text{Percent decomposition} = \frac{\text{change in pressure}}{\text{initial pressure}} \times 100\% = \frac{0.29}{2.00} \times 100\% = 14.5\%$$

**Note:** If $K$ is less than $10^{-3}$ in a problem of this type, the adjustment to equilibrium is small. The pressure of reactant then changes by only a small amount. In this case, an approximate solution that avoids the quadratic equation is possible.

# The Response of Equilibria to Changes in Conditions
## (Sections 9.9–9.13)

## 9.9 Adding and Removing Reagents

- **Qualitative features**
  - → Use Le Chatelier's principle to predict the qualitative adjustments needed to obtain a new equilibrium composition that relieves the stress.
  - → Assessing the effects of the applied stress on $Q$ relative to $K$ is an alternative approach.

    **Example:** Use Le Chatelier's principle and reaction quotient arguments to describe the effect on the equilibrium

$$2\,SO_2\,(g) + O_2\,(g) \rightleftharpoons 2\,SO_3\,(g) \qquad Q = \left( \frac{P_{SO_3}^2}{P_{SO_2}^2 \, P_{O_2}} \right)$$

of (a) addition of $SO_2$, (b) removal of $O_2$, (c) addition of $SO_3$, and (d) removal of $SO_3$. Both reactant and product pressures appear in the equilibrium constant expression. Assuming no change in volume or temperature, adding reactant (or removing product) shifts the equilibrium composition to produce more product molecules, thereby helping to relieve the stress. Adding product (or removing reactant) shifts the equilibrium composition to produce more reactant molecules, thereby helping to relieve the stress.

(a) Adding reactant $SO_2$ results in $Q < K$. The system adjusts by producing more product $SO_3$ (consuming $SO_2$ and $O_2$) until $Q = K$.

(b) Removing reactant $O_2$ results in $Q > K$. The system adjusts by consuming product $SO_3$ (producing more $SO_2$ and $O_2$) until $Q = K$.

(c) Adding $SO_3$ results in $Q > K$. The system adjusts by producing more $SO_2$ and $O_2$ (consuming $SO_3$) until $Q = K$.

(d) Removing $SO_3$ results in $Q < K$. The system adjusts by consuming $SO_2$ and $O_2$ (producing more $SO_3$) until $Q = K$.

**Note:** In this example, the stress changes only the value of $Q$. The composition adjusts to return the value of $Q$ to the original value of $K$. The equilibrium constant does not change. The equilibrium constant has, however, a very weak pressure dependence that is usually safe to ignore.

- **Quantitative features** $\rightarrow$ Use the equilibrium table method to calculate the new composition of an equilibrium system after a stress is applied.

**Example:** Assume that the reaction, $2\,HI\,(g) \rightleftharpoons H_2\,(g) + I_2\,(g)$, has the equilibrium composition determined in Example 9.8a. If HI, $H_2$ and $I_2$ are added to the system in amounts equivalent to 0.008 bar for HI, 0.008 bar for $H_2$, and 0.008 bar for $I_2$, predict the direction of change and calculate the new equilibrium composition.

From the first example in Section 9.8, the equilibrium pressures are $P_{HI} = 0.050$ bar, $P_{H_2} = P_{I_2} = 0.025$ bar; and $K = 0.25$. After adding 0.008 bar to each reactant and product, the *nonequlibrium* pressures are $P_{HI} = 0.058$ bar, $P_{H_2} = 0.033$ bar, and $P_{I_2} = 0.033$ bar. $Q = \dfrac{(0.033)(0.033)}{(0.058)^2} = 0.324 > K$ and the adjustment to the new equilibrium is in the direction toward the reactants.

Quantitative treatment: Set up an equilibrium table, letting $-x$ be the unknown adjustment to the initial *nonequilibrium* pressures of each product $H_2$ and $I_2$. The value of $+2x$ is then the adjustment to the initial *nonequilibrium* pressure of HI. A negative value of $x$ is expected from our qualitative analysis, which suggests that reactants should form from products to relieve the stress on the system.

|  | Species | | |
| --- | --- | --- | --- |
|  | HI | $H_2$ | $I_2$ |
| 1. Initial pressure | 0.058 | 0.033 | 0.033 |
| 2. Change in pressure | $+2x$ | $-x$ | $-x$ |
| 3. Equilibrium pressure | $0.058 + 2x$ | $0.033 - x$ | $0.033 - x$ |

$$K = \frac{P_{H_2}P_{I_2}}{P_{HI}^2} = \frac{(0.033 - x)^2}{(0.058 + 2x)^2} = 0.25$$

This expression is relatively easy to solve for $x$ by taking the square-root of both sides:

$$\sqrt{K} = \frac{(0.033 - x)}{(0.058 + 2x)} = \sqrt{0.25} = 0.50$$

$0.033 - x = 0.029 + x; \quad 2x = 0.004; \quad x = 0.002$

$P_{H_2} = P_{I_2} = 0.033 - 0.002 = 0.031$ bar $\quad and \quad P_{HI} = 0.058 + 0.004 = 0.062$ bar

Equilibrium expression check: $K = \frac{(0.031)^2}{(0.062)^2} = 0.25$ as expected

## 9.10 Compressing a Reaction Mixture

- **Decrease in volume of a reaction mixture at constant temperature**
  - → The partial pressure of each component J of the reaction mixture increases;
  
    recall $P_J = \frac{n_J RT}{V}$ (see Section 4.11)
  
  - → To minimize the increase in pressure, the equilibrium shifts to form fewer gas molecules, if possible.

- **Increase in volume of a reaction mixture at constant temperature**
  - → The partial pressure of each member J of the reaction mixture decreases; recall $P_J = \frac{n_J RT}{V}$
  
  - → To compensate for the decrease in pressure, the equilibrium shifts to form more gas molecules, if possible.

- **Increase in pressure of a reaction mixture by introducing an inert gas**
  - → The partial pressure of each member J of the reaction mixture does not change;
  
    recall $P_J = \frac{n_J RT}{V}$
  
  - → The equilibrium composition is unaffected. The equilibrium constant has a weak pressure dependence that is safe to ignore unless the pressure change is large.

- **Increase in volume of a reaction mixture by introducing an inert gas**
  - → Accomplished by adding an inert gas to a reaction mixture at constant *total* pressure
  - → Increases the *volume* of the system without changing the amount or reactants or products
  - → The partial pressure of each member J of the reaction mixture decreases; recall $P_J = \frac{n_J RT}{V}$
  - → The return to equilibrium is equivalent to that which occurs by increasing the volume or decreasing the pressure of the system (see Example 9.10 above).

## 9.11 Temperature and Equilibrium

- **Qualitative features** → Use Le Chatelier's principle to predict the qualitative adjustments needed to obtain a new equilibrium composition that relieves the stress.

- **For *endothermic* reactions, heat is treated as a *reactant*.**
  - → *Increasing* the temperature supplies heat to the reaction mixture and the composition of the mixture adjusts to form more product molecules.
  - → The value of the equilibrium constant, *K, increases.*

→ *Decreasing* the temperature removes heat from the reaction mixture and the composition of the mixture adjusts to form more reactant molecules.

→ The value of the equilibrium constant, *K, decreases*.

- **For *exothermic* reactions, heat is treated as a *product*.**

  → *Increasing* the temperature supplies heat to the reaction mixture and the composition of the mixture adjusts to form more reactant molecules.

  → The value of the equilibrium constant, *K, decreases*.

  → *Decreasing* the temperature removes heat from the reaction mixture and the composition of the mixture adjusts to form more product molecules.

  → The value of the equilibrium constant, *K, increases*.

- **Quantitative features**

  → van't Hoff equation $\boxed{\ln\dfrac{K_2}{K_1} = \dfrac{\Delta H_r^\circ}{R}\left(\dfrac{1}{T_1} - \dfrac{1}{T_2}\right)}$ (see derivation in the text)

  → The equilibrium constant is $K$, not $K_c$.

  → A major assumption in the derivation is that both $\Delta H_r^\circ$ and $\Delta S_r^\circ$ are independent of temperature between $T_1$ and $T_2$.

## 9.12  Catalysts and Haber's Achievement

- **Catalyst increases the rate at which a reaction approaches equilibrium.**

  → Rate of forward reaction is increased.

  → Rate of reverse reaction is increased.

  → Dynamic equilibrium is unaffected.

  → The identity of the catalyst is unchanged. The catalyst is *neither* a reactant *nor* a product. It does *not* appear in the chemical equation.

- **Haber's achievement**

  → Production of ammonia: $3\,H_2(g) + N_2(g) \rightleftharpoons 2\,NH_3(g)$
  At 298.15 K, $\Delta H_r^\circ = -92.22$ kJ·mol$^{-1}$, $\Delta G_r^\circ = -32.90$ kJ·mol$^{-1}$, and $K = 5.80 \times 10^5$

  → Compression of the gas mixture favors product ammonia.

  → Removal of ammonia as it is formed encourages more to be formed.

  → Increasing the temperature is required to increase the slow rate of the reaction at 298.15 K, but it adversely affects the position of the equilibrium.

  → To circumvent the problem, an appropriate catalyst (a mixture of $Fe_2O_3$ and $Fe_3O_4$) was discovered at about the time of World War I. It is still used today.

## 9.13  Impact on Biology:  Homeostasis

- **Homeostasis**

  → Mechanism similar to chemical equilibrium

  → Maintenance of constant internal conditions

  → Allows living organisms to control biological processes at a constant level

  → An example is the equilibrium involving oxygen and hemoglobin, Hb.

  $$Hb(aq) + O_2(aq) \rightleftharpoons HbO_2(aq) \quad \text{[see text discussion]}$$

# Chapter 10    Acids and Bases

## Proton Transfer Reactions  (Sections 10.1–10.6)

### 10.1  Brønsted–Lowry Acids and Bases

- **Brønsted–Lowry definitions**
    - → *Acid:* proton donor;  *base:* proton acceptor
    - → General theory for *any* solvent (for example, $H_2O$ or $NH_3$) or *no* solvent at all
    - → Acid (*proton donor*) is *deprotonated* when reacting with a base.
    - → Base (*proton acceptor*) is *protonated* when reacting with an acid.
    - → Acids and bases react with each other in proton transfer reactions.
    - → A strong acid is *completely deprotonated.*
        **Example:** $HBr(aq) + H_2O(l) \rightarrow H_3O^+(aq) + Br^-(aq)$  (*strong* acid: reaction goes essentially to completion; proton is transferred from $HBr(aq)$ to $H_2O(l)$).
    - → A strong base is completely *protonated.*
        **Example:** $NH_2^-(aq) + H_2O(l) \rightarrow NH_3(aq) + OH^-(aq)$  (*strong* base: reaction goes essentially to completion; $NH_2^-(aq)$ accepts a proton from $H_2O$).
    - → A weak acid is *partially deprotonated.*
        **Example:** $HF(aq) + H_2O(l) \rightleftharpoons H_3O^+(aq) + F^-(aq)$ (weak acid reaction, a *chemical equilibrium* is established)
    - → A weak base is *partially protonated.*
        **Example:** $NH_3(aq) + H_2O(l) \rightleftharpoons NH_4^+(aq) + OH^-(aq)$ (*weak* base reaction , a *chemical equilibrium* is established).

- **Conjugate base**
    - → Formed from acid after the proton is donated.
    - → Acid $\Rightarrow$ deprotonation $\Rightarrow$ conjugate base
        **Examples:** The conjugate base of $HBr$ is $Br^-$; the conjugate base of $H_3PO_4$ is $H_2PO_4^-$.

- **Conjugate acid**
    - → Formed from base after the proton is accepted.
    - → Base $\Rightarrow$ protonation $\Rightarrow$ conjugate acid
        **Examples:** The conjugate acid of $NH_3$ is $NH_4^+$; the conjugate acid of $CH_3COO^-$ is $CH_3COOH$.

- **Conjugate acid-base pair** → Differ only by the presence of an $H^+$ species:
    $$HCN, CN^-;\ \ H_3O^+, H_2O;\ \ H_2O, OH^-;\ \ H_2S, HS^-$$

- **Solvent**
    - → Need not be water. Other solvents include $HF(l)$, $H_2SO_4(l)$, $CH_3COOH(l)$, $C_2H_5OH(l)$, $NH_3(l)$.
    - → Acid-base reactions can occur with *no* solvent as in: $HCl(g) + NH_3(g) \rightarrow NH_4Cl(s)$.

### 10.2  Lewis Acids and Bases

- **Lewis definitions**
    - → A Lewis acid is an electron pair *acceptor*  **Examples:** $BF_3(g)$, $Fe^{2+}(aq)$, $H_2O(l)$
    - → A Lewis base is an electron pair *donor*  **Examples:** $NH_3(aq)$, $O^{2-}(aq)$
    - → Most general theory  (holds for any solvent and includes metal cations)

- **Comparison of Lewis and Brønsted–Lowry Acids and Bases**
  - → Proton ($H^+$) transfer (Brønsted–Lowry theory) is a special type of Lewis acid-base reaction.
  - → In a proton transfer reaction, $H^+$ acts as a Lewis acid, accepting an electron pair from a Lewis base.
  - → Brønsted acid: supplier of one particular Lewis acid, a proton
  - → Brønsted base: a particular type of Lewis base that can use a lone pair to bond a proton
    - **Example:** The reaction $O^{2-}(aq) + H_2O(l) \rightarrow 2\,OH^-(aq)$ can be viewed as a Brønsted–Lowry reaction ($H_2O$ is the proton donor, $O^{2-}$ is the proton acceptor) or as a Lewis acid-base reaction ($O^{2-}$ is the electron pair donor (Lewis base) , $H^+$ on water is the electron pair acceptor (Lewis acid) as shown in the diagram below.

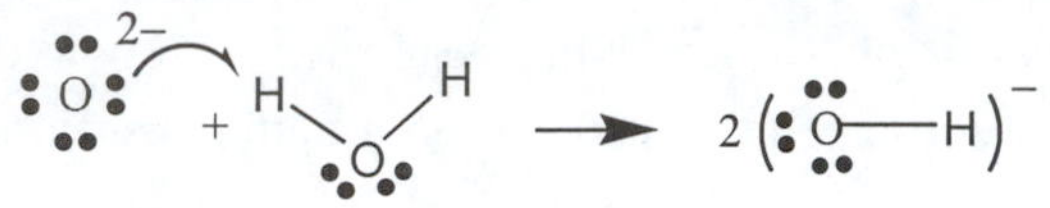

## 10.3 Acidic, Basic, and Amphoteric Oxides

- **Oxides** → React with water as Lewis acids or bases, or both
- **Oxides of nonmetals**
  - → Called *acidic oxides*
  - → Tend to act as Lewis acids  **Examples:** $CO_2$, $SO_3$, $Cl_2O_7$, $N_2O_5$
  - → React with water to form a Brønsted acid  **Examples:**  $CO_2(g) + H_2O(l) \rightarrow H_2CO_3(aq)$
    $$SO_3(g) + H_2O(l) \rightarrow H_2SO_4(aq)$$
    *Lewis acid*          *Brønsted acid*
  - → React with bases to form a salt and water
    - **Example:** $H_2CO_3(aq) + 2\,NaOH(aq) \rightarrow Na_2CO_3(aq) + 2\,H_2O(l)$
- **Oxides of metals**
  - → Called *basic oxides*
  - → Tend to act as Lewis bases  **Examples:** $Na_2O$, $MgO$, $CaO$
  - → React with water to form Brønsted bases  **Example:** $CaO(s) + H_2O(l) \rightarrow Ca(OH)_2(aq)$
    *Lewis base*          *Brønsted base*
  - → React with acids to form a salt and water  **Example:** $CaO(s) + 2\,HBr(aq) \rightarrow CaBr_2(aq) + H_2O(l)$
- **Amphoteric oxides**
  - → Exhibit both acid and base character  **Examples:** $Al_2O_3$, $SnO_2$
  - → React with both acids and bases
    - **Example:** See text for reactions of $Al_2O_3$ as an acid and as a base.
- **Trends in acidity for the oxides of main-group elements**

Increasing acidity ⟶

| 1 | 2 | 13/III | 14/IV | 15/V | 16/VI | 17/VII |
|---|---|---|---|---|---|---|
| $Li_2O$ | **BeO** | $B_2O_3$ | $CO_2$ | $N_2O_5$ | $(O_2)$ | $OF_2$ |
| $Na_2O$ | $MgO$ | **$Al_2O_3$** | $SiO_2$ | $P_4O_{10}$ | $SO_3$ | $Cl_2O_7$ |
| $K_2O$ | $CaO$ | **$Ga_2O_3$** | $GeO_2$ | $As_2O_5$ | $SeO_3$ | $Br_2O_7$ |
| $Rb_2O$ | $SrO$ | $In_2O_3$ | **$SnO_2$** | $Sb_2O_5$ | $TeO_3$ | $I_2O_7$ |
| $Cs_2O$ | $BaO$ | $Tl_2O_3$ | **$PbO_2$** | $Bi_2O_5$ | $PoO_3$ | $At_2O_7$ |

Increasing acidity ⟶

**Note:** Acidity of oxides in water tends to increase as shown in the preceding table. Amphoteric oxides are in bold type, basic oxides are to their left and acidic oxides to their right. Exceptions are $O_2$, which is neutral, and $OF_2$, which is only weakly acidic.

## 10.4  Proton Exchange Between Water Molecules

- **Amphiprotic species** $\rightarrow$ A molecule or ion that can act as either a Brønsted acid ($H^+$ *donor*) or a Brønsted base ($H^+$ *acceptor*).

    **Examples:** $H_2O$: $H^+$ *donor* ($\rightarrow OH^-$) or $H^+$ *acceptor* ($\rightarrow H_3O^+$)

    $NH_3$: $H^+$ *donor* ($\rightarrow NH_2^-$) or $H^+$ *acceptor* ($\rightarrow NH_4^+$)

- **Autoprotolysis** $\rightarrow$ A reaction in which one molecule transfers a proton to another molecule of the same kind

    **Examples:** Autoprotolysis of water: $2\,H_2O(l) \rightleftharpoons H_3O^+(aq) + OH^-(aq)$

    Autoprotolysis of ammonia: $2\,NH_3(l) \rightleftharpoons NH_4^+(am) + NH_2^-(am)$

- **Equilibrium constant** $\rightarrow$ Autoprotolysis of water:
    $$2H_2O(l) \rightleftharpoons H_3O^+(aq) + OH^-(aq); \quad K_w = [H_3O^+][OH^-] = 1 \times 10^{-14} \text{ at } 25°C$$

- **Pure water at 25°C** $\rightarrow [H_3O^+] = [OH^-] = 1.0 \times 10^{-7}$ mol·$L^{-1}$

- **Neutral solution at 25°C** $\rightarrow [H_3O^+] = [OH^-] = 1.0 \times 10^{-7}$ mol·$L^{-1}$

- **Acidic solution 25°C** $\rightarrow [H_3O^+] > 1.0 \times 10^{-7}$ mol·$L^{-1}$ and $[OH^-] < 1.0 \times 10^{-7}$ mol·$L^{-1}$

- **Basic solution 25°C** $\rightarrow [H_3O^+] < 1.0 \times 10^{-7}$ mol·$L^{-1}$ and $[OH^-] > 1.0 \times 10^{-7}$ mol·$L^{-1}$

## 10.5  The pH Scale

- **Concentration variation of $H_3O^+$** $\rightarrow$ Values higher than 1 mol·$L^{-1}$ and less than $10^{-14}$ mol·$L^{-1}$ are not commonly encountered.

- **Logarithmic scale**
    - $\rightarrow$ $pH = -\log[H_3O^+]$ *and* $[H_3O^+] = 10^{-pH}$
      pH values range between 1 and 14 in most cases
    - $\rightarrow$ Pure water at 25°C:  pH = 7.00;  $\Rightarrow$ Neutral solution: pH = 7.00
    - $\rightarrow$ Acidic solution:  pH < 7.00;  $\Rightarrow$ Basic solution:  pH > 7.00

    **Note:** pH has *no* units, but $[H_3O^+]$ has units of molarity.

- **Measurement of pH**
    - $\rightarrow$ *Universal indicator paper* turns different colors at different pH values.
    - $\rightarrow$ *pH meter* measures a difference in electrical potential (proportional to pH) across electrodes that dip into the solution (see Chapter 12).

        **Examples:** The pH of the solutions for which the $[H_3O^+] = 5.4 \times 10^{-4}$ mol·$L^{-1}$ is:
        $$pH = -\log[H_3O^+] = -(-3.27) = 3.27 \quad (\textit{acidic} \text{ solution})$$

        The $[H_3O^+]$ for a solution that has a pH value of 1.70 is: $[H_3O^+] = 10^{-pH} = 10^{-1.70}$
        $= 2.0 \times 10^{-2}$ mol·$L^{-1}$

## 10.6  The pOH of Solutions

- **pX as a generalization of pH**
    - $\rightarrow$ $pX = -\log X$ (X = anything)

$\rightarrow \text{pOH} = -\log[\text{OH}^-] \;\; and \;\; [\text{OH}^-] = 10^{-\text{pOH}}$

$\rightarrow \text{p}K_w = -\log K_w = -\log(1.0 \times 10^{-14}) = 14.00 \text{ at } 25°C$

- **Relationship between pH and pOH of a solution**
  $\rightarrow [\text{H}_3\text{O}^+][\text{OH}^-] = K_w = 1.0 \times 10^{-14} \text{ at } 25°C$
  $\rightarrow \log([\text{H}_3\text{O}^+][\text{OH}^-]) = \log[\text{H}_3\text{O}^+] + \log[\text{OH}^-] = \log K_w$
  $\rightarrow -\log[\text{H}_3\text{O}^+] - \log[\text{OH}^-] = -\log K_w$
  $\rightarrow \text{pH} + \text{pOH} = 14.00 \text{ at } 25°C$

---

# Weak Acids and Bases  (Sections 10.7–10.10)

## 10.7  Acidity and Basicity Constants

- **Weak acid** $\rightarrow$  pH is *larger* than that of a strong acid with the same molarity.
  Incomplete deprotonation

- **General weak acid HA**

  $\rightarrow \text{HA(aq)} + \text{H}_2\text{O(l)} \rightleftharpoons \text{H}_3\text{O}^+(\text{aq}) + \text{A}^-(\text{aq}); \;\; K_a = \dfrac{a_{\text{H}_3\text{O}^+} a_{\text{A}^-}}{a_{\text{HA}} a_{\text{H}_2\text{O}}}$

  But, $a_{\text{H}_2\text{O}} = 1$, $a_{\text{HA}} \approx [\text{HA}]$, $a_{\text{H}_3\text{O}^+} \approx [\text{H}_3\text{O}^+]$, and $a_{\text{A}^-} \approx [\text{A}^-]$.

  $\rightarrow$ Thus, $K_a = \dfrac{[\text{H}_3\text{O}^+][\text{A}^-]}{[\text{HA}]}$  $K_a$ is the *acidity constant;*  $\text{p}K_a = -\log K_a$  *commonly used*

### Some Weak Acids and Their Acidity Constants

|  | Species | Formula | $K_a$ | $\text{p}K_a$ |
|---|---|---|---|---|
| **Neutral acids** | hydrofluoric acid | HF | $3.5 \times 10^{-4}$ | 3.45 |
|  | acetic acid | $CH_3COOH$ | $1.8 \times 10^{-5}$ | 4.75 |
|  | hydrocyanic acid | HCN | $4.9 \times 10^{-10}$ | 9.31 |
| **Cation acids** | pyridinium | $C_5H_5^+$ | $5.6 \times 10^{-6}$ | 5.24 |
|  | ammonium | $NH_4^+$ | $5.6 \times 10^{-10}$ | 9.25 |
|  | ethylammonium | $C_2H_5NH_3^+$ | $1.5 \times 10^{-11}$ | 10.82 |
| **Anion acids** | hydrogen sulfate | $HSO_4^-$ | $1.2 \times 10^{-2}$ | 1.92 |
|  | dihydrogen phosphate | $H_2PO_4^-$ | $6.2 \times 10^{-8}$ | 7.21 |

- **In the preceding table**
  $\rightarrow$ Hydrogen sulfate is the strongest acid  (largest $K_a$, *smallest* $\text{p}K_a$).
  $\rightarrow$ Ethylammonium is the weakest acid  (smallest $K_a$, *largest* $\text{p}K_a$).

- **Weak base** → pH is *smaller* than that of a strong base with the same molarity.
  Incomplete protonation

- **General weak base B**

  → $B(aq) + H_2O(l) \rightleftharpoons BH^+(aq) + OH^-(aq); \quad K_b = \dfrac{a_{BH^+} a_{OH^-}}{a_B a_{H_2O}}$

  But $a_{H_2O} = 1$, $a_B \approx [B]$, $a_{BH^+} \approx [BH^+]$, and $a_{OH^-} \approx [OH^-]$

  → Thus, $K_b = \dfrac{[BH^+][OH^-]}{[B]}$   $K_b$ is the *basicity constant;*   $pK_b = -\log K_b$   *commonly used*

### Weak Bases and Their Basicity Constants

|              | Species       | Formula      | $K_b$                | $pK_b$ |
|--------------|---------------|--------------|----------------------|--------|
| **Neutral bases** | ethylamine    | $C_2H_5NH_2$ | $6.5 \times 10^{-4}$ | 3.19   |
|              | methylamine   | $CH_3NH_2$   | $3.6 \times 10^{-4}$ | 3.44   |
|              | ammonia       | $NH_3$       | $1.8 \times 10^{-5}$ | 4.75   |
|              | hydroxylamine | $NH_2OH$     | $1.1 \times 10^{-8}$ | 7.97   |
| **Anion bases**  | phosphate     | $PO_4^{3-}$  | $4.8 \times 10^{-2}$ | 1.32   |
|              | fluoride      | $F^-$        | $2.9 \times 10^{-11}$| 10.54  |

- **In the preceding table**
  → Phosphate, $PO_4^{3-}$, is the strongest base (largest $K_b$, *smallest* $pK_b$).
  → Fluoride, $F^-$, is the weakest base (smallest $K_b$, *largest* $pK_b$).
- **Acid-base strength**
  → The *larger* the value of $K$, the *stronger* the acid or base.
  → The *smaller* the value of $pK$, the *stronger* the acid or base.

## 10.8  The Conjugate Seesaw

- **Reciprocity of conjugate acid-base pairs**

- **Conjugate acid-base pairs of weak acids**
  → The conjugate base of a *weak* acid is a *weak* base.
  → The conjugate acid of a *weak* base is a *weak* acid.
  → *"Weak"* is defined by $pK$ values between 1 and 14.

- **Conjugate seesaw  (qualitative)**
  → The *stronger* an acid, the *weaker* its conjugate base.
  → The *stronger* a base, the *weaker* its conjugate acid.

- **Conjugate seesaw  (quantitative)**
  → Consider a general *weak* acid HA and its conjugate base A⁻.

  $HA(aq) + H_2O(l) \rightleftharpoons H_3O^+(aq) + A^-(aq) \quad K_a = \dfrac{[H_3O^+][A^-]}{[HA]}$

$$A^-(aq) + H_2O(l) \rightleftharpoons HA(aq) + OH^-(aq) \quad K_b = \frac{[HA][OH^-]}{[A^-]}$$

$$K_a \times K_b = \frac{[H_3O^+][A^-]}{[HA]} \times \frac{[HA][OH^-]}{[A^-]} = [H_3O^+][OH^-] = K_w$$

→ Or, for any conjugate pair, $\boxed{K_a \times K_b = K_w}$ and $pK_a + pK_b = pK_w$

→ If $K_a > 1 \times 10^{-7}$ ($pK_a < 7$), then the acid is *stronger* than its conjugate base.

→ If $K_b > 1 \times 10^{-7}$ ($pK_b < 7$), then the base is *stronger* than its conjugate acid.

See Table 10.3 in the text.

**Example**: Sulfurous acid, $H_2SO_3$, is a stronger acid than nitrous acid, $HNO_2$. Consequently, nitrite, $NO_2^-$, is a stronger base than hydrogen sulfite, $HSO_3^-$.

- **Solvent leveling** → An acid is strong if it is a stronger proton donor than the conjugate acid of the solvent.

- **Strong acid in water**

  → *Stronger* proton donor than $H_3O^+$ ($K_a > 1$; $pK_a < 0$)

  → $H_3O^+(aq) + H_2O(l) \rightleftharpoons H_2O(l) + H_3O^+(aq) \qquad K_a = 1$

  → All acids with $K_a > 1$ appear equally strong in water.

  → Strong acids are *leveled* to the strength of the acid $H_3O^+$.

  **Example:** HCl, HBr, HI, $HNO_3$, and $HClO_4$ are strong acids in water. When dissolved in water, they behave as though they were solutions of the acid $H_3O^+$.

## 10.9 Molecular Structure and Acid Strength

- **Binary acids, $H_nX$**

  → For elements in the *same* period, the more *polar* the H–A bond, the *stronger* the acid.
  **Example:** HF is a stronger acid than $H_2O$, which is stronger than $NH_3$.

  → For elements in the *same* group, the *weaker* the H–A bond, the *stronger* the acid.
  **Example:** HBr is a stronger acid than HCl, which is stronger than HF.

- **Summary**

  → For binary acids, acid strength increases ⇒ from left to right across a period (polarity of H–A bond increases), from top to bottom down a group (H–A bond enthalpy decreases).

  → The more polar or the weaker the H–A bond, the stronger the acid.

## 10.10 The Strengths of Oxoacids (also called oxyacids)

- **Oxoacids**

  → The *greater* the number of oxygen atoms attached to the *same* central atom, the *stronger* the acid.

  → This trend also corresponds to the *increasing* oxidation number of the *same* central atom.
  **Example:** $HClO_3$ is a stronger acid than $HClO_2$, which is stronger than $HClO$.

  → For the *same* number of O atoms attached to the central atom, the *greater* the electronegativity of the central atom, the *stronger* the acid.
  **Example:** HClO is a stronger acid than HBrO, which is stronger than HIO. The halogen is considered the central atom for comparison to other oxoacids.

- **Carboxylic acids** → The *greater* the electronegativities of the groups attached to the carboxyl (COOH) group, the *stronger* the acid.

  **Example:** $CHBr_2COOH$ is a stronger acid than $CH_2BrCOOH$, which is stronger than $CH_3COOH$.

**Note:** The preceding rules are summarized in text Tables 10.4, 10.5, and 10.6.

---

# The pH of Solutions of Weak Acids and Bases
## (Sections 10.11–10.13)

### 10.11 Solutions of Weak Acids

- **Initial concentration** → Initial concentration of an acid ($[HA]_{initial}$) or base ($[B]_{initial}$) as prepared, assuming no proton transfer has occurred.

- **General result for a dilute solution of the *weak* acid HA**

  | Main equilibrium | $HA(aq)$ | $+$ | $H_2O(l)$ | $\rightleftharpoons$ | $H_3O^+(aq)$ | $+$ | $A^-(aq)$ |
  |---|---|---|---|---|---|---|---|
  | Initial molarity | $[HA]_{initial}$ | | — | | 0 | | 0 |
  | Change in molarity | $-x$ | | | | $+x$ | | $+x$ |
  | Equilibrium molarity | $[HA]_{initial} - x$ | | | | $x$ | | $x$ |

  Acidity constant expression: $K_a = \dfrac{[H_3O^+][A^-]}{[HA]} = \dfrac{x^2}{[HA]_{initial} - x}$

  Quadratic equation: $x^2 + (K_a)x - (K_a[HA]_{initial}) = 0$ *and* $x > 0$ (positive root)

  Positive root: $x = \dfrac{-K_a + \sqrt{K_a^2 + 4K_a[HA]_{initial}}}{2} = [H_3O^+] = [A^-] \geq 10^{-6}$ M  (1)

  $[HA] = [HA]_{initial} - x$    $[OH^-] = K_w/[H_3O^+]$ from the secondary equilibrium

  Percentage *deprotonated* $= \dfrac{x}{[HA]_{initial}} \times 100\%$   (2)

- **Approximation to the general result**

  Assume $[HA]_{initial} - x \approx [HA]_{initial}$, then $K_a = \dfrac{x^2}{[HA]_{initial}}$ and $x = \sqrt{K_a[HA]_{initial}}$   (3)

  Assumption is justified if percentage deprotonated < 5%.

  **Example:** For acetic acid, $CH_3COOH$, $K_a = 1.8 \times 10^{-5}$ The value of $[H_3O^+]$ in a 0.100 M solution of $CH_3COOH$, is given by the preceding equation labeled (1). With $[HA]_{initial} = 0.1$, the result is $x = [H_3O^+] = 1.33 \times 10^{-3}$ M. Using the 5% approximation method in equation labeled (3) yields $x = [H_3O^+] =$
  $x = \sqrt{K_a[HA]_{initial}} = [(1.8 \times 10^{-5})(0.100)]^{1/2} = 1.34 \times 10^{-3}$ M
  *The approximation is excellent in this case.*

## 10.12 Solutions of Weak Bases

- **General result for a *weak* base B at *dilute* concentrations**

| Main equilibrium | B(aq) | + | $H_2O(l)$ | $\rightleftharpoons$ | $BH^+$(aq) | + | $OH^-$(aq) |
|---|---|---|---|---|---|---|---|
| Initial molarity | $[B]_{initial}$ | | — | | 0 | | 0 |
| Change in molarity | $-x$ | | | | $+x$ | | $+x$ |
| Equilibrium molarity | $[B]_{initial} - x$ | | | | $x$ | | $x$ |

Basicity constant expression: $\quad K_b = \dfrac{[BH^+][OH^-]}{[B]} = \dfrac{x^2}{[B]_{initial} - x}$

Quadratic equation: $\quad x^2 + (K_b)x - (K_b[B]_{initial}) = 0 \ and \ x > 0$ (positive root)

Positive root: $\quad x = \dfrac{-K_b + \sqrt{K_b{}^2 + 4K_b[B]_{initial}}}{2} = [BH^+] = [OH^-] \geq 10^{-6} \ \text{M} \qquad$ (4)

$[B] = [B]_{initial} - x \quad\quad [H_3O^+] = K_w/[OH^-] \qquad$ secondary equilibrium

Percentage *protonated* $= \dfrac{x}{[B]_{initial}} \times 100\%$ (5)

- **Approximation to the general result**

Assume $[B]_{initial} - x \approx [B]_{initial}$, then $K_b = \dfrac{x^2}{[B]_{initial}}$ and $x = \sqrt{K_b[B]_{initial}}$ (6)

Assumption is justified if percentage protonated < 5%.

> **Example:** $K_b$ for the weak base methyamine, $CH_3NH_2$, is $3.6 \times 10^{-4}$. In a 0.200 M solution of methylamine, the $[OH^-]$ is given by the equation labeled (4) with $[B]_{initial} = 0.200$ M. Substituting the values of $K_b$ and $[B]_{initial}$ yields $x = [OH^-] = [CH_3NH_2{}^+]$ $= 8.31 \times 10^{-3}$ M. Use of the approximate solution in equation labeled (6) $x = [OH^-]$ $= [CH_3NH_2{}^+] = 8.49 \times 10^{-3}$ M. The % difference between the exact and approximate answers is less than 5%, so the approximation is justified.

## 10.13 The pH of Salt Solutions

- **Acidic cations**
  - $\rightarrow$ The conjugate acids of *weak* bases  **Examples:** Ammonium, $NH_4{}^+$; anilinium, $C_6H_5NH_3{}^+$
  - $\rightarrow$ Certain small, highly charged metal cations  **Examples:** Iron(III), $Fe^{3+}$; copper(II), $Cu^{2+}$
- **Neutral cations** $\rightarrow$ Group 1 and 2 cations; cations with +1 charge.  **Examples:** $Li^+$, $Mg^{2+}$, $Ag^+$
- **Basic anions** $\rightarrow$ Conjugate bases of weak acids
  **Examples:** Acetate, $CH_3COO^-$; fluoride, $F^-$; sulfide, $S^{2-}$
- **Neutral anions** $\rightarrow$ Conjugate bases of strong acids  **Examples:** $Cl^-$, $ClO_4{}^-$; $NO_3{}^-$.
- **Acidic anions** $\rightarrow$ Only a few such as $HSO_4{}^-$, $H_2PO_4{}^-$
- **Acid salt**
  - $\rightarrow$ Typically, a salt with an acidic cation
    **Examples:** $FeCl_3$, $NH_4NO_3$ (in these examples, the anions are neutral – see above)
  - $\rightarrow$ Dissolves in water to yield an acidic solution
    **Example:** $NH_4NO_3$(aq) $\rightarrow NH_4{}^+$(aq) $+ NO_3{}^-$(aq) (dissociation of salt in water)
    $\quad\quad\quad\quad NH_4{}^+$(aq) $+ H_2O(l) \rightleftharpoons H_3O^+$(aq) $+ NH_3$(aq) (acidic cation reacts with water)

- **Basic salt**
  - → A salt with a basic anion

    **Examples:** Sodium acetate, $NaO_2CCH_3$; potassium cyanide, KCN

    (in these examples, the cations are neutral – see above)

  - → Dissolves in water to yield a basic solution

    **Example:** $NaO_2CCH_3(aq) \rightarrow Na^+(aq) + CH_3COO^-(aq)$ (dissociation of salt in water)

    $CH_3COO^-(aq) + H_2O(l) \rightleftharpoons OH^-(aq) + CH_3COOH(aq)$

    (basic cation reacts with water to yield a *basic solution*)

- **Neutral salt**
  - → A salt with neutral cations and anions

    **Examples:** Sodium chloride, NaCl; potassium nitrate, $KNO_3$; magnesium iodide, $MgI_2$

  - → Dissolves in water to yield a neutral solution

- **Method of solution (main equilibrium)**
  - → For acid salts, use the same method as for *weak* acids, but replace HA with $BH^+$.
  - → For basic salts, use the same method used for *weak* bases, but replace B with $A^-$. (All are *anions*.)

    **Example:** Iron(III) chloride, $FeCl_3$, is the salt of an acid cation and a neutral anion. It dissolves in water to give an *acidic* solution:

    $FeCl_3(s) + 6H_2O(l) \rightarrow Fe(H_2O)_6^{3+}(aq) + 3Cl^-(aq)$ (dissolution of iron(III) chloride)

    $Fe(H_2O)_6^{3+}(aq) + H_2O(l) \rightleftharpoons Fe(H_2O)_5(OH)^{2+}(aq) + H_3O^+(aq)$;

    (reaction of a weak acid with water)

    For the above equation, $K_a = 3.5 \times 10^{-3}$ (see text Table 10.7)

    The cation is a weak acid and the equation above is the main equilibrium; use the methodology in section 10.11 above to calculate concentration. For example, the $[H_3O^+]$ of a 0.010 M solution of $FeCl_3(s)$ is given by equation (1):

    $$x = \frac{-K_a + \sqrt{K_a^2 + 4K_a[HA]_{initial}}}{2} = [H_3O^+]$$

    $$= \frac{-(3.5\times10^{-3}) + \sqrt{(3.5\times10^{-3})^2 + 4(3.5\times10^{-3})(0.010)}}{2}$$

    $$x = [H_3O^+] = 4.42\times10^{-3} \quad pH = -\log(4.42\times10^{-3}) = 2.35 \approx 2.4$$

    The solution is quite acidic and would be considered corrosive.

## Polyprotic Acids and Bases (Sections 10.14–10.17)

### 10.14 The pH of a Polyprotic Acid Solution

- **Successive deprotonations of a polyprotic acid**
  - → A polyprotic acid is a species that can donate more than one proton.
  - → Stepwise equilibria must be considered if successive $K_a$ values differ by a factor of $\geq 10^3$, which is normally the case. Values of $K_a$ are given in Table 10.9 in the text.
  - → Polyprotic bases (species that can accept more than one proton) also exist.

- **Polyprotic acids and pH**
  - → Use the main equilibrium for a dilute solution of a polyprotic acid to calculate the pH. Neglect any secondary equilibria except for $H_2SO_4$ whose first deprotonation is complete.
  - → The calculation is identical to that for any weak monoprotic acid.

    **Example:** The pH of a 0.025 M solution of diprotic carbonic acid, $H_2CO_3$ is calculated using the first dissociation step and equation (1) in Section 10.11:

    $$H_2CO_3(aq) + H_2O(l) \rightleftharpoons H_3O^+(aq) + HCO_3^-(aq) \quad K_{a1} = 4.3 \times 10^{-7} \text{ (main equilibrium)}$$

    $$x = \frac{-K_{a1} + \sqrt{K_{a1}^2 + 4K_a[HA]_{initial}}}{2} = [H_3O^+]$$

    $$= \frac{-(4.3 \times 10^{-7}) + \sqrt{(4.3 \times 10^{-7})^2 + 4(4.3 \times 10^{-7})(0.025)}}{2}$$

    $$x = [H_3O^+] = 1.03 \times 10^{-4} \quad pH = -\log(1.03 \times 10^{-4}) = 3.9935 \approx 4.0$$

## 10.15 Solutions of Salts of Polyprotic Acids

- **pH of a solution of an amphiprotic anion $HA^-$**

  - → Given by $pH = \frac{1}{2}(pK_{a1} + pK_{a2})$

  - → Independent of the concentration of the anion

- **Solution of basic anion of polyprotic acid : $A^{2-}$ or $A^{3-}$  Examples:** solutions of $Na_2S$ or $K_3PO_4$
  - → Main equilibrium is hydrolysis of the basic anion: $A^{2-}(aq) + H_2O(aq) \rightleftharpoons OH^-(aq) + HA^-(aq)$
  - → Treat using methodology for solutions for basic anions:  [section 10:12, eqs. (4) – (6), this guide) ]

## 10.16 The Concentrations of Solute Species

- **Determination of the concentration of all species for a solution of a weak *diprotic* acid, $H_2A$**
  - → Solve the *main* equilibrium ($K_{a1}$) as for any weak acid.
  - → This yields $[H_2A]$, $[H_3O^+]$ and $[HA^-]$.
  - → Solve the *secondary* equilibrium ($K_{a2}$) for $[A^{2-}]$ using the $[H_3O^+]$ and $[HA^-]$ values determined above.
  - → Determine $[OH^-]$ using $K_w$ and $[H_3O^+]$.

- **Determination of the concentration of all species for a solution of a weak *triprotic* acid, $H_3A$**
  - → Solve the *main* equilibrium ($K_{a1}$) as for any weak acid.
  - → This yields $[H_3A]$, $[H_3O^+]$ and $[H_2A^-]$.
  - → Solve the *secondary* equilibrium ($K_{a2}$) for $[HA^{2-}]$ using the $[H_3O^+]$ and $[H_2A^-]$ values from above.
  - → Solve the *tertiary* equilibrium ($K_{a3}$) for $[A^{3-}]$ using the $[H_3O^+]$ and $[HA^{2-}]$ values from above.
  - → Determine $[OH^-]$ using $K_w$ and $[H_3O^+]$.

## 10.17 Composition and pH

- **Polyprotic acid** → Concentrations of the several solution species vary with the pH of the solution.

- **Composition and pH** → Solution to the stepwise equilibrium expressions for a polyprotic acid as a function of pH

- **Diprotic acid ($H_2A$) – qualitative considerations  (see text Fig. 10.18)**

$$H_2A(aq) + H_2O(l) \rightleftharpoons H_3O^+(aq) + HA^-(aq) \qquad K_{a1} = \frac{[H_3O^+][HA^-]}{[H_2A]} \qquad \text{equation (7)}$$

$$HA^-(aq) + H_2O(l) \rightleftharpoons H_3O^+(aq) + A^{2-}(aq) \qquad K_{a2} = \frac{[H_3O^+][A^{2-}]}{[HA^-]} \qquad \text{equation (8)}$$

→ At low pH, equilibria represented by equation (7) and (8) shift to the left (Le Chatelier) and the solution contains a high concentration of $H_2A$ and very little $A^{2-}$.

→ At high pH, equilibria represented by equation (7) and (8) shift to the right and the solution contains a high concentration of $A^{2-}$ and very little $H_2A$.

→ At intermediate pH values, the solution contains a relatively high concentration of the amphiprotic anion $HA^-$.

- **Diprotic acid ($H_2A$) – quantitative considerations**

  → *Fraction ($\alpha$) of $H_2A$, $HA^-$, and $A^{2-}$ in solution as a function of $[H_3O^+]$*

  → $\alpha(H_2A) = \dfrac{[H_3O^+]^2}{f} \qquad \alpha(HA^-) = \dfrac{[H_3O^+]K_{a1}}{f} \qquad \alpha(A^{2-}) = \dfrac{K_{a1}K_{a2}}{f}$

  → where $f = [H_3O^+]^2 + [H_3O^+]K_{a1} + K_{a1}K_{a2}$

---

## Autoprotolysis and pH (Sections 10.18–10.19)

### 10.18 Very Dilute Solutions of Strong Acids and Bases

- **Very dilute solution** → Concentration of a strong acid or base is less than $10^{-6}$ M.

- **General result for a *strong* acid HA at *very dilute* concentrations**

$$HA(aq) + H_2O(l) \rightarrow H_3O^+(aq) + A^-(aq)$$

*Three* equations in *three* unknowns ($[H_3O^+]$, $[OH^-]$, and $[A^-]$)

Autoprotolysis equilibrium: $\quad K_w = [H_3O^+][OH^-] = 1.0 \times 10^{-14}$

Charge balance: $\qquad\quad [H_3O^+] = [OH^-] + [A^-]$

Material balance: $\qquad\quad [A^-] = [HA]_{initial}$ (complete deprotonation)

Solution: $\qquad\qquad\quad [OH^-] = [H_3O^+] - [HA]_{initial}$ (from charge and material balance)

$\qquad\qquad\qquad\qquad K_w = [H_3O^+]([H_3O^+] - [HA]_{initial})$

$\qquad\qquad\qquad\qquad K_w = [H_3O^+]^2 - [HA]_{initial}[H_3O^+], \quad$ let $x = [H_3O^+]$

Quadratic equation: $\qquad x^2 - [HA]_{initial}\, x - K_w = 0$ *and* $x > 0$ *(positive root)*

Positive root: $\quad x = \dfrac{[HA]_{initial} + \sqrt{[HA]_{initial}^2 + 4K_w}}{2} = [H_3O^+] \quad (9)$

$and\ [OH^-] = K_w/[H_3O^+] \quad (10)$

- **General result for a *strong* base MOH at *very dilute* concentrations**

$$MOH(aq) + H_2O(l) \rightarrow M^+(aq) + OH^-(aq)$$

*Three* equations in *three* unknowns

Autoprotolysis equilibrium: $\quad K_w = [H_3O^+][OH^-] = 1.0 \times 10^{-14}$

Charge balance: $\qquad\quad [M^+] + [H_3O^+] = [OH^-]$

Material balance: $\qquad\quad [M^+] = [MOH]_{initial} \quad$ (complete dissociation)

Solution: $\qquad\qquad\quad [OH^-] = [H_3O^+] + [MOH]_{initial}$

$\qquad\qquad\qquad\qquad K_w = [H_3O^+]\{[H_3O^+] + [MOH]_{initial}\}$

$\qquad\qquad\qquad\qquad K_w = [H_3O^+]^2 + [MOH]_{initial}[H_3O^+], \ \text{let } x = [H_3O^+]$

Quadratic equation: $\qquad x^2 + [MOH]_{initial}\,x - K_w = 0\ and\ x > 0$ (positive root)

Positive root: $\quad x = \dfrac{-[MOH]_{initial} + \sqrt{[MOH]_{initial}^2 + 4K_w}}{2} = [H_3O^+] \quad (11)$

$and\ \ [OH^-] = K_w/[H_3O^+] \quad (10)$

**Note:** Strong bases such as $O^{2-}$ and $CH_3^-$ produce aqueous solutions that are equivalent to MOH.

**Example:** The pH of a $1.5 \times 10^{-8}$ M solution of the strong base KOH can be calculated from equations (10) and (11).

From (11), $[H_3O^+] = \dfrac{-[MOH]_{initial} + \sqrt{[MOH]_{initial}^2 + 4K_w}}{2} = 9.28 \times 10^{-8}$ M.

From (10), pH $= -\log[H_3O^+] = -\log(9.28 \times 10^{-8}) = 7.03$
(slightly *basic* solution as expected)

## 10.19 Very Dilute Solutions of Weak Acids

- **Very dilute solution** $\rightarrow$ Concentration of a weak acid or base is less than about $10^{-3}$ M.

- **General result for a *weak* acid HA at *very dilute* concentrations**
  *Four* equations in *four* unknowns

  1) Weak acid equilibrium: $HA(aq) + H_2O(l) \rightleftharpoons H_3O^+(aq) + A^-(aq);$ $\qquad K_a = \dfrac{[H_3O^+][A^-]}{[HA]}$

  2) Autoprotolysis equilibrium: $\quad K_w = [H_3O^+][OH^-] = 1.0 \times 10^{-14}$

  3) Charge balance: $\qquad\qquad\qquad [H_3O^+] = [OH^-] + [A^-]$

4) Material balance: $\qquad [HA]_{initial} = [HA] + [A^-]$ (incomplete deprotonation)

Solution: $\qquad [A^-] = [H_3O^+] - [OH^-] = [H_3O^+] - \dfrac{K_w}{[H_3O^+]}$ (from charge balance)

$$[HA] = [HA]_{initial} - [A^-] = [HA]_{initial} - [H_3O^+] + \frac{K_w}{[H_3O^+]}$$

Substitute expressions for [HA] and $[A^-]$ into $K_a$.

$$K_a = \frac{[H_3O^+]\left([H_3O^+] - \dfrac{K_w}{[H_3O^+]}\right)}{[HA]_{initial} - [H_3O^+] + \dfrac{K_w}{[H_3O^+]}} \quad (12)$$

Let $x = [H_3O^+]$ and then

$$K_a = \frac{x\left(x - \dfrac{K_w}{x}\right)}{[HA]_{initial} - x + \dfrac{K_w}{x}}$$

Rearrange to a cubic form.

Cubic equation: $\qquad x^3 + K_a x^2 - (K_w + K_a[HA]_{initial})x - K_a K_w = 0$

Solve by using a graphing calculator, trial and error, or the mathematical software present on text web site: www.whfreeman.com/chemicalprinciples/
Only one of the three roots will be physically meaningful.

Alternatively, consider the following approximations:

If $[H_3O^+] > 10^{-6}$ M, then $\dfrac{K_w}{[H_3O^+]} < 10^{-8}$ and this term can be neglected in (12) to yield

$$K_a = \frac{[H_3O^+]^2}{[HA]_{initial} - [H_3O^+]} = \frac{x^2}{[HA]_{initial} - x} \quad (13)$$

Further, if $[H_3O^+] > 10^{-6}$ M $and$ $[H_3O^+] \ll [HA]_{initial}$, then (13) reduces to

$$K_a = \frac{[H_3O^+]^2}{[HA]_{initial}} = \frac{x^2}{[HA]_{initial}} \quad (14)$$

Use equation (12), (13), or (14) as appropriate to solve problems.

**Note:** The software on the text web site does not work when numbers in the polynomial equation differ by many orders of magnitude.

# Chapter 11   Aqueous Equilibria

## Mixed Solutions and Buffers  (Sections 11.1–11.3)

### 11.1  Buffer Action

- **Buffer**
  - → A mixed solution containing weak conjugate acid–base pairs that stabilizes the pH of a solution
  - → Provides a source and a sink for protons

- **Acid buffer**
  - → Consists of a weak acid and its conjugate base provided as a salt
    - **Examples:** Acetic acid ($CH_3COOH$) and sodium acetate ($NaCH_3COO$);
      - phosphorous acid ($H_3PO_3$) and potassium dihydrogenphosphite ($KH_2PO_3$)
  - → Buffers a solution on the acid side of neutrality (pH < 7), if $pK_a < 7$

- **Base buffer**
  - → Consists of a weak base and its conjugate acid provided as a salt
    - **Examples:** Ammonia  ($NH_3$) and ammonium chloride ($NH_4Cl$);
      - sodium dihydrogenphosphate ($NaH_2PO_4$) and
      - sodium hydrogenphosphate ($Na_2HPO_4$)
  - → Buffers a solution on the base side of neutrality (pH > 7), if $pK_b < 7$

### 11.2  Designing a Buffer

- **pH of a buffer**
  - → The pH of a buffer solution is given approximately by the Henderson–Hasselbalch equation

$$pH \approx pK_a + \log\left(\frac{[\text{base}]_{\text{initial}}}{[\text{acid}]_{\text{initial}}}\right)$$

  - → The adjustment from initial concentrations to the equilibrium ones of the conjugate acid-base pair is *usually* negligible.  If it is not, the pH can be adjusted to the desired value by adding additional acid or base.
  - → Use of this equation is equivalent to setting up a table of concentrations and solving an equilibrium problem if adjustment from initial concentrations is negligible.

- **Preparation of an acid buffer**
  - → Use an acid (HA) with a $pK_a$ value close to the desired pH.
  - → Add the acid and its conjugate base ($A^-$) in appropriate amounts to satisfy the Henderson–Hasselbalch equation.
  - → Measure the pH and make any necessary adjustments to the acid or base concentrations to achieve the desired pH.

- **Preparation of a base buffer**
  - → Use a base (B) whose conjugate acid ($BH^+$) has a $pK_a$ value close to the desired pH.
  - → Add the base and its conjugate acid in appropriate amounts to satisfy the Henderson–Hasselbalch equation.

→ Measure the pH and make any necessary adjustments to the acid or base concentrations to achieve the desired pH.

## 11.3  Buffer Capacity

- **Buffer capacity**
  → The amount of acid or base that can be added before the buffer loses its ability to resist a change in pH
  → Determined by its concentration and pH
  → The *greater* the concentrations of the conjugate acid-base pair in the buffer system, the *greater* the buffer capacity

- **Exceeding the capacity of a buffer**
  → Buffers function by providing a weak acid to react with added base and a weak base to react with added acid.
  → Adding sufficient base to react with *all* of the weak acid or adding sufficient acid to react with *all* of the weak base exhausts (completely destroys) the buffer.

- **Limits of buffer action**
  → A buffer has a practical limit to its buffering ability that is reached before the buffer is completely exhausted.
  → As a *rule of thumb,* a buffer acts most effectively in the range

$$10 > \frac{[\text{base}]_{\text{initial}}}{[\text{acid}]_{\text{initial}}} > 0.1$$

  → The molarity ratios above yield the following expression for the *effective pH range of a buffer*

$$\text{pH} = \text{p}K_a \pm 1$$

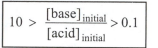

# Titrations  (Sections 11.4–11.7)

## 11.4  Strong Acid–Strong Base Titrations

- **pH curve**
  → Plot of the pH of the *analyte* solution as a function of the volume of the *titrant* added during a titration
  → The stoichiometric point has a pH value of 7 for all strong base–strong acid or strong acid–strong base titrations.

- **Titration of a strong base (analyte) with a strong acid (titrant) (see Fig. 11.4 in the text)**
  → The pH *changes (drops) slowly* with added titrant until the stoichiometric point is approached.
  → The pH then *changes (drops) rapidly* near the stoichiometric point.

→ At the stoichiometric point, the base is neutralized.

→ The pH approaches the value of the titrant solution in the region well beyond the stoichiometric point.

- **Titration of a strong acid (analyte) with a strong base (titrant) (see Fig. 11.5 in the text)**
  → The pH *changes (rises) slowly* with added titrant until the stoichiometric point is approached.
  → The pH then *changes (rises) rapidly* near the stoichiometric point.
  → At the stoichiometric point, all the acid is neutralized.
  → The pH approaches the value of the titrant solution in the region well beyond the stoichiometric point.

- **pH curve calculations**
  → As a titration proceeds, both the volume of the analyte solution and the amount of acid or base present change.
  → To calculate $[H_3O^+]$ in the analyte solution, the amount of acid or base remaining should be determined and divided by the total volume of the solution.
  → Before the stoichiometric point is reached, follow the procedure outlined in Example 11.4 in the text.
  → For a strong acid–strong base titration, the pH at the stoichiometric point is 7.
  → After the stoichiometric point, follow the procedure outlined in Toolbox 11.1 and Example 11.6 in the text.
  → Always assume that the volumes of the titrant and analyte solutions are additive:
  $$(V_{total} = V_{analyte} + V_{titrant\ added})$$
  → If a solid is added to a solution, its volume is usually neglected.

## 11.5 Strong Acid–Weak Base and Weak Acid–Strong Base Titrations

- **Characteristics of the titrations**
  → The titrant is the *strong* acid or base, the analyte is the *weak* base or acid.
  → Stoichiometric point has a pH value different from 7.

- **Strong base–weak acid titration**
  → A strong base *dominates* the weak acid and the pH at the stoichiometric point is *greater* than 7.
  → A strong base converts a weak acid, HA, into its conjugate base form, $A^-$.
  → At the stoichiometric point, the pH is that of the resulting basic salt solution.

- **Strong acid–weak base titration**
  → A strong acid *dominates* the weak base and the pH at the stoichiometric point is *less* than 7.
  → A strong acid converts a weak base, B, into its conjugate acid form, $BH^+$.
  → At the stoichiometric point, the pH is that of the resulting acidic salt solution.

- **Shape of the pH curve**
  → *Similar* to that of a strong acid–strong base titration
  → *Differences* include
    ➤ A buffer region appears between the initial analyte solution and the stoichiometric point.
    ➤ The pH of the stoichiometric point is different from 7.
    ➤ There is a less abrupt pH change in the region of the stoichiometric point.

- **pH curve calculations**
  - $\rightarrow$ The pH is governed by the main species in solution.
  - $\rightarrow$ Calculate amounts and solution volume separately; divide to obtain concentrations.
  - $\rightarrow$ Before the titrant is added, the analyte is a solution of a weak acid or base in water. Determine the pH of the solution as in Sections 10.11 and 10.12 in the text.
  - $\rightarrow$ Before the stoichiometric point is reached, the solution is a buffer with differing amounts of acid/conjugate base. To calculate the pH, follow the procedure outlined in Example 11.6 in the text.
  - $\rightarrow$ At the stoichiometric point, the analyte solution consists of the salt of a weak acid or a weak base. To calculate the pH, follow the procedure outlined in Example 11.5 in the text.
  - $\rightarrow$ Well after the stoichiometric point is passed, the pH of the solution approaches the value of the titrant.
  - $\rightarrow$ Assume that the volumes of the titrant solution added and the analyte solution are additive:

$$(V_{total} = V_{analyte} + V_{titrant\ added})$$

  If the sample is a solid, its volume is usually neglected.

## 11.6 Acid–Base Indicators

- **Acid–base indicator**
  - $\rightarrow$ A water-soluble dye with different *characteristic colors* associated with its acid and base forms
  - $\rightarrow$ Exhibits a color change over a *narrow* pH range
  - $\rightarrow$ Added in low concentration to the analyte so that a titration is primarily that of the analyte, not the indicator
  - $\rightarrow$ Selected to monitor the stoichiometric or *end point* of a titration
  - $\rightarrow$ An indicator is a weak acid. It takes part in the following proton-transfer equilibrium:

$$HIn(aq) + H_2O(l) \rightleftharpoons H_3O^+(aq) + In^-(aq) \qquad K_{In} = \frac{[H_3O^+][In^-]}{[HIn]}$$

  Values of $pK_{In}$ are given in Table 11.3 in the text.

  - $\rightarrow$ The color change is usually most apparent when $[HIn] = [In^-]$ and $pH = pK_{In}$. The color change begins typically within one pH unit before $pK_{In}$ and is essentially complete about 1 pH unit after $pK_{In}$. The $pK_{In}$ value of an indicator is usually chosen to be within 1 pH unit of the stoichiometric point.

$$\boxed{pK_{In} \approx pH(\text{stoichiometric point}) \pm 1}$$

- **End point of a titration** $\rightarrow$ Stage of a titration where the color of the indicator in the analyte is midway between its acid and base colors

## 11.7 Stoichiometry of Polyprotic Acid Titrations

- **Polyprotic acids**
  - $\rightarrow$ Review the list of the common polyprotic acids in Table 10.9 of the text.
  - $\rightarrow$ All the acids in Table 10.9 are *diprotic,* except for phosphoric acid, which is *triprotic.*

→ Phosphorous acid, $H_3PO_3$, is *diprotic*. The major Lewis structures are

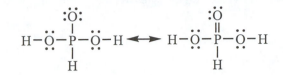

The proton bonded to phosphorus is *not* acidic.

- **Titration curve of a polyprotic acid**
  - → Has stoichiometric points corresponding to the removal of each acidic hydrogen atom
  - → Has buffer regions between successive stoichiometric points
  - → At each stoichiometric point, the analyte is the salt of a conjugate base of the acid or one of its hydrogen-bearing anions.
  - → See the examples of pH curves for $H_3PO_4$ in Fig. 11.13 and $H_2C_2O_4$ in Fig. 11.14 of the text.

- **pH calculations** → The pH curve can be estimated at any point by considering the primary species in solution and the *main* proton-transfer equilibrium that determines the pH.

---

# Solubility Equilibria  (Sections 11.8–11.14)

## 11.8  The Solubility Product

- **Solubility product, $K_{sp}$**
  - → Equilibrium constant for a solubility equilibrium between a solid and its dissolved form

    Example:   $Sb_2S_3(s) \rightleftharpoons 2\,Sb^{3+}(aq) + 3\,S^{2-}(aq)$    $K_{sp} = \dfrac{a_{Sb^{3+}}^2 \, a_{S^{2-}}^3}{a_{Sb_2S_3(s)}}$

  - → Activity of a pure solid is 1, and, for *dilute* solutions (sparingly soluble salts), the activity of a solute species can be replaced by its molarity.
  - → To a fair approximation, $K_{sp} = [Sb^{3+}]^2 [S^{2-}]^3$    (saturated solution)

- **Molar solubility, $s$** → Maximum amount of solute that dissolves in enough water to make one liter of solution

- **Relationship between $s$ and $K_{sp}$**
  - → The stoichiometric relationships are
    1 mol $Sb_2S_3 \simeq$ 2 mol $Sb^{3+}$ and 1 mol $Sb_2S_3 \simeq$ 3 mol $S^{2-}$
  - → The molar solubility, $s$, of the salt $Sb_2S_3(s)$ is related to the ion concentrations at equilibrium as follows:
    $[Sb^{3+}] = 2s$  and $[S^{2-}] = 3s$.
  - → The relationship between $s$ and $K_{sp}$ is then
    $K_{sp} = [Sb^{3+}]^2 [S^{2-}]^3 = (2s)^2(3s)^3 = (4s^2)(27s^3) = 108s^5$ and $s = (K_{sp}/108)^{1/5}$

→ The relationship $s = (K_{sp}/108)^{1/5}$ holds for any salt whose general formula is $X_2Y_3$.

→ Similar relationships can be generated for other type of salts; for example for an AB salt such as AgCl, $s = K_{sp}^{1/2}$.

## 11.9 The Common-Ion Effect

- **Common-ion effect**
  → A decrease in the solubility of a salt caused by the presence of one of the ions of the salt in solution
  → Follows from Le Chatelier's principle
  → Used to help remove unwanted ions from solution

    **Example:** For the equilibrium $Sb_2S_3(s) \rightleftharpoons 2Sb^{3+}(aq) + 3S^{2-}(aq)$

    increasing the concentration of $Sb^{3+}$ ions by addition of $Sb(NO_3)_3$ shifts the equilibrium to the left, causing some precipitation of $Sb_2S_3(s)$. Thus, less $Sb_2S_3(s)$ is dissolved in the presence of an external source of $Sb^{3+}$ (or $S^{2-}$) than in pure water.

- **Quantitative features**
  → Use the solubility product expression to estimate the reduction in solubility resulting from the addition of a common ion.
  → For the equilibrium $Sb_2S_3(s) \rightleftharpoons 2Sb^{3+}(aq) + 3S^{2-}(aq)$, $K_{sp} = [Sb^{3+}]^2 [S^{2-}]^3$
    If $Sb_2S_3(s)$ is added to a solution already containing $Sb^{3+}$ ions with a concentration $[Sb^{3+}]$, then dissolution of this *sparingly soluble salt* will add a negligible amount to the cation concentration. The solubility, *s,* and its relationship to $K_{sp}$ is then
    $$K_{sp} = [Sb^{3+}]^2 (3s)^3 \text{ and } s = (K_{sp}/27[Sb^{3+}]^2)^{1/3}$$

## 11.10 Predicting Precipitation

- **Prediction of whether precipitation occurs**
  → Calculate $Q_{sp}$ using actual or predicted ion concentrations.
  → Compare $Q_{sp}$ to $K_{sp}$.
  → If $Q_{sp} > K_{sp}$, the salt will precipitate from solution.
  → If $Q_{sp} = K_{sp}$, the solution is saturated with respect to the ions in the expression for $K_{sp}$. Any additional amount of these ions will cause precipitation.
  → If $Q_{sp} < K_{sp}$, the solution is unsaturated and no precipitate will form.

## 11.11 Selective Precipitation

- **Separation of different cations in solution**
  → Add a soluble salt containing an *anion* with which the cations form insoluble salts.
  → If the insoluble salts have sufficiently different solubilities, they will precipitate at different anion concentrations and can be collected separately.

## 11.12 Dissolving Precipitates

- **Ion removal**
  → Removal from solution of one of the ions in the solubility equilibrium *increases* the solubility of the solid.

→ A precipitate will dissolve completely if a sufficient number of such ions are removed.

→ In this section, we focus on the removal of *anions*.

- **Removal of basic anions**

    → Addition of a strong acid, such as HCl, neutralizes a basic anion.
      For example, hydroxides react to form water:

    $$H_3O^+(aq) + OH^-(aq) \rightarrow 2H_2O(l)$$

    → In some cases, gases are evolved.
      For example, carbonates react with acid to form carbon dioxide gas:

    $$2H_3O^+(aq) + CO_3^{2-}(aq) \rightarrow CO_2(g) + 3H_2O(l)$$

- **Removal by oxidation of anions**

    → In some cases, gases are evolved.
      For example, $S^{2-}(aq)$ from a metal sulfide is oxidized to S(s) with nitric acid:

    $$3S^{2-}(aq) + 8H_3O^+(aq) + 2NO_3^-(aq) \rightarrow 3S(s) + 2NO(g) + 12H_2O(l)$$

- **Summary** → The solubility of a salt may be increased by removing its anion from solution. An acid can be used to dissolve a hydroxide, sulfide, sulfite, or carbonate salt. Nitric acid can be used to oxidize metal sulfides to sulfur and a soluble salt.

## 11.13 Complex Ion Formation

- **Complex ions**

    → Formed by the reaction of a metal cation acting as a Lewis acid with a Lewis base

    → May disturb a solubility equilibrium by reducing the concentration of metal ions

    → Net result is an increase in the solubility of salts.

    **Example:** $Fe^{2+}(aq) + 6CN^-(aq) \rightleftharpoons Fe(CN)_6^{4-}(aq)$

    In the above reaction, $Fe^{2+}$ acts as a Lewis acid and $CN^-$ acts as a Lewis base. Complexation reduces the concentration of $Fe^{2+}$ in solution thereby causing a shift in any solubility equilibrium in which $Fe^{2+}$ is involved.

    → A number of formation reactions are listed in Table 11.5 in the text.

    → Electron pair acceptors (Lewis acids) that form complex ions include metal cations such as $Fe^{3+}(aq)$, $Cu^{2+}(aq)$, and $Ag^+(aq)$.

    → Electron pair donors (Lewis bases) that form complex ions include basic anions or neutral bases such as $Cl^-(aq)$, $NH_3(aq)$, $CN^-(aq)$, and $S_2O_3^{2-}(aq)$.

**Quantitative features**

    → The equilibrium for a *formation reaction* is characterized by the *formation constant, $K_f$*.

    **Example:** $Fe^{2+}(aq) + 6CN^-(aq) \rightleftharpoons Fe(CN)_6^{4-}(aq)$    $K_f = \dfrac{[Fe(CN)_6^{4-}]}{[Fe^{2+}][CN^-]^6}$

    → Values of $K_f$ at 25°C are given in Table 11.5 in the text.

    → Formation and solubility reactions may be combined to predict the solubility of a salt in the presence of a species that forms a complex with the metal ion.

**Example:** $FeS(s) \rightleftharpoons Fe^{2+}(aq) + S^{2-}(aq)$ $\qquad K_{sp} = [Fe^{2+}][S^{2-}]$ *solubility reaction*

$Fe^{2+}(aq) + 6CN^-(aq) \rightleftharpoons Fe(CN)_6^{2+}(aq)$ $\quad K_f = \dfrac{[Fe(CN)_6^{2+}]}{[Fe^{2+}][CN^-]^6}$ *formation reaction*

Combining the above two reactions produces the following:

$FeS(s) + 6CN^-(aq) \rightleftharpoons Fe(CN)_6^{4-}(aq) + S^{2-}(aq)$

$$K = K_{sp} \times K_f = \dfrac{[Fe(CN)_6^{4-}][S^{2-}]}{[CN^-]^6} = (6.3 \times 10^{-18})(7.7 \times 10^{36}) = 4.9 \times 10^{19} \gg 1$$

The equilibrium lies far to the right indicating a large solubility for FeS(s) in a cyanide solution.

Because 1 mol FeS $\triangleq$ 1 mol $S^{2-}$, the molar solubility, $s$, is equal to the sulfide ion concentration.

## 11.16 Qualitative Analysis

- **Qualitative analysis**
  - → Method used to identify and separate ions present in an unknown solution (for example, seawater)
  - → Utilizes complex formation, selective precipitation, and pH control
  - → Uses solubility equilibria to remove and identify ions selectively
- **Outline of the method**
  - → The method is outlined in Fig. 11.20 in the text and summarized below.

### Partial Qualitative Analysis Scheme

| Step | Possible Precipitate | $K_{sp}$ |
|---|---|---|
| 1) Add HCl(aq) | AgCl | $1.6 \times 10^{-10}$ |
| | $Hg_2Cl_2$ | $1.3 \times 10^{-18}$ |
| | $PbCl_2$ | $1.6 \times 10^{-5}$ |
| 2) Add $H_2S$(aq) | $Bi_2S_3$ | $1.0 \times 10^{-97}$ |
| (in acid solution, there is a | CdS | $4.0 \times 10^{-29}$ |
| low $S^{2-}$ concentration) | CuS | $7.9 \times 10^{-45}$ |
| | HgS | $1.6 \times 10^{-52}$ |
| | $Sb_2S_3$ | $1.6 \times 10^{-93}$ |
| 3) Add base to $H_2S$(aq) | FeS | $6.3 \times 10^{-18}$ |
| (in basic solution, there is a | MnS | $1.3 \times 10^{-15}$ |
| higher $S^{2-}$ concentration) | NiS | $1.3 \times 10^{-24}$ |
| | ZnS | $1.6 \times 10^{-24}$ |

**Note:** In solubility calculations, the fairly strong basic character of $S^{2-}$ must be considered.
$S^{2-}(aq) + H_2O(l) \rightleftharpoons HS^-(aq) + OH^-(aq)$ $\quad K_b = 1.4$

→ A solution may contain any or all of the cations given in the preceding table. We wish to determine which ones are present and which are absent.

- **Procedure 1: Add HCl to the solution.**
  → Most chlorides are soluble and will not precipitate as the $[Cl^-]$ increases.
  → For AgCl and $Hg_2Cl_2$, we expect their $Q_{sp}$ values to be exceeded and for them to precipitate.
  → $PbCl_2$ is slightly more soluble, but it should precipitate, too.
  → The precipitate obtained, if any, using Procedure 1 is collected and the remaining solution tested for additional ions using Procedure 2.

- **Procedure 2: Add $H_2S$ to the solution.**
  → Because the solution is highly acidic after Procedure 1, the concentration of sulfide will be very low:
  $$H_2S(aq) + 2H_2O(l) \rightleftharpoons 2H_3O^+(aq) + S^{2-}(aq)$$
  → Thus, only sulfide salts with very low solubility products, such as CuS or HgS, will precipitate.
  → The resulting precipitate, if any, is collected and the remaining solution tested for additional ions using Procedure 3.

- **Procedure 3: Add base to the solution.**
  → Base reduces $[H_3O^+]$ and, as seen from the equilibrium above, increases $[S^{2-}]$.
  → At higher sulfide concentration, more soluble sulfide salts such as FeS and ZnS precipitate.
  → The resulting precipitate, if any, is collected.

- **Precipitates obtained in Procedures 1, 2, and 3 are analyzed separately for the presence of each cation in the group.**

- **Determining the presence or absence of $Ag^+$, $Hg_2^{2+}$, and $Pb^{2+}$ in Procedure 1**
  → Of the three possible chlorides, $PbCl_2$ is the most soluble and its solubility increases with increasing temperature.
  → Rinse the precipitate with hot water, collect the liquid, and test for $Pb^{2+}$ by adding chromate.
  → A precipitate of insoluble yellow lead(II) chromate indicates that $Pb^{2+}$ is present in the initial solution:
  $$Pb^{2+}(aq) + CrO_4^{2-}(aq) \rightarrow PbCrO_4(s)$$
  → Add aqueous ammonia to the remaining precipitate.
  → Silver(I) will dissolve forming the diammine complex:
  $$Ag^+(aq) + 2NH_3(aq) \rightarrow Ag(NH_3)_2^+(aq)$$
  → In aqueous ammonia, mercury(I) will disproportionate to form a gray mixture of liquid mercury and solid $HgNH_2Cl$:
  $$Hg_2Cl_2(s) + 2NH_3(aq) \rightarrow Hg(l) + HgNH_2Cl(s) + NH_4^+(aq) + Cl^-(aq)$$
  → Formation of the gray mixture confirms the presence of mercury(I) in the original solution.
  → Silver(I) is confirmed by adding HCl to the ammine solution to precipitate white AgCl(s):
  $$Ag(NH_3)_2^+(aq) + Cl^-(aq) + 2H_3O^+(aq) \rightarrow AgCl(s) + 2NH_4^+(aq) + 2H_2O(l)$$

# Chapter 12   Electrochemistry

## Representing Redox Reactions  (Sections 12.1–12.2)

### 12.1  Half-Reactions

- **Half-reaction**
  - → *Conceptual* way of reporting an oxidation or reduction process
  - → Important for understanding oxidation-reduction (*redox*) reactions and electrochemical cells

- **Oxidation**
  - → *Removal* of electrons from a species
  - → Increase in oxidation number of a species
  - → Represented by an oxidation half-reaction

    **Example:** $Zn(s) \rightarrow Zn^{2+}(aq) + 2\,e^-$  Zinc metal loses two electrons to form  zinc(II) ions.

- **Reduction**
  - → *Gain* of electrons by a species
  - → Decrease in oxidation number of a species
  - → Represented by a reduction half-reaction

    **Example:** $Cu^{2+}(aq) + 2\,e^- \rightarrow Cu(s)$  Copper(II) ions gain two electrons each to form copper metal.

- **Redox couple**  →  Oxidized and reduced species in a half-reaction with the oxidized form listed first by convention

    **Example:** The $Zn^{2+}/Zn$  redox couple

- **Half-reactions**
  - → Express separately the oxidation and reduction contributions to a redox reaction
  - → May be added to obtain a redox reaction, if the electrons cancel

    **Example:** From the half reactions above, the following redox reaction is obtained:

    | | |
    |---|---|
    | Oxidation half-reaction: | $Zn(s) \rightarrow Zn^{2+}(aq) + 2\,e^-$ |
    | Reduction half-reaction: | $Cu^{2+}(aq) + 2\,e^- \rightarrow Cu(s)$ |
    | Redox reaction: | $Zn(s) + Cu^{2+}(aq) \rightarrow Zn^{2+}(aq) + Cu(s)$ |

### 12.2  Balancing Redox Equations

- **Balancing methods**
  - → Two methods are commonly used to balance redox equations:
    the *half-reaction method* and the oxidation number method
  - → We will use the half-reaction method.
  - → In either method, mass and charge are balanced separately.
  - → For simplicity, hydronium ions, $H_3O^+(aq)$, are represented as $H^+(aq)$.

- **Half-reaction method**
  - → A six-step procedure (see also Toolbox 12.1 in the text)

    1. Identify all species being oxidized and reduced from changes in their oxidation numbers.
    2. Write unbalanced skeletal equations for each half-reaction.

3. Balance all elements in each half-reaction except O and H.

4. a) For an *acidic* solution, balance O by using $H_2O$, then balance H by adding $H^+$.

   b) For a *basic* solution, balance O by using $H_2O$, then balance H by adding $H_2O$ to the side of each half-reaction that needs H and adding $OH^-$ to the other side.

**Note:** Adding $H_2O$ to one side and $OH^-$ to the other side has the *net* effect of adding H atoms to balance hydrogen, if charge is ignored. Charge is balanced in the next step.

5. Balance charge by adding electrons to the appropriate side of each half-reaction.

6. a) Multiply the half-reactions by factors that give an equal number of electrons in each reaction.

   b) Add the two half-reactions to obtain the balanced equation.

→ An alternative procedure for *basic* solutions is to balance the redox equation for *acid* solutions first. Then, add

$$n\,H^+(aq) + n\,OH^-(aq) \rightarrow n\,H_2O(l) \quad or \quad n\,H_2O(l) \rightarrow n\,H^+(aq) + n\,OH^-(aq)$$

to the redox equation to eliminate $n\,H^+$ from the product *or* reactant side, whichever is necessary.

---

# Galvanic Cells  (Sections 12.3–12.10)

## 12.3  The Structure of Galvanic Cells

- **Galvanic cell**
  → Electrochemical cell in which a spontaneous reaction produces an electric current

- **Galvanic cell components**
  → Two *electrodes* (metallic conductors and/or conducting solids) make electrical contact with the cell contents, the conducting medium.

  → An *electrolyte* (an ionic solution, paste or crystal) is the conducting medium.

  → *Anode*, labeled (−) or ⊖: the electrode at which *oxidation* occurs.

  → *Cathode*, labeled (+) or ⊕: the electrode at which *reduction* occurs.

  → If the anode and cathode do not share a common electrolyte, oxidation reaction and reduction reaction compartments are separated to prevent direct mixing of the electrolyte solutions. Electrical contact between these solutions is maintained by a *salt bridge,* a gel containing the electrolyte $KCl(aq)$, or an equivalent device.

- **The Daniell cell, an example of a galvanic cell**
  → Zinc metal reacts spontaneously with Cu(II) ions in aqueous solution.

  → Copper metal precipitates and Zn(II) ions enter the solution.

  → The reaction is exothermic, as suggested by the standard enthalpy of the cell reaction:
  $$Zn(s) + Cu^{2+}(aq) \rightarrow Zn^{2+}(aq) + Cu(s) \qquad \Delta H_r^\circ = -225.56 \text{ kJ·mol}^{-1}.$$

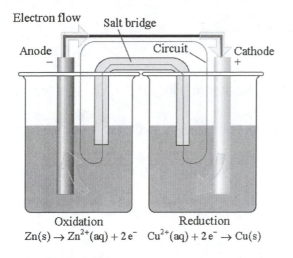

Electron flow   Salt bridge

Anode   Circuit   Cathode
−   +

Oxidation   Reduction
$Zn(s) \rightarrow Zn^{2+}(aq) + 2e^-$   $Cu^{2+}(aq) + 2e^- \rightarrow Cu(s)$

The galvanic cell to the left contains *anode* and *cathode compartments* linked by a salt bridge to prevent the mixing of ions.

The anode compartment contains a zinc electrode in an electrolyte solution such as KCl or $ZnCl_2$.

The cathode compartment contains an electrode (not necessarily Cu) in an electrolyte solution containing $Cu^{2+}(aq)$ ions.

The salt bridge contains an electrolyte such as KCl in an aqueous gelatinous medium to control the flow of $K^+(aq)$ and $Cl^-(aq)$ ions. When current is drawn from the cell, the electrodes are linked by an external wire to a load (examples include a light bulb or motor). Current (electrons) flows through the wire from anode to cathode as indicated and the spontaneous chemical *cell reaction*
$Zn(s) + Cu^{2+}(aq) \rightarrow Cu(s) + Zn^{2+}(aq)$ occurs.

As the reaction proceeds, Zn(s) is oxidized to $Zn^{2+}(aq)$ in the anode compartment (oxidation) and $Cu^{2+}(aq)$ is reduced to Cu(s) in the cathode compartment (reduction). The Cu(s) deposits on the cathode. The closed line labeled *circuit* traces the direction that current flows in the cell. Current is carried by *electrons* in the electrodes and external connections, and by *ions* in the conducting medium.

## 12.4 Cell Potential and Reaction Free Energy

- **Cell potential and emf**
  - → *Cell potential, E:* a measure of a cell reaction's ability to move electrons through the external circuit
  - → *Electromotive force* (emf): cell potential when a cell is operated reversibly
  - → For our purposes, cell potential $E$ will be taken to mean emf unless otherwise noted.
  - → The amount of useful work a cell can perform depends on the cell potential and the amount of the limiting reactant present.

- **Units and definitions**
  - → Current:   *ampere*, A, the SI base unit of electric current
  - → Charge:   *coulomb*, C ≡ A·s, the SI derived unit of charge
  - → Energy:   *joule,* J ≡ kg·m²·s⁻², the SI derived unit of energy
  - → Potential:   *volt*, V ≡ J·C⁻¹, SI derived unit of the cell potential
  - → A charge of one *coulomb* falling through a potential difference of one *volt* produces one *joule* of energy.
  - → Cell potential: Measured using a voltmeter connected to the two electrodes of a cell (see Fig. 12.4 in the text)
  - → Convention:  Write galvanic cells with anode on the left and cathode on the right. This arrangement gives a positive cell potential.

- **Examples of galvanic cell reactions with measured cell potentials**
  - → $Mg(s) + Zn^{2+}(aq, 1\ M) \rightarrow Zn(s) + Mg^{2+}(aq, 1\ M)$        $E° = 1.60$ V
  - → $2H^+(aq, 1\ M) + Zn(s) \rightarrow Zn^{2+}(aq, 1\ M) + H_2(g, 1\ bar)$        $E° = 0.76$ V

**Note:** For the preceding cells, all reactants and products are in their standard states and the potentials measured are called standard potentials, $E°$.

- **Relation between reaction free energy, $\Delta G_r$, and cell potential, $E$**
  - → Reaction free energy, $\Delta G_r$: *maximum nonexpansion* work, $w_e$, obtainable from a reaction at constant $T$ and $P$: $\Delta G_r = w_e$
  - → When $n$ electrons move through a potential difference, $E$, the work done is the total charge times the potential difference: $w_e = QV$.
  - → The charge on one mole of electrons is $N_A$ times the charge of a single electron, $-e = -1.602\,177 \times 10^{-19}$ C: $Q = -neN_A$.
  - → The Faraday constant is the *magnitude* of the charge of *one mole of electrons*:
    $F = eN_A = 9.648\,531 \times 10^4$ C·mol$^{-1}$.
  - → The work done and reaction free energy are given by the product of the total charge, $-nF$, and the cell potential, $E$: $\Delta G_r = w_e = -nFE$.
  - → In the expression above, $n$ is the stoichiometric coefficient of the *electrons* in the oxidation and reduction half-reactions that are combined to determine the cell reaction.
  - → *Maximum* nonexpansion work can only be obtained if the cell operates *reversibly*.
  - → *Reversibility* is only realized when the applied potential equals the potential generated by the cell.
  - → All working cells operate *irreversibly* and produce a lower potentials than the emf.
  - → In sum, the relationship between the free energy and the cell potential for *reversible* electrochemical cells is one of the most important concepts in this chapter and is given by

$$\boxed{\Delta G_r = -nFE}$$

- **Standard cell potential, $E°$**
  - → The standard state of a substance (s, l, or g) is the pure substance at a pressure of 1 bar.
  - → For solutes in solution, we take a molar concentration of 1 mol·L$^{-1}$ as the standard-state activity.
  - → The standard cell potential, $E°$, is the cell potential measured when all reactants and products are in their standard states.
  - → The standard reaction free energy can be obtained from the standard cell potential (emf) and *vice versa:*

$$\boxed{\Delta G_r° = -nFE°}$$

## 12.5  The Notation for Cells

- **Cell diagram**
  - → A symbolic representation of the cell components.
  - → A single vertical line | represents a phase boundary.
  - → A double vertical line ‖ represents a salt bridge.
  - → Reactants and products are represented by chemical symbols.
  - → Components in the same phase are separated by commas.
  - → The anode components are written first, followed by the salt bridge, if present, and the cathode components.
  - → Phase symbols are normally used.
  - → Concentrations of solutes and pressures of gases may also be given.

- **Cell diagrams with concentrations of reactants and products**
  - $\rightarrow$ Often the concentrations of reactants and products are included in the cell diagram to give a more complete description of the cell contents.

    **Example:** $Mg(s) + Zn^{2+}(aq, 1 \text{ M}) \rightarrow Zn(s) + Mg^{2+}(aq, 0.5 \text{ M})$     cell reaction

    $Mg(s) \mid Mg^{2+}(aq, 0.5 \text{ M}) \parallel Zn^{2+}(aq, 1 \text{ M}) \mid Zn(s)$     cell diagram

## 12.6 Standard Potentials

- **Standard cell potential**
  - $\rightarrow$ *Difference* between the two standard electrodes of an electrochemical cell
  - $\rightarrow$ The standard cell potential, $E^\circ$, is the difference between the standard potentials of the electrode on the right side of the diagram and the electrode on the left side of the diagram.

    $$\boxed{\begin{array}{c} E^\circ = E^\circ(\text{electrode on right of cell diagram}) - E^\circ(\text{electode on left of cell diagram}) \\ \text{or} \\ E^\circ = E_\text{R}^\circ - E_\text{L}^\circ \end{array}}$$

  - $\rightarrow$ If $E^\circ > 0$, the cell reaction is spontaneous under standard conditions as written.
  - $\rightarrow$ If $E^\circ < 0$, the reverse cell reaction is spontaneous under standard conditions as written.
  - $\rightarrow$ The standard potential of an electrode is sometimes called the *standard electrode potential* or the *standard reduction potential*.
- **Measuring a cell potential**
  - $\rightarrow$ The potential of a cell is measured using a voltmeter (see text Fig. 12.4).
  - $\rightarrow$ The meter displays a positive voltage when its (+) terminal is connected to the cathode.
- **Standard electrode potentials, $E^\circ$**
  - $\rightarrow$ The potential of a single electrode cannot be measured.
  - $\rightarrow$ A relative scale is required.
  - $\rightarrow$ By convention, the standard potential of the hydrogen electrode is assigned a value of 0 at all temperatures: $E^\circ(\textbf{H}^+/\textbf{H}_2) \equiv \textbf{0}$.
  - $\rightarrow$ For the standard hydrogen electrode (SHE), the half-reaction for reduction is $2\,H^+(aq, 1 \text{ M}) + 2\,e^- \rightarrow H_2(g, 1 \text{ bar})$, and the half-cell diagram is

    $$H^+(aq, 1 \text{ M}) \mid H_2(g, 1 \text{ bar}) \mid Pt(s) \qquad E^\circ(H^+/H_2) \equiv 0$$

  - $\rightarrow$ If an electrode is found to be the anode in combination with the SHE, it is assigned a *negative potential*. If it is the cathode, its standard potential is *positive*.
  - $\rightarrow$ Values of standard electrode potentials measured at 25°C are given in Table 12.1 and Appendix 2B in the text.

- **Standard electrode potentials and free energy**
  - $\rightarrow$ The relation between free energy and potential also applies to standard electrode potentials:

    $$\boxed{\Delta G_\text{r}^\circ = -nFE^\circ}$$

  - $\rightarrow$ $\Delta G_\text{f}^\circ$ values in Appendix 2A can be used to determine an unknown standard potential.

- **Calculating the standard potential of a redox couple from those of two related couples**
  - → If two redox couples are added or subtracted to give a third redox couple:
    - ➤ The standard potentials in general cannot be added to give the unknown potential.
    - ➤ This is because potential is an intensive property.
    - ➤ Free energies are extensive and can be added.
  - → To calculate the unknown potential, we convert potential values to free energy values.

- **Summary of the process of combining half-reactions**
  - → Combining two half-reactions (A and B) to obtain a third half-reaction (C) is accomplished by addition of the associated free energy changes.
  - → The equation, $\Delta G_r° = -nFE°$, applies to each half-reaction and leads to the following equation:

$$E_C° = \frac{n_A E_A° + n_B E_B°}{n_C}$$

- **Standard electrode potentials of half-reactions**
  - → Consider the half-reaction $M^{2+}(aq, 1\ M) + 2\ e^- \rightarrow M(s)$ $\quad E = E°(M^{2+}/M)$
  - → Metals with *negative* standard potentials ($E = E°(M^{2+}/M) < 0$) have a thermodynamic tendency to reduce $H_3O^+$ (aq) to $H_2$(g) under standard conditions (1 M acid).
  - → Metals with *positive* standard potentials (($E = E°(M^{2+}/M) >$) cannot reduce hydrogen ions under standard conditions; in this instance hydrogen gas is the stronger reducing agent.
  - → In general,
    - ➤ The more *negative* the standard electrode potential, the greater the tendency of a metal to *reduce* $H^+$.
    - ➤ The more *positive* the standard electrode potential, the greater the tendency of a metal ion to *oxidize* $H_2$.

## 12.7 The Electrochemical Series

- **Electrochemical series**
  - → Standard half-reactions arranged in order of *decreasing* standard electrode potential (see both Table 12.1 and text Appendix 2B)
  - → Table of relative strengths of oxidizing and reducing agents (see table below)
  - → In going *up the table,* the oxidizing strength of *reactants* increases.
    **Example:** $F_2$(g) is an exceptionally *strong oxidizing agent* with a strong tendency to gain electrons and be reduced.
  - → $F_2$(g) will oxidize any of the *product species* on the right side of the table in the reactions listed below it.
  - → In going *down the table,* the reducing strength of *products* increases.
    **Example:** Li(s) is an exceptionally *strong reducing agent* with a large tendency to lose electrons and be oxidized.

→ Li(s) will reduce any of the *reactant species* on the left side of the table in the reactions listed above it.

→ *Standard* electrode potentials have values ranging from about +3 V to −3 V, a difference of about 6 V.

→ No single *standard* galvanic cell may have a potential larger than about 6 V, because there are no half-reactions that can give a larger potential.

| Reduction Half-Reaction | | $E\,^{\circ}$ (V) |
|---|---|---|
| $F_2(g) + 2e^- \rightarrow 2F^-(aq)$ | reducing strength | +2.87 |
| $Mn^{3+}(aq) + e^- \rightarrow Mn^{2+}(aq)$ | | +1.51 |
| $I_2(s) + 2e^- \rightarrow 2I^-(aq)$ | | +0.54 |
| $2H^+(aq) + 2e^- \rightarrow H_2(g)$ | | 0 |
| $Fe^{2+}(aq) + 2e^- \rightarrow Fe(s)$ | | −0.44 |
| $Al^{3+}(aq) + 3e^- \rightarrow Al(s)$ | | −1.66 |
| $Na^+(aq) + e^- \rightarrow Na(s)$ | | −2.71 |
| $Li^+(aq) + e^- \rightarrow Li(s)$ | | −3.05 |

oxidizing strength

- **Viewing half-reactions as conjugate pairs**
  → For a Brønsted-Lowry acid-base conjugate pair, the stronger the conjugate acid, the weaker the conjugate base.
  → For a given electrochemical half-reaction, the stronger the *conjugate* oxidizing agent, the weaker the *conjugate* reducing agent. $F_2(g)$ is an extremely powerful oxidizing agent, whereas $F^-(aq)$ is an extremely weak reducing agent.

- **Uses of the electrochemical series**
  → Predicting *relative* reducing and oxidizing strengths
  → Predicting which reactants may react spontaneously in a redox reaction
  → Calculating standard cell potentials

## 12.8 Standard Potentials and Equilibrium Constants

- **Equilibrium constants can be obtained from standard cell potentials for**
  → Redox, acid-base, dissolution/precipitation, and dilution (change in concentration) reactions

- **Quantitative aspects**
  → Combining $\Delta G_r^{\circ} = -RT\ln K$ (from Chapter 9) with $\Delta G_r^{\circ} = -nFE^{\circ}$ yields

$$\ln K = \frac{nFE^{\circ}}{RT} = \frac{nE^{\circ}}{0.025\,693\ \text{V}} \quad \text{at } T = 298.15\ \text{K}$$

  → Standard electrode potentials are used to calculate $E^{\circ}$ values.
  → Because $\Delta G_r^{\circ}$ applies to *all* reactions, the above equation may also be used for acid-base reactions, precipitation reactions, and dilution reactions.
  → For non-redox reactions, the overall "cell" reaction will not show the electron transfer that occurs in the half-reactions explicitly.

## 12.9 The Nernst Equation

- **Properties of a galvanic cell**
  - → As a galvanic cell discharges, the cell potential, $E$, decreases and reactants form products.
  - → When $E = 0$, the cell is completely discharged and the cell reaction is at equilibrium.

- **Quantitative aspects**
  - → From Chapter 9, $\Delta G_r = \Delta G_r^\circ + RT \ln Q$, where $Q$ is the reaction quotient. (Note that the *identical* symbol $Q$ is also used for the quantity of electricity.)
  - → From this chapter, $\Delta G_r = -nFE$ and $\Delta G_r^\circ = -nFE^\circ$
  - → Combining the above equations leads to the *Nernst equation:*

$$E = E^\circ - \frac{RT}{nF} \ln Q = E^\circ - \frac{(0.025\ 693\ \text{V})}{n} \ln Q \quad \text{at } 298.15\ \text{K}$$

- **The Nernst equation**
  - → Describes the quantitative relationship between cell potential and the chemical composition of the cell
  - → Is used to estimate the potential for the following:
    - ➤ Cell from its chemical composition
    - ➤ Half-cell *not* in its standard state

## 12.10 Ion-Selective Electrodes

- **Ion-selective electrode**
  - → An electrode sensitive to the concentration of a particular ion
  - → Example: a metal wire in a solution containing the metal ion. The electrode potential for $E\,(Ag^+, Ag)$ or $E\,(Cu^{2+}, Cu)$ is sensitive to the concentration of $Ag^+$ or $Cu^{2+}$, respectively.

- **Measurement of pH**
  - → An important application of the Nernst equation
  - → Utilizes a galvanic cell containing an electrode sensitive to the concentration of the $H^+$ ion
  - → A calomel electrode connected by a salt bridge to a hydrogen electrode:

    Calomel electrode: $Hg_2Cl_2(s) + 2e^- \rightarrow 2Hg(l) + 2Cl^-(aq)$      $E^\circ = +0.27$ V
    Cell reaction: $Hg_2Cl_2(s) + H_2(g) \rightarrow 2H^+(aq) + 2Hg(l) + 2Cl^-(aq)$      $E^\circ = +0.27$ V

  - → $Q$ in the Nernst equation for the cell reaction above will depend only on the unknown concentration of $H^+$ if the $Cl^-$ concentration in the calomel electrode is fixed at a value determined for a *saturated* solution of KCl and the pressure of $H_2$ gas is kept constant at 1 bar.

- **Glass electrode**
  - → A thin-walled glass bulb containing an electrolyte with a potential proportional to pH
  - → Used to replace the (messy) hydrogen electrode in modern pH meters
  - → Modern pH meters use glass and calomel electrodes in a single unit (probe) for convenience. The probe contacts the test solution (unknown pH) through a small salt bridge.

- **pX meters**
  - → Devices sensitive to other ions such as $Na^+$ and $CN^-$
  - → Used in industrial applications and in pollution control

---

# Electrolysis (Sections 12.11–12.12)

## 12.12 Electrolytic Cells

- **Electrolysis**
  - → Process of driving a reaction in a nonspontaneous direction by using an electric current
  - → Conducted in an electrolytic cell

- **Electrolytic cell**
  - → An electrochemical cell in which electrolysis occurs
  - → Different from a galvanic cell in design
  - → Both electrodes normally share the same compartment.
  - → As in a galvanic cell, oxidation occurs at the anode and reduction occurs at the cathode.
  - → Unlike a galvanic cell, the anode has a *positive* charge and the cathode is *negative*.

- **Example of an electrolytic cell**
  - → Sodium metal is produced by the electrolysis of molten rock salt (impure sodium chloride) and calcium chloride. Calcium chloride lowers the melting point of sodium chloride (recall freezing point depression), increasing the energy efficiency of the process.
  - → A schematic diagram of an electrolytic cell for the production of Na(l) and $Cl_2$(g) is shown below (also see text Fig. 12.14).

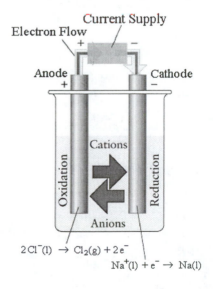

The cell has one compartment that contains both the anode and cathode. The electrode material is usually an inert metal such as platinum. Electrical current is supplied by an external source, such as a battery or power supply to drive this nonspontaneous reaction. This current *forces* reduction to occur at the cathode and oxidation at the anode. The cathode is placed on the right side as in a galvanic cell.

Sodium ions are reduced at the cathode and chloride ions are oxidized at the anode:

cathode: $Na^+(l) + e^- \rightarrow Na(l)$

anode: $2\,Cl^-(l) \rightarrow Cl_2(g) + 2\,e^-$

Note that liquid sodium metal is formed in the electrolysis because the cell temperature of 600°C is above the melting point of sodium (98°C).

**Notes:** In the molten salt mixture, $Na^+$ is reduced in preference to $Ca^{2+}$. Care must be taken in predicting the products of electrolysis in *aqueous* salt solutions because of the possibility of oxidizing or reducing water instead of either the anion or cation, respectively.

- **Overpotential and the products of electrolysis**
  - → An external potential *at least as great as* that of the spontaneous cell reaction must be applied to an electrolytic cell for electrolysis to occur.
  - → The *actual* potential required is often *greater* than this minimum value. The additional voltage required is called *overpotential*.
  - → Overpotential depends on the structure of a solution near an electrode, which differs from that in the bulk solution. Near an electrode, electrons are transferred across an electrode-solution interface that depends on the solution and the condition of the electrode.
  - → Because of overpotential, the observed electrolysis products may differ from those predicted using standard electrode potentials and the Nernst equation.
  - → In a solution with more than one *reducible* species, the *reactant* with the *most positive* reduction potential will be reduced.
  - → In a solution with more than one *oxidizable* species, the *product* with the *most negative* reduction potential will be oxidized.

## 12.14 The Products of Electrolysis

- **Faraday's law of electrolysis**
  - → Number of moles of product formed is *stoichiometrically equivalent* to the number of moles of electrons supplied.
  - → The amount of product depends on the current, the time it is applied, and the number of moles of electrons in the half-reaction.

- **Quantitative aspects of Faraday's law**
  - → Determine the stoichiometry of a half-reaction; for example, $M^{n+} + ne^- \rightarrow M(s)$.
  - → The quantity of electricity, $Q$, passing through the electrolysis cell is determined by the electric current, $I$, and the time, $t$, of current flow. Recall that 1 ampere (A) = 1 coulomb (C) s$^{-1}$.

$$\boxed{\text{Charge supplied (C)} = \text{current (A)} \times \text{time (s)} \quad \text{or} \quad Q = It}$$

  - → Because $Q = nF$, where $n$ is the number of moles of electrons and $F$ is the Faraday constant, then

$$\boxed{\text{Moles of electrons} = \frac{\text{charge supplied (C)}}{F} = \frac{\text{current } (I) \times \text{time } (t)}{F} \quad \text{or} \quad n(e^-) = \frac{It}{F}}$$

  - → The mass of product is determined from $n(e^-)$, the half-reaction, and the molar mass, $M$.

$$\text{Moles of product} = n(p) = \frac{It}{n(e^-)F}$$

$$\text{Mass of product} = m(p) = \left(\frac{It}{n(e^-)F}\right)M(p)$$

# The Impact on Materials (Sections 12.13–12.15)

## 12.13 Applications of Electrolysis

- **Uses of electrolysis**
  - → Extracting metals from their salts. Examples include Al(s), Na (see text Fig. 12.14), and Mg.
  - → Preparation of fluorine, chlorine, and sodium hydroxide
  - → Refining metals such as copper
  - → Electroplating metals
- **Electroplating**
  - → Electrolytic deposition of a thin metal film on an object
  - → Object to be plated is made the cathode of an electrolysis cell.
  - → Cell electrolyte is a salt of the metal to be plated.
  - → Cations are supplied either by the added salt, or from oxidation of the anode, which is then made of the plating metal. For example, silver, gold, and chromium plating (see text Fig. 12.15)

## 12.14 Corrosion

- **Corrosion**
  - → Unwanted oxidation of a metal
  - → An electrochemical process that is destructive and costly
- **Retarding corrosion**
  - → Coating the metal with paint (**painting**) or plastic. Non-uniform coverage or uneven bonding leads to deterioration of the coating and rust.
  - → **Passivation,** the formation of a nonreactive surface layer (Example: $Al_2O_3$ on aluminum metal)
  - → **Galvanization**, coating the metal with an unbroken layer of zinc. Zinc is preferentially oxidized and the zinc oxide that forms provides a protective coating (passivation).
  - → Use of a *sacrificial anode,* a metal more easily oxidized than the one to be protected, attached to it. Used for large structures. A disadvantage is that sacrificial anodes must be replaced when completely oxidized (think of an underground pipeline).
- **Mechanism of corrosion**
  - → Exposed metal surfaces can act as anodes or cathodes, points where oxidation or reduction can occur.
  - → If the metal is wet (from dew or groundwater), a film of water containing dissolved ions may link the anode and cathode, acting as a salt bridge.
  - → The circuit is completed by electrons flowing through the metal, which acts as a wire in the external circuit of a galvanic cell.
- **Mechanism of rust formation on iron metal**
  - → At the *anode,* iron is oxidized to iron(II) ions.
    - 1) $Fe(s) \rightarrow Fe^{2+}(aq) + 2e^-$ $\quad\quad\quad -E° = -(-0.44\text{ V}) = +0.44\text{ V}$
  - → At the *cathode,* oxygen is reduced to water.
    - 2) $O_2(g) + 4H^+(aq) + 4e^- \rightarrow 2H_2O(l)$ $\quad\quad E° = +1.23\text{ V}$
  - → Iron(II) migrates to the cathode where it may be oxidized by oxygen to iron(III).
    - 3) $Fe^{2+}(aq) \rightarrow Fe^{3+}(aq) + e^-$ $\quad\quad\quad -E° = -(+0.77\text{ V}) = -0.77\text{ V}$

→ The overall redox equation is obtained by adding reaction 1) four times, reaction 3) four times, and reaction 2) three times. The resulting cell reaction has $n = 12$.

4)    $4Fe(s) + 3O_2(g) + 12H^+(aq) \rightarrow 4Fe^{3+}(aq) + 6H_2O(l)$        $E° = +1.27$ V

→ Next, the iron(III) ions in reaction 4) precipitate as a hydrated iron(III) oxide or rust. The coefficient, $x$, of the waters of hydration in the oxide is not well defined.

5)    $4Fe^{3+}(aq) + 2(3 + x)H_2O(l) \rightarrow 2Fe_2O_3 \cdot xH_2O(s) + 12H^+(aq)$

→ Finally, the overall reaction for the formation of rust is obtained by adding reaction 4) and 5):

6)    $4Fe(s) + 3O_2(g) + 2xH_2O(l) \rightarrow 2Fe_2O_3 \cdot xH_2O(s)$

→ From this mechanism, summarized in reaction 6), we expect corrosion to be accelerated by moisture, oxygen, and salts, which enhance the rate of ion transport.

## 12.15 Practical Cells

- **Practical galvanic cells**
  - → Commonly called batteries
  - → Should be inexpensive, potable, safe, and environmentally benign
  - → Should have a high *specific energy* (reaction energy per kilogram) and stable current

- **Primary cells**
  - → Cannot be recharged
  - → Some types of cells:
    - ➤ **Dry cells** – (A, AA, D, *etc.*, batteries), used in flashlights, toys, remote controls, etc.
      **Cell diagram:** $Zn(s) \mid ZnCl_2(aq), NH_4Cl(aq) \mid MnO(OH)(s) \mid MnO_2(s) \mid C(gr)$, 1.5 V
      **Cell reaction:** $Zn(s) + 2NH_4^+(aq) + 2MnO_2(s) \rightarrow Zn^{2+}(aq) + 2MnO(OH)(s) + 2NH_3(g)$
    - ➤ **Alkaline cells** – similar to dry cells but with an alkaline electrolyte that increases battery life Used in backup (emergency) power supplies and smoke detectors
      **Cell diagram:** $Zn(s) \mid ZnO(s) \mid OH^-(aq) \mid Mn(OH)_2(s) \mid MnO_2(s) \mid C(gr)$, 1.5 V
      **Cell reaction:** $Zn(s) + MnO_2(s) + H_2O(l) \rightarrow ZnO(s) + Mn(OH)_2(s)$
    - ➤ **Silver cells** – have only solid reactants and products, high emf, long life. Used in pacemakers, hearing aids, and cameras
      **Cell diagram:** $Zn(s) \mid ZnO(s) \mid KOH(aq) \mid Ag_2O(s) \mid Ag(s) \mid$ steel, 1.6 V
      **Cell reaction:** $Zn(s) + Ag_2O(s) \rightarrow ZnO(s) + 2Ag(s)$

- **Secondary cells**
  - → Can be recharged
  - → Must be charged before initial use
  - → Some types of cells:
    - ➤ **Lead-acid cell** (automobile battery) has low specific energy but generates high current over small time period to start vehicles.
      **Cell diagram:** $Pb(s) \mid PbSO_4(s) \mid H^+(aq), HSO_4^-(aq) \mid PbO_2(s) \mid PbSO_4(s) \mid Pb(s)$, 2 V
      **Cell reaction:** $Pb(s) + PbO_2(s) + 2H_2SO_4(aq) \rightarrow 2PbSO_4(s) + 2H_2O(l)$
    - ➤ **Lithium-ion cell** (automobile battery) high energy density, high emf, can be recharged many times. Used in laptop computers.

➤ **Sodium-sulfur cell** used to power electric vehicles, has all *liquid* reactants and products, a solid electrolyte and relatively-high voltage.

**Cell diagram:** $Na(l) \mid Na^+(solution) \parallel S^{2-}(solution) \mid S_8(l)$, 2.2 V

**Cell reaction:** $16\,Na(l) + S_8(l) \rightarrow 16\,Na^+(solution) + 8\,S^{2-}(solution)$

- **Fuel cells**
  - → Galvanic cells designed for continuous reaction
  - → Require a continuous supply of reactants
  - → **Alkali-fuel cell.** Used in the space shuttle.

    **Cell diagram:** $Ni(s) \mid H_2(g) \mid KOH(aq) \mid O_2(g) \mid Ni(s)$, 1.23 V

    **Cell reaction:** $2\,H_2(g) + O_2(g) \rightarrow 2\,H_2O(l)$

# Chapter 13   Chemical Kinetics

## Reaction Rates  (Sections 13.1–13.3)

### 13.1  Concentration and Reaction Rate

- **Definition of rate**

  **General** $\rightarrow$ Rate is the change in a property divided by the time required for the change to occur.

  **Chemical kinetics** $\rightarrow$ Reaction rate is the change in molar concentration of a reactant or product divided by the time required for the change to occur.

- **Average *vs.* instantaneous rates: an analogy to travel**
  - $\rightarrow$ The *average* speed of an automobile trip is the length of the journey divided by the total time for the journey.
  - $\rightarrow$ The *instantaneous* speed is obtained if the car is timed over a very short distance at some point in the journey.

- **Average reaction rate**
  - $\rightarrow$ For a general chemical reaction, $a\mathrm{A} + b\mathrm{B} \rightarrow c\mathrm{C} + d\mathrm{D}$, the average reaction rate is the *change* in molar concentration of a reactant or product divided by the time interval required for the change to occur.
  - $\rightarrow$ The average reaction rate may be determined for *any* reactant or product.
  - $\rightarrow$ Rates for different reactants and products may have different numerical values.

- **Unique average reaction rate**
  - $\rightarrow$ For a general chemical reaction, $a\mathrm{A} + b\mathrm{B} \rightarrow c\mathrm{C} + d\mathrm{D}$, the unique average reaction rate is the average reaction rate of reactant or product divided by its stoichiometric coefficient used as a *pure number*.
  - $\rightarrow$ The unique average reaction rate may be determined from *any* reactant or product.
  - $\rightarrow$ The unique average reaction rate is the same for any reactant or product.

  **Note:** Two critical underlying assumptions for generating a unique reaction rate are that the overall reaction time is slow with respect to the buildup and decay of any intermediate and that the stoichiometry of the reaction is maintained throughout.

- **Reaction rates**
  - $\rightarrow$ The time required to measure the concentration changes of reactants and products varies considerably according to the reaction.
  - $\rightarrow$ Spectroscopic techniques are often used to monitor concentration, particularly for fast reactions. An important example is the *stopped-flow technique* shown in Fig. 13.3 in the text. The fastest reactions occur on a time scale of femtoseconds ($10^{-15}$ s).

### 13.2  The Instantaneous Rate of Reaction

- **Reaction rates**
  - $\rightarrow$ Most reactions slow down as they proceed and reactants are depleted.
  - $\rightarrow$ If equilibrium is reached, the forward and reverse reaction rates are equal.
  - $\rightarrow$ The reaction rate at a specific time is called the *instantaneous reaction rate*.

- **Instantaneous reaction rate**
  - → For a product, the *slope* of a tangent line of a graph of concentration *vs.* time
  - → For a reactant, the *negative of the slope* of a tangent line on a graph of concentration *vs.* time

- **Mathematical form of the instantaneous rate**
  - → The slope of the concentration *vs.* time curve is the derivative of the concentration with respect to time.
  - → For the instantaneous disappearance of a reactant, R,

$$\text{Reaction rate} = -\frac{d[R]}{dt}$$

  - → For the instantaneous appearance of a product, P,

$$\text{Reaction rate} = \frac{d[P]}{dt}$$

  - → For the general reaction, $a\text{A} + b\text{B} \rightarrow c\text{C} + d\text{D}$,

$$\boxed{\text{Unique instantaneous reaction rate} = -\frac{1}{a}\frac{d[A]}{dt} = -\frac{1}{b}\frac{d[B]}{dt} = \frac{1}{c}\frac{d[C]}{dt} = \frac{1}{d}\frac{d[D]}{dt}}$$

  - → In the remainder of the chapter, the term reaction rate is used specifically to mean the *unique* instantaneous reaction rate written here. The two conditions relating to *unique average* rates described in Section 13.1 of the *Study Guide* also apply here.

## 13.3  Rate Laws and Reaction Order

- **Initial reaction rate**
  - → Instantaneous rate at $t = 0$
  - → In what is called the method of initial rates, analysis is simpler at $t = 0$ because products that may affect reaction rates are not present.
  - → The method of initial rates is often used to determine the *rate law* of a reaction.

- **Rate law of a reaction**
  - → Expression for the instantaneous reaction rate in terms of the concentrations of species that affect the rate such as reactants or products
  - → Always determined by experiment, and usually difficult to predict
  - → May have relatively simple or complex mathematical form
  - → The exponents in a rate expression are *not* necessarily the stoichiometric coefficients of a reaction; therefore, we use $m$ and $n$ in place of $a$ and $b$ in rate equations.

- **Simple rate laws**
  - → Form of a simple rate law:  $\boxed{\text{Rate} = k[A]^m}$
  - → $k$ = rate constant, which is the reaction rate when all the concentrations appearing in the rate law are 1 M (recall that the symbol M stands for the units of $\text{mol·L}^{-1}$).
  - → $k$ is *independent* of concentration(s).
  - → $k$ is *dependent* on temperature, $k(T)$.

→ A refers to the reactant, and $m$, the power of $[A]$, is called the order in reactant A and is determined experimentally. Order is *not related to the reaction stoichiometry*. Exceptions include a special class of reactions, which are called *elementary* (see Section 13.7).

→ Typical orders for reactions with one reactant are 0, 1, and 2. Fractional values are also common.

| | | |
|---|---|---|
| Zero-order reaction | $(m = 0)$: | Rate $= k$ |
| First-order reaction | $(m = 1)$: | Rate $= k[A]$ |
| Second-order reaction | $(m = 2)$: | Rate $= k[A]^2$ |

- **More complex rate laws**

  → Many reactions have experimental rate laws containing concentrations of more than one species.

  → Form of a complex rate law:

  $$\text{Rate} = k[A]^m[B]^n \ldots$$

  The overall order is the sum of the powers $m + n + \ldots$.

  → The powers may also be negative values, and forms of even greater complexity with no overall order are found (see Section 13.15 for an example).

  → More complex rate laws are treated in Section 13.8.

  → The following table lists experimental rate laws for a variety of reactions:

| Reaction | Rate law | Order | Units of $k$ |
|---|---|---|---|
| $2NH_3(g) \rightarrow N_2(g) + 3H_2(g)$ | rate $= k$ | zero | $mol \cdot L^{-1} \cdot s^{-1}$ |
| $2N_2O_5(g) \rightarrow 4NO_2(g) + O_2(g)$ | rate $= k[N_2O_5]$ | first | $s^{-1}$ |
| $2NO_2(g) \rightarrow NO(g) + NO_3(g)$ | rate $= k[NO_2]^2$ | second | $L \cdot mol^{-1} \cdot s^{-1}$ |
| $2NO(g) + O_2(g) \rightarrow 2NO_2(g)$ | rate $= k[NO]^2[O_2]$ | first in $O_2$ second in NO third overall | $L^2 \cdot mol^{-2} \cdot s^{-1}$ |
| $I_3^-(aq) + 2N_3^-(aq)$ $\rightarrow 3I^-(aq) + 3N_2(g)$ in the presence of $CS_2(aq)$ | rate $= k[CS_2][N_3^-]$ | first in $CS_2$ first in $N_3^-$ second overall | $L \cdot mol^{-1} \cdot s^{-1}$ |
| $2O_3(g) \rightarrow 3O_2(g)$ | rate $= k\dfrac{[O_3]^2}{[O_2]}$ $= k[O_3]^2[O_2]^{-1}$ | second in $O_3$ minus one in $O_2$ first overall | $s^{-1}$ |

→ In general, the reaction order does not follow from the stoichiometry of the chemical equation. In the case of the *catalytic* decomposition of ammonia on a hot platinum wire, the reaction order is zero initially because the reaction occurs on the surface of the wire, and the surface coverage is independent of concentration. The rate of a zero-order reaction is independent of concentration until the reactant is nearly exhausted or until equilibrium is reached.

→ Reaction rates may depend on species that do not appear in the overall chemical equation. The reaction of triiodide ion with nitride ion, for example, is accelerated by the presence of carbon disulfide, neither a reactant nor product. The concentration of carbon disulfide, a *catalyst*, appears in the rate law. Catalysis is discussed in Section 13.14.

→ In the last example above, the order with respect to the product $O_2$ is −1. Negative orders often arise from reverse reactions in which products reform reactants, thereby slowing the overall reaction rate. Oxygen is present at the beginning of this reaction as commonly studied, because ozone is made by an electrical discharge in oxygen gas. It is possible to study the kinetics of the decomposition of pure ozone. In the absence of oxygen, the initial rate law is quite different from the one listed in the above table.

- **Procedure for determining the reaction orders and rate law**

  → The reaction order for a given species is determined by varying its initial concentration and measuring the initial reaction rate, while keeping the concentrations of all other species constant.

  → The procedure is repeated for the other species until all the reaction orders are obtained.

  → Once the form of the rate law is known, a value of the rate constant is calculated for each set of initial concentrations. The reported rate constant is typically the average value for all the measurements.

- **Pseudo-order reactions**

  → If a species in the rate-law equation is present in great excess, its concentration is effectively constant as the reaction proceeds. The constant concentration may be incorporated into the rate constant and the apparent order of the reaction changes.

  → Reactions in which the solvent appears in the rate law often show pseudo-order behavior, because the solvent is usually present in large excess.

  → A technique called the *isolation method* is often used to determine the order in a particular reactant by keeping the concentration of other reactants in excess. The method is repeated for all reactants and the complete rate law is obtained. A hidden danger is that occasionally the rate law is *not* the same in the extreme ranges of the concentrations of reactants. Different reaction mechanisms that yield different rate laws may occur under the extreme conditions.

---

# Concentration and Time  (Sections 13.4–13.6)

## 13.4  First-Order Integrated Rate Laws

- **Integration of a first-order rate law**

  → First-order rate law:

  $$-\frac{d[A]}{dt} = k[A]$$

  → Integration with limits:

  $$-\int_{[A]_0}^{[A]_t} \frac{d[A]}{[A]} = \int_0^t k\,dt$$

→ Result in logarithmic form: $\boxed{\ln\left(\dfrac{[A]_t}{[A]_0}\right) = -kt}$

→ Result in exponential form: $\boxed{[A]_t = [A]_0 e^{-kt}}$

→ If a plot of $\ln[A]_t$ vs. $t$ gives a straight line, the rate law is first-order in $[A]$.

→ The rearranged logarithmic expression, $\ln[A]_t = \ln[A]_0 - kt$, shows the straight-line behavior, $y = \text{intercept} + (\text{slope} \times x)$, of a plot of $\ln[A]_t$ vs. $t$.

→ Plot of $\ln[A]_t$ vs. $t$: slope $= -k$ and intercept $= \ln[A]_0$

- **Integrated first-order rate law**

    → $[A]$ decays exponentially with time.

    → Used to confirm that a reaction is first order and to measure its rate constant

    → Known $k$ and $[A]_0$ values can be used to predict the value of $[A]$ at any time $t$.

## 13.5  Half-Lives for First-Order Reactions

- **Half-life expression**

    → The half-life, $t_{1/2}$, of a substance is the time required for its concentration to fall to one-half its initial value.

    → At $t = t_{1/2}$, $[A]_{t_{1/2}} = \frac{1}{2}[A]_0$. The ratio of concentrations is $\dfrac{[A]_0}{[A]_{t_{1/2}}} = \dfrac{[A]_0}{\frac{1}{2}[A]_0} = 2$.

    → Solve for $t_{1/2}$ by rearranging the logarithmic form of the integrated rate law to give

    $t_{1/2} = \dfrac{1}{k}\ln\left(\dfrac{[A]_0}{[A]_{t_{1/2}}}\right)$. Then substitute $\dfrac{[A]_0}{[A]_{t_{1/2}}} = 2$.

    → Half-life expression: $\boxed{t_{1/2} = \dfrac{\ln 2}{k}}$

- **Half-life of a first-order reaction**

    → Independent of the initial concentration

    → Characteristic of the reaction

    → Inversely proportional to the rate constant, $k$

    → The logarithmic form of the rate law may be modified as follows:

    $$\boxed{\ln\left(\dfrac{[A]_0}{[A]_t}\right) = kt = (\ln 2)\left(\dfrac{t}{t_{1/2}}\right)}$$

## 13.6 Second-Order Integrated Rate Laws

- **Integration of a second-order rate law**

  → Second-order rate law:
  $$-\frac{d[A]}{dt} = k[A]^2$$

  → Integration with limits:
  $$-\int_{[A]_0}^{[A]_t} \frac{d[A]}{[A]^2} = \int_0^t k\, dt$$

  → Integrated form:
  $$\frac{1}{[A]_t} - \frac{1}{[A]_0} = kt$$

  → Alternative form:
  $$[A]_t = \frac{[A]_0}{1 + [A]_0 kt}$$

  → If a plot of $1/[A]_t$ vs. $t$ gives a straight line, the rate law is second-order in $[A]$.

  → The rearranged expression, $1/[A]_t = 1/[A]_0 + kt$, reveals the linear behavior of a plot of $1/[A]_t$ vs. $t$.

  → Plot of $1/[A]_t$ vs. $t$:   slope $= +k$   *and*   intercept $= 1/[A]_0$

- **Integrated second-order rate law**

  → For the same initial rates, $[A]$ decays more slowly with time for a second-order process than for a first-order one (see Fig. 13.13 in the text).

  → Used to confirm that a reaction is second order and to measure its rate constant

  → If $k$ and $[A]_0$ are known, the value of $[A]$ at any time $t$ can be predicted.

---

# Reaction Mechanisms  (Sections 13.7–13.10)

## 13.7 Elementary Reactions

- **Reactions**

  → A net chemical reaction is assumed to occur at the molecular level in a series of separable steps, called *elementary reactions*.

  → A *reaction mechanism* is a proposed group of elementary reactions that accounts for the reaction's overall stoichiometry and experimental rate law.

- **Elementary reaction**

  → A single-step reaction, written *without* state symbols, that describes the behavior of *individual* atoms and molecules taking part in a chemical reaction

  → The *molecularity* of an elementary reaction specifies the number of *reactant* molecules that take part in it.

  → *Unimolecular:*   Only one reactant species is written (molecularity = 1).
  $$C_3H_6 \text{ (cyclopropane)} \rightarrow CH_2=CH-CH_3 \text{ (propene)}$$
  *Cyclopropane molecule decomposes spontaneously to form a propene molecule.*

→ *Bimolecular:* Two reactant species combine to form products (molecularity = 2).

$H_2 + Br \rightarrow HBr + H$

*A hydrogen molecule and a bromine atom collide to form a hydrogen bromide molecule and a hydrogen atom.*

→ *Termolecular:* Three reactant species combine to form products (molecularity = 3).

$I + I + H_2 \rightarrow HI + HI$

*Two iodine atoms and a hydrogen molecule collide simultaneously to form two hydrogen iodide molecules.*

- **Reaction mechanism**

  → *Sequence* of elementary reactions which, when added together, gives the net chemical reaction and reproduces its rate law

  → *Reaction intermediate,* *not* a reactant or a product, is a species that appears in one or more of the elementary reactions in a proposed mechanism. It usually has a small concentration, and does not appear in the overall rate law.

  → The plausibility of a reaction mechanism may be tested, but it cannot be proved.

  → The presence and behavior of a reaction intermediate are sometimes testable features of a proposed reaction mechanism.

## 13.8 The Rate Laws of Elementary Reactions

- **Forms of rate laws for elementary reactions**

  → Order follows from the stoichiometry of the *reactants* in an *elementary* reaction. Note that products do *not* appear in the rate law of an *elementary* reaction.

  → *Unimolecular* elementary reactions are first-order.

  Reaction: $A \rightarrow$ products    Rate = $k[A]$

  → *Bimolecular* elementary reactions are second-order overall. If there are two different species present, the reaction is first-order in each one.

  Reaction: $A + A \rightarrow$ products    Rate = $k[A]^2$
  Reaction: $A + B \rightarrow$ products    Rate = $k[A][B]$

  → *Termolecular* elementary reactions are third-order overall. If there are two different species, the reaction is second-order in one of them, and first-order in the other. If there are three different species, the reaction is first-order in each one.

  Reaction: $A + A + A \rightarrow$ products    Rate = $k[A]^3$
  Reaction: $A + A + B \rightarrow$ products    Rate = $k[A]^2[B]$
  Reaction: $A + B + C \rightarrow$ products    Rate = $k[A][B][C]$

  → The molecularity of an elementary reaction that is written in the reverse direction also follows from its stoichiometry. The *products* have become *reactants*.

  Forward reaction:    $A + B \rightarrow P + Q + R$    Forward rate = $k[A][B]$
  Reverse reaction:    $P + Q + R \rightarrow A + B$    Reverse rate = $k'[P][Q][R]$

**Note:** The prime symbol, ′, is used here and in the following sections to denote a rate constant written for a *reverse* elementary reaction. The same symbol is used in Section 13.10 to denote a rate constant for the *same* reaction (elementary or otherwise) measured at a *different* temperature. *Do not confuse these two very different meanings.*

- **Rate laws from reaction mechanisms**
  - → The time evolution of each reactant, intermediate, and product may be determined by integrating the rate laws for the elementary reactions in a proposed reaction mechanism. Recall that a successful mechanism must account for the stoichiometry of the overall reaction as well as the observed rate law.
  - → Mathematical solution of kinetic equations often requires numerical methods for solving simultaneous differential equations. In a few cases, exact analytical solutions exist.
  - → For multistep mechanisms, approximate methods are often used to determine a rate law consistent with a proposed reaction mechanism. If the derived rate law is in agreement with the experimental one, the plausibility of the mechanism is tested by further experimentation, if possible.

- **Approximate methods for determining rate laws**
  - → Characteristics of reactions commonly encountered allow for the determination of rate laws with the use of three types of approximations.
  - → The *rate-determining step* approximation is made to determine a rate law for a mechanism in which one step occurs at a rate *substantially* slower than any others. The slow step is a bottleneck, and the overall reaction rate cannot be larger than the slow step. The rate law for the rate-determining step is written first. If a reaction intermediate appears as a reactant in this step, its concentration term must be eliminated from the rate law. The final rate law only has concentration terms for reactants and products.
  - → The *pre-equilibrium condition* describes situations in which reaction intermediates are formed and removed in steps prior to the rate-determining one. If the formation and removal of the intermediate in prior steps is rapid, an equilibrium concentration is established. If the intermediate appears in the rate-determining step, its relatively slow reaction in that step does not change its equilibrium concentration. The pre-equilibrium condition is sometimes called a *fast equilibrium*.
  - → The *steady-state approximation* is a more general method for solving reaction mechanisms. The net rate of formation of any intermediate in the reaction mechanism is set equal to 0. An intermediate is assumed to attain its steady-state concentration instantaneously, decaying slowly as reactants are consumed. An expression is obtained for the steady-state concentration of each intermediate in terms of the rate constants of elementary reactions and the concentrations of reactants and products. The rate law for an elementary step that leads directly to product formation is usually chosen. The concentrations of all intermediates are removed from the chosen rate law, and a final rate law for the formation of product that reflects the concentrations of reactants and products is obtained.
  - → Often a predicted rate law does not quite agree with the experimental one. Reaction conditions that modify the form of the rate law predicted by the mechanism may lead to final agreement (see Section 13.15 for an example). If complete exploration of conditions fails to reproduce the experimental result, the proposed reaction mechanism is rejected.

## 13.9 Chain Reactions

- **Chain reaction**
  - → Series of linked elementary reactions that *propagate* in chain cycles
  - → A reaction intermediate is produced in an *initiation* step.
  - → Reaction intermediates are the *chain carriers,* which are often radicals.

→ In one propagation cycle, a reaction intermediate typically reacts to form a different intermediate, which in turn reacts to regenerate the original one.

→ In each cycle, some product is usually formed.

→ The cycle is eventually broken by a *termination* step that consumes the intermediate.

→ In *chain branching,* a chain carrier reacts to form two or more carriers in a single step.

→ Chain branching often leads to chemical explosions.

## 13.10 Rates and Equilibrium

- **Elementary reaction**

  → At *equilibrium,* the forward and reverse reaction rates for an *elementary* reaction are equal. For a given elementary reaction at equilibrium, $A + B \rightleftharpoons P + Q$
  forward rate $= k[A][B] =$ reverse rate $= k'[P][Q]$

  → The *equilibrium constant* for an *elementary* reaction is equal to the ratio of the rate constants for the forward and reverse reactions.

  In the above example, $K = \dfrac{[P][Q]}{[A][B]} = \dfrac{(\text{reverse rate})/k'}{(\text{forward rate})/k} = \dfrac{k}{k'}$

  $$\boxed{K = \frac{k}{k'}}$$ for any elementary reaction

- **Equilibrium constants from reaction mechanisms**

  → Consider an overall reaction with a multistep mechanism in which the rate constants for the elementary reactions are $k_1$, $k_2$, $k_3$, ... in the forward direction and $k_1'$, $k_2'$, $k_3'$, ... in the reverse direction.

  → If each elementary step is in equilibrium, the overall reaction is obtained by summing the elementary reactions, and the equilibrium constant for the overall reaction is obtained by multiplying the equilibrium constants for the individual steps.

  → For the reaction at equilibrium, the above procedure gives

  $$\boxed{K = \frac{k_1}{k_1'} \times \frac{k_2}{k_2'} \times \frac{k_3}{k_3'} \times \ \cdots}$$

- **Temperature dependence of the equilibrium constant of an elementary reaction**

  → Follows from the temperature dependence of the ratio of the rate constant for the forward reaction to that of the reverse reaction, with both expressed in Arrhenius form

  → If the reaction is *exothermic*, the activation energy in the forward direction is smaller than in the reverse one (see the two upper figures in the *Study Guide* in Section 13.13).

  ➤ The rate constant with the larger activation energy increases more rapidly as temperature increases than the one with smaller energy.

  ➤ As temperature is *increased*, $k'$ is predicted to increase faster than $k$ and the equilibrium constant *decreases*.

  ➤ *Reactants* are favored.

→ If the reaction is *endothermic,* the activation energy is larger in the forward direction than in the reverse one.

> The rate constant with the larger activation energy increases more rapidly as temperature increases than the one with smaller energy.

> As temperature is *increased,* $k$ is predicted to increase faster than $k'$ and the equilibrium constant *increases.*

> *Products* are favored.

---

# Models of Reactions  (Sections 13.11–13.13)

## 13.11  The Effect of Temperature

- **Qualitative aspects**

  → A change in temperature results in a change of the rate constant of a reaction, and, therefore, its rate.

  → For *most* reactions, *increasing* the temperature results in a *larger value* of the rate constant.

  → For *many* reactions in organic chemistry, reaction rates *approximately* double for a 10°C increase in temperature.

- **Quantitative aspects**

  → The temperature dependence of the rate constant of many reactions is given by the Arrhenius equation, shown here in two forms:

  Exponential form:   $$k = A e^{-\frac{E_a}{RT}}$$

  Logarithmic form:   $$\ln k = \ln A - \frac{E_a}{RT}$$

  $R$ is the universal gas constant ($8.31451$ J·K$^{-1}$·mol$^{-1}$)
  $T$ is the absolute temperature (K)

  → The two constants, $A$ and $E_a$, called the *Arrhenius parameters,* are *nearly* independent of temperature.

  → The parameter $A$ is the *pre-exponential factor.* The parameter $E_a$ is the *activation energy.* The two parameters depend on the reaction being studied.

  → The Arrhenius equation applies to all types of reactions, gas phase or in solution.

- **Obtaining Arrhenius parameters**

  → Plot $\ln k$ *vs.* $1/T$. If the plot displays straight-line behavior, the *slope* of the line is $-E_a/R$ and the *intercept* is $\ln A$. Note that the *intercept* at ($1/T = 0$) implies $T = \infty$.

  → The parameter $A$ *always* has a positive value. The *larger* the value of $A$, the *larger* the value of $k$.

→ The parameter $E_a$ *almost always* has a positive value. The *larger* the value of $E_a$, the *smaller* the value of $k$. If $E_a > 0$, the value of $k$ *increases* as temperature *increases*.

→ The rate constant increases more quickly with increasing temperature if $E_a$ is large.

→ If $E_a$ and $k$ are known at a given temperature, $T$, the logarithmic form of the Arrhenius equation can be used to calculate the rate constant, $k'$, at another temperature, $T'$. Or alternatively, a value of $E_a$ can be estimated from values of $k'$ at $T'$ and $k$ at $T$.

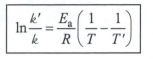

$$\ln\frac{k'}{k} = \frac{E_a}{R}\left(\frac{1}{T} - \frac{1}{T'}\right)$$

- **Units of the Arrhenius parameters**

  → $A$ has the same units as the rate constant of the reaction.

  → $E_a$ has units of energy, and values are usually reported in $kJ \cdot mol^{-1}$.

  **Note:** An equivalent SI unit for the liter is $dm^3$. Rate constants and Arrhenius pre-exponential factors incorporating volume units are sometimes reported using $dm^3$ or $cm^3$ instead of liters. Those incorporating amount units are sometimes reported in units of molecules instead of moles. For example, a second-order rate constant of $1.2 \times 10^{11}$ $L \cdot mol^{-1} \cdot s^{-1}$ is equivalent to $1.2 \times 10^{11}$ $dm^3 \cdot mol^{-1} \cdot s^{-1}$ *or* $2.0 \times 10^{-10}$ $cm^3 \cdot molecule^{-1} \cdot s^{-1}$.

## 13.12 Collision Theory

- **Collision theory**

  → Model for how reactions occur at the molecular level

  → Applies to gases

  → Accounts for rate constants and the exponential form of the Arrhenius equation

  → Reveals the significance of the Arrhenius parameters $A$ and $E_a$

- **Assumptions of collision theory**

  → Molecules must collide in order to react.

  → Only collisions with kinetic energy in excess of some minimum value, $E_{min}$, lead to reaction.

  → Only molecules with the correct orientation with respect to each other may react.

- **Total rate of collisions in a gas mixture**

  → Determined quantitatively given from the kinetic model of a gas

  → Rate of collision = total number of collisions per unit volume per second

$$\text{Rate of collision} = \sigma \bar{v}_{rel} N_A^2 [A][B]$$

  $\sigma =$ collision cross-section (area a molecule presents as a target during collision) Larger molecules are more likely to collide with other molecules than are smaller ones.

  $\bar{v}_{rel} =$ mean speed at which molecules approach each other in a gas

→ For a gas mixture of A and B with molar masses $M_A$ and $M_B$, respectively,

$$\bar{v}_{rel} = \sqrt{\frac{8RT}{\pi\mu}} \qquad \mu = \frac{M_A M_B}{M_A + M_B}$$

Molecules at high temperature collide more often than at low temperature.

$N_A$ = Avogadro constant

[A] and [B] are the molar concentrations of A and B, respectively.

- **Total rate of reaction in a gas mixture**
  - → Rate of reaction = number of collisions that lead to reaction per unit volume per second

  $$\text{Rate of reaction} = P\sigma \bar{v}_{rel} N_A^2 [A][B] \times e^{-E_{min}/RT}$$

  $E_{min}$ = minimum kinetic energy required for a collision to lead to reaction
  Collisions with energy less than the minimum do not lead to reaction.

  $e^{-E_{min}/RT}$ = fraction of collisions with at least the energy $E_{min}$
  Derived from the *Boltzmann distribution* (see Fig. 13.27 in the text)

  $P$ = steric factor < 1
  Reactive collisions often require preferred directions of approach.

- **Reaction rate constant**
  - → Rate constant = $k = \dfrac{\text{rate of reaction}}{[A][B]}$

  $$k = P\sigma \bar{v}_{rel} N_A^2 e^{-E_{min}/RT}$$

  → Comparison with the Arrhenius equation in exponential form, $k = Ae^{-\frac{E_a}{RT}}$ :

  $E_a \approx E_{min}$ and $A \approx P\sigma \bar{v}_{rel} N_A^2$
  Because the speed is proportional to $T^{1/2}$, the model predicts that the pre-exponential factor *and* activation energy both have a *slight* temperature dependence.

- **Summary**
  - → According to collision theory, reaction occurs only if reactant molecules collide with a kinetic energy equal to or greater than the Arrhenius activation energy.
  - → The reaction rate constant increases with the size and speed of the colliding species (molecules).
  - → The steric factor accounts for collisions that lead to no reaction because the orientation of the molecules during such a collision is unfavorable.

## 13.13 Activated Complex Theory

- **Activated complex theory**
  - → Applies to gas phase and solution reactions
  - → Reacting molecules collide and distort.

→ During an encounter, the kinetic energy, $E_K$, of reactants decreases, the potential energy increases, the original chemical bonds lengthen, and new bonds begin to form.

→ Products may form if $E_K$ for the collision is equal to or greater than $E_a$, and if the reactants have the proper orientation.

→ Reactants reform if $E_K$ is less than $E_a$, regardless of orientation.

→ If $E_K$ is equal to or greater than $E_a$, the molecules form either reactants or products. The lowest kinetic energy that may produce products is equal to $E_a$. An *activated complex* is a combination of the two reacting molecules that are at the transition point between reactants and products.

- **Reaction profile**

  → Shows how energy changes as colliding reactants form an *activated complex* and then products along the *minimum energy pathway*. (An analogous process is hiking through a mountain range and choosing the pathway of lowest elevation through it.)

  → Progress of reaction refers to the relevant spatial coordinates of the molecules that show bond breaking as reactants approach the energy maximum and bond formation as products form.

  → Reaction profiles for exothermic and endothermic reactions are shown in the schematic drawings on the next page.

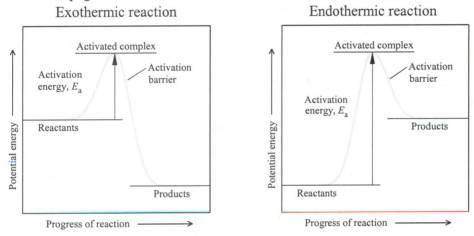

- **Potential energy surface**

  → Three-dimensional plot in which potential energy is plotted on the $z$-axis, and the other two axes correspond to interatomic distances

  → Similar to a mountain pass, the *saddle point* corresponds to an activated complex with the minimum activation energy, $E_a$, required for reactants to form products.

## Accelerating Reactions (Sections 13.14–13.15)

### 13.14 Catalysis

- **Nature of a catalyst**

  → Accelerates a reaction rate without being consumed

  → Does not appear in the overall reaction

  → Lowers the activation energy of a reaction by changing the pathway (mechanism)

→ Accelerates *both* the forward and reverse reaction rates

→ Does *not* change the final equilibrium composition of the reaction mixture

→ May appear in the rate equation, but usually does not

- **Homogeneous catalyst** → Present in the same phase as the *reactants*
- **Heterogeneous catalyst** → Present in a different phase from the *reactants*
- **Poisoning a catalyst**

  → Catalysts can be poisoned or inactivated.

  → For a heterogeneous catalyst, irreversible adsorption of a substance on the catalyst's surface may prevent reactants from reaching the surface and reacting on it.

  → For a homogeneous catalyst, a reactive site on a molecule may be similarly "poisoned" by the binding of a substance directly to the site or nearby either in a permanent or reversible way. The properties of the site are significantly altered to prevent catalytic action.

## 13.15 Living Catalysts: Enzymes

- **Enzyme, E**

  → A biological catalyst

  → Typically, a large molecule (protein) with crevices, pockets, and ridges on its surface (see Fig. 13.39 in the text)

  → Catalyzes the reaction of a reactant known as the *substrate*, S, that produces a *product*, P:

  $$E + S \rightarrow E + P$$

  → Capable of increasing the rate of specific reactions enormously

- **Induced-fit mechanism**

  → Enzymes are highly specific catalysts that function by an *induced-fit mechanism* in which the substrate approaches a correctly-shaped pocket of the enzyme, known as the *active site*.

  → On binding, the enzyme distorts to accommodate the substrate, allowing the reaction to occur.

  → The shape of the pocket determines which substrate can react.

  → The induced-fit mechanism is a more realistic version of the lock and key mechanism, which assumes that the substrate and enzyme fit together like a lock and key.

- **Michaelis–Menten mechanism of enzyme reaction**

  → A two-step mechanism that accounts for the observed dependence of the reaction rate of many enzyme-substrate reactions

  → For an enzyme, E, and a substrate, S, the overall reaction and the Michaelis–Menten mechanism are

  Overall reaction:     $E + S \rightarrow E + P$

  Mechanism:    Step 1  $E + S \rightleftharpoons ES$
  Step 2  $ES \rightarrow E + P$

  → In step 1, the enzyme and substrate form an enzyme-substrate complex, ES, that can dissociate to reform substrate (reverse of step 1) or decay to form product P (step 2). The rate of formation of product is given by step 2.

  Rate of formation of P = $k_2[ES]$

→ The steady-state approximation may be applied to the intermediate, ES, and its concentration is

$$[ES] = \frac{k_1[E][S]}{k_1' + k_2}$$

→ Because the free enzyme concentration may be substantially diminished during the reaction, it is customary to express the rate in terms of the total enzyme concentration, $[E]_0 = [E] + [ES]$. Then, $[E]$ is replaced by $[E]_0 - [ES]$ in the steady-state result, leading to

$$[ES] = \frac{k_1[E]_0[S]}{k_1' + k_2 + k_1[S]}$$

$$\text{Rate of formation of P} = \frac{k_1 k_2 [E]_0 [S]}{k_1' + k_2 + k_1[S]} = \frac{k_2[E]_0[S]}{K_M + [S]}$$

$$K_M = \frac{k_1' + k_2}{k_1} \text{ is known as the Michaelis constant.}$$

→ At low concentration of substrate, $[S] \ll K_M$, and the rate law is given by

$$\text{Rate of formation of P} = \frac{k_2[E]_0[S]}{K_M}$$

The rate varies *linearly* with the concentration of substrate in this region.
*Physically, in this region, most enzyme sites lack substrate molecules, so doubling the substrate concentration doubles the number of occupied sites and the rate as well.*

→ At high concentration of substrate, $[S] \gg K_M$ and the rate law becomes

$$\text{Rate of formation of P} = \frac{k_2[E]_0[S]}{[S]} = k_2[E]_0$$

In this region, the rate is *independent* of the substrate concentration (see Fig. 13.40 in the text).

*Physically, all the enzyme sites are occupied with substrate molecules, so adding substrate has no effect on the rate.*

- **Summary**
  → *Enzymes* are large proteins that function as *biological catalysts*.
  → *Substrate* molecules (reactants) bind to *active sites* on enzyme molecules.
  → The model of enzyme *binding* and *action* is called the *induced-fit mechanism*.
  → Substrate molecules undergo reaction at the active site to form *product* molecules.
  → Product molecules are released and the active site is restored.

# Chapter 14   The Elements: The First Four Main Groups

## Periodic Trends  (Sections 14.1–14.2)

## 14.1  Atomic Properties
- **Atomic properties mainly responsible for properties of an element:**
  - → Atomic radii; first ionization potential; electron affinity; electronegativity; polarizability
- **Atomic and ionic radii**
  - → Typically *decrease* from left to right across a period and *increase* down a group
- **First ionization energy**
  - → Typically *increases* from left to right across a period and *decreases* down a group
  - → Increase across a period: caused by increasing attraction of nuclear charge to electrons in the same valence shell
  - → Trend down a group: concentric shells of electrons are progressively more distant from the nucleus.
- **Electron affinity ($E_{ea}$)**
  - → Measure of the energy *released* when an anion is formed.  The greater the $E_{ea}$, the more stable the anion formed.
  - → *Highest* values of $E_{ea}$ are found at the top right of the periodic table.  Note that Cl has a higher *electron affinity* than F, but F has a higher *electronegativity*.
  - → Elements with high values of $E_{ea}$ are present as *anions* in compounds with *metallic* elements, and commonly have *negative* oxidation states in covalent compounds with other *nonmetallic* elements. For example, F has an oxidation state of −1 in sulfur hexafluoride, $SF_6$.
- **Electronegativity ($\chi$)**
  - → Measure of the tendency of an atom to attract bonding electrons when part of a compound
  - → Typically *increases* from left to right across a period and *decreases* down a group
  - → A useful guide to the type of bond an element may form
  - → A predictor of the *polarity* of chemical bonds between nonmetallic elements
- **Polarizability**
  - → Typically *decreases* across a period (left to right) and *increases* in a group (top to bottom)
  - → A measure of the ease of distorting an atom's electron cloud
  - → Greatest for the more massive atoms in a group
  - → Anions are more polarizable than parent atoms.
  - → Diagonal neighbors in the periodic table have similar polarizability values and a similar amount of covalent character in the bonds they form.
- **Polarizing power**
  - → Associated with ions of small size and high charge
  - → Cations with high polarizing power tend to have covalent character in bonds formed with highly polarizable anions.

## 14.2  Bonding Trends
- **Elements bonded to other elements**
  - → Period 2 elements: valence is determined by the number of valence-shell electrons and the octet rule.
  - → Period 3 or higher ($Z \geq 14$): the valence of an element may be increased through octet expansion allowed by access to empty *d*-orbitals.  Two examples are $PCl_5$ and $SF_6$.

→ *Inert-pair effect:* elements at the bottom of the *p* block may display an oxidation number two less than the group number suggests. For example, lead has an oxidation number of +2 in PbO and +4 in $PbO_2$.

- Pure elements
  → With *low* ionization energies tend to have *metallic* bonds, for example, Li(s) and Mg(s)
  → With *high* ionization energies are typically *molecular* and form *covalent* bonds, for example, $F_2(g)$ and $O_2(g)$
  → In or near the diagonal band of metalloids (intermediate ionization energies and three, four, or five valence electrons) tend to form network structures in which each atom is bonded to three or more other atoms. For example, the *network* structure of *black phosphorus,* in which each P atom is bonded to three others in a network structure.

## *Chemical Properties: Hydrides*

- **Binary hydrides**
  → Binary compounds with hydrogen (hydrides)
  → Formed by all main-group elements — except the noble gases, and possibly In and Tl — and several *d*-block elements
  → Properties show periodic behavior.
  → Formulas directly related to the group number
  **Examples:** $SiH_4$, $NH_3$, $H_2S$, and HCl in Groups 14, 15, 16, and 17, respectively

- **Saline (or salt-like) hydrides**
  → Formed by all members of the *s* block except Be
  → Ionic salts of strongly electropositive metals with hydrogen, present as the hydride ion, $H^-$
  → White, high-melting solids with crystal structures resembling those of the corresponding halides
  → Made by heating the metal in hydrogen, for example, $2 Na(s) + H_2(g) \rightarrow 2 NaH(s)$
  → React readily with water ($H^-$ is a strong base) to form a basic solution and releasing $H_2(g)$:
  $$H^-(aq) + H_2O(l) \rightarrow H_2(g) + OH^-(aq)$$

- **Metallic hydrides**
  → Black, powdery, electrically conducting solids formed by certain *d*-block elements
  → Release hydrogen when heated or when treated with acids. A typical reaction in acid is
  $$CuH(s) + H_3O^+(aq) \rightarrow Cu^+(aq) + H_2(g) + H_2O(l)$$

- **Molecular hydrides**
  → Formed by nonmetals and consist of discrete molecules
  → Volatile, many are Brønsted acids or bases.
  → Gaseous compounds include the base $NH_3(g)$, the hydrogen halides HX(g), X= F, Cl, Br, and I, and the lighter hydrocarbons such as $CH_4$.
  → Liquids include water and heavier hydrocarbons such as benzene, $C_6H_6$, and octane, $C_8H_{18}$.

## *Chemical Properties: Oxides*

- **Binary oxides**
  → Formed by all the main-group elements except the noble gases
  → Main-group *metals* binary oxides are basic and the oxides of *nonmetals* are acidic.
  Exception: $Al_2O_3$ is *amphoteric.* Other *amphoteric oxides* include BeO, $Ga_2O_3$, $SnO_2$, and $PbO_2$.
  → Soluble ionic oxides: formed from elements on the left side of the periodic table
  → Insoluble oxides with high melting points: formed from elements in the left of the *p* block
  → Oxides with low melting points, often gaseous: formed from elements on the right of the *p* block

- **Metallic elements and metalloids**
  - → Ionic oxides: formed from metallic elements with low ionization energies
  - → Ionic oxides: typically yield basic solutions in water
  - → Amphoteric oxides: formed from metallic elements with intermediate ionization energies (Be, B, and Al) and from metalloids
  - → Amphoteric oxides: do not react with water but dissolve in both acidic and basic solutions
- **Nonmetallic elements**
  - → Many oxides of nonmetals are gaseous molecular compounds. Most can act as Lewis acids. Form acidic solutions in water and are called *acid anhydrides*

    **Examples:** The acids $HNO_3$ and $H_2SO_4$ are derived from the oxides $N_2O_5$ and $SO_3$, respectively.
  - → Oxides that do *not* react with water are called *formal anhydrides* of acids. The formal anhydride is obtained by removing the elements of water (H, H, and O) from the molecular formula of the acid, for example, the formal anhydride of acetic acid, $CH_3COOH$, is $CH_2CO$ (ketene), whose Lewis structure is given on the right.

$$\begin{array}{c} H \\ \diagdown \\ \phantom{H}\diagup \\ H \end{array} C=C=\ddot{O}:$$

---

# Hydrogen  (Sections 14.3–14.4)

## 14.3  The Element

- **Hydrogen**
  - → One valence electron but few similarities to the alkali metals
  - → A nonmetal that resembles the halogens but has some different, unique properties
  - → Does not fit clearly into any group and is therefore not assigned to any group
- **Commercial production of hydrogen**
  - → Formed 1) as a by-product of the petroleum *refining* and 2) by the *electrolysis* of water
  - → In the refining process, a Ni-catalyzed *re-forming reaction,* produces products called *synthesis gas*.

    $$CH_4(g) + H_2O(g) \rightarrow CO(g) + 3\,H_2(g)$$

    In a second step, a *shift reaction* employing an Fe/Cu catalyst yields hydrogen gas:

    $$CO(g) + H_2O(g) \rightarrow CO_2(g) + H_2(g)$$

  - → *Electrolysis of water* utilizes a salt solution to improve conductivity: $2\,H_2O(l) \rightarrow 2\,H_2(g) + O_2(g)$
- **Laboratory production and uses of hydrogen**
  - → Produced by reaction of an active metal such as Zn with a strong acid:

    $$Zn(s) + 2\,H_3O^+(aq) \rightarrow Zn^{2+}(aq) + H_2(g) + H_2O(l)$$

  - → Produced by reaction of a hydride with water:

    $$CaH_2(s) + 2\,H_2O(l) \rightarrow Ca^{2+}(aq) + 2\,OH^-(aq) + 2\,H_2(g)$$

## 14.4  Compounds of Hydrogen

- **Unusual properties of hydrogen**
  - → It can form a cation, $H^+$, or an anion, hydride, $H^-$.
  - → It can form covalent bonds with many other elements, because of its intermediate $\chi$ value of 2.2.

- **The hydride ion, H⁻**
  - → Large, with a radius (154 pm) between those of F⁻ (133 pm) and Cl⁻ (181 pm) ions
  - → Highly polarizable, which adds covalent character in its bonds to cations
  - → The two electrons in H⁻ are weakly bound to the single proton, and an electron is easily lost.
  - → Saline hydrides are very strong reducing agents. The reduction half-reaction and standard reduction potential are: $H_2(g) + 2e^- \rightarrow 2H^-(aq)$ and $E° = -2.25$ V, respectively.
  - → The alkali metals are also strong reducing agents with reduction potentials similar to the hydride value.
  - → Hydride ions in saline hydrides reduce water upon contact: $H^-(aq) + H_2O(l) \rightarrow H_2(g) + OH^-(aq)$.
- **Hydrogen bonds**
  - → Relatively strong intermolecular bonds
  - → Formed by hydrides of N, O, and F
  - → Can be understood in terms of the coulombic attraction between the partial positive charge on a hydrogen atom and the partial negative charge of another atom
  - → The H atom is covalently bonded to a very electronegative N, O, or F atom, giving it a partial positive charge.
  - → The partial negative charge is the lone-pair electrons on a different N, O, or F atom.
  - → A hydrogen bond is represented as three dots between atoms *not* covalently bonded to each other: [O–H···:O].

- **Molecular Orbital (MO) theory for hydrogen bonding**
  - → The three atoms in the hydrogen bond each provide one atomic orbital (AO), which combine to give three MOs.
  - → The three MOs are bonding, nonbonding, and antibonding in character.
  - → The four electrons, two in the σ-bond plus the two lone-pair electrons, are placed in the MOs with two electrons in the bonding MO and two in the nonbonding MO.
  - → The antibonding orbital is vacant, and there is net bonding.
  - → A hydrogen bond is typically about 5% as strong as a covalent bond between the same types of atoms. **Example:** Hydrogen bonding in HF

$$H-\ddot{\underset{\cdot\cdot}{F}}: \cdots H-\ddot{\underset{\cdot\cdot}{F}}: \cdots H-\ddot{\underset{\cdot\cdot}{F}}:$$

# Group 1: The Alkali Metals  (Sections 14.5–14.7)

## 14.5  The Group 1 Elements
- **Preparation**
  - → Pure alkali metals are obtained by electrolysis of the molten salts.
  - → Exception: K, prepared by reaction of molten KCl and Na vapor at 750°C:

$$KCl(l) + Na(g) \rightarrow NaCl(s) + K(g)$$

This reaction is driven to form products by the condensation of K(g) because the equilibrium constant is unfavorable.

- **Properties of alkali metals**
  - → Soft, silver-gray metals with low melting points, boiling points, and densities

→ Melting and boiling points decrease down the group. Cs (melting point of 28°C) is very reactive and is transported in sealed ampoules. Fr is very radioactive and little is known about its properties. They are highly reactive.

- **Applications**
  → Li metal is used in the rechargeable lithium-ion battery, which holds a charge for a long time.
  → Li is also used in thermonuclear weapons.

## 14.6 Chemical Properties of the Alkali Metals

- **Alkali metals**
  → Strong reducing agents (low first ionization energies lead to their ease of oxidation)
  **Example:** Reduction of Ti(IV): $TiCl_4 + 4\,Na(l) \rightarrow 4\,NaCl(s) + Ti(s)$
  → Reduce water to form basic solutions, releasing hydrogen gas
  **Example:** $2\,Na(s) + 2\,H_2O(l) \rightarrow 2\,NaOH(aq) + H_2(g)$
  → Dissolve in $NH_3(l)$ to yield solvated electrons, which are used to reduce organic compounds
  → React directly with almost all nonmetals, noble gases excluded
  → However, only one alkali metal, Li, reacts with nitrogen to form the nitride ($Li_3N$).

- **Formation of compounds containing oxygen**
  → Li forms mainly the oxide, $Li_2O$.
  → Na forms a pale yellow peroxide, $Na_2O_2$.
  → K, Rb, and Cs form mainly the superoxide, for example, $KO_2$.

## 14.7 Compounds of Lithium, Sodium, and Potassium

- **Lithium**
  → Differs significantly from the other Group 1 elements. The difference is common for elements at the head of a group and originates in part from the small size of $Li^+$.
  → Cations have strong polarizing power and a tendency toward covalency in their bonding.
  → Has a diagonal relationship with Mg; it is found in the minerals of Mg
  → Found in compounds with oxidation number +1; compounds used in ceramics, lubricants, and medicine
  **Examples:** Lithium carbonate is an effective drug for manic-depressive (bipolar) behavior. Lithium soaps, which have higher melting points than sodium and potassium soaps, are used as thickeners in lubricating greases for high-temperature applications.

- **Sodium compounds**
  → Important in part because they are *plentiful, inexpensive,* and water *soluble*
  → NaCl is mined as *rock salt* and used in the electrolytic production of $Cl_2(g)$ and NaOH.
  → NaOH is an inexpensive starting material for the production of other sodium salts.
  → $Na_2CO_3 \cdot 10\,H_2O$ was once used as *washing soda.*
  → $NaHCO_3$ is also known as *sodium bicarbonate, bicarbonate of soda,* or *baking soda.*
  → Weak acids [lactic acid (milk), citric acid (lemons), acetic acid (vinegar)] react with the hydrogen carbonate anion to release carbon dioxide gas: $HCO_3^-(aq) + HA)(aq) \rightarrow A^-(aq) + H_2O(l) + CO_2(g)$
  → Double-acting *baking powder* contains a solid weak acid as well as the hydrogen carbonate ion.

- **Potassium compounds**
  → Usually more expensive than the corresponding sodium compounds

→ Use may be justified because potassium compounds are generally less hygroscopic than sodium compounds. The $K^+$ cation is larger and is less strongly hydrated by $H_2O$ molecules.

→ $KNO_3$ releases $O_2(g)$ when heated and is used to help matches ignite. $KNO_3$ is also the oxidizing agent in black gunpowder (75% $KNO_3$, 15% charcoal, and 10% sulfur).

→ KCl is used directly in some fertilizers as a source of potassium, but $KNO_3$ is required for some crops that cannot tolerate high chloride ion concentrations.

→ Mineral sources of potassium are *sylvite,* KCl, and *carnallite,* $KCl·MgCl_2·6H_2O$.

---

# Group 2: The Alkaline Earth Metals (Sections 14.8–14.10)

## 14.8   The Group 2 Elements

- **Preparation**
  → Pure metals: obtained by electrolysis of the molten salts or by chemical reduction (except Be)
  → Ca(s), Sr(s), and Ba(s) may be obtained by reduction with Al(s) in a variation of the thermite reaction., for example, $3CaO(s) + 2Al(s) → Al_2O_3(s) + 3Ca(s)$
  → Be occurs in nature as *beryl,* $3BeO·Al_2O_3·6SiO_2$.
  → Mg is found in seawater as $Mg^{2+}$ (aq) and in the mineral dolomite, $CaCO_3·MgCO_3$.

- **Properties**
  → Silver-gray metals with much higher melting points, boiling points, and densities than those of the preceding alkali metals in the same period
  → Melting points decrease down the group with the exception of Mg, which has the lowest value. The trend in boiling point is irregular.
  → All Group 2 elements except Be reduce water.
  **Example:** $Sr(s) + 2H_2O(l) → Sr^{2+}(aq) + 2OH^-(aq) + H_2(g)$
  → Be does not react with water. Mg reacts with hot water, and Ca reacts with cold water.
  → Be and Mg do not dissolve in $HNO_3$ because they develop a protective oxide film.
  → Mg burns vigorously in air because it reacts with $O_2(g)$, $N_2(g)$, and $CO_2(g)$.
  → Reactivity of the Group 2 metals with water and oxygen increases down the group.

- **Applications of beryllium**
  → Has a low density, making it useful for missile and satellite fabrication
  → Used as a "window" for x-ray tubes
  → Added in small amounts to Cu to increase its rigidity. Be/Cu alloys are used in nonsparking tools (oil refineries and grain elevators), and in nonmagnetic parts and contacts that resist deformation and corrosion (electronics industry).

- **Applications of magnesium**
  → A silver-white metal protected from air oxidation by a white oxide film. It appears dull gray for that reason.
  → Light and very soft, but its alloys have great strength and are used in applications where lightness and toughness are desired (airplanes).
  → Limitation in widespread usage is related to its expense, low melting temperature, and difficulty in machining.

- **Applications of calcium, strontium, and barium**
  - → Group 2 metals can be detected in burning compounds by the colors they give to flames. Ca burns orange-red, Sr crimson, and Ba yellow-green. Fireworks are made from their salts.

## 14.9 Compounds of Beryllium and Magnesium

- **Beryllium**
  - → Differs significantly from the other Group 2 elements, as is typical of an element at the head of its group. The difference originates in part from the small size of the $Be^{2+}$ cation, which has strong polarizing power and a tendency toward covalence in its bonding.
  - → Be has a diagonal relationship with Al. It is the only member of the group that, like Al, reacts in aqueous NaOH. In strongly basic solution, it forms the *beryllate ion*, $[Be(OH)_4]^{2-}$.
  - → Compounds are extremely toxic.
  - → The high polarizing power of $Be^{2+}$ produces moderately covalent compounds while its small size allows no more than four groups to be attached, accounting for the prominence of the tetrahedral $BeX_4$ unit.
  - → $BeH_2(g)$ and $BeCl_2(g)$ condense to form solids with chains of tetrahedral $BeH_4$ and $BeCl_4$ units, respectively. In $BeCl_2$, the Be atoms act as Lewis acids and accept electron pairs from the Cl atoms, forming a chain of tetrahedral $BeCl_4$ units in the solid.

- **Magnesium**
  - → More metallic properties than Be
  - → Compounds are primarily ionic with some covalent character.
  - → MgO and $Mg_3N_2$ are formed when magnesium burns in air. MgO dissolves very slowly and only slightly in water. It can withstand high temperatures (*refractory*) because it melts at 2800°C. The oxide conducts heat well but electricity only poorly, and is used as an insulator in electric heaters.
  - → $Mg(OH)_2$ is a base. It is not very soluble in water but forms instead a white colloidal suspension, which is used as a stomach antacid called *milk of magnesia*. A side effect of stomach acid neutralization is the formation of $MgCl_2$, which acts as a purgative. $MgSO_4$, (*Epsom salts*) is also a common purgative.
  - → Chlorophyll is the most important compound of Mg. Magnesium also plays an important role in energy generation in living cells. It is involved in the contraction of muscles.

## 14.10 Compounds of Calcium

- **Carbonate, oxide, and hydroxide**
  - → Ca is more metallic in character than Mg but its compounds share some similar properties.
  - → $CaCO_3$ occurs naturally as chalk and limestone. Marble is a very dense form with colored impurities, most commonly Fe cations.
  - → The most common forms of pure calcium carbonate are *calcite* and *aragonite*. All carbonates are the fossilized remains of marine life.
  - → CaO is formed by heating $CaCO_3$: $CaCO_3(s) \rightarrow CaO(s) + CO_2(g)$

    CaO is called *quicklime* because it reacts rapidly and exothermically with water:

    $$CaO(s) + H_2O(l) \rightarrow Ca^{2+}(aq) + 2OH^-(aq)$$

  - → $Ca(OH)_2$, the product of the equation above, is known as *slaked lime* because in this form the thirst for water has been quenched or slaked.
  - → An aqueous solution of $Ca(OH)_2$, which is only slightly soluble in water, is called *lime water*, which is used as a test for carbon dioxide: $Ca(OH)_2(aq) + CO_2(g) \rightarrow CaCO_3(s) + H_2O(l)$

- **Carbide, sulfate, and phosphate**
    - → CaO may be converted into calcium carbide: $CaO(s) + 3C(s) \rightarrow CaC_2(s) + CO(g)$
    - → Calcium sulfate dihydrate (*gypsum*), $CaSO_4 \cdot 2H_2O$, is used in construction materials.
    - → Ca is found in the rigid structural components of living organisms, either as $CaCO_3$ (shellfish shells) or as $Ca_3(PO_4)_2$ (bone).
    - → Tooth enamel is a *hydroxyapatite*, $Ca_5(PO_4)_3OH$, which is subject to attack by acids produced when bacteria act on food residues. A more resistant coating is formed when the $OH^-$ ions are replaced by $F^-$ ions. The resulting mineral is called *fluoroapatite*:

$$Ca_5(PO_4)_3OH(s) + F^-(aq) \rightarrow Ca_5(PO_4)_3F(s) + OH^-(aq)$$

    Tooth enamel is strengthened by the addition of fluorides (fluoridation) to drinking water, and by the use of fluoridated toothpaste containing tin(II) fluoride, $SnF_2$, or sodium monofluorophosphate (MFP), $Na_2FPO_3$.

# Group 13 / III: The Boron Family (Sections 14.11–14.14)

## 14.11 The Group 13/III Elements

- **Preparation of boron**
    - → Mined as the hydrates *borax* and *kernite*, $Na_2B_4O_7 \cdot xH_2O$, with $x = 10$ and 4, respectively
    - → The ore is converted to the oxide, $B_2O_3$, and extracted in its amorphous elemental form by reaction of the oxide with Mg(s): $B_2O_3(s) + 3Mg(s) \rightarrow 2B(s) + 3MgO(s)$
    - → A purer form of B is obtained by reduction of volatile $BBr_3(g)$ or $BCl_3(g)$ with $H_2(g)$.

- **Preparation of aluminum**
    - → Mined as *bauxite*, $Al_2O_3 \cdot xH_2O$, where $x$ ranges to a maximum of 3
    - → The ore, which contains considerable iron oxide impurity, is processed to obtain alumina, $Al_2O_3$.
    - → Al is then extracted by an electrolytic process (*Hall process*). $Al_2O_3$ is mixed with *cryolite*, $Na_3AlF_6$, to lower the melting point from 2050°C (alumina) to 950°C (mixture). The half-cell and cell reactions are:

| | |
|---|---|
| Cathode: | $Al^{3+}(melt) + 3e^- \rightarrow Al(l)$ |
| Anode: | $2O^{2-}(melt) + C(s,gr) \rightarrow CO_2(g) + 4e^-$ |
| Cell: | $4Al^{3+}(melt) + 6O^{2-}(melt) + 3C(s,gr) \rightarrow 4Al(l) + 3CO_2(g)$ |

- **Properties of boron**
    - → The element is attacked by only the strongest oxidizing agents
    - → A nonmetal in most of its chemical properties
    - → Has acidic oxides and forms a fascinating array of binary molecular hydrides
    - → Exists in several allotropic forms in which B atoms attempt to share eight electrons, despite the small size of B and its ability to contribute only three electrons. In this sense, B is regarded as electron deficient.
    - → One common form of B is a gray-black nonmetallic, high-melting solid.
    - → Another common form is a dark brown powder with a structure (icosahedral with 20 faces) that has clusters of 12 atoms as shown on the right. The bonds create a very hard, three-dimensional structure.

- **Properties of aluminum**
  - → Has a low density, but is a strong metal with excellent electrical conductivity
  - → Easily oxidized, but its surface is passivated by a protective oxide film
  - → If the thickness of the oxide layer is increased electrolytically, *anodized aluminum* results.
  - → Al is amphoteric, reacting with both *nonoxidizing* acids and hot aqueous bases. The reactions are
    $$2\,Al(s) + 6\,H^+(aq) \rightarrow 2\,Al^{3+}(aq) + 3\,H_2(g)$$
    $$2\,Al(s) + 2\,OH^-(aq) + 6\,H_2O(l) \rightarrow 2\,Al(OH)_4^-(aq) + 3\,H_2(g)$$

- **Applications**
  - → B fibers are incorporated into plastics, forming resilient material stiffer than steel and lighter than Al(s).
  - → Boron carbide is similar in hardness to diamond, and boron nitride is similar in structure and mechanical properties to graphite, but, unlike graphite, boron nitride does not conduct electricity.
  - → Al has widespread use in construction and aerospace industries. Because it is a soft metal, its strength is improved by alloy formation with Cu and Si.
  - → Because of its high electrical conductivity, Al is used in overhead power lines. Its high negative electrode potential has led to its use in fuel cells.

## 14.12 Group 13/III Oxides

- **Boron oxide**
  - → Boron, like most nonmetals, has *acidic* oxides. Hydration of $B_2O_3$ yields boric acid, $H_3BO_3$ or $B(OH)_3$, in which B has an incomplete octet and can therefore act as a Lewis acid:
    $$(OH)_3B + :OH_2 \rightarrow (OH)_3B{-}OH_2$$
    The complex formed is a weak *monoprotic* acid ($pK_a = 9.14$).
  - → $H_3BO_3$ is a mild antiseptic and pesticide. It is also used as a fire retardant in home insulation and clothing.
  - → The main use of boric acid, $H_3BO_3$, is as a source of boron oxide, the anhydride of boric acid:
    $$2\,H_3BO_3(s) \xrightarrow{\text{heat}} B_2O_3(s) + 3\,H_2O(l)$$

  $B_2O_3$, which melts 450°C and dissolves many metal oxides, is used as a flux for soldering or welding. It is also used in the manufacture of fiberglass and borosilicate glass, such as Pyrex.

- **Aluminum oxide**
  - → $Al_2O_3$, known as *alumina*, exists in several crystal forms, each of which is used in science and in commerce. In the form of α-alumina, it is a very hard substance, *corundum,* used as an abrasive known as *emery*.
  - → γ-Alumina is absorbent and is used as the stationary phase in chromatography. It is produced by heating $Al(OH)_3$, is moderately reactive, and is amphoteric:
    $$Al_2O_3(s) + 6\,H_3O^+(aq) + 3\,H_2O(l) \rightarrow 2[Al(H_2O)_6^{3+}](aq)$$
    $$Al_2O_3(s) + 2\,OH^-(aq) + 3\,H_2O(l) \rightarrow 2\,Al(OH)_4^-(aq)$$

  The strong polarizing effect of the small, highly charged $Al^{3+}$ ion on the water molecules surrounding it gives the $[Al(H_2O)_6^{3+}]$ ion acidic properties.
  - → Aluminum sulfate is prepared by reaction of $Al_2O_3$ with sulfuric acid:
    $$Al_2O_3(s) + 3\,H_2SO_4(aq) \rightarrow Al_2(SO_4)_3(aq) + 3\,H_2O(l)$$

$Al_2(SO_4)_3$ (*papermaker's alum*) is used in the preparation of paper. True alums are hydrated sulfates containing two metallic cations, a +1 cation and $Al^{3+}$. Example: $KAl(SO_4)_2$.

→ Sodium aluminate, $NaAl(OH)_4$, is used along with $Al_2(SO_4)_3$ in water purification.

### 14.13 Nitrides and Halides

- **Nitrides**
  → When B is heated in $NH_3$, it forms boron nitride: $2B(s) + 2NH_3(g) \rightarrow 2BN(s) + 3H_2(g)$
  The structure BN is similar to graphite, but with the planes of hexagons of C atoms replaced by hexagons of alternating B and N atoms
  → BN is a fluffy, slippery powder which can be pressed into a solid rod that is easy to machine. Unlike graphite, BN is a good insulator.
  → At high pressure, BN is converted to a diamond-like crystalline material called Borazon.
  → BN nanotubes, similar to those formed by C, are semiconductors.
  → Boron trifluoride reacts with $NH_3$ to produce the Lewis acid-base complex, $F_3BNH_3$:

  $$F_3B(g) + :NH_3(g) \rightarrow F_3B-NH_3(s)$$

  Solid $F_3BNH_3$ is stable up to about 125°C, where it decomposes:

  $$4F_3BNH_3(s) \rightarrow BN(s) + 3NH_4BF_4(s)$$

- **Halides**
  → Boron halides are made either by direct reaction of the elements or from the oxide.
  **Example:** $B_2O_3(s) + 3CaF_2(s) + 3H_2SO_4(l) \xrightarrow{heat} 2BF_3(g) + 3CaSO_4(s) + 3H_2O(l)$
  → All boron halides are trigonal-planar, covalently bonded, and electron deficient, with an incomplete octet on B, as a consequence of which, they are strong Lewis acids.
  → $AlCl_3$ is formed as an ionic compound by direct reaction of the elements, or of alumina with chlorine in the presence of carbon: $2Al(s) + 3Cl_2(g) \rightarrow 2AlCl_3(s)$

  $$Al_2O_3(s) + 3Cl_2(g) + 3C(s, gr) \rightarrow 2AlCl_3(s) + 3CO(g)$$

  Ionic $AlCl_3$ melts at 192°C to form a molecular liquid, $Al_2Cl_6$. This dimer of $AlCl_3$, formed by donation of an unshared electron pair on Cl to a vacant orbital on an adjacent Al atom, contains $AlCl_4$ tetrahedral units and two bridging Cl atoms, as shown on the right.

  → Aluminum halides react with water in a highly exothermic reaction.
  → The white solid, lithium aluminum hydride, prepared from $AlCl_3$ is an important reducing agent in organic chemistry: $4LiH + AlCl_3 \rightarrow LiAlH_4 + 3LiCl$.

### 14.14 Boranes, Borohydrides, and Borides

- **Boranes**
  → Compounds of boron and hydrogen, such as $B_2H_6$ and $B_{10}H_{14}$
  → Electron-deficient compounds with three-center, two-electron B–H–B bonds. Valid Lewis structures cannot be written. Molecular Orbital theory provides a framework for understanding the delocalized behavior of an electron pair associated with three atoms, as shown on the right.

  → Diborane, $B_2H_6$, is highly reactive. When heated, it decomposes to hydrogen and boron:

  $$B_2H_6(g) \rightarrow 2B(s) + 3H_2(g)$$

It reacts with water, reducing hydrogen and forming boric acid:

$$B_2H_6(g) + 6H_2O(l) \rightarrow 2B(OH)_3(aq) + 6H_2(g)$$

- **Borohydrides**
  - → Anionic versions of boranes. The most important borohydride is $BH_4^-$, as in sodium borohydride, $NaBH_4$, which is produced by the reaction of NaH with $BCl_3$:

$$4NaH + BCl_3 \rightarrow NaBH_4 + 3NaCl$$

  The borohydride anion $BH_4^-$ is an effective and important reducing agent.
  - → Diborane is produced by the reaction of sodium borohydride with boron trifluoride:

$$3BH_4^- + 4BF_3 \rightarrow 3BF_4^- + 4B_2H_6$$

- **Borides**
  - → Many borides of the metals and nonmetals are known.
  - → Formulas of borides are often unrelated to the position of the two elements in the periodic table. Examples include $AlB_2$, $CaB_6$, $B_{13}C_2$, $B_{12}S_2$, $Ti_3B_4$, TiB, and $TiB_2$.
  - → B atoms in borides commonly form extended structures such as zigzag chains, branched chains, or networks of hexagonal rings.

---

# Group 14/IV: The Carbon Family (Sections 14.16–14.20)

## 14.15 The Group 14/IV Elements

- **General properties**
  - → Carbon forms covalent compounds with nonmetals and ionic compounds with metals.
  - → The oxides of C and Si are acidic.
  - → Ge, a metalloid, exhibits metallic or nonmetallic properties, depending upon the compound.
  - → Sn and Pb are classified as metals, but tin has some intermediate properties.
  - → Sn is amphoteric and reacts with hot concentrated hydrochloric acid and with hot alkali:

$$Sn(s) + 2H_3O^+(aq) \rightarrow Sn^{2+}(aq) + H_2(g) + 2H_2O(l)$$
$$Sn(s) + 2OH^-(aq) + 2H_2O(l) \rightarrow Sn(OH)_4^{2-}(aq) + H_2(g)$$

  - → Carbon, at the top of the group, differs greatly in its properties from the other members.
  - → Carbon tends to form multiple bonds, whereas silicon does not. $CO_2$ is a molecular gas, whereas $SiO_2$ (silica) contains networks of –O–Si–O– groups and is a mineral. Si compounds can act as Lewis acids, whereas C compounds usually cannot. Si may expand its octet, whereas C does not.
  - → Sn and Pb are commonly found with an oxidation number of +2 because of the inert-pair effect.

## 14.16 The Different Forms of Carbon

- **Graphite**
  - → The thermodynamically stable allotrope of carbon (Soot contains small crystals of graphite).
  - → Produced pure commercially by heating C rods in an electric furnace for several days
  - → Contains $sp^2$ hybridized C atoms
  - → Consists structurally of large sheets of fused benzene-like hexagonal units. A π-bonding network of delocalized electrons accounts for the high electrical conductivity of graphite.

When certain impurities are present, graphitic sheets can slip past one another, and graphite becomes an excellent dry lubricant.

→ Only soluble in a few liquid metals

→ Soot and carbon black have commercial applications in rubber and inks. Activated charcoal is an important, versatile purifier. Unwanted compounds are adsorbed onto its microcrystalline surface.

- **Diamond**

  → It is the hardest substance known and an excellent conductor of heat. Its hardness makes it an excellent abrasive, and the heat generated by friction is rapidly conducted away.

  → Natural diamonds are uncommon. Synthetic diamonds are produced from graphite at high pressures (> 80 kbar) and temperatures (> 1500°C), and by thermal decomposition of methane.

  → It is soluble in liquid metals, such as Cr and Fe, but less soluble than graphite. This solubility difference is utilized in the synthesis of diamond at high pressure and high temperature.

  → Carbon in diamond is $sp^3$ hybridized, and each C atom in a diamond crystal is directly bonded to four other C atoms and indirectly interconnected to all of the other carbon atoms in the crystal through C–C single σ-bonds. The σ-bonding network of localized electrons accounts for the electrical insulating properties of diamond.

- **Fullerenes**

  → Buckminsterfullerene, $C_{60}$, is a soccer-ball-like molecule first identified in 1985. Numerous fullerenes, containing 44 to 84 carbon atoms, were discovered quickly thereafter.

  → Fullerenes are molecular and consequently electrical insulators, also soluble in organic solvents.

  → They have relatively low ionization energies *and* large electron affinities, so they readily lose or gain electrons to form ions.

  → Fullerenes contain an even number of carbon atoms, differ by a $C_2$ unit, and have pentagonal and hexagonal ring structures. Solid samples of fullerenes are called *fullerites*.

  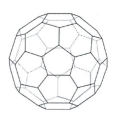

  → $C_{60}$ has the atomic arrangement of a *truncated icosahedron*, with 32 faces, 12 pentagons, and 20 hexagons. The interior of the molecule is large, and other atoms may be inserted to produce compounds with unusual properties. The crystal structure of $C_{60}$ is face-centered cubic.

  → The anion $C_{60}{}^{3-}$ is very stable, and $K_3C_{60}$ is a superconductor below 18 K.   Buckminsterfullerene, $C_{60}$

## 14.17 Silicon, Germanium, Tin, and Lead

- **Silicon**

  → Occurs naturally as silicon dioxide and as silicates, which contain $SiO_4{}^{4-}$

  → Forms of $SiO_2$ include quartz, quartzite, and sand.

  → Pure Si, which is widely used in semiconductors, is obtained from quartzite in a three-step process. Silicon produced in the first step is crude and requires further purification:

$$SiO_2(s) + 2C(s) \rightarrow Si(s) + 2CO_2(g)$$
$$Si(s) + 2Cl_2(g) \rightarrow SiCl_4(l)$$
$$SiCl_4(l) + 2H_2(g) \rightarrow Si(s) + 4HCl(g)$$

- **Germanium**

  → Mendeleev predicted the existence of Ge, "eka-silicon," before its discovery.

  → Occurs as an impurity in Zn ores

  → Used mainly in the semiconductor industry

- **Tin**
  - → Reduced from its ore *cassiterite,* $SnO_2$, through reaction with C at 1200°C in a process that requires great care in order to avoid contamination with iron
  $$SnO_2(s) + C(s) \rightarrow Sn(l) + CO_2(g)$$
  - → Used to plate other materials and for the production of alloys
- **Lead**
  - → The principal ore of lead is *galena,* PbS, which is converted to the oxide in air. The oxide in turn is reduced with coke to produce the metal:
  $$2\,PbS(s) + 3\,O_2(g) \rightarrow 2\,PbO(s) + 2\,SO_2(g)$$
  $$PbO(s) + C(s) \rightarrow Pb(s) + CO(g)$$
  - → Pb is very dense, malleable, and chemically inert. It is used as the electrode material in car batteries.

## 14.18 Oxides of Carbon

- **Carbon dioxide, $CO_2$**
  - → A nonmetal oxide and the acid anhydride of carbonic acid, $H_2CO_3$ (The equilibrium between $CO_2$ and $H_2CO_3$ favors $CO_2$.)
  - → $CO_2(aq)$ is an equilibrium mixture of $CO_2$, $H_2CO_3$, $HCO_3^-$, and a very small amount of $CO_3^{2-}$.
  - → Solid $CO_2$ (*dry ice*), sublimes to the gas phase, making it useful as a refrigerant and cold pack.
- **Carbon monoxide, CO**
  - → A colorless, odorless, flammable gas, which is nearly insoluble in water, and highly toxic [It has an extremely large bond enthalpy ($1074$ kJ·mol$^{-1}$).]
  - → CO is produced when an organic compound or C itself is burned in a limited amount of oxygen. It is commercially produced as *synthesis gas:*
  $$CH_4(g) + H_2O(g) \rightarrow CO(g) + 3\,H_2(g)$$
  - → CO is the *formal* anhydride of formic acid and is produced in the laboratory by dehydration of formic acid with hot, concentrated sulfuric acid:
  $$HCOOH(l) \rightarrow CO(g) + H_2O(l)$$
  - → CO is a moderately good Lewis base. One example is the reaction of CO with a *d*-block metal or ion: $\quad Ni(s) + 4\,CO(g) \rightarrow Ni(CO)_4(l)$
  - → Complex formation is responsible for the toxicity of CO, which attaches more strongly than $O_2$ to iron in hemoglobin, thereby blocking the acceptance of $O_2$ by red blood cells.

## 14.19 Oxides of Silicon: The Silicates

- **Silica or silicon dioxide, $SiO_2$**
  - → Derives its strength from its covalently bonded network structure
  - → Occurs naturally in pure form only as quartz and sand
- **Silicates, compounds containing $SiO_4^{4-}$ units**
  - → *Orthosilicates*, the simplest silicates, contain $SiO_4^{4-}$ ions. Examples: $Na_4SiO_4$ and $ZrSiO_4$, *zircon*
  - → *Pyroxenes* contain chains of tetrahedral $SiO_4^{4-}$ *units* in which one O atom bridges two Si atoms (Si–O–Si) from adjacent tetrahedra, such that the "average" unit is $SiO_3^{2-}$ unit. Cations placed along the chains provide electrical neutrality. Examples are $Al(SiO_3)_2$ and $CaMg(SiO_3)_2$.

→ *Amphiboles* contain chains of silicate ions linked to form a cross-linked ladderlike structure containing $Si_4O_{11}^{6-}$ units. One example is the fibrous mineral *tremolite*, $Ca_2Mg_5(Si_4O_{11})_2(OH)_2$, one form of *asbestos*, a material that withstands extreme heat but is a known carcinogen. Almost all amphiboles contain hydroxide ions attached to the metal.

→ Other types of silicates have sheets containing $Si_2O_5^{2-}$ units, with cations lying between the sheets and linking them together. One example is *talc*, $Mg_3(Si_2O_5)_2(OH)_2$. Additional structural types of silicon oxides exist.

- **Aluminosilicates and cements**

  → *Aluminosilicates* form when a $Si^{4+}$ ion is replaced by $Al^{3+}$ plus additional cations for charge balance. One example is *mica*, $KMg_3(Si_3AlO_{10})_2(OH)_2$. The extra cations hold the sheets of tetrahedra together, accounting for the hardness of these materials. *Feldspar* is a silicate material in which more than half the silicon is replaced by aluminum.

  → *Cements* are produced by melting aluminosilicates with limestone and other materials, and allowing them to solidify. Cement is one of the materials in concrete.

- **Silicones**

  → Contain long $-O-Si-O-$ chains with the remaining two silicon bonding positions occupied by organic groups, such as the methyl group, $-CH_3$

  → Possess both hydrophobic and hydrophilic properties, which makes them useful in waterproofing fabrics and in biological applications, such as surgical and cosmetic implants

## 14.20 Other Important Group 14/IV Compounds

- **Carbides**

  → *Saline carbides* are most commonly formed from Group 1 and 2 metals, Al, and a few other metals. *s*-Block metals form saline carbides when their oxides are heated with C.

  → Anions present in carbides are normally *methanide*, $C^{4-}$, or *acetylide*, $C_2^{2-}$, both very strong Brønsted bases. Methanides react with water to produce a basic solution and methane gas:

  $$Al_4C_3(s) + 12H_2O(l) \rightarrow 4Al(OH)_3(s) + 3CH_4(g)$$

  Acetylides release acetylene, $HC{\equiv}CH$, upon reaction with water:

  $$CaC_2(s) + 2H_2O(l) \rightarrow Ca(OH)_2(s) + C_2H_2(g)$$

  → *Covalent carbides* are formed by metals with $\chi$ values similar to that of C, such as Si and B.

  → Silicon carbide, SiC (*carborundum*), an extremely hard, covalent carbide, is an excellent abrasive. It is produced by the reaction of silicon dioxide with carbon at 2000°C:

  $$SiO_2(s) + 3C(s) \rightarrow SiC(s) + 2CO(g)$$

  → *Interstitial carbides* are compounds formed by direct reaction of a *d*-block metal with C at temperatures above 2000°C. The C atoms lie in holes in close-packed arrays of metal atoms. Bonds between the C and metal atoms stabilize the lattice and help hold the metal atoms together. Examples are WC, $Cr_3C$, and $Fe_3C$, an important component of steel.

- **Tetrachlorides**

  → All Group 14/IV elements form *liquid* molecular tetrachlorides, of which $PbCl_4$ is the least stable. Carbon tetrachloride is formed by the reaction of chlorine and methane:

  $$CH_4(g) + 4Cl_2(g) \rightarrow CCl_4(l) + 4HCl(g)$$

  → Si reacts directly with chlorine to form silicon tetrachloride. Unlike $CCl_4$, $SiCl_4$ reacts with water as a Lewis acid, accepting a lone pair of electrons from $H_2O$:

  $$SiCl_4(l) + 2H_2O(l) \rightarrow SiO_2(s) + 4HCl(aq)$$

- **Cyanides**
  - → Cyanides are strong Lewis bases which form a range of complexes with *d*-block metal ions. They are extremely poisonous, combining with cytochromes involved with the transfer of electrons and the supply of energy in cells.
  - → The cyanide ion, $CN^-$, is the conjugate base of the acid HCN, which is made by heating methane, ammonia, and air in the presence of a Pt catalyst at 1100°C:

$$2CH_4(g) + 2NH_3(g) + 3O_2(g) \rightarrow 2HCN(g) + 6H_2O(g)$$

  - → *Acrylonitrile*, $H_2C{=}CHCN$, the starting point for the production of the synthetic fibers Acrilan and Orlon, is produced by the reaction of HCN with $HC{\equiv}CH$, using a copper(I) chloride catalyst.

- **Methane and silane**
  - → C bonds readily with itself to form chains and rings, and forms a multitude of compounds with hydrogen (hydrocarbons).
  - → Si forms a much smaller number of compounds with hydrogen; the simplest is the analogue of methane, *silane*, $SiH_4$, formed by the reaction of the reducing agent $LiAlH_4$ with silicon halides:

$$SiCl_4 + LiAlH_4 \rightarrow SiH_4 + LiCl + AlCl_3$$

  - → $SiH_4$ is more reactive than $CH_4$ and ignites spontaneously in air. Higher silanes are unstable and decompose on standing.

---

# The Impact on Materials (Sections 14.21–14.22)

## 14.21 Glasses

- **Preparation and properties**
  - → A *glass* is an ionic solid with an amorphous structure similar to a liquid. It is characterized by a network structure based on a nonmetal oxide, commonly silica, $SiO_2$, melted with metal oxides that function as *network modifiers*.
  - → Heating $SiO_2$ ruptures Si–O bonds, enabling metal ions to form ionic bonds with some of the O atoms. The nature of the glass depends upon the metal added.
  - → *Soda-lime glass*, used for windows and bottles, is made by adding $Na^+$ (12% $Na_2O$) and $Ca^{2+}$ (12% CaO). Heating $Na_2CO_3$ (soda) produces the $Na_2O$; heating $CaCO_3$ (lime) produces the CaO.
  - → *Borosilicate glass*, such as Pyrex, is produced using less soda and lime, and adding 16% $B_2O_3$. These glasses expand very little on heating and are used for ovenware and laboratory glassware.

- **Reaction with acids and bases**
  - → Glass is resistant to attack by most chemicals.
  - → But, the silica in glass can react with HF:

$$SiO_2(s) + 6HF(aq) \rightarrow SiF_6{}^{2-}(aq) + 2H_3O^+(aq)$$

  - → Silica also reacts with the Lewis base $OH^-$ in molten NaOH and with the carbonate anion in molten $Na_2CO_3$ at 1400°C:

$$SiO_2(s) + 2Na_2CO_3(l) \rightarrow Na_4SiO_4(l) + 2CO_2(g)$$

  - → Removal of silica from glass by the ions $F^-$ (from HF), $OH^-$, and $CO_3{}^{2-}$ is called *etching*.

## 14.22 Ceramics

- **Preparation and properties**

*Ceramics*

→ Inorganic materials hardened by heating to a high temperature

→ Typically very hard, insoluble in water, and stable to corrosion and high temperatures

→ Can be used at high temperatures without failing, and resist deformation

→ Tend to brittle, however

→ Often are oxides of elements on the border between metals and nonmetals

→ Used in many automobile parts including spark plugs, pressure and vibration sensors, brake linings, and catalytic converters

→ Some $d$-metal oxides and compounds of B and Si with C and N are also ceramic materials.

→ Most ceramics are electrical insulators, but some are semiconductors and superconductors.

*Aluminosilicate ceramics*

→ Formed in many cases by heating clays to expel water trapped between sheets of tetrahedral aluminosilicate units

→ The result is a rigid heterogeneous mass of small interlocking crystals bound together by glassy silica.

→ One form is *china clay,* used to make porcelain and china. It is free of iron impurities that tend to make clays reddish brown.

→ Methods developed to make ceramics less brittle are the *sol-gel process* and the *composite material technique*.

# Chapter 15  The Elements:
# The Last Four Main Groups

## Group 15 / V: The Nitrogen Family
## (Sections 15.1–15.4)

### 15.1  The Group 15 / V Elements

- **Nitrogen**
  - → *Nitrogen* differs greatly from other group members owing to its high electronegativity ($\chi = 3.0$), small size (radius of 74 pm), ability to form multiple bonds, and lack of available *d*-orbitals. It is found with oxidation numbers from −3 to +5.
  - → *Elemental nitrogen*, $N_2$, is found in air (78.1% by volume) and is prepared by fractional distillation of liquid air. The strong triple bond in $N_2$ ($\Delta H_B = 944$ kJ·mol$^{-1}$) makes it very stable.
  - → *Nitrogen fixation* is the process in which $N_2$ is converted into compounds usable by plants. The *Haber synthesis* of ammonia is the major industrial method for fixing nitrogen, but it requires high temperatures and pressures and is, therefore, expensive. Lightning produces some oxides of nitrogen, which are washed into soil by rain. Bacteria found in the root nodules of legumes also fix $N_2$. An important area of current research is the search for catalysts that can mimic bacteria and fix nitrogen under ordinary temperatures.

- **Phosphorus**
  - → *Phosphorus* differs from nitrogen in its lower electronegativity ($\chi = 2.2$), larger size (radius of 110 pm), and the presence of available *d*-orbitals. Phosphorus may form as many as six bonds, whereas nitrogen can form a maximum of four.
  - → *Phosphorus* is prepared from the *apatites*, mineral forms of *calcium phosphate*, $Ca_3(PO_4)_2$. The rocks are heated in an electric furnace with carbon and sand. The phosphorus vapor formed condenses as *white phosphorus*, a soft, white, poisonous molecular solid consisting of tetrahedral $P_4$ molecules. It bursts into flame upon exposure to air and is normally stored under water.
  - → *Red phosphorus* is less reactive than white phosphorus and is used in match heads because it can be ignited by friction. It is a network solid consisting, most likely, of linked $P_4$ tetrahedral units.
  - → *Red phosphorus* is the most thermodynamically stable form of phosphorus, yet *white phosphorus* crystallizes from the vapor or from the liquid phase. It is the form chosen to have an enthalpy and free energy of formation equal to 0. For *red phosphorus*, $\Delta H_f° = -17.6$ kJ·mol$^{-1}$ and $\Delta G_f° = -12.1$ kJ·mol$^{-1}$ at 25°C.
  - → *Black phosphorus* is a third form of the element; it has a network structure in which each P atom is bonded to three others at approximately right angles.

- **Arsenic and antimony**
  - → *Arsenic* and *antimony* are metalloids. They are elements known since ancient times because they are easily reduced from their ores.
  - → As elements, *arsenic* and *antimony* are used in lead alloys in the electrodes of storage batteries and in the semiconductor industry. *Gallium arsenide* is used as a light detecting material with near infrared response and in lasers, including ones used for CD players.

- **Bismuth**
  - → *Bismuth* is a metal with large, weakly bonded atoms. Bismuth has a low melting point, and it is used in alloys that serve as fire detectors in sprinkler systems.

→ *Bismuth* is also used to make low temperature castings. Like ice, solid bismuth is less dense than the liquid. Bismuth ($Z = 83$) is the last element in the periodic table with a stable isotope, $^{209}$Bi.

## 15.2 Compounds with Hydrogen and the Halogens

- **Ammonia, $NH_3$, and phosphine, $PH_3$**

  → *Ammonia* is prepared in large amounts by the Haber process. Ammonia is a pungent, toxic gas that condenses to a colorless liquid at $-33°C$. It acts as a solvent for a wide range of substances. Autoprotolysis occurs to a much smaller extent in liquid ammonia (am) than in pure water.

  $$2NH_3(am) \rightleftharpoons NH_4^+(am) + NH_2^-(am) \qquad K_{am} = 1 \times 10^{-33} \text{ at } -35°C$$

  Very strong bases that are protonated by water survive in liquid ammonia.

  → *Ammonia* is a weak Brønsted base in water; it is also a fairly strong Lewis base, particularly toward *d*-block elements. One reaction is $Cu^{2+}(aq) + 4NH_3(aq) \rightarrow Cu(NH_3)_4^{2+}(aq)$.
  In this reaction, unshared electron pairs on nitrogen interact with the Lewis acid $Cu^{2+}$ to form four $Cu–NH_3$ bonds.

  → *Ammonium salts* decompose when heated; if the salt contains a nonoxidizing anion, $NH_3$ is produced. The reaction is $(NH_4)_2SO_4(s) \rightarrow 2NH_3(g) + H_2SO_4(l)$.
  Decomposing ammonium carbonate has a pungent odor and is used as a "smelling salt" to revive people who have fainted.

  → *Phosphine* is a poisonous gas that smells faintly of garlic and bursts into flame in air if it is impure. It is much less stable than ammonia. Because it does not form hydrogen bonds, it is not very soluble in water and is a very weak base ($pK_b = 27.4$). Its aqueous solution is nearly neutral.

- **Hydrazine, $N_2H_4$**

  → *Hydrazine* is an oily, colorless liquid. It is prepared by gentle oxidation of ammonia with alkaline hypochlorite solution. $2NH_3(aq) + ClO^-(aq) \rightarrow N_2H_4(aq) + Cl^-(aq) + H_2O(l)$
  Its physical properties are similar to water; its melting point is $1.5°C$ and its boiling point is $113°C$. It is dangerously explosive and is stored in aqueous solution.

  → A mixture of *methylhydrazine* and liquid $N_2O_4$ is used as a rocket fuel because these two liquids ignite on contact and produce a large volume of gas.

  $$4CH_3NHNH_2(l) + 5N_2O_4(l) \rightarrow 9N_2(g) + 12H_2O(g) + 4CO_2(g)$$

- **Nitrides, $N^{3-}$, and phosphides, $P^{3-}$**

  → *Nitrides* are solids that contain the nitride ion. Nitrides are stable only for small cations such as lithium, magnesium, or aluminum. Nitrides dissolve in water to form ammonia and the corresponding hydroxide. $Zn_3N_2(s) + 6H_2O(l) \rightarrow 2NH_3(g) + 3Zn(OH)_2(aq)$

  → *Phosphides* are solids that contain the phosphide ion. Because $PH_3$ is a very weak parent acid of the strong Brønsted base $P^{3-}$, water is a sufficiently strong proton donor to react with phosphides to form phosphine. $2P^{3-}(s) + 6H_2O(l) \rightarrow 2PH_3(g) + 6OH^-(aq)$

- **Azides, $N_3^-$**

  → The *azide ion* is a highly reactive polyatomic anion of nitrogen. Sodium azide is prepared by the reaction of dinitrogen oxide with molten sodium amide, $NaNH_2$.

  → Some *azides*, such as $AgN_3$, $Cu(N_3)_2$, and $Pb(N_3)_2$, are shock sensitive. The azide ion is a weak base, and its conjugate acid, hydrazoic acid, $HN_3$, is a weak acid similar in strength to acetic acid.

- **Halides**

  → *Nitrogen* has an oxidation number of +3 in the *nitrogen halides*. Nitrogen trifluoride, $NF_3$, is the most stable and does not react with water. Nitrogen trichloride, however, reacts with water to form ammonia and hypochlorous acid, HClO. Nitrogen triodide is so unstable that it decomposes explosively upon contact with light.

→ *Phosphorus forms chlorides with an oxidation number of +3 in PCl₃ and +5 in PCl₅.* *Phosphorus trichloride*, a liquid, is formed by direct chlorination of phosphorus. It is a major intermediate in the production of pesticides, oil additives, and flame retardants. *Phosphorus pentachloride*, a solid, is made by allowing the trichloride to react with additional chlorine. It exists in the solid as tetrahedral $PCl_4^+$ cations and octahedral $PCl_6^-$ anions, but it vaporizes to a gas of trigonal bipyramidal $PCl_5$ molecules. *Phosphorus pentabromide* is also molecular in the vapor and ionic as a solid; but in the solid, the anions are $Br^-$ anions, presumably because of the difficulty in fitting six large Br atoms around a central P atom.

→ Both *phosphorus trichloride* and *phosphorus pentachloride* react with water in a *hydrolysis reaction*, a reaction with water in which new element-oxygen bonds are formed and there is no change in oxidation state. An *oxoacid* and *hydrogen chloride* gas are formed in the reaction.

$$PCl_3(l) + 3H_2O(l) \rightarrow H_3PO_3(s) + 3HCl(g)$$

$$PCl_5(s) + 4H_2O(l) \rightarrow H_3PO_4(l) + 5HCl(g)$$

## 15.3 Nitrogen Oxides and Oxoacids

- **Dinitrogen oxide, $N_2O$**

  → *Dinitrogen oxide* is commonly called *nitrous oxide* (laughing gas), and it is the oxide of nitrogen with the lowest oxidation number for nitrogen (+1). It is formed by gentle heating of ammonium nitrate. $NH_4NO_3(s) \rightarrow N_2O(g) + 2H_2O(g)$

  → $N_2O$ is fairly unreactive, but it is toxic if inhaled in large amounts.

- **Nitrogen oxide, NO**

  → *Nitrogen oxide* is commonly called *nitric oxide*; the nitrogen atom has an oxidation number of +2. Nitrogen oxide, a colorless gas, is produced industrially by the catalytic oxidation of ammonia at 1000°C. $4NH_3(g) + 5O_2(g) \rightarrow 4NO(g) + 6H_2O(g)$

  → The endothermic formation of NO from the oxidation of $N_2$ occurs readily at the high temperatures that exist in automobile engines and turbine engine exhausts.
  $$N_2(g) + O_2(g) \rightarrow 2NO(g)$$

  → NO is further oxidized to $NO_2$ upon exposure to air. In this form, it contributes to smog, acid rain, and the destruction of the stratospheric ozone layer.

- **Nitrogen dioxide, $NO_2$**

  → *Nitrogen dioxide* is a choking, poisonous, brown gas that contributes to the color and odor of smog. The oxidation number of nitrogen in $NO_2$ is +4. $NO_2$, like NO, is paramagnetic. It disproportionates in water to form nitric acid and nitrogen oxide.
  $$3NO_2(g) + H_2O(l) \rightarrow 2HNO_3(aq) + NO(g)$$

  → *Nitrogen dioxide* is prepared in the laboratory by heating lead(II) nitrate:
  $$2Pb(NO_3)_2(s) \rightarrow 4NO_2(g) + 2PbO(s) + O_2(g)$$
  and exists in equilibrium with its colorless dimer, dinitrogen tetroxide:
  $$2NO_2(g) \rightleftharpoons N_2O_4(g)$$
  Only the dimer exists in the solid, so the brown gas condenses to a colorless solid.

- **Nitrous acid, $HNO_2$**
  → *Nitrous acid* can be produced in aqueous solution by mixing its anhydride, dinitrogen trioxide, with water. The oxidation number of nitrogen in $HNO_2$ is +3. $N_2O_3(g) + H_2O(l) \rightarrow 2HNO_2(aq)$

→ *Nitrous a*cid has not been isolated as a pure compound, but it has many uses in aqueous solution. It is a weak acid with $pK_a = 3.4$. The conjugate base, nitrite ion, $NO_2^-$, forms ionic solids that are soluble in water and mildly toxic.

- **Nitric acid, $HNO_3$**

  → *Nitric acid,* a widely used industrial and laboratory acid, is produced by the three-step *Ostwald process*. The oxidation number of nitrogen in $HNO_3$ is +5.

  $$4NH_3(g) + 2O_2(g) \rightarrow 4NO(g) + 6H_2(g)$$
  $$2NO(g) + O_2(g) \rightarrow 2NO_2(g)$$
  $$3NO_2(g) + H_2O(l) \rightarrow 3HNO_3(aq) + NO(g)$$

  → Nitric acid has been isolated as a pure compound, and it is a colorless liquid with a boiling point of 83°C. It is usually used in aqueous solution. It is an excellent oxidizing agent as well as a strong acid.

## 15.4 Phosphorus Oxides and Oxoacids

- **Phosphorus(III) oxide, $P_4O_6$, and phosphorous acid, $H_3PO_3$**

  → *Phosphorus(III) oxide* is made by heating white phosphorus in a limited supply of air. $P_4O_6$ molecules consist of $P_4$ tetrahedral units, with each oxygen atom lying between two phosphorus corner atoms. Six oxygen atoms lie along the six edges of the tetrahedron, one per edge. The edge bonds are represented as $-P-O-P-$.

  → $P_4O_6$ is the anhydride of *phosphorous acid:* $\quad 4H_3PO_3(aq) \rightarrow P_4O_6(s) + 6H_2O(l)$
  Phosphorous acid is a *diprotic* acid.

- **Phosphorus(V) oxide, $P_4O_{10}$, and phosphoric acid, $H_3PO_4$**

  → *Phosphorus(V) oxide* is made by heating white phosphorus in an excess supply of air. $P_4O_{10}$ molecules are similar to $P_4O_6$, but have an additional oxygen attached to each phosphorus at the apices of the tetrahedron.

  → $P_4O_{10}$ is the *anhydride* of *phosphoric acid*, $H_3PO_4$. It traps and reacts with water very efficiently and is widely used as a drying agent. Phosphoric acid is a *triprotic acid*. It is only mildly oxidizing, despite the fact that phosphorus is in a high oxidation state. Phosphoric acid is the parent acid of phosphate salts that contain the tetrahedral phosphate ion, $PO_4^{3-}$. Phosphates are generally not very soluble, which makes them suitable structural material for bones and teeth.

- **Polyphosphates**

  → *Polyphosphates* are compounds made up of linked $PO_4^{3-}$ tetrahedral units. The simplest structure is the *pyrophosphate ion*, $P_2O_7^{4-}$, in which two phosphate ions are linked by an oxygen atom through $-O-P-O-$ bonds.

  → More complicated structures that form with longer chains or rings also exist. The biochemically most important polyphosphate is *adenosine triphosphate*, ATP, which contains three phosphorus tetrahedral units linked by $-O-P-O-$ bonds. The hydrolysis of ATP to *adenosine diphosphate*, ADP, by the rupture of an O-P bond releases energy that is used by cells to drive biochemical reactions within the cell. $\text{ATP} + H_2O \rightarrow \text{ADP} + HPO_4^{2-} \quad \Delta H = -41 \text{ kJ}$

  ATP is often called the "fuel of life," because it provides the energy for many of the biochemical reactions occurring in cells.

# Group 16 / VI: The Oxygen Family
## (Sections 15.5–15.8)

### 15.5 The Group 16 / VI Elements

- **Oxygen**

  → *Elemental oxygen*, $O_2$, is found in air (21.0% by volume) and is prepared by fractional distillation of liquid air. It is a colorless, odorless, paramagnetic gas.

  → A satisfactory Lewis structure for oxygen cannot be drawn. The Molecular Orbital theory accounts for the paramagnetic properties (see Chapter 3).

  → *Elemental oxygen* exists in two allotropic forms, $O_2$ and $O_3$ (*ozone*). Inorganic chemists often use the term *dioxygen* to refer to molecular oxygen. Ozone is produced by an electric discharge in oxygen gas; its pungent odor is apparent near sparking electrical equipment and lightning strikes.

- **Sulfur**

  → *Sulfur* is found in ores and as elemental sulfur. It is widely distributed as sulfide ores, which include *galena*, $PbS$; *cinnabar*, $HgS$; *iron pyrite*, $FeS_2$ (fool's gold), and *spahlerite*, $ZnS$. Sulfur is found in elemental form as deposits, called *brimstone*, which are created by bacterial action on $H_2S$.

  → *Elemental sulfur* is a yellow, tasteless, almost odorless, insoluble, nonmetallic molecular solid of crownlike $S_8$ rings. The more stable allotrope, *rhombic sulfur*, forms beautiful yellow crystals, whereas the less stable allotrope, *monoclinic sulfur*, forms needlelike crystals. The two allotropes differ in the manner in which the $S_8$ rings are stacked together. For simplicity, the elemental form of sulfur is often represented as $S(s)$ rather than $S_8(s)$.

  → One method of recovering sulfur is the *Claus process*, in which some of the $H_2S$ that occurs in oil and natural gas is first oxidized to sulfur dioxide. $2H_2S(g) + 3O_2(g) \rightarrow 2SO_2(g) + 2H_2O(l)$ Sulfur dioxide oxidizes the remaining hydrogen sulfide and both are converted to elemental sulfur at 300°C in the presence of an alumina catalyst. $2H_2S(g) + SO_2(g) \rightarrow 3S(s) + 2H_2O(l)$

  → *Sulfur* is used primarily to produce *sulfuric acid* and to vulcanize rubber. Vulcanization increases the toughness of rubber by introducing cross-links between the polymer chains of natural rubber.

  → An important property of sulfur is its ability to form chains of atoms, *catenation*. The $-S-S-$ links that connect different parts of the chains of amino acids in proteins are an important example. These *disulfide links* contribute to the shapes of proteins (see Chapter 19).

- **Selenium and tellurium**

  → *Selenium* and *tellurium* are found in sulfide ores; they are also recovered from the anode sludge formed during the electrolytic refining of copper.

  → Both elements have several allotropic forms; the most stable one consists of long *zigzag chains* of atoms. The appearance of the allotropes is that of a silver-white metal, but electrical conductivity is poor. Exposing selenium to light increases its electrical conductivity, so it is used in solar cells.

- **Polonium**

  → *Polonium* is a radioactive, low-melting metalloid. It was identified in uranium ores in 1898 by the Curies. Its most stable isotope, $^{209}Po$, has a half-life of 103 y.

  → *Polonium* decays by emission of an alpha particle, $^4He^{2+}$. It is used in antistatic devices in textile mills to counteract the buildup of negative charges on fast moving fabric.

## 15.6 Compounds with Hydrogen

- **Water, $H_2O$, and hydrogen sulfide, $H_2S$**

  → *Water* is a remarkable compound possessing a unique set of physical and chemical properties. Water is purified in domestic water supplies in a multi-step procedure that includes *aeration*, addition of *slaked lime*, removal of particles by *coagulation* and *flocculation*, addition of $CO_2$, *filtration*, reduction of pH, and addition of chlorine *disinfectant* to kill bacteria. High purity water is obtained by *distillation*.

  → *Water* has a higher boiling point than expected on the basis of its molar mass because of the extensive hydrogen bonding between $H_2O$ molecules. Hydrogen bonding also causes a more open structure and, therefore, a lower density for the solid than for the liquid. Because of its high polarity, water is an excellent solvent for ionic compounds.

  → Chemically, water is *amphiprotic,* and it can both donate and accept protons; so it is both a *Brønsted acid* and *Brønsted base*.

  $$H_2O(l, \textit{acid}) + NH_3(aq) \rightarrow NH_4^+(aq) + OH^-(aq)$$

  $$CH_3COOH(aq) + H_2O(l, \textit{base}) \rightarrow H_3O^+(aq) + CH_3COO^-(aq)$$

  → *Water* can act as a *Lewis base* by donating its unshared pair of electrons on the oxygen atom.

  $$Fe^{3+}(aq) + 6H_2O(l) \rightarrow Fe(H_2O)_6^{3+}(aq)$$

  → *Water* can act as an *oxidizing agent* and a *reducing agent*.

  $$2Na(s) + 2H_2O(l, \textit{ox. agent}) \rightarrow 2NaOH(aq) + H_2(g)$$

  $$2H_2O(l, \textit{red. agent}) + 2F_2(g) \rightarrow 4HF(aq) + O_2(g)$$

  → A *hydrolysis reaction* is one in which water as a reactant is used to form a bond between oxygen and another element. Hydrolysis can occur *with* or *without* a change in oxidation number.

  $$PCl_5(s) + 4H_2O(l) \rightarrow H_3PO_4(aq) + 5HCl(aq)$$

  $$Cl_2(g) + H_2O(l) \rightarrow ClOH(aq) + HCl(aq)$$

  The first example given above has *no* change in oxidation number. The second is an example of a *disproportionation reaction,* in which the oxidation number of chlorine changes from 0 in $Cl_2$ to +1 in HClO and −1 in HCl.

  → *Hydrogen sulfide* is an example of a Group 16/VI binary compound with hydrogen. It is a toxic gas with an offensive odor (rotten eggs). Egg proteins contain sulfur and release $H_2S$ when they decompose. Iron(II) sulfide sometimes appears as a pale green discoloration at the boundary between the white and the yolk.

  → *Hydrogen sulfide* is prepared by the direct reaction of hydrogen and sulfur at 600°C or by protonation of the sulfide ion, which is a Brønsted base.

  $$FeS(s) + 2HCl(aq) \rightarrow FeCl_2(aq) + H_2S(g)$$

  → *Hydrogen sulfide* dissolves in water and is slowly oxidized by dissolved air to form a colloidal dispersion of particles of sulfur. Hydrogen sulfide is a *weak diprotic acid,* and the parent acid of the hydrogen sulfides, $HS^-$, and the sulfides, $S^{2-}$. Sulfides of *s*-block elements are moderately soluble in water, whereas the sulfides of the heavy *p*- and *d*-block metals are generally very insoluble.

- **Hydrogen peroxide, $H_2O_2$, and polysulfanes, $HS-S_n-SH$**

  → *Hydrogen peroxide* is a highly reactive, pale blue liquid that is appreciably more dense (1.44 g·mL$^{-1}$) than water. Its melting point is −0.4°C and its boiling point is 152°C.

→ Chemically, *hydrogen peroxide* is more acidic than water ($pK_a = 11.75$). It is a strong oxidizing agent, but it can also function as a reducing agent.

$$2\,Fe^{2+}(aq) + H_2O_2(aq, \textit{ox. agent}) + 2\,H^+(aq) \rightarrow 2\,Fe^{3+}(aq) + 2\,H_2O(l)$$
$$Cl_2(g) + H_2O_2(aq, \textit{red. agent}) + 2\,OH^-(aq) \rightarrow 2\,Cl^-(aq) + 2\,H_2O(l) + O_2(g)$$

→ The O–O bond in *hydrogen peroxide* is very weak. The oxidation state of oxygen in hydrogen peroxide is –1.

→ A *polysulfane* is a *catenated* molecular compound of composition $HS–S_n–SH$, where $n$ can take values from 0 through 6. The sulfur analogue of hydrogen peroxide occurs with $n = 0$. Two polysulfide ions obtained from polysulfanes are found to occur in the mineral *lapis lazuli* (see Fig. 15.16 in the text); its color derives from impurities of $S_2^-$ (blue) and $S_3^-$ (hint of green). The $S_2^-$ ion is an analogue of the *superoxide* ion, $O_2^-$; and $S_3^-$ is an analogue of the *ozonide* ion, $O_3^-$.

## 15.7 Sulfur Oxides and Oxoacids

- **Sulfur dioxide, $SO_2$, sulfurous acid, $H_2SO_3$, and sulfite, $SO_3^{2-}$**

  → *Sulfur dioxide,* a poisonous gas, is made by burning sulfur in air. Volcanic activity, the combustion of fuels contaminated with sulfur, and the oxidation of hydrogen sulfide are the major sources of $SO_2$ in the atmosphere. The chemical equation for the oxidation of hydrogen sulfide in the atmosphere is

  $$2\,H_2S(g) + 3\,O_2(g) \rightarrow 2\,SO_2(g) + 2\,H_2O(g)$$

  → *Sulfurous acid* is formed by the reaction of its acid anhydride, $SO_2$, and water.

  $$SO_2(g) + H_2O(l) \rightarrow H_2SO_3(aq)$$

  Sulfurous acid is an equilibrium mixture of two molecules, $(HO)SHO_2$ and $(HO)SO(OH)$. These molecules are also in equilibrium with molecules of $SO_2$, each of which is surrounded by a cage of water molecules. When the solution is cooled, crystals with a composition of roughly $SO_2 \cdot 7H_2O$ form. This substance is an example of a *clathrate,* in which a molecule sits in a cage of other molecules. Methane, carbon dioxide, and the noble gases also form clathrates with water.

  → In *sulfite ions* and *sulfur dioxide,* the oxidation number of the sulfur atom is +4. Because this value is intermediate in sulfur's range of –2 to +6, these compounds can act as either oxidizing agents or reducing agents. The most important reaction of sulfur dioxide is its slow oxidation to sulfur trioxide, in which the oxidation number of the sulfur atom is +6.

  $$2\,SO_2(g) + O_2(g) \rightarrow 2\,SO_3(g)$$

  This reaction is catalyzed by the presence of metal cations in droplets of water or by certain surfaces. Indirect pathways also exist in the atmosphere for the conversion of $SO_2$ to $SO_3$.

- **Sulfur trioxide, $SO_3$, sulfuric acid, $H_2SO_4$, and sulfate, $SO_4^{2-}$**

  → At normal temperatures, *sulfur trioxide* is a volatile liquid with a boiling point of 45°C. Its shape is trigonal planar with bond angles of 120°. In the solid, and to some extent in the liquid, the molecules aggregate to form *trimers,* $S_3O_9$, as well as larger groupings.

  → *Sulfuric acid* is produced by the *contact process,* in which sulfur is first burned in oxygen at 1000°C to form sulfur dioxide, which then reacts with oxygen to form sulfur trioxide over a $V_2O_5$ catalyst at 500°C. Because sulfur trioxide forms a corrosive acid with water, it is absorbed in 98% concentrated sulfuric acid to give a dense, oily liquid called *oleum,* $H_2S_2O_7$.

  $$SO_3(g) + H_2SO_4(l) \rightarrow H_2S_2O_7(l)$$

  → *Sulfuric acid* is a colorless, corrosive, oily liquid that boils (decomposes) at 300°C. It is a strong *Brønsted acid,* a powerful *dehydrating agent,* and a mild *oxidizing agent.* In water, its first ionization is complete, but $HSO_4^-$ is a *weak acid* with $pK_a = 1.92$. As a *dehydrating agent,* sulfuric acid is able to extract water from a variety of compounds, including sucrose, $C_{12}H_{22}O_{11}$, and formic acid, HCOOH.

$$C_{12}H_{22}O_{11}(s) \rightarrow 12\,C(s) + 11\,H_2O(l)$$

$$HCOOH(l) \rightarrow CO(g) + H_2O(l)$$

Because of the large exothermicity associated with dilution, mixing sulfuric acid with water can cause violent splashing, so for reasons of safety, sulfuric acid (which is more dense than water) is always carefully added to water, not the water to acid. The *sulfate ion* is a mild *oxidizing agent*.

$$4H^+(aq) + SO_4^{2-}(aq) + 2e^- \rightarrow SO_2(g) + 2H_2O(l) \quad E° = 0.17\ V$$

## 15.8 Sulfur Halides

- **Fluorides**
  - → Sulfur ignites in fluorine to produce *sulfur hexafluoride*, $SF_6$, a dense, colorless, odorless, nontoxic, thermally stable ($\Delta G_f° = -1105.3\ kJ \cdot mol^{-1}$), insoluble gas.
  - → The oxidation number of sulfur in *sulfur hexafluoride* attains its maximum value of +6, but the large number of strongly electronegative fluorine atoms around the sulfur atom protects the compound from attack. Because it has a high ionization energy, sulfur hexafluoride is a good gas phase electrical insulator.
  - → The predicted shape of $SF_6$ is *octahedral* and the molecule is *nonpolar*, $\mu = 0$ (see Section 3.3).

- **Chlorides**
  - → *Disulfur dichloride,* $S_2Cl_2$, is one of the products of the reaction of sulfur with chlorine. A yellow, toxic liquid with a nauseating odor, *disulfur dichloride* is used mainly for the vulcanization of rubber.
  - → The reaction of *disulfur dichloride* with excess chlorine in the presence of a catalyst, iron(III) chloride, produces a foul smelling, red liquid of *sulfur dichloride*, $SCl_2$. It reacts with ethene, $C_2H_4$, to produce *mustard gas*, $S(CH_2CH_2Cl)_2$, which has been used in chemical warfare. *Mustard gas* causes symptoms of severe discomfort, skin blisters, nasal discharges, and vomiting. It also destroys the cornea of the eye, leading to blindness.

---

# Group 17 / VII: The Halogens
## (Sections 15.9–15.10)

## 15.9 The Group 17 / VII Elements

- **Fluorine**
  - → *Fluorine* is a reactive, almost colorless gas of $F_2$ molecules. Its properties follow from its high electronegativity, small size, and lack of available *d*-orbitals for bonding. It is the most electronegative element and has an oxidation number of −1 in all its compounds. Elements combined with fluorine are often found with their highest oxidation numbers, such as +7 for I in $IF_7$. The small size of the fluoride ion results in high lattice enthalpies for fluorides, a property that makes them less soluble than other halides.
  - → *Fluorine* occurs widely in many minerals including *fluorspar*, $CaF_2$; *cryolite*, $Na_3AlF_6$; and the *fluoroapatites*, $Ca_5F(PO_4)_3$.

→ *Elemental fluorine* is obtained by the electrolysis of an anhydrous, molten KF–HF mixture at 75°C, using a carbon electrode. Fluorine is highly reactive and highly oxidizing; it is used in the production of $SF_6$ for electrical equipment, Teflon (polytetrafluoroethylene), and $UF_6$ for isotope enrichment.

→ The strong bonds formed by *fluorine* make many of its compounds relatively inert. Fluorine's ability to form *hydrogen bonds* results in relatively high melting points, boiling points, and enthalpies of vaporization for many of its compounds.

- **Chlorine**

  → *Chlorine* is a reactive, pale yellow-green gas of $Cl_2$ molecules that condenses at –34°C. It reacts directly with all the elements except carbon, nitrogen, oxygen, and the noble gases. It is used in a large number of industrial processes, including the manufacture of plastics, solvents, and pesticides. It is also used as a disinfectant to treat water supplies.

  → *Elemental chlorine* is obtained by the electrolysis of molten or aqueous NaCl. It is a strong oxidizing agent and oxidizes metals to high oxidation states.
  $$2\,Fe(s) + 3\,Cl_2(g) \rightarrow 2\,FeCl_3(s)$$

- **Bromine**

  → *Bromine* is a corrosive, reddish brown liquid of $Br_2$ molecules. It has a penetrating odor. Organic bromides are used in textiles as fire retardants and in pesticides. Inorganic bromides, particularly silver bromide, are used in photographic emulsions.

  → *Elemental bromine* is produced from brine wells by the oxidation of $Br^-$ by $Cl_2$.
  $$2\,Br^-(aq) + Cl_2(g) \rightarrow Br_2(l) + 2\,Cl^-(aq)$$

- **Iodine**

  → *Iodine* is a blue-black, lustrous solid of $I_2$ molecules that readily sublimes to form a purple vapor. It occurs as $I^-$ in seawater and as an impurity in Chile saltpeter, $KNO_3$. The best source is the brine from oil wells. *Elemental iodine* is produced from brine wells by the oxidation of $I^-$ by $Cl_2$.

  → *Elemental iodine* is slightly soluble in water, but it dissolves well in iodide solutions because it reacts with aqueous $I^-$ to form the triodide ion, $I_3^-$.

  → *Iodine* is an essential trace element in human nutrition, and *iodides* are often added to table salt. This "iodized salt" prevents iodine deficiency, which leads to the enlargement of the thyroid gland in the neck (a condition called *goiter*).

- **Astatine**

  → *Astatine* is a radioactive element that occurs in uranium ores, but only to a tiny extent. Its most stable isotope, $^{210}At$, has a half-life of 8.3 h. The isotopes formed in uranium ores have much shorter lifetimes. The properties of astatine are surmised from spectroscopic measurements.

  → *Astatine* is created by bombarding bismuth with alpha particles in a cyclotron, which accelerates particles to high speed.

## 15.10 Compounds of the Halogens

- **Interhalogens**

  → *Interhalogen* compounds consist of a heavier halogen atom at the center of the compound and an odd number of lighter ones bonded to it on the periphery. An exception is $I_2Cl_6$, which contains two (–I–Cl–I–) bridges between the two central iodine atoms. The two bridging chlorine atoms act as Lewis bases (electron pair donors), in a manner similar to $Al_2Cl_6$.

  → *Interhalogens* are typically formed by direct reaction between stoichiometric amounts of the elements. $I_2(g) + 7F_2(g) \rightarrow 2IF_7(g)$

→ As the size of the central atom in an *interhalogen* compound increases, the bond enthalpies also increase, *usually* resulting in lower reactivity of the compound.

   (X–F):    $ClF_5$ (154 kJ·mol$^{-1}$) < $BrF_5$ (187 kJ·mol$^{-1}$) < $IF_5$ (269 kJ·mol$^{-1}$)

→ Table 15.5 in the text provides a compilation of the known *interhalogens*.

- **Hydrogen halides and metal halides**

   → *Hydrogen halides* are prepared by direct reaction of the elements or by the action of nonvolatile acids on *metal halides:*

$$Cl_2(g) + H_2(g) \rightarrow 2HCl(g)$$
$$CaF_2(s) + H_2SO_4(aq, conc.) \rightarrow Ca(HSO_4)_2(aq) + 2HF(g)$$

   Because Br$^-$ and I$^-$ are oxidized by sulfuric acid, phosphoric acid is used in the preparation of HBr and HI.

   → *Hydrogen halides* are pungent, colorless gases, but *hydrogen fluoride* is a liquid at temperatures below 20°C. Its low volatility is a sign of extensive hydrogen bonding, and short zigzag chains up to about $(HF)_5$ persist in the vapor.

   → *Hydrogen halides* dissolve readily in water to produce acidic solutions. *Hydrogen fluoride* has the unusual property of attacking and dissolving silica glasses, and it is used for glass etching (see Section 14.21).

   → Anhydrous *metal halides* may be formed by direct reaction:

$$2Fe(s) + 3Cl_2(g) \rightarrow 2FeCl_3(s)$$

   *Halides* of metals tend to be ionic unless the metal has an oxidation number greater than +2. For example, sodium chloride and copper(II) chloride are ionic compounds with high melting points, whereas titanium(IV) chloride and iron(III) chloride sublime as molecules.

- **Halogen oxides and oxoacids**

   → *Hypohalous acids,* HXO *or* HOX, consist of the halogen atom as a *terminal* atom with an oxidation number of +1. They are prepared by the direct reaction of a halogen with water; the corresponding *hypohalite* (XO$^-$) salts are prepared by the reaction of a halogen with aqueous alkali solution.

$$Cl_2(g) + H_2O(l) \rightarrow HOCl(aq) + HCl(aq)$$
$$Cl_2(g) + NaOH(aq) \rightarrow NaOCl(aq) + HCl(aq)$$

   → *Hypochlorites* cause the oxidation of organic material in water by producing oxygen in aqueous solution, which readily oxidizes organic materials. The production of $O_2$ occurs in two steps; the net result is   $2OCl^-(aq) \rightarrow 2Cl^-(aq) + O_2(g)$

   → *Chlorates,* $ClO_3^-$, contain a *central* chlorine atom with an oxidation number of +5. They are prepared by the reaction of chlorine with hot aqueous alkali:

$$3Cl_2(g) + 6OH^-(aq) \rightarrow ClO_3^-(aq) + 5Cl^-(aq) + 3H_2O(l)$$

   *Chlorates* decompose upon heating; the identity of the final product is determined by the presence of a catalyst:   $4KClO_3(l) \rightarrow 3KClO_4(s) + KCl(s)$        *no catalyst*
   $$2KClO_3(l) \rightarrow 2KCl(s) + 3O_2(g) \qquad \textit{MnO}_2 \textit{ catalyst}$$

   *Chlorates* are good oxidizing agents and are also used to produce the important compound, $ClO_2$. Sulfur dioxide is a convenient reducing agent for this reaction:

$$2NaClO_3(l) + SO_2(g) + H_2SO_4(aq, dilute) \rightarrow 2NaHSO_4(aq) + 2ClO_2(g)$$

→ *Chlorine dioxide*, $ClO_2$, contains a *central* chlorine atom with an oxidation number of +4. It has an odd number of electrons and is a paramagnetic yellow gas that is used to bleach paper pulp. It is highly reactive and may explode violently under the right conditions.

→ *Perchlorates*, $ClO_4^-$, contain a *central* chlorine atom with an oxidation number of +7. They are prepared by the electrolysis of aqueous chlorates. The half-reaction is

$$ClO_3^-(aq) + H_2O(l) \rightarrow ClO_4^-(aq) + 2H^+(aq) + 2e^-$$

*Perchlorates* and *perchloric acid*, $HClO_4$, are powerful oxidizing agents; *perchloric acid* in contact with small amounts of organic materials may explode.

→ The *oxidizing strength* and *acidity* of oxoacids both increase as the oxidation number of the halogen increases. For oxoacids with the same number of oxygen atoms, the oxidizing and acid strength both increase as the electronegativity value of the halogen increases. The general rules for acid strength of oxoacids are summarized in Chapter 10.

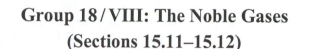

# Group 18 / VIII: The Noble Gases
## (Sections 15.11–15.12)

## 15.11 The Group 18 / VIII Elements

- **Helium** → *Helium* derives its name from the Greek word *helios,* sun. The first evidence for helium was discovered in the solar spectrum taken during a solar eclipse in 1868. *Helium* is the second most abundant element in the universe after hydrogen. Helium atoms are light and travel at velocities sufficient to escape from the earth's atmosphere. Alpha particles, $^4He^{2+}$, are released by nuclear decay of naturally occurring uranium and thorium ores. They are high-energy nuclei of helium-4, which capture two electrons apiece to become helium atoms. The low density and lack of flammability of helium make it an appropriate gas for lighter-than-air airships such as blimps. Helium is found in the atmosphere with a concentration of 5.24 ppm (parts per million by volume).

- **Neon**
  → *Neon* derives its name from the Greek word *neos*, new. It is widely used in advertising display signs because it emits an orange-red glow when an electric current passes through it. Of all the noble gases, the discharge of neon is the most intense at ordinary voltages and currents. The neon atom is the source of laser light in the helium-neon laser.

  → *Neon* has over 40 times more refrigerating capacity per unit volume than liquid helium and more than three times that of liquid hydrogen. It is compact, inert, and less expensive than helium when used as a *cryogenic* coolant. Neon is found in the atmosphere with a concentration of 18.2 ppm.

- **Argon**
  → *Argon* derives its name from the Greek word *argos,* inactive. Its main use is used to fill electric light bulbs at a pressure of about 400 Pa, where it conducts heat away from the filament. It is also used to fill fluorescent tubes.

  → *Argon* is also used as an inert gas shield (to prevent oxidation) for arc welding and cutting, and as a protective atmosphere for growing silicon and germanium crystals. Argon is the most abundant noble gas and is found in the atmosphere with a concentration of about 0.93% (or about 1 part per hundred by volume).

- **Krypton**
  - → *Krypton* derives its name from the Greek word *kryptos,* hidden. It gives an intense white light when an electric discharge is passed through it, and it is used in airport runway lighting.
  - → *Krypton* is produced in nuclear fission, and its atmospheric abundance is a measure of worldwide nuclear activity. Krypton is found in the atmosphere with a concentration of about 1 ppm.

- **Xenon**
  - → *Xenon* derives its name from the Greek word *xenos,* stranger. It is used in halogen lamps for automobile headlights and in high-speed photographic flash tubes.
  - → *Xenon* is used in the nuclear energy area in bubble chambers, probes, and other applications where a high atomic mass is desirable. Currently, the anesthetic properties of xenon are being explored. Xenon is found in the atmosphere with a concentration of about 0.087 ppm.

- **Radon**
  - → *Radon* derives its name from the element *radium*. The gas is radioactive and is formed by radioactive decay processes deep in the earth. Uranium-238 decays very slowly to radium-226, which further decays by alpha particle emission to radon-222 (see Chapter 17). The half-life of radon-222 is 3.825 days. Other shorter-lived isotopes are formed from the decay of thorium-232 and uranium-235. Every square mile of soil to a depth of 6 inches is estimated to contain about 1 g of radium, which releases radon in tiny amounts into the atmosphere.
  - → *Radon* is used in implant seeds for the therapeutic treatment of localized tumors. Radon is found in the atmosphere with a concentration estimated to be 1 in $10^{21}$ parts of air.

## 15.12 Compounds of the Noble Gases

- **Helium, neon, and argon**
  - → The ionization energies of helium, neon, and argon are very high. They form no stable *neutral* compounds.
  - → If electrons are removed from noble gas atoms in ionization processes, the resulting noble gas ions are radicals, and they may form stable molecular ions such as $NeAr^+$, $ArH^+$, and $HeNe^+$ in the gas phase. If an electron is added to them, these ions rapidly dissociate into neutral atoms.

- **Krypton**
  - → *Krypton* forms a thermodynamically unstable neutral compound, $KrF_2$ (recall that diamond is also thermodynamically unstable). It is a volatile white solid, which decomposes slowly at room temperature.
  - → In 1988, a compound with a Kr–N bond was discovered, but it is stable only at temperatures below $-50°C$.

- **Xenon**
  - → *Xenon* is the noble gas element with the richest chemistry. It forms several compounds with fluorine and oxygen, and even compounds with Xe–N and Xe–C bonds. The compound, $XeF_2$, is thermodynamically stable. Xenon is found in compounds with oxidation numbers of +2, +4, +6, and +8.
  - → Direct reaction of *fluorine* with *xenon* at high temperature results in the formation of compounds with oxidation numbers of +2 ($XeF_2$), +4 ($XeF_4$), and +6 ($XeF_6$). In the case of $XeF_6$, high pressure is required. All three fluorides are crystalline solids. In the gas phase, all are molecular compounds. Solid *xenon hexafluoride* is an ionic compound with a complex structure of $XeF_5^+$ cations bridged by $F^-$ anions.

→ *Xenon fluorides* are powerful *fluorinating agents*, reagents that attach fluorine to other substances. *Xenon trioxide* is synthesized by the hydrolysis of *xenon tetrafluoride:*

$$6\,XeF_4(s) + 12\,H_2O(l) \rightarrow 2\,XeO_3(aq) + 4\,Xe(g) + 3\,O_2(g) + 24\,HF(aq)$$

The trioxide is the anhydride of *xenic acid*, $H_2XeO_4$. In aqueous basic solution, the acid forms the *hydrogen xenate ion*, $HXeO_4^-$, which further disproportionates to the *perxenate ion*, $XeO_6^{4-}$, in which xenon attains its highest oxidation number of +8.

$$2\,HXeO_4^-(aq) + 2\,OH^-(aq) \rightarrow XeO_6^{4-}(aq) + Xe(g) + O_2(g) + 2\,H_2O(l)$$

- **Radon**
  - → *Radon* chemistry is difficult to study because all of its isotopes are radioactive.
  - → A *radon fluoride* of unknown composition is found to form readily, but it decomposes rapidly during attempts to vaporize it for further study.

---

# The Impact on Materials
# (Sections 15.13–15.14)

## 15.13 Soft Materials: Colloids, Gels, and Biomaterials

- **Colloids**
  - → Particles (*dispersed phase*) with lengths or diameters between 1 nm and 1 μm *dispersed* or *suspended* in a *dispersion medium* (gas, liquid, or solid solvent)
  - → *Classification of colloids* is given in Table 15.7 in the text.
  - → *Aerosols* are solids dispersed in a gas (smoke) *or* liquids dispersed in a gas (hairspray, mist, fog).
  - → *Sols* or *gels* are solids dispersed in a liquid (printing ink, paint).
  - → *Emulsions* are liquids dispersed in a liquid (milk, mayonnaise).
  - → *Solid emulsions* are liquids dispersed in a solid (ice cream).
  - → *Foams* are gases dispersed in a liquid (soapsuds). *Solid foams* are gases dispersed in a solid (Styrofoam).
  - → *Solid dispersions* are solids dispersed in a solid (ruby glass, some alloys).
- **Aqueous colloids**
  - → *Hydrophilic colloids* contain molecules with *polar groups* that are strongly attracted to water. Examples include proteins that form gels and puddings.
  - → *Hydrophobic colloids* contain molecules with *nonpolar groups* that are only weakly attracted to water. Examples include fats that form emulsions (milk and mayonnaise).
  - → Rapid mixing of silver nitrate and sodium bromide may produce a *hydrophobic colloidal suspension* rather than a precipitate of silver bromide. The tiny silver bromide particles are kept from further aggregation by *Brownian motion*, the motion of small particles resulting from constant collisions with solvent molecules. The *sol* is further stabilized by *adsorption* of ions on the surfaces of the particles. The adsorbed ions are hydrated by surrounding water molecules and help prevent further aggregation.

## 15.14 Phosphors and Other Luminescent Materials

- **Light emission from materials**

  → *Incandescence* is light emission from a heated object, such as the filament in a lamp or the particles of hot soot in a candle flame. See the discussion of *black-body radiation* in Section 1.2.

  → *Luminescence* is light emission from materials caused by other processes, such as light absorption, chemical reaction, impact with electrons, radioactivity, or mechanical shock.

  → *Chemiluminescence* is light emission from materials caused by chemical reaction. The products of reaction are formed in energetically excited states that decay by light emission. The highly exothermic reaction of hydrogen atoms and fluorine molecules provides an example of *infrared chemiluminescence*. Hydrogen fluoride is produced with considerable *vibrational* energy, and it decays by *emission* in the *infrared* region of the spectrum (see Major Technique 1 for a discussion of *infrared absorption*). *Visible chemiluminescence* is usually produced by *electronic* excitation of product molecules.

  → *Bioluminescence* is a form of chemiluminescence produced by living organisms, such as fireflies and certain bacteria.

- **Light emission from molecules**

  → *Fluorescence* is the *prompt* emission of light from molecules excited by radiation of higher frequency. Normally, absorption of ultraviolet radiation by a molecule leads to emission in the visible region; higher frequency *absorption* leads to lower frequency *emission*.

    [**Note:** With high-power lasers, simultaneous absorption of two or more photons by a molecule may lead to emission of a single photon with a *greater* frequency than that of the individual photons absorbed.]

  → *Phosphorescence* is the *slow* or *delayed* emission of light from molecules excited by radiation of higher frequency. In this case, the initially excited molecule undergoes a transition to a state that decays more slowly.

  → *Triboluminescence* (from the Greek word, *tribos,* a rubbing) is luminescence that is produced by a mechanical shock to a crystal. It is readily observed in striking or grinding sugar crystals. Trapped nitrogen gas escapes while in an excited state and produces the radiation.

- **Phosphors**

  → *Phosphorescent materials* that glow when activated by the impact of fast electrons, as well as high frequency radiation such as ultraviolet or x-ray

  → Clusters of three *phosphors* are used for each dot in a color television or computer display screen. Commonly used phosphors for this purpose are europium-activated yttrium orthovanadate, $YVO_4$, for the red color, silver-activated zinc sulfide for blue, and copper-activated zinc sulfide for green.

  → *Fluorescent lamps* make use of fluorescent materials that are activated by ultraviolet light. Mercury atoms in the lamp are excited by electrons in a discharge. The atoms emit light at 254 and 185 nm, which is absorbed by a phosphor thinly coated on the surface of the lamp. A commonly used phosphor is *calcium halophosphate,* $Ca_5(PO_4)_3F_{1-x}Cl_x$, doped with manganese(II) and antimony(III) ions. An antimony(III)-activated phosphor emits blue light and a manganese(II)-activated phosphor emits yellow light. The net result is light with a spectral range that is approximately white.

  → *Fluorescent materials* have important applications in medical research. Dyes such as *fluorescein* are attached to protein molecules to probe biological reactions. *Fluorescent materials*, such as sodium iodide and zinc sulfide, can be activated by radioactivity and are used in scintillation counters to measure radiation (see Chapter 17).

# Chapter 16  The Elements: The *d* Block

## The *d*-Block Elements and Their Compounds (Sections 16.1–16.2)

### 16.1  Trends in Physical Properties

- **d-Block elements**
  - → All are metals, most are good electrical conductors, particularly Ag, Cu, and Au.
  - → Most are malleable, ductile, lustrous, and silver-white in color.
    Exceptions are Cu (red-brown) and Au (yellow).
  - → Most have higher melting and boiling points than main group elements.
    A major exception is Hg, which is a liquid at room temperature.

- **Shapes of the *d*-orbitals**
  - → Some properties of the *d*-block elements follow from the shapes of the *d*-orbitals.
  - → *d*-Orbital *lobes* are relatively far apart from each other, so electrons in different *d*-orbitals on the same atom repel each other weakly.
  - → Electron density in *d*-orbitals is low near the nucleus.  Because *d*-electrons are relatively far from the nucleus, they are not very effective in shielding other electrons from its positive charge.

- **Trends in atomic radii**

  *Across a Period*

  - → *Nuclear charge* and the *number* of *d*-electrons both increase across a row from left to right.
  - → The first five electrons added to a *d*-subshell are placed in different orbitals (Hund's rule).
  - → Since repulsion between *d*-electrons in different orbitals is small, increasing *effective nuclear charge* is the dominant factor, so atoms become smaller as Z increases.  Exception: Mn
  - → Further across the block, radii begin to *increase* slightly with Z because *electron-electron repulsion* outweighs the effect of increasing *effective nuclear charge*.
  - → Attractions and repulsions are finely balanced, and the range of atomic *d*-metal radii is small.
  - → *d*-Metals form many alloys, because atoms of one metal can easily replace  atoms of another metal.

  *Down a group – Periods 4 and 5*

  - → *d*-Metals in *Period 5* are typically larger than those in *Period 4*.
  - → The effective nuclear charge on the outer electrons is roughly the same, whereas the *n* quantum number increases down the group.  The usual pattern is an increase in size for group members from top to bottom.  This pattern is followed in *Periods 4* and *5* for the *d*-metals.  Exception: Mn

  *Across Period 6 – the lanthanide contraction.*

  - → *d*-Metal radii in *Period 6* are approximately the same as in *Period 4*.  They are smaller than expected because of the *lanthanide contraction*, the decrease in radius along the first row of the *f* block.
  - → There are 14 *f*-block elements before the first 6*d*-block element in *Period 6,* Lu.  At Lu, the atomic radius has fallen from 224 pm for Ba to 172 pm for Lu.
  - → *Period 6 d*-block elements are substantially more dense than those in *Period 5* owing to the *lanthanide* contraction.  The radii are about equal, but the masses are almost twice as large.  Ir and Os are the two most dense elements (approximately 22.6 g·cm$^{-3}$).
  - → A second effect of the *contraction* is the *low reactivity* of Pt and Au, whose valence electrons are so tightly bound that they are not readily available for chemical reaction.

## 16.2  Trends in Chemical Properties

- **Oxidation numbers (states)**

  → We will use *oxidation state* and *oxidation number* interchangeably.

  → Most *d*-block elements have more than one common oxidation number and one or more less common oxidation numbers.

  → Oxidation numbers range from negative values to +8.  Example: in $[Fe(CO)_4]^{2-}$, Fe has an oxidation number of –2, whereas in $OsO_4$, Os has an oxidation number of +8.

  → Oxidation numbers show greatest variability for elements in the middle of the *d* block.  Example: Mn, at the center of its row, has 7 oxidation numbers, whereas Sc, at the beginning of the same row, has only one.

  → *Period 5* and *Period 6* *d*-block elements tend to have higher oxidation numbers than those in *Period 4*.  Example: the highest oxidation number of Ni (Group 10, *Period 4*) is +4, whereas that of Pt (Group 10, *Period 6*) is +6.

- **Chemical properties and oxidation states**

  → Species containing *d*-block elements with *high oxidation numbers* tend to be good *oxidizing agents* ($MnO_4^-$ with Mn +7 and $CrO_4^{2-}$ with Cr +6 are examples); to exhibit *covalent bonding* ($Mn_2O_7$ with Mn +7 is a covalent liquid at room temperature); and to have *oxides* that are *acid anhydrides*.

  → Species containing *d*-block elements with *low oxidation numbers* tend to be good *reducing agents* (CrO with Cr +2 and $FeCl_2$ with Fe +2 are examples); to exhibit *ionic bonding* ($Mn_3O_4$, with Mn +2 and Mn +3 is an ionic solid); and to have *oxides* that are *basic anhydrides*.

# Selected Elements: A Survey  (Sections 16.3–16.4)

## 16.3  Scandium Through Nickel

- **Properties of the metals**

  → Densities increase from Sc (2.99 g·cm$^{-3}$) to Ni (8.91 g·cm$^{-3}$).

  → Melting points increase from Sc (1540 K) to V (1920 K), then decrease uniformly to Ni (1455 K), with the exception of the very low value of Mn (1250 K).

  → Boiling points increase from Sc (2800 K) to V (3400K), then show irregular behavior to Ni (2150 K).

  → Valence electron configurations follow the filling rule: $4s$ fills before $3d$ from Sc, $[Ar]\,3d^1\,4s^2$, to Ni, $[Ar]\,3d^8\,4s^2$, with the exception of Cr, $[Ar]\,3d^5\,4s^1$.

- **Scandium, Sc, $[Ar]\,3d^1\,4s^2$**

  → A highly reactive metal which reacts with water as vigorously as Ca

  → Has only a few commercial uses and is *not* essential to life

  → Has one oxidation state, +3, in compounds

  → Small, highly charged $Sc^{3+}$ is strongly hydrated in water, forming the $Sc(H_2O)_6^{3+}$ ion, which is a weak acid (about as strong as acetic acid).

- **Titanium, Ti, $[Ar]\,3d^2\,4s^2$**

  → A light, strong metal passivated by an oxide coating, which masks its inherent reactivity

  → Commonly found with the +4 oxidation number in its ores: *rutile,* $TiO_2$, and *ilmenite,* $FeTiO_2$

- → Requires strong reducing agents to extract from its ores
- → Metal used in jet engines, dental appliances; the oxide, $TiO_2$, used as a paint pigment
- → Forms a series of oxides called *titanates*
- → *Barium titanate*, $BaTiO_3$, is a *piezoelectric* material (develops an electrical signal when stressed).

- **Vanadium, V, [Ar] $3d^3 4s^2$**
  - → A soft, silver-gray metal
  - → Has many commercial uses and it is thought to be essential for life
  - → Used in *ferroalloys* (Fe, V, C) to make tough steels for automotive springs
- **Compounds of V**
  - → Exhibit positive oxidation numbers ranging from +1 to +5
  - → Common oxidation numbers are +4 and +5.
  - → Many colored compounds containing vanadium species are used as ceramic glazes.
  - → Orange-yellow *Vanadium(V) oxide*, $V_2O_5$, is used as an oxidizing agent and as an oxidizing catalyst in the contact process for the manufacture of sulfuric acid.

- **Chromium, Cr, [Ar] $3d^5 4s^1$**
  - → A bright, corrosion-resistant metal
  - → Used to make stainless steel and for chromium plating
  - → Obtained from *chromite*, $FeCr_2O_4$, by reduction with C in an electric furnace:
  $$FeCr_2O_4(s) + 4\,C(s) \rightarrow Fe(l) + 2\,Cr(l) + 4\,CO(g)$$
  - → An exception to the normal filling order of electrons ($3d^5 4s^1$, not $3d^4 4s^2$)
- **Compounds of Cr**
  - → Positive oxidation numbers range from +1 to +6.
  - → Common oxidation numbers are +3 and +6.
  - → *Chromium(IV) oxide*, $CrO_2$, is used as a *ferromagnetic* coating for "chrome" magnetic tapes.
  - → *Sodium chromate*, $Na_2CrO_4(s)$, is used to prepare many other Cr compounds. In acid solution, $CrO_4^{2-}$ forms the *dichromate ion*, $Cr_2O_7^{2-}$; in both ions, the oxidation number of Cr is +6.
  $$2\,CrO_4^{2-}(aq) + 2\,H_3O^+(aq) \rightarrow Cr_2O_7^{2-}(aq) + 3\,H_2O(l)$$
  - → Cr compounds are used as pigments, corrosion inhibitors, fungicides, and ceramic glazes.
  - → *Cr(III)* is essential to human health, possibly playing a role in glucose metabolism regulation.

- **Manganese, Mn, [Ar] $3d^5 4s^2$**
  - → A gray metal that corrodes easily and is rarely used alone
  - → Important in alloys such as steel and bronze
  - → Obtained from the ore, *pyrolusite*, mostly $MnO_2$
- **Compounds of Mn**
  - → Positive oxidation numbers range from +1 to +7.
  - → The most stable oxidation number is +2, but +4, +7, and, to some extent, +3 are also common.
  - → The most important compound of Mn, *manganese(IV) oxide* ($MnO_2$ or *manganese dioxide*), is a brown-black solid used as a decolorizer in glass, to prepare other Mn compounds, and in dry cells.
  - → The *permanganate ion*, $MnO_4^-$, with Mn in the +7 oxidation state, is an important oxidizing agent and a mild disinfectant.

- **Iron, Fe, [Ar] $3d^6 4s^2$**
  - $\rightarrow$ The most abundant element on earth and the most widely used $d$-metal
  - $\rightarrow$ Otained from the principal ores *hematite*, $Fe_2O_3$, and *magnetite*, $Fe_3O_4$
  - $\rightarrow$ Reactive metal which readily corrodes in moist air, forming rust
  - $\rightarrow$ Forms corrosion-resistant alloys and steels see text Table 16.2
  - $\rightarrow$ A *ferromagnetic* metal, as are its oxides and many of its alloys

- **Compounds of Fe**
  - $\rightarrow$ Positive oxidation numbers range from +2 to +6.
  - $\rightarrow$ Common oxidation numbers are +2 and +3.
  - $\rightarrow$ Salts of iron vary in color from pale yellow to dark green.
  - $\rightarrow$ Color in aqueous solution is dominated by $[FeOH(H_2O)_5]^{2+}$, the conjugate base of $Fe(H_2O)_6]^{3+}$.
  - $\rightarrow$ *Fe(II)* is readily oxidized to *Fe(III)*; the reaction is slow in acid solution and rapid in base.
  - $\rightarrow$ Iron is essential for human health. It is present in ionic form in hemoglobin and in iron-containing proteins. Iron deficiency, *anemia*, results in reduced transport of oxygen to the brain and muscles.

- **Cobalt, Co, [Ar] $3d^7 4s^2$**
  - $\rightarrow$ A silver-gray metal used mainly in Fe alloys
  - $\rightarrow$ Permanent magnets like those used in loudspeakers are made of *alnico steel,* an alloy of Fe, Ni, Co, and Al.
  - $\rightarrow$ Co steels are hard and are used for drill bits and surgical tools.

- **Compounds of Co**
  - $\rightarrow$ Positive oxidation numbers range from +1 to +4.
  - $\rightarrow$ Common oxidation numbers are +2 and +3.
  - $\rightarrow$ *Co(II) oxide*, CoO(s), is a deep blue salt used to color glass and ceramic glazes.
  - $\rightarrow$ Co is present in the coenzyme Vitamin $B_{12}$, an essential nutrient in human diets.

- **Nickel, Ni, [Ar] $3d^8 4s^2$**
  - $\rightarrow$ A hard, silver-white metal used mainly in stainless steel and for alloying with Cu to produce *cupronickels,* the alloys used for nickel coins
  - $\rightarrow$ Used as a catalyst for the *hydrogenation* of organic molecules
  - $\rightarrow$ Obtained as a pure metal by the *Mond process* (the first step requires heat):

    $$NiO(s,ore) + H_2(g) \rightarrow Ni(s, impure) + H_2O(g)$$

    $$Ni(s, impure) + 4CO(g) \rightarrow Ni(CO)_4(l, pure)$$

    $$Ni(CO)_4(l, pure) \rightarrow Ni(s, pure) + 4CO(g)$$

- **Compounds of Ni**
  - $\rightarrow$ Positive oxidation numbers range from +1 to +4.
  - $\rightarrow$ Common oxidation numbers are +2 and +3; +2 is the most stable.
  - $\rightarrow$ The green color of aqueous solutions of nickel salts arises from $[Ni(H_2O)_6]^{2+}$ ions.
  - $\rightarrow$ In nickel-cadmium (nicad) batteries, *Ni(III)* is reduced to *Ni(II)*.

- **Carbonyl compounds**
  - $\rightarrow$ Many transition metals form compounds with CO.
  - $\rightarrow$ In *carbonyl compounds,* both the metal and carbonyl (CO) groups are regarded as *nearly* neutral species, and the metal is assigned an oxidation state of 0.

→ The CO group is not very electronegative and the number of electrons surrounding the metal in the compound is *approximately* the same as in the free metal.

→ When Fe is heated in CO, it reacts to form trigonal-bipyramidal *iron pentacarbonyl*, $Fe(CO)_5$, a yellow molecular liquid which melts at $-20°C$, boils at $103°C$, and decomposes in visible light.

→ Other examples of transition metal carbonyls are *nickel tetracarbonyl*, $Ni(CO)_4$, a colorless, toxic, flammable liquid that boils at $43°C$ and *chromium hexacarbonyl*, $Cr(CO)_6$, a colorless crystal that sublimes readily. $Ni(CO)_4$ is *tetrahedral*; and $Cr(CO)_6$ is *octahedral*.

→ Some metal carbonyl compounds can be reduced to produce anions in which the metal has a *negative* oxidation number. Example: in the $[Fe(CO)_4]^{2-}$ anion, produced by reduction of $Fe(CO)_5$, Fe has an oxidation number of $-2$:

$$Fe(CO)_5 \xrightarrow[\text{tetrahydrofuran}]{Na} [Fe(CO)_4]^{2-} + CO$$

## 16.4 Groups 11 and 12

- **Properties of the metals (see Text Table 16.3)**
  → All the elements of Groups 11 and 12 are metals with completely filled *d*-subshells.

  → The Group 11 metals, *copper, silver, and gold,* are called the *coinage metals,* and have valence electron configurations of $(n-1)d^{10}ns^1$. The low reactivity of the coinage metals derives partly from the poor shielding abilities of the *d*-electrons, and hence the strong attraction of the nucleus on the outermost electrons. The effect is enhanced in Period 6 by lanthanide contraction, accounting for the inertness of gold.

  → The Group 12 metals, *zinc, cadmium, and mercury,* have valence electron configurations of $(n-1)d^{10}ns^2$. Zinc and, to a lesser extent, cadmium show some resemblance to beryllium or magnesium in their chemistry.

- **Copper, Cu, [Ar] $3d^{10}4s^1$**
  → Found as the metal and in sulfide ores such as *chalcopyrite,* $CuFeS_2$

  → Used as an electrical conductor in wires

  → Forms alloys, such as brass and bronze, which are used in plumbing and casting

  → Corrodes in moist air forming green *basic copper carbonate:*
  $$2\,Cu(s) + H_2O(l) + O_2(g) + CO_2(g) \rightarrow Cu_2(OH)_2CO_3(s)$$

  → Oxidized by *oxidizing* acids such as $HNO_3$

- **Compounds of Cu**
  → Common oxidation numbers are $+1$ and $+2$; $+2$ is more stable.

  → *Cu(I)* disproportionates in water to metallic copper and copper(II).

  → *Cu(II)* forms a pale-blue hydrated ion in water, $[Cu(H_2O)_6]^{2+}$

  → Cu is essential in animal metabolism. In humans, Cu-containing enzymes are required for healthy nerve and connective tissue. In species such as octopi and lobsters, Cu (not Fe) is used to transport oxygen in the blood.

- **Silver, Ag, [Kr] $4d^{10}5s^1$**
  → Rarely found as the free metal and obtained as a byproduct of the refining of Cu and Pb

  → Reacts with sulfur to produce a black tarnish on silver dishes and cutlery

  → Like Cu, Ag is oxidized by *oxidizing* acids.
  $$3\,Ag(s) + 4\,H_3O^+(aq) + NO_3^-(aq) \rightarrow 3\,Ag^+(aq) + NO(g) + 6\,H_2O(l)$$

- **Compounds of Ag**
  - → Oxidation numbers are +1, +2, and +3, but +1 is most stable.
  - → *Ag(I)* does *not* disproportionate in water.
  - → Except for $AgNO_3$ and $AgF$, Ag salts are insoluble in water.
  - → Silver halides are used in photographic film.

- **Gold, Au, [Xe] $4f^{14}5d^{10}6s^{1}$**
  - → So inert that it is usually found as the free metal
  - → Highly malleable, easily made into thin foil (gold leaf)
  - → *Cannot* be oxidized by nitric acid, but it dissolves in *aqua regia*, a mixture of sulfuric and hydrochloric acids:
    $$Au(s) + 6H^+(aq) + 3NO_3^-(aq) + 4Cl^-(aq) \rightarrow AuCl_4^-(aq) + 3NO_2(g) + 3H_2O(l)$$
  - → Is used in coins, jewelry, dental fillings, and for decorative ornamentation.

- **Compounds of Au**
  - → Oxidation numbers are +1, +2, and +3, but +1 and +3 are most common.
  - → Compounds are used to treat arthritis.
  - → Au has no known role in human health.

- **Zinc, Zn, [Ar] $3d^{10}4s^{2}$**
  - → A silvery, reactive metal, found mainly in the sulfide ore, *sphalerite,* ZnS
  - → Obtained from the ore by *froth flotation,* followed by *smelting* with coke
  - → Used primarily for *galvanizing* Fe
  - → Like Cu, Zn is protected by a hard film of *basic carbonate*, $Zn_2(OH)_2CO_3$.

- **Compounds of Zn**
  - → Common oxidation number is +2.
  - → An *amphoteric* metal, Zn dissolves in both acidic and basic solutions:
    Acid:   $Zn(s) + 2H_3O^+(aq) \rightarrow Zn^{2+}(aq) + H_2(g) + 2H_2O(l)$
    Base:   $Zn(s) + 2OH^-(aq) + 2H_2O(l) \rightarrow [Zn(OH)_4]^{2-} + H_2(g)$

    $Zn(OH)_4^{2-}$ is called the *zincate ion.* The reactions above suggest that galvanized containers should not be used to transport either acids or alkalis.
  - → Zn is an essential element for human health; it occurs in many enzymes.
  - → Zn is toxic only in very large amounts.

- **Cadmium, Cd, [Kr] $4d^{10}5s^{2}$**
  - → A silvery, reactive metal
  - → Like Zn, obtained from its ore by *froth flotation,* followed by *smelting* with coke

- **Compounds of Cd**
  - → Common oxidation number is +2.
  - → *Cadmiate ions* (analogous to *zincate ions*) are known, but Cd does not react with strong bases. Like Zn, Cd reacts with nonoxidizing acids.
  - → Unlike Zn, Cd salts are *deadly poisons*.
  - → Cd disrupts human metabolism by replacing other essential metals in the body, such as Zn and Ca, leading to soft bones, and to kidney and lung disorders.
  - → Cd and Zn are alike chemically in many ways, but both are quite different from Hg.

- **Mercury, Hg, [Xe] $4f^{14} 5d^{10} 6s^2$**
  - → A volatile, silvery metal, and a liquid at room temperature (Ga and Cs are liquid on *warm* days)
  - → Found mainly in the sulfide ore, *cinnebar,* HgS
  - → Obtained from the ore by *froth flotation,* followed by *roasting* in air; *vapor is poisonous.*
  - → Combines with many metals to form mercury alloys called *amalgams*
  - → The liquid temperature range (from $-39°C$ to $357°C$) is unusually wide and makes Hg well suited for use in thermometers, silent electrical switches, and high-vacuum pumps.
- **Compounds of Hg**
  - → Oxidation numbers are +1, +2, and +3 (most common are +1 and +2).
  - → The *Hg(I)* cation, $[Hg–Hg]^{2+}$, is diatomic, with a covalent bond. It is usually written as $Hg_2^{2+}$.
  - → Hg does not react with nonoxidizing acids or with bases, but does react with oxidizing acids in a reaction similar to those for Cu and Ag:

$$3\,Hg(l) + 8\,H^+(aq) + 2\,NO_3^-(aq) \rightarrow 3\,Hg^{2+}(aq) + 2\,NO(g) + 4\,H_2O(l)$$

  - → Hg compounds, particularly organic ones, are acutely poisonous. Frequent exposure to low levels of Hg(g) results in the accumulation of Hg in the body. Damaging effects after such chronic exposure include impaired neurological function and hearing loss.

---

# Coordination Compounds  (Sections 16.5–16.7)

## 16.5 Coordination Complexes

- **Understanding complexes (terminology)**
  - → *Coordination number* is the number of bonds formed by a central metal atom or ion.
  - → *Ligand* is a molecule or ion attached to a metal atom or ion by a coordinate-covalent bond.
  - → A ligand *coordinates* to the metal when it attaches.
  - → For ionic complexes, ligands *directly* attached to a central atom are enclosed in *brackets*.
  - → The *charge* of the ion is displayed in the upper right corner as usual.
  - → Ligands attached to a metal define the *coordination sphere* of the central atom or ion.
  - → Common coordination spheres include octahedral (almost all 6-coordinate complexes), tetrahedral (4-coordinate complexes) or square planar (also 4-coordinate).
  - → The *isocyano* ligand is attached to the metal by the electron pair on the nitrogen end.
- **Naming *d*-metal complexes and coordination compounds**
  - → Use the formulas and names for the common ligands given in Text Table 16.4.
  - → Follow the procedure outlined in Text Toolbox 16.1.
- **Ligand substitution reaction**
  - → A reaction in which one Lewis base takes the place of another
  - → Sometimes all the ligands are replaced, but often the substitution is incomplete.
    **Example:** $Ni(H_2O)_6^{2+}(aq) + 6\,NH_3(g) \rightarrow Ni(NH_3)_6^{2+}(aq) + 6\,H_2O(l)$

## 16.6 The Shapes of Complexes

- **Complexes** → Exhibit a variety of shapes (coordination geometries) depending on the *coordination number*

- **Coordination number 6**
  - → Complexes are usually *octahedral*, with ligands at the vertices of the octahedron and the metal at the center.
  - → Some examples are hexamminecobalt(III), $[Co(NH_3)_6]^{3+}$, on the right, and hexacyanoferrate(II), $[Fe(CN)_6]^{4-}$.

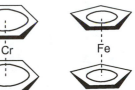

- **Coordination number 4**
  - → Two shapes are common, *tetrahedral* and *square planar*.
    **Examples:** $[Zn(OH)_4]^{2-}$ and $TiCl_4$ are *tetrahedral* complexes.
  - → *Square planar* complexes are most common for metal species with a $d^8$ electronic configuration, such as $Ni^{2+}$, $Pt^{2+}$, and $Au^{3+}$.
    **Examples:** $[PtCl_4]^{2-}$ and $[Ni(CN)_4]^{2-}$ are *square planar* complexes.

- **Coordination number 2** → Complexes are *linear* in shape.
    **Examples:** Dimethylmercury(0), $Hg(CH_3)_2$, and diamminesilver(I), $[Ag(NH_3)_2]^+$

- **Metallocenes** → "Sandwich" compounds with aromatic rings bonded by $2p\pi$-electrons to a metal atom or ion. The aromatic rings act as ligands.
    **Examples:** Dibenzenechromium, $Cr(C_6H_6)_2$, and ferrocene, $[Fe(C_5H_5)_2]$, both shown on the right.

- **Other coordination geometries**
  - → Complexes with coordination number 5 (trigonal-bipyramidal or square- pyramidal)
  - → Coordination number 8 (square-antiprism or dodecahedral) is also known (see Text).

- **Monodentate and polydentate ligands**
  - → *Monodentate* ligand can bind to a metal at only one site.
    **Examples:** Chloro, $Cl^-$; cyano, $CN^-$; carbonyl, CO; and methyl, $CH_3$.
  - → *Polydentate* ligand can bind to a metal at more than one site simultaneously.
  - → *Bidentate* ligand can bind to a metal at two sites.
    **Examples:** Oxalato (ox), $C_2O_4^{2-}$, and ethylenediamine (en), $NH_2CH_2CH_2NH_2$

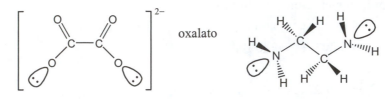

oxalato                    ethylenediamine

  - → *Tridentate* ligands may bind at three sites.
    **Example:** Diethylenetriamine (dien), $NH_2CH_2CH_2NHCH_2CH_2NH_2$

## 16.7 Isomers

- **Isomers**
  - → *Isomers* are compounds that contain the same numbers of the same atoms but in different arrangements.
  - → Two major classes of isomers are *structural isomers,* in which *some* atoms have different partners (different connectivity), and *stereoisomers,* in which *all* atoms have the same partners (same connectivity) but *some* atoms are arranged differently in space.
  - → *Structural isomers* are subdivided into *ionization isomers, hydrate isomers, linkage isomers,* and *coordination isomers.*
  - → *Stereoisomers* are subdivided into *geometrical isomers* and *optical isomers.*

- **Ionization isomers**
  - → Differ by the exchange of a ligand with an anion or molecule *outside* the coordination sphere
  - → Form *different* ions in solution
  - **Example:** $[CoBr(NH_3)_5]SO_4$ and $[CoSO_4(NH_3)_5]Br$ are ionization isomers.

- **Hydrate isomers**
  - → Special type of ionization isomer
  - → Differ by the exchange of a $H_2O$ molecule and another ligand in the coordination sphere
  - **Example:** The solid hexahydrate of chromium(III) chloride, $CrCl_3 \cdot 6H_2O$, is known in three forms: $[Cr(H_2O)_6]Cl_3$, $[CrCl(H_2O)_5]Cl_2 \cdot H_2O$, and $[CrCl_2(H_2O)_4]Cl \cdot 2H_2O$.

- **Linkage isomers**
  - → Differ in the identity of the atom in a given ligand attached to the metal
  - → A ligand may have more than one atom with a lone pair that can bond to the metal, but because of its size or shape, only one atom from the ligand may bond to the metal at one time.
  - **Examples:** $NO_2^-$ (M–$NO_2$ *nitro* and M–ONO *nitrito*); $CN^-$ (M–CN *cyano* and M–NC *isocyano*) The Lewis structure for $CN^-$ clearly shows the lone pairs on C and N that can bond to a metal ion. $[:C\equiv N:]^-$

- **Coordination isomers**
  - → Occur when one or more ligands are exchanged between a complex that is a cation and one that is an anion in a coordination compound
  - **Example:** $[Cu(NH_3)_4][PtCl_4]$ and $[Pt(NH_3)_4][CuCl_4]$

- **Geometrical isomers**
  - → Atoms bonded to the same neighbors, but with different orientations
  - → *Geometrical isomers* are also *stereoisomers.*
  - **Example:** Cis (adjacent) and trans (across) complexes of $[Pt(NH)_3Cl_2]$

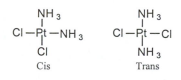

- **Optical isomers**
  - → Special type of *stereoisomers:* they are *nonsuperimposable* mirror images of each other.
  - **Example:** CHBrClF has a *tetrahedral shape* and *two* optical isomers:

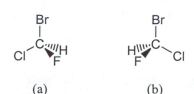

(a)        (b)

  - → Molecules (a) and (b) are not superimposable; therefore, they are different compounds. They are related to each other as the left hand is related to the right hand (*mirror images*).

- **Chirality and optical activity**
  - $\rightarrow$ A *chiral species* is *not* identical to its mirror image. All *optical isomers* are *chiral*.
  - $\rightarrow$ An *achiral species* is identical to its mirror image.
  - $\rightarrow$ A *chiral species* and its mirror image form a pair of *enantiomers*. Molecules (a) and (b) shown above are *enantiomers*.
  - $\rightarrow$ *Chiral molecules* display *optical activity*, the ability to rotate the plane of polarization of light. For a given pair, one enantiomer rotates the plane clockwise, whereas its mirror image rotates it by the same amount counterclockwise.
  - $\rightarrow$ A *racemic mixture* is a mixture of enantiomers in equal molar proportions.
  - $\rightarrow$ Because enantiomers rotate light equally in opposite directions, a racemic sample is *not* optically active.

- **Shapes of complexes** $\rightarrow$ Geometrical and optical isomers are related to the shape of the complex and the types of ligands that bind to the metal.

---

# The Electronic Structures of Complexes   (Sections 16.8–16.12)

## 16.8 Crystal Field Theory *(accounts for the optical and magnetic properties of complexes)*

- ***d*-Orbital splitting in an octahedral complex, $ML_6$**
  - $\rightarrow$ Ligands are considered to be point *negative* charges.
  - $\rightarrow$ Ligands approach the *positively charged* metal ion along the $x$-, $y$-, and $z$-axes. *d*-Orbital valence electrons repel the negative charges on the ligands as they approach.
  - $\rightarrow$ The *d*-orbitals are split into two sets: $t_{2g}$ (three orbitals) and $e_g$ (two orbitals).
  - $\rightarrow$ The $d_{xy}$-, $d_{yz}$-, and $d_{zx}$-orbitals form the $t_{2g}$-orbitals and are *lowered* in energy by $(2/5)\Delta_O$ with respect to the *d*-orbitals in the free metal ion.
  - $\rightarrow$ The $d_{x^2-y^2}$- and $d_{z^2}$-orbitals form the $e_g$-orbitals and are *raised* in energy by $(3/5)\Delta_O$.
  - $\rightarrow$ The separation in energy between the $t_{2g}$- and $e_g$-orbitals is the *octahedral ligand field splitting*, $\Delta_O$, where the O denotes octahedral.

- ***d*-Orbital splitting in a tetrahedral complex, $ML_4$**
  - $\rightarrow$ Ligands are again approximated by point *negative* charges.
  - $\rightarrow$ Ligands approach the *positively charged* metal ion along the corner directions of a tetrahedron.
  - $\rightarrow$ The *d*-orbitals are split into two sets: $t_2$ (three orbitals) and $e$ (two orbitals).
  - $\rightarrow$ The $d_{xy}$-, $d_{yz}$-, and $d_{zx}$-orbitals form the $t_2$-orbitals *raised* in energy by $(2/5)\Delta_T$ with respect to the *d*-orbitals in the free metal ion.
  - $\rightarrow$ The $d_{x^2-y^2}$- and $d_{z^2}$-orbitals form the $e$-orbitals *lowered* in energy by $(3/5)\Delta_T$.
  - $\rightarrow$ The separation in energy between the $t_2$- and $e$-orbitals is the *tetrahedral ligand field splitting*, $\Delta_T$, where the T denotes tetrahedral.

- **Comparison of octahedral and tetrahedral complexes**
  - $\rightarrow$ Tetrahedral ligand-field splitting values tends to be *smaller* than corresponding octahedral splitting values, in part because there are fewer repelling ligands in the tetrahedral case.

→ The relative energies of the $t_2$ and $e$ sets of orbitals are reversed in octahedral and tetrahedral complexes.

→ In tetrahedral complexes, the subscript g is *not* used to describe the orbitals.

## 16.9 The Spectrochemical Series

- **Relative strength of the ligand field splitting, $\Delta$**

  → For a given $d$-metal ion in a given oxidation state, $\Delta$ depends on the ligand.

  → Some ligands produce larger values of $\Delta$ than others.

  → The *relative* values of $\Delta$ are *approximately* in the same order for all $d$-metal ions.

- **Spectrochemical series**

  → Ligands are arranged according to the magnitude of the ligand field splitting they produce.

  → *Weak field ligands* produce small values of $\Delta$:  $I^- < Br^- < Cl^- \approx SCN^- < F^- < OH^- < (ox) < H_2O$

  → *Strong field ligands* produce large values of $\Delta$:  $NH_3 < (en) < NO_2^- < CN^- \approx CO$

- **Electron configurations in isolated atoms**

  → In an isolated atom, the five $d$-orbitals in a subshell have the same energy.

  → According to Hund's rule, electrons occupy each degenerate orbital separately with spins *parallel* until each has one electron.

- **Octahedral complexes**

  → Similar rules apply to $d$-metal complexes; however, the five $d$-orbitals in a subshell are divided into two groups.

  → The lower energy $t_{2g}$-orbitals fill first; degenerate orbitals ($d_{xy}$, $d_{yz}$, $d_{zx}$) fill with spins parallel until the $t_{2g}$-orbitals are half full ($d^3$ complex).

  → The electron configurations of $d^4$ through $d^7$ complexes depend on the magnitude of $\Delta_O$ compared with the energy required to pair an electron in a $t_{2g}$-orbital.

  → If $\Delta_O$ is *large* compared with the spin-pairing energy (*strong field ligands*), electrons will enter the lower-energy $t_{2g}$-orbitals with spins paired before filling the $e_g$-orbitals.

  → If $\Delta_O$ is *small* compared with the spin-pairing energy (*weak field ligands*), electrons will enter the higher-energy $e_g$-orbitals with spins parallel before pairing begins in the $t_{2g}$-orbitals.

  → For $d^8$ through $d^{10}$ complexes, the orbitals filled are the same for strong field and weak field ligands.  Both procedures predict the same final configuration.

  → A similar procedure is followed for *tetrahedral complexes,* but the two groups of $d$-orbitals are reversed in energy, and the $e$-orbitals fill first.

  → Electron configurations of $d^n$ complexes are given in Text Table 16.5.

- **High- and low-spin complexes**

  → A $d^n$ complex with the maximum number of unpaired electrons is called a *high-spin complex*.  High-spin complexes are expected for *weak-field ligands*.

  → A $d^n$ complex with the minimum number of unpaired electrons is called a *low-spin complex*.  Low-spin complexes are expected for *strong-field ligands*.

  → The $d^4$ through $d^7$ octahedral complexes may be high- or low-spin.

  → Because $\Delta_T$ is small with respect to the energy required to pair an electron in the $e$-orbitals, $t_2$-orbitals are always accessible and *tetrahedral complexes* are *almost always* high-spin.

## 16.10 The Colors of Complexes

- **Absorption of light and transmission or reflection**
  - → Visible light contains all wavelengths from about 400 (blue) to 800 nm (red).
  - → Transition metal complexes often *absorb* some visible light and *transmit* or *reflect* the rest.
  - → *Transmitted* or *reflected* light (the *complementary color* of the light absorbed) is the light we see and determines the color of an object.
  - → The color wheel in Major Technique 2 of the text provides a simple way of relating color *absorbed* to color *seen* in some systems.

- **Electronic transitions in complexes**
  - → Light absorption in complexes is often associated with *d–d transitions*. An electron is excited from a $t_{2g}$-to an $e_g$-orbital in an *octahedral complex* or from an $e$- to a $t_2$-orbital in a *tetrahedral complex*.
  - → *Charge-transfer transitions* may also occur in which an electron is excited from a ligand-centered orbital to a metal-centered orbital or *vice versa*. These transitions are often quite intense. An example is the deep purple color of the permanganate ion, $MnO_4^-$.
  - → Color arises when a substance absorbs light from a portion of the visible spectrum and transmits the rest. Both *d–d* and charge-transfer transitions can occur in the visible region and either or both processes can contribute to the color of a transition metal complex.

- **Color, *d–d* transitions, and the spectrochemical series**
  - → *Weak-field* complexes have *small* Δ values and absorb *low-energy* (*red* or *infrared*) radiation.
  - → Absent strong charge-transfer transitions, solutions of these complexes appear *green* (the complementary color of *red*) or *colorless* (if the absorption is in the *infrared*).
  - → *Strong-field* complexes have *large* Δ values and absorb *high-energy* (*blue, violet,* or *ultraviolet*) radiation.
  - → In the absence of complicating charge-transfer transitions, solutions of these complexes appear *orange* or *yellow* (the complementary colors of *blue* and *violet,* respectively) or *colorless* (if absorption is in the *ultraviolet*).
  - → The ligand field splitting Δ can be determined from the electronic spectrum of a given complex, but, in most cases, it is not related simply to the color of its solution.

## 16.11 Magnetic Properties of Complexes

- **Complexes may be diamagnetic or paramagnetic**
  - → *Diamagnetic complexes* have no unpaired electrons and are repelled by a magnetic field.
  - → *Paramagnetic complexes* have one or more unpaired electrons and are attracted into a magnetic field. The greater the number of unpaired electrons, the greater the attraction.

- **Magnetic properties of complexes**
  - → Determined from the electronic configuration of the complex
  - → Strong-field ligands produce weakly paramagnetic low-spin complexes.
  - → Weak-field ligands produce strongly paramagnetic high-spin complexes.

## 16.12 Ligand Field Theory

- **Ligand field theory**
  - → Molecular Orbitals (MOs) are generated from the available Atomic Orbitals (AOs) in a complex.

→ Only one Atomic Orbital (AO) is used for each monodentate ligand.

**Examples:** For a chloride ligand, a Cl $3p$-orbital, directed toward the metal, is used. For an ammonia ligand, the $sp^3$ lone-pair orbital is chosen.

→ The treatment of $\pi$-*bonding* in complexes provides further insight into the *spectrochemical series*.

→ Ligand field theory explains why uncharged species such as CO are strong-field ligands, whereas negatively charged species such as Cl⁻ are weak-field ligands.

→ Based on electrostatic considerations, we would expect the *opposite* to be true.

- **Octahedral complexes**

→ For first-row transition elements, the $4s$-, $4p$-, and $3d$-orbitals of the metal ion are chosen, because they have similar energies.

→ The nine metal and six ligand orbitals yield a total of 15 AOs, which overlap to form 15 MOs (see text Fig. 16.39).

→ The final result is that six MOs are bonding, six are antibonding, and three are nonbonding.

→ The 12 ligand electrons and the available $d$-valence electrons are placed in the MOs in accordance with the building-up principle to determine the ground-state electron configuration of the complex.

→ In transition metal ions, the valence $s$- and $p$-orbitals are typically vacant, whereas the $d$-orbitals are partially occupied.

→ The first 12 electrons form six metal-ligand sigma bonds. The $d$-electrons of the metal enter the $t_{2g}$- and $e_g$-orbitals in the same way they did in the crystal field theory. Ligand field theory, however, identifies the $t_{2g}$-orbitals as *nonbonding* and the $e_g$-orbitals as *antibonding*.

→ $\pi$-*Bonding* in octahedral complexes arises from overlap of the metal $t_{2g}$-orbitals with $p$- or $\pi$-ligand orbitals to form *bonding* and *antibonding* MOs, $\pi$ and $\pi^*$, respectively.

→ The details of this interaction are shown in Fig. 16.40 in the text.

- **Weak-field ligand (Fig. 16.40a in the text)**

→ The *antibonding* ligand $\pi^*$-orbital is too high in energy to take part in bonding.

→ The *bonding* ligand $\pi$-orbital combines with a metal $t_{2g}$-orbital and the combination *raises* the energy of the MO that the metal $d$-electron occupies.

→ The electron from the metal enters an *antibonding* MO. The energy of the $e_g$-orbitals is unchanged, but the splitting between the $t_{2g}$- and $e_g$-orbitals is reduced. Therefore, the ligand field splitting, $\Delta_O$, is *decreased* by $\pi$-bonding.

→ Some examples of weak-field ligands are Cl⁻ and Br⁻, which have a valence shell $p$-orbital with two electrons oriented perpendicular to the metal-ligand sigma bond.

- **Strong-field ligand (Fig. 16.40b in the text)**

→ The bonding $\pi$-orbital of the ligand is too low in energy to take part in the bonding. The *empty* antibonding $\pi^*$-orbital of the ligand combines with a $t_{2g}$-orbital of the metal, and the combination *lowers* the energy of the MO that the metal $d$-electron occupies. The electron from the metal enters a *bonding* MO.

→ The energy of the $e_g$-orbitals is unchanged, and the splitting between the $t_{2g}$- and $e_g$-orbitals is increased. The ligand field splitting, $\Delta_O$, is, therefore, *increased* by $\pi$-bonding.

→ Some examples of strong-field ligands are CO and CN⁻, which have an unoccupied valence shell $\pi^*$-orbital oriented perpendicular to the metal-ligand sigma bond.

# The Impact on Materials (Sections 16.13–16.15)

## 16.13 Steel

- **Iron**
    - → *Pig iron*, which contains 3-5% C, ~2% Si, and lesser amounts of other impurities, is produced by reduction of iron ores in a blast furnace. *Cast iron* is similar to pig iron, but with some impurities removed. Both forms are hard and brittle, whereas *pure iron* is malleable and more flexible.
    - → In the blast furnace, limestone is used to remove impurities, such as $SiO_2$, $Al_2O_3$, and $P_4O_{10}$.

- **Steels**
    - → Steels are homogeneous alloys (solid solutions) made from *pig iron*. First, *pig iron* is purified to *lower* the carbon content and to remove other impurities. Then, appropriate metals are *added* to form the desired steel.
    - → Steels typically contain 2% or less carbon; their hardness, tensile strength, and ductility depend on the carbon content. Higher carbon content produces harder steel. The strength of steel can be increased greatly by heat treatment.
    - → The corrosion resistance of steel is significantly improved by alloying with other metals. Highly corrosion-resistant *stainless steel* is an alloy containing about 15% Cr by mass.

## 16.14 Nonferrous Alloys

- **Alloys**
    - → Alloys are solid mixtures of metals, and are *homogeneous* or *heterogeneous* solutions.
    - → *Homogeneous alloys* include brass, bronze, and coins containing gold or silver.
    - → *Heterogeneous alloys* include tin-lead solder and mercury amalgams for dental fillings.

- **Substitutional alloys**
    - → Homogeneous alloys with metals of similar radius: brass, consisting of Cu and Zn
    - → Typically harder than the pure metals but with lower electrical and thermal conductivity

## 16.15 Magnetic Materials

- **Magnetism**
    - → Arises from unpaired electrons in substances (three general types)
    - → *Paramagnetic:* Unpaired electron spins oriented *randomly*
    - → *Ferromagnetic:* Unpaired electron spins all aligned *parallel* in *domains*
    - → *Antiferromagnetic:* Unpaired electron spins aligned *antiparallel,* as in manganese
    - → Many *d*-block elements have unpaired *d*-electrons and are *paramagnetic* or *ferromagnetic.*

- **Ferromagnets** → Permanent magnets used for coating cassette tapes and computer disks

- **Ceramic magnets**
    - → Made from barium ferrite, $BaO \cdot n Fe_2O_3$, or strontium ferrite, $SrO \cdot n Fe_2O_3$
    - → Used as refrigerator magnets; brittle, relatively lightweight, and inexpensive

- **Ferrofluids**
    - → Magnetic liquids; suspensions of powdered magnetite, $Fe_3O_4$, in a viscous oil/detergent mixture
    - → The $Fe_3O_4$ particles are attracted to the polar ends of the detergent molecules, which form micelles that are distributed throughout the oil. When a magnet approaches the fluid, the particles try to align, but are inhibited from doing so by the viscous oil. As a result, it is possible to control the flow and position of the *ferrofluid* by means of an applied magnetic field.

# Chapter 17  Nuclear Chemistry

## Nuclear Decay  (Sections 17.1–17.5)

### 17.1  The Evidence for Spontaneous Nuclear Decay

- **Radioactivity**
  - → Spontaneous emission of radiation or particles by a nucleus undergoing decay
  - → Radioactivity includes the emission of $\alpha$ particles, $\beta$ particles, and $\gamma$ radiation.
  - → These three types, the most common forms of radioactivity, were the first ones to be discovered.

- **Nuclear radiation (see Table 17.1 in the text):**

  **$\alpha$ particles**
  - → Helium-4 nuclei, $_2^4He^{2+}$, traveling at speeds equal to about 10% of the speed of light, $c$
  - → Deflected by electric and magnetic fields
  - → When $\alpha$ particles interact with matter, their speed is reduced and they are neutralized.
  - → Denoted by $_2^4\alpha$ or $\alpha$

  **$\beta$ particles**
  - → Rapidly moving *electrons,* $e^-$, emitted by nuclei at speeds less than 90% of the speed of light
  - → Deflected by electric and magnetic fields, sometimes called *negatrons*
  - → When traveling between positively and negatively charged plates, $\alpha$ and $\beta$ particles are deflected in opposite directions.
  - → Denoted by $_{-1}^0e$, $\beta^-$, or $\beta$

  **$\gamma$ radiation**
  - → High-energy *photons* (electromagnetic radiation) traveling at the speed of light
  - → Uncharged and not deflected by electric and magnetic fields
  - → Denoted by $\gamma$ or $_0^0\gamma$

  **$\beta^+$ particles**
  - → Called *positrons*
  - → Have the mass of an electron but carry a positive charge
  - → When an electron and a positron, its *antiparticle*, meet, they are annihilated and are completely transformed into energy, mostly $\gamma$ radiation.
  - → Denoted by $_{+1}^0e$ or $\beta^+$

  **p particles**
  - → *Protons* are positively charged subatomic particles found in nuclei.  When emitted in nuclear decay, protons typically travel at speeds equal to about 10% of the speed of light.
  - → Denoted by $_1^1H^+$, $_1^1p$, or p

  **n particles**
  - → *Neutrons* are uncharged subatomic particles found in nuclei.  The mass of a neutron is slightly larger than the mass of a proton (see Table B.1 in the text).  When emitted in nuclear decay, neutrons typically travel at speeds less than about 10% of the speed of light.
  - → Denoted by $_0^1n$ or n

- **Penetrating power of nuclear radiation**
  - → Penetrating power into matter decreases with increased charge and mass of the particle.
  - → Uncharged γ radiation and neutrons are more penetrating than charged positrons or alpha particles.
  - → Singly charged β particles (electrons) are more penetrating than the more massive, doubly charged α particles, which are virtually nonpenetrating.
- **Antiparticles**
  - → Subatomic particles with equal mass and opposite charge
  - → Annihilate each other upon encounter, producing energy

## 17.2 Nuclear Reactions

- **Nucleons and nuclides**
  - → A *nucleon* is a proton or a neutron in the nucleus of an atom.
  - → A *nuclide* is an atom characterized by its atomic number, mass number, and nuclear energy state.
  - → Nuclides are often denoted by $_Z^A E$, where $Z$ is the atomic number, $A$ is the mass number (protons plus neutrons), and E is the symbol for the element. Note that $A - Z$ equals the number of neutrons in the nuclide.
  - → *A nuclide is a neutral atom and therefore is uncharged.*
  - → *Radioactive nuclei* can change their structure spontaneously by *nuclear decay,* the partial breakup of a nucleus.
- **Nuclear reactions**
  - → Any transformations that a nucleus undergoes
  - → Differ from chemical reactions in three important ways:
    1) Isotopes of a given element often have different nuclear properties but always have similar chemical properties.
       Example: $^{12}C$ and $^{14}C$ are difficult to separate chemically because they have similar chemical properties, yet the two isotopes have different nuclear stability. Carbon-12 is a stable nuclide, whereas carbon-14 decays with a half-life of 5730 y (see Section 17.7).
    2) Nuclear reactions often produce a different element.
       Example: Titanium-44 captures one of its electrons to produce the nuclide scandium-44.
    3) Nuclear reactions involve enormous energies relative to chemical reactions.
- **Balancing nuclear reactions**
  - → Like chemical reactions, *nuclear reactions* must be balanced with respect to both *charge* and *mass*.
  - → **Charge balance:** For the several nuclides $_Z^A E$ in a nuclear reaction, the sum of the atomic numbers $Z$ of the reactants must equal the sum of the $Z$ values of the products.
  - → **Mass balance:** The sum of the mass numbers $A$ of the reactants must equal the sum of the mass numbers $A$ of the products.
  - → The result of *most* nuclear decay reactions is *nuclear transmutation,* the conversion of one element into another.
  - → For nuclear decay reactions
    - ➤ A reactant nuclide is called a *parent nuclide.*
    - ➤ A product nuclide is called a *daughter nuclide.*

## 17.3 The Pattern of Nuclear Stability

- **Characterization of nuclei:**

  **Even-even**

  → Nuclei with an even number of protons and of neutrons, consequently $A$ and $Z$ are even numbers

  → 157 even-even nuclides are stable.

  **Examples:** helium-4, $^4_2\text{He}$; carbon-12, $^{12}_6\text{C}$; oxygen-16, $^{16}_8\text{O}$; and neon-22, $^{12}_{10}\text{Ne}$

  **Even-odd**

  → Nuclei with an even number of protons and an odd number of neutrons, consequently $Z$ is even and $A$ is odd

  → 53 even-odd nuclides are stable.

  **Examples:** helium-3, $^3_2\text{He}$; berylium-9, $^9_4\text{Be}$; carbon-13, $^{13}_6\text{C}$; and neon-21, $^{21}_{10}\text{Ne}$

  **Odd-even**

  → Nuclei with an odd number of protons and an even number of neutrons, consequently both $Z$ and $A$ are odd numbers

  → 50 odd-even nuclides are stable.

  **Examples:** hydrogen-1, $^1_1\text{H}$; lithium-7, $^7_3\text{Li}$; boron-11, $^{11}_5\text{B}$; and nitrogen-15, $^{15}_7\text{N}$

  **Odd-odd**

  → Nuclei with an odd number of protons and of neutrons, consequently $Z$ is odd and $A$ is even

  → *Only 4 odd-odd nuclides are stable.*

  → The stable odd-odd nuclides are hydrogen-2 (deuterium), $^2_1\text{H}$; lithium-6, $^6_3\text{Li}$; boron-10, $^{10}_5\text{B}$; and nitrogen-14, $^{14}_7\text{N}$.

  **Summary**

  → Even-even nuclides are the most stable and most abundant nuclides.

  → Odd-odd nuclides are the least stable and least abundant nuclides.

- **Strong force**

  → An attractive force that holds nucleons together in the nucleus

  → Overcomes the coulomb repulsion of protons

  → Acts only over a very short distance, approximately the diameter of the nucleus

- **Magic numbers**

  → Are 2, 8, 20, 50, 82, 114, 126, and 184 for *either* protons or neutrons

  → Nuclei with *magic numbers* of protons and/or neutrons are more likely to be stable.

  → Analogous to the pattern of electronic stability in atoms associated with electrons in filled subshells

  → *Doubly magic* nuclides are very stable.

  **Examples:** helium-4, $^4_2\text{He}$; oxygen-16, $^{16}_8\text{O}$; calcium-40, $^{40}_{20}\text{Ca}$; and lead-208, $^{208}_{82}\text{Pb}$

- **Band of stability and sea of instability**

  → Stable nuclides are found in a narrow band, the *band of stability*, that ends at $Z = 83$ (bismuth).

  → All nuclides with $Z > 83$ are unstable.

  → Unstable nuclides are found in the *sea of instability,* a region above and below the *band of stability*.

→ Nuclides *above* the band of stability are *neutron rich.*

→ Neutron rich nuclei are likely to emit β particles (neutron → proton + β).

→ Nuclides *below* the band of stability are *proton rich.*

→ Proton rich nuclei are likely to emit $\beta^+$ particles (proton → neutron + $\beta^+$) or to capture electrons (proton + β → neutron).

→ Heavier nuclides below the band of stability and those with $Z > 83$ may also decay by emitting α particles.

- **Models of nuclear structure**
  → Three common models of nuclear structure
  → Description and order of sophistication
    1) The *liquid drop model,* in which nucleons are considered to be packed together in the nucleus like molecules in a liquid.
    2) The *independent particle model,* in which nucleons are described by quantum numbers and assigned to shells. Nucleons in the outermost shells are most easily lost as a result of radioactive decay.
    3) The *collective model,* in which nucleons are considered to occupy quantized energy levels and to interact with each other by the strong force and the electrostatic (coulomb) force.

## 17.4 Predicting the Type of Nuclear Decay

- **Massive nuclei with $Z > 83$**
  → Lose protons to reduce their atomic number and generally lose neutrons as well
  → Stepwise decay gives rise to a *radioactive series,* a characteristic sequence of nuclides

  **Examples of radioactive series:**
  1) The uranium-238 series (see Fig. 17.16 in the text) is one in which alpha and beta particles are successively ejected. The final step is the formation of a stable isotope of lead (magic number 82), lead-206.
  2) The uranium-235 series proceeds similarly and ends at lead-207. This series is sometimes called the actinium series.
  3) The thorium-232 series ends at lead-208.

- **Neutron rich nuclides**
  → Nuclides that have high neutron to proton (n/p) ratios with respect to the band of stability
  → Tend to decay by reducing the number of neutrons
  → Commonly undergo β decay, which decreases the n/p ratio

- **Proton rich nuclides**
  → Nuclides that have high proton to neutron (p/n) ratios with respect to the band of stability
  → Tend to decay by reducing the number of protons
  → Commonly undergo electron capture, positron emission, or proton emission (least likely), which decreases the p/n ratio

## 17.5 Nucleosynthesis

- **Formation of the elements**
  → Elements are formed by *nucleosynthesis.*

- → Nucleosynthesis occurs when particles collide vigorously, as in stars.
- → *Transmutation* is the conversion of one element into another.
- → The first artificial transmutation was observed in 1919 by Rutherford who converted $^{14}_{7}N$ to $^{17}_{8}O$ by bombarding the parent nuclide with high-speed ↦ particles:

$$^{14}_{7}N + ^{4}_{2}\alpha \rightarrow ^{17}_{8}O + ^{1}_{1}p$$

The reactants must collide with enough energy to overcome the repulsive coulomb force between positively charged nuclei.

- → Nucleosynthesis reactions are commonly written in shorthand as

target (incoming species, ejected species) product

In the Rutherford example above, the shorthand notation is $^{14}_{7}N\,(\alpha,p)\,^{17}_{8}O$

- → Nucleosynthesis can be effected using a variety of projectile species including γ radiation, protons, neutrons, deuterium nuclides, α particles, and heavier nuclides. Because they are uncharged, neutrons can travel relatively slowly and still react with nuclides.

- **Transuranium elements**
  - → Elements following uranium ($Z = 92$) in the periodic table
  - → Elements from rutherfordium (Rf, $Z = 104$) to meitnerium (Mt, $Z = 109$) were given official names in 1997.

- **Transmeitnerium elements**
  - → Elements following meitnerium; six have been reported as of this writing.
  - → Temporary names are based on short hand numbers derived from Latin (see Table 17.2 in the text).

---

# Nuclear Radiation  (Sections 17.6–17.8)

## 17.6  The Biological Effects of Radiation

- **Penetrating power**
  - → Alpha, beta, and gamma particles are forms of *ionizing radiation* with different *penetration power*.
  - → Different amount of shielding is required to prevent harmful effects (see Table 17.3 in the text).
  - → Alpha particles have the least penetrating power and are absorbed by paper and the outer layers of skin. They are extremely dangerous if inhaled or ingested.
  - → Beta particles are 100 times more penetrating than alpha particles and are absorbed by 1 cm of flesh or 3 mm of aluminum, for example.
  - → Gamma radiation is 100 times more penetrating than beta particles. It can pass through buildings and bodies, leaving a trail of ionized or damaged molecules, and is absorbed by lead bricks or thick concrete.
  - → The order of *increasing* penetrating power is $\alpha < \beta < \gamma$.

- **Radiation damage to tissue**
  - → Depends on the *type* and *strength* of the radiation, the *length* of exposure, and the *extent* to which the radiation can reach sensitive tissue
  - → For example, $Pu^{4+}$ is an $\alpha$ emitter and, if ingested, replaces $Fe^{3+}$ in the body resulting in inhibition of the production of red blood cells.

- **Absorbed dose of radiation**
  - → Energy deposited in a sample, such as a human body, when exposed to radiation
  - → Units of the absorbed dose of radiation are
    - 1 rad = amount of radiation that deposits 0.01 J of energy per kg of tissue
    - 1 gray (Gy) = an energy deposit of 1 $J \cdot kg^{-1}$   (SI unit)
    - 1 Gy = 100 rad

    **Note:** The name *rad* stands for "radiation absorbed dose."

- **Dose equivalent**
  - → Dose modified to account for different destructive powers of radiation
  - → Relative destructive power is called the *relative biological effectiveness*, $Q$.
    For $\beta^-$ and $\gamma$ radiation, $Q \equiv 1$.  For $\alpha$ radiation, $Q \approx 20$.
  - → The natural unit of dose equivalent is the rem (*roentgen equivalent man*).
  - → Definition:   Dose equivalent in rem = $Q \times$ absorbed dose in rad
  - → SI Units:   Dose equivalent in sievert (Sv) = $Q \times$ absorbed dose in gray (Gy)
    1 Sv = 100 rem

- **Common radiation dose and harmful effects**
  - → Typical annual human dose equivalent from natural sources is 0.2 $rem \cdot y^{-1}$, a number with a wide range depending upon habitat and lifestyle.
  - → A typical chest X ray gives a dose equivalent of about 7 millirem.
  - → 30 rad (30 rem) of $\gamma$ radiation may cause a reduction in white blood cell count.
  - → 30 rad (600 rem) of $\alpha$ radiation causes death.

## 17.7  Measuring the Rate of Nuclear Decay

- **Activity of a sample**
  - → Number of nuclear disintegrations per second
  - → One nuclear disintegration per second is called 1 becquerel, Bq (SI unit).
  - → The curie, Ci, an older unit for activity, is equal to $3.7 \times 10^{10}$ Bq, which is equal to the radioactive output of 1 g of radium-226.
  - → Because the curie represents a large value of decay, the activity of typical samples is given in millicuries (mCi) or microcuries ($\mu$Ci).

    **Note:**   See Table 17.4 in the text for a list of the radiation units.

- **Nuclear decay**
  - → The decay of a nucleus can be written as
    - Parent nucleus → daughter nucleus + radiation
  - → This form is the same as a *unimolecular elementary reaction,* with an unstable nucleus taking the place of an excited molecule.

→ The rate of nuclear decay depends only on the identity of the nucleus (isotope), not on its chemical form or temperature.

- **Law of radioactive decay**
  - → The rate of nuclear decay is proportional to the number of radioactive nuclei $N$:

  $$\boxed{\text{Activity} = \text{rate of decay} = k \times N}$$

  - → $k$ is the *decay constant* (or rate constant for the reaction)
  - → The units of the *decay constant* are (time)$^{-1}$, the same as for any *first order* rate constant.

- **Integrated rate law for nuclear decay and half-life**

  - → Exponential form: $\boxed{N = N_0 e^{-kt}}$ The number of radioactive nuclei N decays exponentially with time. Large values of $k$ correspond to more rapid decay.

  - → Logarithmic form: $\boxed{\ln\left(\dfrac{N}{N_0}\right) = -kt}$

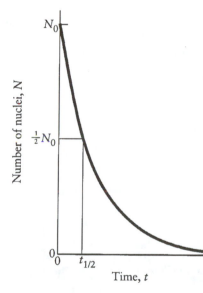

On the left is a plot of $N$ vs. $t$ for nuclear decay. $N_0$ is the number of radioactive nuclei present at $t = 0$. $N$ is the number of radioactive nuclei present at a later time $t$. The time required for the nuclei to decay to half their initial number is equal to the half-life, $t_{1/2}$. At this time,

$$N = \frac{1}{2}N_0$$

Substituting this value into the integrated rate law yields

$$\boxed{t_{1/2} = \frac{\ln 2}{k}}$$

Half-lives of radioactive nuclides span an enormous range, from picoseconds ($^{215}$Fr, 120 ps) to billions of years ($^{238}$U, $4.5 \times 10^9$ y). See Table 17.5 in the text for a list of half-lives of common radioactive isotopes.

- **Isotopic dating**
  - → Used to determine the ages of rocks and of archeological artifacts
  - → Carried out by measuring the activity of a radioactive isotope in the sample
  - → Useful isotopes for dating are uranium-238, potassium-40, tritium, $^3_1$H, and carbon-14.

- **Radiocarbon dating**
  - → Most important example of isotopic dating
  - → Uses the beta decay of carbon-14, for which the half-life is 5730 y

- **Radioactive $^{14}$C**
  - → Formed in the atmosphere by neutron bombardment of $^{14}$N

  $$^{14}_{7}\text{N} + ^1_0\text{n} \rightarrow ^{14}_{6}\text{C} + ^1_1\text{p}$$

→ Enters living organisms as $^{14}CO_2$ through photosynthesis and digestion

→ Leaves living organisms by excretion and respiration

→ Achieves a steady-state concentration in living organism with an activity of 15 disintegrations per minute per gram of *total* carbon

→ Shows decreasing activity in dead organisms because they no longer ingest $^{14}C$ from the atmosphere

→ The time of death of an organism can be estimated by measuring the activity of the sample and using the integrated rate law.

### 17.8  Uses of Radioisotopes

- **Radioisotopes**
  → Used to cure disease, preserve food, and trace the mechanisms of chemical reactions

  → Used to power spacecraft, locate sources of water, and determine the age of nonliving materials including wood ($^{14}C$ dating), rocks ($^{238}U/^{206}Pb$ ratio), and ground water (tritium dating using the $^{1}H/^{3}H$ ratio)

- **Radioactive tracers**
  → Radioactive isotopes that are used to track changes and locations

  **Example:** Phosphorus-32, a $\beta$ emitter with $t_{1/2}$ = 14.28 d, can be incorporated into phosphate-containing fertilizer to follow the mechanism of plant growth.

  → In chemical reactions, tracers can help determine the mechanism of the reaction. Nonradioactive isotopes are also used for this purpose.

---

# Nuclear Energy  (Sections 17.9–17.12)

### 17.9  Mass-Energy Conversion

- **Nuclear binding energy**
  → Energy *released* when protons and neutrons join together to form a *nucleus*
  $$(Z)^{1}_{1}p + (A - Z)^{1}_{0}n \rightarrow ^{A}_{Z}E^{Z+}$$

  → Since *nuclides* are *neutral* atoms, the energy released is also given *approximately* by
  $$(Z)^{1}_{1}H + (A - Z)^{1}_{0}n \rightarrow ^{A}_{Z}E$$

  → The binding energy is then given by applying the Einstein mass-energy equation to each reactant and product species:
  $$E_{bind} = |\Delta E| = |\Sigma E \,(\text{products}) - \Sigma E \,(\text{reactants})|$$
  $$= | E \,(\text{product}) - \Sigma E \,(\text{reactants})| = | mc^2 \,(\text{product}) - \Sigma mc^2 \,(\text{reactants})|$$
  $$= | m \,(\text{product}) - \Sigma m \,(\text{reactants})| \times c^2$$

  $$\boxed{E_{bind} = |\Delta m| \times c^2}$$

  → *Binding energy* is a measure of the stability of the nucleus.

→ The *binding energy per nucleon* is defined as $E_{bind}/A$, and is a better measure of the stability of the nucleus.

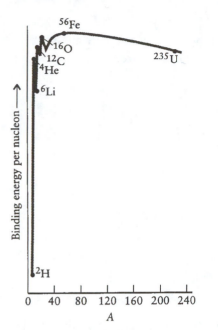

A plot of the binding energy per nucleon *vs.* atomic mass number is shown on the left (see Fig. 17.20 in the text). Nucleons are bound together most strongly in the elements near iron and nickel, which accounts for the high abundance of iron and nickel in meteorites and on planets similar to the earth. From the figure, it appears that light nuclei become more stable when they "fuse" together to form heavier ones. Heavy nuclei become more stable when they undergo "fission" and split into lighter ones. Also, note the increased stability of the nuclei with magic numbers, particularly oxygen-16, which is *doubly magic*.

- **Units used in nuclear calculations**

    → *Mass* is usually reported in *atomic mass units,* u.

    → One atomic mass unit is defined as *exactly* $\frac{1}{12}$ the mass of one atom (nuclide) of $^{12}C$:

    $$1\ u = 1.660\ 54 \times 10^{-27}\ kg$$

    → *Binding energy* is usually reported in electronvolts, eV, or millions (mega) of electronvolts, MeV.

    → An electronvolt is the change in potential energy of an electron (charge of $1.602\ 18 \times 10^{-19}$ C) when it is moved through a potential difference of 1 V:

    $$1\ eV = 1.602\ 18 \times 10^{-19}\ J$$

## 17.10 Nuclear Fission

- **Nuclear fission**

    → Occurs when a nucleus breaks into two or more smaller nuclei

    → Releases a large amount of energy

    → May be *spontaneous* or *induced*

- **Spontaneous nuclear fission**

    → Occurs when oscillations in heavy nuclei lead them to break into two smaller nuclei of similar mass

    → Yields a variety of products for a given nuclide (see Fig. 17.22 in the text)

    **Example:** Consider the spontaneous fission of americium-244. Two of the daughter nuclides formed are iodine-134 and molybdenum-107. The reaction is

    $$^{244}_{95}Am \rightarrow\ ^{134}_{53}I\ +\ ^{107}_{42}Mo\ +\ 3\ ^{1}_{0}n$$

- **Induced nuclear fission**
  - $\rightarrow$ Caused by bombarding a heavy nucleus with neutrons
  - $\rightarrow$ Yields a variety of daughter nuclides for a given parent

    **Example:** Consider the induced fission of plutonium-239 by neutron bombardment. A number of products are formed. Two important reactions are

    $$^{239}_{94}\text{Pu} + {}^{1}_{0}\text{n} \rightarrow {}^{98}_{42}\text{Mo} + {}^{138}_{52}\text{Te} + 4\,{}^{1}_{0}\text{n}$$

    $$^{239}_{94}\text{Pu} + {}^{1}_{0}\text{n} \rightarrow {}^{100}_{43}\text{Tc} + {}^{135}_{51}\text{Sb} + 5\,{}^{1}_{0}\text{n}$$

- **Energy changes in fission reactions**
  - $\rightarrow$ Energy released during fission is calculated by using Einstein's equation.
  - $\rightarrow$ For a balanced nuclear reaction,

    $$\boxed{\Delta m = \sum m\,(\text{products}) - \sum m\,(\text{reactants})} \quad \text{and} \quad \boxed{\Delta E = \Delta m \times c^2}$$

- **Fissionable and fissile nuclei**
  - $\rightarrow$ *Fissionable nuclei* are nuclei that can undergo induced fission.
  - $\rightarrow$ *Fissile nuclei* are fissionable nuclei that can undergo induced fission *with slow-moving neutrons*.

    Examples of fissile nuclei: $^{235}_{92}\text{U}$, $^{233}_{92}\text{U}$, and $^{239}_{94}\text{Pu}$, which are nuclear power plant fuels.

    The nuclide, $^{238}_{92}\text{U}$, is fissionable by fast-moving neutrons but is *not* fissile.

- **Nuclear branched chain reactions**
  - $\rightarrow$ Neutrons are *chain carriers* in a branched chain reaction (see Section 13.12).
  - $\rightarrow$ Chain branching occurs when an induced fission reaction produces two or more neutrons.

    Example: $^{235}_{92}\text{U} + {}^{1}_{0}\text{n} \rightarrow {}^{97}_{40}\text{Zr} + {}^{137}_{52}\text{Te} + 2\,{}^{1}_{0}\text{n}$, in which two neutrons are produced for each uranium-235 nuclide that reacts. The two neutrons can either escape into the surroundings or be captured by (react with) other uranium-235 nuclides.
  - $\rightarrow$ A *critical mass* is the mass of fissionable material above which so few neutrons escape from the sample that the fission chain reaction is sustained.
  - $\rightarrow$ The critical mass for pure plutonium of normal density is about 15 kg.
  - $\rightarrow$ A sample of plutonium with mass greater than 15 kg is *supercritical;* the reaction is self-sustaining and may result in an explosion.
  - $\rightarrow$ A *subcritical* sample is one that has less than the critical mass for its density.

- **Controlled and uncontrolled (explosive) nuclear reactions**
  - $\rightarrow$ Explosive nuclear reactions occur when a *subcritical* amount of fissile material is made *supercritical* so rapidly that the chain reaction occurs uniformly throughout the material.
  - $\rightarrow$ Controlled nuclear reactions occur in nuclear reactors.
  - $\rightarrow$ Controlled reactions are not explosive because there is a subcritical amount of fuel.
  - $\rightarrow$ The rate of reaction is controlled by *moderators,* such as graphite rods.
  - $\rightarrow$ Moderators slow the emitted neutrons so that a greater fraction can induce fission.

→ The difference between a fission explosion and a controlled fission is shown in the figure below. Each symbol ⊗ represents a fission event. The neutrons from the fission are shown but products are not.

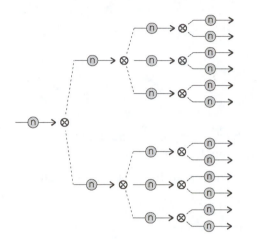

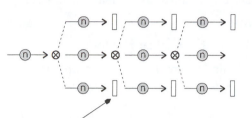

Control rod and reactor geometry prevent some neutrons from sustaining fission.

Uncontrolled fission (explosion)                     Controlled fission

**Note:** The number of neutrons released in an event is variable as is the identity of the products (see Fig. 17.24 in the text).

- **Breeder reactors**

  → Nuclear reactors used to synthesize fissile nuclides for fuel and weapon use.

  → Breeder reactors run very hot and fast because no moderator is present. As a result, they are more dangerous than nuclear reactors used for power generation.

## 17.11 Nuclear Fusion

- **Nuclear fusion**

  → Occurs when lighter nuclei fuse together to form a heavier nucleus

  → Is accompanied by a large release of energy because heavier nuclei have larger binding energies per nucleon

  → Occurs in stars and in the hydrogen bomb, in which the fusion reaction is uncontrolled

  → Is difficult to achieve because charged nuclei must collide with tremendous kinetic energies to fuse

  → Fusion is essentially free of radioactive waste.

  → Fusion has not yet been sustained for an appreciable time in a controlled reaction on earth.

  → A safe, sustained fusion reaction could provide an almost unlimited source of energy.

- **Energy changes in fusion reactions**

  → Energy released during a fusion reaction is calculated by using Einstein's equation.

  → The procedure is the same as for fission reactions.

## 17.12 The Chemistry of Nuclear Power

- **Chemistry**
  - → The key to the safe use of nuclear power
  - → Used to prepare nuclear fuel
  - → Used to recover important fission products
  - → Used to dispose of nuclear waste safely

- **Uranium**
  - → The fuel of nuclear reactors
  - → Obtained primarily from the ore pitchblende, $UO_2$, by reducing the oxide to the metal and then *enriching* or increasing the fraction of the fissile nuclide, uranium-235
  - → To be useful as a fuel, the percentage of $^{235}U$ must be increased from its natural abundance of 0.7% to about 3%.

- **Enrichment**
  - → Exploits the mass difference between $^{235}U$ and $^{238}U$
  - → Separation is accomplished by repeated effusion of $^{235}UF_6$ and $^{238}UF_6$ vapor (see Graham's law of effusion in Section 4.12).

$$\frac{\text{Rate of effusion of } ^{235}UF_6}{\text{Rate of effusion of } ^{238}UF_6} = \sqrt{\frac{M_{238}}{M_{235}}} = \sqrt{\frac{352.1}{349.0}} = 1.004$$

  **Note:** Because the ratio is close to one, repeated effusion steps (hundreds) are required to get the desired separation.

- **Nuclear waste**
  - → Spent nuclear fuel called *nuclear waste* is still radioactive.
  - → Waste is a mixture of uranium and fission products.
  - → Nuclear waste must be stored safely for about 10 half-lives, and is generally buried underground.
  - → For underground burial, incorporation of the highly radioactive fission (HRF) products into a glass or ceramic material is better than placing it in metal storage drums.
  - → Storage drums can corrode, allowing the waste to seep into aquifers and/or contaminate large areas of soil.

# Chapter 18  Organic Chemistry I:  The Hydrocarbons

## Aliphatic Hydrocarbons  (Sections 18.1–18.6)

### 18.1  Types of Aliphatic Hydrocarbons

- **Types of hydrocarbons (compounds containing only C and H)**
  - → *Aromatic* hydrocarbon: hydrocarbon with one or more benzene rings.
    **Examples:** Benzene, $C_6H_6$; toluene, $C_6H_5CH_3$; and naphthalene, $C_{10}H_8$; all shown on the right
    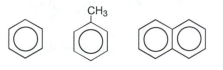
  - → *Aliphatic* hydrocarbon: hydrocarbon without a benzene ring
  - → *Saturated hydrocarbon:* an aliphatic hydrocarbon with no carbon-carbon multiple bonds
    **Example:** Butane, $CH_3CH_2CH_2CH_3$
  - → *Unsaturated hydrocarbon:* an aliphatic hydrocarbon with at least one carbon-carbon multiple bond.
    **Examples:** Butene, $CH_2=CHCH_2CH_3$, and butyne, $CH\equiv CCH_2CH_3$

- **Aliphatic hydrocarbons: structural formula** → Shows chemical connectivity, how many atoms are attached to one another

    **Example:**  Butane and methylpropane have the same *empirical* ($C_2H_5$) and *molecular* ($C_4H_{10}$) formulas, respectively, but different *structural formulas*.

    In the structure of methylpropane, an abbreviated notation for the methyl group ($CH_3-$) attached to the propane chain is used (the three C–H bonds are *not* shown).  A structure with all three terminal methyl groups in abbreviated form is shown on the right.

- **Condensed structural formula**
  - → Simple way to represent structures of complicated organic molecules
  - → Shows in an abbreviated form how the atoms are grouped together
  - → The *condensed structural formula* for butane is $CH_3CH_2CH_2CH_3$ or $CH_3(CH_2)_2CH_3$; the latter is often written when the chain is lengthy.
  - → The *condensed structural formula* for methylpropane is $CH_3CH(CH_3)CH_3$ or $(CH_3)_3CH$, because the methyl groups are equivalent.  The first formula is the one chosen to name the compound (methylpropane *not* trimethylmethane).

- **Line structure**
  - → A *line structure* represents a chain of carbon atoms as a zigzag line.
  - → The end of each short line represents a carbon atom.
  - → C–H bonds and H atoms are not shown, but are added "mentally" to the structure.
  - → Other atoms, such as O and N atoms, are written explicitly, along with any H atoms to which they are bonded.
    **Example:** The zigzag *structural formula* and the *line structure* of methylpropane are shown on the right.

  - → A benzene ring is represented by a circle inside a hexagon or by a Kekulé structure.  The presence of a H atom attached to each C atom is implied.  These forms of the benzene ring are also used in structural formulas. The respective *structural formulas* and *line structures* of ethylbenzene, $C_6H_5CH_2CH_3$, are shown below.

## Nomenclature of Hydrocarbons

- **Alkanes**
  - → Alkanes are saturated hydrocarbons.
  - → In naming single chain *alkanes,* add a prefix denoting the number of C atoms in the chain to the suffix *-ane* (see Table 18.1 in the text).
  - → The first three members of the alkane family are methane, $CH_4$, ethane, $CH_3CH_3$, and propane, $CH_3CH_2CH_3$.
  - → If a *hydrogen atom* is removed from an *alkane,* the name of the *alkyl* group that results is derived by replacing the suffix *-ane* with *-yl.*
  - → The first three members of the alkyl family are the methyl, $CH_3-$, ethyl, $CH_3CH_2-$, and propyl, $CH_3CH_2CH_2-$, groups.
  - → Methylpropane, $CH_3CH(CH_3)CH_3$, is a propane molecule in which a H atom on the middle C atom is replaced by a methyl $(CH_3)$ group.
  - → *Cycloalkanes* are alkanes that contain *rings* of $-CH_2-$ units. The first three members are cyclopropane, $C_3H_6$; cyclobutane, $C_4H_8$; and cyclopentane, $C_5H_{10}$; all shown on the right.

- **Properties of alkanes**
  - → Structures may be written as linear, but they are actually linked tetrahedral units. Rotation about C–C bonds is fairly unrestricted, so the chains often roll up into a ball in gases and liquids to maximize the attraction between different parts of the chain.
  - → Electronegativity values of C and H are similar, so hydrocarbon molecules can be regarded as nonpolar.
  - → The dominant interaction between the molecules is therefore the London interaction, whose strength increases with the number of electrons in the molecule.
  - → Thus, alkanes should become less volatile with increasing molar mass (see Text Fig. 18.4).
  - → Many alkane molecules appear in forms with the same atoms bonded together in different arrangements such as the isomers butane and methylpropane, shown on the previous page.
  - → In general, *isomers* are different compounds with the same molecular formula.

- **Alkenes**
  - → *Alkenes* are a series of compounds with formulas derived from $H_2C=CH_2$ by inserting $CH_2$ groups. The next member of the family is propene, $H-CH_2-CH=CH_2$ ($CH_3CH=CH_2$).
  - → Ethene or ethylene, $C_2H_4$ or $H_2C=CH_2$, is the simplest *unsaturated* hydrocarbon.
  - → Alkenes are named in a manner analogous to that of alkanes except the ending is *-ene.*
  - → The double bond is located by numbering the C atoms in the chain and writing the lower of the two numbers of the two C atoms joined by the double bond (see Text Toolbox 18.1).

- **Alkynes**
  - → *Alkynes* are aliphatic hydrocarbons with at least one carbon-carbon triple bond.
  - → Ethyne or acetylene, $HC\equiv CH$, is the simplest alkyne.
  - → Alkynes are named like the alkenes but with the suffix *-yne.*

- **Nomenclature rules** for aliphatic hydrocarbons are given in Text Toolbox 18.1
- **Alkane nomenclature**
  - → If the C atoms in an alkane are bonded in a row with a formula of the form $CH_3(CH_2)_mCH_3$, the compound, called an *unbranched alkane,* is named according to the number of C atoms present. All are named with a stem name and the suffix *-ane.* A cyclic (ring) alkane has the general formula $(CH_2)_m$ and is designated by the prefix *cyclo-.*

    **Examples:** $CH_3(CH_2)_3CH_3$, pentane;  cyclopentane, $C_5H_{10}$

  - → A *branched alkane* has one or more C atoms not in a single row, and these carbons constitute *side chains*.

  - → Butane is an *unbranched alkane,* and methylpropane is a *branched alkane* with a $CH_3$ *side chain.* Side chains are treated as *substituents,* species that have been substituted for a H atom.

  - → A systematic procedure for naming alkanes is

    *Step 1.*   Identify the *longest* chain of C atoms in the molecule.

    *Step 2.*   Number the C atoms on the longest chain so that the carbon atoms with substituents have the *lowest* possible numbers.

    *Step 3.*   Identify each substituent and the number of the C atom in the chain on which it is located.

    *Step 4.*   Name the compound by listing the substituents in *alphabetical order*, with the numbered location of each substituent preceding its name.  The name(s) of the substituent(s) are followed by the parent name of the alkane, which is determined by the number of C atoms in the longest chain.

    Example:  heptane with a methyl group at the number 3 carbon is named 3-methylheptane.

    *Step 5.*   If two or more of the same substituents are present (such as two methyl groups), a Greek prefix such as *di-*, *tri-*, or *tetra-* is attached to the name of the group.  Numbers of the C atoms to which groups are attached, separated by commas, are included in the name.

    Example: heptane, with methyl groups at C atoms 3 and 4, is named 3,4-dimethylheptane.

    *Step 6.*   As in the example above, numbers in a name are always separated from letters by a hyphen, while numbers are separated from each other by commas.

- **Alkene nomenclature**
  - → Alkenes are named by using the stem name of the corresponding alkane with a number that specifies the location of the double bond.  The number is obtained by numbering the longest carbon chain so that the double bond is associated with the lowest numbered carbon possible and assigning to the double bond the lower number of the two double-bonded carbons.

    $$\overset{6}{C}H_3\overset{5}{C}H_2\overset{4}{C}H_2\overset{3}{C}H=\overset{2}{C}H\overset{1}{C}H_3$$

    2-Hexene (not 3-hexene)

  - → Other substituents are named as for alkanes, with a number specifying the location of the group and an appropriate prefix denoting how many groups of each kind are present.

  - → Numbering the longest chain so that the double bond has the lowest number takes precedence over keeping the substituent numbers low.

  - → If more than one double bond is present, the number of these bonds is indicated by a Greek prefix. The suffix is *-ene.*

    **Example:** 1,3-hexadiene, $H_2C=CHCH=CHCH_2CH_3$, or

- **Alkyne nomenclature**
  - → Alkynes are named by using the stem name of the corresponding alkane with a number specifying the location of the triple bond. The numbering convention is the same as that used for double bonds. If more than one triple bond is present, the number of these bonds is indicated by a Greek prefix. The suffix is *-yne*.
  - → When numbering atoms in the chain, the lowest numbers are given preferentially to (a) functional groups named by suffixes (see Toolbox 19.1), (b) double bonds, (c) triple bonds, and (d) groups named by prefixes.

    **Example:** $CH_3CH_2CH(Cl)C\equiv CCH_3$, 4-chloro-2-hexyne

## 18.2 Isomers

- **Structural isomers**
  - → Molecules constructed from the same atoms connected differently
  - → Have the *same* molecular formula, but *different* structural formulas

    **Example:** Butane, $CH_3(CH_2)_2CH_3$, and methylpropane, $CH_3CH(CH_3)CH_3$, are structural isomers with the same molecular formula, $C_4H_{10}$. The molecules have a different *connectivity*.

- **Stereoisomers**
  - → Molecules with the same connectivity but with some of their atoms arranged differently in space
  - → *Geometrical isomers* are *stereoisomers* with different arrangements in space on either side of a double bond, or above and below the ring of a cycloalkane.
  - → *Optical isomers* are *stereoisomers* in which each isomer is the mirror image of the other, *and* the images are nonsuperimposable.

- **Geometrical isomers**
  - → Compounds with the same molecular formula and with atoms bonded to the same neighbors, but with a different arrangement of atoms in space
  - → For molecules with a double bond, effective lack of rotation about the double bond makes this type of isomer possible.
  - → *Geometrical isomers* of organic compounds are distinguished by the italicized prefixes: *cis* (from the Latin word for *on this side*) and *trans* (from the Latin word for *across*).
  - → An example is 2-butene, which exists as *cis* and *trans* geometrical isomers.

    *cis*-2-butene    *trans*-2-butene

    In the trans isomer, the two methyl groups lie across the double bond from each other. In the cis isomer, they are on the same side of the double bond.

    Rotation about the double bond does not normally occur because the $\pi$-bond is rigid, and the two isomers are distinct compounds with different chemical and physical properties.

    If sufficient energy is added to either compound, a *cis–trans isomerization* reaction may occur in which the $\pi$-bond is partially broken and part of the sample is converted to the other isomer.

- **Optical isomers**
  - → *Nonsuperimposable* mirror images
  - → A *chiral* molecule has a mirror image that is nonsuperimposable. Your hand is a chiral object; its mirror image is not superimposable.
  - → An organic molecule is chiral if at least one of its central C atoms has four *different* groups attached to it. Such a C atom is called a *stereogenic center*.
  - → A pair of *enantiomers* consists of a chiral molecule and its mirror image.
  - → Enantiomers have identical chemical properties, except when they react with other chiral species.

→ Enantiomers differ in only one physical property; chiral molecules display *optical activity,* the ability to rotate the plane of polarization of light.

→ Mixtures of enantiomers in equal proportions are *racemic mixtures,* which are not optically active.

→ A molecule that has a superimposable mirror image is said to be *achiral.*

- **Summary** → The three different types of isomerism are displayed in Text Fig. 18.1.

## 18.3 Properties of Alkanes

- **Physical properties**

  → *Alkanes* are best regarded as nonpolar molecules held together primarily by London forces which increase with the number of electrons in a molecule.

  → Consequently, for unbranched *alkanes,* melting and boiling points increase with chain length.

  → Unbranched *alkanes* tend to have higher melting points, boiling points, and heats of vaporization than their branched structural isomers.

  → Molecules with unbranched chains can get closer together than molecules with branched chains. As a result, molecules with branched chains have *weaker* intermolecular forces than their isomers with unbranched chains.

    **Example:** Unbranched butane has a higher boiling point (–0.5°C) than its branched-chain structural isomer, methylpropane (–11.6°C).

- **Chemical properties**

  → *Alkanes* are not very reactive chemically. They were once called *paraffins,* derived from the Latin "little affinity."

  → *Alkanes* are unaffected by concentrated sulfuric acid, boiling nitric acid, strong oxidizing agents such as $KMnO_4$, and by boiling aqueous NaOH.

  → The C–C bond enthalpy, 348 kJ·mol$^{-1}$, and the C–H bond enthalpy, 412 kJ·mol$^{-1}$, are large, so there is little energy advantage in replacing them with most other bonds. Notable exceptions are C=O, 743 kJ·mol$^{-1}$; C–OH, 360 kJ·mol$^{-1}$; and C–F, 484 kJ·mol$^{-1}$.

- **Combustion (oxidation) reactions**

  → *Alkanes* are used as fuels, because their enthalpies of combustion are high.

  → The products of combustion are carbon dioxide and water. Strong C–H bonds are replaced by even stronger O–H bonds in $H_2O$, and the O=O bonds are replaced by two strong C=O bonds in $CO_2$.

    **Example:** Combustion of one mole of octane releases 5471 kJ of heat:
    $$C_8H_{18}(l) + 12.5\ O_2(g) \rightarrow 8\ CO_2(g) + 9\ H_2O(l) \qquad \Delta H_r° = -5471\ kJ\cdot mol^{-1}$$

- **Substitution reactions**

  → *Alkanes* are used as raw materials for the synthesis of many reactive organic compounds.

  → Organic chemists introduce reactive groups into alkane molecules in a process called *functionalization*.

  → Functionalization of alkanes is achieved by a *substitution reaction,* in which an atom or group of atoms replaces an atom in the original molecule (hydrogen, in the case of alkanes).

  → Reaction of methane, $CH_4$, and chlorine, $Cl_2$, is an example of a substitution reaction. In the presence of ultraviolet light or temperatures above 300°C, the gases react explosively:

    $$CH_4(g) + Cl_2(g) \xrightarrow{\text{light or heat}} CH_3Cl(g) + HCl(g)$$

    Chloromethane, $CH_3Cl$, is only one of four products; the others are dichloromethane, $CH_2Cl_2$, trichloromethane, $CHCl_3$, and tetrachloromethane, $CCl_4$, which is carcinogenic.

## 18.4 Alkane Substitution Reactions

- **Radical chain mechanism**

  → Kinetic studies suggest that alkane substitution reactions proceed by a *radical chain* mechanism.

  → The *initiation step* is the dissociation of chlorine:

  $$Cl_2 \xrightarrow{\text{light or heat}} 2\,Cl\cdot$$

  → Chlorine atoms proceed to attack methane molecules and abstract a hydrogen atom:

  $$Cl\cdot + CH_4 \rightarrow HCl + \cdot CH_3$$

  Because one of the products is a radical, this reaction is a *propagation step*.

  → In a second propagation step, the methyl radical may react with a chlorine molecule:

  $$Cl_2 + \cdot CH_3 \rightarrow CH_3Cl + Cl\cdot$$

  The chlorine atom formed may take part in the other *propagation step* or attack a $CH_3Cl$ molecule to eventually form $CH_2Cl_2$. $CHCl_3$ and $CCl_4$ can also be formed by a continuation of this process.

  → A termination step occurs when two radicals combine to form a nonradical product, as in the

  reaction:   $Cl\cdot + \cdot CH_3 \rightarrow CH_3Cl$

  The substitution reaction is not very clean and the product is usually a mixture of compounds. One may limit the production of the more highly substituted alkanes by using a large excess of the alkane.

## 18.5 Properties of Alkenes

- **Double bond**

  → The carbon-carbon double bond, C=C, consists of a $\sigma$-bond and a $\pi$-bond.

  → Each carbon atom is $sp^2$ hybridized and one of the hybrid orbitals is used to form the $\sigma$-bond.

  → The unhybridized $p$-orbitals on each atom overlap with each other to form a $\pi$-bond.

  → All four atoms attached to the C=C group lie in the same plane and are fixed in that arrangement by resistance to twisting of the $\pi$-bond (see Text Fig. 18.7).

  → Alkenes can not roll up into a compact arrangement as alkanes can, so alkenes have lower melting points.

  → In C=C, the $\pi$-bond is weaker than the $\sigma$-bond. A consequence of this weakness is the reaction most common in alkenes: replacement of the $\pi$-bond by two new $\sigma$-bonds.

- **Formation of alkenes by elimination reactions**

  → In the petrochemical industry, alkanes are converted to more reactive alkenes by a catalytic process called *dehydrogenation:*

  $$CH_3CH_3(g) \xrightarrow{Cr_2O_3} CH_2{=}CH_2(g) + H_2(g)$$

  This is an example of an *elimination reaction,* one in which two groups on neighboring C atoms are removed from a molecule, leaving a multiple bond (see Text Fig. 18.8).

  → In the laboratory, alkenes are produced by *dehydrohalogenation* of haloalkanes, the removal of a hydrogen atom and a halogen atom from neighboring carbon atoms:

  $$CH_3CH_2Br \xrightarrow{CH_3CH_2O^- \text{ in ethanol at } 70°C} CH_2{=}CH_2 + HBr$$

  This elimination reaction is carried out in hot ethanol containing sodium ethoxide, $CH_3CH_2ONa$.

  → *Dehydrohalogenation* occurs by attack of the ethoxide ion on a hydrogen atom of the methyl group. A H atom is removed as a proton and $CH_3CH_2O{-}H$ is formed. When the methyl C atom forms a second bond to its neighbor, the $Br^-$ ion departs.

  States of reactants and products in organic reactions are often not give because the reaction may take place on a catalyst surface or in a non-aqueous solvent, as in the reaction above.

## 18.6 Electrophilic Addition

- **Addition reaction**
  - → A characteristic reaction of alkenes, in which atoms supplied by the reactant form σ-bonds to the two C atoms joined by the π-bond, which is broken (see Text Fig. 18.9).
  - → Almost all addition reactions are exothermic.
  - → A *hydrogenation* reaction is the addition of two hydrogen atoms at a double bond:
  $$CH_3CH=CHCH_3 + H_2 \rightarrow CH_3CH_2-CH_2CH_3$$
  - → A *halogenation* reaction is the addition of two halogen atoms at a double bond:
  $$CH_3CH=CHCH_3 + Cl_2 \rightarrow CH_3CHCl-CHClCH_3$$
  - → A *hydrohalogenation* reaction is the addition of a hydrogen atom and a halogen atom at a double bond:    $CH_3CH=CHCH_3 + HCl \rightarrow CH_3CH_2-CHClCH_3$

- **Estimating the reaction enthalpy of an addition reaction**
  - → Use bond enthalpies for the bonds that are broken and formed. Note that bond enthalpies strictly apply to gas-phase reactions.
  - → Recall:
  $$\Delta H_r° \approx \sum_{\text{reactants}} n\Delta H_B(\text{bonds broken}) - \sum_{\text{products}} n\Delta H_B(\text{bonds formed})$$

- **Mechanism of addition reactions**
  - → Double bonds contain a high density of high energy electrons associated with the π-bond. This region of high electron density is attractive to positively charged reactants.
  - → An *electrophile* is a reactant that is attracted to a region of high electron density. It may be a positively charged species or one that has or can acquire a *partial* positive charge during the reaction.
  - → An example treated in the text is the bromination of ethene to form dibromoethane:
  $$H_2C=CH_2 + Br_2 \rightarrow CH_2BrCH_2Br$$
  - → Bromine molecules are polarizable; a partial positive charge builds up on the bromine atom closest to the double bond. Bromine acts as an *electrophile*. The partial charge becomes a full charge and a bromine cation attaches to the double bond, leaving a Br⁻ ion behind. The cyclic intermediate is called a *bromonium ion*.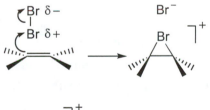
  - → A Br⁻ ion is attracted by the positive charge of the bromonium ion. It forms a bond to one C atom, and the bromine atom in the cyclic ion forms another bond, giving 1,2-dibromoethane.

- **Hydrogenation reaction**
  - → Hydrogen can be added across a double bond with the help of a solid-state catalyst:
  $$H_2(g) + \bullet\bullet\bullet \underset{}{=\!=\!=} \bullet\bullet\bullet \xrightarrow{\text{catalyst}} \bullet\bullet\bullet CH\!-\!\!-\!\!CH \bullet\bullet\bullet$$
  - → This reaction is used in the food industry, for example, to convert liquid oils to solids. Recall that alkenes have lower melting points than alkanes, other things equal.

# Aromatic Compounds (Sections 18.7–18.8)

## 18.7 Nomenclature of Arenes

- **Definition**

  → Aromatic hydrocarbons are called *arenes*.

  → *Benzene*, $C_6H_6$, is the parent compound. The benzene ring is also called the *phenyl* group, as in 2-phenyl-*trans*-2-butene, $CH_3C(C_6H_5)=CHCH_3$, shown on the right.

  → Aromatic compounds include those with fused benzene rings, such as naphthalene, $C_{10}H_8$; anthracene, $C_{14}H_{10}$; and phenanthrene, $C_{14}H_{10}$.

  naphthalene      anthracene      phenanthrene

- **Benzene ring numbering**

  → Benzene ring substituents are designated by numbering the C atoms from 1 to 6 around the ring. Compounds are named by counting around the ring in the direction that gives the smallest numbers to the substituents. The prefixes, *ortho-*, *meta-*, and *para-* are commonly used to denote substituents at C atoms 2, 3, and 4, respectively, relative to another substituent at C atom 1.

  **Example:** The three possible dichlorobenzene isomers are shown in the diagram on the right.

  → Numbering carbon atoms in fused ring systems is somewhat more complicated and is not covered in the text

  1,2-Dichlorobenzene     1,3-Dichlorobenzene     1,4-Dichlorobenzene
  ortho-Dichlorobenzene   meta-Dichlorobenzene    para-Dichlorobenzene

## 18.8 Electrophilic Substitution

- **Substitution reactions** → *Arenes* have delocalized π-electrons; but, unlike alkenes, arenes undergo predominantly *substitution* reactions, with the π-bonds of the ring unaffected.

  **Example:** The reaction of benzene, $C_6H_6$, with chlorine produces chlorobenzene when one chlorine atom substitutes for a hydrogen atom:   $C_6H_6 + Cl_2 \xrightarrow{Fe} C_6H_5Cl + HCl$

- **Electrophilic substitution** → If the mechanism of substitution of a benzene ring involves electrophilic attack, the reaction is called *electrophilic substitution*.

- **Halogenation**

  → In the halogenation of benzene, an iron catalyst is used. The iron is converted to iron(III) halide, $FeX_3$. Iron(III) halide reacts further to polarize the $X_2$ molecule.

  → The commonly accepted mechanism for aromatic halogenation, $X_2$, is then

  $$FeX_3 + X–X \rightleftharpoons X_3Fe^{\delta-}\cdots X–X^{\delta+} \qquad \textit{fast equilibrium}$$

  $$X_3Fe^{\delta-}\cdots X–X^{\delta+} + C_6H_6 \rightarrow FeX_4^- + C_6H_6X^+ \qquad \textit{slow}$$

  $$C_6H_6X^+ + FeX_4^- \rightarrow C_6H_5X + HX + FeX_3 \qquad \textit{fast}$$

  → In the last step, the hydrogen atom is easily removed from the ring, for in that way the π-electron delocalization, lost in the slow step, is regained.

- **Nitration**

  → A mixture of $HNO_3$ and concentrated $H_2SO_4$ slowly converts benzene into nitrobenzene.

→ The nitronium ion, $NO_2^+$, is the electrophile and the nitrating agent.

→ The commonly accepted mechanism for aromatic nitration is

$$HNO_3 + 2\,H_2SO_4 \rightleftharpoons NO_2^+ + H_3O^+ + 2\,HSO_4^-$$

$$NO_2^+ + C_6H_6 \rightarrow C_6H_6NO_2^+ \qquad\qquad slow$$

$$C_6H_6NO_2^+ + HSO_4^- \rightarrow C_6H_5NO_2 + H_2SO_4 \quad fast$$

→ In the last step, the hydrogen atom is removed from the ring, restoring aromaticity.

- **Ortho- and para-directing activators**

  → Electrons are donated into the delocalized Molecular Orbitals of the ring (*resonance effect*).

  → The reaction rate is *much* faster than in unsubstituted benzene, and the substituent is called an *activator*.

  → Products of substitution reactions favor the *ortho* and *para* positions of the ring, because these locations have more electron density than the *meta* positions.

  → Examples of ortho- and para-directing activators include $-OH$, $-NH_2$, and substituted amines.

  → The atom bonded to the benzene ring has a *nonbonding* pair of electrons.

  **Note:** Alkyl groups have no nonbonding electron pair, yet they are also ortho- and para-directing activators. The mechanism in this case is a different one (see any organic chemistry text).

- **Ortho- and para-directing deactivators**

  → Electrons are donated into the delocalized Molecular Orbitals of the ring (*resonance effect*), but the substituent is very electronegative.

  → The reaction rate is *slightly* slower than in unsubstituted benzene, and the substituent is called a *deactivator*. Deactivation occurs when the substituent is *highly* electronegative and withdraws some electron density from the ring.

  → Products of a substitution reaction still favor the *ortho* and *para* positions of the ring, because these positions have *relatively* more electron density than the *meta* positions.

  → The *only* examples are $-F$, $-Cl$, $-Br$, and $-I$.

  → The atom bonded to the benzene ring has a *nonbonding* pair of electrons.

- **Meta-directing deactivators**

  → Highly electronegative substances that can withdraw electrons partially and/or a substituent that *removes* electrons by resonance

  → The reaction rate is *much* slower than in unsubstituted benzene, and the substituent is called a *deactivator*.

  → Products of a substitution reaction favor the *meta* position, because the *ortho* and *para* positions have greatly decreased electron density.

  → Examples include $-COOH$, $-NO_2$, $-CF_3$, and $-C{\equiv}N$.

  → *All* electron pairs on the atom bonded to the benzene ring are *bonding* pairs.

# Impact on Materials: Fuels  (Sections 18.9–18.10)

## 18.9  Gasoline

- **Gasoline**
  - → Derived from petroleum by fractional distillation and further processing
  - → Contains primarily $C_5$ to $C_{11}$ hydrocarbons
  - → Petroleum is refined to increase the quantity and quality of gasoline.

- **Quantity** → The quantity of gasoline is increased by *cracking* (breaking down long hydrocarbon chains) and by *alkylation* (combining small molecules to make larger ones).

  **Example:** Cracking fuel oil to obtain an octane/octane isomeric mixture

  **Example:** Alkylation converts a butane/butene mixture to octane.

  $$C_4H_{10} + C_4H_8 \xrightarrow{\text{catalyst}} C_8H_{18}$$

- **Quality**
  - → Octane rating: used to measure the quality of gasoline
  - → Other things equal, branched hydrocarbons have higher octane ratings than unbranched ones.
  - → *Isomerization* and *aromatization* are used to improve octane rating of gasoline.
  - → *Isomerization* is used to convert straight chain hydrocarbons to branched ones.

  $$CH_3(CH_2)_6CH_3 \xrightarrow{\text{AlCl}_3} CH_3C(CH_3)_2CH_2CH(CH_3)CH_3$$

  - → *Aromatization* converts alkanes to arenes, as in the conversion of heptane to methylbenzene.

  $$CH_3(CH_2)_5CH_3 \xrightarrow{\text{AlCl}_3,\ \text{Cr}_2\text{O}_3} CH_3C_6H_5 + 4\,H_2$$

  - → Ethanol, a renewable fuel, also increases the octane rating of gasoline.

## 18.10  Coal

- → End product of anaerobic decay of vegetable matter
- → Less environmentally friendly than gasoline, in part because it releases ash when burned
- → Contains many aromatic rings and is primarily aromatic in nature
- → Coal fragments, when heated in the absence of oxygen, to yield coal tar which contains many aromatic hydrocarbons and their derivatives. Consequently, coal is used as a raw material for other chemicals.
- → Many pharmaceuticals, fertilizers, and dyes are derived from coal tar.

  **Examples:** Naphthalene for making indigo dyes (for blue jeans) and benzene, used to make nylon, detergents and pesticides

# Chapter 19  Organic Chemistry II:
# Polymers and Biological Compounds

## Common Functional Groups  (Sections 19.1–19.8)

### 19.1  Haloalkanes

- **Haloalkanes**
  - → Alkanes in which a *halogen atom,* X, replaces one or more H atoms
  - → Insoluble in water
  - → Some are highly toxic and environmentally unfriendly.

    **Example:** The chlorofluorocarbon (CFC) 1,2-dichloro-1-fluoroethane, $HFClC–CClH_2$, is partly responsible for depletion of the ozone layer.

  - → The C–X bonds in haloalkanes are *polar;* C carries a partial positive charge and the halogen, X, a partial negative one.
  - → C–X bond polarity governs much of the chemistry of haloalkanes.
  - → Haloalkanes are susceptible to *nucleophilic substitution,* in which a reactant that seeks out centers of positive charge in a molecule (a *nucleophile*) replaces a halogen atom.
  - → Examples of *nucleophiles:* Anions such as $OH^-$ and Lewis bases with lone pairs such as $NH_3$.

    **Example:** Water acts as a nucleophile in the following *hydrolysis* reaction:

$$CH_3Br + H_2O \rightarrow CH_3OH + HBr$$

### 19.2  Alcohols

- **Definitions**
  - → *Hydroxyl group:* An –OH group covalently bonded to a C atom
  - → *Alcohol:* An organic compound that contains a *hydroxyl* group not directly bonded to an aromatic ring or to a carbonyl group, $\overset{\diagdown}{\underset{\diagup}{C}}=O$

  - → Alcohols are named by adding the suffix *–ol* to the stem of the parent hydrocarbon.

    **Examples:** Methanol for $CH_3OH$; ethanol for $CH_3CH_2OH$
  - → To identify the location of the –OH group, the number of the C atom attached to it is given.

    **Examples:** 1-Propanol for $CH_3CH_2CH_2OH$; 2-propanol for $CH_3CH(OH)CH_3$
  - → A *diol* is an organic compound with two hydroxyl groups.

    **Example:** 1,2-Ethanediol for $HOCH_2CH_2OH$ (ethylene glycol)

- **Classes of alcohols**
  - → *Primary alcohol:* $RCH_2$–OH, where R can be any group  **Examples:** Methanol and 1-propanol
  - → *Secondary alcohol:* $R_2CH$–OH, where the R groups can be the same or different

    **Examples:** 2-Propanol, $(CH_3)_2CH–OH$ (same R groups); 2-butanol, $CH_3CH_2CH(–OH)CH_3$ (different R groups)
  - → *Tertiary alcohol*: $R_3C$–OH, where the R groups can be the same or different

    **Examples:** 2-Methyl-2-propanol, $(CH_3)_3C–OH$ (*same* R groups, $CH_3$–); 3-methyl-3-pentanol, $CH_3CH_2C(CH_3)(–OH)CH_2CH_3$ (*different* R groups, $CH_3CH_2$– and $CH_3$–).  A structure for the latter is:

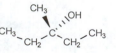

- **Properties of alcohols**
  - → Alcohols are polar molecules that can lose the −OH proton in certain solvents, but typically not in water.
  - → Alcohols have relatively high boiling points and low volatility because of hydrogen bond formation through the −OH group. In this way, they are similar to water.
  - → Compare the normal boiling point of ethanol (78.2°C) with a molar mass of 46.07 g·mol$^{-1}$ to that of pentane (36.0°C) with a molar mass of 72.14 g·mol$^{-1}$.

## 19.3 Ethers

- **Ethers**
  - → Organic compounds of the form R−O−R, where R is any alkyl group and the two R groups may be the same or different
  - → More volatile than alcohols with the same molar mass because ethers do not form hydrogen bonds with each other
  - → Act, however, as hydrogen-bond acceptors by using the lone pairs of electrons on the O atom
  - → Useful solvents for other organic compounds, because they have low polarity and low reactivity
  - → Quite flammable and must be handled with care
  **Examples:** Diethyl ether, $CH_3CH_2-O-CH_2CH_3$; 1-butyl methyl ether, $CH_3CH_2CH_2CH_2-O-CH_3$

- **Crown ethers**
  - → Cyclic polyethers of formula $+(CH_2CH_2-O)_n$
  - → Name reflects the crownlike shape of the molecules.
  - → Bind strongly to alkali metal ions such as $Na^+$ and $K^+$, allowing inorganic salts to be dissolved in organic solvents
  **Example:** The common oxidizing agent potassium permanganate, $KMnO_4$, is insoluble in nonpolar solvents such as benzene, $C_6H_6$. In the presence of [18]-crown-6, it dissolves in benzene according to the reaction on the right:

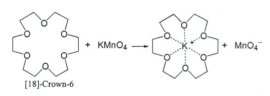

[18]-Crown-6

## 19.4 Phenols

- **Phenols**
  - → Organic compounds with a *hydroxyl group* attached *directly* to an aromatic ring
  - → The parent compound, *phenol,* is a white, crystalline, molecular solid.

Phenol, $C_6H_5OH$:  or  Melting point: 40.9°C

  - → Substituted phenols occur naturally, and some are responsible for the fragrances of plants.
  **Examples:** *Thymol* is the active ingredient in oil of thyme, and *eugenol* provides the scent and flavor in oil of cloves.

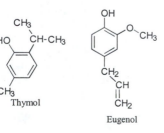

Thymol

Eugenol

- **Acid-base properties of phenols**

  → Phenols are generally *weak* acids in contrast to alcohols, which typically are *not* acidic.

  → The acidity of phenols can be understood on the basis of resonance stabilization of the negative charge of the conjugate base of phenol, $C_6H_5O^-$.

  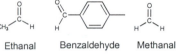

  → Because of resonance stabilization, the phenoxide ($C_6H_5O^-$) anion is a weaker conjugate base than the corresponding conjugate bases of typical alcohols.

  **Example:** The ethoxide ($CH_3CH_2O^-$) anion is a stronger base than the phenoxide anion.

## 19.5 Aldehydes and Ketones

- **Aldehydes**

  → Organic compounds of the form shown on the right, where R is a H atom, an aliphatic group, or an aromatic group

  → The group characteristic of aldehydes is written as –CHO, as in formaldehyde, HCHO, the first member of the family.

  → Named systematically by replacing the ending *–e* by *–al,* but many common names exist, such as formaldehyde for methanal and acetaldehyde for ethanal

  → The C atom on the carbonyl group is *included* in the count of C atoms when determining the alkane from which the aldehyde is derived.

  → A few common aldehydes are shown to the right.

  Ethanal    Benzaldehyde    Methanal

  → Aldehydes generally contribute to the flavor of fruits and nuts, and to the odors of plants.

  **Example:** Benzaldehyde provides part of the aroma of almonds and cherries.

  → Aldehydes can be prepared by the mild oxidation of *primary* alcohols.

  **Example:** Formaldehyde is prepared industrially by oxidizing methanol with a Ag catalyst. Ag acts as a mild oxidizing agent preventing further oxidation of formaldehyde to formic acid, HCOOH. The overall reaction is

  $$2\,CH_3OH(g) + O_2(g) \xrightarrow{\text{600°C, Ag}} 2\,HCHO(g) + 2\,H_2O(g)$$

- **Ketones**

  → Organic compounds of the form shown on the right, where the R groups (alkyl or aryl) may be the same or different

  → The *carbonyl group,* –C=O, characteristic of ketones, is written –CO, as in propanone, $CH_3COCH_3$ (acetone), the first member of the family.

  → Named systematically by replacing the ending *-e* by *-one,* but many common names exist such as acetone or dimethyl ketone for propanone, and methyl ethyl ketone for butanone

  → The C atom on the carbonyl group is *included* in the count of C atoms when determining the alkane from which the ketone is derived.

  → A number is used to locate the C atom in the carbonyl group to avoid ambiguity.

→ A few common ketones are shown on the right: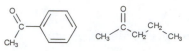

Propanone    Benzophenone    2-Pentanone

→ *Ketones* contribute to flavors and fragrances.

**Example:** *Carvone,* shown on the right, is the essential oil in spearmint.

→ Ketones can be prepared by the oxidation of *secondary* alcohols.

→ There is less risk of further oxidation than with aldehydes, so stronger oxidizing agents are used. Dichromate oxidation of secondary alcohols produces ketones with little excess oxidation and in good yields.

**Example:** The oxidation of 2-propanol to propanone is shown schematically on the right: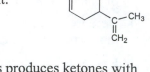

## 19.6 Carboxylic Acids

• **Carboxylic acids**

→ Organic compounds of the form shown on the right, where R is a H atom, an alkyl, or an aryl group

→ The *carboxyl group* characteristic of carboxylic acids is written –COOH as in formic acid, HCOOH, the first member of the family.

→ The *carboxyl group* contains a *hydroxyl group,* –OH, attached to a *carbonyl group,* –C=O.

→ Named systematically by replacing the ending *-e* by *-oic acid,* but many common names are used such as formic acid for methanoic acid, and acetic acid for ethanoic acid

→ The carbonyl C atom is included in the C atom count to determine the parent hydrocarbon molecule.

→ Some common carboxylic acids are shown on the right: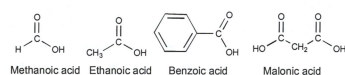

Methanoic acid  Ethanoic acid   Benzoic acid   Malonic acid

**Note:** Malonic acid contains two –COOH groups; it is called a *diacid.*

→ Carboxylic acids contain hydroxyl groups, –OH, which can take part in hydrogen bonding.

→ Carboxylic acids can be prepared by oxidation of aldehydes or of *primary* alcohols in an acidified solution containing a strong oxidizing agent such as $KMnO_4$, or $Na_2Cr_2O_7$:

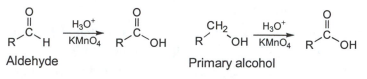

Aldehyde              Primary alcohol

→ In some cases, alkyl groups can be oxidized *directly* to carboxyl groups.

→ An industrial example is the oxidation of the methyl groups on *p*-xylene by a cobalt(III) catalyst to form *terephthalic acid,* which is used in the production of artificial fibers.

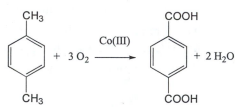

# 19.7 Esters

- **Esters**

  → An *ester* is the product of a *condensation* reaction between a carboxylic acid and an alcohol. A water molecule is also produced.

  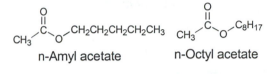

  → *Esterification* reactions can be acid catalyzed as in the example above.

  → *Condensation* reaction: one in which two molecules combine to form a large one, and a small molecule is eliminated.

  → Many *esters* have fragrant aromas and contribute to flavors of fruits.

    **Examples:** The esters *n*-amyl acetate and *n*-octyl acetate are responsible for the aromas of bananas and oranges, respectively.

# 19.8 Amines, Amino Acids, and Amides

- **Amines**

  → Derivatives of $NH_3$ formed by replacing one or more H atoms with organic groups, R

  → The *amino group*, $-NH_2$, is the functional group of *amines*.

  → Named by specifying the groups attached to the nitrogen atom, N, alphabetically, followed by the suffix *amine*

  → Designated as *primary, secondary,* or *tertiary,* depending on the number of R groups attached to the N atom

  → Ammonia and several representative amines are shown on the right:

- **Properties of amines**

  → Characterized by four $sp^3$ hybrid orbitals on the N atom; three participate in three single bonds and the fourth has a lone pair of electrons

  → *Amines* are widespread in nature.

  → They often have disagreeable odors (for example, *putrescine*, $NH_2(CH_2)_4NH_2$) and, similar to ammonia itself, are weak bases.

- **Quaternary ammonium ions**

  → Tetrahedral ions of formula $R_4N^+$

  → Four groups bonded to the central nitrogen atom, which may be the same or different

  → Negligible acid or base properties and little effect on pH

  → Isolated as quaternary ammonium salts

    **Example:** Dimethylethylpropylammonium chloride, with four R groups, is a quaternary ammonium salt.

- **Amino acids**

  → Carboxylic acids with an *amino* group and the *carboxyl* group separated by *one* C atom

  → A more precise name is an *aminocarboxylic acid*.

  → Have the general formula R(NH$_2$)CH(COOH)
    **Examples:** *Glycine, methionine,* and
    *phenylalanine*

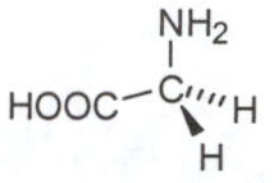

  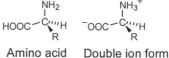

  Glycine 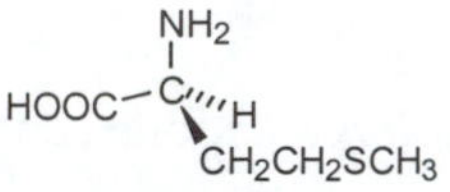     Methionine 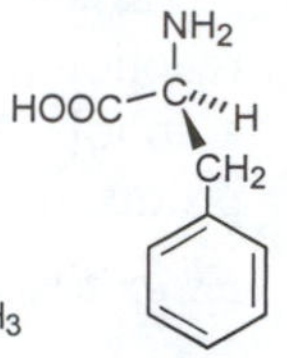     Phenylalanine

  **Note:** In amino acids, the central C atom is *stereogenic* and the molecule is *chiral* except for *glycine*.

- **Properties of amino acids**

  → Building blocks of proteins (see Section 19.13)

  → Essential for human health

  → Form *hydrogen bonds* with *both* the *amino* and *carboxyl* groups

  → Form *double ions* (*zwitterions*) by transferring a proton from the *carboxyl* to the *amino* group as shown on the right:

  Amino acid     Double ion form

  → In aqueous solution, amino acids exist in four forms whose concentrations are pH dependent.

  → The concentrations of these species can be determined as with polyprotic acids (see Section. 10.16).

  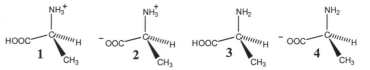

  → Form 1 predominates at low pH while form 4 predominates at high pH. At neutral pH, form 3 predominates.

- **Amides**

  → Molecules with the general formula

  → Formed by the condensation reaction of carboxylic acids with amines, eliminating water

  → Near room temperature, the reaction mixture forms an ammonium salt that reacts further to form the amide at higher temperature.

  → The general reaction is

  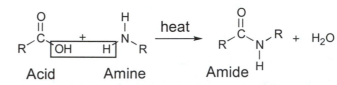

  Acid     Amine     Amide

  **Note:** The amide shown above has an N–H group that may participate in *intermolecular* hydrogen bonding. The atoms that form the product water molecule are shown in a box.

# The Impact on Materials (Sections 19.9–19.12)

## 19.9 Addition Polymerization

- **Addition polymers**
  - → Form when *alkene* monomers react with themselves with no net loss of atoms to form *polymers*
  - → Prepared by *radical polymerization,* a radical chain reaction
  - → Radical polymerization reactions are started by using an *initiator* such as an organic peroxide, R–O–O–R, which decomposes when heated to form two free radicals.
  - → A radical polymerization mechanism is shown for ROOR and $CH_2=CHX$:

  Initiation:     $R—O—O—R \xrightarrow{heat} R—O\bullet + \bullet O—R$

  Propagation:

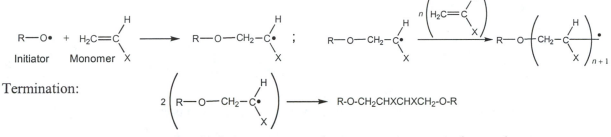

  Termination:

  → Radical polymerization reactions include *initiation* and *propagation* steps as shown above.

  → The reaction terminates when all the monomer is consumed or when two radical chains of any length react to form a single diamagnetic (nonradical) species.

  → One possible *termination* step is shown in the scheme above.

- **Stereoregular polymer**
  - → Each unit or pair of repeating units in the polymer has the same relative orientation.
  - → These polymers pack well together, are relatively strong, and are impact resistant.

- **Isotactic polymer**
  - → A *stereoregular* polymer with all substituents on the same side of the extended carbon chain

    **Example:** A portion of an isotactic polypropylene chain:

    Isotactic configuration - same side

- **Syndiotactic polymer**
  - → A *stereoregular* polymer with substituents alternating regularly on either side of the extended carbon chain

    **Example:** A portion of a syndiotactic polypropylene chain with alternating methyl groups:

    Syndiotactic configuration - alternating sides

- **Atactic polymer**
  - → A polymer with substituents randomly oriented with respect to the extended carbon chain
  - → *Atactic* polymers are *not* stereoregular. They tend to be amorphous and less well suited for many applications.

**Example:** An example of an *atactic* polypropylene chain with randomly-oriented methyl side groups:

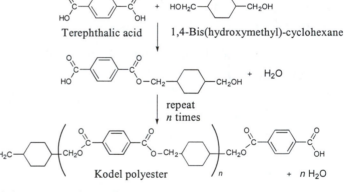

Atactic configuration - random sides

- **Ziegler–Natta catalysts**

  → Consist of aluminum- and titanium-containing compounds, such as titanium tetrachloride, $TiCl_4$, and triethyl aluminum, $(CH_3CH_2)_3Al$

  → Used in the production of *stereoregular* polymers, including synthetic rubber

  → Chemists were unable to synthesize rubber with useful mechanical properties until the discovery of Ziegler–Natta catalysts in 1953.

## 19.10 Condensation Polymerization

- **Condensation polymers**

  → Formed

  ➤ By a series of *condensation* reactions

  ➤ From reactants, each of which has *two* functional groups

  ➤ From stoichiometric amounts of the reactants

  → Encompass the classes of polymers known as *polyesters* and *polyamides*

  → Characterized by reactions proceeding without the need for an initiator

  → Typically have shorter chain lengths than addition polymers because each monomer can initiate the reaction

- **Polyesters**

  → Formed from the condensation of a *diacid* and a *diol*

  → For polyester polymerization, it is necessary to have two functional groups on each monomer and to mix stoichiometric amounts of the reactants.

  **Example:** Kodel polyester is formed from the esterification of terephthalic acid and 1,4-bis(hydroxymethyl)-cyclohexane as shown by the reactions above:

- **Polyamides**

  → Are formed from the condensation of a *diacid* with a *diamine*

  → For polyamide polymerization, it is necessary to have two functional groups on each monomer and to mix stoichiometric amounts of the reactants.

**Example:** Nylon-66 forms by condensation of 1,6-hexanedioic acid and 1,6-hexanediamine. The reactants first form a salt, which then condenses to the polyamide.

1,6-Hexanedioic acid + 1,6-Hexanediamine

repeat *n* times

Nylon-66 + $n$ H$_2$O

## 19.11 Copolymers and Composites

- **Copolymers**
  - → Polymers with more than one type of repeating unit
  - → Produced from more than one type of monomer
  - → Four different forms: *alternating, block, random,* and *graft* copolymers (see Text Fig. 19.14)

- **Alternating copolymers**
  - → Follow the pattern: −A-B-A-B-A-B-A-B−, where A and B are monomer units
  - → Nylon-66 (see above) is an *alternating* copolymer. The index (66) indicates the number of C atoms (6) in each type of monomer.

- **Block copolymers**
  - → Follow the pattern: −A-A-A-A-B-B-B-B-A-A-A−
  - → Long segments of one monomer, A, are followed by long segments of monomer B.

    **Example:** *High-impact polystyrene* is a block copolymer of *styrene* and *butadiene:*

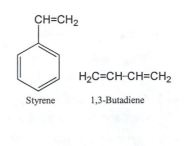

Styrene     1,3-Butadiene

- **Random copolymers**
  - → Follow no particular pattern:

    −A-A-B-A-B-A-B-B-A-B-A-B-B-B-A−B-AA-BBB-

    **Example:** Radical polymerization of the monomers styrene and 3-methylstyrene is expected to form a random copolymer:

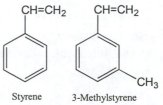

Styrene     3-Methylstyrene

- **Graft copolymers**
  → Consist of long chains of one monomer, A, with pendant chains of the second monomer, B:

  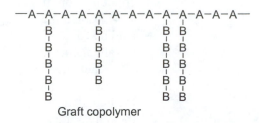
  Graft copolymer

  **Example:** Soft contact lenses are composed of a graft copolymer that has a backbone of nonpolar monomers but side groups of a different water-absorbing monomer.

- **Properties of copolymers**
  → The different types of copolymers extend the range of physical properties obtainable for materials.

  **Example:** Soft polyurethane foams formed from *diisocyanates* and *glycols* are used for insulation and for furniture stuffing.  The formation of a *polyurethane* is shown below:

  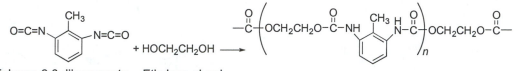

  Toluene-2,6-diisocyanate    Ethylene glycol                                     A  polyurethane

- **Composite materials**
  → Consist of two or more materials solidified together
  → Combine the advantages of the component materials
  → Can exhibit properties superior to those of the component materials

  **Example:** Fiberglass, a material of great strength and flexibility, consists of inorganic materials in a polymer matrix.

## 19.12 Physical Properties of Polymers

- **Synthetic polymers**
  → No definite molar mass, only an *average* value
  → Tend to soften gradually upon heating and have no definite melting point
  → Contain chains of various lengths mixed together
  → For a given polymer, longer *average* chain length leads to higher viscosity and a higher softening point.

- **Properties of polymers**
  → Depend upon the *average* chain length
  → Depend upon the *polarity* of the functional groups
  → Polar groups are associated with stronger intermolecular forces and tend to increase both softening points and mechanical strength.
  → Depend on the manner in which chains pack.
  → Long unbranched chains form crystalline regions that lead to strong, dense materials.
  → Branched-chain polymers exhibit more tangled arrangements; they are less likely to form crystalline regions and tend to be weaker, less dense materials.

- **Elasticity**
  - → Ability of a polymer to return to its original shape after being stretched
  - → *Elastomers* are materials that return easily to their original shape after stretching.
  - → *Elasticity* of natural rubber is improved by vulcanization (heating with S), which forms disulfide (–S–S-) links between chains, increasing the resilience of the polymer (see Text Fig. 19.17).
  - → Extensive cross-linking provides a rigid network of interlinked polymer chains and leads to very hard materials.
  - → Most polymers are electrical insulators, but conducting polymers are known (see Box 19.1).

---

# The Impact on Biology  (Sections 19.13–19.15)

## 19.13 Proteins

- **Proteins**
  - → Condensation copolymers of *up to* 20 naturally occurring amino acids (see Text Table 19.4)
  - → Nine of the 20 amino acids, known as *essential amino acids,* cannot be produced by the human body.
  - → Perform highly specific functions in the human body
  - → Enzymes (globular proteins) act as specific and efficient catalysts.

    **Example:** The enzyme *alcohol dehydrogenase,* a globular protein, oxidizes ethanol to ethanal.

- **Peptides**
  - → Molecules formed from two or more amino acids
  - → Named starting with the amino acid on the left (N-terminus)
  - → The –CO–NH– link is called a *peptide bond,* and each amino acid in a peptide is called a *residue.*
  - → Typical proteins contain *polypeptide chains* of more than a hundred residues joined through peptide bonds and arranged in a particular order.
  - → *Oligopeptides* are molecules with only a few amino acid residues.

    **Example:** Reaction of the naturally occurring amino acids, aspartic acid (Asp) and phenylalanine (Phe), produces the artificial sweetener aspartame (Asp-Phe), a *dipeptide* which contains two *residues*. Notice the peptide link, –CO–NH–, shown in the box.

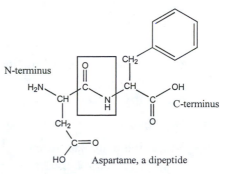

Aspartame, a dipeptide

- **Structure of proteins**
  - → *Primary structure*
    - ➤ Sequence of residues in the peptide chain

    **Example:** The *primary* structure of *aspartame* is Asp-Phe. The *dipeptide* with the N- and C-termini interchanged is Phe-Asp, a *different* oligopeptide.

→ *Secondary structure*

➤ Describes the shape of the polypeptide chain

➤ Controlled by *intramolecular* interactions between different parts of the peptide chain

➤ Common *secondary* structures include the α *helix* and the β *sheet* (see Text Figs. 19.19 and 19.20).

➤ The α *helix* is a helical portion of a polypeptide held together by hydrogen bonding.

➤ The β-*pleated sheet* is a portion of a polypeptide in which the segments of the chain lie side by side to form nearly flat sheets.

→ *Tertiary structure*

➤ *Tertiary* structure describes the overall three-dimensional shape of the polypeptide, including the way in which α helix and β sheet regions fold together to shape the macromolecule.

➤ Folding is a consequence of the hydrophobic and hydrophilic interactions between residues lying in different parts of the primary structure. One important link responsible for *tertiary* structure is the *disulfide link* (–S–S–) between amino acids containing sulfur.

➤ Proteins are classified as *globular* (ball shaped) or *fibrous* (hair shaped with long chains of polypeptides that occur in bundles).

➤ Globular proteins are soluble in water; fibrous proteins are not. Essentially all enzymes are globular proteins.

**Example:** *Hemoglobin* (responsible for oxygen transport in blood) and *cytochrome c* (a component of electron transport chains in mitochondria and bacteria) are globular proteins; *fibroin* (the protein of silk) is a fibrous protein.

→ *Quaternary structure*

➤ Describes the arrangement of *subunits* in proteins with more than one polypeptide chain

➤ Not all proteins contain more than one subunit, so not all have a *quaternary* structure.

**Example:** *Hemoglobin* (see Text Fig. 19.20) contains four polypeptide units and has a *quaternary* structure.

- **Denaturation**

  → Loss of structure of proteins

  → Occurs when a protein loses quaternary, tertiary, or secondary structure (disruption of *noncovalent* interactions)

  → *Denaturation* is caused by heating and other means, is often irreversible, and is accompanied by loss of function of the protein.

## 19.14 Carbohydrates

- **Carbohydrates**

  → Most abundant class of naturally occurring organic compounds

  → Constitute more than 50% of the Earth's biomass

  → Members include starches, cellulose, and sugars.

  → Often have the empirical formula $CH_2O$ (hence the name *carbohydrates*)

- **D-Glucose**

  → Most abundant carbohydrate

→ Forms starch and cellulose by polymerization reactions

→ Has the molecular formula $C_6H_{12}O_6$

→ Is classified as both an alcohol *and* an aldehyde (*pentahydroxyl aldehyde*)

→ Exists in an acyclic (open-chain) and in two cyclic (closed-chain) forms called *anomers*

→ In water, the cyclic structures, also called *glucopyranoses*, are favored.

- **Structure of D-glucose**

  → The straight-chain structure of D-glucose and the interconversion to the two cyclic *anomers* are on the right.

  → The percentages indicate proportions of the several species present in water solution.

  → The anomers, α-D-glucose and β-D-glucose, differ in the stereochemistry at the carbon atom derived from the aldehyde C atom.

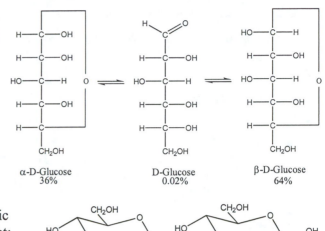

  → More realistic representations of the cyclic forms of D-glucose are shown on the right:

  **Notes:** 1) L-Glucose, not found in nature, is the mirror image of D-glucose, and together they form a pair of *enantiomers*, which differ in the direction of rotation of polarized light.

   2) The symbols D and L refer to a configuration relationship to the compound D-glyceraldehyde, and not necessarily to the sign of rotation of light.

- **Polysaccharides**

  → Polymers of glucose

  → Include starch (digestible by humans) and cellulose (not digestible by humans)

   **Example:** Cellulose, the structural material of plants and wood, is a condensation polymer of β-D-glucose. A three subunit chain of cellulose is shown below. The polymer is formed formally by elimination of $H_2O$ from glucose monomers.

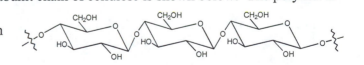

## 19.15 Nucleic Acids

- **DNA**

  → Abbreviation for *deoxyribonucleic acid*

  → DNA molecules

   ➤ Carry genetic information from generation to generation

   ➤ Control the production of proteins

   ➤ Serve as the template for the synthesis of RNA (see RNA subsection)

- **Structure of DNA**

  → DNA molecules are condensation copolymers of enormous size.

  → DNA is composed of a sugar phosphate backbone and pendent bases as shown on the right:

  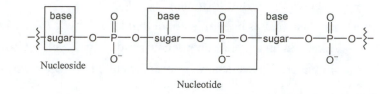

  → A base-sugar unit is called a *nucleoside*.

  → A base-sugar-phosphate grouping is called a *nucleotide*.

- **Sugar and bases**

  → In DNA, the sugar is *deoxyribose*.

  → There are four possible bases: *cytosine* (C), *guanine* (G), *adenine* (A), and *thymine* (T).

  → The sugar and four bases are shown on the right:

  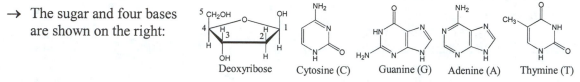

  Deoxyribose    Cytosine (C)    Guanine (G)    Adenine (A)    Thymine (T)

- **Nucleosides**

  → Formed in DNA by condensation involving the OH group at carbon 1 in deoxyribose and an appropriate amine hydrogen atom of a base

  → Cytosine, guanine, adenine, and thymine nucleosides are shown here:

  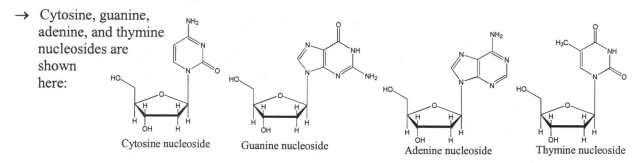

  Cytosine nucleoside    Guanine nucleoside    Adenine nucleoside    Thymine nucleoside

- **Nucleotides**

  → Condense at carbon 3 and carbon 5, eliminating water, to form the DNA molecule (nucleic acid), a *polynucleotide*

  → A trinucleotide segment of a polynucleotide is shown on the right:

  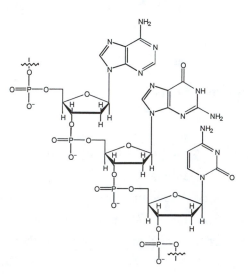

- **DNA double helix**
  - → Polynucleotide (nucleic acid) strands link to each other in pairs to form the well-known double helix structure (see Text Fig. 19.30).
  - → Association between strands of two polynucleotides occurs when hydrogen bonds between bases on *different* strands are formed.
  - → In DNA, only two types of base pairs (G with C and A with T) occur.
  - → These are shown below with the hydrogen bonds indicated by dashed lines:

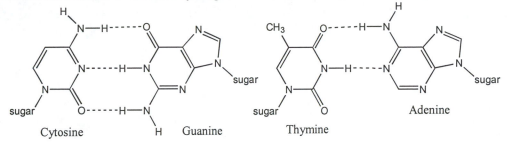

Cytosine    Guanine    Thymine    Adenine

**Note:** CG and AT base pairs have approximately the same size and shape, which reduces distortions in the double helix structure.

- **RNA**
  - → The abbreviation for *ribonucleic acid*
  - → Is a polynucleotide, like DNA
  - → Is shorter than DNA, and generally single-stranded

- **Three types of RNA**
  - → *Messenger RNA* (mRNA) carries genetic information from a cell nucleus to the cytoplasm where translation to protein takes place.
  - → *Ribosomal RNA* (rRNA) appears to act as a catalyst in the biosynthesis of proteins.
  - → *Transfer RNA* (tRNA) is used to transport amino acids during protein synthesis.

- **Structure of RNA**
  - → RNA contains four base pairs as in DNA, but *uracil* (U) substitutes for *thymine* (T).
  - → *Uracil* lacks the methyl group of *thymine* at the carbon-5 position. The structures of thymine and uracil are

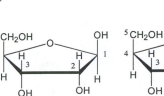

Thymine (T)    Uracil (U)

  **Note:** The position of the methyl group on thymine does not interfere with its hydrogen bonding with adenine (A) (see the preceding figure showing an AT base pair).

  - → The sugar in RNA is ribose (deoxyribose is shown for comparison):
  - → Although RNA molecules are generally single-stranded, they contain regions of a double helix produced by the formation of *hairpin loops*.

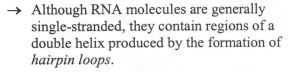

Ribose    Deoxyribose

  - → In these regions, the usual base pairing is A with U, and G with C.
  - → Imperfections in base pairing are common, however, in RNA.

# SOLUTIONS MANUAL

# FUNDAMENTALS

**A.1** (a) chemical;  (b) physical;  (c) physical

**A.3** The temperature, the humidity, and the evaporation of water are physical properties. The ripening of oranges is a chemical change.

**A.5** (a) intensive;  (b) intensive;  (c) extensive;  (d) extensive

**A.7** $d = \dfrac{m}{V}$

$$= \left( \frac{112.32 \text{ g}}{29.27 \text{ mL} - 23.45 \text{ mL}} \right) \left( \frac{1 \text{ mL}}{1 \text{ cm}^3} \right)$$

$$= 19.3 \text{ g} \cdot \text{cm}^{-3}$$

**A.9** $d = \dfrac{m}{V}$, rearranging gives $V = \dfrac{m}{d}$

$$= \left( \frac{0.750 \text{ carat}}{3.51 \text{ g} \cdot \text{cm}^{-3}} \right) \left( \frac{200 \text{ mg}}{1 \text{ carat}} \right) \left( \frac{1 \text{ g}}{1000 \text{ mg}} \right)$$

$$= 0.0427 \text{ cm}^3$$

**A.11** $d = \dfrac{m}{V}$

$$= \left( \frac{3.95 \times 10^{-22} \text{ g}}{\frac{4}{3} \pi (138 \text{ pm})^3} \right) \left( \frac{1 \text{ pm}}{1 \times 10^{-10} \text{ cm}} \right)^3$$

$$= 35.9 \text{ g} \cdot \text{cm}^{-3}$$

Because the density of metallic uranium is much less than the density of a uranium atom, the metallic form of uranium must contain considerable empty space.

**A.13** (a) $d = \dfrac{m}{V}$

$$= \left( \frac{0.213 \text{ g}}{1.100 \text{ cm} \times 0.531 \text{ cm} \times 0.212 \text{ cm}} \right)$$

$$= \left( \frac{0.213 \text{ g}}{0.1238 \text{ cm}^3} \right)$$

$$= 1.72 \text{ g} \cdot \text{cm}^{-3}$$

This determination is more precise because the volume is not limited to 2 significant figures as it is in part (b).

(b) $d = \dfrac{m}{V}$

$$= \left( \frac{41.003 \text{ g} - 39.753 \text{ g}}{20.37 \text{ mL} - 19.65 \text{ mL}} \right)$$

$$= \left( \frac{1.250 \text{ g}}{0.72 \text{ mL}} \right) \left( \frac{1 \text{ mL}}{1 \text{ cm}^3} \right)$$

$$= 1.7 \text{ g} \cdot \text{cm}^{-3}$$

**A.15** $E_K = \dfrac{1}{2} m v^2$

$$= \frac{1}{2} (4.2 \text{ kg})(14 \text{ km} \cdot \text{h}^{-1})^2 \left( \frac{1 \text{ h}}{3600 \text{ s}} \right)^2 \left( \frac{1000 \text{ m}}{1 \text{ km}} \right)^2$$

$$= 32 \text{ kg} \cdot \text{m}^2 \cdot \text{s}^{-2}$$

$$= 32 \text{ J}$$

**A.17** $m = 2.8$ metric tons, $v_i = 100 \text{ km} \cdot \text{hr}^{-1}$, $v_f = 50 \text{ km} \cdot \text{hr}^{-1}$

$$E_K = \frac{1}{2} m v^2$$

$$E_{K(init)} = \frac{1}{2} (2.8 \text{ metric tons}) \left( \frac{10^3 \text{ kg}}{1 \text{ metric ton}} \right)$$

$$\left[ \left( \frac{100 \text{ km}}{1 \text{ hr}} \right) \left( \frac{1 \text{ hr}}{60 \text{ min.}} \right) \left( \frac{1 \text{ min.}}{60 \text{ sec.}} \right) \right]^2$$

$$= 4.32 \text{ kg} \cdot \text{km}^2 \cdot \text{s}^{-2} = 4.32 \times 10^6 \text{ kg} \cdot \text{m}^2 \cdot \text{s}^{-2} = 4,320 \text{ kJ}$$

$$E_{K(final)} = \frac{1}{2}(2.8 \text{ metric tons})\left(\frac{10^3 \text{ kg}}{1 \text{ metric ton}}\right)$$

$$\left[\left(\frac{50 \text{ km}}{1 \text{ hr}}\right)\left(\frac{1 \text{ hr}}{60 \text{ min.}}\right)\left(\frac{1 \text{ min.}}{60 \text{ sec.}}\right)\right]^2$$

$$= 0.27 \text{ kg} \cdot \text{km}^2 \cdot \text{s}^{-2} = 0.27 \times 10^6 \text{ kg} \cdot \text{m}^2 \cdot \text{s}^{-2} = 270 \text{ kJ}$$

$$E_{K(init)} - E_{K(final)} = (4{,}320 - 270) \text{ kJ} = 4{,}050 \text{ kJ} = 4.0 \times 10^3 \text{ kJ (2 SF)}$$

This amount of energy could have been recovered, neglecting friction and other losses, or used to drive the vehicle up a hill.

$$E_P = mgh \qquad g = 9.81 \text{ ms}^{-2}$$

Setting potential energy equal to 4,050 kJ=4.05 kg m$^2$ s$^{-2}$ and solving for height gives

$$h = \frac{E_P}{mg} = \left(\frac{4.05 \times 10^6 \text{ kg} \cdot \text{m}^2 \cdot \text{s}^{-2}}{(2800 \text{ kg})(9.81 \text{ ms}^{-2})}\right) = 147 \text{ m} = 150 \text{ m (2 SF)}$$

**A.19** $E_P = mgh$

$$= (40.0 \text{ g})(9.81 \text{ m} \cdot \text{s}^{-2})(0.50 \text{ m})\left(\frac{1 \text{ kg}}{1000 \text{ g}}\right)$$

$$= 0.20 \text{ kg} \cdot \text{m}^2 \cdot \text{s}^{-2} \text{ for one raise of a fork.}$$

For 30 raises, $(30)(0.20 \text{ kg} \cdot \text{m}^2 \cdot \text{s}^{-2}) = 6.0 \text{ J}$

**A.21** (a) The energy is all potential energy before the ball is dropped. After the ball has fallen halfway, half of the energy has been converted to kinetic energy.

$$E_K = \frac{1}{2}mgh$$

$$= \frac{1}{2}(0.95 \text{ kg})(9.81 \text{ m} \cdot \text{s}^{-2})(13.9 \text{ m})$$

$$= 65 \text{ kg} \cdot \text{m}^2 \cdot \text{s}^{-2}$$

$$= 65 \text{ J}$$

(b) When the ball hits the floor, all of the energy has been converted to kinetic energy.

$$E_K = mgh$$
$$= (0.95 \text{ kg})(9.81 \text{ m} \cdot \text{s}^{-2})(13.9 \text{ m})$$
$$= 1.3 \times 10^2 \text{ kg} \cdot \text{m}^2 \cdot \text{s}^{-2}$$
$$= 1.3 \times 10^2 \text{ J}$$

**A.23** We need to use the expansion given in Exercise A.22 to help solve this problem. We also need to recognize that $E_p = egh$ for the small difference in distance, $h$, can be represented by subtracting $E_p$ at distance $r$ between the proton and electron from $E_p$ at distance $r+h$.

$$E_p = \Delta E_p = \frac{q_1 q_2}{4\pi\varepsilon_0 (r+h)} - \frac{q_1 q_2}{4\pi\varepsilon_0 r}$$

Since $q_1 = e$ and $q_2 = -e$,

$$= -\frac{e^2}{4\pi\varepsilon_0 (r+h)} + \frac{e^2}{4\pi\varepsilon_0 r}$$

$$= -\frac{e^2}{4\pi\varepsilon_0 r} \left( \frac{1}{1 + \dfrac{h}{r}} - \frac{1}{1} \right)$$

If $x = \dfrac{h}{r}$, then $\dfrac{1}{1 + \dfrac{h}{r}} = 1 - \dfrac{h}{r}$ from the expansion in $x$ (see A.22). Then

$$= -\frac{e^2}{4\pi\varepsilon_0 r} \left( 1 - \frac{h}{r} - 1 \right) = \frac{e^2}{4\pi\varepsilon_0 r} \left( \frac{h}{r} \right) = \frac{e^2 h}{4\pi\varepsilon_0 r^2}$$

$$E_p = \frac{e^2 h}{4\pi\varepsilon_0 r^2} = egh$$

So $g = \dfrac{e^2 h}{4\pi\varepsilon_0 r^2} \left( \dfrac{1}{eh} \right) = \dfrac{e}{4\pi\varepsilon_0 r^2}$ when $E_p = egh$

**A.25** The relationship between distance of separation and potential energy for charged particles is given in section A.2, equation 4.

$$E_p = V(r) = \frac{q_1 q_2}{4\pi\varepsilon_0 r} = \frac{(-e)(+e)}{4\pi\varepsilon_0 r}$$

$$= \frac{-(1.602\times10^{-19}\,\text{C})^2}{4\pi(8.85419\times10^{-12}\,\text{C}^2\cdot\text{J}^{-1}\cdot\text{m}^{-1})(53\times10^{-12}\,\text{m})}$$

$$= -4.352\times10^{-18}\,\text{J}\left(\frac{1\,\text{eV}}{1.602\times10^{-19}\,\text{J}}\right) = -27.17\,\text{eV} = -27\,\text{eV (2 SF)}$$

Considering the proton and electron beginning at rest and at infinite separation sets the initial total energy to 0. Since the electron is not at rest in a hydrogen atom, its total energy is represented by equation 5:

$$E = E_K + E_P$$

We have only calculated the potential energy. The discrepancy between the calculated value of the potential energy, -27 eV, and the measured amount released, -13.6 eV, is the kinetic energy of the electron, 13.6 eV.

**A.27**  SI unit of pressure $= \text{Pa} = \text{N}\cdot\text{m}^{-2} = \left(\frac{\text{kg}\cdot\text{m}}{\text{s}^2}\right)\left(\frac{1}{\text{m}^2}\right) = \frac{\text{kg}}{\text{m}\cdot\text{s}^2}$

SI unit of volume $= \text{m}^3$ or $(\text{L} = \text{dm}^3 = 1\times10^{-3}\,\text{m}^3)$

SI unit of energy $= \text{J} = \dfrac{\text{kg}\cdot\text{m}^2}{\text{s}^2}$

The product of pressure and volume is

$$1\,\text{Pa}\times1\,\text{m}^3 = \left(\frac{\text{kg}}{\text{m}\cdot\text{s}^2}\right)(\text{m}^3) = \frac{\text{kg}\cdot\text{m}^2}{\text{s}^2} = 1\,\text{J}$$

(Note: If one uses L instead of $\text{m}^3$ for volume, the same overall units result.)

**A.29**  $105.50\,\text{g} - 43.50\,\text{g} = 62.00\,\text{g} = m_{\text{H}_2\text{O}}$

$$d = \frac{m}{V} \qquad V = \frac{m}{d} = \frac{m_{\text{H}_2\text{O}}}{d_{\text{H}_2\text{O}}} = \frac{62.00\,\text{g}}{0.9999\,\text{g}\cdot\text{cm}^{-3}} = 62.00\,\text{cm}^3$$

$$d_{liquid} = \left(\frac{96.75\,\text{g} - 43.50\,\text{g}}{62.00\,\text{cm}^3}\right) = 0.8589\,\text{g}\cdot\text{cm}^{-3}$$

**B.1** number of beryllium atoms $= \dfrac{\text{mass of sample}}{\text{mass of one atom}}$

$$= \left( \dfrac{0.210\ \text{g}}{1.50 \times 10^{-26}\ \text{kg} \cdot \text{atom}^{-1}} \right) \left( \dfrac{1\ \text{kg}}{1000\ \text{g}} \right)$$

$$= 1.40 \times 10^{22}\ \text{atoms}$$

**B.3** (a) Radiation may pass through a metal foil. (b) All light (electromagnetic radiation) travels at the same speed; the slower speed supports the particle model. (c) This observation supports the radiation model. (d) This observation supports the particle model; electromagnetic radiation has no mass and no charge.

**B.5** (a) $4.80 \times 10^{-10}$ esu; (b) 14 electrons

**B.7** (a) 5p, 6n, 5e; (b) 5p, 5n, 5e; (c) 15p, 16n, 15e; (d) 92p, 146n, 92e

**B.9** (a) $^{194}\text{Ir}$; (b) $^{22}\text{Ne}$; (c) $^{51}\text{V}$

**B.11** (a) they all have the same mass; (b) they have differing numbers of protons, neutrons, and electrons

**B.13** bromine-79, bromine-81

**B.15** (a) 0.5359; (b) 0.4639; (c) 0.0002526; (d) 463.9 kg

**B.17** (a) Rubidium, Group 1 metal; (b) Radium, Group 2 metal; (c) Ruthenium, Group 8 metal; (d) Radon, Group 18 nonmetal.

**B.19** (a) Os, metal; (b) Tl, metal; (c) At, nonmetal

**B.21** Fluorine, F, Z=9, gas; Chlorine, Cl, Z=17, gas; Bromine, Br, Z=35, liquid; Iodine, I, Z=53, solid

**B.23** (a) $d$ block; (b) $s$ block; (c) $p$ block; (d) $d$ block; (e) $p$ block; (f) $d$ block

**C.1** (a) An ionic compound is made up of ions. Sodium chloride is an example of an ionic compound. (b) A molecular compound is made up of molecules. Sucrose (sugar) is an example of a molecular compound. Ionic compounds have higher melting points than solids made of molecules. Ionic compounds dissolve only in polar solvents if at all. Molecules are more likely to dissolve in nonpolar solvents. See Chapter 5.

**C.3** The empirical formula can be represented as $(C_6H_{10}O_1)_n$ . Since each molecule has 2 oxygen atoms, n=2 and the molecular formula is $C_{12}H_{20}O_2$ .

**C.5** (a) Cesium is a metal in Group 1; it will form $Cs^+$ ions. (b) Iodine is a nonmetal in Group 17/VII and will form $I^-$ ions. (c) Selenium is a Group 16/VI nonmetal and will form $Se^{2-}$ ions. (d) Calcium is a Group 2 metal and will form $Ca^{2+}$ ions.

**C.7** (a) $^4He^{2+}$ has 2 protons, 2 neutrons, and no electrons. (b) $^{15}N^{3-}$ has 7 protons, 8 neutrons, and 10 electrons. (c) $^{127}I^-$ has 53 protons, 74 neutrons, and 54 electrons. (d) $^{80}Se^{2-}$ has 34 protons, 46 neutrons, and 36 electrons.

**C.9** (a) $^{19}F^-$; (b) $^{24}Mg^{2+}$; (c) $^{128}Te^{2-}$; (d) $^{86}Rb^+$

**C.11** (a) Aluminum forms $Al^{3+}$ ions; tellurium forms $Te^{2-}$ ions. Two aluminum atoms produce a charge of $2 \times +3 = +6$. Three tellurium atoms produce a charge of $3 \times -2 = -6$. The formula for aluminum telluride is $Al_2Te_3$. (b) Magnesium forms $Mg^{2+}$ ions and oxygen forms $O^{2-}$ ions. A magnesium ion produces a charge of $+2$, which is required to balance the charge on one $O^{2-}$ ion. The formula for magnesium oxide is MgO. (c) Sodium forms $+1$ ions; sulfur forms $-2$ ions. The formula for sodium sulfide is $Na_2S$. (d) Rubidium forms $+1$ ions and iodine forms $-1$ ions. One iodide ions are required to balance the charge of one rubidium ion, so the formula is RbI.

**C.13** (a) HCl, molecular compound (in the gas phase); (b) $S_8$, element (molecular substance); (c) CoS, ionic compound; (d) Ar, element; (e) $CS_2$, molecular compound; (f) $SrBr_2$, ionic compound

**C.15** (a) Group 13; (b) aluminum, Al

**C.17** The formula is $Al_2O_3$.

$$d = \frac{m}{V}$$
$$= \left( \frac{102 \text{ g}}{2.5 \text{ cm} \times 3.0 \text{ cm} \times 4.0 \text{ cm}} \right)$$
$$= 3.4 \text{ g} \cdot \text{cm}^{-3}$$

**D.1** (a) $Al_2O_3$. Aluminum forms $3+$ ions and oxygen forms $2-$ ions. (b) Strontium forms $+2$ ions and the phosphate ion is $PO_4^{3-}$, so the formula is $Sr_3(PO_4)_2$. (c) Aluminum forms $+3$ ions and the carbonate ion is $CO_3^{2-}$, giving a formula of $Al_2(CO_3)_3$. (d) Lithium forms $Li^+$ ions and the nitride ion is $N^{3-}$. The formula of lithium nitride is $Li_3N$.

**D.3** (a) calcium phosphate; (b) tin(II) fluoride, stannous fluoride; (c) vanadium(V) oxide; (d) copper(I) oxide, cuprous oxide

**D.5** (a) $TiO_2$; (b) $SiCl_4$; (c) $CS_2$; (d) $SF_4$; (e) $Li_2S$; (f) $SbF_5$; (g) $N_2O_5$; (h) $IF_7$

**D.7** (a) sulfur hexafluoride; (b) dinitrogen pentoxide; (c) nitrogen triiodide; (d) xenon tetrafluoride; (e) arsenic tribromide; (f) chlorine dioxide

**D.9** (a) hydrochloric acid; (b) sulfuric acid; (c) nitric acid; (d) acetic acid; (e) sulfurous acid; (f) phosphoric acid

**D.11** (a) $Na_2O$; (b) $K_2SO_4$; (c) $AgF$; (d) $Zn(NO_3)_2$; (e) $Al_2S_3$

**D.13** (a) sodium sulfite; (b) iron(III) oxide or ferric oxide; (c) iron(II) oxide or ferrous oxide; (d) magnesium hydroxide; (e) nickel(II) sulfate hexahydrate; (f) phosphorus(V) chloride or phosphorus pentachloride; (g) chromium(III) dihydrogen phosphate; (h) diarsenic trioxide; (i) ruthenium(II) chloride

**D.15** (a) heptane; (b) propane; (c) pentane; (d) butane

**D.17** (a) cobalt (III) oxide monohydrate; $Co_2O_3 \cdot H_2O$; (b) cobalt (II) hydroxide; $Co(OH)_2$

**D.19** E=C; methane; sodium carbide

**D.21** (a) lithium aluminum hydride, ionic (with a molecular anion); (b) sodium hydride, ionic

**D.23** (a) telluric acid; (b) sodium arsenate; (c) calcium selenite; (d) barium antimonate; (e) arsenic acid; (f) cobalt(III) tellurate

**D.25** (a) butanoic acid; (b) 2,3-dimethylbutane; (c) 1,1-dichloropentane

**E.1** (a) $\text{moles of people} = \dfrac{6.0 \times 10^9 \text{ people}}{6.022 \times 10^{23} \text{ people} \cdot \text{mol}^{-1}}$

$$= 1.0 \times 10^{-14} \text{ mol}$$

(b) $\text{time} = \dfrac{1 \text{ mol peas}}{1.0 \times 10^{-14} \text{ mol} \cdot \text{s}^{-1}} = 1.0 \times 10^{14} \text{ s}$

$$(1.0 \times 10^{14} \text{ s}) \left( \frac{1 \text{ h}}{3600 \text{ s}} \right) \left( \frac{1 \text{ day}}{24 \text{ h}} \right) \left( \frac{1 \text{ yr}}{365 \text{ days}} \right) = 3.2 \times 10^6 \text{ years}$$

**E.3** (a) mass of average Li atom

$$= \left( \frac{7.42}{100} \right) (9.988 \times 10^{-24} \text{ g}) + \left( \frac{92.58}{100} \right) (1.165 \times 10^{-23} \text{ g})$$

$$= 1.153 \times 10^{-23} \text{ g} \cdot \text{atom}^{-1}$$

$$\text{molar mass} = (1.153 \times 10^{-23} \text{ g} \cdot \text{atom}^{-1}) (6.022 \times 10^{23} \text{ atoms} \cdot \text{mol}^{-1})$$

$$= 6.94 \text{ g} \cdot \text{mol}^{-1}$$

(b) mass of average Li atom

$$= \left( \frac{5.67}{100} \right) (9.988 \times 10^{-24} \text{ g}) + \left( \frac{100 - 5.67}{100} \right) (1.165 \times 10^{-23} \text{ g})$$

$$= 1.1556 \times 10^{-23} \text{ g} \cdot \text{atom}^{-1}$$

$$\text{molar mass} = (1.1556 \times 10^{-23} \text{ g} \cdot \text{atom}^{-1}) (6.022 \times 10^{23} \text{ atoms} \cdot \text{mol}^{-1})$$

$$= 6.96 \text{ g} \cdot \text{mol}^{-1}$$

**E.5** (a) $MgSO_4 \cdot 7 H_2O$ formula mass $= 246.48 \text{ g} \cdot \text{mol}^{-1}$

$$\text{atoms of O} = \left( \frac{5.15 \text{ g}}{246.48 \text{ g} \cdot \text{mol}^{-1}} \right) \left( \frac{11 \text{ mol O atoms}}{\text{mol } MgSO_4 \cdot 7 H_2O} \right)$$

$$(6.022 \times 10^{23} \text{ atoms} \cdot \text{mol}^{-1}) = 1.38 \times 10^{23}$$

(b) formula units $= \left( \dfrac{5.15 \text{ g}}{246.48 \text{ g} \cdot \text{mol}^{-1}} \right) (6.022 \times 10^{23} \text{ atoms} \cdot \text{mol}^{-1})$

$$= 1.26 \times 10^{22}$$

(c) moles of $H_2O = 7 \left( \dfrac{5.15 \text{ g}}{246.48 \text{ g} \cdot \text{mol}^{-1}} \right) = 0.146 \text{ mol}$

**E.7**   The percentage $^{10}B = 100 -$ percentage $^{11}B$

$$\text{molar mass} = \left( \dfrac{\% \ ^{10}B}{100\%} \right) (\text{mass } ^{10}B) + \left( \dfrac{\% \ ^{11}B}{100\%} \right) (\text{mass } ^{11}B)$$

$$= \left( \dfrac{100\% - \% \ ^{11}B}{100\%} \right) (\text{mass } ^{10}B) + \left( \dfrac{\% \ ^{11}B}{100\%} \right) (\text{mass } ^{11}B)$$

Rearranging gives

$$\% \ ^{11}B = \dfrac{100 \cdot \text{molar mass} - 100 \cdot \text{mass } ^{10}B}{\text{mass } ^{11}B - \text{mass } ^{10}B}$$

$$= \dfrac{100(10.81 \text{ g} \cdot \text{mol}^{-1}) - 100(10.013 \text{ g} \cdot \text{mol}^{-1})}{11.093 \text{ g} \cdot \text{mol}^{-1} - 10.013 \text{ g} \cdot \text{mol}^{-1}}$$

$$= 73.8 \ \%$$

$$\% \ ^{10}B = 26.2 \ \%$$

**E.9**   (a) $\dfrac{75 \text{ g}}{114.82 \text{ g} \cdot \text{mol}^{-1}} = 0.65 \text{ mol In}$

$\dfrac{80 \text{ g}}{127.60 \text{ g} \cdot \text{mol}^{-1}} = 0.63 \text{ mol Te}$

80. g of tellurium contains more moles of atoms than does 75 g of indium.

(b) $\dfrac{15.0 \text{ g}}{30.97 \text{ g} \cdot \text{mol}^{-1}} = 0.484 \text{ mol P}$

$\dfrac{15.0 \text{ g}}{32.07 \text{ g} \cdot \text{mol}^{-1}} = 0.468 \text{ mol S}$

15.0 g of P has slightly more atoms than 15.0 g of S.

(c) Because the two samples have the same number of atoms, they will have the same number of moles, which is given by

$$\frac{2.49 \times 10^{22} \text{ atoms}}{6.022 \times 10^{23} \text{ atoms} \cdot \text{mol}^{-1}} = 0.0413 \text{ mol}$$

**E.11** (a) $m_{\text{Rh}} = \left( \dfrac{57 \text{ g N}}{14.01 \text{ g} \cdot \text{mol}^{-1} \text{ N}} \right) (102.91 \text{ g} \cdot \text{mol}^{-1} \text{ Rh})$

$= 4.1 \times 10^2 \text{ g Rh}$

(b) $m_{\text{Rh}} = \left( \dfrac{57 \text{ g Zr}}{91.22 \text{ g} \cdot \text{mol}^{-1} \text{ Zr}} \right) (102.91 \text{ g} \cdot \text{mol}^{-1} \text{ Rh})$

$= 63 \text{ g Rh}$

**E.13** (a) molar mass of $Al_2O_3 = 101.96 \text{ g} \cdot \text{mol}^{-1}$

$$n_{Al_2O_3} = \frac{10.0 \text{ g}}{101.96 \text{ g} \cdot \text{mol}^{-1}} = 0.0981 \text{ mol}$$

$N_{Al_2O_3} = (0.0981 \text{ mol})(6.022 \times 10^{23} \text{ atoms} \cdot \text{mol}^{-1}) = 5.91 \times 10^{22} \text{ molecules}$

(b) molar mass of HF $= 20.01 \text{ g} \cdot \text{mol}^{-1}$

$$n_{\text{HF}} = \frac{25.92 \times 10^{-3} \text{ g}}{20.01 \text{ g} \cdot \text{mol}^{-1}} = 1.30 \times 10^{-3} \text{ mol}$$

$N_{\text{HF}} = (1.30 \times 10^{-3} \text{ mol})(6.022 \times 10^{23} \text{ atoms} \cdot \text{mol}^{-1})$

$= 7.83 \times 10^{20} \text{ molecules}$

(c) molar mass of hydrogen peroxide $= 34.02 \text{ g} \cdot \text{mol}^{-1}$

$$n_{H_2O_2} = \frac{1.55 \times 10^{-3} \text{ g}}{34.02 \text{ g} \cdot \text{mol}^{-1}} = 4.56 \times 10^{-5} \text{ mol}$$

$N_{H_2O_2} = (4.56 \times 10^{-5} \text{ mol})(6.022 \times 10^{23} \text{ atoms} \cdot \text{mol}^{-1})$

$= 2.75 \times 10^{19} \text{ molecules}$

(d) molar mass of glucose $= 180.15 \text{ g} \cdot \text{mol}^{-1}$

$$n_{\text{glucose}} = \frac{1250 \text{ g}}{180.15 \text{ g} \cdot \text{mol}^{-1}} = 6.94 \text{ mol}$$

$N_{\text{glucose}} = (6.94 \text{ mol})(6.022 \times 10^{23} \text{ atoms} \cdot \text{mol}^{-1}) = 4.18 \times 10^{24} \text{ molecules}$

(e) molar mass of N atoms $= 14.01 \ \text{g} \cdot \text{mol}^{-1}$

$$n_N = \frac{4.37 \ \text{g}}{14.01 \ \text{g} \cdot \text{mol}^{-1}} = 0.312 \ \text{mol}$$

$$N_N = (0.312 \ \text{mol})(6.022 \times 10^{23} \ \text{atoms} \cdot \text{mol}^{-1}) = 1.88 \times 10^{23} \ \text{atoms}$$

molar mass of $N_2$ molecules $= 28.02 \ \text{g} \cdot \text{mol}^{-1}$

$$n_{N_2} = \frac{4.37 \ \text{g}}{28.02 \ \text{g} \cdot \text{mol}^{-1}} = 0.156 \ \text{mol}$$

$$N_{N_2} = (0.156 \ \text{mol})(6.022 \times 10^{23} \ \text{atoms} \cdot \text{mol}^{-1}) = 9.39 \times 10^{22} \ \text{molecules}$$

**E.15** (a) molar mass of AgCl $= 143.32 \ \text{g} \cdot \text{mol}^{-1}$

$$n_{AgCl} = \frac{2.00 \ \text{g}}{143.32 \ \text{g} \cdot \text{mol}^{-1}} = 0.0140 \ \text{mol}$$

The number of moles of $Ag^+$ ions equals the number of moles of AgCl.

(b) molar mass of $UO_3 = 286.03 \ \text{g} \cdot \text{mol}^{-1}$

$$n_{UO_3} = \frac{600 \ \text{g}}{286.03 \ \text{g} \cdot \text{mol}^{-1}} = 2.10 \ \text{mol}$$

(c) molar mass of $FeCl_3 = 162.20$

$$n_{FeCl_3} = \left( \frac{4.19 \ \text{mg}}{162.20 \ \text{g} \cdot \text{mol}^{-1}} \right) \left( \frac{1 \ \text{g}}{1000 \ \text{mg}} \right) = 2.58 \times 10^{-5} \ \text{mol}$$

The number of moles of $Cl^-$ ions equals 3 times the number of moles of $FeCl_3$.

$$n_{Cl^-} = 7.75 \times 10^{-5} \ \text{mol}$$

(d) molar mass of $AuCl_3 \cdot 2 \ H_2O = 339.35 \ \text{g} \cdot \text{mol}^{-1}$

$$\left( \frac{2 \ \text{mol} \ H_2O}{1 \ \text{mol} \ AuCl_3 \cdot 2 \ H_2O} \right) (n_{AuCl_3 \cdot 2 \ H_2O}) = 2 \left( \frac{1.00 \ \text{g}}{339.35 \ \text{g} \cdot \text{mol}^{-1}} \right)$$
$$= 5.89 \times 10^{-3} \ \text{mol} \ H_2O$$

**E.17** (a) number of formula units $= (0.750 \ \text{mol})(6.022 \times 10^{23} \ \text{formula units} \cdot \text{mol}^{-1})$

$= 4.52 \times 10^{23}$ formula units

(b) molar mass of $Ag_2SO_4 = 311.80 \text{ g} \cdot \text{mol}^{-1}$

$$\left( \frac{2.39 \times 10^{20} \text{ formula units}}{6.022 \times 10^{23} \text{ formula units} \cdot \text{mol}^{-1}} \right) (311.80 \text{ g} \cdot \text{mol}^{-1}) \left( \frac{1000 \text{ mg}}{1 \text{ g}} \right)$$

$= 124 \text{ mg}$

(c) molar mass of $NaHCO_2 = 68.01 \text{ g} \cdot \text{mol}^{-1}$

$$\left( \frac{3.429 \text{ g}}{68.01 \text{ g} \cdot \text{mol}^{-1}} \right) (6.022 \times 10^{23} \text{ formula units} \cdot \text{mol}^{-1})$$

$= 3.036 \times 10^{22}$ formula units

**E.19** (a) molar mass of $H_2O = 18.02 \text{ g} \cdot \text{mol}^{-1}$

$$\left( \frac{18.02 \text{ g} \cdot \text{mol}^{-1}}{6.022 \times 10^{23} \text{ molecules} \cdot \text{mol}^{-1}} \right) = 2.992 \times 10^{-23} \text{ g} \cdot \text{molecule}^{-1}$$

(b) $N_{H_2O} = \left( \frac{1000 \text{ g}}{18.02 \text{ g} \cdot \text{mol}^{-1}} \right) (6.022 \times 10^{23} \text{ molecules} \cdot \text{mol}^{-1})$

$= 3.34 \times 10^{25}$ molecules

**E.21** (a) molar mass of $CuBr_2 \cdot 4 H_2O = 295.42 \text{ g} \cdot \text{mol}^{-1}$

$$\left( \frac{7.35 \text{ g}}{295.42 \text{ g} \cdot \text{mol}^{-1}} \right) = 2.49 \times 10^{-2} \text{ mol}$$

(b) Because there are 2 mol $Br^-$ per mole of compound, the number of moles will be twice the amount in part (a), $4.98 \times 10^{-2}$ mol.

(c)

$$\left( \frac{4 \text{ mol } H_2O}{1 \text{ mol } CuBr_2 \cdot 4 H_2O} \right) (2.49 \times 10^{-2} \text{ mol}) (6.022 \times 10^{23} \text{ molecules} \cdot \text{mol}^{-1})$$

$= 6.00 \times 10^{22}$ molecules $H_2O$

(d) fraction of mass due to $O = \dfrac{4(16.00 \text{ g} \cdot \text{mol}^{-1})}{295.42 \text{ g} \cdot \text{mol}^{-1}} = 0.2166$

**E.23** (a)

$$\left(\frac{10 \text{ mol H}_2\text{O}}{1 \text{ mol hydrated cpd}}\right)\left(\frac{18.01 \text{ g H}_2\text{O}}{1 \text{ mol H}_2\text{O}}\right)\left(\frac{1 \text{ mol hydrated cpd}}{264.166 \text{ g hydrated cpd}}\right) \times 100$$

$$= 68.2\% \text{ H}_2\text{O}$$

Therefore 6.82 kg out of 10 kg was water.

$$\frac{? \$}{\text{L H}_2\text{O}} = \left(\frac{\$72.00}{10 \text{ kg hydrated cpd}}\right)\left(\frac{10 \text{ kg hydrated cpd}}{6.82 \text{ kg H}_2\text{O}}\right)\left(\frac{1 \text{ kg H}_2\text{O}}{1 \text{ L H}_2\text{O}}\right)$$

$$= \$10.60 \text{ per L H}_2\text{O}$$

(b) Since the anhydrous compound costs $80.00 for 10 kg, or $8.00/kg, and only 10-6.82=3.18 kg is $NaHCO_3$, a fair price would have been

$$\frac{\$8.00}{\text{kg}} \times 3.18 \text{ kg} = \$25.44 \, .$$

**E.25** (a) moles of Cu $= \dfrac{43.4 \text{ g}}{63.54 \text{ g} \cdot \text{mol}^{-1}} = 0.683$ mol

S atoms required $= (0.683 \text{ mol})(6.022 \times 10^{23} \text{ atoms} \cdot \text{mol}^{-1}) = 4.11 \times 10^{23}$

(b) $S_8$ molecules required $= \dfrac{4.11 \times 10^{23}}{8} = 5.14 \times 10^{22}$

(c) mass of sulfur needed $= (0.683 \text{ mol})(32.06 \text{ g} \cdot \text{mol}^{-1}) = 21.9$ g

**E.27** Solve by factor label (dimensional analysis).

$$? \text{ mole H atoms} = 28.0 \text{ cm}^3 \text{ NaBH}_4 \left(\frac{1.074 \text{ g}}{1 \text{ cm}^3}\right)\left(\frac{2.50 \times 10^{23} \text{ H atoms}}{3.93 \text{ g}}\right)$$

$$\times \left(\frac{1 \text{ mol}}{6.022 \times 10^{23} \text{ H atoms}}\right)$$

$$= 3.18 \text{ moles H atoms}$$

**E.29** For example, Br has $m_1 = 80.9163$ u, $m_2 = 78.9183$ u

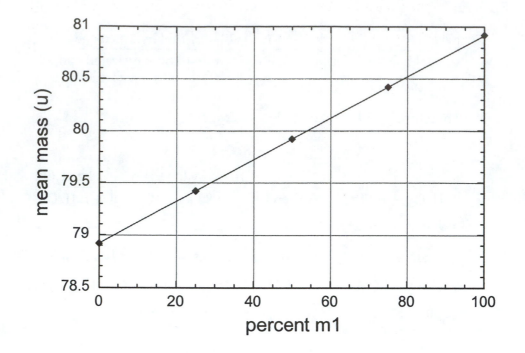

**F.1**    $C_7H_{15}NO_3$

$$7 \times 12.01 \text{ g} = 84.07 \text{ g} \qquad \times \left( \frac{100}{161.20 \text{ g}} \right) = 52.15\% \text{ C}$$

$$15 \times 1.0079 \text{ g} = 15.1185 \text{ g} \quad \times \left( \frac{100}{161.20 \text{ g}} \right) = 9.3787\% \text{ H}$$

$$1 \times 14.01 \text{ g} = 14.01 \text{ g} \qquad \times \left( \frac{100}{161.20 \text{ g}} \right) = 8.691\% \text{ N}$$

$$3 \times 16.00 \text{ g} = \underline{48.00 \text{ g}} \qquad \times \left( \frac{100}{161.20 \text{ g}} \right) = \underline{29.78\%} \text{ O}$$

$$\qquad\qquad\qquad 161.20 \text{ g} \qquad\qquad\qquad\qquad 100.00\%$$

**F.3**    (a) $M_2O$    88.8% M    For 100 g of compound, 88.8 g is M, 11.2 g is O.

$$\frac{? \text{ g M}_2\text{O}}{\text{mole M}_2\text{O}} = \left( \frac{100 \text{ g M}_2\text{O}}{11.2 \text{ g O}} \right)\left( \frac{16.00 \text{ g O}}{1 \text{ mol O}} \right)\left( \frac{1 \text{ mol O}}{1 \text{ mol M}_2\text{O}} \right)$$

$$= 143 \text{ g} \cdot \text{mol}^{-1} \text{ M}_2\text{O}$$

Therefore, 143-16=127 g/mol are due to M. Since there are 2 moles of M per mole of $M_2O$,

the molar mass of M=63.4 g/mol. That molar mass matches Cu.  (b)
copper (I) oxide

**F.5**    (a)  For 100 g of compound,

$$\text{moles of Na} = \frac{32.79 \text{ g}}{22.99 \text{ g} \cdot \text{mol}^{-1}} = 1.426 \text{ mol}$$

$$\text{moles of Al} = \frac{13.02 \text{ g}}{26.98 \text{ g} \cdot \text{mol}^{-1}} = 0.4826 \text{ mol}$$

$$\text{moles of F} = \frac{54.19 \text{ g}}{19.00 \text{ g} \cdot \text{mol}^{-1}} = 2.852 \text{ mol}$$

Dividing each number by 0.4826 gives a ratio of 1 Al : 2.95 Na : 5.91 F.
The formula is $Na_3AlF_6$.

(b)  For 100 g of compound,

$$\text{moles of K} = \frac{31.91 \text{ g}}{39.10 \text{ g} \cdot \text{mol}^{-1}} = 0.8161 \text{ mol}$$

$$\text{moles of Cl} = \frac{28.93 \text{ g}}{35.45 \text{ g} \cdot \text{mol}^{-1}} = 0.8161 \text{ mol}$$

mass of O is obtained by difference:

$$\text{moles of O} = \frac{100 \text{ g} - 31.91 \text{ g} - 28.93 \text{ g}}{16.00 \text{ g} \cdot \text{mol}^{-1}} = 2.448 \text{ mol}$$

Dividing each number by 0.8161 gives a ratio of 1.00 K : 1 Cl : 3.00 O.
The formula is $KClO_3$.

(c)  For 100 g of compound,

$$\text{moles of N} = \frac{12.2 \text{ g}}{14.01 \text{ g} \cdot \text{mol}^{-1}} = 0.871 \text{ mol}$$

$$\text{moles of H} = \frac{5.26 \text{ g}}{1.0079 \text{ g} \cdot \text{mol}^{-1}} = 5.22 \text{ mol}$$

$$\text{moles of P} = \frac{26.9 \text{ g}}{30.97 \text{ g} \cdot \text{mol}^{-1}} = 0.869 \text{ mol}$$

$$\text{moles of O} = \frac{55.6 \text{ g}}{16.00 \text{ g} \cdot \text{mol}^{-1}} = 3.475 \text{ mol}$$

Dividing each number by 0.869 gives a ratio of 1.00 N : 6.01 H : 1.00 P : 4.00 O. The formula is $NH_6PO_4$ or $[NH_4][H_2PO_4]$, ammonium dihydrogen phosphate.

**F.7** $$\text{moles of P} = \frac{4.14 \text{ g}}{30.97 \text{ g} \cdot \text{mol}^{-1}} = 0.134 \text{ mol}$$

$$\text{moles of Cl} = \frac{27.8 \text{ g} - 4.14 \text{ g}}{35.45 \text{ g} \cdot \text{mol}^{-1}} = 0.667 \text{ mol}$$

Dividing each number by 0.134 mol gives a ratio of 4.98 Cl : 1 P. The formula is $PCl_5$.

**F.9** For 100 g of compound,

$$\text{moles of C} = \frac{54.82 \text{ g}}{12.01 \text{ g} \cdot \text{mol}^{-1}} = 4.565 \text{ mol}$$

$$\text{moles of H} = \frac{5.62 \text{ g}}{1.0079 \text{ g} \cdot \text{mol}^{-1}} = 5.58 \text{ mol}$$

$$\text{moles of N} = \frac{7.10 \text{ g}}{14.01 \text{ g} \cdot \text{mol}^{-1}} = 0.507 \text{ mol}$$

$$\text{moles of O} = \frac{32.46 \text{ g}}{16.00 \text{ g} \cdot \text{mol}^{-1}} = 2.029 \text{ mol}$$

Dividing each number by 0.507 mol gives a ratio of 9.00 C : 11.01 H : 1.00 N : 4.00 O. The formula is $C_9H_{11}NO_4$.

**F.11** For 100 g of the osmium carbonyl compound,

$$\text{moles of C} = \frac{15.89 \text{ g}}{12.01 \text{ g} \cdot \text{mol}^{-1}} = 1.323 \text{ mol}$$

$$\text{moles of O} = \frac{21.18 \text{ g}}{16.00 \text{ g} \cdot \text{mol}^{-1}} = 1.324 \text{ mol}$$

$$\text{moles of Os} = \frac{62.93 \text{ g}}{190.2 \text{ g} \cdot \text{mol}^{-1}} = 0.3309 \text{ mol}$$

Dividing each number by 0.3309 mol gives a ratio of 4.00 C : 4.00 O : 1.00 Os. (a) The empirical formula is $OsC_4O_4$. (b) The formula mass of $OsC_4O_4$ is 302.24 $g \cdot mol^{-1}$. The molar mass is 907 $g \cdot mol^{-1}$ which is 3 times the formula mass, so the molecular formula is $Os_3C_{12}O_{12}$.

**F.13** For 100 g of caffeine,

$$\text{moles of C} = \frac{49.48 \text{ g}}{12.01 \text{ g} \cdot mol^{-1}} = 4.12 \text{ mol}$$

$$\text{moles of H} = \frac{5.19 \text{ g}}{1.0079 \text{ g} \cdot mol^{-1}} = 5.15 \text{ mol}$$

$$\text{moles of N} = \frac{28.85 \text{ g}}{14.01 \text{ g} \cdot mol^{-1}} = 2.059 \text{ mol}$$

$$\text{moles of O} = \frac{16.48 \text{ g}}{16.00 \text{ g} \cdot mol^{-1}} = 1.03 \text{ mol}$$

Dividing each number by 1.03 mol gives a ratio of 4.00 C : 5.00 H : 2.00 N : 1.00 O. The formula is $C_4H_5N_2O$ with a molar formula mass of 97.10 $g \cdot mol^{-1}$. Because the molecular molar mass is twice this value, the actual formula will be $C_8H_{10}N_4O_2$.

**F.15** Glucose ($C_6H_{12}O_6$) has a molar mass of 180.15 $g \cdot mol^{-1}$ and will have the following composition:

$$\%C = \frac{6(12.01 \text{ g} \cdot mol^{-1})}{180.15 \text{ g} \cdot mol^{-1}} = 40.00\%$$

$$\%H = \frac{12(1.0079 \text{ g} \cdot mol^{-1})}{180.15 \text{ g} \cdot mol^{-1}} = 6.71\%$$

$$\%O = \frac{6(16.00 \text{ g} \cdot mol^{-1})}{180.15 \text{ g} \cdot mol^{-1}} = 53.29\%$$

Sucrose ($C_{12}H_{22}O_{11}$) has a molar mass of 342.29 $g \cdot mol^{-1}$ and will have the following composition:

$$\%C = \frac{12(12.01 \text{ g} \cdot \text{mol}^{-1})}{342.29 \text{ g} \cdot \text{mol}^{-1}} = 42.10\%$$

$$\%H = \frac{22(1.0079 \text{ g} \cdot \text{mol}^{-1})}{342.29 \text{ g} \cdot \text{mol}^{-1}} = 6.48\%$$

$$\%O = \frac{11(16.00 \text{ g} \cdot \text{mol}^{-1})}{342.29 \text{ g} \cdot \text{mol}^{-1}} = 51.42\%$$

While the %H values for glucose and sucrose are too close to allow us to distinguish between them by this value alone, %C (40.00 versus 42.10%) and %O (53.29 versus 51.42%) values are sufficient that, when taken together, can give us a reasonable amount of confidence in distinguishing between them.

**F.17**  Calculate the mass percent carbon for each fuel from its formula.

ethene, $C_2H_4$ $\qquad \frac{2(12.01)}{2(12.01)+4(1.0079)} \times 100 = 85.63\% \text{ C}$

propanol, $C_3H_7OH$ $\qquad \frac{3(12.01)}{3(12.01)+8(1.0079)+16.00} \times 100 = 59.96\% \text{ C}$

heptane, $C_7H_{16}$ $\qquad \frac{7(12.01)}{7(12.01)+16(1.0079)} \times 100 = 83.91\% \text{ C}$

ethene (85.63%) > heptane (83.91%) > propanol (59.96%)

**F.19**  This problem requires that we relate unknowns to each other appropriately by writing a balanced chemical equation and using other information in the problem.

$$x \text{ NaNO}_3 + y \text{ Na}_2\text{SO}_4 \rightarrow (x + 2y) \text{ Na}^+ + x \text{ NO}_3^- + y \text{ SO}_4^{2-}$$

$$\frac{1.61 \text{ g Na}^+}{22.99 \text{ g} \cdot \text{mol}^{-1}\text{Na}^+} = 0.07003 \text{ mol Na}^+ = x + 2y$$

$$5.37 \text{ g total} - 1.61 \text{ g Na}^+ = 3.76 \text{ g} = (62.01 \text{ g} \cdot \text{mol}^{-1})x + (96.07 \text{ g} \cdot \text{mol}^{-1})y$$

Rearrange and substitute:

$3.76 = 62.01(0.07003 - 2y) + 96.07y$

$0.06065 = 0.07003 - 2y + 1.548y$

$0.009385 = 0.4516y$

$y = 0.02078$ moles sulfate, $x = 0.02847$ moles nitrate

Therefore, the mass of sodium nitrate in the mixture was

$$0.02847 \text{ mol} \left( \frac{85.00 \text{ g NaNO}_3}{1 \text{ mol}} \right) = 2.42 \text{ g NaNO}_3$$

$$\frac{2.42 \text{ g NaNO}_3}{5.37 \text{ g total}} \times 100 = 45.1\% \text{ NaNO}_3$$

**G.1**  (a) solubility;   (b) the abilities of the components to adsorb;   (c) boiling points

**G.3**  (a) homogeneous, distillation;   (b) heterogeneous, dissolving followed by filtration and distillation;   (c) homogeneous, distillation

**G.5**  mass of $AgNO_3 = (0.179 \text{ mol} \cdot L^{-1})(0.5000 \text{ L})(169.88 \text{ g} \cdot \text{mol}^{-1}) = 15.2 \text{ g}$

**G.7**  (a) molarity of $Na_2CO_3 = \dfrac{2.111 \text{ g}}{(105.99 \text{ g} \cdot \text{mol}^{-1})(0.2500 \text{ L})} = 0.079\,67 \text{ M}$

$Na_2CO_3$

$$V = \frac{(2.15 \times 10^{-3} \text{ mol Na}^+ (1 \text{ mol Na}_2CO_2)}{(0.079\,67 \text{ mol} \cdot L^{-1} \text{ Na}_2CO_3)(2 \text{ mol Na}^+)} = 1.35 \times 10^{-2} \text{ L or 13.5 mL}$$

(b)  $$V = \frac{(4.98 \times 10^{-3} \text{ mol CO}_3^{2-})(1 \text{ mol Na}_2CO_2)}{(0.079\,67 \text{ mol} \cdot L^{-1} \text{ Na}_2CO_3)(1 \text{ mol CO}_3^{2-})}$$

$$= 6.25 \times 10^{-2} \text{ L or 62.5 mL}$$

(c)  $$V = \frac{(50.0 \times 10^{-3} \text{ g Na}_2CO_3)}{(105.99 \text{ g} \cdot \text{mol}^{-1})(0.079\,67 \text{ mol} \cdot L^{-1} \text{ Na}_2CO_3)}$$

$$= 5.92 \times 10^{-3} \text{ L or 5.92 mL}$$

**G.9**   (a)  Weigh 1.6 g (0.010 mol, molar mass of $KMnO_4 = 158.04 \ g \cdot mol^{-1}$) into a 1.0-L volumetric flask and add water to give a total volume of 1.0 L. Smaller (or larger) volumes could also be prepared by using a proportionally smaller (or larger) mass of $KMnO_4$.

(b)  Starting with $0.050 \ mol \cdot L^{-1} \ KMnO_4$, add four volumes of water to one volume of starting solution, because the concentration desired is one-fifth of the starting solution. This relation can be derived from the expression

$$V_i \times molarity_i = V_f \times molarity_f$$

where i represents the initial solution and f the final solution. But $V_f = V_i + V_d$ where $V_d$ represents the volume of solvent that must be added to dilute the initial solution. Rearranging the first equation gives

$$\frac{V_i}{V_f} = \frac{molarity_f}{molarity_i}$$

$$\frac{V_i}{V_i + V_d} = \frac{molarity_f}{molarity_i}$$

So if the ratio of final molarity to initial molarity is 1 : 5, we can write

$$\frac{V_i}{V_i + V_d} = \frac{1}{5}$$
$$5V_i = V_i + V_d$$
$$4V_i = V_d$$

For example, to prepare 50 mL of solution, you would add 40 mL of water to 10 mL of $0.050 \ mol \cdot L^{-1} \ KMnO_4$.


**G.11**   (a)  $V(0.778 \ mol \cdot L^{-1}) = (0.1500 \ L)(0.0234 \ mol \cdot L^{-1})$

$V = 4.51 \times 10^{-3}$ L or 4.51 mL

(b)  The concentration desired is one-fifth of the starting NaOH solution, so the stockroom attendant will need to add four volumes of water to one volume of the $2.5 \ mol \cdot L^{-1}$ solution. To prepare 60.0 mL of solution, divide 60.0 by 5; so 12.0 mL of 2.5 mol·L$^{-1}$ NaOH solution are added to 48.0 mL of water. (See the solution to G.9.)

**G.13** (a) mass of $CuSO_4 = (0.20 \text{ mol·L}^{-1})(0.250 \text{ L})(159.60 \text{ g·mol}^{-1})$
$$= 8.0 \text{ g}$$

(b) mass of $CuSO_4 \cdot 5 H_2O = (0.20 \text{ mol·L}^{-1})(0.250 \text{ L})(249.68 \text{ g·mol}^{-1})$
$$= 12 \text{ g}$$

**G.15** (a) Chloride ions are supplied only by the $NiCl_2 \cdot 6 H_2O$ complex:

$$[Cl^-] = \frac{(2 \text{ mol } Cl^-)(0.129 \text{ g } NiCl_2 \cdot 6 H_2O)}{(1 \text{ mol } NiCl_2 \cdot 6 H_2O)(237.70 \text{ g·mol}^{-1} \text{ } NiCl_2 \cdot 6 H_2O)(0.250 \text{ L})}$$

$$= 0.00434 \text{ M}$$

(b) $Ni^{2+}$ ions are present in both the $NiSO_4 \cdot 6 H_2O$ and the $NiCl_2 \cdot 6 H_2O$, so the final concentration will be the sum of the ions provided from the two sources:

$Ni^{2+}$ from $NiCl_2 \cdot 6 H_2O$

$$m_{Ni^{2+} \text{ from nickel chloride}} = \frac{(1 \text{ mol } Ni^{2+})(0.129 \text{ g } NiCl_2 \cdot 6 H_2O)}{(1 \text{ mol } NiCl_2 \cdot 6 H_2O)(237.70 \text{ g·mol}^{-1} \text{ } NiCl_2 \cdot 6 H_2O)}$$

$$= 0.0005 \text{ 43 mol}$$

$Ni^{2+}$ from $NiSO_4 \cdot 6 H_2O$

$$m_{Ni^{2+} \text{ from nickel sulfate}}$$
$$= \frac{(1 \text{ mol } Ni^{2+})(0.376 \text{ g } NiSO_4 \cdot 6 H_2O)}{(1 \text{ mol } NiSO_4 \cdot 6 H_2O)(262.86 \text{ g·mol}^{-1} \text{ } NiSO_4 \cdot 6 H_2O)}$$

$$= 0.00143 \text{ mol}$$

total moles of $Ni^{2+} = 0.0005 \text{ 43 mol} + 0.001 \text{ 43 mol} = 0.001 \text{ 97 mol}$

$$[Ni^{2+}] = \frac{0.001 \text{ 97 mol}}{0.2500 \text{ L}} = 0.00789 \text{ M}$$

**G.17** (a) mass of $K_2SO_4 = (0.125 \text{ mol·L}^{-1})(1.00 \text{ L})(174.26 \text{ g·mol}^{-1}) = 21.8 \text{ g}$

(b)   mass of $NaF = (0.015 \text{ mol} \cdot L^{-1})(0.375 \text{ L})(41.99 \text{ g} \cdot \text{mol}^{-1}) = 0.24 \text{ g}$

(c)   mass of $C_{12}H_{22}O_{11} = (0.35 \text{ mol} \cdot L^{-1})(0.500 \text{ L})(342.29 \text{ g} \cdot \text{mol}^{-1})$
$= 60. \text{ g}$

**G.19**   We can show that fewer than 1 molecule of X would be left after only 70 doublings.

$$10 \text{ mL} \times \frac{0.10 \text{ mol}}{1000 \text{ mL}} \times \frac{6.02214 \times 10^{23} \text{ molecules}}{1 \text{ mol}} = 6.0 \times 10^{20} \text{ molecules are}$$

in the first 10 mL aliquot of the solution.  In order to find the number of times the volume must be doubled to get to one molecule, we can solve for the number of times this amount of molecules must be cut in half until it equals 1.

$$6.0 \times 10^{20} \text{ molecules} \times \left(\frac{1}{2}\right)^{n} = 1 \text{ molecule}$$

$$\log(6.0 \times 10^{20}) + n\log\left(\frac{1}{2}\right) = \log(1)$$

$$20.8 + n(-0.301) = 0$$

$$20.8 = 0.301n$$

$$n = 69$$

The other 21 additional doublings involve solutions with no X remaining. There can be no health benefits if there are no molecules of the active substance, X, left in the solution.

**G.21**   Find the volume of concentrated HCl solution that is equivalent to 10.0 L of the dilute HCl solution with respect to the number of moles of solute present in each volume.

? mL con. HCl(aq)

$$=10.0 \text{ L dilute HCl(aq)} \left(\frac{0.7436 \text{ mol HCl}}{1 \text{ L dilute HCl(aq)}}\right)\left(\frac{36.46 \text{ g HCl}}{1 \text{ mol HCl}}\right)$$

$$\times \left(\frac{100 \text{ g con. HCl(aq)}}{37.50 \text{ g HCl}}\right)\left(\frac{1 \text{ cm}^3}{1.205 \text{ g}}\right)$$

$= 600. \text{ mL con. HCl}$

Therefore 600. mL of concentrated HCl(aq) must be diluted up to a final volume of 10.0 L by adding water in order to form 0.7436 M HCl(aq).

**H.1**    (a)   $BCl_3(g) + 3 H_2O(l) \rightarrow B(OH)_3(aq) + 3 HCl(aq)$

        (b)   $2 NaNO_3(s) \rightarrow 2 NaNO_2(s) + O_2(g)$

        (c)

        $2 Ca_3(PO_4)_2(s) + 6 SiO_2(s) + 10 C(s) \rightarrow 6 CaSiO_3(s) + 10 CO(g) + P_4(s)$

        (d)   $4 Fe_2P(s) + 18 S(s) \rightarrow P_4S_{10}(s) + 8 FeS(s)$

**H.3**    (a)   $2 K(s) + 2 H_2O(l) \rightarrow H_2(g) + 2 KOH(aq)$

        (b)   $Na_2O(s) + H_2O(l) \rightarrow 2 NaOH(aq)$

        (c)   $6 Li(s) + N_2(g) \rightarrow 2 Li_3N(s)$

        (d)   $Ca(s) + 2 H_2O(l) \rightarrow H_2(g) + Ca(OH)_2(aq)$

**H.5**    (I)   $3 Fe_2O_3(s) + CO(g) \rightarrow 2 Fe_3O_4(s) + CO_2(g)$

        (II)   $Fe_3O_4(s) + 4 CO(g) \rightarrow 3 Fe(s) + 4 CO_2(g)$

**H.7**    (I)   $N_2(g) + O_2(g) \rightarrow 2 NO(g)$

        (II)   $2 NO(g) + O_2(g) \rightarrow 2 NO_2(g)$

**H.9**    $4 HF(aq) + SiO_2(s) \rightarrow SiF_4(aq) + 2 H_2O(l)$

**H.11**   $2 C_8H_{18}(l) + 25 O_2(g) \rightarrow 16 CO_2(g) + 18 H_2O(g)$

**H.13**   $4 C_{10}H_{15}N(s) + 55 O_2(g) \rightarrow 40 CO_2(g) + 30 H_2O(l) + 2 N_2(g)$

**H.15**   (I)   $H_2S(g) + 2 NaOH(s) \rightarrow Na_2S(aq) + 2 H_2O(l)$

         (II)   $4 H_2S(g) + Na_2S(alc) \rightarrow Na_2S_5(alc) + 4 H_2(g)$

(III)

$$2 \, Na_2S_5(alc) + 9 \, O_2(g) + 10 \, H_2O(l) \rightarrow 2 \, Na_2S_2O_3 \cdot 5 \, H_2O(s) + 6 \, SO_2(g)$$

**H.17** We can find the empirical formulas from the percent compositions.

First oxide:

P   43.64 g ÷ 30.97 g/mol = 1.409 mol

O   56.36 g ÷ 16.00 g/mol = 3.523 mol

$$\frac{3.523 \text{ mol}}{1.409 \text{ mol}} = 2.500 \qquad 1:2.5 \text{ or } 2:5 \qquad P_2O_5$$

Second oxide:

P   56.34 g ÷ 30.97 g/mol = 1.819 mol

O   43.66 g ÷ 16.00 g/mol = 2.729 mol

$$\frac{2.729 \text{ mol}}{1.819 \text{ mol}} = 1.500 \qquad 1:1.5 \text{ or } 2:3 \qquad P_2O_3$$

These empirical formulas could be named diphosphorus pentoxide and diphosphorus trioxide. The names according to the Stock system are given below with the formulas of the actual compounds.

(a) $P_2O_5$ (phosphorus (V) oxide), $P_2O_3$ (phosphorus (III) oxide);

(b) Since the molar masses of the empirical formulas are 142 g/mol and 110 g/mol respectively, and these masses are both half as big as the molar masses of the compounds, both molecular formulas are twice the empirical formulas: $P_4O_{10}$ (phosphorous (V) oxide), $P_4O_6$ (phosphorous (III) oxide).

(c) $P_4(s) + 3 \, O_2(g) \rightarrow P_4O_6(s)$; $P_4(s) + 5 \, O_2(g) \rightarrow P_4O_{10}(s)$

**I.1** (a) $CH_3OH$, nonelectrolyte; (b) $CaBr_2$, strong electrolyte; (c) KI, strong electrolyte

**I.3** (a) soluble; (b) slightly soluble; (c) insoluble; (d) insoluble

**I.5** (a) $Na^+(aq)$ and $I^-(aq)$; (b)

$Ag^+(aq)$ and $CO_3^{2-}(aq)$, $Ag_2CO_3$ is insoluble. The very small amount that does go into solution will be present as $Ag^+$ and $CO_3^{2-}$ ions.

(c) $NH_4^+(aq)$ and $PO_4^{3-}(aq)$   (d) $Fe^{2+}(aq)$ and $SO_4^{2-}(aq)$

**I.7**   (a) $Fe(OH)_3$, precipitate; (b) $Ag_2CO_3$, precipitate forms; (c) No precipitate will form because all possible products are soluble in water.

**I.9**   (a) net ionic equation: $Fe^{2+}(aq) + S^{2-}(aq) \rightarrow FeS(s)$ ;

spectator ions: $Na^+$, $Cl^-$

(b) net ionic equation: $Pb^{2+}(aq) + 2\,I^-(aq) \rightarrow PbI_2(s)$ ;

spectator ions: $K^+$, $NO_3^-$

(c) net ionic equation: $Ca^{2+}(aq) + SO_4^{2-}(aq) \rightarrow CaSO_4(s)$ ;

spectator ions: $NO_3^-$, $K^+$

(d) net ionic equation: $Pb^{2+}(aq) + CrO_4^{2-}(aq) \rightarrow PbCrO_4(s)$ ;

spectator ions: $Na^+$, $NO_3^-$

(e) net ionic equation: $Hg_2^{2+}(aq) + SO_4^{2-}(aq) \rightarrow Hg_2SO_4(s)$ ;

spectator ions: $K^+$, $NO_3^-$

**I.11**   (a)   overall equation: $(NH_4)_2CrO_4(aq) + BaCl_2(aq)$
$\rightarrow BaCrO_4(s) + 2\,NH_4Cl(aq)$

complete ionic equation:

$2\,NH_4^+(aq) + CrO_4^{2-}(aq) + Ba^{2+}(aq) + 2\,Cl^-(aq)$
$\rightarrow BaCrO_4(s) + 2\,NH_4^+(aq) + 2\,Cl^-(aq)$

net ionic equation: $Ba^{2+}(aq) + CrO_4^{2-}(aq) \rightarrow BaCrO_4(s)$ ;

spectator ions: $NH_4^+$, $Cl^-$

(b) $CuSO_4(aq) + Na_2S(aq) \rightarrow CuS(s) + Na_2SO_4(aq)$

complete ionic equation:

$$Cu^{2+}(aq) + SO_4^{2-}(aq) + 2\ Na^+(aq) + S^{2-}(aq)$$
$$\rightarrow CuS(s) + 2\ Na^+(aq) + SO_4^{2-}(aq)$$

net ionic equation: $Cu^{2+}(aq) + S^{2-}(aq) \rightarrow CuS(s)$ ;

spectator ions: $Na^+$, $SO_4^{2-}$

(c)
$$3\ FeCl_2(aq) + 2\ (NH_4)_3PO_4(aq)$$
$$\rightarrow Fe_3(PO_4)_2(s) + 6\ NH_4^+(aq) + 6\ Cl^-(aq)$$

complete ionic equation:

$$3\ Fe^{2+}(aq) + 6\ Cl^-(aq) + 6\ NH_4^+(aq) + 2\ PO_4^{3-}(aq)$$
$$\rightarrow Fe_3(PO_4)_2(s) + 6\ NH_4^+(aq) + 6\ Cl^-(aq)$$

net ionic equation: $3\ Fe^{2+}(aq) + 2\ PO_4^{3-}(aq) \rightarrow Fe_3(PO_4)_2(s)$ ;

spectator ions: $Cl^-$, $NH_4^+$

(d) $K_2C_2O_4(aq) + Ca(NO_3)_2(aq) \rightarrow CaC_2O_4(s) + 2\ KNO_3(aq)$

complete ionic equation:

$$2\ K^+(aq) + C_2O_4^{2-}(aq) + Ca^{2+}(aq) + 2\ NO_3^-(aq)$$
$$\rightarrow CaC_2O_4(s) + 2\ K^+(aq) + 2\ NO_3^-(aq)$$

net ionic equation: $Ca^{2+}(aq) + C_2O_4^{2-}(aq) \rightarrow CaC_2O_4(s)$ ;

spectator ions: $K^+$, $NO_3^-$

(e) $NiSO_4(aq) + Ba(NO_3)_2(aq) \rightarrow Ni(NO_3)_2(aq) + BaSO_4(s)$

complete ionic equation:

$$Ni^{2+}(aq) + SO_4^{2-}(aq) + Ba^{2+}(aq) + 2\ NO_3^-(aq)$$
$$\rightarrow Ni^{2+}(aq) + 2\ NO_3^-(aq) + BaSO_4(s)$$

net ionic equation: $Ba^{2+}(aq) + SO_4^{2-}(aq) \rightarrow BaSO_4(s)$ ;

spectator ions: $Ni^{2+}$, $NO_3^-$

I.13    (a) $Ba(CH_3CO_2)_2(aq) + Li_2CO_3(aq) \rightarrow BaCO_3(s) + 2\ LiCH_3CO_2(aq)$

$$Ba^{2+}(aq) + 2\ CH_3CO_2^-(aq) + 2\ Li^+(aq) + CO_3^{2-}(aq) \rightarrow$$

$$BaCO_3(s) + 2\ Li^+(aq) + 2CH_3CO_2^-(aq)$$

net ionic equation: $Ba^{2+}(aq) + CO_3^{2-}(aq) \rightarrow BaCO_3(s)$

(b) $2\,NH_4Cl(aq) + Hg_2(NO_3)_2(aq) \rightarrow 2\,NH_4NO_3(aq) + Hg_2Cl_2(s)$

$2\,NH_4^+(aq) + 2\,Cl^-(aq) + Hg_2^{2+} + 2\,NO_3^-(aq) \rightarrow$

$Hg_2Cl_2(s) + 2\,NH_4^+(aq) + 2\,NO_3^-(aq)$

net ionic equation: $Hg_2^{2+}(aq) + 2\,Cl^-(aq) \rightarrow Hg_2Cl_2(s)$

(c) $Cu(NO_3)_2(aq) + Ba(OH)_2(aq) \rightarrow Cu(OH)_2(s) + Ba(NO_3)_2(aq)$

$Cu^{2+}(aq) + 2\,NO_3^-(aq) + Ba^{2+}(aq) + 2\,OH^-(aq) \rightarrow$

$Cu(OH)_2(s) + Ba^{2+}(aq) + 2\,NO_3^-(aq)$

net ionic equation: $Cu^{2+}(aq) + 2\,OH^-(aq) \rightarrow Cu(OH)_2(s)$

**I.15** (a) $AgNO_3$ and $Na_2CrO_4$

(b) $CaCl_2$ and $Na_2CO_3$

(c) $Cd(ClO_4)_2$ and $(NH_4)_2S$

**I.17** (a) $2\,Ag^+(aq) + SO_4^{2-}(aq) \rightarrow Ag_2SO_4(s)$

(b) $Hg^{2+}(aq) + S^{2-}(aq) \rightarrow HgS(s)$

(c) $3\,Ca^{2+}(aq) + 2\,PO_4^{3-}(aq) \rightarrow Ca_3(PO_4)_2(s)$

(d) $AgNO_3$ and $Na_2SO_4$; $Na^+$, $NO_3^-$;

$Hg(CH_3CO_2)_2$ and $Li_2S$; $Li^+$, $CH_3CO_2^-$

$CaCl_2$ and $K_3PO_4$; $K^+$, $Cl^-$

**I.19** white ppt.=$AgCl(s)$, $Ag^+$; no ppt. with $H_2SO_4$, no $Ca^{2+}$; black ppt.=ZnS, $Zn^{2+}$

**I.21** (a) $2\,NaOH(aq) + Cu(NO_3)_2(aq) \rightarrow Cu(OH)_2(s) + 2\,NaNO_3(aq)$

complete ionic equation:

$$2\ Na^+(aq) + 2\ OH^-(aq) + Cu^{2+}(aq) + 2\ NO_3^-(aq)$$
$$\rightarrow Cu(OH)_2(s) + 2\ Na^+(aq) + 2\ NO_3^-(aq)$$

net ionic equation: $\quad Cu^{2+}(aq) + 2\ OH^-(aq) \rightarrow Cu(OH)_2(s)$

(b) $\dfrac{?\ mol\ Na^+}{L\ solution} = \dfrac{20.00\ mL\ NaOH(aq)\left(\dfrac{0.100\ mol\ Na^+}{1000\ mL\ NaOH(aq)}\right)}{(20.00+40.00)\ mL\ solution}$

$\qquad\qquad \cdot\left(\dfrac{1000\ mL}{L}\right)$

$\qquad\qquad = 0.0333\ M\ Na^+$

**I.23** (a) Find the number of moles of potassium chromate per liter of solution.

$\dfrac{?\ mol\ K_2CrO_4}{L\ solution} = \left(\dfrac{3.50\ g\ K_2CrO_4}{75.0\ mL\ solution}\right)\left(\dfrac{1\ mol\ K_2CrO_4}{194.20\ g\ K_2CrO_4}\right)\left(\dfrac{1000\ mL}{1\ L}\right)$

$\qquad\qquad = 0.240\ M\ K_2CrO_4(aq)$

(b) Find the number of grams of potassium in the amount of potassium chromate that was dissolved.

$?\ g\ K^+ = 3.50\ g\ K_2CrO_4\left(\dfrac{1\ mol\ K_2CrO_4}{194.20\ g\ K_2CrO_4}\right)\left(\dfrac{2\ mol\ K^+}{1\ mol\ K_2CrO_4}\right)$

$\qquad \times\left(\dfrac{39.0983\ g\ K^+}{1\ mol\ K^+}\right)$

$\qquad = 1.41\ g\ K^+$

(c) $MgCrO_4$

**I.25** $2\ Ag^+ + CrO_4^{2-} \rightarrow Ag_2CrO_4(s)$

$?\ g\ Ag_2CrO_4 = 25.0\ mL\ K_2CrO_4(aq)\left(\dfrac{5.0\ mol\ K_2CrO_4}{1000\ mL\ K_2CrO_4(aq)}\right)$

$\qquad \times\left(\dfrac{1\ mol\ Ag_2CrO_4}{1\ mol\ K_2CrO_4}\right)\left(\dfrac{331.74\ g\ Ag_2CrO_4}{1\ mol\ Ag_2CrO_4}\right)$

$\qquad = 41.5\ g\ Ag_2CrO_4$ solid precipitate

**J.1** (a) base; (b) acid; (c) base; (d) acid; (e) base

**J.3** (a) overall equation: $HF(aq) + NaOH(aq) \rightarrow NaF(aq) + H_2O(l)$

total ionic equation:

$HF(aq) + Na^+(aq) + OH^-(aq) \rightarrow Na^+(aq) + F^-(aq) + H_2O(l)$

net ionic equation: $\quad HF(aq) + OH^-(aq) \rightarrow F^-(aq) + H_2O(l)$

(b) overall equation: $(CH_3)_3N(aq) + HNO_3(aq) \rightarrow [(CH_3)_3NH]NO_3(aq)$

total ionic equation:
$(CH_3)_3N(aq) + H_3O^+(aq) + NO_3^-(aq)$
$\rightarrow [(CH_3)_3NH]^+(aq) + NO_3^-(aq) + H_2O(l)$

net ionic equation:

$(CH_3)_3N(aq) + H_3O^+(aq) \rightarrow [(CH_3)_3NH]^+(aq) + H_2O(l)$

(c) overall equation: $\quad LiOH(aq) + HI(aq) \rightarrow LiI(aq) + H_2O(l)$

complete ionic equation:

$Li^+(aq) + OH^-(aq) + H_3O^+(aq) + I^-(aq) \rightarrow Li^+(aq) + I^-(aq) + 2\,H_2O(l)$

**J.5** (a) $HBr(aq) + KOH(aq) \rightarrow KBr(aq) + H_2O(l)$

(b) $Zn(OH)_2(aq) + 2\,HNO_2(aq) \rightarrow Zn(NO_2)_2(aq) + 2\,H_2O(l)$

(c) $Ca(OH)_2(aq) + 2\,HCN(aq) \rightarrow Ca(CN)_2(aq) + 2\,H_2O(l)$

(d) $3\,KOH(aq) + H_3PO_4(aq) \rightarrow K_3PO_4(aq) + 3\,H_2O(l)$

**J.7** (a) acid: $H_3O^+(aq)$; base: $CH_3NH_2(aq)$; (b) acid: $HCl(aq)$;

base: $C_2H_5NH_2(aq)$;

(c) acid: $HI(aq)$; base: $CaO(s)$

**J.9** Since X turns litmus red and conducts electricity poorly, it is a weak acid.

We can find the empirical formula from the percent composition.

C $26.68\text{ g} \div 12.01\text{ g/mol} = 2.221$ mol

H $2.239\text{ g} \div 1.0079\text{ g/mol} = 2.221$ mol

O $71.081\text{ g} \div 16.00\text{ g/mol} = 4.443$ mol

So the subscripts are 1:1:2 on the empirical formula.

(a) $CHO_2$;

(b) Since the molar mass of the empirical formula is 45.0 g/mol while the molar mass of X is 90.0 g/mol, the molecular formula is twice the empirical formula or $C_2H_2O_4$ ;.

(c) The weak acid whose formula matches the one given in part (b) is oxalic acid.

$(COOH)_2(aq) + 2\ NaOH(aq) \rightarrow Na_2C_2O_4(aq) + 2\ H_2O(l)$

net ionic equation:   $(COOH)_2(aq) + 2\ OH^- \rightarrow C_2O_4^{2-}\ (aq) +\ 2\ H_2O(l)$

**J.11** (a) $C_6H_5O^-\ (aq) + H_2O(l) \rightarrow C_6H_5OH\ (aq) + OH^-(aq)$

(b) $ClO^-\ (aq) + H_2O(l) \rightarrow HClO(aq) + OH^-(aq)$

(c) $C_5H_5NH^+\ (aq) + H_2O(l) \rightarrow C_5H_5N(aq) + H_3O^+(aq)$

(d) $NH_4^+\ (aq) + H_2O(l) \rightarrow NH_3(aq) + H_3O^+\ (aq)$

**J.13** (a) $AsO_4^{3-}(aq) + H_2O(l) \rightarrow HAsO_4^{2-}(aq) + OH^-(aq)$

$HAsO_4^{2-}(aq) + H_2O(l) \rightarrow H_2AsO_4^-(aq) + OH^-(aq)$

$H_2AsO_4^-(aq) + H_2O(l) \rightarrow H_3AsO_4(aq) + OH^-(aq)$

(b)

$$? \text{ mol Na}^+ = 35.0 \text{ g Na}_3\text{AsO}_4 \left( \frac{1 \text{ mol Na}_3\text{AsO}_4}{207.89 \text{ g Na}_3\text{AsO}_4} \right)$$

$$\times \left( \frac{3 \text{ mol Na}^+}{1 \text{ mol Na}_3\text{AsO}_4} \right)$$

$$= 0.505 \text{ mol Na}^+$$

**K.1** (a) $2\ NO_2(g) + O_3(g) \rightarrow N_2O_5(g) + O_2(g)$

(b) $S_8(s) + 16\ Na(s) \rightarrow 8\ Na_2S(s)$

(c) $2\ Cr^{2+}(aq) + Sn^{4+}(aq) \rightarrow 2\ Cr^{3+}(aq) + Sn^{2+}(aq)$

(d) $2\ As(s) + 3\ Cl_2(g) \rightarrow 2\ AsCl_3(l)$

**K.3** (a) $Mg^0(s) + Cu^{2+}(aq) \rightarrow Mg^{2+}(aq) + Cu^0(s)$

(b) $Fe^{2+}(aq) + Ce^{4+}(aq) \rightarrow Fe^{3+}(aq) + Ce^{3+}(aq)$

(c) $H_2(g) + Cl_2(g) \rightarrow 2\,HCl(g)$

(d) $4\,Fe(s) + 3\,O_2(g) \rightarrow 2\,Fe_2O_3(s)$

**K.5** (a) +4; (b) +4; (c) $-2$; (d) +5; (e) +1; (f) 0

**K.7** (a) +2; (b) +2; (c) +6; (d) +4; (e) +1

**K.9** (a) Methanol $CH_3OH\,(aq)$ is oxidized to formic acid (the carbon atom goes from an oxidation number of +2 to +4). The $O_2(g)$ is reduced to $O^{2-}$ present in water. (b) Mo is reduced from +5 to +4, while *some* sulfur (that which ends up as $S(s)$) is oxidized from $-2$ to 0. The sulfur present in $MoS_2(s)$ remains in the $-2$ oxidation state.

(c) $Tl^+$ is both oxidized and reduced. The product $Tl(s)$ is a reduction of $Tl^+$ (from +1 to 0) while the $Tl^{3+}$ is produced via an oxidation of $Tl^+$. A reaction in which a single substance is both oxidized and reduced is known as a *disproportionation reaction.*

**K.11** (a) $Cl_2$ will be reduced more easily and is therefore a stronger oxidizing agent than $Cl^-$.

(b) $N_2O_5$ will be a stronger oxidizing agent because it will be readily reduced. $N^{5+}$ will accept $e^-$ more readily than will $N^+$.

**K.13** (a) oxidizing agent: $H^+$ in HCl(aq); reducing agent: Zn(s)

(b) oxidizing agent: $SO_2(g)$; reducing agent: $H_2S(g)$

(c) oxidizing agent: $B_2O_3(s)$; reducing agent: Mg(s)

**K.15** (a) $ClO_3^- \rightarrow ClO_2$, Cl goes from +5 to +4; reducing agent

(b) $SO_4^{2-} \rightarrow S^{2-}$, S goes from +6 to $-2$; reducing agent

(c) $Mn^{2+} \rightarrow MnO_2$, Mn goes from +2 to +4; oxidizing agent

(d) $HCHO \rightarrow HCOOH$, C goes from 0 to +2; oxidizing agent

**K.17** (a) oxidizing agent: $WO_3(s)$ ; reducing agent: $H_2(g)$

(b) oxidizing agent: HCl reducing agent: Mg(s)

(c) oxidizing agent: $SnO_2(s)$; reducing agent: C(s)

(d) oxidizing agent: $N_2O_4(g)$ ; reducing agent: $N_2H_4(g)$

**K.19** (a) $3 N_2H_4(l) \rightarrow 4 NH_3(g) + N_2(g)$; (b) $2^-$ in $N_2H_4$; $3^-$ in $NH_3$; (c) $N_2H_4$ is both oxidizing and reducing agent;

(d) Factor label (dimensional analysis) can be used to find the volume of nitrogen.

$$? \text{ L } N_2(g) = 1.0 \text{ L } N_2H_4(l) \left( \frac{1000 \text{ cm}^3}{1 \text{ L}} \right) \left( \frac{1.004 \text{ } g}{1 \text{ cm}^3} \right) \left( \frac{1 \text{ mol}}{32.0 \text{ g}} \right)$$

$$\times \left( \frac{1 \text{ mol } N_2}{3 \text{ mol } N_2H_4} \right) \left( \frac{28.0 \text{ g}}{1 \text{ mol}} \right) \left( \frac{24 \text{ L}}{28 \text{ g}} \right)$$

$$= 2.5 \times 10^2 \text{ L } N_2(g)$$

**K.21** (a) $2 Cr^{2+} + Cu^{2+} \rightarrow 2 Cr^{3+} + Cu(s)$; (b) 2 e- transferred;

(c) Use factor label (dimensional analysis):

$$\frac{? \text{ mol } NO_3^-}{\text{L solution}} = \left( \frac{50.5 \text{ g } Cr(NO_3)_2}{250.0 \text{ mL solution}} \right) \left( \frac{1 \text{ mol } Cr(NO_3)_2}{176.0 \text{ g } Cr(NO_3)_2} \right)$$

$$\times \left( \frac{2 \text{ mol } NO_3^-}{1 \text{ mol } Cr(NO_3)_2} \right) \left( \frac{1000 \text{ mL}}{1 \text{ L}} \right)$$

$$= 2.27 \text{ M } NO_3^-$$

$$\frac{? \text{ mol } SO_4^{2-}}{\text{L solution}} = \left( \frac{60.0 \text{ g } CuSO_4}{250.0 \text{ mL solution}} \right) \left( \frac{1 \text{ mol } CuSO_4}{159.62 \text{ g } CuSO_4} \right)$$

$$\times \left( \frac{1 \text{ mol } SO_4^{2-}}{1 \text{ mol } CuSO_4} \right) \left( \frac{1000 \text{ mL}}{1 \text{ L}} \right)$$

$$= 1.50 \text{ M } SO_4^{2-}$$

**L.1**  $2 Na_2S_2O_3(aq) + AgBr(s) \rightarrow NaBr(aq) + Na_3[Ag(S_2O_3)_2](aq)$

(a) moles of $Na_2S_2O_3$ needed to dissolve 1.0 mg AgBr

$$= 1.0 \text{ mg AgBr} \left( \frac{1 \text{ g AgBr}}{1000 \text{ mg AgBr}} \right) \left( \frac{1 \text{ mol AgBr}}{187.78 \text{ g AgBr}} \right) \left( \frac{2 \text{ mol } Na_2S_2O_3}{1 \text{ mol AgBr}} \right)$$

$$= 1.1 \times 10^{-5} \text{ mol } Na_2S_2O_3$$

(b) mass of AgBr to produce 0.033 mol $Na_3[Ag(S_2O_3)_2]$

$$= 0.033 \text{ mol } Na_3[Ag(S_2O_3)_2] \left( \frac{1 \text{ mol AgBr}}{1 \text{ mol } Na_3[Ag(S_2O_3)_2]} \right) \left( \frac{187.78 \text{ g AgBr}}{1 \text{ mol AgBr}} \right)$$

$$= 6.2 \text{ g AgBr}$$

**L.3**  $6 NH_4ClO_4(s) + 10 Al(s) \rightarrow 5 Al_2O_3(s) + 3 N_2(g) + 6 HCl(g) + 9 H_2O(g)$

(a) $(1.325 \text{ kg } NH_4ClO_4) \left( \frac{1 \text{ mol } NH_4ClO_4}{117.49 \text{ g } NH_4ClO_4} \right) \left( \frac{1000 \text{ g}}{1 \text{ kg}} \right)$

$\left( \frac{10 \text{ mol Al}}{6 \text{ mol } NH_4ClO_4} \right) \left( \frac{26.98 \text{ g Al}}{1 \text{ mol Al}} \right) = 507.1 \text{ g Al}$

(b) $3500 \text{ kg Al} \left( \frac{1000 \text{ g Al}}{1 \text{ kg Al}} \right) \left( \frac{1 \text{ mol Al}}{26.98 \text{ g Al}} \right)$

$\times \left( \frac{5 \text{ mol } Al_2O_3}{10 \text{ mol Al}} \right) \left( \frac{101.96 \text{ g } Al_2O_3}{1 \text{ mol } Al_2 O_3} \right)$

$= 6.613 \times 10^6 \text{ g } Al_2O_3 \text{ or } 6.613 \times 10^3 \text{ kg } Al_2O_3$

**L.5**  $2 C_{57}H_{110}O_6(s) + 163 O_2(g) \rightarrow 114 CO_2(g) + 110 H_2O(l)$

(a) $(454 \text{ g fat}) \left( \frac{1 \text{ mol fat}}{891.44 \text{ g fat}} \right) \left( \frac{110 \text{ mol } H_2O}{2 \text{ mol fat}} \right) \left( \frac{18.02 \text{ g } H_2O}{1 \text{ mol } H_2O} \right)$

$= 505 \text{ g } H_2O$

(b) $(454 \text{ g fat}) \left( \frac{1 \text{ mol fat}}{891.44 \text{ g}} \right) \left( \frac{163 \text{ mol } O_2}{2 \text{ mol fat}} \right) \left( \frac{32.00 \text{ g } O_2}{1 \text{ mol } O_2} \right)$

$= 1.33 \times 10^3 \text{ g } O_2$

**L.7**   $2\,C_8H_{18}(l) + 25\,O_2(g) \rightarrow 16\,CO_2(g) + 18\,H_2O(l)$

$d = 0.79\ \text{g}\cdot\text{mL}^{-1}$, density of gasoline

$$(3.785\ \text{L gas})\left(\frac{1000\ \text{mL}}{1\ \text{L}}\right)\left(\frac{0.79\ \text{g gas}}{1\ \text{mL}}\right)\left(\frac{1\ \text{mol gas}}{114.22\ \text{g gas}}\right)\left(\frac{18\ \text{mol H}_2\text{O}}{2\ \text{mol C}_8\text{H}_{18}}\right)$$

$$\left(\frac{18.02\ \text{g H}_2\text{O}}{1\ \text{mol H}_2\text{O}}\right) = 4.246 \times 10^3\ \text{g H}_2\text{O or 4.2 kg H}_2\text{O}$$

**L.9**   (a)  $HCl + NaOH \rightarrow NaCl + H_2O$

$$17.40\ \text{mL}\left(\frac{0.234\ \text{mol HCl}}{1000\ \text{mL}}\right)\left(\frac{1\ \text{mol NaOH}}{1\ \text{mol HCl}}\right) = 0.004\,07\ \text{mol}$$

$$\text{concentration of NaOH} = \frac{0.004\,07\ \text{mol}}{15.00 \times 10^{-3}\ \text{L}} = 0.271\ \text{M}$$

(b)  $(0.271\ \text{mol}\cdot\text{L}^{-1})(0.01500\ \text{L})\,(40.00\ \text{g}\cdot\text{mol}^{-1}) = 0.163\ \text{g NaOH}$

**L.11**   (a)  $Ba(OH)_2(aq) + 2\,HNO_3(aq) \rightarrow Ba(NO_3)_2(aq) + 2\,H_2O(l)$

$$\frac{?\ \text{mol HNO}_3}{\text{L HNO}_3(aq)} = (11.56\ \text{mL Ba(OH)}_2(aq)\,)\left(\frac{9.670\ \text{g Ba(OH)}_2}{250.\ \text{mL Ba(OH)}_2(aq)}\right)$$

$$\times\frac{\left(\dfrac{1\ \text{mol Ba(OH)}_2}{171.36\ \text{g Ba(OH)}_2}\right)\left(\dfrac{2\ \text{mol HNO}_3}{1\ \text{mol Ba(OH)}_2}\right)}{(25.0\ \text{mL HNO}_3(aq))\left(\dfrac{1\ \text{L}}{1000\ \text{mL}}\right)}$$

$$= 0.209\ \text{mol}\cdot\text{L}^{-1}$$

(b)  mass of $HNO_3$ in solution:

$$\left(\frac{0.209\ \text{mol}}{1000\ \text{mL}}\right)(25.0\ \text{mL})\left(\frac{63.02\ \text{g HNO}_3}{1\ \text{mol HNO}_3}\right) = 0.329\ \text{g}$$

**L.13**   $HX(aq) + NaOH(aq) \rightarrow NaX(aq) + H_2O(l)$

$$(68.8\ \text{mL})\left(\frac{0.750\ \text{mol NaOH}}{1000\ \text{mL NaOH}}\right) = 0.0516\ \text{mol NaOH}$$

3.25 g HX corresponds to 0.0516 mol NaOH used

$$\frac{3.25 \text{ g}}{0.0516 \text{ mol}} = 63.0 \text{ g} \cdot \text{mol}^{-1} = \text{molar mass of acid}$$

**L.15** (a) $Na_2CO_3(aq) + 2\,HCl(aq) \rightarrow 2\,NaCl(aq) + H_2CO_3(aq)$

(b) First find the concentration of the diluted acid.

$$? \text{ M HCl(aq) dilute} = \left( \frac{0.832 \text{ g Na}_2\text{CO}_3}{0.100 \text{ L base solution}} \right) \left( \frac{1 \text{ mol Na}_2\text{CO}_3}{105.99 \text{ g Na}_2\text{CO}_3} \right)$$

$$\times \left( \frac{0.025 \text{ L base solution}}{0.031\,25 \text{ L acid solution}} \right)$$

$$\times \left( \frac{2 \text{ mol HCl}}{1 \text{ mol Na}_2\text{CO}_3} \right) = 0.126 \text{ M HCl(aq) dilute}$$

The original HCl solution is 100 times more concentrated than the solution used for titration (diluted 10.00 mL to 1000 mL), so the original concentration of the HCl solution is $12.6 \text{ mol} \cdot \text{L}^{-1}$.

**L.17** $I_3^-(aq) + SnCl_x(aq) + (y - x)Cl^-(aq) \rightarrow 3\,I^-(aq) + SnCl_y(aq)$

The information given can be used to find the molar mass of the reactant in order to identify it.

$$25.00 \text{ mL} \left( \frac{0.120 \text{ mol I}_3^-}{1000 \text{ mL}} \right) = 3.00 \times 10^{-3} \text{ mol I}_3^-$$

$$30.00 \text{ mL} \left( \frac{19.0 \text{ g tin chloride}}{1000 \text{ mL}} \right) = 0.570 \text{ g tin chloride}$$

If the reaction is 1:1 then the # moles of $I_3^-$ is the same as the number of moles of $SnCl_x$. In that case, the molar mass of the tin chloride reactant is

$\dfrac{0.570 \text{ g}}{3.00 \times 10^{-3} \text{ mol}} = 190.\ \text{g} \cdot \text{mol}^{-1}$. This molar mass matches that of $SnCl_2$,

$189.61 \text{ g} \cdot \text{mol}^{-1}$. Tin (II) chloride also has the correct mass percent tin.

$$\frac{118.71 \text{ g} \cdot \text{mol}^{-1} \text{ Sn}}{189.61 \text{ g} \cdot \text{mol}^{-1} \text{ SnCl}_2} \times 100 = 62.6\%$$

Since the product compound is oxidized relative to the reactant, we can expect it to be Sn(IV). The net ionic equation for the reaction is $I_3^-(aq) + Sn^{2+}(aq) \rightarrow 3\,I^-(aq) + Sn^{4+}(aq)$. Another way to write a balanced reaction would be

$I_3^-(aq) + SnCl_2(aq) + 2\,Cl^-(aq) \rightarrow 3\,I^-(aq) + SnCl_4(aq)$.

**L.19** (a) $S_2O_3^{2-}$ is both oxidized and reduced.

(b) Find the number of grams of thiosulfate ion in 10.1 mL of solution.

$$? \text{ g } S_2O_3^{2-} = 10.1 \text{ mL } HSO_3^-(aq) \left( \frac{1.45 \text{ g } HSO_3^-(aq)}{1 \text{ mL } HSO_3^-(aq)} \right) \left( \frac{55.0 \text{ g } HSO_3^-}{100 \text{ g } HSO_3^-(aq)} \right)$$

$$\times \left( \frac{1 \text{ mol } HSO_3^-}{81.0 \text{ g } HSO_3^-} \right) \left( \frac{1 \text{ mol } S_2O_3^{2-}}{1 \text{ mol } HSO_3^-} \right) \left( \frac{112.0 \text{ g } S_2O_3^{2-}}{1 \text{ mol } S_2O_3^{2-}} \right)$$

$$= 11.1 \text{ g } S_2O_3^{2-} \text{ present initially}$$

**L.21** $XCl_4 + 2\,NH_3 \rightarrow XCl_2(NH_3)_2 + Cl_2$

The reactant and product that contain X are in a 1:1 ratio, so 3.571 g of the reactant is equivalent to 3.180 g of the product. The molar mass of the reactant is $x + 4(35.453 \text{ g/mol})$ while that of the product is $x + 2(35.453 \text{ g/mol}) + 2(14.01 \text{ g/mol}) + 6(1.0079 \text{ g/mol})$. Therefore we can set up the following proportion in order to solve for $x$ in g/mol:

$$\frac{x + 141.8}{x + 104.97} = \frac{3.571}{3.180}$$

$$1.1230x - x = 23.923$$

$$x = 194.6 \text{ g} \cdot \text{mol}^{-1}, \text{ or Pt}$$

**L.23** The number of moles of product is

$2.27 \text{ g} \div 208.23 \text{ g} \cdot \text{mol}^{-1} = 0.0109 \text{ moles } BaCl_2$. An equivalent number of moles is represented by 3.25 g of $BaBr_x$, so its molar mass is

3.25 g ÷ 0.0109 moles=298 g·mol⁻¹ BaBr$_x$. Since 137.33 g is attributable to Ba, 161 g must be Br. Each Br has a mass of 80.4 g/mol so there must be 2 moles of Br for each mole of Ba in the reactant.

$x = 2$, $BaBr_2 + Cl_2 \rightarrow BaCl_2 + Br_2$

**M.1** $CaCO_3(s) \rightarrow CaO(s) + CO_2(g)$

theoretical yield:

$$(42.73 \text{ g CaCO}_3)\left(\frac{1 \text{ mol CaCO}_3}{100.09 \text{ g CaCO}_3}\right)\left(\frac{1 \text{ mol CO}_2}{1 \text{ mol CaCO}_3}\right)\left(\frac{44.01 \text{ g CO}_2}{1 \text{ mol CO}_2}\right)$$

$= 18.79$ g $CO_2$

actual yield:

$$\frac{17.5 \text{ g}}{18.79 \text{ g}} \times 100\% = 93.1\% \text{ yield}$$

**M.3** $C_xH_yCl_z + \left(x + \frac{y}{4}\right)O_2 \rightarrow x\,CO_2 + \left(\frac{y}{2}\right)H_2O + \left(\frac{z}{2}\right)Cl_2$

$$\frac{1.52 \text{ g}}{360.88 \text{ g·mol}^{-1}} = 4.21\times10^{-3} \text{ mol Arochlor yields } \frac{2.224 \text{ g}}{44.0 \text{ g·mol}^{-1}}$$

$= 5.055\times10^{-2}$ mol $CO_2$

$$\frac{2.53 \text{ g}}{360.88 \text{ g·mol}^{-1}} = 7.01\times10^{-3} \text{ mol Arochlor yields } \frac{0.2530 \text{ g}}{18.01 \text{ g·mol}^{-1}}$$

$= 1.405\times10^{-2}$ mol $H_2O$

Therefore, $x = \dfrac{5.055\times10^{-2}}{4.21\times10^{-3}} = 12.0$ and $y = \dfrac{1.405\times10^{-2}}{7.01\times10^{-3}} = 2.00$

$$12.011x + 1.0079y + 35.453z = 360.88$$
$$12.011(12.0) + 1.0079(2.00) + 35.453z = 360.88$$
$$35.453z = 214.7$$
$$z = 6.06$$

Since the number or Cl atoms per Arochlor 1254 molecule must be a whole number, the number of chlorine atoms is 6.

**M.5** (a) $P_4(s) + 3\,O_2(g) \rightarrow P_4O_6(s)$

$P_4O_6(s) + 2\,O_2(g) \rightarrow P_4O_{10}(s)$

In the first reaction, 5.77 g $P_4$ uses

$$(5.77\text{ g }P_4)\left(\frac{1\text{ mol }P_4}{123.88\text{ g }P_4}\right)\left(\frac{3\text{ mol }O_2}{1\text{ mol }P_4}\right)\left(\frac{32.00\text{ g }O_2}{1\text{ mol }O_2}\right) = 4.47\text{ g }O_2\text{ (g)}$$

excess $O_2 = 5.77\text{ g} - 4.47\text{ g }O_2 = 1.30\text{ g }O_2$

In the second reaction, 5.77 g $P_4$ uses

$$\left(\frac{5.77\text{ g }P_4}{123.88\text{ g}\cdot\text{mol}^{-1}\text{ }P_4}\right)\left(\frac{1\text{ mol }P_4O_6}{1\text{ mol }P_4}\right)\left(\frac{2\text{ mol }O_2}{1\text{ mol }P_4O_6}\right)\left(\frac{32.00\text{ g }O_2}{1\text{ mol }O_2}\right)$$

$= 2.98\text{ g }O_2$

limiting reagent: $O_2$

(b) $$\left(\frac{1.30\text{ g }O_2}{32.00\text{ g}\cdot\text{mol}^{-1}\text{ }O_2}\right)\left(\frac{1\text{ mol }P_4O_{10}}{2\text{ mol }O_2}\right)\left(\frac{283.88\text{ g }P_4O_{10}}{1\text{ mol }P_4O_{10}}\right) = 5.77\text{ g }P_4O_{10}$$

(c) $$\left(\frac{1.30\text{ g }O_2}{32.00\text{ g}\cdot\text{mol}^{-1}\text{ }O_2}\right)\left(\frac{1\text{ mol }P_4O_6}{2\text{ mol }O_2}\right)\left(\frac{219.88\text{ g }P_4O_6}{1\text{ mol }P_4O_6}\right)$$

$= 4.47\text{ g }P_4O_6\text{ used}$

In the first reaction, 5.77 g $P_4$ produces

$$\left(\frac{5.77\text{ g }P_4}{123.88\text{ g}\cdot\text{mol}^{-1}}\right)\left(\frac{219.88\text{ g }P_4O_6}{1\text{ mol }P_4O_6}\right)\left(\frac{1\text{ mol }P_4O_6}{1\text{ mol }P_4}\right) = 10.2\text{ g }P_4O_6$$

excess reagent: $10.2\text{ g} - 4.47\text{ g} = 5.7\text{ g }P_4O_6$

**M.7** $C_{63}H_{88}CoN_{14}O_{14}P$. The molar mass of cobalamin is $1355.37\text{ g}\cdot\text{mol}^{-1}$.

$$n_{\text{cobalamin}} = \frac{0.1674\text{ g}}{1355.37\text{ g}\cdot\text{mol}^{-1}} = 1.235\times 10^{-4}\text{ mol}$$

1 mole of cobalamin will produce 63 moles of $CO_2$ and 44 moles of $H_2O$.

$$m_{CO_2} = (1.235\times 10^{-4}\text{ mol cobalamin})\left(\frac{63\text{ mol }CO_2}{1\text{ mol cobalamin}}\right)$$

$$\times\,(44.01\text{ g}\cdot\text{mol}^{-1}\text{ }CO_2)$$

$$= 0.3424\text{ g }CO_2$$

$$m_{H_2O} = (1.235 \times 10^{-4} \text{ mol cobalamin}) \left( \frac{44 \text{ mol H}_2\text{O}}{1 \text{ mol cobalamin}} \right)$$

$$\times (18.02 \text{ g} \cdot \text{mol}^{-1} \text{ CO}_2)$$

$$= 0.097\,92 \text{ g H}_2\text{O}$$

**M.9** $(0.682 \text{ g CO}_2) \left( \frac{1 \text{ mol CO}_2}{44.01 \text{ g CO}_2} \right) \left( \frac{1 \text{ mol C}}{1 \text{ mol CO}_2} \right) = 0.0155 \text{ mol C}$

$(0.0155 \text{ mol C})(12.01 \text{ g} \cdot \text{mol}^{-1} \text{ C}) = 0.186 \text{ g C}$

$(0.174 \text{ g H}_2\text{O}) \left( \frac{1 \text{ mol H}_2\text{O}}{18.02 \text{ g H}_2\text{O}} \right) \left( \frac{2 \text{ mol H}}{1 \text{ mol H}_2\text{O}} \right) = 0.0193 \text{ mol H}$

$(0.0193 \text{ mol H})(1.0079 \text{ g} \cdot \text{mol}^{-1} \text{ H}) = 0.0195 \text{ g H}$

$(0.110 \text{ g N}_2) \left( \frac{1 \text{ mol N}_2}{28.02 \text{ g N}_2} \right) \left( \frac{2 \text{ mol N}}{1 \text{ mol N}_2} \right) = 0.007\,85 \text{ mol N}$

$(0.007\,85 \text{ mol N})(14.01 \text{ g} \cdot \text{mol}^{-1} \text{ N}) = 0.110 \text{ g N}$

mass of O = 0.376 g − (0.186 g + 0.0193 g + 0.110 g) = 0.061 g O

$$\frac{0.061 \text{ g O}}{16.00 \text{ g O}} = 0.0038 \text{ mol O}$$

Dividing each amount by 0.0038 gives C : H : N : O ratios = 4.1 : 5.1 : 2.1

: 1. The empirical formula is $C_4H_5N_2O$.

The molecular mass of caffeine is 194 g · mol$^{-1}$. Its empirical mass is

97.10 g · mol$^{-1}$.

molecular formula = 2 × empirical formula = $C_8H_{10}N_4O_2$

$2 \text{ C}_8\text{H}_{10}\text{N}_4\text{O}_2(\text{s}) + 19 \text{ O}_2(\text{g}) \rightarrow 16 \text{ CO}_2(\text{g}) + 10 \text{ H}_2\text{O(l)} + 4 \text{ N}_2(\text{g})$

**M.11** $3 \text{ Ca(NO}_3)_2(\text{aq}) + 2 \text{ H}_3\text{PO}_4(\text{aq}) \rightarrow \text{Ca}_3(\text{PO}_4)_2(\text{s}) + 6 \text{ HNO}_3(\text{aq})$

(a) The solid is calcium phosphate, $\text{Ca}_3(\text{PO}_4)_2$.

(b) $(206 \text{ g Ca(NO}_3)_2) \left( \frac{1 \text{ mol Ca(NO}_3)_2}{164.10 \text{ g Ca(NO}_3)_2} \right) \left( \frac{2 \text{ mol H}_3\text{PO}_4}{3 \text{ mol Ca(NO}_3)_2} \right)$

$\left( \frac{97.99 \text{ g H}_3\text{PO}_4}{1 \text{ mol H}_3\text{PO}_4} \right) = 82.01 \text{ g H}_3\text{PO}_4$

Therefore $Ca(NO_3)_2$ is the limiting reagent.

$$(206 \text{ g } Ca(NO_3)_2)\left(\frac{1 \text{ mol } Ca(NO_3)_2}{164.10 \text{ g } Ca(NO_3)_2}\right)\left(\frac{1 \text{ mol } Ca_3(PO_4)_2}{3 \text{ mol } Ca(NO_3)_2}\right)$$

$$\left(\frac{310.18 \text{ g } Ca_3(PO_4)_2}{1 \text{ mol } Ca_3(PO_4)_2}\right) = 130. \text{ g } Ca_3(PO_4)_2$$

**M.13** If the 2-naphthol $(144.16 \text{ g} \cdot mol^{-1})$ were pure, it would give the following combustion analysis:

$$\%C = \frac{10(12.01 \text{ g} \cdot mol^{-1})}{(144.16 \text{ g} \cdot mol^{-1} \text{ naphthol})} \times 100\% = 83.31\% \text{ C}$$

$$\%H = \frac{8(1.0079 \text{ g} \cdot mol^{-1})}{(144.16 \text{ g} \cdot mol^{-1} \text{ naphthol})} \times 100\% = 5.59\% \text{ H}$$

The observed percentages are low as is expected for a sample contaminated with a substance that contains no C or H. Because the sample does not contain C or H, the percent purity can be easily obtained by

$$\%\text{purity (based on C)} = \frac{\% \text{ found}}{\% \text{ theoretical}} = \frac{77.48\% \text{ mixture}}{83.31\% \text{ pure naphthol}} \times 100\%$$

$$= 93.00\%$$

$$\%\text{purity (based on H)} = \frac{\% \text{ found}}{\% \text{ theoretical}} = \frac{5.20\% \text{ mixture}}{5.59\% \text{ pure naphthol}} \times 100\%$$

$$= 93.0\%$$

**M.15** (a) $C_xH_yO_z + (x + \frac{y}{2} - \frac{z}{2}) O_2 \rightarrow x \text{ CO}_2 + y \text{ H}_2\text{O}$

$$\frac{2.492 \text{ g } CO_2}{44.0 \text{ g} \cdot mol^{-1}} = 5.664 \times 10^{-2} \text{ mol CO}_2 = \text{mol C}$$

$$\frac{0.6495 \text{ g } H_2O}{18.01 \text{ g} \cdot mol^{-1}} = 3.608 \times 10^{-2} \text{ mol H}_2\text{O} = 7.216 \times 10^{-2} \text{ mol H}$$

$$(5.664 \times 10^{-2} \text{ mol C})\left(\frac{12.011 \text{ g}}{\text{mol}}\right) = 0.6803 \text{ g C}$$

$$(7.216 \times 10^{-2} \text{ mol H}) \left( \frac{1.0079 \text{ g}}{\text{mol}} \right) = 0.07273 \text{ g H}$$

1.000 g compound−(0.6803 g+0.07273 g)=0.2470 g

$$0.2470 \text{ g O} \div 16.00 \text{ g} \cdot \text{mol}^{-1} = 1.544 \times 10^{-2} \text{ mol O}$$

$$\frac{5.664 \times 10^{-2}}{1.544 \times 10^{-2}} = 3.67 \qquad \frac{7.216 \times 10^{-2}}{1.544 \times 10^{-2}} = 4.67$$

The mole ratio of C:H:O is 3.67:4.67:1 or 11:14:3, so the empirical

formula is $C_{11}H_{14}O_3$.

(b) The molar mass of the empirical formula is 194 g/mol, which is half of

388.46 g/mol. Therefore, the molecular formula of the compound is

$C_{22}H_{28}O_6$.

**M.17** Determine the number of moles of each element present in the compound

then find their ratios to get the subscripts for the empirical formula.

$$\frac{0.055 \text{ g Cl}}{35.453 \text{ g} \cdot \text{mol}^{-1}} = 1.55 \times 10^{-3} \text{ mol Cl}$$

$$\frac{0.0682 \text{ g CO}_2}{44.0 \text{ g} \cdot \text{mol}^{-1}} = 1.55 \times 10^{-3} \text{ mol CO}_2 = \text{mol C}$$

$$\frac{0.0140 \text{ g H}_2\text{O}}{18.01 \text{ g} \cdot \text{mol}^{-1}} = 7.78 \times 10^{-4} \text{ mol H}_2\text{O} = 1.56 \times 10^{-3} \text{ mol H}$$

0.100 g compound

$$-\left( 0.055 \text{ g Cl} + 1.55 \times 10^{-3} \text{ mol} \cdot \frac{12.0 \text{ g C}}{\text{mol}} + 1.55 \times 10^{-3} \text{ mol} \cdot \frac{1.0079 \text{ g H}}{\text{mol}} \right)$$

$$= 0.0247 \text{ g O}$$

$$\times \left( \frac{1 \text{ mol}}{16.00 \text{ g}} \right) = 1.55 \times 10^{-3} \text{ mol O}$$

So the mole ratio is 1:1:1:1 and the empirical formula is CHOCl.

There is not sufficient information to allow the determination of the

molecular formula.

# CHAPTER 1

# ATOMS: THE QUANTUM WORLD

**1.1** microwaves < visible light < ultraviolet light < x-rays < $\gamma$-rays

**1.3** (a) If wavelength is known, the frequency can be obtained from the relation

$c = \nu\lambda$:

$2.997\,92 \times 10^{8}\ \mathrm{m \cdot s^{-1}} = (\nu)\,(925 \times 10^{-9}\ \mathrm{m})$

$$\nu = \frac{2.997\,92 \times 10^{8}\ \mathrm{m \cdot s^{-1}}}{925 \times 10^{-9}\ \mathrm{m}}$$

$$= 3.24 \times 10^{14}\ \mathrm{s^{-1}}$$

(b) $2.997\,92 \times 10^{8}\ \mathrm{m \cdot s^{-1}} = (\nu)\,(4.15 \times 10^{-3}\ \mathrm{m})$

$$\nu = \frac{2.997\,92 \times 10^{8}\ \mathrm{m \cdot s^{-1}}}{4.15 \times 10^{-3}\ \mathrm{m}}$$

$$= 7.22 \times 10^{10}\ \mathrm{s^{-1}}$$

**1.5** Wien's law states that $T\lambda_{max} = \text{constant} = 2.88 \times 10^{-3}\ \mathrm{K \cdot m}$.

If $T/\mathrm{K} = 1540°\mathrm{C} + 273°\mathrm{C} = 1813\ \mathrm{K}$, then $\lambda_{max} = \dfrac{2.88 \times 10^{-3}\ \mathrm{K \cdot m}}{1813\ \mathrm{K}}$

$\lambda_{max} = 1.59 \times 10^{-6}\ \mathrm{m}$, or 1590 nm

**1.7** (a) From $c = n\lambda$ and $E = hn$, we can write

$E = hc\lambda^{-1}$

$= (6.626\,08 \times 10^{-34}\ \mathrm{J \cdot s})\,(2.997\,92 \times 10^{8})\,(589 \times 10^{-9}\ \mathrm{m})^{-1}$

$= 3.37 \times 10^{-19}\ \mathrm{J}$

(b) $E = \left( \dfrac{5.00 \times 10^{-3} \text{ g Na}}{22.99 \text{ g} \cdot \text{mol}^{-1} \text{ Na}} \right) (6.022 \times 10^{23} \text{ atoms} \cdot \text{mol}^{-1})$

$\times (3.37 \times 10^{-19} \text{ J} \cdot \text{atom}^{-1})$

$= 44.1 \text{ J}$

(c) $E = (6.022 \times 10^{23} \text{ atoms} \cdot \text{mol}^{-1})(3.37 \times 10^{-19} \text{ J} \cdot \text{atom}^{-1})$

$= 2.03 \times 10^{5} \text{ J or 203 KJ}$

**1.9** The energy is first converted from eV to joules:

$E = (140.511 \times 10^{3} \text{ eV}) (1.6022 \times 10^{-19} \text{ J} \cdot \text{eV}^{-1}) = 2.2513 \times 10^{-14} \text{ J}$

From $E = h\nu$ and $c = \nu\lambda$ we can write

$\lambda = \dfrac{hc}{E}$

$= \dfrac{(6.626\ 09 \times 10^{-34} \text{ J} \cdot \text{s}) (2.997\ 92 \times 10^{8} \text{ m} \cdot \text{s}^{-1})}{2.2513 \times 10^{-14} \text{ J}}$

$= 8.8236 \times 10^{-12} \text{ m or 8.8236 pm}$

**1.11** (a) false. The total intensity is proportional to $T^4$. (Stefan-Boltzmann Law)   (b) true;   (c) false. Photons of radio-frequency radiation are lower in energy than photons of ultraviolet radiation.

$\lambda = m/s$

**1.13** (a) Use the de Broglie relationship, $\lambda = hp^{-1} = h(mv)^{-1}$.

$m_e = (9.109\ 39 \times 10^{-28} \text{ g}) (1 \text{ kg}/1000 \text{ g}) = 9.109\ 39 \times 10^{-31} \text{ kg}$

$(3.6 \times 10^{3} \text{ km} \cdot \text{s}^{-1}) (1000 \text{ m} \cdot \text{km}^{-1}) = 3.6 \times 10^{6} \text{ m} \cdot \text{s}^{-1}$

$\lambda = h(mv)^{-1}$

$= \dfrac{6.626\ 08 \times 10^{-34} \text{ J} \cdot \text{s}}{(9.109\ 39 \times 10^{-31} \text{ kg}) (3.6 \times 10^{6} \text{ m} \cdot \text{s}^{-1})}$

$= 2.0 \times 10^{-10} \text{ m}$

(b) $E = h\nu$

$= (6.626\ 08 \times 10^{-34} \text{ J} \cdot \text{s}) (2.50 \times 10^{16} \text{ s}^{-1})$

$= 1.66 \times 10^{-17} \text{ J}$

(c) The photon needs to contain enough energy to eject the electron from the surface as well as to cause it to move at $3.6 \times 10^3$ km$\cdot$s$^{-1}$. The energy involved is the kinetic energy of the electron, which equals $\frac{1}{2}mv^2$.

$$E_{photon} = 1.66 \times 10^{-17} \text{ J} + \frac{1}{2}mv^2$$
$$= 1.66 \times 10^{-17} \text{ J} + \frac{1}{2}(9.109\ 39 \times 10^{-31} \text{ kg})(3.6 \times 10^6 \text{ m}\cdot\text{s}^{-1})^2$$
$$= 1.66 \times 10^{-17} \text{ J} + 5.9 \times 10^{-18} \text{ J}$$
$$= 2.3 \times 10^{-17} \text{ J}$$

But we are asked for the wavelength of the photon, which we can get from $E = hv$ and $c = v\lambda$ or $E = hc\lambda^{-1}$.

$$2.3 \times 10^{-17} \text{ kg}\cdot\text{m}^2\cdot\text{s}^{-2} = (6.626\ 08 \times 10^{-34} \text{ kg}\cdot\text{m}^2\cdot\text{s}^{-1})$$
$$\times (2.997\ 92 \cdot 10^8 \text{ m}\cdot\text{s}^{-1})\lambda^{-1}$$
$$\lambda = 6.8 \times 10^{-8} \text{ m}$$
$$= 8.8 \text{ nm}$$

(d)  8.8 nm is in the x-ray/gamma ray region.

**1.15**  To answer this question, we need to convert the quantities to a consistent set of units, in this case, SI units.

(5.15 ounce) (28.3 g$\cdot$ounce$^{-1}$) (1 kg/1000 g) = 0.146 kg

$$\left(\frac{92 \text{ mi}}{\text{h}}\right)\left(\frac{1 \text{ h}}{3600 \text{ s}}\right)\left(\frac{1 \text{ km}}{0.6214 \text{ mi}}\right)\left(\frac{1000 \text{ m}}{1 \text{ km}}\right) = 41 \text{ m}\cdot\text{s}^{-1}$$

Use the de Broglie relationship.

$$\lambda = hp^{-1} = h(mv)^{-1}$$
$$= h(mv)^{-1}$$
$$= \frac{6.626\ 08 \times 10^{-34} \text{ J}\cdot\text{s}}{(0.146 \text{ kg})(0.041 \text{ km}\cdot\text{s}^{-1})}$$
$$= \frac{6.626\ 08 \times 10^{-34} \text{ kg}\cdot\text{m}^2\cdot\text{s}^{-1}}{(0.146 \text{ kg})(41 \text{ m}\cdot\text{s}^{-1})}$$
$$= 1.1 \times 10^{-34} \text{ m}$$

**1.17** From the de Broglie relationship, $p = h\lambda^{-1}$ or $h = mv\lambda$, we can calculate the velocity of the neutron:

$$v = \frac{h}{m\lambda}$$

$$= \frac{(6.626\ 08 \times 10^{-34}\ \text{kg} \cdot \text{m}^2 \cdot \text{s}^{-1})}{(1.674\ 93 \times 10^{-27}\ \text{kg})(100 \times 10^{-12}\ \text{m})} \quad (\text{remember that } 1\ \text{J} = 1\ \text{kg} \cdot \text{m}^2 \cdot \text{s}^{-2})$$

$$= 3.96 \times 10^3\ \text{m} \cdot \text{s}^{-1}$$

**1.19** Yes there are degenerate levels. The first three cases of degenerate levels are:

$n_1 = 1, n_2 = 2$ is degenerate with $n_1 = 2, n_2 = 1$

$n_1 = 1, n_2 = 3$ is degenerate with $n_1 = 3, n_2 = 1$

$n_1 = 2, n_2 = 3$ is degenerate with $n_1 = 3, n_2 = 2$

**1.21** (a) Integrate over the "left half of the box" or from 0 to ½ L:

$$\int_0^{\frac{L}{2}} \Psi^2 = \frac{2}{L} \int_0^{\frac{L}{2}} \left(\sin \frac{n\pi x}{L}\right)^2 dx$$

$$= \frac{2}{L}\left[\left(\frac{-1}{2n\pi} \cdot \cos \frac{n\pi x}{L} \cdot \sin \frac{n\pi x}{L} + \frac{x}{2}\right)\Big|_0^{\frac{L}{2}}\right]$$

given $n$ is an integer:

$$= \frac{2}{L}\left[\left(\frac{L/2}{2}\right) - 0\right] = \frac{1}{2}$$

(b) Integrate over the "left third of the box" or from 0 to 1/3 L:

$$\int_0^{\frac{L}{3}} \Psi^2 = \frac{2}{L} \int_0^{\frac{L}{3}} \left(\sin \frac{n\pi x}{L}\right)^2 dx$$

$$= \frac{2}{L}\left[\left(\frac{-1}{2n\pi} \cdot \cos \frac{n\pi x}{L} \cdot \sin \frac{n\pi x}{L} + \frac{x}{2}\right)\Big|_0^{\frac{L}{3}}\right]$$

$$= \frac{-1}{L \cdot n \cdot \pi} \cos \frac{n\pi}{3} \cdot \sin \frac{n\pi}{3} + \frac{1}{3}$$

if $n$ is a multiple of 3, the first term in this sum is zero and the probability of finding an electron in the left third of the box is $1/3$. Also, as $n$ becomes large the probability of finding the electron in the left third of the box approaches $1/3$.

**1.23** (a) The Rydberg equation gives $\nu$ when $\mathfrak{R} = 3.29 \times 10^{15}$ s$^{-1}$, from which one can calculate $\lambda$ from the relationship $c = \nu\lambda$.

$$\nu = \mathfrak{R}\left(\frac{1}{n_1^2} - \frac{1}{n_2^2}\right)$$

and $c = \nu\lambda = 2.99792 \times 10^8$ m$\cdot$s$^{-1}$

$$c = \mathfrak{R}\left(\frac{1}{n_1^2} - \frac{1}{n_2^2}\right)\lambda$$

$$2.99792 \times 10^8 \text{ m}\cdot\text{s}^{-1} = (3.29 \times 10^{15} \text{ s}^{-1})\left(\frac{1}{4} - \frac{1}{16}\right)\lambda$$

$$\lambda = 4.86 \times 10^{-7} \text{ m} = 486 \text{ nm}$$

(b) Balmer series

(c) visible, blue

**1.25** For hydrogen-like one-electron ions, we use the $Z$-dependent Rydberg relation with the relationship $c = \lambda\nu$ to determine the transition wavelength. For He$^+$, $Z = 2$.

$$\nu = Z^2\mathfrak{R}\left(\frac{1}{n_1^2} - \frac{1}{n_2^2}\right) = (2^2)(3.29 \times 10^{15} \text{ s}^{-1})\left(\frac{1}{1^2} - \frac{1}{2^2}\right) = 9.87 \times 10^{15} \text{ Hz}$$

$$\lambda = \frac{c}{\nu} = \frac{2.99792 \times 10^8 \text{ m}\cdot\text{s}^{-1}}{9.87 \times 10^{15} \text{ s}^{-1}} = 3.04 \times 10^{-8} \text{ m} = 30.4 \text{ nm}$$

**1.27** (a) This problem is the same as that solved in Example 1.5, but the electron is moving between different energy levels. For movement between energy levels separated by a difference of 1 in principal quantum number, the expression is

$$\Delta E = E_{n+1} - E_n = \frac{(n+1)^2 h^2}{8mL^2} - \frac{n^2 h^2}{8mL^2} = \frac{(2n+1)h^2}{8mL^2}$$

For $n = 2$ and $n + 1 = 3$, $\Delta E = \dfrac{5h^2}{8mL^2}$

Then $\lambda_{3,2} = \dfrac{hc}{E} = \dfrac{8mhcL^2}{5h^2} = \dfrac{8mcL^2}{5h}$

For an electron in a 150-pm box, the expression becomes

$$\lambda_{3,2} = \frac{8(9.109\,39 \times 10^{-31}\ \text{kg})\,(2.997\,92 \times 10^8\ \text{m} \cdot \text{s}^{-1})\,(150 \times 10^{-12}\ \text{m})^2}{5(6.626\,08 \times 10^{-34}\ \text{J} \cdot \text{s})}$$

$$= 1.48 \times 10^{-8}\ \text{m}$$

(b) We need to remember that the equation for $\Delta E$ was originally determined for energy separations between successive energy levels, so the expression needs to be altered to make it general for energy levels two units apart:

$$\Delta E = E_{n+2} - E_n = \frac{(n+2)^2 h^2}{8mL^2} - \frac{n^2 h^2}{8mL^2} = \frac{(n^2 + 4n + 4 - n^2)h^2}{8mL^2} = \frac{(4n+4)h^2}{8mL^2}$$

$$\lambda = \frac{hc}{\Delta E} = \frac{hc(8mL^2)}{h^2[4n+4]} = \frac{8mcL^2}{h^2[4n+4]}$$

For $n = 2$, the expression becomes

$$\lambda = \frac{8mcL^2}{h[(4 \times 2) + 4]} = \frac{8mcL^2}{12h}$$

$$= \frac{8(9.10939 \times 10^{-31}\ \text{kg})\,(2.99792 \times 10^8\ \text{m} \cdot \text{s}^{-1})\,(150 \times 10^{-12}\ \text{m})^2}{12(6.62608 \times 10^{-34}\ \text{kg} \cdot \text{m}^2 \cdot \text{s}^{-1})}$$

$$= 6.18 \times 10^{-9}\ \text{m}$$

**1.29** In each of these series, the principal quantum number for the lower energy level involved is the same for each absorption line. Thus, for the Lyman series, the lower energy level is $n = 1$; for the Balmer series, $n = 2$; for Paschen series, $n = 3$; and for the Brackett series, $n = 4$.

**1.31** (a) 1s        2p        3d

(b) A node is a region in space where the wavefunction $\psi$ passes through 0.   (c) The simplest $s$-orbital has 0 nodes, the simplest $p$-orbital has 1 nodal plane, and the simplest $d$-orbital has 2 nodal planes.   (d) Given the increase in number of nodes, an $f$-orbital would be expected to have 3 nodal planes.

**1.33** The $p_x$ orbital will have its lobes oriented along the x axis, the $p_y$ orbital will have its lobes oriented along the y axis, and the $p_z$ orbital will have its lobes oriented along the z axis.

**1.35** The equation derived in Illustration 1.4 can be used:

$$\frac{\psi^2(r=0.55a_0,\theta,\phi)}{\psi^2(0,\theta,\phi)} = \frac{\dfrac{e^{-2(0.55a_0)/a_0}}{\pi a_0^{\,3}}}{\left(\dfrac{1}{\pi a_0^{\,3}}\right)} = 0.33$$

**1.37** To show that three p orbitals taken together are spherically symmetric, sum the three probability distributions (the wavefunctions squared) and show that the magnitude of the sum is not a function of $\theta$ or $\phi$.

$$p_x = R(r)C\sin\theta\cos\phi$$
$$p_y = R(r)C\sin\theta\sin\phi$$
$$p_z = R(r)C\cos\theta$$

where $C = \left(\dfrac{3}{4\pi}\right)^{\frac{1}{2}}$

Squaring the three wavefunctions and summing them:

$$R(r)^2 C^2 \sin^2 \theta \cos^2 \phi + R(r)^2 C^2 \sin^2 \theta \sin^2 \phi + R(r)^2 C^2 \cos^2 \theta$$
$$= R(r)^2 C^2 \left( \sin^2 \theta \cos^2 \phi + \sin^2 \theta \sin^2 \phi + \cos^2 \theta \right)$$
$$= R(r)^2 C^2 \left( \sin^2 \theta \left( \cos^2 \phi + \sin^2 \phi \right) + \cos^2 \theta \right)$$

Using the identity $\cos^2 x + \sin^2 x = 1$ this becomes

$$R(r)^2 C^2 \left( \sin^2 \theta + \cos^2 \theta \right) = R(r)^2 C^2$$

With one electron in each p orbital, the electron distribution is not a fuction of $\theta$ or $\phi$ and is, therefore, spherically symmetric.

**1.39** (a) The probability ($P$) of finding an electron within a sphere of radius $a_o$ may be determined by integrating the appropriate wavefunction squared from 0 to $a_o$ :

$$P = \frac{4}{a_o^2} \int_0^{a_o} r^2 \exp\left( -\frac{2r}{a_o} \right) dr$$

This integral is easier to evaluate if we allow the following change of variables:

$$z = \frac{2r}{a_o} \qquad \therefore \ z = 2 \text{ when } r = a_o, z = 0 \text{ when } r = 0, \text{ and } dr = \left( \frac{a_o}{2} \right) dz$$

$$P = \frac{1}{2} \int_0^2 z^2 \exp(-z) dz = -\frac{1}{2}(z^2 + 2z + 2)\exp(-z) \Big|_0^2$$

$$= -\frac{1}{2}\left[ \left( (4 + 4 + 2)\exp(-2) \right) - 2 \right]$$

$$= 0.323 \text{ or } 32.3\%$$

(b) Following the answer developed in (a) changing the integration limits to 0 to 2 $a_o$:

$$z = 4 \text{ when } r = 2a_o, z = 0 \text{ when } r = 0, \text{ and } dr = \left( \frac{a_o}{2} \right) dz$$

$$P = \frac{1}{2} \int_0^4 z^2 \exp(-z) dz = -\frac{1}{2}(z^2 + 2z + 2)\exp(-z) \Big|_0^4$$

$$= -\frac{1}{2}\left[ \left( 26\exp(-4) \right) - 4 \right]$$

$$= 0.761 \text{ or } 76.1\%$$

**1.41**   (a) 1 orbital;   (b) 5 orbitals;   (c) 3 orbitals;   (d) 7 orbitals

**1.43**   (a) 7 values: 0, 1, 2, 3, 4, 5, 6;   (b) 5 values; $-2, -1, 0, 1, 2$;   (c) 3 values: $-1, 0, 1$; (d) 4 subshells: $4s, 4p, 4d$, and $4f$

**1.45**   (a) $n = 6; l = 1$;   (b) $n = 3; l = 2$;   (c) $n = 2; l = 1$;   (d) $n = 5; l = 3$

**1.47**   (a) $-1, 0, +1$;   (b) $-2, -1, 0, +1, +2$;   (c) $-1, 0, +1$;
(d) $-3, -2, -1, 0, +1, +2, +3$.

**1.49**   (a) 6 electrons;   (b) 10 electrons;   (c) 2 electrons;   (d) 14 electrons

**1.51**   (a) $5d$, five;   (b) $1s$, one;   (c) $6f$, seven;   (d) $2p$, three

**1.53**   (a) six;   (b) two;   (c) eight;   (d) two

**1.55**   (a) cannot exist;   (b) exists;   (c) cannot exist;   (d) exists

**1.57**   (a) The total Coulomb potential energy $V(r)$ is the sum of the individual coulombic attractions and repulsions. There will be one attraction between the nucleus and each electron plus a repulsive term to represent the interaction between each pair of electrons. For lithium, there are three protons in the nucleus and three electrons. Each attractive Coulomb potential will be equal to

$$\frac{(-e)(+3e)}{4\pi\varepsilon_0 r} = \frac{-3e^2}{4\pi\varepsilon_0 r}$$

where $-e$ is the charge on the electron and $+3e$ is the charge on the nucleus, $e_0$ is the vacuum permittivity, and $r$ is the distance from the electron to the nucleus. The total attractive potential will thus be

$$\left(\frac{-3e^2}{4\pi\varepsilon_0 r_1}\right) + \left(\frac{-3e^2}{4\pi\varepsilon_0 r_2}\right) + \left(\frac{-3e^2}{4\pi\varepsilon_0 r_3}\right) = \left(\frac{-3e^2}{4\pi\varepsilon_0}\right)\left(\frac{1}{r_1} + \frac{1}{r_2} + \frac{1}{r_3}\right)$$

The repulsive terms will have the form

$$\frac{(-e)(-e)}{4\pi\varepsilon_0 r_{ab}} = \frac{e^2}{4\pi\varepsilon_0 r_{ab}}$$

where $r_{ab}$ represents the distance between two electrons a and b. The total repulsive term will thus be

$$\frac{e^2}{4\pi\varepsilon_0 r_{12}} + \frac{e^2}{4\pi\varepsilon_0 r_{13}} + \frac{e^2}{4\pi\varepsilon_0 r_{23}} = \frac{e^2}{4\pi\varepsilon_0}\left(\frac{1}{r_{12}} + \frac{1}{r_{13}} + \frac{1}{r_{23}}\right)$$

This gives

$$V(r) = \left(\frac{-3e^2}{4\pi\varepsilon_0}\right)\left(\frac{1}{r_2} + \frac{1}{r_2} + \frac{1}{r_3}\right) + \frac{e^2}{4\pi\varepsilon_0}\left(\frac{1}{r_{12}} + \frac{1}{r_{13}} + \frac{1}{r_{23}}\right)$$

(b)  The first term represents the coulombic attractions between the nucleus and each electron, and the second term represents the coulombic repulsions between each pair of electrons.

1.59   (a) false. $Z_{eff}$ is considerably affected by the total number of electrons present in the atom because the electrons in the lower energy orbitals will "shield" the electrons in the higher energy orbitals from the nucleus. This effect arises because the e-e repulsions tend to offset the attraction of the electron to the nucleus.   (b) true;   (c)  false. The electrons are increasingly less able to penetrate to the nucleus as $l$ increases.   (d) true.

1.61   Only (d) is the configuration expected for a ground-state atom; the others all represent excited-state configurations.

1.63   (a) This configuration is possible.   (b)  This configuration is not possible because $l = 0$ here,  so $m_l$ must also equal 0.   (c)  This configuration is not possible because the maximum value $l$ can have is $n - 1$; $n = 4$, so $l_{max} = 3$.

**1.65**  (a) silver  $[Kr]4d^{10}5s^1$

(b) beryllium  $[He]2s^2$

(c) antimony  $[Kr]4d^{10}5s^25p^3$

(d) gallium  $[Ar]3d^{10}4s^24p^1$

(e) tungsten  $[Xe]4f^{14}5d^46s^2$

(f) iodine  $[Kr]4d^{10}5s^25p^5$

**1.67**  (a) tellurium;  (b) vanadium;  (c) carbon;  (d) thorium

**1.69**  (a) $4p$;  (b) $4s$;  (c) $6s$;  (d) $6s$

**1.71**  (a) 5;  (b) 11;  (c) 5;  (d) 20

**1.73**  (a) 3;  (b) 2;  (c) 3;  (d) 2

**1.75**  (a) $ns^1$;  (b) $ns^2np^1$;  (c) $(n-1)d^5ns^2$;  (d) $(n-1)d^{10}ns^1$

**1.77**  (a) oxygen $(1310\,kJ\cdot mol^{-1}) >$ selenium $(941\,kJ\cdot mol^{-1}) >$ tellurium $(870\,kJ\cdot mol^{-1})$; ionization energies generally decrease as one goes down a group.  (b) gold $(890\,kJ\cdot mol^{-1}) >$ osmium $(840\,kJ\cdot mol^{-1}) >$ tantalum $(761\,kJ\cdot mol^{-1})$; ionization energies generally decrease as one goes from right to left in the periodic table.  (c) lead $(716\,kJ\cdot mol^{-1}) >$ barium $(502\,kJ\cdot mol^{-1}) >$ cesium $(376\,kJ\cdot mol^{-1})$; ionization energies generally decrease as one goes from right to left in the periodic table.

**1.79**  The atomic radii (in pm) are

Sc    161    Fe    124

| | | | |
|---|---|---|---|
| Ti | 145 | Co | 125 |
| V | 132 | Ni | 125 |
| Cr | 125 | Cu | 128 |
| Mn | 137 | Zn | 133 |

The major trend is for decreasing radius as the nuclear charge increases, with the exception that Cu and Zn begin to show the effects of electron-electron repulsions and become larger as the $d$-subshell becomes filled. Mn is also an exception as found for other properties; this may be attributed to having the $d$-shell half-filled.

**1.81** (a) $Sb^{3+}$, $Sb^{5+}$; (c) $Tl^+$, $Tl^{3+}$; (b) and (d) only form one positive ion each.

**1.83** $P^{3-} > S^{2-} > Cl^-$

**1.85** (a) fluorine; (b) carbon; (c) chlorine; (d) lithium.

**1.87** (a) A diagonal relationship is a similarity in chemical properties between an element in the periodic table and one lying one period lower and one group to the right. (b) It is caused by the similarity in size of the ions. The lower-right element in the pair would generally be larger because it lies in a higher period, but it also will have a higher oxidation state, which will cause the ion to be smaller. (c) For example, $Al^{3+}$ and $Ge^{4+}$ compounds show the diagonal relationship, as do $Li^+$ and $Mg^{2+}$.

**1.89** (a) N and S; (b) Li and Mg

**1.91** The ionization energies of the $s$-block metals are considerably lower, thus making it easier for them to lose electrons in chemical reactions.

**1.93** (a) metal; (b) nonmetal; (c) metal; (d) metalloid; (e) metalloid; (f) metal

**1.95** (a) $\dfrac{\nu}{c} = 3600 \text{ cm}^{-1}$

$\nu = c(3600 \text{ cm}^{-1})$

$\nu = (2.997\,92 \times 10^8 \text{ m} \cdot \text{s}^{-1})(3600 \text{ cm}^{-1})$

$\nu = (2.997\,92 \times 10^{10} \text{ cm} \cdot \text{s}^{-1})(3600 \text{ cm}^{-1})$

$\nu = 1.1 \times 10^{14} \text{ s}^{-1}$

(b) From $E = h\nu$: $E = (6.626\,08 \times 10^{-34} \text{ J} \cdot \text{s})(1.079 \times 10^{14} \text{ s}^{-1})$

$= 7.2 \times 10^{-20}$ J.

(c) 1.00 mol of molecules $= 6.022 \times 10^{23}$ molecules, so the energy absorbed by 1.00 mol will be

$(7.151 \times 10^{-20} \text{ J} \cdot \text{molecule}^{-1})(6.022 \times 10^{23} \text{ molecules} \cdot \text{mol}^{-1})$

$= 4.3 \times 10^4 \text{ J} \cdot \text{mol}^{-1}$ or $43 \text{ kJ} \cdot \text{mol}^{-1}$.

**1.97** In copper it is energetically favorable for an electron to be promoted from the 4s orbital to a 3d orbital, giving a completely filled 3d subshell. In the case of Cr, it is energetically favorable for an electron to be promoted from the 4s orbital to a 3d orbital to exactly ½ fill the 3d subshell.

**1.99** This trend is attributed to the inert-pair effect, which states that the *s*-electrons are less available for bonding in the heavier elements. Thus, there is an increasing trend as we descend the periodic table for the preferred oxidation number to be 2 units lower than the maximum one. As one descends the periodic table, ionization energies tend to decrease. For Tl, however, the values are slightly higher than those of its lighter analogues.

**1.101** (a) The relation is derived as follows: the energy of the photon entering, $E_{\text{total}}$, must be equal to the energy to eject the electron, $E_{\text{ejection}}$, plus the energy that ends up as kinetic energy, $E_{kinetic}$, in the movement of the electron:

$$E_{total} = E_{ejection} + E_{kinetic}$$

But $E_{total}$ for the photon $= h\nu$ and $E_{kinetic} = \left(\frac{1}{2}\right)mv^2$ where $m$ is the mass of the object and $v$ is its velocity. $E_{ejection}$ corresponds to the ionization energy, $I$, so we arrive at the final relationship desired.

(b)

$$E_{total} = h\nu = hc\lambda^{-1}$$

$$= (6.62608 \times 10^{-34} \text{ J} \cdot \text{s}) (2.99792 \times 10^8 \text{ m} \cdot \text{s}^{-1}) (58.4 \times 10^{-9} \text{ m})^{-1}$$

$$= 3.401 \times 10^{-18} \text{ J}$$

$$= E_{ejection} + E_{kinetic}$$

$$E_{kinetic} = \left(\frac{1}{2}mv^2\right) = \left(\frac{1}{2}\right)(9.10939 \times 10^{-28} \text{ g}) (2450 \text{ km} \cdot \text{s}^{-1})^2$$

$$= \left(\frac{1}{2}\right)(9.10939 \times 10^{-31} \text{ kg}) (2.450 \times 10^6 \text{ m} \cdot \text{s}^{-1})^2$$

$$= 2.734 \times 10^{-18} \text{ kg} \cdot \text{m}^2 \cdot \text{s}^{-2} = 2.734 \times 10^{-18} \text{ J}$$

$$3.401 \times 10^{-18} \text{ J} = E_{ejection} + 2.734 \times 10^{-18} \text{ J}$$

$$E_{ejection} = 6.67 \times 10^{-19} \text{ J}$$

**1.103** By the time we get to the lanthanides and actinides— the two series of *f*-orbital filling elements—the energy levels become very close together and minor changes in environment cause the different types of orbitals to switch in energy-level ordering. For the elements mentioned, the electronic configurations are

| La | $[\text{Xe}] 6s^2 5d^1$ | Lu | $[\text{Xe}] 6s^2 4f^{14} 5d^1$ |
|---|---|---|---|
| Ac | $[\text{Rn}] 7s^2 6d^1$ | Lr | $[\text{Rn}] 7s^2 5f^{14} 6d^1$ |

As can be seen, all these elements have one electron in a *d*-orbital, so placement in the third column of the periodic table could be considered appropriate for either, depending on what aspects of the chemistry of these elements we are comparing. The choice is not without argument, and it is discussed by W. B. Jensen (1982), *J. Chem. Ed.* **59**, 634.

**1.105** (a)—(c) We can use the hydrogen $2s$ wavefunction found in Table 1.2. Remember that the probability of locating an electron at a small region in space is proportional to $\psi^2$, not $\psi$.

$$\psi_{2s} = \frac{1}{4}\left(\frac{1}{2\pi a_0^3}\right)^{\frac{1}{2}}\left(2 - \frac{r}{a_0}\right)e^{-\frac{r}{2a_0}}$$

$$\psi_{2s}^2 = \left(\frac{1}{32\pi a_0^3}\right)\left(2 - \frac{r}{a_0}\right)^2 e^{-\frac{r}{a_0}}$$

$$\frac{\psi_{2s}^2(r,\theta,\phi)}{\psi_{2s}^2(0,\theta,\phi)} = \frac{\left(\frac{1}{32\pi a_0^3}\right)\left(2 - \frac{r}{a_0}\right)^2 e^{-\frac{r}{a_0}}}{\left(\frac{1}{32\pi a_0^3}\right)2^2 e^{-\frac{0}{a_0}}}$$

$$= \frac{\left(2 - \frac{r}{a_0}\right)^2 e^{-\frac{r}{a_0}}}{4}$$

Because $r$ will be equal to some fraction $x$ times $a_0$, the expression will simplify further:

$$\frac{\psi_{2s}^2(r,\theta,\phi)}{\psi_{2s}^2(0,\theta,\phi)} = \frac{\left(2 - \frac{xa_0}{a_0}\right)^2 e^{-\frac{xa_0}{a_0}}}{4} = \frac{(2-x)^2 e^{-x}}{4}$$

Carrying out this calculation for the other points, we obtain:

| x | relative probability |
|---|---|
| 0.1 | 0.82 |
| 0.2 | 0.66 |
| 0.3 | 0.54 |
| 0.4 | 0.43 |
| 0.5 | 0.34 |
| 0.6 | 0.27 |
| 0.7 | 0.21 |
| 0.8 | 0.16 |

| | |
|---|---|
| 0.9 | 0.12 |
| 1 | 0.092 |
| 1.1 | 0.067 |
| 1.2 | 0.048 |
| 1.3 | 0.033 |
| 1.4 | 0.022 |
| 1.5 | 0.014 |
| 1.6 | 0.0081 |
| 1.7 | 0.0041 |
| 1.8 | 0.0017 |
| 1.9 | 0.0004 |
| 2 | 0.0000 |
| 2.1 | 0.00031 |
| 2.2 | 0.0011 |
| 2.3 | 0.0023 |
| 2.4 | 0.0036 |
| 2.5 | 0.0051 |
| 2.6 | 0.0067 |
| 2.7 | 0.0082 |
| 2.8 | 0.0097 |
| 2.9 | 0.011 |
| 3 | 0.012 |

This can be most easily carried out graphically by simply plotting the

function $f(x) = \dfrac{(2-x)^2 e^{-x}}{4}$ from 0 to 3.

The node occurs when x = 2, or when $r = 2a_0$. This is exactly what is

obtained by setting the radial part of the equation equal to 0.

**1.107** The approach to showing that this is true involves integrating the

probability function over all space. The probability function is given by

the square of the wave function, so that for the particle in the box we have

$$\psi = \left(\frac{2}{L}\right)^{\frac{1}{2}} \sin\left(\frac{n\pi x}{L}\right)$$

and the probability function will be given by

$$\psi^2 = \left(\frac{2}{L}\right) \sin^2\left(\frac{n\pi x}{L}\right)$$

Because $x$ can range from 0 to $L$ (the length of the box), we can write the integration as

$$\int_0^x \psi^2\, dx = \int_0^x \left(\frac{2}{L}\right) \sin^2\left(\frac{n\pi x}{L}\right) dx$$

for the entire box, we write

probability of finding the particle somewhere in the box =

$$\int_0^L \left(\frac{2}{L}\right) \sin^2\left(\frac{n\pi x}{L}\right) dx$$

$$probability = \left(\frac{2}{L}\right) \int_0^x \sin^2\left(\frac{n\pi x}{L}\right) dx$$

$$\int_0^{\frac{L}{2}} \Psi^2 = \frac{2}{L} \int_0^{\frac{L}{2}} \left(\sin\frac{n\pi x}{L}\right)^2 dx$$

$$= \frac{2}{L}\left[\left(\frac{-1}{2n\pi}\cdot\cos\frac{n\pi x}{L}\cdot\sin\frac{n\pi x}{L} + \frac{x}{2}\right)\Big|_0^L\right]$$

$$= \frac{2}{L}\left[\left(\frac{-1}{2n\pi}\cdot\cos n\pi \cdot\sin n\pi + \frac{L}{2}\right) - 0\right]$$

if $n$ is an integer, $\sin n\pi$ will always be zero and

$$probability = \frac{2}{L}\left[\frac{L}{2}\right] = 1$$

# CHAPTER 2
# CHEMICAL BONDS

**2.1**   The coulombic attraction is directly proportional to the charge on each ion (Equation 1) so the ions with the higher charges will give the greater coulombic attraction. The answer is therefore (b) $Ga^{3+}$, $O^{2-}$.

**2.3**   The $Li^+$ ion is smaller than the $Rb^+$ ion (58 vs 149 pm). Because the lattice energy is related to the coulombic attraction between the ions, it will be inversely proportional to the distance between the ions (see Equation 2). Hence the larger rubidium ion will have the lower lattice energy for a given anion.

**2.5**   (a) 5;   (b) 4;   (c) 7;   (d) 3

**2.7**   (a) [Ar];   (b) $[Ar]3d^{10}4s^2$;   (c) $[Kr]4d^5$;   (d) $[Ar]3d^{10}4s^2$

**2.9**   (a) $[Ar]3d^{10}$;   (b) $[Xe]4f^{14}5d^{10}6s^2$;   (c) $[Ar]3d^{10}$;   (d) $[Xe]4f^{14}5d^{10}$

**2.11**   (a) $[Kr]4d^{10}5s^2$; same   (b) none   (c) $[Kr]4d^{10}$; Pd

**2.13**   (a) $Co^{2+}$;   (b) $Fe^{2+}$;   (c) $Mo^{2+}$;   (d) $Nb^{2+}$

**2.15**   (a) $Co^{3+}$;   (b) $Fe^{3+}$;   (c) $Ru^{3+}$;   (d) $Mo^{3+}$

**2.17**   (a) $4s$;   (b) $3p$;   (c) $3p$;   (d) $4s$

**2.19** (a) $-1$;  (b) $-2$;  (c) $+1$;  (d) $+3$ ($+1$ sometimes observed);
(e) $+2$

**2.21** (a) 3;  (b) 6;  (c) 6;  (d) 2

**2.23** (a) $[Kr]4d^{10}5s^2$; no unpaired electrons;  (b) $[Kr]4d^{10}$; no unpaired electrons;  (c) $[Xe]4f^{14}5d^4$; four unpaired electrons;  (d) $[Kr]$; no unpaired electrons;  (e) $[Ar]3d^8$; two unpaired electrons

**2.25** (a) $3p$;  (b) $5s$;  (c) $5p$;  (d) $4d$

**2.27** (a) $+7$;  (b) $-1$;  (c) $[Ne]$ for $+7$, $[Ar]$ for $-1$;  (d) electrons are lost or added to give noble-gas configuration.

**2.29** (a) $Mg_3As_2$;  (b) $In_2S_3$;  (c) $AlH_3$;  (d) $H_2Te$;  (e) $BiF_3$

**2.31** (a) $Bi_2O_3$;  (b) $PbO_2$;  (c) $Tl_2O_3$

**2.33**

(a) 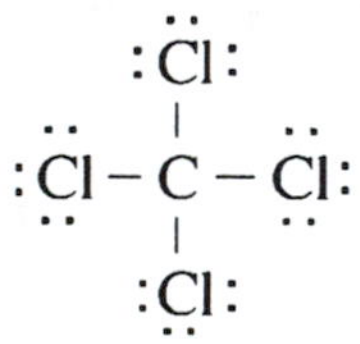
$$
\begin{array}{c}
:\!\ddot{C}l\!: \\
| \\
:\!\ddot{C}l - C - \ddot{C}l\!: \\
| \\
:\!\ddot{C}l\!:
\end{array}
$$

(b)
$$
\begin{array}{c}
:\!O\!: \\
\| \\
:\!\ddot{C}l - C - \ddot{C}l\!:
\end{array}
$$

(c)
$$
\ddot{O} = \ddot{N} - \ddot{F}:
$$

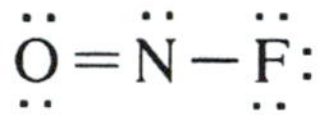

(d) 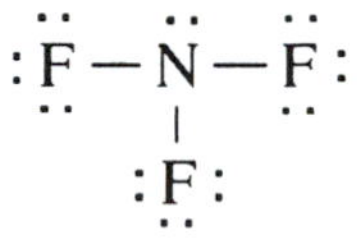
$$
\begin{array}{c}
:\!\ddot{F} - \ddot{N} - \ddot{F}\!: \\
| \\
:\!\ddot{F}\!:
\end{array}
$$

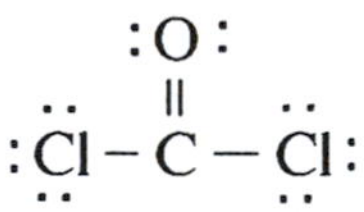

**2.35**

(a) 
$$\left[\; H-\underset{\underset{H}{|}}{\overset{\overset{H}{|}}{B}}-H \;\right]^{-}$$

(b) $\left[\; :\ddot{B}r-\ddot{O}: \;\right]^{-}$

(c) $\left[\; :\ddot{N}-\underset{|}{H} \atop H \;\right]^{-}$

**2.37**

(a) $\left[\; H-\underset{\underset{H}{|}}{\overset{\overset{H}{|}}{N}}-H \;\right]^{+}$ $\;:\ddot{C}l:^{-}$

(b) $K^{+}\left[\; :\ddot{P}: \;\right]^{3-}$
$\qquad K^{+}$
$\qquad K^{+}$

(c) $Na^{+}\left[\; :\ddot{C}l-\ddot{O}: \;\right]^{-}$

**2.39**

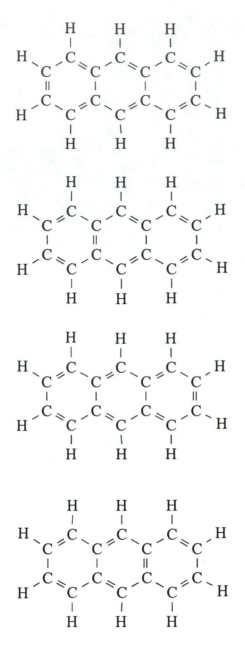

**2.41**

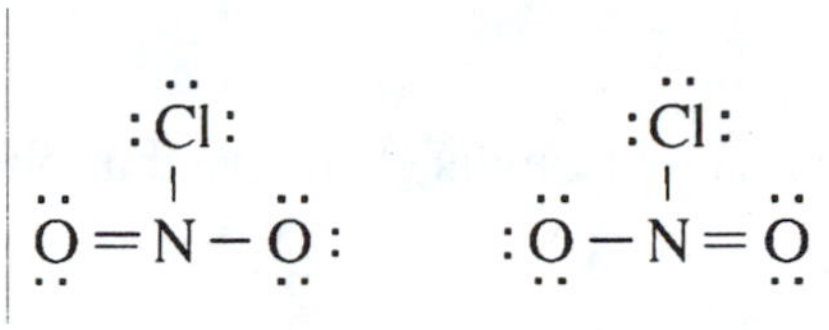

```
        :Cl:              :Cl:
         |                 |
    ..   |   ..        ..  |   ..
    O = N - O :     : O - N = O
    ..       ..      ..        ..
```

**2.43**

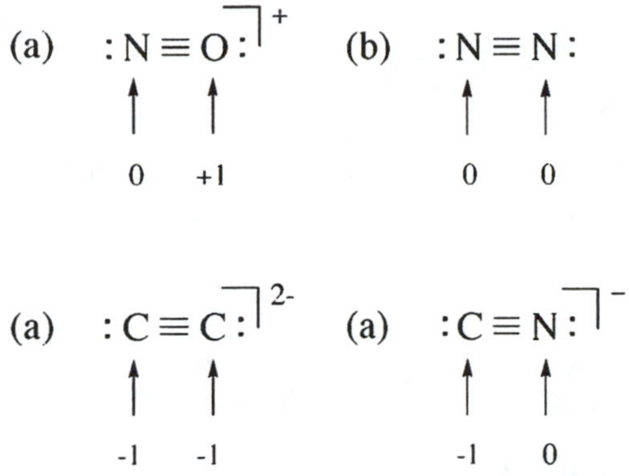

(a)  :N≡O:⁺        (b)  :N≡N:        (a)  :C≡O:
     ↑   ↑              ↑   ↑              ↑   ↑
     0  +1              0   0             -1  +1

(a)  :C≡C:²⁻        (a)  :C≡N:⁻
     ↑   ↑              ↑   ↑
    -1  -1             -1   0

**2.45**

(a)
```
      ..      ..      ..
   0  O = Cl - O : 0
           || 0  |
          :O:  H
           0
       lower energy
```
```
     ..     ..     ..
   :O - Cl - O : 0
    ..   | +2  |
    -1  :O:  H
         ..
        -1
```

(b)
```
        0
     ..         ..
  0  O = C = S  0
     ..         ..
    lower energy
```
```
        0
     ..         ..
  -1 :O - C = S -1
     ..         ..
```

(c)
```
   H - C ≡ N:        H - C = N:
   0   0   0         0   -1  +1
    lower energy
```

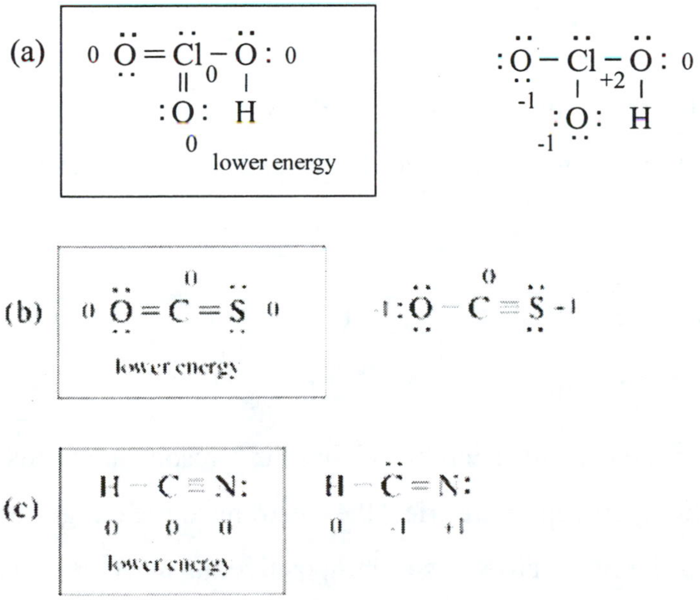

**2.47**  (a) In the first structure, the formal charges at Xe and F are 0, whereas, in the second structure, Xe is −1, one F is 0, and the other F is +1. The first structure is favored on the basis of formal charges.   (b) In the first structure, all of the atoms have formal charges of 0, whereas, in the second structure, one O atom has a formal charge of +1 and the other O has a formal charge of −1. The first structure is thus preferred.

**2.49**  (a)  The sulfite ion has one Lewis structure that obeys the octet rule:

$$\left[\,:\!\ddot{O}-\overset{\displaystyle .\,.}{S}-\ddot{O}\!:\,\right]^{2-}$$
$$\underset{:\ddot{O}:}{|}$$

and three with an expanded octet:

$$\left[\,\ddot{O}=\overset{\displaystyle .\,.}{S}-\ddot{O}\!:\,\right]^{2-}\qquad\left[\,:\!\ddot{O}-\overset{\displaystyle .\,.}{S}=\ddot{O}\,\right]^{2-}\qquad\left[\,:\!\ddot{O}-\overset{\displaystyle .\,.}{S}-\ddot{O}\!:\,\right]^{2-}$$
$$\underset{:\ddot{O}:}{|}\qquad\qquad\underset{:\ddot{O}:}{|}\qquad\qquad\underset{:\ddot{O}:}{\parallel}$$

The structures with expanded octets have lower formal charges.

(b)  There is one Lewis structure that obeys the octet rule:

$$\left[\,:\!\ddot{O}-\overset{\displaystyle .\,.}{S}-\ddot{O}\!:\,\right]^{-}$$
$$\underset{:\ddot{O}-H}{|}$$

The formal charge at sulfur can be reduced to 0 by including one double bond contribution. This change gives rise to two expanded octet structures.

$$\left[\,:\!\ddot{O}-\overset{\displaystyle .\,.}{S}=\ddot{O}\,\right]^{-}\qquad\qquad\left[\,\ddot{O}=\overset{\displaystyle .\,.}{S}-\ddot{O}\!:\,\right]^{-}$$
$$\underset{:\ddot{O}-H}{|}\qquad\qquad\qquad\underset{:\ddot{O}-H}{|}$$

Notice that, unlike the sulfite ion, which has three resonance forms, the presence of the hydrogen ion restricts the electrons to the oxygen atom to which it is attached. Because H is electropositive, its placement near an oxygen atom makes it less likely for that oxygen atom to donate a lone pair to an adjacent atom.

(c)  The perchlorate ion has one Lewis structure that obeys the octet rule:

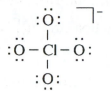

The formal charge at Cl can be reduced to 0 by including three double-bond contributions, thereby giving rise to four resonance forms.

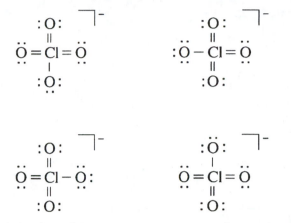

(d) For the nitrite ion, there are two resonance forms, both of which obey the octet rule:

**2.51** Radicals are species with an unpaired electron, therefore only (b) and (c) are radicals since they have an odd number of electrons while (a) and (d) have an even number of electrons allowing Lewis structures to be drawn with all electrons paired.

**2.53**

(a)  :Cl – O:
    radical

(b)  :Cl – O – O – Cl:
    not a radical

(c)  O = N – O:  with  :O – Cl:
    not a radical

(d)  :Cl – O – O·
    radical

**2.55** (a)            (b)

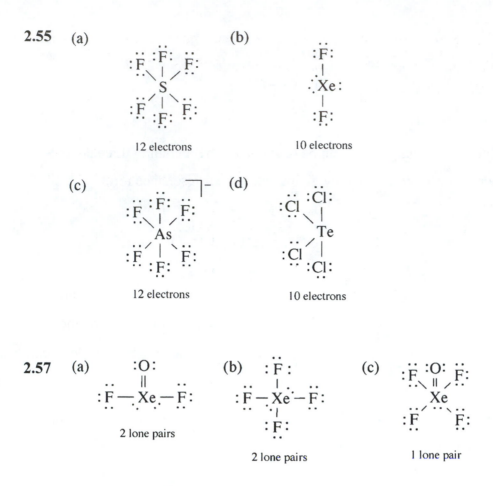

12 electrons          10 electrons

(c)            (d)

12 electrons          10 electrons

**2.57** (a)      (b)      (c)

2 lone pairs       2 lone pairs       1 lone pair

**2.59**     I (2.7) < Br (3.0) < Cl (3.2) < F (4.0)

**2.61**     In (1.8) < Sn (2.0) < Sb (2.1) < Se (2.6)

**2.63**     (a) Iodide is more polarizable than $Cl^-$, so HI would be more covalent than HCl.

(b) The bonds in $CF_4$ would be more ionic. The electronegativity difference is greater between C and F than between C and H, making the C—F bonds more ionic.    (c) C and S have nearly identical electronegativities, so the C—S bonds would be expected to be almost completely covalent, whereas the C—O bonds would be more ionic.

**2.65** $Rb^+ < Sr^{2+} < Be^{2+}$; smaller, more highly charged cations have greater polarizing power. The ionic radii are 149 pm, 116 pm, 27 pm, respectively.

**2.67** $O^{2-} < N^{3-} < Cl^- < Br^-$; the polarizability increases as the ion gets larger and less electronegative. The ionic radii for these species are 140 pm, 171 pm, 181 pm, 196 pm, respectively.

**2.69** (a) $CO_3^{2-} > CO_2 > CO$

$CO_3^{2-}$ will have the longest C—O bond length. In CO there is a triple bond and in $CO_2$ the C—O bonds are double bonds. In carbonate, the bond is an average of three Lewis structures in which the bond is double in one form and single in two of the forms. We would thus expect the bond order to be approximately 1.3. Because the bond length is inversely related to the number of bonds between the atoms, we expect the bond length to be longest in carbonate.

(b) $SO_3^{2-} > SO_2 \sim SO_3$

Similar arguments can be used for these molecules as in part (a). In $SO_2$ and $SO_3$, the Lewis structures with the lowest formal charge at S have double bonds between S and each O. In the sulfite ion, however, there are three Lewis structures that have a 0 formal charge at S. Each has one S—O double bond and two S—O single bonds. Because these S—O bonds would have a substantial amount of single bond character, they would be expected to be longer than those in $SO_2$ or $SO_3$. This is consistent with the experimental data that show the S—O bond lengths in $SO_2$ and $SO_3$ to be 143 pm, whereas those in $SO_3^{2-}$ range from about 145 pm to 152 pm depending on the compound.

(c) $CH_3NH_2 > CH_2NH > HCN$

The C—N bond in HCN is a triple bond, in $CH_2NH$ it is a double bond, and in $CH_3NH_2$ it is a single bond. The C—N bond in the last molecule would, therefore, be expected to be the longest.

**2.71** (a) The covalent radius of N is 75 pm, so the N—N single bond in hydrazine would be expected to be ca. 150 pm. The experimental value is 145 pm.   (b) The C—O bonds in carbon dioxide are double bonds. The covalent radius for doubly bonded carbon is 67 pm and that of O is 60 pm. Thus we predict the CRO in $CO_2$ to be ca. 127 pm. The experimental bond length is 116.3 pm.   (c) The C—O bond is a double bond so it would be expected to be the same as in (b), 127 ppm. This is the experimentally found value. The C—N bonds are single bonds and so one might expect the bond distance to be the sum of the single bond C radius and the single bond N radius (77 plus 75 pm) which is 152 pm. However, because the C atom is involved in a multiple bond, its radius is actually smaller. The sum of that radius (67 pm) and the N single bond radius gives 132 pm, which is close to the experimental value of 133 pm.   (d) The N—N bond is a double bond so we expect the bond distance to be two times the double bond covalent radius of N, which is 2 × (60 pm) or 120 pm. The experimental value is 123.0 pm.

**2.73** (a) 77 pm + 72 pm = 149 pm   (b) 111 pm + 72 pm = 183 pm
(c) 141 pm + 72 pm = 213 pm.  Bond distance increases with size going down Group 14/IV.

**2.75**

$$:N \equiv N - \overset{\cdot\cdot}{\underset{\cdot\cdot}{N}} - N \equiv N:$$

$$\overset{\cdot\cdot}{\underset{\cdot\cdot}{N}} = N = N - N \equiv N:$$

$$:N \equiv N - N = N = \overset{\cdot\cdot}{\underset{\cdot\cdot}{N}}$$

**2.77**

(a)

$$
\begin{array}{cc}
\overset{0}{:\!O\!:} & \overset{-1}{:\!\ddot{O}\!:} \\
\parallel & \parallel \\
\overset{0}{C} \!-\! \overset{0}{C} \\
\mid & \parallel \\
:\!O\!: & :\!O\!: \\
\overset{-1}{} & \overset{0}{}
\end{array}
\Biggr]^{2-}
\qquad
\begin{array}{cc}
:\!\ddot{O}\!: & :\!O\!: \\
\mid & \parallel \\
C \!-\! C \\
\parallel & \mid \\
:\!O\!: & :\!\ddot{O}\!:
\end{array}
\Biggr]^{2-}
$$

$$
\begin{array}{cc}
:\!O\!: & :\!O\!: \\
\parallel & \parallel \\
C \!-\! C \\
\mid & \mid \\
:\!\ddot{O}\!: & :\!\ddot{O}\!:
\end{array}
\Biggr]^{2-}
\qquad
\begin{array}{cc}
:\!\ddot{O}\!: & :\!\ddot{O}\!: \\
\mid & \mid \\
C \!-\! C \\
\parallel & \parallel \\
:\!O\!: & :\!O\!:
\end{array}
\Biggr]^{2-}
$$

(b) $\overset{+1}{\ddot{B}r} \!=\! \overset{0}{\ddot{O}}$ ]$^{+}$  

(c) $:\!\overset{-1}{C} \!=\! \overset{-1}{C}\!:$ ]$^{2-}$

**2.79** (a)

$$
CH_3 \!-\! \underset{}{\overset{\displaystyle :O:}{\overset{\parallel}{C}}} \!-\! \ddot{O}:
\Bigr]^{-}
\qquad
CH_3 \!-\! \underset{}{\overset{\displaystyle :\ddot{O}:}{\overset{\mid}{C}}} \!=\! \ddot{O}
\Bigr]^{-}
$$

(b)

$$
\underset{\displaystyle CH_3}{\overset{\displaystyle H}{H \!-\! C \!=\! C \!-\! \ddot{O}:}}
\Bigr]^{-}
\qquad
\underset{\displaystyle CH_3}{\overset{\displaystyle H}{H \!-\! \ddot{C} \!-\! C \!=\! \ddot{O}}}
\Bigr]^{-}
$$

(c)

$$
\overset{H \quad H \quad H}{H \!-\! C \!=\! C \!-\! C \!-\! H}
\Bigr]^{+}
\qquad
\overset{H \quad H \quad H}{H \!-\! C \!-\! C \!=\! C \!-\! H}
\Bigr]^{+}
$$

(d)

$$
CH_3 \!-\! \overset{\displaystyle :O:}{\overset{\parallel}{C}} \!-\! \overset{\displaystyle H}{N}:
\Bigr]^{-}
\qquad
CH_3 \!-\! \overset{\displaystyle :\ddot{O}:}{\overset{\mid}{C}} \!=\! \overset{\displaystyle H}{N}:
\Bigr]^{-}
$$

**2.81** P and S are larger atoms that are less able to form multiple bonds to themselves, unlike the small N and O atoms. All bonds in $P_4$ and $S_8$ are single bonds, whereas $N_2$ has a triple bond and $O_2$ a double bond.

**2.83** **(a)**

$$H-C\equiv C-H \qquad H-C\equiv Si-H$$
$$H-Si\equiv Si-H \qquad H-C\equiv N: \qquad :N\equiv N:$$

**(b)**

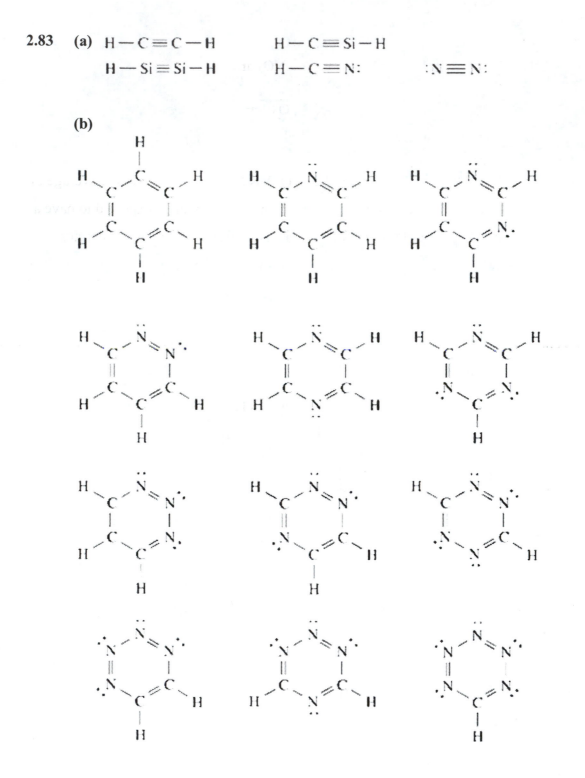

 + suitable resonance forms

**2.85** The Lewis structures for NO and $NO_2$ are

$$\ddot{N}=\ddot{O} \qquad \ddot{O}=N-\ddot{O}: \longleftrightarrow :\ddot{O}-N=\ddot{O}$$

Both compounds are radicals.

(a) NO has a double bond, but $NO_2$ has N-O bonds that are the average of a single bond and a double bond. Thus, NO would be expected to have a shorter, stronger bond, and this fact is indicated by the bond energies.

(b) The fact that the two N-O bonds in $NO_2$ are equal is a result of the two available resonance forms.

**2.87** (a)

| Metal Iodide | d(M − I), pm | Lattice Energy, kJ/mol |
|---|---|---|
| LiI | 278 | 761 |
| NaI | 322 | 705 |
| KI | 358 | 649 |
| RbI | 369 | 632 |
| CsI | 390 | 601 |
| AgI | 353 | 886 |

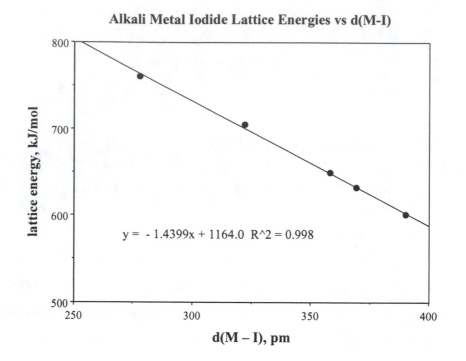

**Alkali Metal Iodide Lattice Energies vs d(M-I)**

$y = -1.4399x + 1164.0 \quad R^2 = 0.998$

The correlation is excellent with an agreement coefficient of greater than 99%.

(b)  From the equation of this line

$$\text{Lattice Energy} = -1.440\, d_{M-X} + 1164$$

and the Ag-I distance of 353 pm, we can estimate the AgI lattice energy to be $656\ \text{kJ} \cdot \text{mol}^{-1}$. (c) This is not very good agreement with the experimental value of $886\ \text{kJ} \cdot \text{mol}^{-1}$. One possible explanation is that the structure of silver iodide is different from that of the alkali metal iodides and therefore would have a different Madelung constant. This is the case as AgI crystallizes in a type of lattice known as the Wurtzite structure. However, the Madelung constant for the Wurtzite structure is 1.641 versus 1.748 for the rock salt structure of all the alkali metal iodides (except CsI, which adopts the CsCl structure with a very similar Madelung constant of 1.763). One would therefore predict that the lattice energy of AgI would be less than the calculated value based upon the alkali metal series, rather than greater. A better explanation is that the $Ag^+$ ion is much more polarizable than the alkali metal cations of similar size and therefore the bonding in AgI is much more covalent.

**2.89** (a)

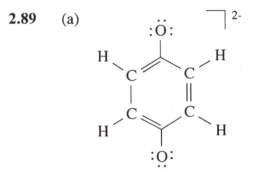

(b) All the atoms have formal charge 0 except the two oxygen atoms, which are -1. The negative charge is most likely to be concentrated at the oxygen atoms.

(c) Because the oxygen atoms are the most negative sites and the lone pairs on the oxygen atoms are good Lewis bases, the protons will bond to the oxygen atoms. This compound is known as hydroquinone.

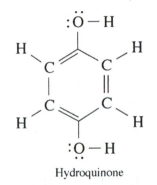

Hydroquinone

**2.91**

| Z | Configuration | Number of unpaired $e^-$ | Element | Charge | Energy state |
|---|---|---|---|---|---|
| 26 | $[Ar]3d^6$ | 4 | Fe | 2+ | ground |
| 52 | $[Kr]5s^2 4d^{10} 5p^5 6s^1$ | 2 | Te | 2- | excited |
| 16 | $[Ne]3s^2 3p^6$ | 0 | S | 2- | ground |
| 39 | $[Kr]4d^1$ | 1 | Y | 2+ | ground |
| 30 | $[Ar]4s^2 3d^8$ | 2 | Zn | 2+ | excited |

**2.93** (a)

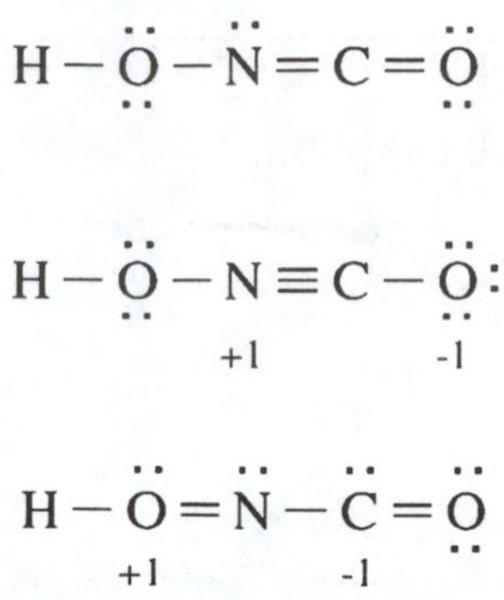

$$H-\ddot{O}-N=C=\ddot{O}$$

$$\underset{+1 \qquad -1}{H-\ddot{O}-N\equiv C-\ddot{O}:}$$

$$\underset{+1 \qquad -1}{H-\ddot{O}=\ddot{N}-\ddot{C}=\ddot{O}}$$

(b)

$$\underset{H}{\overset{H}{>}}C=\ddot{S}=\ddot{O} \qquad \underset{H}{\overset{H}{>}}\underset{+1 \quad -1}{C=\ddot{S}-\ddot{O}:}$$

(c)

$$\underset{H}{\overset{H}{>}}\underset{+1 \quad -1}{C=N=\ddot{N}}$$

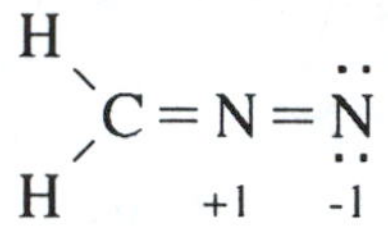

(d)

$$\underset{+1 \qquad -1}{\ddot{O}=N=C=\ddot{N}} \qquad \ddot{O}=\ddot{N}-C\equiv N:$$

**2.95**

$$\underset{H-C=C-C-C=C-H}{\overset{R \quad H \quad H \quad H \quad R'}{|\quad |\quad |\quad |\quad |}}$$

$$\underset{H-C-C=C-C=C-H}{\overset{R \quad H \quad H \quad H \quad R'}{|\quad |\quad |\quad |\quad |}}$$

$$\underset{H-C=C-C=C-C-H}{\overset{R \quad H \quad H \quad H \quad R'}{|\quad |\quad |\quad |\quad |}}$$

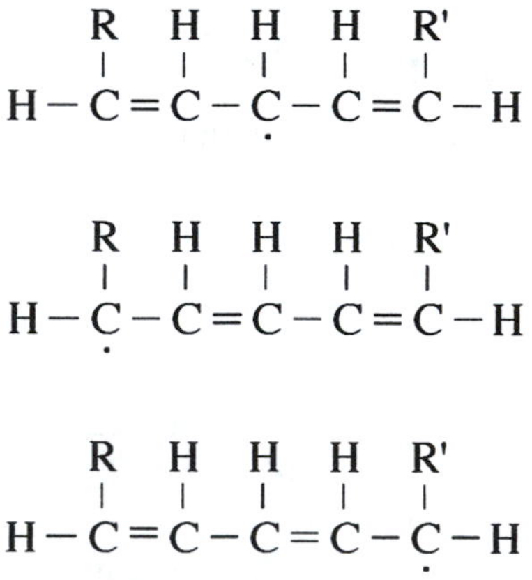

**2.97**

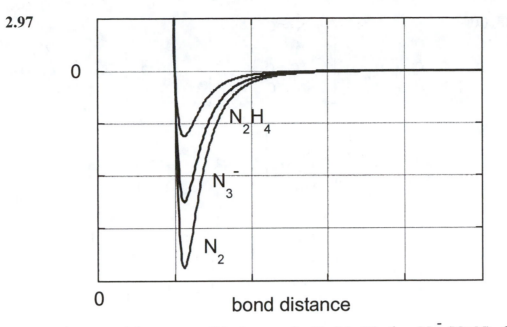

The potential energy well is deepest for $N_2$ ($N\equiv N$), then $N_3^-$ ($N=N$), then $N_2H_4$ (N-N).

**2.99** (a) I: $Tl_2O_3$; II: $Tl_2O$; (b) +3; +1; (c) $[Xe]4f^{14}5d^{10}$; $[Xe]4f^{14}5d^{10}6s^2$; (d) Because compound II has a lower melting point, it is probably more covalent which is consistent with the fact that the +3 ion is more polarizing.

**2.101** The alkyne group has the stiffer C—H bond because a large force constant, $k$, results in a higher-frequency absorption.

**2.103** (a)

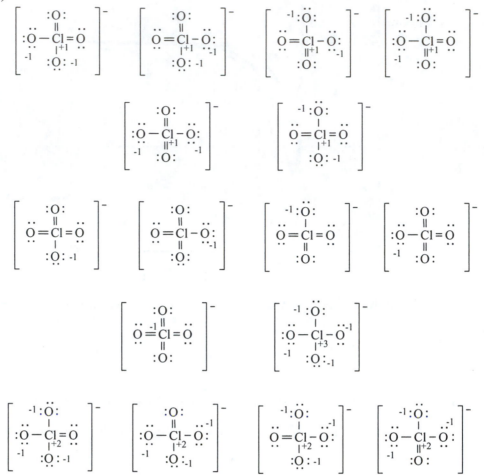

The four structures with three double bonds (third row) and the one with four double bonds are the most plausible Lewis structures according to formal charge arguments because these five structures minimize the formal charges. (b) The structure with four double bonds fits these observations best since its bond lengths would all be 140 pm, or only 4 pm shorter than the observed length. However, the four structures with three double bonds also fit because, if the double bonds are delocalized by resonance, we can estimate the average bond length to be $\frac{1}{4}(170 \text{ pm}) + \frac{3}{4}(140 \text{ pm}) = 147.5 \text{ pm}$, or just 3.5 pm longer than observed. (c) +7; The structure with all single bonds fits this criterion best. (d) Approaches (a) and (b) are consistent but approach (c) is not. This result is reasonable because oxidation numbers are assigned by assuming ionic bonding.

# CHAPTER 3

# MOLECULAR SHAPE AND STRUCTURE

**3.1**  (a) The shape of the thionyl chloride molecule is trigonal pyramidal.
(b) The O—S—Cl angles are identical. The lone electron pair repels the
bonded electron pairs equally and thus, all O—S—Cl bond angles are
compressed equally.  (c) The expected bond angle is slightly less
than 109.5°.

**3.3**  (a) angular, the electron pair on the central atom results in a trigonal
planar arrangement .  (b) The bond angle will be slightly less than 120° .

**3.5**  (a) linear;  (b) slightly less than 180°

**3.7**

(a)

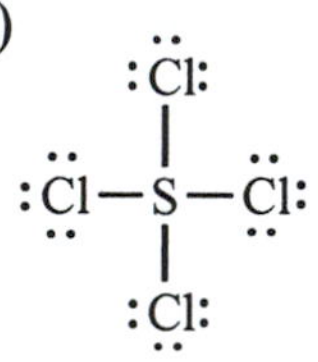

(b)

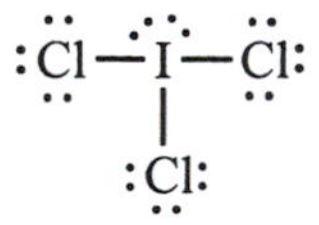

(c)

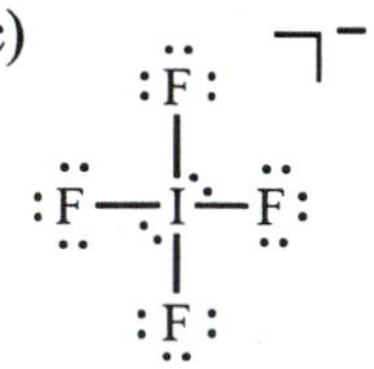

(d)

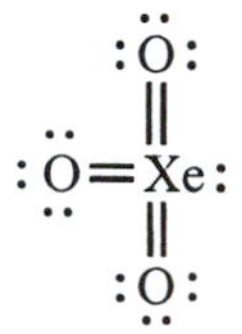

(a) The sulfur atom will have five pairs of electrons about it: one
nonbonding pair and four bonding pairs to chlorine atoms. The
arrangement of electron pairs will be trigonal bipyramidal; the nonbonding
pair of electrons will prefer to lie in an equatorial position, because in that

location the e-e repulsions will be lowest. The actual structure is described as a seesaw. $AX_4E$

(b) Like the sulfur atom in (a), the iodine in iodine trichloride has five pairs of electrons about it, but here there are two lone pairs and three bonding pairs. The arrangement of electron pairs will be the same as in (a), and again the lone pairs will occupy the equatorial positions. Because the name of the molecule ignores the lone pairs, it will be classified as T-shaped. $AX_3E_2$

(c) There are six pairs of electrons about the central iodine atom in $IF_4^-$. Of these, two are lone pairs and four are bonding pairs. The pairs will be placed about the central atom in an octahedral arrangement with the lone pairs opposite each other. This will minimize repulsions between them. The name given to the structure is square planar. $AX_4E_2$

(d) In determining the shape of a molecule, double bonds count the same as single bonds. The $XeO_3$ structure has four "objects" about the central Xe atom: three bonds and one lone pair. These will be placed in a tetrahedral arrangement. Because the lone pair is ignored in naming the molecule, it will be classified as trigonal pyramidal. $AX_3E$

**3.9** (a) The $I_3^-$ molecule is predicted to be linear, so the $<I$—$I$—$I$ should equal 180°. $AX_2E_3$

(b) The $SbCl_5$ molecule is trigonal bipyramidal. There should be three Cl—Sb—Cl angles of 120°, and two of 90°. $AX_5$

(c) The structure of $IO_4^-$ will be tetrahedral, so the O—I—O bond angles should be 109.5°. $AX_4$

(d) The structure of $NO_2^-$ is angular with a bond angle of around 120°. $AX_2E$

**3.11** The Lewis structures are

(a) $\overset{\displaystyle :\!\overset{..}{F}\!:}{\underset{\displaystyle :\!\overset{..}{F}\!:}{:\!\overset{..}{Cl}\!-\!C\!-\!\overset{..}{F}\!:}}$ 　(b) $\overset{\displaystyle :\!\overset{..}{Cl}\!:}{\underset{\displaystyle :\!\overset{..}{Cl}\!:}{:\!\overset{..}{Cl}\!-\!\overset{..}{Te}\!-\!\overset{..}{Cl}\!:}}$ 　(c) $\underset{\displaystyle :\!\overset{..}{F}\!:}{:\!\overset{..}{F}\!-\!C\!=\!\overset{..}{O}\!:}$ 　d) $\left[\,H\!-\!\overset{\displaystyle ..}{\underset{\displaystyle H}{C}}\!-\!H\,\right]^{-}$

(a) The shape of $CF_3Cl$ is tetrahedral; all halogen—C—halogen angles should be approximately 109.5°. $AX_4$;

(b) $TeCl_4$ molecules will be see saw shaped with Cl—Te—Cl bond angles of approximately 90° and 120°. $AX_4E$;

(c) $COF_2$ molecules will be trigonal planar with F—C—F and O—C—F angles of 120°. $AX_3$;

(d) $CH_3^-$ ions will be trigonal pyramidal with H—C—H angles of slightly less than 109.5°. $AX_3E$

**3.13** (a) a and b are expected to be about 120°, c is expected to be about 109.5° in 2, 4-pentanedione. All of the angles are expected to be about 120° in the acetylacetonate ion.

(b) The major difference arises at the C of the original $sp^3$-hybridized $CH_2$ group, which upon deprotonation goes to $sp^2$ hybridization with only three groups attached.

**3.15** (a) slightly less than 120°;  (b) 180°;  (c) 180°;  (d) slightly less then 109.5°

**3.17** The Lewis structures are

(a) $\overset{\displaystyle H}{\underset{\displaystyle :\!\overset{..}{Cl}\!:}{H\!-\!C\!-\!\overset{..}{Cl}\!:}}$ 　(b) $\overset{\displaystyle :\!\overset{..}{Cl}\!:}{\underset{\displaystyle :\!\overset{..}{Cl}\!:}{:\!\overset{..}{Cl}\!-\!C\!-\!\overset{..}{Cl}\!:}}$ 　(c) $:\!\overset{..}{S}\!=\!C\!=\!\overset{..}{S}\!:$ 　(d) $\overset{\displaystyle :\!\overset{..}{F}\!:}{\underset{\displaystyle :\!\overset{..}{F}\!:}{:\!\overset{..}{F}\!-\!S\!-\!\overset{..}{F}\!:}}$

Molecules (a) and (d) are polar; (b) and (c) are nonpolar.

**3.19**  (a) pyridine: polar

(b) ethane: nonpolar

(c) trichloromethane: polar

**3.21**  **(a)** Of the three forms, **1** and **2** are polar; only **3** is nonpolar. This is because the C—Cl bond dipoles are pointing in exactly opposite directions in **3**. (b) The dipole moment for **1** would be the largest because the C—Cl bond vectors are pointing most nearly in the same direction in **1** (60° apart) whereas in **2** the C—Cl vectors point more away from each other (120°), giving a larger cancellation of dipole.

**3.23**

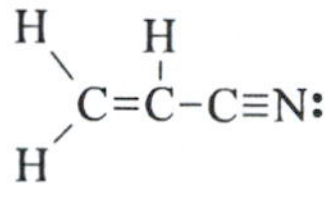

$$H\underset{H}{\overset{H}{\diagdown}}C=C-C\equiv N\ddot{}$$

The first two carbons ($CH_2$ and CH) are $sp^2$ hybridized with H—C—H and C—C—H angles of 120°. The third carbon (bonded to N) is $sp$ hybridized with a C—C—N angle of 180°.

**3.25**  (a) tetrahedral, bond angle of 109.5°

(b) Tetrahedral about the carbon atoms (109.5°) C—Be—C angle of 180°.

(c) angular, H—B—H angle slightly less than 120°

(d) angular, Cl—Sn—Cl angle slightly less than 120°

**3.27**  (a) H—C—H and H—C—C angles of 120°.

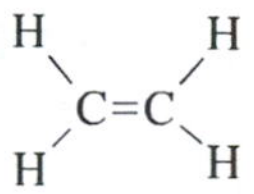

$$H\underset{H}{\overset{H}{\diagdown}}C=C\underset{H}{\overset{H}{\diagup}}$$

(b) 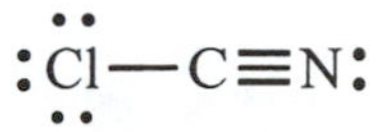 $\ddot{}\ddot{Cl}\ddot{}$—C≡N$\ddot{}$   linear, 180 °

(c) Tetrahedral, 109.5°.

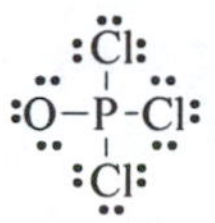

$$\ddot{}\ddot{O}-\overset{\ddot{Cl}\ddot{}}{\underset{\ddot{Cl}\ddot{}}{P}}-\ddot{Cl}\ddot{}$$

(d) The arrangement of atoms about each N is trigonal pyramidal giving H—N—H and H—N—N bond angles of approximately 107°.   $H-\overset{..}{N}-\overset{..}{N}-H$ with H and H

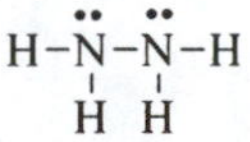

**3.29** (a) tetrahedral:

$$:\overset{\overset{\displaystyle ::}{\cdot\cdot}}{\underset{\underset{\displaystyle ::}{}}{O-Sb-Cl:}}$$

(b) tetrahedral:

$$:\overset{\overset{\displaystyle ::}{\cdot\cdot}}{\underset{\underset{\displaystyle ::}{}}{O-S-Cl:}}$$

(c) seesaw:

$$\left[ :\overset{\overset{\displaystyle :F:}{}}{\underset{\underset{\displaystyle :O:}{}}{O-I-F:}} \right]^{-}$$

**3.31** (a) $sp^3$, orbitals oriented toward corners of a tetrahedron (109.5° apart); (b) $sp$, orbitals oriented directly opposite to each other (180° apart); (c) $sp^3d^2$, orbitals oriented toward the corners of an octahedron (interorbital angles of 90° and 180°); (d) $sp^2$, orbitals oriented toward the corners of an equilateral triangle trigonal planar array (angles $=120°$); trigonal planar.

**3.33** (a) $sp^3d$; (b) $sp^2$; (c) $sp^3$; (d) $sp$

**3.35** (a) $sp^2$; (b) $sp^3$; (c) $sp^3d$; (d) $sp^3$

**3.37** (a) $sp^3$; (b) $sp^3d^2$; (c) $sp^3d$; (d) $sp^3$

**3.39** As the s-character of a hybrid orbital increases, the bond angle increases.

**3.41** Atomic orbitals a and b are mutually orthogonal

if $\int a \cdot b \, d\tau = 0$ (assuming $a \neq b$) where the integration is over all space.

Furthermore, an orbital, a, is normalized if $\int a^2 \, d\tau = 1$.

In this problem, the two hybrid orbitals

are: $h_1 = s + p_x + p_y + p_z$ and $h_2 = s - p_x + p_y - p_z$. Therefore, to show

these two orbitals are orthogonal we must show $\int h_1 h_2 \, d\tau = 0$.

$$\int h_1 h_2 \, d\tau = \int (s + p_x + p_y + p_z)(s - p_x + p_y - p_z) d\tau =$$

$$\int (s^2 - sp_x + sp_y - sp_z + sp_x - p_x^2 + p_x p_y - p_x p_z + sp_y -$$

$$p_x p_y + p_y^2 - p_y p_z + sp_z - p_z p_x + p_z p_y - p_z^2) d\tau$$

Of course, this integral of a sum may be written as a sum of integrals:

$$\int s^2 d\tau - \int sp_x d\tau + \int sp_y d\tau - \int sp_z d\tau + ...$$

Because the hydrogen wavefunctions are mutually orthogonal, the members of this sum which are integrals of a product of two different wavefunctions are zero. Therefore, this sum of integrals simplifies to:

$$\int s^2 d\tau - \int p_x^2 d\tau + \int p_y^2 d\tau - \int p_z^2 d\tau = 1 - 1 + 1 - 1 = 0$$

(recall that the integral of the square of a normalized wavefunction is one.)

**3.43** We are given: $\lambda = -\dfrac{\cos\theta}{\cos^2(\frac{1}{2}\theta)}$. In the $H_2O$ molecule, the bond angle is

104.5°. Therefore, $\lambda = 0.67$ and the hybridization is $sp^{0.67}$.

**3.45**

(a)   $Li_2$   $BO = \frac{1}{2}(2 + 2 - 2) = 1$

diamagnetic, no unpaired electrons

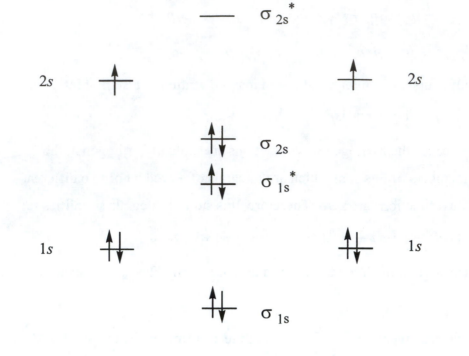

$2s$             $2s$

$\sigma_{2s}$

$\sigma_{1s}^{*}$

$1s$             $1s$

$\sigma_{1s}$

(b)    $Li_2^{+}$    $BO = \frac{1}{2}(2 + 2 - 2 - 1) = \frac{1}{2}$

paramagnetic, one unpaired electron

$\sigma_{2s}^{*}$

$2s$             $2s$

$\sigma_{2s}$

$\sigma_{1s}^{*}$

$1s$             $1s$

$\sigma_{1s}$

(c)  $Li_2^-$  $BO = \frac{1}{2}(2 + 2 - 2 - 1) = \frac{1}{2}$

paramagnetic, one unpaired electron

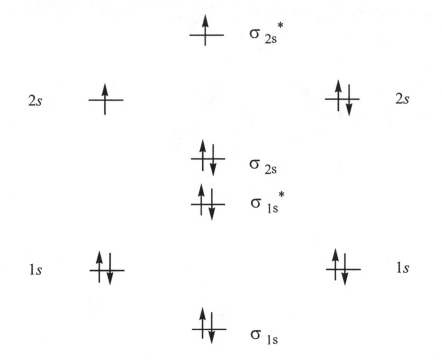

**3.47** (a)  (1) $\left(\sigma_{2s}\right)^2 \left(\sigma_{2s}^*\right)^2 \left(\sigma_{2p}\right)^2 \left(\pi_{2p_x}\right)^2 \left(\pi_{2p_y}\right)^2 \left(\pi_{2p_x}^*\right)^2 \left(\pi_{2p_y}^*\right)^1$

(2) $\left(\sigma_{2s}\right)^2 \left(\sigma_{2s}^*\right)^2 \left(\sigma_{2p}\right)^2 \left(\pi_{2p_x}\right)^2 \left(\pi_{2p_y}\right)^2 \left(\pi_{2p}^*\right)^1$

(3) $\left(\sigma_{2s}\right)^2 \left(\sigma_{2s}^*\right)^2 \left(\sigma_{2p}\right)^2 \left(\pi_{2p_x}\right)^2 \left(\pi_{2p_y}\right)^2 \left(\pi_{2p_x}^*\right)^2 \left(\pi_{2p_y}^*\right)^2$

(b) (1) 1.5;  (2) 2.5;  (3) 1

(c) (1) and  (2) are paramagnetic with one unpaired electron each

(d) $\pi$ in all three cases

**3.49** (a) See Figure 3.34 for the energy level diagram for $N_2$. (b) The nitrogen atom is more electronegative, which will make its orbitals lower in energy than those of C. The revised energy-level diagram is shown below. This will make all of the bonding orbitals closer to N than to C in energy and will make all the antibonding orbitals closer to C than to N in energy.

Energy level diagram for $CN^-$

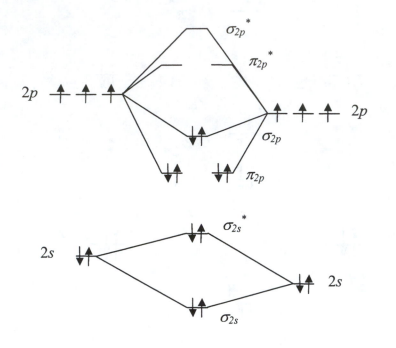

energy levels on C                    energy levels on N

(c)  The electrons in the bonding orbitals will have a higher probability of being at N because N is more electronegative and its orbitals are lower in energy.

**3.51**  (a)  $B_2$ (6 valence electrons): $(\sigma_{2s})^2 \, (\sigma_{2s}^*)^2 \, (\pi_{2p_x})^1 \, (\pi_{2p_y})^1$,

bond order = 1.

(b)  $Be_2$ (4 valence electrons): $(\sigma_{2s})^2 \, (\sigma_{2s}^*)^2$, bond order = 0.

(c)  $F_2$ (14 valence electrons):

$(\sigma_{2s})^2 \, (\sigma_{2s}^*)^2 \, (\sigma_{2p})^2 \, (\pi_{2p_x})^2 \, (\pi_{2p_y})^2 \, (\pi_{2p_x}^*)^2 \, (\pi_{2p_y}^*)^2$, bond order = 1.

**3.53**   (a) – (c) All of these molecules are paramagnetic. $O_2^-$ and $O_2^+$ have an odd number of electrons and must, therefore, have at least one unpaired electron. $O_2$ has an even number of electrons, but in its molecular orbital energy level diagram, the HOMO is a degenerate set of orbitals that are each singly occupied, giving this molecule two unpaired electrons. For $O_2^-$, one more electron will be placed in this degenerate set of orbitals, causing one of the original unpaired electrons to now be paired. $O_2^-$ will therefore have one unpaired electron. Likewise, $O_2^+$ will have one less electron than $O_2$; thus one of the originally unpaired electrons will be removed, leaving one unpaired electron in this molecule.

**3.55**   (a) $F_2$ with 14 valence electrons has a valence electron configuration of $(\sigma_{2s})^2 \, (\sigma_{2s}{}^*)^2 \, (\sigma_{2p})^2 (\pi_{2px})^2 (\pi_{2py})^2 (\pi_{2px}{}^*)^2 (\pi_{2py}{}^*)^2$ with a bond order of 1. After forming $F_2^-$ from $F_2$, an electron is added into a $\sigma_{2p}{}^*$ orbital. The addition of an electron to this antibonding orbital will result in a reduction of the bond order to 1/2 (See 51). $F_2$ will have the stronger bond.    (b) $B_2$ will have an electron configuration of $(\sigma_{2s})^2 \, (\sigma_{2s}{}^*)^2 \, (\pi_{2px})^1 \, (\pi_{2py})^1$ with a bond order of 1. Removing one electron to form $B_2^+$ will eliminate one electron in the bonding orbitals, creating a bond order of 1/2. $B_2$ will have the stronger bond.

**3.57**   The conductivity of a semiconductor increases with temperature as increasing numbers of electrons are promoted into the conduction band, whereas the conductivity of a metal will decrease as the motion of the atoms will slow down the migration of electrons.

Note: The electron due to the charge has arbitrarily been placed in a $2p$ orbital on C on the left-hand side of the diagram.

**3.59** (a) In and Ga;   (b) P and Sb

**3.61** Given the overlap integral $S = \int \Psi_{A1s} \Psi_{B1s} d\tau$, the bonding orbital $\Psi = \Psi_{A1s} + \Psi_{B1s}$, and the fact that the individual atomic orbitals are normalized, we are asked to find the normalization constant N which will normalize the bonding orbital $\Psi$ such that:

$$\int N^2 \Psi^2 d\tau = N^2 \int (\Psi_{A1s} + \Psi_{B1s})^2 d\tau = 1$$

$$N^2 \int (\Psi_{A1s} + \Psi_{B1s})^2 d\tau = N^2 \int (\Psi_{A1s}^2 + 2\Psi_{A1s}\Psi_{B1s} + \Psi_{B1s}^2) d\tau$$

$$= N^2 \left( \int \Psi_{A1s}^2 d\tau + 2 \int \Psi_{A1s}\Psi_{B1s} d\tau + \int \Psi_{B1s}^2 d\tau \right)$$

Given the definition of the overlap integral above and the fact that the individual orbitals are normalized, this expression simplifies to:

$$N^2(1 + 2S + 1) = 1$$

Therefore,    $N = \sqrt{\dfrac{1}{2 + 2S}}$

**3.63** The antibonding molecular orbital is obtained by taking the difference between two atomic orbitals that are proportional to $e^{-r/a_0}$. Halfway between the two nuclei, the distance from the first nucleus, $r_1$, is equal to the distance to the second nucleus, $r_2$, and the antibonding orbital is proportional to:    $\Psi \propto e^{-r/a_0} - e^{-r/a_0} = 0$

**3.65** (a) $\begin{bmatrix} \phantom{:}\overset{\displaystyle ..}{\underset{\displaystyle ..}{:\ddot{C}l:}} \\ :\ddot{C}l - In - \ddot{C}l: \\ \phantom{:}:\ddot{C}l: \end{bmatrix}^{-}$ tetrahedral, $sp^3$, all Cl—In—Cl bond angles $= 109.5°$, nonpolar

(b)

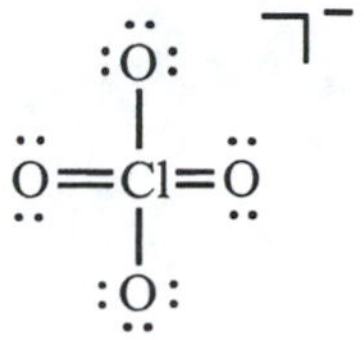

$$\left[\begin{array}{c} :\ddot{O}: \\ | \\ \ddot{O}=\overset{..}{Cl}=\ddot{O} \\ | \\ :\ddot{O}: \end{array}\right]^{-}$$

tetrahedral, $sp^3$, all O—Cl—O bond angles

$= 109.5°$, nonpolar

(c) 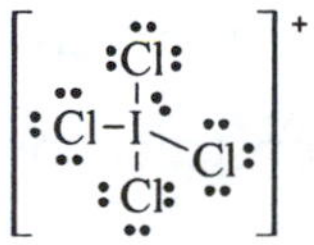 $\left[\begin{array}{c} :\ddot{Cl}: \\ :\ddot{Cl}-\overset{..}{I}\overset{..}{\diagdown}\ddot{Cl}: \\ :\ddot{Cl}: \end{array}\right]^{+}$  seesaw, $sp^3d$, Cl—I—Cl bond angles

$= 90°, 120°$ and $180°$, polar

(d)  $:\ddot{O}—\overset{..}{N}=\ddot{O}\phantom{x}\Bigg]^{-} \longleftrightarrow \ddot{O}=\overset{..}{N}—\ddot{O}:\phantom{x}\Bigg]^{-}$

bent, O—N—O bond angle slightly less than 120°, polar

**3.67** (a)  $SiF_4 : SiF_4$ is nonpolar (tetrahedral $AX_4$ structure) but $PF_3$ is polar

(trigonal pyramidal $AX_3E$ structure);

b)  $SF_6 : SF_6$ is nonpolar (octahedral $AX_6$ structure) whereas $SF_4$ is polar

(seesaw, $AX_4E$ structure);

(c)  $AsF_5 : IF_5$ is polar (square pyramidal, $AX_5E$ structure) whereas $AsF_5$ is

nonpolar ($AX_5$, trigonal bipyramidal structure).

**3.69**  (a) The elemental composition gives an empirical formula of $CH_4O$,

which agrees with the molar mass. There is only one reasonable Lewis

structure; this corresponds to the compound methanol. All of the bond

angles about carbon should be 109.5°. The bond angles about oxygen

should be close to 109.5° but will be somewhat less, due to the repulsions

by the lone pairs. (b) Both carbon and oxygen are $sp^3$ hybridized. (c) The

molecule is polar.

**3.71**

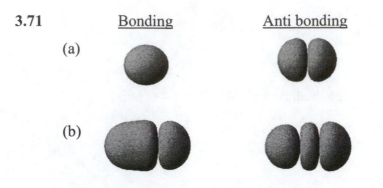

(c) The bonding and antibonding orbitals for HF appear different due to the fact that a *p*-orbital from the F atom is used to construct bonding and antibonding orbitals whereas in the $H_2$ molecule s orbitals on each atom are used to construct bonding and antibonding orbitals.

**3.73**    The expected molecular orbital diagram for CF is

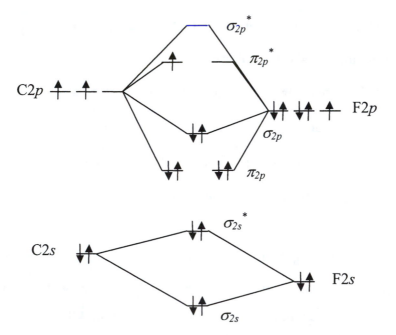

The bond order for the neutral species is 2.5 because one electron occupies a $\pi_{2p}{}^*$ orbital. Adding an electron to form $CF^-$ will reduce the bond order by 1/2 to 2, while removing an electron from form $CF^+$ will increase the bond order to 3. The bond lengths will increase as the bond

order decreases: $CF^- < CF < CF^+$. The $CF^+$ ion will be diamagnetic but both CF and $CF^-$ will have unpaired electrons (one in the case of CF and two in the case of $CF^-$).

**3.75** The Lewis structure of borazine is nearly identical to that of benzene. It is obtained by replacing alternating C atoms in the benzene structure with B and N, as shown. The orbitals at each B and N atom will be $sp^2$ hybridized.

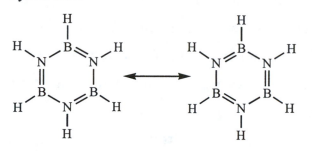

**3.77** (a) The Lewis structures are:

$$\left[\begin{array}{c} H \\ H-C \\ H \end{array}\right]^+ \quad \begin{array}{c} H \\ H-C-H \\ H \end{array} \quad \left[\begin{array}{c} H \\ H-C: \\ H \end{array}\right]^- \quad \begin{array}{c} H \\ C: \\ H \end{array} \quad \left[\begin{array}{c} H \\ C \\ H \end{array}\right]^{2+} \quad \left[\begin{array}{c} H \\ :C: \\ H \end{array}\right]^{2-}$$

(b) All of these species are expected to be diamagnetic. None are radicals.

(c) The predicted bond angles in each species based upon the Lewis structure and VSEPR theory will be

| | | | |
|---|---|---|---|
| $CH_3^+$ | $AX_3$ | trigonal planar 120° | |
| $CH_4$ | $AX_4$ | tetrahedral | 109.5° |
| $CH_3^-$ | $AX_3E$ | pyramidal | slightly less than 109.5° |
| $CH_2$ | $AX_2E$ | angular | slightly less than 120° |
| $CH_2^{2+}$ | $AX_2$ | linear | 180° |
| $CH_2^{2-}$ | $AX_2E_2$ | angular | less than 109.5°, more so than $CH_3^-$ due to the presence of two lone pairs |

The order of increasing H—C—H bond angle will be

$$CH_2^{2-} < CH_3^- < CH_4 < CH_2 < CH_3^+ < CH_2^{2+}$$

**3.79** Acetylene : $H-C \equiv C-H$    Polymer:

$$\left(\begin{matrix} H & & H \\ | & & | \\ C = C - C = C \\ | & & | \\ H & & H \end{matrix}\right)_n$$

Polyacetylene retains multiple bonds along the chain. It is through the series of orbitals that electrons can be conducted. A resonance form of the Lewis structure can be drawn showing that the electrons may be delocalized along the polyacetylene chain. No such resonance form is possible for polyethylene.

Note: The dark color of the material results from the formation of a large number of molecular orbitals that are not very different in energy. These molecular orbitals are made up of combinations of the $p$-orbitals on the carbons that make up the double bonds. Because the orbitals are closely spaced in energy, electrons in them can readily absorb visible light to be promoted to a higher energy orbital. See section 3.13.

**3.81** The energy of an electron in the $n^{th}$ quantum state of a one-dimensional box is given by: $E = n^2 h^2 / (8mL^2)$, where $h$ is Plank's constant, $m$ is the mass of the electron and $L$ is the length of the box (recall $n$ must be an integer). The lowest energy transition for this system will be from the highest occupied quantum state, which we will identify with the quantum number $n_{HO}$, to the next highest state or the lowest unoccupied quantum state, $n_{LU}$. Since each quantum state can hold 2 electrons and each carbon atom contributes one electron, $n_{HO} = N/2$ and $n_{LU} = (N/2) + 1$. (In the case $N$ is odd, $N/2$ must be rounded up to the nearest integer.) Also, the length of the box is given by $L = NR$ where $R$ is the average C—C bond length. Therefore, the lowest energy transition is given by:

$$\Delta E = E_{LU} - E_{HO} = \frac{n_{LU}h^2}{8mN^2R^2} - \frac{n_{HO}h^2}{8mN^2R^2} = \frac{h^2}{8mN^2R^2}\left[\left(\frac{N}{2}+1\right)^2 - \left(\frac{N}{2}\right)^2\right]$$

$$= \frac{h^2(N+1)}{8mN^2R^2}$$

To shift the wavelength of the absorption to longer wavelengths (lower energies) the length of the carbon chain, $N$, must increase.

**3.83** (a)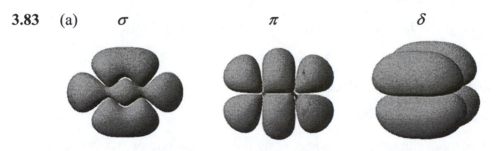

(b) The overlap between the orbitals will decrease as one goes from $\sigma$ to $\pi$ to $\delta$, so we expect the bond strengths to decline in the same order.

**3.85** The effect of changes (a) and (b) will be similar. The overall bond order will change. In the first case, electrons will be removed from $\pi$ orbitals so that the net $\pi$-bond order will drop from three to two. The same thing will happen in (b), but because two electrons are added to antibonding orbitals, a net total of one $\pi$-bond will be broken. Based on this simple model, the ions formed should be paramagnetic because the electrons are added to or taken from doubly degenerate orbitals.

**3.87** (a) The Lewis structure of benzyne is

(b) Benzyne would be highly reactive because the two carbon atoms that are $sp$ hybridized are constrained to have a very strained structure compared to what their hybridization would like to adopt—namely a linear arrangement. Instead of 100° angles at these carbon atoms, the angles by necessity of being in a six-membered ring are constrained to be close to

120°. A possibility that allows the carbon atoms to adopt more reasonable angles is the formation of a diradical:

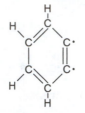

**3.89** (a) The carbon atoms are all $sp^3$ hybridized.  (b) The C—C—C, H—C—H and H—C—C bond angles should be 109.5° based upon the answer to (a).  (c) Because of the ring structure, however, the C—C—C bond angles must be 60°.  (d) The $\sigma$-bond will have the electron density of the bond located on a line between the two atoms that it joins. (e) If the C atoms are truly $sp^3$ hybridized, then the bonding orbitals will not necessarily point directly between the C atoms.  (f) The $sp^3$ hybridized orbitals can still overlap even if they do not point directly between the atoms as shown. Such bonds are sometimes called "bent" bonds, or "banana" bonds. As a result of the situation in the C—C—C bond angles, the H—C—H bond angles are also distorted from 109.5°.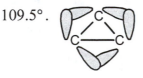

**3.91** (a) and (d) have the possibility of $n$-to-$\pi^*$ transitions because these molecules possess both an atom with a lone pair of electrons (on O in HCOOH and on N in HCN) and a $\pi$-bond to that atom. The other molecules have either a lone pair or a $\pi$-bond, but not both.

**3.93** (a)

(b)

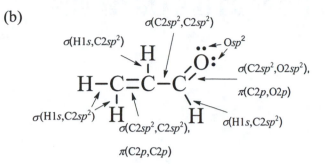

(c)

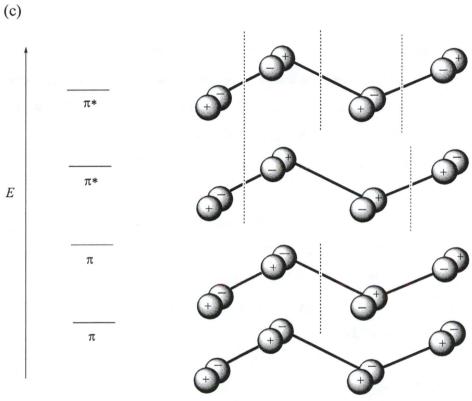

(d) $(C1sp^2 + H1s)^2 (C1sp^2 + H1s)^2 (C1sp^2 + C2sp^2)^2 (C2sp^2 + H1s)^2$
$(C2sp^2 + H1s)^2 (C2sp^2 + C3sp^2)^2 (C3sp^2 + H1s)^2 (C3sp^2 + Osp^2)^2 (Osp^2)^2$
$(Osp^2)^2 (\pi)^2 (\pi)^2$

# CHAPTER 4

# THE PROPERTIES OF GASES

**4.1**   (a)  $8 \times 10^9$ Pa;   (b)  80 kilobars;   (c)  $6 \times 10^7$ Torr;

(d)  $1 \times 10^6$ lb $\cdot$ in$^{-2}$

**4.3**   (a)  The difference in column height will be equal to the difference in pressure between atmospheric pressure and pressure in the gas bulb. If the pressures were equal, the height of the mercury column on the air side and on the apparatus side would be the same. The pressure in the gas bulb is 0.890 atm or $0.890 \times 760$  Torr $\cdot$ atm$^{-1}$ = 676 Torr. The difference would be 762 Torr – 676 Torr = 86 Torr = 86 mm Hg.   (b)  The side attached to the bulb will be higher because the neon pressure is less than the pressure of the atmosphere.   (c)  If the student had recorded the level in the atmosphere arm to be higher than the level in the bulb arm by 86 mm Hg then the pressure in the bulb would have been reported as 762 Torr + 86 Torr = 848 Torr.

**4.5**   $d_1 h_1 = d_2 h_2$

$$73.5 \text{ cm} \times \frac{13.6 \text{ g} \cdot \text{cm}^{-3}}{1.10 \text{ g} \cdot \text{cm}^{-3}} = 909 \text{ cm or } 9.09 \text{ m}$$

**4.7**   $(20. \text{ in})(10. \text{ in})(14.7 \text{ lb} \cdot \text{in}^2) = 2.9 \times 10^3$ lb

**4.9** (a) Volume, L   $\dfrac{nR}{V}$, atm$\cdot$K$^{-1}$

| | |
|---|---|
| 0.01 | 8.21 |
| 0.02 | 4.10 |
| 0.03 | 2.74 |
| 0.04 | 2.05 |
| 0.05 | 1.64 |

(b) The slope is equal to $\dfrac{nR}{V}$

(c) The intercept is equal to 0.00 for all the plots.

**4.11** (a) from $P_1V_1 = P_2V_2$, we have

$(2.0 \times 10^5 \text{ kPa}) (7.50 \text{ mL}) = (P_2) (1000 \text{ mL})$; solving for $P_2$ we get

$1.5 \times 10^3 \text{ kPa};$  (b) similar to (a),

$P_1V_1 = P_2V_2$ or $(643 \text{ Torr}) (54.2 \text{ cm}^3) = (P_2)(7.8 \text{ cm}^3)$, $P_2 = 4.5 \times 10^3 \text{ Torr}$

**4.13** Using

$\dfrac{P_1}{T_1} = \dfrac{P_2}{T_2}$ and expressing $T$ in Kelvins, $\dfrac{1.10 \text{ atm}}{298 \text{ K}} = \dfrac{P_2}{898 \text{ K}}$; $P_2 = 3.31 \text{ atm}$

**4.15** Using

$\dfrac{P_1}{T_1} = \dfrac{P_2}{T_2}$ and expressing $T$ in Kelvins, $\dfrac{1.5 \text{ atm}}{283 \text{ K}} = \dfrac{P_2}{303 \text{ K}}$; $P_2 = 1.6 \text{ atm}$

**4.17** If $P$ and $T$ are constant, then $\dfrac{V_1}{n_1} = \dfrac{V_2}{n_2}$, or $\dfrac{V_1}{0.100 \text{ mol}} = \dfrac{V_2}{0.110 \text{ mol}}$.

Solving for $V_2$ in terms of $V_1$, we obtain $V_2 = \dfrac{n_2 V_1}{n_1} = \dfrac{0.110 V_1}{0.10} = 1.10 V_1$.

So the volume must be increased by 10.% to keep $P$ and $T$ constant.

**4.19**   (a)  Because $P$, $V$, and $T$ all change, we use the relation $\dfrac{P_1V_1}{T_1} = \dfrac{P_2V_2}{T_2}$.

Substituting for the appropriate values we get

$$\frac{(0.255\ \text{atm})(35.5\ \text{mL})}{228\ \text{K}} = \frac{(1.00\ \text{atm})(V_2)}{298\ \text{K}}; V_2 = 11.8\ \text{mL}$$

(b)  The same relation holds as in (a) but here the final temperature and volume are known:

$$\frac{(0.255\ \text{atm})\ (35.5\ \text{mL})}{228\ \text{K}} = \frac{(P_2)(12.0\ \text{mL})}{293\ \text{K}} \cdot P_2 = 0.969\ \text{atm} \quad \text{(c)  Similarly,}$$

we can use the same expression, with $P$ and $V$ known and $T$ wanted.

$$\frac{(0.255\ \text{atm})(35.5\ \text{mL})}{228\ \text{K}} = \frac{\left(\dfrac{(500\ \text{Torr})}{760\ \text{Torr}\cdot\text{atm}^{-1}}\right)(12.0\ \text{mL})}{T_2}; T_2 = 199\ \text{K}$$

**4.21**   (a)  Using the ideal gas law with the gas constant $R$ expressed in kPa:

$$P(0.3500\ \text{L}) = (0.1500\ \text{mol})\,(8.314\,51\ \text{L}\cdot\text{kPa}\cdot\text{K}^{-1}\cdot\text{mol}^{-1})\,(297\ \text{K});$$
$$P = 1.06 \times 10^3\ \text{kPa}$$

(b)  $BrF_3$ has a molar mass of $136.91\ \text{g}\cdot\text{mol}^{-1}$. We then substitute into the ideal gas equation:

$$\left(\frac{10.0\ \text{Torr}}{760\ \text{Torr}\cdot\text{atm}^{-1}}\right)V$$
$$= \left(\frac{23.9 \times 10^{-3}\ \text{g}}{136.91\ \text{g}\cdot\text{mol}^{-1}}\right)(0.082\,06\ \text{L}\cdot\text{atm}\cdot\text{K}^{-1}\cdot\text{mol}^{-1})\,(373\ \text{K})$$
$$V = 4.06 \times 10^2\ \text{mL}$$

(c)
$$\frac{(0.77\ \text{atm})(0.1000\ \text{L})}{} = \left(\frac{m}{64.06\ \text{g}\cdot\text{mol}^{-1}}\right)(0.082\,06\ \text{L}\cdot\text{atm}\cdot\text{K}^{-1}\cdot\text{mol}^{-1})\,(303\ \text{K})$$

$$m = 0.20\ \text{g}$$

(d)
$$(129\ \text{kPa})(6.00 \times 10^3\ \text{m}^3)\left(\frac{1 \times 10^6\ \text{cm}^3}{\text{m}^3}\right)\left(\frac{1\ \text{L}}{1000\ \text{cm}^3}\right)$$
$$= n(8.314\,51\ \text{L}\cdot\text{kPa}\cdot\text{K}^{-1}\cdot\text{mol}^{-1})(287\ \text{K})$$

$n = 3.24 \times 10^5 \, \text{mol CH}_4$

(e) The number of He atoms is the Avogadro constant $N_A$ multiplied by the number of moles. The number of moles is obtained from the ideal gas equation:

$$PV = nRT$$

$$n = \frac{PV}{RT}; \; n = \frac{N}{N_A}$$

so the number of atoms $N$ will be given by

$$N = N_A \left( \frac{PV}{RT} \right)$$

$$= (6.022 \times 10^{23} \, \text{atoms} \cdot \text{mol}^{-1}) \left( \frac{(2.00 \, \text{kPa}) \, (1.0 \times 10^{-6} \, \text{L})}{(8.314 \, 51 \, \text{L} \cdot \text{kPa} \cdot \text{K}^{-1} \cdot \text{mol}^{-1}) \, (158 \, \text{K})} \right)$$

$$= 9.2 \times 10^{14} \, \text{atoms}$$

**4.23** (a) $V = \dfrac{nRT}{P} = \dfrac{(1 \, \text{mol})(0.082 \, 06 \, \text{L} \cdot \text{atm} \cdot \text{K}^{-1} \cdot \text{mol}^{-1})(773 \, \text{K})}{1 \, \text{atm}} = 63.4 \, \text{L}$

(b) $V = \dfrac{nRT}{P} = \dfrac{(1 \, \text{mol})(0.082 \, 06 \, \text{L} \cdot \text{atm} \cdot \text{K}^{-1} \cdot \text{mol}^{-1})(77 \, \text{K})}{1 \, \text{atm}} = 6.32 \, \text{L}$

**4.25** Because $P, V$, and $T$ are state functions, the intermediate conditions are irrelevant to the final states. We can simply use the ideal gas law in the form

$$\frac{P_1 V_1}{T_1} = \frac{P_2 V_2}{T_2}$$

$$\frac{\left( \dfrac{759 \, \text{Torr}}{760 \, \text{Torr} \cdot \text{atm}^{-1}} \right)(1.00 \, \text{L})}{253 \, \text{K}} = \frac{\left( \dfrac{252 \, \text{Torr}}{760 \, \text{Torr} \cdot \text{atm}^{-1}} \right)(V_2)}{1523 \, \text{K}}$$

$$V_2 = 18.1 \, \text{L}$$

**4.27**   Because $T$ is constant, we can use

$$P_1V_1 = P_2V_2$$

$$(1.00 \text{ atm}) (1.00 \text{ L}) = P_2 (0.239 \text{ L})$$

$$P_2 = 4.18 \text{ atm}$$

**4.29**   $PV = nRT$

$$\left( \frac{24.5 \text{ kPa}}{101.325 \text{ kPa} \cdot \text{atm}^{-1}} \right)(0.2500 \text{ L})$$

$$= n(0.082 \, 06 \text{ L} \cdot \text{atm} \cdot \text{K}^{-1} \cdot \text{mol}^{-1})(292.7 \text{ K})$$

$$n = 2.52 \times 10^{-3} \text{ mol}$$

**4.31**   (a)  $\dfrac{P_1V_1}{T_1} = \dfrac{P_2V_2}{T_2}$

$$\frac{(104 \text{ kPa})(2.0 \text{ m}^3)}{294.3 \text{ K}} = \frac{(52 \text{ kPa}) V_2}{268.2 \text{ K}}$$

$$V_2 = 3.6 \text{ m}^3$$

(b)  $\dfrac{P_1V_1}{T_1} = \dfrac{P_2V_2}{T_2}$

$$\frac{(104 \text{ kPa})(2.0 \text{ m}^3)}{294.3 \text{ K}} = \frac{(0.880 \text{ kPa}) V_2}{221.2 \text{ K}}$$

$$V_2 = 1.8 \times 10^2 \text{ m}^3$$

**4.33**   The pressure of the Ar sample will be given by

$$P_{\text{Ar}} = \frac{nRT}{V} = \frac{\left( \dfrac{2.00 \times 10^{-3} \text{g}}{39.95 \text{ g} \cdot \text{mol}^{-1}} \right)(0.082 \, 06 \text{ L} \cdot \text{atm} \cdot \text{K}^{-1} \cdot \text{mol}^{-1})(293 \text{ K})}{0.050 \, 0 \text{ L}}$$

$$P_{\text{Kr}} = \frac{\left( \dfrac{2.00 \times 10^{-3} \text{ g}}{83.80 \text{ g} \cdot \text{mol}^{-1}} \right)(0.082 \, 06 \text{ L} \cdot \text{atm} \cdot \text{K}^{-1} \cdot \text{mol}^{-1})(T_2)}{0.050 \, 0 \text{ L}}$$

Because we want the pressure to be the same, we can set these two equal to each other. Because volume, mass of the gases, and the gas constant $R$ are the same on both sides of the equation, they will cancel.

$$\left(\frac{1}{83.80 \text{ g} \cdot \text{mol}^{-1}}\right)(T_2) = \left(\frac{1}{39.95 \text{ g} \cdot \text{mol}^{-1}}\right)(293 \text{ K})$$

Solving for $T_2$, we obtain temperature = 615 K or 342°C.

**4.35** Density is proportional to the molar mass of the gas as seen from the ideal gas law:

$$PV = nRT$$

$$PV = \frac{m}{M} RT$$

$$\text{density} = \text{mass per unit volume} = \frac{m}{V} = \frac{MP}{RT}$$

The molar masses of the gases in question are

$28.01 \text{ g} \cdot \text{mol}^{-1}$ for $CO(g)$, $44.01 \text{ g} \cdot \text{mol}^{-1}$ for $CO_2(g)$,

and $34.01 \text{ g} \cdot \text{mol}^{-1}$ for $H_2S(g)$. The most dense will be the one with the highest molar mass, which in this case is $CO_2$. The order of increasing density will be $CO < H_2S < CO_2$.

**4.37** (a) Density is proportional to the molar mass of the gas as seen from the ideal gas law. See Section 4.9.

$$d = \frac{(119.37 \text{ g} \cdot \text{mol}^{-1})\left(\dfrac{200. \text{ Torr}}{760 \text{ Torr} \cdot \text{atm}^{-1}}\right)}{(0.082\ 06 \text{ L} \cdot \text{atm} \cdot \text{K}^{-1} \cdot \text{mol}^{-1})(298 \text{ K})} = 1.28 \text{ g} \cdot \text{L}^{-1}$$

(b) $d = \dfrac{(119.37 \text{ g} \cdot \text{mol}^{-1})(1.00 \text{ atm})}{(0.082\ 06 \text{ L} \cdot \text{atm} \cdot \text{K}^{-1} \cdot \text{mol}^{-1})(373 \text{ K})} = 3.90 \text{ g} \cdot \text{L}^{-1}$

**4.39** (a) $M = \dfrac{dRT}{P} = \dfrac{(8.0 \text{ g} \cdot \text{L}^{-1})(0.082\,06 \text{ L} \cdot \text{atm} \cdot \text{K}^{-1} \cdot \text{mol}^{-1})(300.\text{ K})}{2.81 \text{ atm}}$

$= 70.\text{ g} \cdot \text{mol}^{-1}$

(b) The compound is most likely $CHF_3$, for which $M = 70 \text{ g} \cdot \text{mol}^{-1}$. It might also be $C_2H_4F_2$, for which $M = 66 \text{ g} \cdot \text{mol}^{-1}$.

(c) You can use the relationship in (a) to calculate the new density, or you can apply the proportionality changes expected from the change in pressure and temperature to the original density:

$$d_2 = (8.0 \text{ g} \cdot \text{L}^{-1})\left(\frac{1.00 \text{ atm}}{2.81 \text{ atm}}\right)\left(\frac{300.\text{ K}}{298 \text{ K}}\right) = 2.9 \text{ g} \cdot \text{L}^{-1}$$

**4.41** From the analytical data, an empirical formula of CHCl is calculated. The empirical formula mass is $48.47 \text{ g} \cdot \text{mol}^{-1}$. The problem may be solved using the ideal gas law:

$PV = nRT$

$PV = \dfrac{m}{M}RT$

$M = \dfrac{mRT}{PV}$

$M = \dfrac{(3.557 \text{ g})(0.082\,06 \text{ L} \cdot \text{atm} \cdot \text{K}^{-1} \cdot \text{mol}^{-1})(273 \text{ K})}{(1.10 \text{ atm})(0.755 \text{ L})} = 95.9 \text{ g} \cdot \text{mol}^{-1}$

$n$ in the formula $(CHCl)_n$ is equal to

$95.9 \text{g} \cdot \text{mol}^{-1} \div 48.47 \text{g} \cdot \text{mol}^{-1} = 1.98$. The formula is $C_2H_2Cl_2$.

**4.43** Density is proportional to the molar mass of the gas as seen from the ideal gas law:

$PV = nRT$

$PV = \dfrac{m}{M}RT$

$$\text{density} = \text{mass per unit volume} = \frac{m}{V} = \frac{MP}{RT}$$

$$0.943 \text{ g} \cdot \text{L}^{-1} = \frac{M\left(\dfrac{53.1 \text{ kPa}}{101.325 \text{ kPa} \cdot \text{atm}^{-1}}\right)}{(0.082\ 06 \text{ L} \cdot \text{atm} \cdot \text{K}^{-1} \cdot \text{mol}^{-1})(298 \text{ K})}$$

$$M = 44.0 \text{ g} \cdot \text{mol}^{-1}$$

**4.45** (a) The number of moles of $H_2$ needed will be 1.5 times the amount of $NH_3$ produced, as seen from the balanced equation:

$$\tfrac{1}{2} N_2(g) + \tfrac{3}{2} H_2(g) \rightarrow NH_3(g)$$

or

$$N_2(g) + 3 H_2(g) \rightarrow 2 NH_3(g)$$

Once the number of moles is known, the volume can be obtained from the ideal gas law.

$$V = \frac{n_{H_2} RT}{P} = \frac{\left(\tfrac{3}{2} n_{NH_3}\right) RT}{P} = \frac{\left(\left(\dfrac{3 \text{ mol } H_2}{2 \text{ mol } NH_3}\right) \dfrac{(1.0 \times 10^3 \text{ kg})(10^3 \text{ g} \cdot \text{kg}^{-1})}{17.03 \text{ g} \cdot \text{mol}^{-1}}\right) RT}{P}$$

$$= \frac{\left(\dfrac{3}{2}\right)\left(\dfrac{10^6 \text{ g}}{17.03 \text{ g} \cdot \text{mol}^{-1}}\right)(0.082\ 06 \text{ L} \cdot \text{atm} \cdot \text{K}^{-1} \cdot \text{mol}^{-1})(623 \text{ K})}{15.00 \text{ atm}}$$

$$= 3.0 \times 10^5 \text{ L}$$

(b) The ideal gas equation

$$\frac{P_1 V_1}{n_1 R T_1} = \frac{P_2 V_2}{n_2 R T_2} \quad \text{simplifies to} \quad \frac{P_1 V_1}{T_1} = \frac{P_2 V_2}{T_2}$$

because $R$ and $n$ are constant for this problem.

$$\frac{(15.00 \text{ atm})(3.0 \times 10^5 \text{ L})}{623 \text{ K}} = \frac{(376 \text{ atm})V_2}{(523 \text{ K})}$$

$$V_2 = 1.0 \times 10^4 \text{ L}$$

**4.47** We need to find the number of moles of $CH_4(g)$ present in each case. Because the combustion reaction is the same in both cases, as are the temperature and pressure, the larger number of moles of $CH_4(g)$ should produce the larger volume of $CO_2(g)$. We will use the ideal gas equation to solve for $n$ in the first case:

$$n = \frac{PV}{RT} = \frac{(1.00 \text{ atm})(2.00 \text{ L})}{(0.082\ 06 \text{ L} \cdot \text{atm} \cdot \text{K}^{-1} \cdot \text{mol}^{-1})(348 \text{ K})} = 0.0700 \text{ mol CH}_4$$

$2.00$ g of $CH_4$ will be $\dfrac{2.00 \text{ g}}{16.04 \text{ g} \cdot \text{mol}^{-1}} = 0.124 \text{ mol}$

The latter case will have the greater number of moles of $CH_4$ and should produce the larger amount of $CO_2(g)$.

**4.49** The molar mass of glucose is $180.15 \text{ g} \cdot \text{mol}^{-1}$. From this, we can calculate the number of moles of glucose formed and, using the reaction stoichiometry, determine the number of moles of $CO_2$ needed. With that information and the other information provided in the problem, we can use the ideal gas law to calculate the volume of air that is needed:

$$PV = nRT$$

$$V = \frac{\left[\left(\dfrac{10.0 \text{ g glucose}}{180.15 \text{ g glucose} \cdot \text{mol}^{-1} \text{ glucose}}\right)\left(\dfrac{6 \text{ mol CO}_2}{1 \text{ mol glucose}}\right)\right]}{\left(\dfrac{0.26 \text{ Torr}}{760 \text{ Torr} \cdot \text{atm}^{-1}}\right)}$$

$$\times (0.082\ 06 \text{ L} \cdot \text{atm} \cdot \text{K}^{-1} \cdot \text{mol}^{-1})(298 \text{ K})$$

$$= 2.4 \times 10^4 \text{ L}$$

**4.51** (a) This is a limiting reactant problem. Our first task is to determine the number of moles of $NH_3$ and $HCl$ that are present to start with. This can be done from the ideal gas equation:

$$PV = nRT$$

$$n_{NH_3} = \frac{PV}{RT} = \frac{\left(\dfrac{100\ \text{Torr}}{760\ \text{Torr} \cdot \text{atm}^{-1}}\right)(0.0150\ \text{L})}{(0.082\ 06\ \text{L} \cdot \text{atm} \cdot \text{K}^{-1} \cdot \text{mol}^{-1})(303\ \text{K})} = 7.94 \times 10^{-5}\ \text{mol}$$

$$n_{HCl} = \frac{PV}{RT} = \frac{\left(\dfrac{150\ \text{Torr}}{760\ \text{Torr} \cdot \text{atm}^{-1}}\right)(0.0250\ \text{L})}{(0.082\ 06\ \text{L} \cdot \text{atm} \cdot \text{K}^{-1} \cdot \text{mol}^{-1})(298\ \text{K})} = 2.02 \times 10^{-4}\ \text{mol}$$

The ammonia is the limiting reactant. The number of moles of $NH_4Cl(s)$ that form will be equal to the number of moles of $NH_3$ that react. From the molar mass of $NH_4Cl$ ($53.49\ \text{g} \cdot \text{mol}^{-1}$) and the number of moles, we can calculate the mass of $NH_4Cl$ that forms:

$$(7.94 \times 10^{-5}\ \text{mol}\ NH_4Cl(s))(53.49\ \text{g} \cdot \text{mol}^{-1}) = 4.25 \times 10^{-3}\ \text{g}$$

(b) There will be $(2.02 \times 10^{-4}\ \text{mol} - 7.94 \times 10^{-5}\ \text{mol}) = 1.23 \times 10^{-4}\ \text{mol}\ HCl$ left after the reaction. This quantity will exist in a total volume after mixing of 40.0 mL or 0.0400 L. Again, we use the ideal gas law to determine the final pressure:

$$PV = nRT$$

$$P = \frac{(1.23 \times 10^{-4}\ \text{mol})(0.082\ 06\ \text{L} \cdot \text{atm} \cdot \text{K}^{-1} \cdot \text{mol}^{-1})(300\ \text{K})}{0.0400\ \text{L}} = 0.0757\ \text{atm}$$

**4.53** (a) The molar volume of an ideal gas is 22.4 L at 273.15 K. 1.0 mol of ideal gas will exert a pressure of 1.0 atm under those conditions. The partial pressure of $N_2(g)$ will be 1.0 atm. Because there are 2.0 mol of $H_2(g)$, the partial pressure of $H_2(g)$ will be 2.0 atm. (b) The total pressure will be 1.0 atm + 2.0 atm = 3.0 atm.

**4.55** (a) We find the pressure of $SO_2(g)$ originally present by difference. The initial data gives us the total number of moles present, whereas the data for the gas sample after being passed over $CaSO_3(s)$ represents the number of moles of $N_2(g)$.

$$PV = nRT$$

$$n_{total} = \frac{PV}{RT} = \frac{(1.09\ atm)(0.500\ L)}{(0.082\ 06\ L \cdot atm \cdot K^{-1} \cdot mol^{-1})(298\ K)} = 0.0223\ mol$$

$$n_{N_2} = \frac{PV}{RT} = \frac{(1.09\ atm)(0.150\ L)}{(0.082\ 06\ L \cdot atm \cdot K^{-1} \cdot mol^{-1})(323\ K)} = 0.006\ 17\ mol$$

The number of moles of $SO_2$ gas is $0.0223 - 0.006\ 17\ mol = 0.0161\ mol$.

The partial pressure will be given by the mole fraction multiplied by the total pressure. The mole fraction of $SO_2\,(g)$ will be

$0.0161\ mol \div 0.0223\ mol = 0.722$. The pressure due to $SO_2$ in the original mixture is $(0.722)(1.09\ atm) = 0.787\ atm$.

(b) The mass of $SO_2$ will be obtained by multiplying the number of moles of $SO_2$ by the molar mass of $SO_2$ :

$$m_{SO_2} = (0.0161\ mol)(64.06\ g \cdot mol^{-1}) = 1.03\ g$$

**4.57** (a) Of the 756.7 Torr measured, 17.54 Torr will be due to water vapor. The pressure due to $H_2\,(g)$ will, therefore, be

756.7 Torr $- 17.54$ Torr $= 739.2$ Torr. (b) $H_2O(l) \rightarrow H_2\,(g) + \frac{1}{2}O_2\,(g)$;

(c) To answer this question, we must determine the number of moles of $H_2$ produced in the reaction. Using the partial pressure of $H_2$ calculated in part (a) and the ideal gas equation, we can set up the following:

$$\left(\frac{739.2\ Torr}{760\ Torr \cdot atm^{-1}}\right)(0.220\ L) = n(0.082\ 06\ L \cdot atm \cdot K^{-1} \cdot mol^{-1})(293\ K)$$

Solving for $n$, we obtain $n = 0.008\ 90\ mol$. According to the stoichiometry of the reaction, half as much oxygen as hydrogen should be produced, so the number of moles of $O_2 = 0.004\ 45\ mol$. The mass of $O_2$ will be given by $(0.004\ 45\ mol)(32.00\ g \cdot mol^{-1}) = 0.142\ g$.

**4.59** Graham's law of effusion states that the rate of effusion of a gas is inversely proportional to the square root of its molar mass:

$$\text{rate of effusion} = \frac{1}{\sqrt{M}}$$

Diffusion also follows this relationship. If we have two different gases whose rates of diffusion are measured under identical conditions, we can take the ratio

$$\frac{\text{rate}_1}{\text{rate}_2} = \frac{\dfrac{1}{\sqrt{M_1}}}{\dfrac{1}{\sqrt{M_2}}} = \sqrt{\frac{M_2}{M_1}}$$

If a compound takes 1.24 times as long to diffuse as Kr gas, the rate of diffusion of Kr is 1.24 times that of the unknown. We can now use the expression to calculate the molar mass of the unknown, given the mass of Kr:

$$\frac{1.24}{1} = \sqrt{\frac{M_2}{83.80 \text{ g} \cdot \text{mol}^{-1}}}$$

$$M_2 = 129 \text{ g} \cdot \text{mol}^{-1}$$

A mass of $129 \text{ g} \cdot \text{mol}^{-1}$ corresponds to a molecular formula of $C_{10}H_{10}$.

**4.61** The rate of effusion is inversely proportional to the square root of the molar mass. Using a ratio as follows allows us to calculate the time of effusion without knowing the exact conditions of pressure and temperature:

$$\frac{\text{rate}_1}{\text{rate}_2} = \frac{\dfrac{1}{\sqrt{M_1}}}{\dfrac{1}{\sqrt{M_2}}} = \sqrt{\frac{M_2}{M_1}}$$

The rate will be equal to the number of molecules $N$ that effuse in a given time interval. For the conditions given, $N$ will be the same for argon and for the second gas chosen.

$$\frac{\dfrac{N}{time}}{\dfrac{N}{147 \text{ s}}} = \frac{\dfrac{1}{time}}{\dfrac{1}{147 \text{ s}}} = \sqrt{\frac{39.95 \text{ g} \cdot \text{mol}^{-1}}{M_1}}$$

In order to calculate the time of effusion, we need to know only the molar mass of the gases.

(a)  For $CO_2$ with a molar mass of

$$44.01 \text{ g} \cdot \text{mol}^{-1} : \frac{\dfrac{1}{time_{CO_2}}}{\dfrac{1}{147 \text{ s}}} = \sqrt{\frac{39.95 \text{ g} \cdot \text{mol}^{-1}}{44.01 \text{ g} \cdot \text{mol}^{-1}}}$$

time = 154 s

(b)  For $C_2H_4$ with a molar mass of

$$28.05 \text{ g} \cdot \text{mol}^{-1} : \frac{\dfrac{1}{time_{C_2H_4}}}{\dfrac{1}{147 \text{ s}}} = \sqrt{\frac{39.95 \text{ g} \cdot \text{mol}^{-1}}{28.05 \text{ g} \cdot \text{mol}^{-1}}}$$

time = 123 s

(c)  For $H_2$ with a molar mass of

$$2.01 \text{ g} \cdot \text{mol}^{-1} : \frac{\dfrac{1}{time_{CO_2}}}{\dfrac{1}{147 \text{ s}}} = \sqrt{\frac{39.95 \text{ g} \cdot \text{mol}^{-1}}{2.01 \text{ g} \cdot \text{mol}^{-1}}}$$

time = 33.0 s

(d)  For $SO_2$ with a molar mass of

$$64.06 \text{ g} \cdot \text{mol}^{-1} : \frac{\dfrac{1}{time_{CO_2}}}{\dfrac{1}{147 \text{ s}}} = \sqrt{\frac{39.95 \text{ g} \cdot \text{mol}^{-1}}{64.06 \text{ g} \cdot \text{mol}^{-1}}}$$

time = 186 s

**4.63** The formula mass of $C_2H_3$ is 27.04 g·mol$^{-1}$. From the effusion data, we can calculate the molar mass of the sample.

$$\frac{\text{rate}_1}{\text{rate}_2} = \frac{\frac{1}{\sqrt{M_1}}}{\frac{1}{\sqrt{M_2}}} = \sqrt{\frac{M_2}{M_1}}$$

Because time is inversely proportional to rate, we can write alternatively

$$\frac{\frac{1}{349\ \text{s}}}{\frac{1}{210\ \text{s}}} = \sqrt{\frac{39.95\ \text{g·mol}^{-1}}{M_1}}$$

$$\frac{210}{349} = \sqrt{\frac{39.95\ \text{g·mol}^{-1}}{M_1}}$$

$$M_1 = 110\ \text{g·mol}^{-1}$$

The molar mass is 4.1 times that of the empirical formula mass, so the molecular formula is $C_8H_{12}$.

**4.65** (a) The average kinetic energy is obtained from the expression: average kinetic energy $= \frac{3}{2}RT$. The value is independent of the nature of the monatomic ideal gas. The numerical values are:

(a) 4103.2 J·mol$^{-1}$;   (b) 4090.7 J·mol$^{-1}$;

(c) 4103.2 J·mol$^{-1}$ − 4090.7 J·mol$^{-1}$ = 12.5 J·mol$^{-1}$

**4.67** The root mean square speed is calculated from the following equation:

$$c = \sqrt{\frac{3\ RT}{M}}$$

(a) methane, $CH_4$, $M = 16.04$ g·mol$^{-1}$

$$c = \sqrt{\frac{3(8.314\ \text{kg·m}^2 \cdot \text{s}^{-2} \cdot \text{K}^{-1} \cdot \text{mol}^{-1})(253\ \text{K})}{1.604 \times 10^{-2}\ \text{kg·mol}^{-1}}}$$

$$= 627\ \text{m·s}^{-1}$$

(b) ethane, $C_2H_6$, $M = 30.07\ g \cdot mol^{-1}$

$$c = \sqrt{\frac{3(8.314\ kg \cdot m^2 \cdot s^{-2} \cdot K^{-1} \cdot mol^{-1})(253\ K)}{3.007 \times 10^{-2}\ kg \cdot mol^{-1}}}$$

$$= 458\ m \cdot s^{-1}$$

(c) propane, $C_3H_8$, $M = 44.09\ g \cdot mol^{-1}$

$$c = \sqrt{\frac{3(8.314\ kg \cdot m^2 \cdot s^{-2} \cdot K^{-1} \cdot mol^{-1})(253\ K)}{4.409 \times 10^{-2}\ kg \cdot mol^{-1}}}$$

$$= 378\ m \cdot s^{-1}$$

**4.69** Use the expression for the root mean square speed to determine the temperature.

$$v_{rms} = \sqrt{\frac{3RT}{M}}$$

$$T = \frac{v_{rms}^2 \cdot M}{3R} = \frac{(1477\ m \cdot s^{-1})^2 \cdot (4.00 \times 10^{-3}\ kg \cdot mol^{-1})}{3(8.314\ J \cdot K^{-1} \cdot mol^{-1})}$$

$$= 349.9\ K$$

Use the Maxwell Distribution of Speeds (Equation 4.29) appropriately for both gases:

$$\frac{f(v_{He})}{f(v_{Ar})} = \frac{4\pi N \left(\dfrac{M_{He}}{2RT}\right)^{\frac{3}{2}} (v_{He})^2 e^{-M_{He}(v_{He})^2/(2RT)}}{4\pi N \left(\dfrac{M_{Ar}}{2RT}\right)^{\frac{3}{2}} (v_{Ar})^2 e^{-M_{Ar}(v_{Ar})^2/(2RT)}}$$

$$= \frac{(M_{He})^{\frac{3}{2}} (v_{He})^2 e^{-M_{He}(v_{He})^2/(2RT)}}{(M_{Ar})^{\frac{3}{2}} (v_{Ar})^2 e^{-M_{Ar}(v_{Ar})^2/(2RT)}}$$

$$= \frac{(4.00\ g \cdot mol^{-1})^{\frac{3}{2}} (1477\ m \cdot s^{-1})^2}{(39.95\ g \cdot mol^{-1})^{\frac{3}{2}} (467\ m \cdot s^{-1})^2}$$

$$\times \frac{e^{-(4.00 \times 10^{-3}\,kg/mol)(1477\,m/s)^2/(2(8.314\,J/(K \cdot mol))(349.9\,K))}}{e^{-(39.95 \times 10^{-3}\,kg/mol)(467\,m/s)^2/(2(8.314\,J/(K \cdot mol))(349.9\,K))}}$$

$$= \left(\frac{4.00}{39.95}\right)^{\frac{3}{2}} \left(\frac{1477}{467}\right)^2 (0.9977)$$

$$= 0.316$$

**4.71** (a) The most probable speed is the one that corresponds to the maximum on the distribution curve. (b) The percentage of molecules having the most probable speed decreases as the temperature is raised (the distribution spreads out).

**4.73** Hydrogen bonding is important in HF. At low temperatures, this hydrogen bonding causes the molecules of HF to be attracted to each other more strongly, thus lowering the pressure. As the temperature is increased, the hydrogen bonds are broken and the pressure rises more quickly than for an ideal gas. Dimers (2 HF molecules bonded to each other) and chains of HF molecules are known to form.

**4.75** The pressures are calculated very simply from the ideal gas law:

$$P = \frac{nRT}{V} = \frac{(1.00 \text{ mol})(0.082\,06 \text{ L} \cdot \text{atm} \cdot \text{K}^{-1} \cdot \text{mol}^{-1})(298 \text{ K})}{V}$$

Calculating for the volumes requested, we obtain $P =$ (a) 1.63 atm; (b) 48.9 atm; (c) 489 atm. The calculations can now be repeated using the van der Waals equation:

$$\left(P + \frac{an^2}{V^2}\right)(V - nb) = nRT$$

We can rearrange this to solve for $P$:

$$P = \left(\frac{nRT}{V - nb}\right) - \left(\frac{an^2}{V^2}\right)$$

$$= \left(\frac{(1.00 \text{ mol})(0.082\,06 \text{ L} \cdot \text{atm} \cdot \text{K}^{-1} \cdot \text{mol}^{-1})(298 \text{ K})}{V - (1.00 \text{ mol})(0.04267 \text{ L} \cdot \text{mol}^{-1})}\right)$$

$$- \left(\frac{(3.640 \text{ L}^2 \cdot \text{atm} \cdot \text{mol}^{-2})(1.00^2)}{V^2}\right)$$

Using the three values for $V$, we calculate for $P =$ (a) 1.62; (b) 38.9; (c) $1.88 \times 10^3$ atm. Note that at low pressures, the ideal gas law gives essentially the same values as the van der Waals equation, but at high pressures there is a very significant differences.

**4.77** (a) and (b) The values for the pressures of gas with varying numbers of moles of $CO_2$ present are calculated as follows:

The ideal gas law values are calculated from

$$P = \frac{nRT}{V}$$

values for the van der Waals equation can be obtained by rearranging the equation:

$$\left(P + \frac{an^2}{V^2}\right)(V - nb) = nRT$$

$$P = \left(\frac{nRT}{V - nb}\right) - \left(\frac{an^2}{V^2}\right)$$

$$= \left(\frac{(n)\,(0.082\ 06\ \text{L} \cdot \text{atm} \cdot \text{K}^{-1} \cdot \text{mol}^{-1})(300\ \text{K})}{1.00\ \text{L} - (n)(0.042\ 67\ \text{L} \cdot \text{mol}^{-1})}\right)$$

$$- \left(\frac{(3.640\ \text{L}^2 \cdot \text{atm} \cdot \text{mol}^{-1})\,(n^2)}{(1.00\ \text{L})^2}\right)$$

The resulting values are

| $n$ | $P_{ideal}$ | $P_{\text{van der Waals}}$ | % deviation* |
|---|---|---|---|
| 0.100 | 2.46 | 2.44 | 0.8 |
| 0.200 | 4.92 | 4.82 | 2.1 |
| 0.300 | 7.38 | 7.15 | 3.2 |
| 0.400 | 9.85 | 9.44 | 4.3 |
| 0.500 | 12.31 | 11.67 | 5.5 |

$$*\% \text{ deviation} = \frac{P\ \text{ideal} - P\ \text{van der Waals}}{P\ \text{van der Waals}} \times 100$$

(c) Consider one point, for example, the case for $n = 0.400$ mol. The term

$V - nb$ will increase the ideal value $\frac{nRT}{V}$ by 1.7% of the ideal value

(10.02 atm versus 9.85 atm) whereas the correction from $\frac{an^2}{V^2}$ will

decrease the value by 0.58 atm, a change of 5.9% over the ideal gas value.

The second effect, which is due to the intermolecular attractions, dominates in this case.

(d) The gas starts to deviate from ideality by more than 5% at pressures above about 10 atm.

**4.79** Ammonia: $a = 4.225 \ \text{L}^2 \cdot \text{atm} \cdot \text{mol}^{-2}$; $b = 0.037\,07 \ \text{L} \cdot \text{atm}^{-1}$

Oxygen: $a = 1.378 \ \text{L}^2 \cdot \text{atm} \cdot \text{mol}^{-2}$; $b = 0.0318\,83 \ \text{L} \cdot \text{atm}^{-1}$

| Volume | $P$, ammonia | $P$, oxygen | $P$, ideal |
|---|---|---|---|
| 0.05 | 3581 | 1897 | 489 |
| 0.1 | 811 | 497 | 245 |
| 0.2 | 256 | 180 | 122 |
| 0.3 | 140 | 106 | 82 |
| 0.4 | 94 | 75 | 61 |
| 0.5 | 70 | 58 | 49 |
| 0.6 | 55 | 47 | 41 |
| 0.7 | 46 | 39 | 35 |
| 0.8 | 39 | 34 | 31 |
| 0.9 | 34 | 30 | 27 |
| 1 | 30 | 27 | 24 |

Clearly, the greater deviation from the ideal gas law values occurs at low volumes or high pressures. Ammonia deviates more strongly and its van der Waals constants are larger than those for oxygen. This may likely arise because ammonia is more polar and will have stronger intermolecular interactions.

**4.81** (a)

$$\ddot{\text{O}} = \ddot{\text{N}} - \ddot{\ddot{\text{O}}}: \quad \longleftrightarrow \quad :\ddot{\ddot{\text{O}}} - \ddot{\text{N}} = \ddot{\text{O}}$$

(b) Since the wavelength of the absorbed photons is 197 nm, we can find the energy per photon.

$$E_{photon} = \frac{hc}{\lambda} = \frac{(6.626 \times 10^{-34} \text{ J} \cdot \text{s})(2.998 \times 10^8 \text{m} \cdot \text{s}^{-1})}{197 \text{ nm}} \left( \frac{10^9 \text{ nm}}{\text{m}} \right)$$

$$= 1.006 \times 10^{-18} \text{ J}$$

The number of photons in 1.07 mJ must be equal to the number of $NO_2$ molecules.

$$? \text{ photons} = 1.07 \text{ mJ} \left( \frac{1 \text{ J}}{1000 \text{ mJ}} \right) \left( \frac{1 \text{ photon}}{1.006 \times 10^{-18} \text{ J}} \right)$$

$$= 1.064 \times 10^{15} \text{ photons} = 1.064 \times 10^{15} \text{ } NO_2 \text{ molecules}$$

The pressure is created by all the molecules in the sample, so the ideal gas law can be used to find the total molecules.

$$? \text{ molecules total} = N_{AV} \times n_{tot} = N_{AV} \times \frac{P_{tot}V}{RT}$$

$$= (6.022 \times 10^{23} \text{ molecules} \cdot \text{mol}^{-1})$$

$$\times \left( \frac{(0.85 \text{ atm})(2.5 \text{ L})}{(0.08206 \text{ L} \cdot \text{atm} \cdot \text{K}^{-1} \cdot \text{mol}^{-1})(293 \text{ K})} \right)$$

$$= 5.32 \times 10^{22} \text{ molecules total}$$

Therefore, the proportion of $NO_2$ molecules in the sample is

$$\frac{1.064 \times 10^{15} \text{ } NO_2 \text{ molecules}}{5.32 \times 10^{22} \text{ molecules total}} = 0.020 \text{ ppm}$$

**4.83** Use the ideal gas law to calculate the number of moles of HCl.

$$n = \frac{PV}{RT} = \frac{(690. \text{ Torr})(200. \text{ mL})}{(0.082 \, 06 \text{ L} \cdot \text{atm} \cdot \text{K}^{-1} \cdot \text{mL}^{-1})(293 \text{ K})} \cdot \frac{(1 \text{ atm})}{(760 \text{ Torr})} \cdot \frac{(1 \text{ L})}{(1000 \text{ mL})}$$

$$= 7.55 \times 10^{-3} \text{ mol HCl}$$

Since the reaction between HCl and NaOH occurs in a 1:1 mole ratio, this number of moles of NaOH is also present in the volume of NaOH(aq) required to reach the stoichiometric point of the titration. Therefore, the molarity of the NaOH solution is

$$\frac{\text{moles NaOH}}{\text{1 L of solution}} = \frac{(7.55 \times 10^{-3} \text{ mol NaOH})}{(15.7 \text{ mL})} \cdot \frac{1000 \text{ mL}}{1 \text{ L}}$$

$$= 0.481 \text{ M}$$

**4.85** (a) $N_2O_4(g) \rightarrow 2 NO_2(g)$; (b) If all the gas were $N_2O_4(g)$, then the moles can be calculated from the ideal gas equation:

$$P = \frac{nRT}{V}$$

$$= \frac{\left(\dfrac{43.78 \text{ g}}{92.02 \text{ g} \cdot \text{mol}^{-1}}\right)(0.082\,06 \text{ L} \cdot \text{atm} \cdot \text{K}^{-1} \cdot \text{mol}^{-1})(298 \text{ K})}{5.00 \text{ L}}$$

$$= 2.33 \text{ atm}$$

(c) The only difference in the calculation between part (b) and part (c) is that the molar mass of $NO_2$ is half that of $N_2O_4$.

$$P = \frac{nRT}{V}$$

$$= \frac{\left(\dfrac{43.78 \text{ g}}{46.01 \text{ g} \cdot \text{mol}^{-1}}\right)(0.082\,06 \text{ L} \cdot \text{atm} \cdot \text{K}^{-1} \cdot \text{mol}^{-1})(298 \text{ K})}{5.00 \text{ L}}$$

$$= 4.65 \text{ atm}$$

(d) Because both $N_2O_4$ and $NO_2$ are present, we need to determine some way of calculating the relative amounts of each present. This can be done by taking advantage of the gas law relationships. The total pressure at the end of the reaction will give us the total number of moles present:

$$P_{\text{total}} = 2.96 \text{ atm}$$

$$(2.96 \text{ atm})(5.00 \text{ L}) = n_{\text{total}}(0.082\,06 \text{ L} \cdot \text{atm} \cdot \text{K}^{-1} \cdot \text{mol}^{-1})(298 \text{ K})$$

$$n_{\text{total}} = 0.605 \text{ mol}$$

$$\therefore n_{N_2O4} + n_{NO_2} = 0.605 \text{ mol}$$

This gives us one equation, but we have two unknowns, so another relationship is needed. We can take advantage of knowing the stoichiometry of the reaction. If we assume that all of the gas begins at $N_2O_4$ and we allow some to react, we can write the following:

Initial amount of $N_2O_4$  0.476 mol

Amount of $N_2O_4$ that reacts  $x$ (mol)

Amount of $NO_2$ formed  $2x$ (mol)

When the reaction is completed, there will be $0.476 - x$ mole of $N_2O_4$ and

$2x$ mole $NO_2$. The total number of moles will be given by:

$(0.476 - x) + 2x = n_{total}$

$0.605 \text{ mol} = 0.476 \text{ mol} + x$

$$x = 0.129 \text{ mol}$$

$$n_{NO_2} = 2x$$
$$= 2(0.129 \text{ mol})$$
$$= 0.258 \text{ mol}$$

$$n_{N_2O_4} = 0.476 \text{ mol} - x$$
$$= 0.347 \text{ mol}$$

$$X_{NO_2} = \frac{0.258 \text{ mol}}{0.605 \text{ mol}} = 0.426$$

$$X_{N_2O_4} = \frac{0.347 \text{ mol}}{0.605 \text{ mol}} = 0.574$$

**4.87** (a) The elemental analyses yield an empirical formula of $NH_2$. The

formula unit has a mass of $16.02 \text{ g} \cdot \text{mol}^{-1}$. The mass, volume, pressure,

and temperature data will allow us to calculate the molar mass, using the

ideal gas equation:

$$PV = nRT$$

$$PV = \frac{m}{M} RT$$

$$M = \frac{mRT}{PV} = \frac{(0.473 \text{ g})(0.082 \text{ 06 L} \cdot \text{atm} \cdot \text{K}^{-1} \cdot \text{mol}^{-1})(298 \text{ K})}{(1.81 \text{ atm})(0.200 \text{ L})} = 31.9 \text{ g} \cdot \text{mol}^{-1}$$

The molar mass divided by the mass of the empirical formula mass will

give the value of n in the formula

$(NH_2)_n$. $31.9 \text{ g} \cdot \text{mol}^{-1} \div 16.02 \text{ g} \cdot \text{mol}^{-1} = 1.99$, so the molecular formula is $N_2H_4$, which corresponds to the molecule known as hydrazine.

(b)

$$H-\overset{\cdot\cdot}{N}-\overset{\cdot\cdot}{N}-H$$
$$\overset{|}{H} \quad \overset{|}{H}$$

(c)

$$\frac{rate_A}{rate_B} = \sqrt{\frac{M_B}{M_A}}$$

$$\frac{\dfrac{3.5 \times 10^{-4} \text{ mol}}{15.0 \text{ min}}}{\dfrac{X}{25.0 \text{ min}}} = \sqrt{\frac{32.05 \text{ g} \cdot \text{mol}^{-1}}{17.03 \text{ g} \cdot \text{mol}^{-1}}}$$

$$X = \left(\frac{3.5 \times 10^{-4} \text{ mol}}{15.0 \text{ min}}\right)(25.0 \text{ min}) \sqrt{\frac{17.03 \text{ g} \cdot \text{mol}^{-1}}{32.05 \text{ g} \cdot \text{mol}^{-1}}}$$

$$= 4.2 \times 10^{-4} \text{ mol}$$

**4.89** Using $v_{rms}$ = root mean square speed and Equation 4.29 (the Maxwell Distribution of Speeds):

$$\frac{f(10v_{rms})}{f(v_{rms})} = \frac{4\pi N\left(\dfrac{M}{2RT}\right)^{\frac{3}{2}}(10v_{rms})^2 e^{-M(10v_{rms})^2/2RT}}{4\pi N\left(\dfrac{M}{2RT}\right)^{\frac{3}{2}}(v_{rms})^2 e^{-M(v_{rms})^2/2RT}}$$

$$= \frac{100v_{rms}^2 e^{-M100v_{rms}^2/2RT}}{v_{rms}^2 e^{-Mv_{rms}^2/2RT}} = 100e^{\left(-M100v_{rms}^2/2RT + Mv_{rms}^2/2RT\right)}$$

$$= 100e^{-99Mv_{rms}^2/2RT}$$

The ratio is not independent of temperature since the variable $T$ appears in the denominator of a negative exponent on $e$. It makes sense that the ratio should become bigger at higher temperatures as the distribution spreads out such that the number of molecules with higher speeds increases while the number with lower speeds decreases.

**4.91**   The gases react according to the equation:

$$CO(g) + Cl_2(g) \rightarrow COCl_2(g)$$

(a)  We can write the following relationship based upon the stoichiometry of the reaction:

$$P_{final} = P_{final,\,CO} + P_{final,\,chlorine} + P_{final,\,phosgene}$$

By the stoichiometry, we can write

$$P_{final,\,phosgene} = x$$
$$P_{final,\,CO} = P_{initial,\,CO} - x$$
$$P_{final,\,chlorine} = P_{initial,\,chlorine} - x$$
$$P_{final} = P_{initial,\,CO} - x + P_{initial,\,CO} - x + x = P_{initial,\,CO} + P_{initial,\,CO} - x$$

The initial pressures, however, must be adjusted to the new temperature:

$$P_{initial,\,CO,\,223°C} = (3.59\ atm)\left(\frac{500\ K}{298\ K}\right) = 6.02\ atm$$

$$P_{initial,\,Cl_2,\,223°C} = (2.75\ atm)\left(\frac{500\ K}{298\ K}\right) = 4.61\ atm$$

$$P_{final} = 6.02\ atm + 4.61\ atm - x = 9.75\ atm$$
$$x = 0.88\ atm$$
$$P_{final,\,CO} = 6.02\ atm - 0.88\ atm = 5.14\ atm$$
$$P_{final,\,chlorine} = 4.61\ atm - 0.88\ atm = 3.73\ atm$$

The mole fractions are proportional to the pressure so we can write

$$X_{COCl_2} = \frac{0.88}{9.75} = 0.090$$

$$X_{CO} = \frac{5.14}{9.75} = 0.527$$

$$X_{Cl_2} = \frac{3.73}{9.75} = 0.383$$

(b)  The gas density will not change over the course of the reaction because the steel cylinder is a fixed size. The density can be calculated from the relationship

$$d = \frac{m}{V}$$
$$= \frac{PM}{RT}$$

We do not know the mass of samples added, nor the volume of the container, but we can calculate the density from the individual densities of the gases put into the cylinder initially. Because no mass is added or subtracted from the cylinder and its volume does not change, the density will be the same at the end of the reaction as at the beginning.

$$d_{COCl_2} = \frac{PM}{RT}$$

$$= \frac{(3.51 \text{ atm})(28.01 \text{ g} \cdot \text{mol}^{-1})}{(0.082\,06 \text{ L} \cdot \text{atm} \cdot \text{K}^{-1} \cdot \text{mol}^{-1})(298 \text{ K})}$$

$$= 2.45 \text{ g} \cdot \text{L}^{-1}$$

$$d_{Cl_2} = \frac{(2.75 \text{ atm})(70.90 \text{ g} \cdot \text{mol}^{-1})}{(0.082\,06 \text{ L} \cdot \text{atm} \cdot \text{K}^{-1} \cdot \text{mol}^{-1})(298 \text{ K})}$$

$$= 4.75 \text{ g} \cdot \text{L}^{-1}$$

$$d_{total} = 2.45 \text{ g} \cdot \text{L}^{-1} + 4.75 \text{ g} \cdot \text{L}^{-1} = 7.20 \text{ g} \cdot \text{L}^{-1}$$

One could do a similar calculation for all three gases at 500 K to obtain the same answer.

**4.93** The two scents will diffuse according to Eqn 4.19.

$$\frac{\text{rate}_{fruity}}{\text{rate}_{minty}} = \sqrt{\frac{M_{C_8H_8O_2}}{M_{C_{10}H_{20}O_2}}} = \sqrt{\frac{136}{172}} = 0.889$$

Let $x$ = the distance traveled by ethyl octanoate (fruity) and $y$ = the distance traveled by p-anisaldehyde (minty) in the same amount of time. Then

$$x + y = 5 \text{ m} \qquad \text{and} \qquad \frac{x}{y} = \frac{0.889 \text{ m}}{1 \text{ m}}, \text{ or } x = 0.889y$$

Substituting for $x$ gives

$$0.889y + y = 5$$
$$1.889y = 5$$
$$y = 2.65$$

So the fruity smell will travel 5-2.65=2.35 m in the same time that the minty smell will travel 2.65 m. A person must stand more than 2.35 m away from the north end of the room where the fruity smell originates in order to smell the minty scent first. (Note: the problem only gives 1 SF for the length of the room. If we round off to 1 SF for each distance, the answers are 2 m and 3 m respectively.)

**4.95** The molar mass calculation follows from the ideal gas law:

$$PV = nRT$$

$$PV = \frac{m}{M} RT$$

$$M = \frac{mRT}{PV} = \frac{(1.509 \text{ g})(0.082\ 06 \text{ L} \cdot \text{atm} \cdot \text{K}^{-1} \cdot \text{mol}^{-1})(473 \text{ K})}{\left(\dfrac{745 \text{ Torr}}{760 \text{ Torr} \cdot \text{atm}^{-1}}\right)(0.235 \text{ L})}$$

$$= 254 \text{ g} \cdot \text{mol}^{-1}$$

If the molecular formula is $OsO_x$, then the molar mass will be given by:

$$190.2 \text{ g} \cdot \text{mol}^{-1} + x(16.00 \text{ g} \cdot \text{mol}^{-1}) = 254 \text{ g} \cdot \text{mol}^{-1})$$

$$x = 3.99$$

The formula is $OsO_4$.

**4.97** In this problem, the volume, pressure, and molar mass of the substance stay constant. In order to calculate the new mass with the same conditions, we can resort to using the ideal gas equation rearranged to group the constant terms on one side of the equation:

$$PV = nRT$$

$$PV = \frac{m}{M} RT$$

but $M$, $P$, and $V$ are constants, so we can write

$$\frac{MPV}{R} = mT$$

Now we have two sets of conditions, 1 and 2, for which $\dfrac{MPV}{R}$ is constant

so we can set them equal:

$$m_1 T_1 = m_2 T_2$$

$(46.2 \text{ g})(300. \text{ K}) = (m_2)(600. \text{ K})$; therefore $m_2 = 23.1 \text{ g}$

The mass of gas released must therefore be $46.2 \text{ g} - 23.1 \text{ g} = 23.1 \text{ g}$.

**4.99** (a) volume of one atom = molar volume ÷ Avogadro's number

$$2.370 \times 10^{-2} \text{ L} \cdot \text{mol}^{-1} \div 6.022 \times 10^{23} \text{ atoms} \cdot \text{mol}^{-1} = 3.936 \times 10^{-26} \text{ L} \cdot \text{atom}^{-1}$$
$$3.936 \times 10^{-26} \text{ L} \cdot \text{atom}^{-1} \times 1000 \text{ cm}^3 \cdot \text{L}^{-1} = 3.936 \times 10^{-23} \text{ cm}^3 \cdot \text{atm}^{-1}$$
$$3.936 \times 10^{-23} \text{ cm}^3 \cdot \text{atm}^{-1} \times (10^{10} \text{ pm} \cdot \text{cm}^{-1})^3 = 3.936 \times 10^7 \text{ pm}^3$$
$$3.936 \times 10^7 \text{ pm}^3 = \frac{4}{3}\pi r^3$$
$$r = 211 \text{ pm}$$

(b) The atomic radius of He is 128 pm (Appendix 2D).

The volume of the He atom, based upon this radius, is

$$V = \frac{4}{3}\pi r^3$$
$$= \frac{4}{3}\pi (128 \text{ pm})^3$$
$$= 8.78 \times 10^6 \text{ pm}^3$$

(c) The difference in these values illustrates that there is no easy definition for the boundaries of an atom. The van der Waals value obtained from the correction for molar volume is considerably larger than the atomic radius, owing perhaps to longer range and weak interactions between atoms. One should also bear in mind that the value for the van der Waals $b$ is a parameter used to obtain a good fit to a curve, and its interpretation is more complicated than a simple molar volume.

**4.101** (a) $ClNO_2$

(b) $ClNO_2$

(c)

(d) trigonal planar

**4.103** (a) Substitute the van der Waals parameters for ammonia as well as the given values of $n$, $R$, $P$, and $T$ into the vdW equation, then solve for V. Since the equation is cubic, solve graphically for the three roots or use an appropriate program such as Math Cad. Only one of the three roots is physically possible:

$$V^3 + n\left(\frac{RT + bP}{P}\right)V^2 + \left(\frac{n^2 a}{P}\right)V - \frac{n^3 ab}{P} = 0$$

$$V^3 + (0.505 \text{ mol})$$

$$\times \left(\frac{(0.08206 \text{ L} \cdot \text{atm} \cdot \text{K}^{-1} \cdot \text{mol}^{-1})(298 \text{ K}) + (3.707 \text{ L} \cdot \text{mol}^{-1})(95.0 \text{ atm})}{(95.0 \text{ atm})}\right)V^2$$

$$+ \left(\frac{(0.505 \text{ mol})^2 (4.225 \text{ L}^2 \cdot \text{atm} \cdot \text{mol}^{-2})}{(95.0 \text{ atm})}\right)V$$

$$- \frac{(0.505 \text{ mol})^3 (4.225 \text{ L}^2 \cdot \text{atm} \cdot \text{mol}^{-2})(3.707 \text{ L} \cdot \text{mol}^{-1})}{(95.0 \text{ atm})} = 0$$

$$V^3 + (2.002 \text{ L})V^2 + (0.01134 \text{ L}^2)V - 0.02123 \text{ L}^3 = 0$$

$$V = -1.991, \text{ or } -0.1089, \text{ or } 0.09789 \text{ L}$$

but only the positive root is physically possible, so

$$V = 0.0979 \text{ L}$$

(b) Compare the volume calculated in part (a) to that of an ideal gas under the same conditions.

$$PV = nRT$$

$$V = \frac{(0.505 \text{ mol})(0.08206 \text{ L} \cdot \text{atm} \cdot \text{K}^{-1} \cdot \text{mol}^{-1})(298 \text{ K})}{95.0 \text{ atm}} = 0.130 \text{ L}$$

$V_{\text{ideal}} = 0.130 \text{ L} < V_{\text{vdW}}, = 0.0979 \text{ L}$. Attractive forces dominate because the van der Waals, or "real", gas occupies less volume than the "ideal" gas. If the molecules are attracted to one another they will behave less independently, reducing the effective number of moles of gas.

# CHAPTER 5

# LIQUIDS AND SOLIDS

**5.1** (a) London forces, dipole-dipole; (b) London forces, dipole-dipole; (c) London forces, dipole-dipole, hydrogen bonding; (d) London forces

**5.3** Only (b) $CH_3Cl$, (c) $CH_2Cl_2$, and (d) $CHCl_3$ will have dipole-dipole interactions. The molecules $CH_4$ and $CCl_4$ do not have dipole moments.

**5.5** The interaction energies can be ordered based on the relationship the energy has to the distance separating the interacting species. Thus ion-ion interactions are the strongest and are inversely proportional to the distance separating the two interacting species. Ion-dipole energies are inversely proportional to $r^2$, whereas dipole-dipole for constrained molecules (i.e., solid state) is inversely proportional to $r^3$. Dipole-dipole interactions where the molecules are free to rotate become comparable to induced dipole-induced dipole interactions, which are both inversely related to $r^6$. The order thus derived is: (b) dipole-induced dipole $\cong$ (c) dipole-dipole in the gas phase < (e) dipole-dipole in the solid phase < (a) ion-dipole < (d) ion-ion.

**5.7** Only molecules with H attached to the electronegative atoms F, N, and O can hydrogen bond. Additionally, there must be lone pairs available for the H's to bond to. This is true only of (c) $H_2SO_3$.

**5.9**   II, because the dipole-dipole interactions are maximized along the length of the chain.

**5.11**   (a) NaCl (801°C vs. −114.8°C) because it is an ionic compound as opposed to a molecular compound;   (b) butanol (−90°C vs. −116°C) due to hydrogen bonding in butanol that is not possible in diethyl ether; (c) triiodomethane because it will have much stronger London dispersion forces (−82.2°C for trifluoromethane vs. 219°C for triiodomethane);   (d) $H_2O$ (0°C vs. −94°C) because the number of possible hydrogen bonds is greater than the number in methanol.

**5.13**   (a) $PF_3$ and $PBr_3$ are both trigonal pyramidal and should have similar intermolecular forces, but $PBr_3$ has the greater number of electrons and should have the higher boiling point. The boiling point of $PF_3$ is −101.5°C and that of $PBr_3$ is 173.2°C.   (b) $SO_2$ is bent and has a dipole moment whereas $CO_2$ is linear and will be nonpolar. $SO_2$ should have the higher boiling point. $SO_2$ boils at −10°C, whereas $CO_2$ sublimes at −78°C.   (c) $BF_3$ and $BCl_3$ are both trigonal planar, so the choice of higher boiling point depends on the difference in total number of electrons. $BCl_3$ should have the higher boiling point (12.5°C vs. −99.9°C).

**5.15**   The ionic radius of $Al^{3+}$ is 53 pm and that of $Be^{2+}$ is 27 pm. The ratio of energies will be given by

$$E_p \propto \frac{-|z|\mu}{r^2}$$

$$E_{p_{Al^{3+}}} \propto \frac{-|z|\mu}{r^2} = \frac{-|3|\mu}{(53)^2}$$

$$E_{p_{Be^{2+}}} \propto \frac{-|z|\mu}{r^2} = \frac{-|2|\mu}{(27)^2}.$$

The electric dipole moment of the water molecule ($\mu$) will cancel:

$$\text{ratio}\left(\frac{E_{p_{Al^{3+}}}}{E_{p_{Be^{2+}}}}\right) = \frac{-|3|\mu/(53)^2}{-|2|\mu/(27)^2} = \frac{3(27)^2}{2(53)^2} = 0.39$$

The attraction of the $Be^{2+}$ ion will be greater than that of the $Al^{3+}$ ion. Even though the $Be^{2+}$ ion has a lower charge, its radius is much smaller than that of $Al^{3+}$, making the attraction greater.

**5.17** The ionic radius of $Al^{3+}$ is 53 pm and that of $Ga^{3+}$ is 62 pm. The ratio of energies will be given by

$$E_p \propto \frac{-|z|\mu}{r^2}$$

$$E_{p_{Al^{3+}}} \propto \frac{-|3|\mu}{(53 \text{ pm})^2}$$

$$E_{p_{Ga^{3+}}} \propto \frac{-|3|\mu}{(62 \text{ pm})^2}$$

The electric dipole moment of water ($\mu$) will cancel:

$$\text{ratio}\left(\frac{E_{p_{Al^{3+}}}}{E_{p_{Ga^{3+}}}}\right) = \frac{-|3|\mu/(53 \text{ pm})^2}{-|3|\mu/(62 \text{ pm})^2} = \frac{(62 \text{ pm})^2}{(53 \text{ pm})^2} = 1.4$$

The water molecule will be more strongly attracted to the $Al^{3+}$ ion because of its smaller radius.

**5.19** (a) Xenon is larger, with more electrons, giving rise to larger London forces that increase the melting point. (b) Hydrogen bonding in water causes the molecules to be held together more tightly than in diethyl ether.

(c) Both molecules have the same molar mass, but pentane is a linear molecule compared to dimethylpropane, which is a compact, spherical molecule. The compactness of the dimethylpropane gives it a lower surface area. That means that the intermolecular attractive forces, which are of the same type (London forces) for both molecules, will have a larger effect for pentane.

**5.21** $F = \dfrac{-dE_P}{dr} = \dfrac{-d}{dr}\left(\dfrac{1}{r^6}\right) = -\left(\dfrac{-6}{r^7}\right) \propto \dfrac{1}{r^7}$

**5.23** (a) *cis*-Dichloroethene is polar, whereas *trans*-dichloroethene, whose individual bond dipole moments cancel, is nonpolar. Therefore, *cis*-dichloroethene has the greater intermolecular forces and the greater surface tension. (b) Benzene at 20°C, because surface tension of liquids decreases with increasing temperature as a result of thermal motion as temperature rises. Increased thermal motion allows the molecules to more easily break away from each other, which manifests itself as decreased surface tension.

**5.25** At 50°C all three compounds are liquids. $C_6H_6$ (nonpolar) $< C_6H_5SH$ (polar, but no hydrogen bonding) $< C_6H_5OH$ (polar and with hydrogen bonding). The viscosity will show the same ordering as the boiling points, which are 80°C for $C_6H_6$, 169°C for $C_6H_5SH$, 182° for $C_6H_5OH$.

**5.27** $CH_4$, -162°C; $CH_3CH_3$, -88.5°C; $(CH_3)_2CHCH_2CH_3$, 28°C; $CH_3(CH_2)_3CH_3$, 36°C; $CH_3OH$, 64.5°C; $CH_3CH_2OH$, 78.3 °C; $CH_3CHOHCH_3$, 82.5°C; $C_5H_9OH$ (cyclic, but not aromatic), 140°C; $C_6H_5CH_3OH$ (aromatic ring), 205°C; $HOCH_2CHOHCH_2OH$, 290°C

**5.29** (a) hydrogen bonding; (b) London dispersion forces increase

**5.31**　Using $h = \dfrac{2\gamma}{gdr}$ we can calculate the height. For water:

$$r = \frac{1}{2}\,diameter = \frac{1}{2}\,(0.15\ \text{mm})\left(\frac{1\ \text{m}}{1000\ \text{mm}}\right) = 7.5 \times 10^{-5}\ \text{m}$$

$$d = 0.997\ \text{g}\cdot\text{cm}^{-3}\left(\frac{1\ \text{kg}}{1000\ \text{g}}\right)\left(\frac{10^6\ \text{cm}^3}{\text{m}^3}\right) = 9.97 \times 10^2\ \text{kg}\cdot\text{m}^{-3}$$

$$h = \frac{2(72.75 \times 10^{-3}\ \text{N}\cdot\text{m}^{-1})}{(9.81\ \text{m}\cdot\text{s}^{-1})(9.97 \times 10^2\ \text{kg}\cdot\text{m}^{-3})(7.5 \times 10^{-5}\ \text{m})} = 0.20\ \text{m or } 200\ \text{mm}$$

Remember that $1\ \text{N} = 1\ \text{kg}\cdot\text{m}^{-1}\cdot\text{s}^{-2}$

For ethanol:

$$d = 0.79\ \text{g}\cdot\text{cm}^{-3}\left(\frac{1\ \text{kg}}{1000\ \text{g}}\right)\left(\frac{10^6\ \text{cm}^3}{\text{m}^3}\right) = 7.9 \times 10^2\ \text{kg}\cdot\text{m}^{-3}$$

$$h = \frac{2(22.8 \times 10^{-3}\ \text{N}\cdot\text{m}^{-1})}{(9.81\ \text{m}\cdot\text{s}^{-1})(7.9 \times 10^2\ \text{kg}\cdot\text{m}^{-3})(7.5 \times 10^{-5}\ \text{m})} = 0.078\ \text{m or } 78\ \text{mm}$$

Water will rise to a higher level than ethanol. There are two opposing effects to consider. While the greater density of water, as compared to ethanol, acts against it rising as high, it has a much higher surface tension.

**5.33**　(a) At center: 1 center $\times$ 1 atom $\cdot$ center$^{-1}$ $= 1$ atom; at 8 corners, 8 corners $\times$ $\frac{1}{8}$ atom $\cdot$ corner$^{-1}$ $= 1$ atom; total $= 2$ atoms;　(b) There are eight nearest neighbors, hence a coordination number of 8;　(c) The direction along which atoms touch each other is the body diagonal of the unit cell. This body diagonal will be composed of four times the radius of the atom. In terms of the unit cell edge length *a*, the body diagonal will be $\sqrt{3}\,a$. The unit cell edge length will, therefore, be given by

$$4r = \sqrt{3}\,a \text{ or } a = \frac{4r}{\sqrt{3}} = \frac{4\cdot(124\ \text{pm})}{\sqrt{3}} = 286\ \text{pm}$$

**5.35** (a) $a$ = length of side for a unit cell; for an fcc unit cell, $a = \sqrt{8}\ r$ or $2\sqrt{2}\ r = 404$ pm.

$$V = a^3 = (404\ \text{pm} \times 10^{-12}\ \text{m} \cdot \text{pm}^{-1})^3 = 6.59 \times 10^{-29}\ \text{m}^3 = 6.59 \times 10^{-23}\ \text{cm}^3.$$

Because for a fcc unit cell there are 4 atoms per unit cell, we have

$$\text{mass(g)} = 4\ \text{Al atoms} \times \frac{1\ \text{mol Al atoms}}{6.022 \times 10^{23}\ \text{atoms} \cdot \text{mol}^{-1}} \times \frac{26.98\ \text{g}}{\text{mol Al atoms}}$$

$$= 1.79 \times 10^{-22}\ \text{g}$$

$$d = \frac{1.79 \times 10^{-22}\ \text{g}}{6.59 \times 10^{-23}\ \text{cm}^3} = 2.72\ \text{g} \cdot \text{cm}^{-3}$$

(b) $a = \dfrac{4r}{\sqrt{3}} = \dfrac{4 \times 235\ \text{pm}}{\sqrt{3}} = 543$ pm

$$V = (543 \times 10^{-12}\ \text{m})^3 = 1.60 \times 10^{-28}\ \text{m}^3 = 1.60 \times 10^{-22}\ \text{cm}^3$$

There are 2 atoms per bcc unit cell:

$$\text{mass(g)} = 2\ \text{K atoms} \times \frac{1\ \text{mol K atoms}}{6.022 \times 10^{23}\ \text{atoms} \cdot \text{mol}^{-1}} \times \frac{39.10\ \text{g}}{\text{mol K atoms}}$$

$$= 1.30 \times 10^{-22}\ \text{g}$$

$$d = \frac{1.30 \times 10^{-22}\ \text{g}}{1.60 \times 10^{-22}\ \text{cm}^3} = 0.813\ \text{g} \cdot \text{cm}^{-3}$$

**5.37** $a$ = length of unit cell edge

$$V = \frac{\text{mass of unit cell}}{d}$$

(a)

$$V = a^3 = \frac{(1\ \text{unit cell})\left(\dfrac{195.09\ \text{g Pt}}{\text{mol Pt}}\right)\left(\dfrac{1\ \text{mol Pt}}{6.022 \times 10^{23}\ \text{atoms Pt}}\right)\left(\dfrac{4\ \text{atoms}}{1\ \text{unit cell}}\right)}{21.450\ \text{g} \cdot \text{cm}^3}$$

$$a = 3.92 \times 10^{-8}\ \text{cm}$$

Because for an fcc cell, $a = \sqrt{8}\ r$, $r = \dfrac{\sqrt{2}\ a}{4} = \dfrac{\sqrt{2}\ (3.92 \times 10^{-8}\ \text{cm})}{4}$

$$= 1.39 \times 10^{-8}\ \text{cm} = 139\ \text{pm}$$

(b) $V = a^3$

$$= \frac{(1 \text{ unit cell})\left(\dfrac{180.95 \text{ g Ta}}{1 \text{ mol Ta}}\right)\left(\dfrac{1 \text{ mol Ta}}{6.022 \times 10^{23} \text{ atoms Ta}}\right)\left(\dfrac{2 \text{ atoms}}{1 \text{ unit cell}}\right)}{16.654 \text{ g} \cdot \text{cm}^3}$$

$$= 3.61 \times 10^{-23} \text{ cm}^3$$

$a = 3.30 \times 10^{-8} \text{ cm}$

$$r = \frac{\sqrt{3}\,a}{4} = \frac{\sqrt{3}\left(3.30 \times 10^{-8} \text{ cm}\right)}{4} = 1.43 \times 10^{-8} \text{ cm} = 143 \text{ pm}$$

**5.39**  (a)  The volume of the unit cell is

$$V_{\text{unit cell}} = \left(543 \text{ pm} \times \frac{10^{-12} \text{ m}}{\text{pm}} \times \frac{100 \text{ cm}}{\text{m}}\right)^3 = 1.601 \times 10^{-22} \text{ cm}^3$$

mass in unit cell $= (1.601 \times 10^{-22} \text{ cm}^3) \times (2.33 \text{ g} \cdot \text{cm}^{-3}) = 3.73 \times 10^{-22} \text{ g}$

(b)  The mass of a Si atom is

$28.09 \text{ g} \cdot \text{mol}^{-1} \div (6.022 \times 10^{23}) \text{ atoms} \cdot \text{mol}^{-1} = 4.665 \times 10^{-23} \text{ g} \cdot \text{atom}^{-1}$

Therefore, there are $3.73 \times 10^{-22} \text{ g} \div 4.665 \times 10^{-23} \text{ g} \cdot \text{atom}^{-1} = 8$ atoms per

unit cell.

**5.41**  volume of a cylinder = base area x length $= \pi r^2 l$

volume of a triangular prism = base area x length = $(bhl)/2 = \sqrt{3}r^2 l$ for a

triangle inscribed in the centers of three touching cylinder bases

Each of the inscribed triangle's 60° angles accounts for 1/6 of the volume

of each of the three touching cylinders, or a total cylinder volume of (3/6)

$\pi r^2 l$. So the percent of the space occupied by the cylinders is (0.5) $\pi r^2 l$

*100/ $\sqrt{3}r^2 l = 0.5\pi/\sqrt{3} = 90.7\%$

**5.43**  (a)  There are eight chloride ions at the eight corners, giving a total of

$8 \text{ corners} \times \frac{1}{8} \text{ atom} \cdot \text{corner}^{-1} = 1 \text{ Cl}^- \text{ ion}$

There is one $Cs^+$ that lies at the center of the unit cell. All of this ion belongs to the unit cell. The ratio is thus 1 : 1 for an empirical formula of CsCl, with one formula unit per unit cell.

(b) The titanium atoms lie at the corners of the unit cell and at the body center:  8 corners $\times \frac{1}{8}$ atom $\cdot$ corner$^{-1}$ + 1 at body center = 2 atoms per unit cell

Four Oxygen atoms lie on the faces of the unit cell and two lie completely within the unit cell, giving:

4 atoms in faces $\times \frac{1}{2}$ atom $\cdot$ face$^{-1}$ + 2 atoms wholly within cell = 4 atoms

The ratio is thus two Ti per four O, or an empirical formula of $TiO_2$ with two formula units per unit cell (c) The Ti atoms are 6-coordinate and the O atoms are 3-coordinate.

**5.45**  Y:  8 atoms $\times \frac{1}{8}$ atom $\cdot$ corner$^{-1}$ = 1 Y atom

Ba:  8 atoms $\times \frac{1}{4}$ atom $\cdot$ edge$^{-1}$ = 2 Ba atoms

Cu:  3 Cu atoms completely inside unit cell = 3 Cu atoms

O:  10 atoms on faces $\times \frac{1}{2}$ atom $\cdot$ face$^{-1}$ + 2 atoms  completely inside unit cell = 7 O atoms

Formula = $YBa_2Cu_3O_7$

**5.47**  (a)  ratio $= \dfrac{149 \text{ pm}}{133 \text{ pm}} = 1.12$, predict cesium-chloride structure with (8,8) coordination; however, rubidium fluoride actually adopts the rock-salt structure

(b)  ratio $= \dfrac{72 \text{ pm}}{140 \text{ pm}} = 0.51$, predict rock-salt structure with (6,6) coordination

(c) ratio $= \dfrac{102 \text{ pm}}{196 \text{ pm}} = 0.520,$ predict rock-salt structure with (6,6)

coordination

**5.49** (a) In the rock-salt structure, the unit cell edge length is equal to two times the radius of the cation plus two times the radius of the anion. Thus for CaO, $a = 2(100 \text{ pm}) + 2(140 \text{ pm}) = 480 \text{ pm}.$ The volume of the unit cell will be given by (converting to $cm^3$ because density is normally given in terms of $g \cdot cm^{-3}$)

$$V = \left( 480 \text{ pm} \times \frac{10^{-12} \text{ m}}{\text{pm}} \times \frac{100 \text{ cm}}{\text{m}} \right)^3 = 1.11 \times 10^{-22} \text{ cm}^3$$

There are four formula units in the unit cell, so the mass in the unit cell will be given by

$$\text{mass in unit cell} = \frac{\left( \dfrac{4 \text{ formula units}}{1 \text{ unit cell}} \right) \times \left( \dfrac{56.08 \text{ g CaO}}{1 \text{ mol CaO}} \right)}{6.022 \times 10^{23} \text{ molecules} \cdot \text{mol}^{-1}} = 3.725 \times 10^{-22} \text{ g}$$

The density will be given by the mass in the unit cell divided by the volume of the unit cell:

$$d = \frac{3.725 \times 10^{-22} \text{ g}}{1.11 \times 10^{-22} \text{ cm}^3} = 3.36 \text{ g} \cdot \text{cm}^{-3}$$

(b) For a cesium chloride-like structure, it is the body diagonal that represents two times the radius of the cation and plus two times the radius of the anion. Thus the body diagonal for CsBr is equal to $2(170 \text{ pm}) + 2(196) = 732 \text{ pm}.$

For a cubic cell, the body diagonal $= \sqrt{3} \, a = 732 \text{ pm}$
$$a = 423 \text{ pm}$$

$$a^3 = V = \left( 423 \text{ pm} \times \frac{10^{-12} \text{ m}}{\text{pm}} \times \frac{100 \text{ cm}}{\text{m}} \right)^3 = 7.57 \times 10^{-23} \text{ cm}^3$$

There is one formula unit of CsBr in the unit cell, so the mass in the unit cell will be given by

$$\text{mass in unit cell} = \frac{\left(\dfrac{1 \text{ formula units}}{1 \text{ unit cell}}\right)\left(\dfrac{212.82 \text{ g CsBr}}{1 \text{ mol CsBr}}\right)}{6.022 \times 10^{23} \text{ molecules} \cdot \text{mol}^{-1}} = 3.534 \times 10^{-22} \text{ g}$$

$$d = \frac{3.534 \times 10^{-22} \text{ g}}{7.57 \times 10^{-23} \text{ cm}^3} = 4.67 \text{ g} \cdot \text{cm}^3$$

**5.51** (a) Glucose will be held in the solid by London forces, dipole-dipole interactions, and hydrogen bonds; benzophenone will be held in the solid by dipole-dipole interactions and London forces; methane will be held together by London forces only. London forces are strongest in benzophenone, but glucose can experience hydrogen bonding, which is a strong interaction and dominates intermolecular forces. Methane has few electrons so experiences only weak London forces. (b) We would expect the melting points to increase in the order $CH_4$ (m.p. $= -182°C$) $<$ benzophenone (m.p. $= 48°C$) $<$ glucose (m.p. $= 148 - 155°C$).

**5.53** One form of boron nitride, silicon dioxide, plus many others

**5.55** (a) The alloy is undoubtedly interstitial, because the atomic radius of nitrogen is much smaller (74 pm vs. 124 pm) than that of iron. The rule of thumb is that the solute atom be less than 60% the solvent atom in radius, in order for an interstitial alloy to form. That criterion is met here. (b) We expect that nitriding will make iron harder and stronger, with a lower electrical conductivity.

**5.57** Graphite is a metallic conductor parallel to the planes; the electrons are quite free to move within them. Between planes, however, there is an energy barrier to conduction, though this barrier can be partially overcome by raising the temperature. Thus, graphite is a semiconductor perpendicular to the planes and a conductor parallel to the planes.

**5.59**   (a)  These problems are most easily solved by assuming 100 g of substance. In 100 g of Ni-Cu alloy there will be 25 g Ni and 75 g Cu, corresponding to 0.43 mol Ni and 1.2 mol Cu. The atom ratio will be the same as the mole ratio:

$$\frac{1.2 \text{ mol Cu}}{0.43 \text{ mol Ni}} = 2.8 \text{ Cu per Ni}$$

(b)  Pewter, which is 7% Sb, 3% Cu, and 90% Sn, will contain 7 g Sb, 3 g Cu, and 90 g Sn per 100 g of alloy. This will correspond to 0.06 mol Sb, 0.05 mol Cu, and 0.76 mol Sn per 100 g. The atom ratio will be 15 Sn : 1.2 Sb : 1 Cu.

**5.61**   There are too many ways that these molecules can rotate and twist so that they do not remain rod-like. The molecular backbone when the molecule is stretched out is rod-like, but the molecules tend to curl up on themselves, destroying any possibility of long range order with neighboring molecules. This is partly due to the fact that the molecules have only single bonds that allow rotation about the bonds, so that each molecule can adopt many configurations. If multiple bonds are present, the bonding is more rigid.

**5.63**   Use of a nonpolar solvent such as hexane or benzene (etc.) in place of water should give rise to the formation of inverse micelles.

**5.65**   (a)  Pentane and 2,2-dimethylpropane are isomers; both have the chemical formula $C_5H_{12}$. We will assume that 2,2-dimethylpropane is roughly spherical and that all the hydrogen atoms lie on this sphere. The surface area of a sphere is given by $A = 4\pi r^2$. For this particular sphere, $A = 4\pi(254 \text{ pm})^2 = 8.11 \times 10^5 \text{ pm}^2$.

(b) For pentane, the surface area of rectangular prism is

$2(295 \text{ pm} \times 766 \text{ pm}) +$

$2(295 \text{ pm} \times 254 \text{ pm}) + 2(254 \text{ pm} \times 766 \text{ pm}) = 9.91 \times 10^5 \text{ pm}^2$.

(c) The pentane should have the higher boiling point. It has a significantly larger surface area and should have stronger intermolecular forces between the molecules.

**5.67** (a) anthracene, $C_{14}H_{10}$

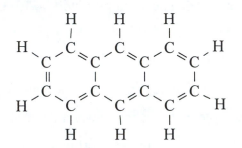

London forces

(b) phosgene, $COCl_2$

Dipole-dipole forces, London forces

(c) glutamic acid, $C_5H_9NO_4$

Hydrogen bonding, dipole-dipole forces, London forces

**5.69** (a) tetrachloromethane; (b) water

**5.71** The unit cell for a cubic close-packed lattice is the fcc unit cell. For this cell, the relation between the radius of the atom $r$ and the unit cell edge length a is

$$4r = \sqrt{2}\,a$$

$$a = \frac{4r}{\sqrt{2}}$$

The volume of the unit cell is given by

$$V = a^3 = \left(\frac{4r}{\sqrt{2}}\right)^3$$

If $r$ is given in pm, then a conversion factor to cm is required:

$$V = a^3 = \left(\frac{4r}{\sqrt{2}} \times \frac{10^{-12}\ \text{m}}{\text{pm}} \times \frac{100\ \text{cm}}{\text{m}}\right)^3$$

Because there are four atoms per fcc unit cell, the mass in the unit cell will be given by

$$\text{mass} = \left(\frac{4\ \text{atoms}}{\text{unit cell}}\right)\left(\frac{M}{6.022 \times 10^{23}\ \text{atoms} \cdot \text{mol}^{-1}}\right)$$

The density will be given by

$$d = \frac{\text{mass of unit cell}}{\text{volume of unit cell}} = \frac{\left(\dfrac{4\ \text{atoms}}{\text{unit cell}}\right) \times \left(\dfrac{M}{6.022 \times 10^{23}\ \text{atoms} \cdot \text{mol}^{-1}}\right)}{\left(\dfrac{4r}{\sqrt{2}} \times \dfrac{10^{-12}\ \text{m}}{\text{pm}} \times \dfrac{100\ \text{cm}}{\text{m}}\right)^3}$$

$$= \frac{(2.936 \times 10^5)M}{r^3}$$

or

$$r = \sqrt[3]{\frac{(2.936 \times 10^5)M}{d}}$$

where $M$ is the atomic mass in $\text{g} \cdot \text{mol}^{-1}$ and $r$ is the radius in pm.

For the different gases we calculate the results given in the following table:

| Gas | Density (g·cm³) | Molar mass (g·mol⁻¹) | Radius (pm) |
|---|---|---|---|
| Neon | 1.20 | 20.18 | 170 |
| Argon | 1.40 | 39.95 | 203 |
| Krypton | 2.16 | 83.80 | 225 |
| Xenon | 2.83 | 131.30 | 239 |
| Radon | 4.4 | 222 | 246 |

**5.73**  There are two approaches to this problem. The information given does not specify the radius of the tungsten atom. This value can be looked up in Appendix 2D. We can calculate an answer, however, based simply upon the fact that the density is $19.3 \text{ g} \cdot \text{cm}^{-3}$ for the bcc cell, by taking the ratio between the expected densities, based upon the assumption that the atomic radius of tungsten will be the same for both. The unit cell for a cubic close-packed lattice is the fcc unit cell. For this cell, the relation between the radius of the atom $r$ and the unit cell edge length a is

$$4r = \sqrt{2}\, a$$

$$a = \frac{4r}{\sqrt{2}}$$

The volume of the unit cell is given by

$$V = a^3 = \left(\frac{4r}{\sqrt{2}}\right)^3$$

If $r$ is given in pm, then a conversion factor to cm is required:

$$V = a^3 = \left(\frac{4r}{\sqrt{2}} \times \frac{10^{-12} \text{ m}}{\text{pm}} \times \frac{100 \text{ cm}}{\text{m}}\right)^3$$

Because there are four atoms per fcc unit cell, the mass in the unit cell is given by

$$\text{mass} = \left(\frac{4 \text{ atoms}}{\text{unit cell}}\right)\left(\frac{M}{6.022 \times 10^{23} \text{ atoms} \cdot \text{mol}^{-1}}\right)$$

The density is given by

$$d = \frac{\text{mass of unit cell}}{\text{volume of unit cell}} = \frac{\left(\dfrac{4 \text{ atoms}}{\text{unit cell}}\right)\left(\dfrac{M}{6.022 \times 10^{23} \text{ atoms} \cdot \text{mol}^{-1}}\right)}{\left(\dfrac{4r}{\sqrt{2}} \times \dfrac{10^{-12} \text{ m}}{\text{pm}} \times \dfrac{100 \text{ cm}}{\text{m}}\right)^3}$$

$$= \frac{(2.936 \times 10^5)M}{r^3}$$

or

$$r = \sqrt[3]{\frac{(2.936 \times 10^5)M}{d}}$$

where $M$ is the atomic mass in $g \cdot mol^{-1}$ and $r$ is the radius in pm.

Likewise, for a body-centered cubic lattice there will be two atoms per unit cell. For this cell, the relationship between the radius of the atom $r$ and the unit cell edge length $a$ is derived from the body diagonal of the cell, which is equal to 4 times the radius of the atom. The body diagonal is found from the Pythagorean theorem to be equal to the $\sqrt{3}\, a$.

$$4r = \sqrt{3}\, a$$
$$a = \frac{4r}{\sqrt{3}}$$

The volume of the unit cell is given by

$$V = a^3 = \left(\frac{4r}{\sqrt{3}}\right)^3$$

If $r$ is given in pm, then a conversion factor to cm is required:

$$V = a^3 = \left(\frac{4r}{\sqrt{3}} \times \frac{10^{-12} \text{ m}}{\text{pm}} \times \frac{100 \text{ cm}}{\text{m}}\right)^3$$

Because there are two atoms per bcc unit cell, the mass in the unit cell will be given by

$$\text{mass} = \left(\frac{2 \text{ atoms}}{\text{unit cell}}\right) \times \left(\frac{M}{6.022 \times 10^{23} \text{ atoms} \cdot \text{mol}^{-1}}\right)$$

The density will be given by

$$d = \frac{\text{mass of unit cell}}{\text{volume of unit cell}} = \frac{\left(\dfrac{2 \text{ atoms}}{\text{unit cell}}\right)\left(\dfrac{M}{6.022 \times 10^{23} \text{ atoms} \cdot \text{mol}^{-1}}\right)}{\left(\dfrac{4\,r}{\sqrt{3}} \times \dfrac{10^{-12} \text{ m}}{\text{pm}} \times \dfrac{100 \text{ cm}}{\text{m}}\right)^3}$$

$$= \frac{\left(\dfrac{2 \text{ atoms}}{\text{unit cell}}\right)\left(\dfrac{M}{6.022 \times 10^{23} \text{ atoms} \cdot \text{mol}^{-1}}\right)}{(2.309 \times 10^{-10}\,r)^3}$$

$$= \frac{(2.698 \times 10^5)M}{r^3}$$

or

$$r = \sqrt[3]{\frac{(2.698 \times 10^5)M}{d}}$$

Setting these two expressions equal and cubing both sides, we obtain

$$\frac{(2.936 \times 10^5)M}{d_{\text{fcc}}} = \frac{(2.698 \times 10^5)M}{d_{\text{bcc}}}$$

The molar mass of tungsten $M$ is the same for both ratios and will cancel from the equation.

$$\frac{(2.936 \times 10^5)}{d_{\text{fcc}}} = \frac{(2.698 \times 10^5)}{d_{\text{bcc}}}$$

Rearranging, we get

$$d_{\text{fcc}} = \frac{(2.936 \times 10^5)}{(2.698 \times 10^5)}\, d_{\text{bcc}}$$

$$= 1.088\, d_{\text{bcc}}$$

For W, $d_{\text{fcc}} = 1.088 \times 19.6 \text{ g} \cdot \text{cm}^3$

$$= 21.3 \text{ g} \cdot \text{cm}^{-3}$$

**5.75** (a) The oxidation state of the titanium atoms must balance the charge on the oxide ions, $O^{2-}$. The presence of 1.18 $O^{2-}$ ions means that the Ti present must have a charge to compensate the $-2.36$ charge on the oxide ions. The average oxidation state of Ti is thus $+2.36$. (b) This is most easily solved by setting up a set of two equations in two unknowns. We know that the total charge on the titanium atoms present must equal 2.36, so if we multiply the charge on each type of titanium by the fraction of titanium present in that oxidation state and sum the values, we should get 2.36:

let x = fraction of $Ti^{2+}$, y = fraction of $Ti^{3+}$, then
$$2x + 3y = 2.36$$

Also, because we are assuming all the titanium is either +2 or +3, the fractions of each present must add up to 1:
$$x + y = 1$$

Solving these two equations simultaneously, we obtain y = 0.36, x = 0.64.

**5.77**   (a)  true. If this is not the case, the unit cell will not match with other unit cells of the same type when stacked to form the entire lattice.

(b)  false. Unit cells do not have to have atoms at the corners.

(c)  true. In order for the unit cell to repeat properly, opposite faces must have the same composition.

(d)  false. If one face is centered, the opposing face must be centered, but the other faces do not necessarily have to be centered.

**5.79**   There are several ways to draw unit cells that will repeat to generate the entire lattice. Some examples are shown below. The choice of unit cell is determined by conventions that are beyond the scope of this text (the smallest unit cell that indicates all of the symmetry present in the lattice is typically the one of choice).

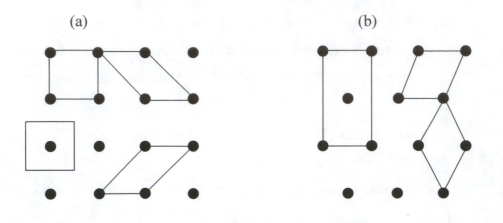

(a)          (b)

**5.81** In a salt such as sodium chloride or sodium bromide, all the interactions will be ionic. In sodium acetate or sodium methoxide, however, the organic anions $CH_3COO^-$ and $CH_3O^-$ will have the negative charges localized largely on the oxygen atoms, which are the most electronegative atoms. The carbon part of the molecule will interact with other parts of the crystal lattice via London forces rather than ionic interactions. In general, solid organic salts will not be as hard as purely ionic substances and will also have lower melting points.

**5.83** Fused silica, also known as fused quartz, is predominantly $SiO_2$ with very few impurities. This glass is the most refractory, which means it can be used at the highest temperatures. Quartz vessels are routinely used for reactions that must be carried out at temperatures up to 1000°C. Vycor, which is 96% $SiO_2$, can be used at up to ca. 900°C, and normal borosilicate glasses at up to approximately 550°C. This is due to the fact that the glasses melt at lower temperatures, as the amount of materials other than $SiO_2$ increases. The borosilicate glasses (Pyrex or Kimax) are commonly used because the lower softening points allow them to be more easily molded and shaped into different types of glass objects, such as reaction flasks, beakers, and other types of laboratory and technical glassware.

**5.85**  For a bcc unit cell (see Example 5.3),

$$d = \frac{3^{3/2}(55.85\ \text{g}\cdot\text{mol}^{-1})}{32(6.022\times10^{23}\ \text{mol}^{-1})(1.24\times10^{-8}\ \text{cm})^3} = 7.90\ \text{g}\cdot\text{cm}^{-3};$$

$$\frac{(1.5\ \text{cm})^3(7.90\ \text{g}\cdot\text{cm}^{-3})}{(55.85\ \text{g}\cdot\text{mol}^{-1})} = 0.477\ \text{mol Fe};$$

$$\frac{(1.00\ \text{atm})(15.5\ \text{L})}{(0.08206\ \text{L}\cdot\text{atm}\cdot\text{K}^{-1}\cdot\text{mol}^{-1})(298\ \text{K})} = 0.634\ \text{mol O}_2;$$

Fe is the limiting reactant; so

$(0.5)(0.477\ \text{mol})(159.70\ \text{g}\cdot\text{mol}^{-1}) = 38.1\ \text{g Fe}_2\text{O}_3$ is the theoretical yield,

or maximum mass that can be produced.

**5.87**  For M, there is total of one atom in the unit cell from the corners;

cations: $8\ \text{corners}\times\frac{1}{8}\ \text{atom}\cdot\text{corner}^{-1} + 6\ \text{faces} + \frac{1}{2}\ \text{atom}\cdot\text{face}^{-1} = 4\ \text{atoms}$

anions: $8\ \text{tetrahedral holes}\times1\ \text{atom}\cdot\text{tetrahedral hole}^{-1} = 8\ \text{atoms}$

The cation to anion ratio is thus $4:8$ or $1:2$; the empirical formula is

$MA_2$.

**5.89**  The cesium chloride lattice is a simple cubic lattice of $Cl^-$ ions with a

$Cs^+$ at the center of the unit cell (See 5.43). In the unit cell, there is a total

of one $Cl^-$ ion and one $Cs^+$ ion. If the density is $3.988\ \text{g}\cdot\text{cm}^{-3}$, then we

can determine the volume and unit cell edge length. The molar mass of

CsCl is $168.36\ \text{g}\cdot\text{mol}^{-1}$.

$$3.988\ \text{g}\cdot\text{cm}^{-3} = \frac{\left(\dfrac{168.36\ \text{g}\cdot\text{mol}^{-1}}{6.022\times10^{23}\ \text{formula units}\cdot\text{mol}^{-1}}\right)}{a^3}$$

$$a^3 = \frac{\left(\dfrac{168.36\ \text{g}\cdot\text{mol}^{-1}}{6.022\times10^{23}\ \text{formula units}\cdot\text{mol}^{-1}}\right)}{3.988\ \text{g}\cdot\text{cm}^{-3}}$$

$a = 4.12\times10^{-8}\ \text{cm}$

$\quad = 412\ \text{pm}$

The volume of the unit cell is $(412 \text{ pm})^3 = 6.99 \times 10^7 \text{ pm}^3$.

We will determine the size of the $Cs^+$ and $Cl^-$ ions from ionic radii given in Appendix 2D, but we can check these values against the unit cell dimensions. For this type of unit cell, the body diagonal will be equal to $2\, r(Cs^+) + 2\, r(Cl^-) = a\sqrt{3} = 714 \text{ pm}$. The sum of the ionic radii gives us $2\,(170 \text{ pm}) + 2\,(181 \text{ pm}) = 702 \text{ pm}$, which is in very good agreement.

Note that we cannot calculate the size of these ions independently from the unit cell data without more information, because this lattice is not close-packed. We will assume that the ions are spherical. The volume occupied in the unit cell will be

$$V_{Cs^+} = \frac{4}{3}\pi r^3 = \frac{4}{3}\pi (170 \text{ pm})^3 = 2.06 \times 10^7 \text{ pm}^3$$

$$V_{Cl^-} = \frac{4}{3}\pi r^3 = \frac{4}{3}\pi (181 \text{ pm})^3 = 2.48 \times 10^7 \text{ pm}^3$$

The total occupied volume in the cell is

$2.06 \times 10^7 \text{ pm}^3 + 2.48 \times 10^7 \text{ pm}^3 = 4.54 \times 10^7 \text{ pm}^3$. The empty space is

$6.99 \times 10^7 \text{ pm}^3 - 4.54 \times 10^7 \text{ pm}^3 = 2.45 \times 10^7 \text{ pm}^3$. The percent empty

space is $2.45 \times 10^7 \text{ pm}^3 \div 6.99 \times 10^7 \text{ pm}^3 \times 100 = 35\%$.

**5.91**  (a)  In a simple cubic unit cell there is a total of one atom. The volume of the atom is given by $\frac{4}{3}\pi r^3$. The volume of the unit cell is given by $a^3$ and $a = 2r$, so the volume of the unit cell is $8r^3$. The fraction of occupied space in the unit cell is given by

$$\frac{\frac{4}{3}\pi r^3}{8r^3} = \frac{\frac{4}{3}\pi}{8} = 0.52$$

52% of the space is occupied, so 48% of this unit cell would be empty.

(b)  The percentage of empty space in an fcc unit cell is 26%, so the fcc cell is much more efficient at occupying the space available.

**5.93** The lattice layers from which constructive x-ray diffraction occurs are parallel. First draw perpendicular lines from the point of intersection of the top x-ray with the lattice plane to the lower x-ray for both the incident and diffracted rays. The x-rays are in phase and parallel at point A. If we want them to be still in phase and parallel when they exit the crystal, then they must still be in phase when they reach point C. In order for this to be true, the extra distance that the second beam travels with respect to the first must be equal to some integral number of wave-lengths. The total extra distance traveled, $A \rightarrow B \rightarrow C$, is equal to $2x$.

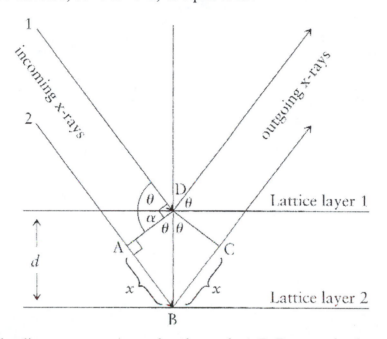

From the diagram, we can see that the angle A-D-B must also be equal to $\theta$. The angles $\theta$ and $\alpha$ sum to 90°, as do the angles $\alpha$ and A-D-B.

We can then write $\sin \theta = \dfrac{x}{d}$ and $x = d \sin \theta$. The total distance traveled is $2x = 2d \sin \theta$. So, for the two x-rays to be in phase as they exit the crystal, $2d \sin \theta$ must be equal to an integral number of wave-lengths.

**5.95** The answer to this problem is obtained from Bragg's law: $\lambda = 2d \sin \theta$ where $\lambda$ is the wavelength of radiation, $d$ is the interplanar spacing, and $\theta$

is the angle of incidence of the x-ray beam. Here $d = 401.8$ pm and
$\lambda = 71.107$ pm.

$71.07 \text{ pm} = 2(401.8 \text{ pm}) \sin \theta$
$\theta = 5.074°$

**5.97**  $3/2 \, kT = 1/2 \, mv^2$   $\lambda = h/mv$   $3kT = h^2/m\lambda^2$   $T = 633$ K

# CHAPTER 6

# THERMODYNAMICS: THE FIRST LAW

**6.1** (a) isolated;  (b) closed;  (c) isolated;  (d) open;  (e) closed;  (f) open

**6.3** (a) Work is given by $w = -P_{ext}\Delta V$. The applied external pressure is known, but we must calculate the change in volume given the physical dimensions of the pump and the distance, $d$, the piston in the pump moves:

$$\Delta V = -\pi r^2 d = \pi(1.5\text{ cm})^2(20\text{ cm})\left(\frac{1\text{ L}}{1000\text{ cm}^3}\right) = -0.141\text{ L}$$

$\Delta V$ is negative because the air in the pump is compressed to a smaller volume; work is then:

$$w = -(2\text{ atm})(-0.141\text{ L})\left(\frac{101.325\text{ L}}{\text{L}\cdot\text{atm}}\right) = 29\text{ J}$$

(b) Work on the air is positive by convention as work is done on the air, it is compressed.

**6.5** The change in internal energy $\Delta U$ is given simply by summing the two energy terms involved in this process. We must be careful, however, that the signs on the energy changes are appropriate. In this case, internal energy will be added to the gas sample by heating and through compression. Therefore the change in internal energy is:

$\Delta U = 524\text{ kJ} + 340\text{ kJ} = 864\text{ kJ}$

**6.7**  (a) The internal energy increased by more than the amount of heat added. Therefore, the extra energy must have come from work done on the system.

(b)  $w = \Delta U - q = 982\ \text{J} - 492\ \text{J} = +4.90 \times 10^2\ \text{J}.$

**6.9**  To get the entire internal energy change, we must sum the changes due to heat and work. In this problem, $q = +5500$ kJ. Work will be given by $w = -P_{\text{ext}} \Delta V$ because it is an expansion against a constant opposing pressure:

$$w = -\left( \frac{750\ \text{Torr}}{760\ \text{Torr} \cdot \text{atm}^{-1}} \right)\left( \frac{1846\ \text{mL} - 345\ \text{mL}}{1000\ \text{mL} \cdot \text{L}^{-1}} \right) = -1.48\ \text{L} \cdot \text{atm}$$

To convert to J we use the equivalency of the ideal gas constants:

$$w = -(1.48\ \text{L} \cdot \text{atm})\left( \frac{8.314\ \text{J} \cdot \text{K}^{-1} \cdot \text{mol}^{-1}}{0.082\,06\ \text{L} \cdot \text{atm} \cdot \text{K}^{-1} \cdot \text{mol}^{-1}} \right) = -1.50 \times 10^2\ \text{J}$$

$\Delta U = q + w = 5500\ \text{kJ} - 0.150\ \text{kJ} = 5500\ \text{kJ}$

The energy change due to the work term turns out to be negligible in this problem.

**6.11**  Using $\Delta U = q + w$ , where $\Delta U = $ -2573 kJ and $q = $ -947 kJ ,

- 2573 kJ $= -947$ kJ $+ w.$

Therefore, $w = -1626$ kJ.

1626 kJ of work can be done by the system on its surroundings.

**6.13**  (a) true if no work is done;   (b) always true;   (c) always false;   (d) true only if $w = 0$ (in which case $\Delta U = q = 0$);   (e) always true

**6.15**  (a) The heat change will be made up of two terms: one term to raise the temperature of the copper and the other to raise the temperature of the water:

$$q = (750.0 \text{ g})(4.18 \text{ J} \cdot (°C)^{-1} \cdot g^{-1})(100.0°C - 23.0°C)$$
$$+ (500.0 \text{ g})(0.38 \text{ J} \cdot (°C)^{-1} \cdot g^{-1})(100.0°C - 23.0°C)$$
$$= 2.4 \times 10^5 \text{ J} + 1.5 \times 10^4 \text{ J} = 2.6 \times 10^5 \text{ J} = 2.6 \times 10^2 \text{ kJ}$$

(b) The percentage of heat attributable to raising the temperature of water will be

$$\left( \frac{241 \text{ kJ}}{256 \text{ kJ}} \right)(100) = 94.3\%$$

**6.17**  heat lost by metal $= -$ heat gained by water

$$(20.0 \text{ g})(T_{final} - 100.0°C)(0.38 \text{ J} \cdot (°C)^{-1} \cdot g^{-1})$$
$$= -(50.7 \text{ g})(4.18 \text{ J} \cdot (°C)^{-1} \cdot g^{-1})(T_{final} - 22.0°C)$$
$$(T_{final} - 100.0°C)(7.6 \text{ J} \cdot (°C)^{-1}) = -(212 \text{ J} \cdot (°C)^{-1})(T_{final} - 22.0°C)$$
$$T_{final} - 100.0°C = -28(T_{final} - 22.0°C)$$
$$T_{final} + 28 \, T_{final} = 100.0°C + 616°C$$
$$29 \, T_{final} = 716°C$$
$$T_{final} = 25°C$$

**6.19**  $C_{cal} = \dfrac{22.5 \text{ kJ}}{23.97°C - 22.45°C} = 14.8 \text{ kJ} \cdot (°C)^{-1}$

**6.21**  (a) The irreversible work of expansion against a constant opposing pressure is given by

$$w = -P_{ex}\Delta V$$
$$w = -(1.00 \text{ atm})(6.52 \text{ L} - 4.29 \text{ L})$$
$$= -2.23 \text{ L} \cdot \text{atm}$$
$$= -2.23 \text{ L} \cdot \text{atm} \times 101.325 \text{ J} \cdot \text{L}^{-1} \cdot \text{atm}^{-1} = -226 \text{ J}$$

(b) An isothermal expansion will be given by

$$w = -nRT\frac{V_2}{V_1}$$

$n$ is calculated from the ideal gas law:

$$n = \frac{PV}{RT} = \frac{(1.79 \text{ atm})(4.29 \text{ L})}{(0.082\ 06 \text{ L} \cdot \text{atm} \cdot \text{K}^{-1} \cdot \text{mol}^{-1})(305 \text{ K})} = 0.307 \text{ mol}$$

$$w = -(0.307 \text{ mol})(8.314 \text{ J} \cdot \text{K}^{-1} \cdot \text{mol}^{-1})(305 \text{ K}) \ln\frac{6.52}{4.29}$$

$$= -326 \text{ J}$$

Note that the work done is greater when the process is carried out reversibly.

**6.23**  $NO_2$. The heat capacity increases with molecular complexity—as more atoms are present in the molecule, there are more possible bond vibrations that can absorb added energy.

**6.25**  (a)  The molar heat capacity of a monatomic ideal gas at constant pressure is $C_{P,m} = \frac{5}{2}R$. The heat released will be given by

$$q = \left(\frac{5.025 \text{ g}}{83.80 \text{ g} \cdot \text{mol}^{-1}}\right)(25.0°C - 97.6°C)(20.8 \text{ J} \cdot \text{mol}^{-1} \cdot (°C)^{-1}) = -90.6 \text{ J}$$

(b)  Similarly, the molar heat capacity of a monatomic ideal gas at constant volume is $C_{V,m} = \frac{3}{2}R$. The heat released will be given by

$$q = \left(\frac{5.025 \text{ g}}{83.80 \text{ g} \cdot \text{mol}^{-1}}\right)(25.0°C - 97.6°C)(12.5 \text{ J} \cdot \text{mol}^{-1} \cdot (°C)^{-1}) = -54.4 \text{ J}$$

**6.27**  (a)  HCN is a linear molecule. The contribution from molecular motions will be $5/2\ R$.

(b)  $C_2H_6$ is a polyatomic, nonlinear molecule. The contribution from molecular motions will be $3R$.

(c)  Ar is a monoatomic ideal gas. The contribution from molecular motions to the heat capacity will be $3/2\ R$.

(d)  HBr is a diatomic, linear molecule. The contribution from molecular motions will be $5/2\ R$.

**6.29** The strategy here is to determine the amount of energy per photon and the amount of energy needed to heat the water. Dividing the latter by the former will give the number of photons needed. Energy per photon is given by:

$$E = \frac{hc}{\lambda} = \frac{(6.626 \times 10^{-34} \text{ J} \cdot \text{s})(2.9979 \times 10^{8} \text{ m} \cdot \text{s}^{-1})}{4.50 \times 10^{-3} \text{ m}} = 4.41 \times 10^{-23} \text{ J} \cdot \text{photon}^{-1}$$

The energy needed to heat the water is:

$$350 \text{ g } (4.184 \text{ J} \cdot \text{g}^{-1} \cdot {}^{\circ}\text{C}^{-1}) (100.0\,{}^{\circ}\text{C} - 25.0\,{}^{\circ}\text{C}) = 1.10 \times 10^{5} \text{ J}$$

The number of photons needed is therefore:

$$\frac{1.10 \times 10^{5} \text{ J}}{4.41 \times 10^{-23} \text{ J} \cdot \text{photon}^{-1}} = 2.49 \times 10^{27} \text{ photons}$$

**6.31** (a) Using the estimation that $3R = C$:

$$3R = C = (0.392 \text{ J} \cdot \text{K}^{-1} \cdot \text{g}^{-1})(M)$$

$$M = \frac{3R}{0.392 \text{ J} \cdot \text{K}^{-1} \cdot \text{g}^{-1}} = 63.6 \text{ g mol}^{-1}$$

This molar mass indicates that the atomic solid is Cu(s)

(b) From Example 5.3 we find that the density of a substance which forms a face-centered cubic unit cell is given by: $d = \dfrac{4M}{8^{3/2} N_A r^3}$. Therefore, we expect the density of copper to be:

$$d = \frac{4 \, (63.55 \text{ g} \cdot \text{mol}^{-1})}{8^{3/2} (6.022 \times 10^{23} \text{ mol}^{-1})(1.28 \times 10^{-8} \text{ cm})^3} = 8.90 \text{ g} \cdot \text{cm}^{-3}$$

**6.33** (a) $\Delta H_{\text{vap}} = \dfrac{4.76 \text{ kJ}}{0.579 \text{ mol}} = 8.22 \text{ kJ} \cdot \text{mol}^{-1}$

(b) $\Delta H_{\text{vap}} = \dfrac{21.2 \text{ kJ}}{\left(\dfrac{22.45 \text{ g}}{46.07 \text{ g} \cdot \text{mol}^{-1}}\right)} = 43.5 \text{ kJ} \cdot \text{mol}^{-1}$

**6.35**  This process is composed of two steps: melting the ice at 0°C and then raising the temperature of the liquid water from 0°C to 25°C:

Step 1: $\Delta H = \left(\dfrac{80.0\ \text{g}}{18.02\ \text{g}\cdot\text{mol}^{-1}}\right)(6.01\ \text{kJ}\cdot\text{mol}^{-1}) = 26.7\ \text{kJ}$

Step 2: $\Delta H = (80.0\ \text{g})(4.18\ \text{J}\cdot(°\text{C})^{-1}\cdot\text{g}^{-1})(20.0°\text{C} - 0.0°\text{C}) = 6.69\ \text{kJ}$

Total heat required $= 26.7\ \text{kJ} + 6.69\ \text{kJ} = 33.4\ \text{kJ}$

**6.37**  The heat gained by the water in the ice cube will be equal to the heat lost by the initial sample of hot water. The enthalpy change for the water in the ice cube will be composed of two terms: the heat to melt the ice at 0°C and the heat required to raise the ice from 0°C to the final temperature.

$\text{heat (ice cube)} = \left(\dfrac{50.0\ \text{g}}{18.02\ \text{g}\cdot\text{mol}^{-1}}\right)(6.01\times10^{3}\ \text{J}\cdot\text{mol}^{-1})$
$+ (50.0\ \text{g})(4.184\ \text{J}\cdot(°\text{C})^{-1}\cdot\text{g}^{-1})(T_f - 0°)$
$= 1.67\times10^{4}\ \text{J} + (209\ \text{J}\cdot(°\text{C})^{-1})(T_f - 0°)$

$\text{heat (water)} = (400\ \text{g})(4.184\ \text{J}\cdot(°\text{C})^{-1}\cdot\text{g}^{-1})(T_f - 45°)$
$= (1.67\times10^{3}\ \text{J}\cdot(°\text{C})^{-1})(T_f - 45°)$

Setting these equal:

$-(1.67\times10^{3}\ \text{J}\cdot(°\text{C})^{-1})T_f + 7.5\times10^{4}\ \text{J} = 1.67\times10^{4}\ \text{J} + (209\ \text{J}\cdot(°\text{C})^{-1})T_f$

Solving for $T_f$:

$T_f = \dfrac{5.8\times10^{4}\ \text{J}}{1.88\times10^{3}\ \text{J}\cdot(°\text{C})^{-1}} = 31\ °\text{C}$

**6.39**  (a)  $\Delta H = (1.25\ \text{mol})(+358.8\ \text{kJ}\cdot\text{mol}^{-1}) = 448\ \text{kJ}$

(b)  $\Delta H = \left(\dfrac{197\ \text{g C}}{12.01\ \text{g}\cdot\text{mol}^{-1}\ \text{C}}\right)\left(\dfrac{358.8\ \text{kJ}}{4\ \text{mol C}}\right) = 1.47\times10^{3}\ \text{kJ}$

(c)  $\Delta H = 415\ \text{kJ} = (n_{CS_2})\left(\dfrac{358.8\ \text{kJ}\cdot\text{mol}^{-1}}{4\ \text{mol CS}_2}\right)$

$n_{CS_2} = 4.63\ \text{mol CS}_2$  or  $(4.63\ \text{mol})(76.13\ \text{g}\cdot\text{mol}^{-1}) = 352\ \text{g CS}_2$

**6.41**    (a)   $(12 \text{ ft} \times 12 \text{ ft} \times 8 \text{ ft})\left(\dfrac{30.48 \text{ cm}}{1 \text{ ft}}\right)^3 = 3.26 \times 10^7 \text{ cm}^3$

The specific heat capacity of air is $1.01 \text{ J} \cdot (°\text{C}) \cdot \text{g}^{-1}$ and the average molar mass of air is $28.97 \text{ g} \cdot \text{mol}^{-1}$ (see Table 4.4). The density of air can be calculated from the ideal gas law:

$$d = \frac{PM}{RT} = \frac{(1.00 \text{ atm})(28.97 \text{ g} \cdot \text{mol}^{-1})(10^{-3} \text{ L} \cdot \text{cm}^{-3})}{(0.08206 \text{ L} \cdot \text{atm} \cdot \text{K}^{-1} \cdot \text{mol}^{-1})(277.55 \text{ K})}$$

$$d = 0.00127 \text{ g} \cdot \text{cm}^{-3}$$

where the temperature of the air is initially $40° \text{ F} = 4.4 \text{ °C} = 277.55 \text{ K}$.

The heat required to raise the room temperature by

$\Delta T = 78 \text{ °F} - 40 \text{ °F} = 26 \text{ °C} - 4 \text{ °C} = 22 \text{ °C}$ is

$$(3.26 \times 10^7 \text{ cm}^3)(0.00127 \text{ g} \cdot \text{cm}^{-3})(1.01 \text{ J} \cdot (°\text{C})^{-1} \cdot \text{g}^{-1})(22 \text{ °C})$$

$$= 9.2 \times 10^5 \text{ J} = 9.2 \times 10^2 \text{ kJ}.$$

Finally, the mass of octane required to produce this much heat is

$$\left(\frac{-9.2 \times 10^2 \text{ kJ}}{-5471 \text{ kJ} \cdot \text{mol}^{-1}}\right)(114.22 \text{ g} \cdot \text{mol}^{-1}) = 19.2 \text{ g}.$$

(b)   $\Delta H = \left(\dfrac{(1.0 \text{ gal})(3.785 \times 10^3 \text{ mL} \cdot \text{gal}^{-1})(0.70 \text{ g} \cdot \text{mL}^{-1})}{114.22 \text{ g} \cdot \text{mol}^{-1}}\right)$

$$\times \left(\frac{-10\,942 \text{ kJ}}{2 \text{ mol octane}}\right)$$

$$= -1.3 \times 10^5 \text{ kJ}$$

**6.43**    (a)   $\left(1250 \dfrac{\text{kJ}}{\text{hr}}\right)\left(1 \dfrac{\text{hr}}{\text{day}}\right)\left(150 \dfrac{\text{days}}{\text{year}}\right) = 1.9 \times 10^5 \dfrac{\text{kJ}}{\text{year}}$

(b)   $\left(150 \dfrac{\text{trips}}{\text{year}}\right)\left(0.40 \dfrac{\text{gal.}}{\text{trip}}\right)\left(3.785 \dfrac{\text{L}}{\text{gal.}}\right)\left(1000 \dfrac{\text{mL}}{\text{L}}\right)\left(0.702 \dfrac{\text{g}}{\text{mL}}\right)$

$$\left(\frac{1 \text{ mol}}{114.23 \text{ g}}\right)\left(5471 \frac{\text{kJ}}{\text{mol}}\right) = 7.6 \times 10^6 \frac{\text{kJ}}{\text{year}}$$

**6.45**   From $\Delta H = \Delta U + P\Delta V$ at constant pressure, or $\Delta U = \Delta H - P\Delta V$.

Because $w = -P\Delta V = +22$ kJ, we get $-15$ kJ $+ 22$ kJ $= \Delta U = +7$ kJ.

**6.47**   To determine the enthalpy of the reaction we must start with a balanced chemical reaction and determine the limiting reagent:

$2HCl(aq) + Zn(s) \rightleftharpoons H_2(g) + ZnCl_2(aq)$.

$0.800$ L $\cdot 0.500$ M HCl $= 0.400$ mol HCl

$\dfrac{8.5\ g}{65.37 g\cdot mol^{-1}} = 0.130$ mol Zn

Examining the reaction stoichiometry and the initial quantities of HCl and Zn, we note that Zn is the limiting reagent (0.260 mol of HCl is needed to completely react with 0.130 moles of Zn). The enthalpy of reaction may be obtained using tabulated enthalpies of formation:

$$\Delta H_r = -153.89\frac{kJ}{mol} + 2\left(-167.16\frac{kJ}{mol}\right) - 2\left(-167.16\frac{kJ}{mol}\right) - 0$$

$$= -153.89\frac{kJ}{mol}$$

This is the enthalpy per mole of Zinc consumed. Therefore, the energy released by the reaction of 8.5 g of Zinc is:

$$\left(-153.89\frac{kJ}{mol}\right)(0.130\ mol) = -20.0\ kJ$$

The change in the temperature of the water is then:

$$-20000\ J = \left(-4.184\ \frac{J}{°C\ g}\right)(800\ g)\Delta T$$

$\Delta T = 5.98°C$ and $T_f = 25°C + 5.98°C = 31°C$

**6.49**   The enthalpy of reaction for the reaction

$4\ C_7H_5N_3O_6(s) + 21\ O_2(g) \rightarrow 28\ CO_2(g) + 10\ H_2O(g) + 6\ N_2(g)$

may be found using enthalpies of formation:

$$28\left(-393.51\frac{kJ}{mol}\right) + 10\left(-241.82\frac{kJ}{mol}\right) - 4\left(-67\frac{kJ}{mol}\right) = -13168\frac{kJ}{mol}$$

This is the energy released per mole of reaction as written. One fourth of this amount of energy or $3292\dfrac{kJ}{mol}$ will be *released* per mole of TNT consumed. The energy density in kJ per L may be found by dividing this amount of energy with the mass of one mole of TNT and then by multiplying with the density of TNT:

$$\dfrac{3292\dfrac{kJ}{mol}}{227.14\dfrac{g}{mol}}\left(1.65\dfrac{g}{cm^3}\right)\left(\dfrac{10^3\,cm^3}{1\,L}\right)=+23.9\times10^3\,\dfrac{kJ}{L}$$

**6.51** The combustion reaction of diamond is reversed and added to the combustion reaction of graphite to give the desired reaction:

$$C(gr)+O_2(g)\longrightarrow CO_2(g) \qquad\qquad \Delta H^\circ=-393.51\,kJ$$

$$CO_2(g)\longrightarrow C(dia)+O_2(g) \qquad\qquad \Delta H^\circ=+395.41\,kJ$$

$$C(gr)\longrightarrow C(dia) \qquad\qquad \Delta H^\circ=+1.90\,kJ$$

**6.53** The first reaction is doubled, reversed, and added to the second to give the desired total reaction:

$$2[SO_2(g)\longrightarrow S(s)+O_2(g)] \qquad (2)[+296.83\,kJ]$$

$$2S(s)+3\,O_2(g)\longrightarrow 2\,SO_3(g) \qquad -791.44\,kJ$$

$$2\,SO_2(g)+O_2(g)\longrightarrow 2\,SO_3(g)$$

$$\Delta H^\circ=(2)(+296.83\,kJ\cdot mol^{-1})-(791.44\,kJ\cdot mol^{-1})=-197.78\,kJ$$

**6.55** First, write the balanced equations for the reaction given:

$$C_2H_2(g)+\tfrac{5}{2}O_2(g)\longrightarrow 2\,CO_2(g)+H_2O(l) \qquad \Delta H^\circ=-1300\,kJ$$

$$C_2H_6(g)+\tfrac{7}{2}O_2(g)\longrightarrow 2\,CO_2(g)+3\,H_2O(l) \qquad \Delta H^\circ=-1560\,kJ$$

$$H_2(g)+\tfrac{1}{2}O_2(g)\longrightarrow H_2O(l) \qquad \Delta H^\circ=-286\,kJ$$

The second equation is reversed and added to the first, plus two times the third:

$$C_2H_2(g) + \tfrac{5}{2}O_2(g) \longrightarrow 2\,CO_2(g) + H_2O(l) \qquad \Delta H° = -1300\,kJ$$

$$2\,CO_2(g) + 3\,H_2O(l) \longrightarrow C_2H_6(g) + \tfrac{7}{2}O_2(g) \qquad \Delta H° = +1560\,kJ$$

$$2[H_2(g) + \tfrac{1}{2}O_2(g) \longrightarrow H_2O(l)] \qquad 2[\Delta H° = -286\,kJ]$$

$$C_2H_2(g) + 2\,H_2(g) \longrightarrow C_2H_6(g)$$
$$\Delta H° = -1300\,kJ + 1560\,kJ + 2(-286\,kJ) = -312\,kJ \cdot mol^{-1}$$

**6.57**  The reaction enthalpy for this reaction is given by:

$$\Delta H° = 12\,(\Delta H°_f(H_2O,l))$$
$$\qquad - [4\,(\Delta H°_f(HNO_3,l)) + 5\,(\Delta H°_f(N_2H_4,l))]$$
$$= 12(-285.83\,kJ \cdot mol^{-1})$$
$$\qquad - [4(-174.10\,kJ \cdot mol^{-1}) + 5(+50.63\,kJ \cdot mol^{-1})]$$
$$= -2986.71\,kJ \cdot mol^{-1}$$

**6.59**  The desired reaction may be obtained by reversing the first reaction and multiplying it by 2, reversing the second reaction, and adding these to the third:

$$2[NH_4Cl(s) \longrightarrow NH_3(g) + HCl(g)] \qquad 2[\Delta H° = +176.0\,kJ]$$

$$2\,NH_3(g) \longrightarrow N_2(g) + 3\,H_2(g) \qquad \Delta H° = +92.22\,kJ$$

$$N_2(g) + 4\,H_2(g) + Cl_2(g) \longrightarrow 2\,NH_4Cl(s) \quad \Delta H° = -628.86\,kJ$$

$$H_2(g) + Cl_2(g) \longrightarrow 2\,HCl(g)$$

$$\Delta H° = 2(+176.0\,kJ) + 92.22\,kJ - 628.86\,kJ = -184.6\,kJ$$

**6.61**  From Appendix 2A, $\Delta H°_f$ (NO) = +90.25 kJ

The reaction we want is

$$N_2(g) + \tfrac{5}{2}O_2(g) \longrightarrow N_2O_5(g)$$

Adding the first reaction to half of the second gives

$$2\,NO(g) + O_2(g) \longrightarrow 2\,NO_2(g) \qquad \Delta H° = -114.1\,kJ$$

$$2\,NO_2(g) + \tfrac{1}{2}O_2(g) \longrightarrow N_2O_5(g) \qquad \Delta H° = -55.1\,kJ$$

$$2\,NO(g) + \tfrac{3}{2}O_2(g) \longrightarrow N_2O_5(g) \qquad -169.2\ kJ$$

The enthalpy of this reaction equals the enthalpy of formation of $N_2O_5\,(g)$ minus twice the enthalpy of formation of NO, so we can write

$$-169.2\ kJ = \Delta H^\circ_f\,(N_2O_5) - 2(+90.25\ kJ)$$
$$\Delta H^\circ_f\,(N_2O_5) = +11.3\ kJ$$

**6.63** The enthalpy of the reaction

$$PCl_3\,(l) + Cl_2\,(g) \longrightarrow PCl_5\,(s) \qquad \Delta H^\circ = -124\ kJ$$

is $\Delta H^\circ_r = \Sigma\,\Delta H^\circ_f\,(products) - \Sigma\,\Delta H^\circ_f\,(reactants)$

$$-124\ kJ = \Delta H^\circ_f\,(PCl_5,\,s) - \Delta H^\circ_f\,(PCl_3,\,l)$$

Remember that the standard enthalpy of formation of $Cl_2\,(g)$ will be 0 by definition because this is an element in its reference state. From the Appendix we find that

$$\Delta H^\circ_f\,(PCl_3,l) = -319.7\ kJ\cdot mol^{-1}$$
$$-124\ kJ = \Delta H^\circ_f\,(PCl_5,s) - (-319.7\ kJ)$$
$$\Delta H^\circ_f\,(PCl_5,s) = -444\ kJ\cdot mol^{-1}$$

**6.65** (a) For $H_2O(l)$, we want to find the enthalpy of the reaction

$$H_2(g) + \tfrac{1}{2}O_2(g) \longrightarrow H_2O(l)$$

The enthalpy change can be estimated from bond enthalpies. We will need to put in $(1\ mol)(436\ kJ\cdot mol^{-1})$ to break the H—H bonds in $1\ mol\,H_2\,(g)$, $(\tfrac{1}{2}mol)\,(496\ kJ\cdot mol^{-1})$ to break the O—O bonds in $\tfrac{1}{2}\,mol\,O_2\,(g)$; we will get back $(2\ mol)\,(463\,kJ\cdot mol^{-1})$ for the formation of 2 mol O—H bonds. This will give $\Delta H = -242\ kJ\cdot mol^{-1}$. This value, however, will be to produce water in the gas phase. In order to get the value for the liquid, we will need to take into account the amount of heat given off when the gaseous water condenses to the liquid phase. This is $44.0\ kJ\cdot mol^{-1}$ at 298 K:

$$\Delta H^\circ_{\text{f,water(l)}} = \Delta H^\circ_{\text{f,water(g)}} - \Delta H^\circ_{\text{vap}} = -242 \text{ kJ} \cdot \text{mol}^{-1} - 44.0 \text{ kJ} \cdot \text{mol}^{-1}$$

$$= -286 \text{ kJ} \cdot \text{mol}^{-1}$$

(b) The calculation for methanol is done similarly:

$$C(gr) + 2 H_2(g) + \tfrac{1}{2} O_2(g) \longrightarrow CH_3OH(l)$$

$\Delta H$ for individual bond contributions:

| | |
|---|---|
| atomize 1 mol C(gr) | $(1 \text{ mol})(717 \text{ kJ} \cdot \text{mol}^{-1})$ |
| break 2 mol H—H bonds | $(2 \text{ mol})(436 \text{ kJ} \cdot \text{mol}^{-1})$ |
| break $\tfrac{1}{2}$ mol $O_2$ bonds | $(\tfrac{1}{2} \text{ mol})(496 \text{ kJ} \cdot \text{mol}^{-1})$ |
| form 3 mol C—H bonds | $-(3 \text{ mol})(412 \text{ kJ} \cdot \text{mol}^{-1})$ |
| form 1 mol C—O bonds | $-(1 \text{ mol})(360 \text{ kJ} \cdot \text{mol}^{-1})$ |
| form 1 mol O—H bonds | $-(1 \text{ mol})(463 \text{ kJ} \cdot \text{mol}^{-1})$ |

Total                                                    $-222$ kJ

$$\Delta H^\circ_{\text{f,methanol(l)}} = \Delta H^\circ_{\text{f,methanol(g)}} - \Delta H^\circ_{\text{vap}}$$

$$= -222 \text{ kJ} \cdot \text{mol}^{-1} - 35.3 \text{ kJ} \cdot \text{mol}^{-1}$$

$$= -257 \text{ kJ} \cdot \text{mol}^{-1}$$

(c)  $6 C(gr) + 3 H_2(g) \longrightarrow C_6H_6(l)$

Without resonance, we do the calculation considering benzene to have three double and three single C—C bonds:

| | | |
|---|---|---|
| atomize: | 6 mol C(gr) | $(6 \text{ mol})(717 \text{ kJ} \cdot \text{mol}^{-1})$ |
| break: | 3 mol H—H bonds | $(3 \text{ mol})(436 \text{ kJ} \cdot \text{mol}^{-1})$ |
| form: | 3 mol CRC bonds | $-(3 \text{ mol})(612 \text{ kJ} \cdot \text{mol}^{-1})$ |
| form: | 3 mol C—C bonds | $-(3 \text{ mol})(348 \text{ kJ} \cdot \text{mol}^{-1})$ |
| form: | 6 mol C—H bonds | $-(6 \text{ mol})(412 \text{ kJ} \cdot \text{mol}^{-1})$ |

Total                                                    $+258$ kJ

$$\Delta H^{\circ}_{f,benzene(l)} = \Delta H^{\circ}_{f,benzene(g)} - \Delta H^{\circ}_{vap} = +258 \text{ kJ} \cdot \text{mol}^{-1} - 30.8 \text{ kJ} \cdot \text{mol}^{-1}$$
$$= +227 \text{ kJ} \cdot \text{mol}^{-1}$$

(d)  $6 \text{ C(gr)} + 3 \text{ H}_2(g) \longrightarrow C_6H_6(l)$

With resonance, we repeat the calculation considering benzene to have six resonance-stabilized C—C bonds:

| | | |
|---|---|---|
| atomize: | 6 mol C(gr) | $(6 \text{ mol})(717 \text{ kJ} \cdot \text{mol}^{-1})$ |
| break: | 3 mol  H—H bonds | $(3 \text{ mol})(436 \text{ kJ} \cdot \text{mol}^{-1})$ |
| form: | 6 mol C—C bonds, resonance | $-(6 \text{ mol})(518 \text{ kJ} \cdot \text{mol}^{-1})$ |
| form: | 6 mol C—H bonds | $-(6 \text{ mol})(412 \text{ kJ} \cdot \text{mol}^{-1})$ |

---

Total                                                           +30 kJ

$$\Delta H^{\circ}_{f,benzene(l)} = \Delta H^{\circ}_{f,benzene(g)} - \Delta H^{\circ}_{vap} = +30 \text{ kJ} \cdot \text{mol}^{-1} - 30.8 \text{ kJ} \cdot \text{mol}^{-1}$$
$$= -1 \text{ kJ} \cdot \text{mol}^{-1}$$

**6.67**  For the reaction  $Na_2O(s) \longrightarrow 2 \text{ Na}^+(g) + O^{2-}(g)$

$$\Delta H_L = 2 \Delta H^{\circ}_f(Na, g) + \Delta H^{\circ}_f(O, g) + 2 I_1(Na)$$
$$-E_{ea1}(O) - E_{ea2}(O) - \Delta H_f(Na_2O(s)$$
$$\Delta H_L = 2(107.32 \text{ kJ} \cdot \text{mol}^{-1}) + 249 \text{ kJ} \cdot \text{mol}^{-1} + 2(494 \text{ kJ} \cdot \text{mol}^{-1})$$
$$-141 \text{ kJ} \cdot \text{mol}^{-1} + 844 \text{ kJ} \cdot \text{mol}^{-1} + 409 \text{ kJ} \cdot \text{mol}^{-1}$$
$$\Delta H_L = 2564 \text{ kJ} \cdot \text{mol}^{-1}$$

**6.69**  (a)    $\Delta H_L = \Delta H^{\circ}_f(Na, g) + \Delta H^{\circ}_f(Cl, g) + I_1(Na)$
$$- E_{ea} \text{ of } Cl - \Delta H_f(NaCl(s))$$
$$787 \text{ kJ} \cdot \text{mol}^{-1} = 108 \text{ kJ} \cdot \text{mol}^{-1} + 122 \text{ kJ} \cdot \text{mol}^{-1} + 494 \text{ kJ} \cdot \text{mol}^{-1}$$
$$- 349 \text{ kJ} \cdot \text{mol}^{-1} - \Delta H_f(NaCl(s))$$
$$\Delta H_f(NaCl(s)) = -412 \text{ kJ} \cdot \text{mol}^{-1}$$

(b)   $\Delta H_L = \Delta H^{\circ}_f(K, g) + \Delta H^{\circ}_f(Br, g) + I_1(K)$
$$- E_{ea}(Br) - \Delta H_f(KBr(s))$$

$$\Delta H_L = 89 \text{ kJ} \cdot \text{mol}^{-1} + 97 \text{ kJ} \cdot \text{mol}^{-1} + 418 \text{ kJ} \cdot \text{mol}^{-1}$$
$$- 325 \text{ kJ} \cdot \text{mol}^{-1} + 394 \text{ kJ} \cdot \text{mol}^{-1}$$
$$= 673 \text{ kJ} \cdot \text{mol}^{-1}$$

(c) $\Delta H_L = \Delta H°_f(\text{Rb, g}) + \Delta H°_f(\text{F, g}) + I_1(\text{Rb}) - E_{ea}(\text{F}) - \Delta H_f(\text{RbF(s)})$

$774 \text{ kJ} \cdot \text{mol}^{-1} = \Delta H°_f(\text{Rb, g}) + 79 \text{ kJ} \cdot \text{mol}^{-1}$
$$+ 402 \text{ kJ} \cdot \text{mol}^{-1} - 328 \text{ kJ} \cdot \text{mol}^{-1} + 558 \text{ kJ} \cdot \text{mol}^{-1}$$

$\Delta H°_f(\text{Rb, g}) = 63 \text{ kJ} \cdot \text{mol}^{-1}$

**6.71** (a) break:  3 mol C≡C bonds 3(837) kJ·mol$^{-1}$

form:  6 mol C=C bonds −6(518) kJ·mol$^{-1}$

---

Total  − 597 kJ·mol$^{-1}$

(b) break:  4 mol C—H bonds 4(412) kJ·mol$^{-1}$

4 mol Cl—Cl bonds 4(242) kJ·mol$^{-1}$

form:  4 mol C—Cl bonds −4(338) kJ·mol$^{-1}$

4 mol H—Cl bonds −4(431) kJ·mol$^{-1}$

---

Total  −460 kJ·mol$^{-1}$

(c) The number and types of bonds on both sides of the equations are equal, so we expect the enthalpy of the reaction to be essentially 0.

**6.73** (a) break:  1 mol N—N triple bonds  (1 mol)(944 kJ·mol$^{-1}$)

3 mol F—F bonds  (3 mol)(158 kJ·mol$^{-1}$)

form:  6 mol N—F bonds  (6 mol)(−195 kJ·mol$^{-1}$)

---

Total  + 248 kJ·mol$^{-1}$

(b) break:  1 mol C=C bonds  (1 mol)(612 kJ·mol$^{-1}$)

1 mol O—H bonds  (1 mol)(463 kJ·mol$^{-1}$)

| form: | 1 mol C—C bonds | $-(1\text{ mol})(348\text{ kJ}\cdot\text{mol}^{-1})$ |
|---|---|---|
| | 1 mol C—O bonds | $-(1\text{ mol})(360\text{ kJ}\cdot\text{mol}^{-1})$ |
| | 1 mol C—H bonds | $-(1\text{ mol})(412\text{ kJ}\cdot\text{mol}^{-1})$ |
| | Total | $-45\text{ kJ}\cdot\text{mol}^{-1}$ |
| (c) break: | 1 mol C—H bonds | $(1\text{ mol})(412\text{ kJ}\cdot\text{mol}^{-1})$ |
| | 1 mol Cl—Cl bonds | $(1\text{ mol})(242\text{ kJ}\cdot\text{mol}^{-1})$ |
| form: | 1 mol C—Cl bonds | $-(1\text{ mol})(338\text{ kJ}\cdot\text{mol}^{-1})$ |
| | 1 mol H—Cl bonds | $-(1\text{ mol})(431\text{ kJ}\cdot\text{mol}^{-1})$ |
| | Total | $-115\text{ kJ}\cdot\text{mol}^{-1}$ |

**6.75** The value that we want is given simply by the difference between three isolated C=C bonds and three isolated C—C single bonds, versus six resonance-stabilized bonds:

3 C=C bonds + 3 C—C bonds = 3(348 kJ) + 3(612 kJ) = 2880 kJ

6 resonance-stabilized bonds = 6(518 kJ) = 3108 kJ

As can be seen, the six resonance-stabilized bonds are more stable by ca. 228 kJ.

**6.77** (a) The enthalpy of vaporization is the enthalpy change associated with the conversion $C_6H_6(l) \longrightarrow C_6H_6(g)$ at constant pressure. The value at 298.2 K will be given by

$$\Delta H^{\circ}_{\text{vaporization at 298 K}} = \Delta H^{\circ}_{f}\ (C_6H_6, g) - \Delta H^{\circ}_{f}\ (C_6H_6, l)$$
$$= 82.93\text{ kJ}\cdot\text{mol}^{-1} - (49.0\text{ kJ}\cdot\text{mol}^{-1})$$
$$= 33.93\text{ kJ}\cdot\text{mol}^{-1}$$

(b) In order to take into account the difference in temperature, we need to use the heat capacities of the reactants and products in order to raise the temperature of the system to 353.2 K. We can rewrite the reactions as follows, to emphasize temperature, and then combine them according to Hess's law:

$$C_6H_6(l)_{\text{at 298 K}} \longrightarrow C_6H_6(g)_{\text{at 298 K}} \quad \Delta H^\circ = 33.93 \text{ kJ}$$

$$C_6H_6(l)_{\text{at 298 K}} \longrightarrow C_6H_6(l)_{\text{at 353.2 K}} \quad \Delta H^\circ = (1 \text{ mol})(353.2 \text{ K} - 298.2 \text{ K})$$
$$(136.1 \text{ J} \cdot \text{mol}^{-1} \cdot \text{K}^{-1})$$
$$= 7.48 \text{ kJ}$$

$$C_6H_6(g)_{\text{at 298 K}} \longrightarrow C_6H_6(g)_{\text{at 353.2 K}} \quad \Delta H^\circ = (1 \text{ mol})(353.2 \text{ K} - 298.2 \text{ K})$$
$$(81.67 \text{ J} \cdot \text{mol}^{-1} \cdot \text{K}^{-1})$$
$$= 4.49 \text{ kJ}$$

To add these together to get the overall equation at 353.2 K, we must reverse the second equation:

$$C_6H_6(l)_{\text{at 298 K}} \longrightarrow C_6H_6(g)_{\text{at 298 K}} \quad \Delta H^\circ = 33.93 \text{ kJ}$$

$$C_6H_6(l)_{\text{at 353.2 K}} \longrightarrow C_6H_6(l)_{\text{at 298 K}} \quad \Delta H^\circ = -7.48 \text{ kJ}$$

$$C_6H_6(g)_{\text{at 298 K}} \longrightarrow C_6H_6(g)_{\text{at 353.2 K}} \quad \Delta H^\circ = 4.49 \text{ kJ}$$

$$C_6H_6(l)_{\text{at 353.2 K}} \longrightarrow C_6H_6(g)_{\text{at 353.2 K}}$$
$$\Delta H^\circ = 33.93 \text{ kJ} - 7.48 \text{ kJ} + 4.49 \text{ kJ} = 30.94 \text{ kJ} \cdot \text{mol}^{-1}$$

(c) The value in the table is $30.8 \text{ kJ} \cdot \text{mol}^{-1}$ for the enthalpy of vaporization of benzene. The value is close to that calculated as corrected by heat capacities. At least part of the error can be attributed to the fact that heat capacities are not strictly constant with temperature.

**6.79** For the reaction: $A + 2B \rightarrow 3C + D$ the molar enthalpy of reaction at temperature 2 is given by:

$$\Delta H^\circ_{r,2} = H^\circ_{m,2}(\text{products}) - H^\circ_{m,2}(\text{reactants})$$

$$= 3H^\circ_{m,2}(C) + H^\circ_{m,2}(D) - H^\circ_{m,2}(A) - 2H^\circ_{m,2}(B)$$

$$= 3[H^\circ_{m,1}(C) + C_{p,m}(C)(T_2 - T_1)] + [H^\circ_{m,1}(D) + C_{p,m}(D)(T_2 - T_1)]$$

$$- [H^\circ_{m,1}(A) + C_{p,m}(A)(T_2 - T_1)] - 2[H^\circ_{m,1}(B) + C_{p,m}(B)(T_2 - T_1)]$$

$$= 3H^\circ_{m,1}(C) + H^\circ_{m,1}(D) - H^\circ_{m,1}(A) - 2H^\circ_{m,1}(B)$$

$$+ [3C_{p,m}(C) + C_{p,m}(D) - C_{p,m}(A) - 2C_{p,m}(B)](T_2 - T_1)$$

$$= \Delta H^\circ_{r,1} + [3C_{p,m}(C) + C_{p,m}(D) - C_{p,m}(A) - 2C_{p,m}(B)](T_2 - T_1)$$

Finally, $\Delta H_{r,2}^\circ = \Delta H_{r,1}^\circ + \Delta C_p(T_2 - T_1)$, which is Kirchhoff's law.

**6.81** This process involves five separate steps: (1) raising the temperature of the ice from −5.042 °C to 0.00 °C. (2) melting the ice at 0.00°C, (3) raising the temperature of the liquid water from 0.00°C to 100.00°C, (4) vaporizing the water at 100.00°C, and (5) raising the temperature of the water vapor from 100.00°C to 150.35°C.

Step 1:

$$\Delta H = (42.30 \text{ g})(2.03 \text{ J} \cdot (°C)^{-1} \cdot g^{-1})(0.00 \text{ °C} - (-5.042°C)) = 0.433 \text{ kJ}$$

Step 2: $\Delta H = \left(\dfrac{42.30 \text{ g}}{18.02 \text{ g} \cdot \text{mol}^{-1}}\right)(6.01 \text{ kJ} \cdot \text{mol}^{-1}) = 14.1 \text{ kJ}$

Step 3:

$$\Delta H = (42.30 \text{ g})(4.18 \text{ J} \cdot (°C)^{-1} \cdot g^{-1})(100.00 \text{ °C} - 0.00°C) = 17.7 \text{ kJ}$$

Step 4: $\Delta H = \left(\dfrac{42.30 \text{ g}}{18.02 \text{ g} \cdot \text{mol}^{-1}}\right)(40.7 \text{ kJ} \cdot \text{mol}^{-1}) = 95.5 \text{ kJ}$

Step 5:

$$\Delta H = (42.30 \text{ g})(2.01 \text{ J} \cdot (°C)^{-1} \cdot g^{-1})(150.35 \text{ °C} - 100.00°C) = 4.3 \text{ kJ}$$

The total heat required

$$= 0.4 \text{ kJ} + 14.1 \text{ kJ} + 17.7 \text{ kJ} + 95.5 \text{ kJ} + 4.3 \text{ kJ}$$

$$= 132.0 \text{ kJ}$$

**6.83** Appendix 2A provides us with the heat of formation of $I_2(g)$ at 298K ($+62.44$ kJ $\cdot$ mol$^{-1}$) and the heat capacities of

$I_2(g)$ $(36.90 \text{ J} \cdot \text{K}^{-1} \cdot \text{mol}^{-1})$ and $I_2(s)$ $(54.44 \text{ J} \cdot \text{K}^{-1} \cdot \text{mol}^{-1})$. We can calculate the $\Delta H_{sub}{}^0$ at 298K:

$$I_2(s) \longrightarrow I_2(g) \qquad\qquad \Delta H_{sub}{}^0 = +62.44 \text{ kJ} \cdot \text{mol}^{-1}$$

We can calculate the enthalpy of fusion from the relationship

$$\Delta H_{sub}{}^0 = \Delta H_{fus}{}^0 + \Delta H_{vap}{}^0$$

but these values need to be at the same temperature. To correct the value for the fact that we want all the numbers for 298K, we need to alter the heat of vaporization, using the heat capacities for liquid and gaseous iodine.

$$I_2(l) \text{ at } 184.3°C \longrightarrow I_2(g) \text{ at } 184.3°C \qquad \Delta H_{vap}{}^0 = +41.96 \text{ kJ} \cdot \text{mol}^{-1}$$

From Section 6.22, we find the following relationship

$$\Delta H_{r,2}{}^0 = \Delta H_{r,1}{}^0 + \Delta C_{P,m}{}^0 (T_2 - T_1)$$

$$\Delta H_{vap, 298K}{}^0 = \Delta H_{vap, 475.5K}{}^0 + (C_{P,m}{}^0 (I_2, g) - C_{P,m}{}^0 (I_2, l)) (T_2 - T_1)$$

$$\Delta H_{vap, 298K}{}^0 = +41.96 \text{ kJ} \cdot \text{mol}^{-1}$$
$$+ (36.90 \text{ J} \cdot \text{K}^{-1} \cdot \text{mol}^{-1} - 80.7 \text{ J} \cdot \text{K}^{-1} \cdot \text{mol}^{-1})(298K - 475.5K)$$
$$= +49.73 \text{ kJ} \cdot \text{mol}^{-1}$$

So, at 298K:

$$+62.44 \text{ kJ} \cdot \text{mol}^{-1} = \Delta H_{fus}{}^0 + 49.73 \text{ kJ} \cdot \text{mol}^{-1}$$
$$\Delta H_{fus}{}^0 = +12.71 \text{ kJ} \cdot \text{mol}^{-1}$$

**6.85** (a)

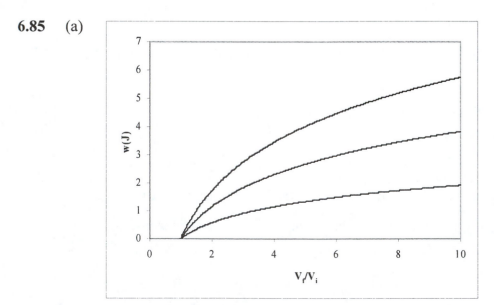

(b) The amount of work done is greater at the higher temperature. This can be seen from the equation:

$$w = -nRT \ln \frac{V_{final}}{V_{initial}}$$

The amount of work done is directly proportional to the temperature at which the expansion takes place.

(c) The comparison requested is the comparison of the terms

$$\ln \frac{V_{final}}{V_{initial}}$$

for the two processes. Even though in both cases the gas expands by 4 L, the relative amount of work done is different. We can get a numerical comparison by taking the ratio of this term for the two conditions:

$$\frac{\ln \left( \dfrac{9.00\ L}{5.00\ L} \right)}{\ln \left( \dfrac{5.00\ L}{1.00\ L} \right)} = \frac{0.588}{1.61} = 0.365$$

The second expansion by 4.00 L produces only about one third the amount of work that the first expansion did.

**6.87** First, we need to calculate how much energy from the sunshine will be hitting the surface of the ethanol, so we convert the rate $kJ \cdot cm^{-2} \cdot s^{-1}$ :

$$1\ kJ \cdot m^{-2} \cdot s^{-1} \left( \frac{1\ m}{100\ cm} \right)^2 = 1 \times 10^{-4}\ kJ \cdot cm^{-2} \cdot s^{-1}$$

$$(1 \times 10^{-4}\ kJ \cdot cm^{-2} \cdot s^{-1})(50.0\ cm^2) \left( 10\ min \times \frac{60\ s}{min} \right) = 3\ kJ$$

The enthalpy of vaporization of ethanol is $43.5\ kJ \cdot mol^{-1}$ (see ~~Table 6.2~~). *Table 6.3*

We will assume that the enthalpy of vaporization is approximately the same at ambient conditions as it would be at the boiling point of ethanol.

$$\left( \frac{3\ kJ}{43.5\ kJ \cdot mol^{-1}} \right) (46.07\ g \cdot mol^{-1}) = 3\ g$$

**6.89** (a) $C_6H_5NH_2(l) + \frac{31}{4}O_2(g) \longrightarrow 6\,CO_2(g) + \frac{7}{2}H_2O(l) + \frac{1}{2}N_2(g)$

(b)

$$m_{CO_2} = \left(\frac{0.1754 \text{ g aniline}}{93.12 \text{ g} \cdot \text{mol}^{-1} \text{ anline}}\right)\left(\frac{6 \text{ mol CO}_2}{1 \text{ mol aniline}}\right)(28.01 \text{ g} \cdot \text{mol}^{-1} \text{ CO}_2)$$

$$= 0.4873 \text{ g CO}_2(g)$$

$$m_{H_2O} = \left(\frac{0.1754 \text{ g aniline}}{93.12 \text{ g} \cdot \text{mol}^{-1} \text{ aniline}}\right)\left(\frac{3.5 \text{ mol H}_2O}{1 \text{ mol aniline}}\right)(18.02 \text{ g} \cdot \text{mol}^{-1} \text{ H}_2O)$$

$$= 0.1188 \text{ g H}_2O(l)$$

$$m_{N_2} = \left(\frac{0.1754 \text{ g aniline}}{93.12 \text{ g} \cdot \text{mol}^{-1} \text{ aniline}}\right)\left(\frac{0.5 \text{ mol N}_2}{1 \text{ mol aniline}}\right)(28.02 \text{ g} \cdot \text{mol}^{-1} \text{ N}_2)$$

$$= 0.026\,39 \text{ g N}_2(g)$$

(c) $n_{O_2} = \left(\frac{0.1754 \text{ g aniline}}{93.12 \text{ g} \cdot \text{mol}^{-1} \text{ aniline}}\right)\left(\frac{\frac{31}{4} \text{ mol O}_2}{1 \text{ mol aniline}}\right)$

$$= 0.014\,60 \text{ g O}_2(g)$$

$$P = \frac{nRT}{V} = \frac{(0.014\,60 \text{ mol O}_2)(0.082\,06 \text{ L} \cdot \text{atm} \cdot \text{K}^{-1} \cdot \text{mol}^{-1})(296K)}{0.355 \text{ L}}$$

$$= 0.999 \text{ atm}$$

**6.91** (a) The reaction enthalpy is obtained by Hess's law:

$\Delta H°_r = \Delta H°_f (CO, g) - \Delta H°_f (H_2O, g)$

$\Delta H°_r = (1)(-110.53 \text{ kJ} \cdot \text{mol}^{-1}) - (1)(-241.82 \text{ kJ} \cdot \text{mol}^{-1})$

$\Delta H°_r = +131.29 \text{ kJ} \cdot \text{mol}^{-1}$

endothermic

(b) The number of moles of $H_2$ produced is obtained from the ideal gas law:

$$n = \frac{PV}{RT} = \frac{\left(\dfrac{500 \text{ Torr}}{760 \text{ Torr} \cdot \text{atm}^{-1}}\right)(200 \text{ L})}{(0.082\,06 \text{ L} \cdot \text{atm} \cdot \text{K}^{-1} \cdot \text{mol}^{-1})(338 \text{ K})} = 4.74 \text{ mol}$$

The enthalpy change accompanying the production of this amount of hydrogen will be given by

$\Delta H = (4.74 \text{ mol})(131.29 \text{ kJ} \cdot \text{mol}^{-1}) = 623 \text{ kJ}$

**6.93** (a) The number of moles burned may be obtained by taking the difference in the number of moles of gas present in the tank before and after the drive using the ideal gas equation:

$$n_1 - n_2 = \frac{P_1 V}{RT} - \frac{P_2 V}{RT} = (P_1 - P_2)\left(\frac{V}{RT}\right)$$

$$= (16.0 \text{ atm} - 4.0 \text{ atm})\left(\frac{30.0 \text{ L}}{(0.0820574 \text{ L} \cdot \text{atm} \cdot \text{K}^{-1} \cdot \text{mol}^{-1})(298 \text{ K})}\right)$$

$$= 14.7 \text{ mol}$$

(b) From a table of enthalpies of combustion, the enthalpy of combustion of $H_2$ is found to be $-286 \text{ kJ} \cdot \text{mol}^{-1}$. The energy change is, therefore,

$$(14.7 \text{ mol})(-286 \text{ kJ} \cdot \text{mol}^{-1}) = -4.20 \times 10^3 \text{ kJ}.$$

**6.95** (a) First we must balance the chemical reaction:

$$C_6H_6(l) + \tfrac{15}{2} O_2(g) \longrightarrow 6 CO_2(g) + 3 H_2O(g)$$

For 1 mol $C_6H_6(l)$ burned, the change in the number of moles of gas is

$$\Delta n = (9.00 - 7.50) \text{ mol} = +1.50 \text{ mol}$$

$$w = -P\Delta V = -P\left(\frac{\Delta nRT}{P}\right) = -\Delta nRT$$

$$w = -(+1.50 \text{ mol})(8.314 \text{ J} \cdot \text{mol}^{-1} \cdot \text{K}^{-1})(298 \text{ K}) = -3.72 \times 10^3 \text{ J} = -3.72 \text{ kJ}.$$

(b)  $\Delta H_c = 6(-393.51 \text{ kJ} \cdot \text{mol}^{-1}) + 3(-241.82 \text{ kJ} \cdot \text{mol}^{-1}) - (+49.0 \text{ kJ} \cdot \text{mol}^{-1})$

$\qquad = -3135.5 \text{ kJ}$

(c)  $\Delta U° = \Delta H° + w = (-3135.5 - 6.19) \text{ kJ} = -3141.7 \text{ kJ}$

**6.97** (a) The heat given off by the reaction, which was absorbed by the calorimeter, is given by

$$\Delta H = -(525.0 \text{ J} \cdot (°C)^{-1})(20.0°C - 18.6°C) = -0.74 \text{ kJ}$$

This, however, is not all the heat produced, as the 100.0 mL of solution resulting from mixing also absorbed some heat. If we assume that the volume of NaOH and $HNO_3$ are negligible compared to the volume of water present and that the density of the solution is $1.00 \text{ g} \cdot \text{mL}^{-1}$, then the change in heat of the solution is given by

$$\Delta H = -(20.0°C - 18.6°C)(4.18 \text{ J} \cdot (°C)^{-1} \cdot g^{-1})(100.0 \text{ g}) = -0.59 \text{ kJ}$$

The total heat given off will be $-0.74 \text{ kJ} + (-0.59 \text{ kJ}) = -1.33 \text{ kJ}$

(b) This heat is for the reaction of $(0.500 \text{ M})(0.0500\text{L}) =$ $0.0250 \text{ mol HNO}_3$, so the amount of heat produced per mole of $HNO_3$ will be given by

$$\frac{-1.33 \text{ kJ}}{0.0250 \text{ mol}} = -53.2 \text{ kJ} \cdot \text{mol}^{-1}$$

**6.99**

(a)  (1)    (2)    (3)

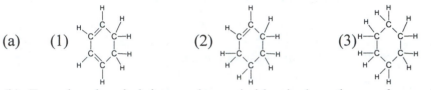

(b) From bond enthalpies, each step is identical, as the number and types of bonds broken and formed are the same:

| | | |
|---|---|---|
| break: | 1 mol C=C bonds | 612 kJ |
| | 1 mol H—H bonds | 436 kJ |
| form: | 1 mol C—C bonds | −348 kJ |
| | 2 mol C—H bonds | 2(−412 kJ) |

| | |
|---|---|
| Total: | −124 kJ |

The total energy change should be equal to the sum of the three steps or $3(-124 \text{ kJ}) = -372 \text{ kJ}$.

(c) The Hess's law calculation using standard enthalpies of formation is easily performed on the composite reaction:

$$C_6H_6(l) + 3 H_2(g) \longrightarrow C_6H_{12}(l)$$
$$\Delta H°_r = \Sigma \Delta H°_f (\text{products}) - \Sigma \Delta H°_f (\text{reactants})$$

$$= \Delta H°_f (\text{cyclohexane}) - \Delta H°_f (\text{benzene})$$
$$= -156.4 \text{ kJ} \cdot \text{mol}^{-1} - (+49.0 \text{ kJ} \cdot \text{mol}^{-1})$$
$$= -205.4 \text{ kJ} \cdot \text{mol}^{-1}$$

(d) The hydrogenation of benzene is much less exothermic than predicted by bond enthalpy estimations. Part of this difference can be due to the inherent inaccuracy of using average values, but the difference is so large that this cannot be the complete explanation. As may be expected, the resonance energy of benzene makes it more stable than would be expected by treating it as a set of three isolated double and three isolated single bonds. The difference in these two values [−205 kJ − (−372 kJ) = 167 kJ] is a measure of how much more stable benzene is than the Kekulé structure would predict.

**6.101** (a) The combustion reaction is

$$C_{60}(s) + 60\ O_2(g) \longrightarrow 60\ CO_2(g)$$

The enthalpy of formation of $C_{60}(s)$ will be given by

$$\Delta H^\circ_c = 60\ \Delta H^\circ_f(CO_2, g) - \Delta H^\circ_f(C_{60}, s)$$
$$-25\ 937\ kJ = 60\ mol \times (-393.51\ kJ \cdot mol^{-1}) - \Delta H^\circ_f(C_{60}, s)$$
$$\Delta H^\circ_f(C_{60}, s) = +2326\ kJ \cdot mol^{-1}$$

(b) The bond enthalpy calculation is

| | |
|---|---|
| $60\ C(gr) \longrightarrow 60\ C(g)$ | $(60)(+717\ kJ \cdot mol^{-1})$ |
| Form 60 mol C—C bonds | $-60(348\ kJ \cdot mol^{-1})$ |
| Form 30 mol C=C bonds | $-30(612\ kJ \cdot mol^{-1})$ |
| $C_{60}(g) \longrightarrow C_{60}(s)$ | $-233\ kJ$ |
| $60\ C(gr) \longrightarrow C_{60}(s)$ | $+3547\ kJ$ |

(c) From the experimental data, the enthalpy of formation of $C_{60}$ shows that it is *more* stable by $(3547\ kJ - 2326\ kJ) = 1221\ kJ$ than predicted by the isolated bond model.

(d) $1221\ kJ \div 60 = 20\ kJ$ per carbon atom

(e) $150\ kJ \div 6 = 25\ kJ$ per carbon atom

(f) Although the comparison of the stabilization of benzene with that of $C_{60}$ should be treated with caution, it does appear that there is slightly less stabilization per carbon atom in $C_{60}$ than in benzene. This fits with expectations, as the $C_{60}$ molecule is forced by its geometry to be curved. This means that the overlap of the $p$-orbitals, which gives rise to the delocalization that results in resonance, will not be as favorable as in the planar benzene molecule. Another perspective on this is obtained by noting that the C atoms in $C_{60}$ are forced to be partially $sp^3$ hybridized because they cannot be rigorously planar as required by $sp^2$ hybridization.

**6.103** The balanced combustion reactions are

$$C_6H_3(NO_2)_3(s) + \tfrac{15}{4}O_2(g) \longrightarrow 6\,CO_2(g) + \tfrac{3}{2}H_2O(l) + \tfrac{3}{2}N_2(g)$$
$$C_6H_3(NH_2)_3(s) + \tfrac{33}{4}O_2(g) \longrightarrow 6\,CO_2(g) + \tfrac{9}{2}H_2O(l) + \tfrac{3}{2}N_2(g)$$

Because the fundamental structures of the two molecules are the same, we need only look at the differences between the two, which in this case are concerned with the groups attached to nitrogen. From the combustion equations we can see that the differences are (1) the consumption of $\tfrac{18}{4}$ more moles of $O_2(g)$ and (2) the production of three more moles of $H_2O(l)$ for the combustion of aniline. Because the $\Delta H°_f$ of $O_2(g)$ is 0, the net difference will be the production of 3 more moles of $H_2O(l)$ or $3 \times (-285.83\ kJ \cdot mol^{-1}) = -857.49\ kJ$.

**6.105** (a) Start by calculating the amount of energy generated by the decomposition of lauric acid:

$$C_{12}H_{24}O_2(s) + 17\,O_2(g) \rightarrow 12\,CO_2(g) + 12\,H_2O(l)$$
$$\Delta H_r = 12(-393.51\ kJ \cdot mol^{-1}) + 12(-285.83\ kJ \cdot mol^{-1}) - (-774.6\ kJ \cdot mol^{-1})$$
$$= -7377.5\ kJ \cdot mol^{-1}$$

if 15.0 g are consumed, then the energy released

is: $-7377.48\ kJ \cdot mol^{-1} \dfrac{15.0\ g}{200.32\ g \cdot mol^{-1}} = -552\ kJ$

The enthalpy for the decomposition of sucrose may be calculated given standard enthalpies of formation:

$$C_{12}H_{21}O_{11}(s) + 12\,O_2(g) \rightarrow 12\,CO_2(g) + 11\,H_2O(l)$$

$$\Delta H_r = 12(-393.51\,kJ \cdot mol^{-1}) + 11(-285.83\,kJ \cdot mol^{-1}) - (-2222\,kJ \cdot mol^{-1})$$

$$= -5644\,kJ \cdot mol^{-1}$$

The amount of sucrose needed to produce the same amount of energy as 15.0 g of lauric acid is:

$$= \frac{552\,kJ}{5644\,kJ \cdot mol^{-1}} = 0.0979\,mol$$

or $(0.0979\,mol)(342.3\,g \cdot mol^{-1}) = 33.5\,g$.

(b) It is more efficient to store energy in the form of fat instead of carbohydrates since, given the same amount of lauric acid (fat) and sucrose (carbohydrate), the lauric acid will produce more energy.

**6.107** (a) Given the enthalpy of combustion, the enthalpy of formation of each compound may be found using:

$$C_4H_4O_4(l) + 3\,O_2(g) \rightarrow 4\,CO_2(g) + 2\,H_2O(l)$$

$$\Delta H_c(X) = 4(-393.51\,kJ \cdot mol^{-1}) + 2(-285.83\,kJ \cdot mol^{-1}) - \Delta H_f(X)$$

$$\Delta H_f(X) = 4(-393.51\,kJ \cdot mol^{-1}) + 2(-285.83\,kJ \cdot mol^{-1}) - \Delta H_c(X)$$

where X is either maleic or fumaric acid. Plugging in the appropriate enthalpies of combustion, the enthalpies of formation are found to be:

$\Delta H_f(\text{maleic acid}) = -790.5\,kJ \cdot mol^{-1}$ and $\Delta H_f(\text{fumaric acid}) = -811.0\,kJ \cdot mol^{-1}$
The enthalpy of the cis-trans isomerization reaction: maleic acid → fumaric acid is:

$$\Delta H_f = \Delta H_f(\text{fumaric acid}) - \Delta H_f(\text{maleic acid})$$

$$= (-811.0\,kJ \cdot mol^{-1}) - (-790.5\,kJ \cdot mol^{-1}) = -20.5\,kJ \cdot mol^{-1}.$$

(b) Fumaric acid has the lower enthalpy of formation.

(c) Because one mole of gas is produced for each mole of acid consumed, work is done on the surroundings by the system under constant pressure conditions, *i.e.* work for the system will be negative. Therefore, $U_c$ will be less negative than $H_c$.

**6.109** (a) $V_{init} = \dfrac{nRT}{P} = \dfrac{(0.060 \text{ mol})(0.0820578 \text{ L} \cdot \text{atm} \cdot \text{K}^{-1} \cdot \text{mol}^{-1})(298.15 \text{ K})}{1.00 \text{ atm}}$

$= 1.5 \text{ L}$

(b) The combustion reaction is: $2 \text{ SO}_2(g) + \text{O}_2(g) \rightarrow 2 \text{ SO}_3(g)$. If equal molar amounts of $SO_2$ and $O_2$ are mixed, as in this case, $SO_2$ is the limiting reagent.

(c) The total number of moles remaining in the container will be: $0.030$ mol $SO_3(g)$ + $0.015$ mol $O_2(g)$ = $0.045$ mol of gas at the end of the reaction. The final volume will, therefore, be:

$V_f = \dfrac{nRT}{P} = \dfrac{(0.045 \text{ mol})(0.0820578 \text{ L} \cdot \text{atm} \cdot \text{K}^{-1} \cdot \text{mol}^{-1})(298.15 \text{ K})}{1.00 \text{ atm}}$

$= 1.1 \text{ L}$

$\Delta V = 1.1 \text{ L} - 1.5 \text{ L} = -0.4 \text{ L}$

(d) $w = -P\Delta V = (1.00 \text{ atm})(-0.4 \text{ L})(101.325 \text{ J} \cdot \text{L}^{-1} \cdot \text{atm}^{-1})$

$= 40 \text{ J}$ of work done on the system (work is positive)

(e) The enthalpy of reaction may be found using standard enthalpies of formation and the balanced equation given above:

$\Delta H_r = 2(-395.72 \text{ kJ} \cdot \text{mol}^{-1}) - 2(-296.83 \text{ kJ} \cdot \text{mol}^{-1}) = -197.78 \text{ kJ} \cdot \text{mol}^{-1}$.
If $0.030$ mol of $SO_2$ are consumed, then enthalpy change is:

$(0.030 \text{ mol SO}_2)\left(\dfrac{-197.78 \text{ kJ}}{2 \text{ mol SO}_2}\right) = -3.0 \text{ kJ} = -3.0 \times 10^3 \text{ J}$.

(f) $\Delta U_r = q + w = -2970 \text{ J} + 37.2 \text{ J} = 2930 \text{ J}$

# CHAPTER 7

# THERMODYNAMICS: THE SECOND AND THIRD LAWS

**7.1** (a) rate of entropy generation $= \dfrac{\Delta S_{surr}}{time} = -\dfrac{q_{rev}}{time \cdot T}$

$$= -\dfrac{\text{rate of heat generation}}{T}$$

$$= \dfrac{-(100. \, \text{J} \cdot \text{s}^{-1})}{293 \, \text{K}} = 0.341 \, \text{J} \cdot \text{K}^{-1} \cdot \text{s}^{-1}$$

(b) $\Delta S_{day} = (0.341 \, \text{J} \cdot \text{K}^{-1} \cdot \text{s}^{-1})(60 \, \sec \cdot \min^{-1})(60 \, \min \cdot \text{hr}^{-1})(24 \, \text{hr} \cdot \text{day}^{-1})$

$$= 29.5 \, \text{kJ} \cdot \text{K}^{-1} \cdot \text{day}^{-1}$$

(c) Less, because in the equation $\Delta S = \dfrac{-\Delta H}{T}$, if $T$ is larger, $\Delta S$ is smaller.

**7.3** (a) $\Delta S = \dfrac{q_{rev}}{T} = \dfrac{65 \, \text{J}}{298 \, \text{K}} = 0.22 \, \text{J} \cdot \text{K}^{-1}$

(b) $\Delta S = \dfrac{65 \, \text{J}}{373 \, \text{K}} = 0.17 \, \text{J} \cdot \text{K}^{-1}$

(c) The entropy change is smaller at higher temperatures, because the matter is already more chaotic. The same amount of heat has a greater effect on entropy changes when transferred at lower temperatures.

**7.5** (a) The relationship to use is $dS = \dfrac{dq}{T}$. At constant pressure, we can

substitute

$dq = n \, C_P dT:$

$dS = \dfrac{n \, C_p \, dT}{T}$

Upon integration, this gives $\Delta S = n\,C_p\,\ln\dfrac{T_2}{T_1}$. The answer is calculated by simply plugging in the known quantities. Remember that for an ideal monatomic gas

$C_P = \frac{5}{2}R$ :

$$\Delta S = (1.00\ \text{mol})\left(\tfrac{5}{2}\times 8.314\ \text{J}\cdot\text{K}^{-1}\cdot\text{mol}^{-1}\right)\ln\dfrac{431.0\ \text{K}}{310.8\ \text{K}} = 6.80\ \text{J}\cdot\text{K}^{-1}$$

(b) A similar analysis using $C_V$ gives $\Delta S = n\,C_V\,\ln\dfrac{T_2}{T_1}$, where $C_V$ for a monatomic ideal gas is $\frac{3}{2}R$ :

$$\Delta S = (1.00\ \text{mol})\left(\tfrac{3}{2}\times 8.314\ \text{J}\cdot\text{K}^{-1}\cdot\text{mol}^{-1}\right)\ln\dfrac{431.0\ \text{K}}{310.8\ \text{K}} = 4.08\ \text{J}\cdot\text{K}^{-1}$$

**7.7** Because the process is isothermal and reversible, the relationship $dS = \dfrac{dq}{T}$ can be used. Because the process is isothermal, $\Delta U = 0$ and hence $q = -w$, where $w = -P\,dV$. Making this substitution, we obtain

$$dS = \dfrac{P\,dV}{T} = \dfrac{nRT}{TV}\,dV = \dfrac{nR}{V}\,dV$$

$$\therefore \Delta S = nR\,\ln\dfrac{V_2}{V_1}$$

Substituting the known quantities, we obtain

$$\Delta S = (5.25\ \text{mol})(8.314\ \text{J}\cdot\text{K}^{-1}\cdot\text{mol}^{-1})\ln\dfrac{34.058\ \text{L}}{24.252\ \text{L}}$$

$$= 14.8\ \text{J}\cdot\text{K}^{-1}$$

**7.9** The change in entropy for each block (block 1 and block 2) are:

$$\Delta S_1 = \dfrac{q_1}{T_1} \quad \text{and} \quad \Delta S_2 = \dfrac{q_2}{T_2}$$ energy is transferred, then the change in entropy for the system is:

$$\Delta S = \Delta S_1 + \Delta S_2 = \dfrac{q_1}{T_1} + \dfrac{q_2}{T_2}$$

If 1 J of energy is transferred from block 2 to block 1,

$$q_1 = +1 \text{ J}, q_2 = -1 \text{ J}, \text{ and } \Delta S = \frac{1}{T_1} - \frac{1}{T_2}.$$

For the transfer of heat from block 2 to block 1 to be spontaneous, $\Delta S$ must be positive and, therefore, $T_2$ would have to be greater than $T_1$.

**7.11** (a) $\Delta S° = \dfrac{q}{T} = \dfrac{\Delta H°}{T} = \dfrac{1.00 \text{ mol} \times (-6.01 \text{ kJ} \cdot \text{mol}^{-1})}{273.2 \text{ K}} = -22.0 \text{ J} \cdot \text{K}^{-1}$

(b) $\Delta S = \dfrac{q}{T} = \dfrac{\Delta H}{T} = \dfrac{\dfrac{50.0 \text{ g}}{46.07 \text{ g} \cdot \text{mol}^{-1}} \times 43.5 \text{ kJ} \cdot \text{mol}^{-1}}{351.5 \text{ K}} = +134 \text{ J} \cdot \text{K}^{-1}$

**7.13** (a) The boiling point of a liquid may be obtained from the relationship

$\Delta S_{\text{vap}} = \dfrac{\Delta H_{\text{vap}}}{T_B}$, or $T_B = \dfrac{\Delta H_{\text{vap}}}{\Delta S_{\text{vap}}}$. This relationship should be rigorously true if we have the actual enthalpy and entropy of vaporization. The data in the Appendix, however, are for 298 K. Thus, calculation of $\Delta H°_{\text{vap}}$ or $\Delta S°_{\text{vap}}$, using the enthalpy and entropy differences between the gas and liquid forms at 298 K, give a good approximation of these quantities but the values are not exact. For ethanal(l) $\longrightarrow$ ethanal(g), the data in the appendix give

$\Delta H_{\text{vap}} \cong -166.19 \text{ kJ} \cdot \text{mol}^{-1} - (-192.30) \text{ kJ} \cdot \text{mol}^{-1} = 26.11 \text{ kJ} \cdot \text{mol}^{-1}$

$\Delta S_{\text{vap}} \cong 250.3 \text{ J} \cdot \text{K}^{-1} \cdot \text{mol}^{-1} - 160.2 \text{ J} \cdot \text{K}^{-1} \cdot \text{mol}^{-1} = 90.1 \text{ J} \cdot \text{K}^{-1} \cdot \text{mol}^{-1}$

$T_B = \dfrac{26.11 \times 10^3 \text{ J} \cdot \text{mol}^{-1}}{90.1 \text{ J} \cdot \text{K}^{-1} \cdot \text{mol}^{-1}} = 290. \text{ K}$

(b) The boiling point of ethanal is 20.8°C or 293.9 K.

(c) These numbers are in very good agreement.

(d) Differences arise partly because the enthalpy and entropy of vaporization are slightly different from the values calculated at 298 K, but the boiling point of ethanal is not 298 K.

**7.15**  (a) Trouton s rule indicates that the entropy of vaporization for a number of organic liquids is approximately $85\,\text{J}\cdot\text{K}^{-1}\cdot\text{mol}^{-1}$. Using this information and the relationship

$$T_{\text{B}} = \frac{\Delta H^{\circ}_{\text{vap}}}{\Delta S^{\circ}_{\text{vap}}} = \frac{21.51\times10^{3}\,\text{J}\cdot\text{mol}^{-1}}{85\,\text{J}\cdot\text{K}^{-1}\cdot\text{mol}^{-1}} = 253\text{ K}.$$

(b) The experimental boiling point of dimethyl ether is 248 K, which is in reasonably close agreement, given the nature of the approximation.

**7.17**  (a)  The value can be estimated from

$$\Delta H^{\circ}_{\text{vap}} = T\Delta S^{\circ}_{\text{vap}}$$

$$\Delta H^{\circ}_{\text{vap}} = (353\text{ K})(85\,\text{J}\cdot\text{mol}^{-1}\cdot\text{K}^{-1})$$

$$= +30.\,\text{kJ}\cdot\text{mol}^{-1}$$

(b)  $\Delta S^{\circ}_{\text{surr}} = -\dfrac{\Delta H^{\circ}_{\text{system}}}{T}$

$$\Delta S_{\text{surr}} = -\left(\frac{10\text{ g}}{78.11\,\text{g}\cdot\text{mol}^{-1}}\right)\left(\frac{30\,\text{kJ}\cdot\text{mol}^{-1}}{353\text{ K}}\right) = -11\,\text{J}\cdot\text{K}^{-1}$$

**7.19**  $COF_2$. $COF_2$ and $BF_3$ are both trigonal planar molecules, but it would be possible for the molecule to be disordered with the fluorine and oxygen atoms occupying the same locations. Because all the groups attached to boron are identical, such disorder is not possible.

**7.21**  There are six orientations of an $SO_2F_2$ molecule as shown below:

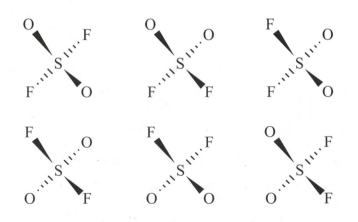

The Boltzmann expression for one mole of $SO_2F_2$ molecules having six possible orientations is

$$S = k \ln 6^{6.02 \times 10^{23}} = (1.38 \times 10^{-23} \; J \cdot K^{-1}) \ln 6^{6.02 \times 10^{23}}$$
$$S = 14.9 \; J \cdot K^{-1}$$

**7.23** (a) HBr(g), because Br is more massive and contains more elementary particles than F in HF; (b) $NH_3$(g), because it has greater complexity, being a molecule rather than a single atom; (c) $I_2$ (l), because molecules in liquids are more randomly oriented than molecules in solids; (d) 1.0 mol Ar(g) at 1.00 atm, because it will occupy a larger volume than 1.0 mol of Ar(g) at 2.00 atm.

**7.25** It is easy to order $H_2O$ in its various phases because entropy will increase when going from a solid to a liquid to a gas. The main question concerns where to place C(s, diamond) in this order, and that will essentially become a question of whether C(s, diamond) should have more or less entropy than $H_2O$(s), because we would automatically expect C(s, diamond) to have less entropy than any liquid. Because water is a molecular substance held together in the solid phase by weak hydrogen bonds, and in C(s, diamond) the carbon is more rigidly held in place and will have less entropy.
In summary, C(s, diamond) < $H_2O$(s) < $H_2O$(l) < $H_2O$(g).

**7.27** (a) In the standard state, bromine is a liquid and iodine is a solid. If the two substances had the same state (i.e., both were gases or both liquids) we would expect iodine to have the higher entropy due to its larger mass and consequently larger number of fundamental particles. However, because the compounds are in different states, we would expect the liquid to have a higher entropy than the solid.

(b) When we consider the two structures, it is clear that 1-pentene will have more flexibility in its framework than cyclopentane, which will be comparatively rigid. Therefore, we predict 1-pentene to have a higher entropy.

(c) Ethene (or ethylene) is a gas and polyethylene is a solid, so we automatically expect ethene to have a higher entropy. Also, for the same mass, a sample of ethene will be composed of many small molecules, whereas polyethylene will be made up of fewer but larger molecules.

**7.29** (a) Entropy should decrease because the number of moles of gas is less on the product side of the reaction.

(b) Entropy should increase because the dissolution of the solid copper phosphate will increase the randomness of the copper and phosphate ions.

(c) Entropy should decrease as the total number of moles decreases.

**7.31** $\Delta S_A > \Delta S_C > \Delta S_B$. The change in entropy for container A is greater than that for container B or C due to the greater number of particles. The change in entropy in container C is greater than that of container B because of the disorder due to the vibrational motion of the molecules in container C.

**7.33** (a) $H_2(g) + \frac{1}{2}O_2(g) \longrightarrow H_2O(l)$

$\Delta S°_f = \sum S°_{products} - \sum S°_{reactants}$

$= 69.91 \text{ J} \cdot \text{K}^{-1} \cdot \text{mol}^{-1} - [130.68 \text{ J} \cdot \text{K}^{-1} \cdot \text{mol}^{-1} + \frac{1}{2}(205.14 \text{ J} \cdot \text{K}^{-1} \cdot \text{mol}^{-1})]$

$= -163.34 \text{ J} \cdot \text{K}^{-1} \cdot \text{mol}^{-1}$

The entropy change is negative because the number of moles of gas has decreased by 1.5. Note that the absolute entropies of the elements are not 0, and that the entropy change for the reaction in which a compound is formed from the elements is also not 0.

(b) $CO(g) + \frac{1}{2}O_2(g) \longrightarrow CO_2(g)$

$$\Delta S_r^\circ = 213.74 \text{ J} \cdot \text{K}^{-1} \cdot \text{mol}^{-1}$$

$$- [197.67 \text{ J} \cdot \text{K}^{-1} \cdot \text{mol}^{-1} + \tfrac{1}{2}(205.14 \text{ J} \cdot \text{K}^{-1} \cdot \text{mol}^{-1})]$$

$$= -86.5 \text{ J} \cdot \text{K}^{-1} \cdot \text{mol}^{-1}$$

The entropy change is negative because the number of moles of gas has decreased by 0.5.

(c) $CaCO_3(s) \longrightarrow CaO(s) + CO_2(g)$

$$\Delta S_r^\circ = 39.75 \text{ J} \cdot \text{K}^{-1} \cdot \text{mol}^{-1} + 213.74 \text{ J} \cdot \text{K}^{-1} \cdot \text{mol}^{-1} - [+92.9 \text{ J} \cdot \text{K}^{-1} \cdot \text{mol}^{-1}]$$

$$= +160.6 \text{ J} \cdot \text{K}^{-1} \cdot \text{mol}^{-1}$$

The entropy change is positive because the number of moles of gas has increased by 1.

(d) $4 \text{ KClO}_3(s) \longrightarrow 3 \text{ KClO}_4(s) + \text{KCl}(s)$

$$\Delta S_r^\circ = 3(151.0 \text{ J} \cdot \text{K}^{-1} \cdot \text{mol}^{-1}) + 82.59 \text{ J} \cdot \text{K}^{-1} \cdot \text{mol}^{-1}$$

$$- [4(143.1 \text{ J} \cdot \text{K}^{-1} \cdot \text{mol}^{-1})]$$

$$= -36.8 \text{ J} \cdot \text{K}^{-1} \cdot \text{mol}^{-1}$$

It is not immediately obvious, but the four moles of solid products are more ordered than the four moles of solid reactants.

**7.35**  $dS = \dfrac{dq_{rev}}{T} = \dfrac{C_{p,m} \cdot dT}{T}$ and, therefore, $\Delta S = \displaystyle\int_{T_1}^{T_2} \dfrac{C_{p,m}}{T} \, dT$.

if $C_{p,m} = a + bT + c/T^2$ then

$$\Delta S = \int_{T_1}^{T_2} \frac{a + bT + c/T^2}{T} \, dT$$

$$= \int_{T_1}^{T_2} \left( \frac{a}{T} + b + \frac{c}{T^3} \right) dT = a \ln(T) + bT - \frac{c}{2T^2} \Bigg|_{T_1}^{T_2}$$

$$= a \ln\left( \frac{T_2}{T_1} \right) + b(T_2 - T_1) - \frac{c}{2}\left( \frac{1}{T_2^2} - \frac{1}{T_1^2} \right)$$

$\Delta S$(true) for heating graphite from 298 K to 400 K is

$$= (16.86 \text{ J} \cdot \text{K}^{-1} \cdot \text{mol}^{-1}) \ln\left(\frac{400 \text{ K}}{298 \text{ K}}\right)$$

$$+ (0.00477 \text{ J} \cdot \text{K}^{-2} \cdot \text{mol}^{-1})(400 \text{ K} - 298 \text{ K})$$

$$- \frac{(-8.54 \times 10^5 \text{ J} \cdot \text{K} \cdot \text{mol}^{-1})}{2}\left(\frac{1}{(400 \text{ K})^2} - \frac{1}{(298 \text{ K})^2}\right)$$

$$= 3.31 \text{ J} \cdot \text{K}^{-1} \cdot \text{mol}^{-1}$$

If we assume a constant heat capacity at the mean temperature of 350 K:

$$C_{p,m} = (16.86 \text{ J} \cdot \text{K}^{-1} \cdot \text{mol}^{-1}) + (0.00477 \text{ J} \cdot \text{K}^{-2} \cdot \text{mol}^{-1})(350 \text{ K})$$

$$+ \frac{(-8.54 \times 10^5 \text{ J} \cdot \text{K} \cdot \text{mol}^{-1})}{(350 \text{ K})^2}$$

$$= 11.6 \text{ J} \cdot \text{K}^{-1} \cdot \text{mol}^{-1}$$

and

$$\Delta S(\text{mean}) = C_{p,m} \ln\frac{T_2}{T_1} = (11.6 \text{ J} \cdot \text{K}^{-1} \cdot \text{mol}^{-1}) \ln\left(\frac{400 \text{ K}}{298 \text{ K}}\right) = 3.41 \text{ J} \cdot \text{K}^{-1} \cdot \text{mol}^{-1}$$

Not integrating the heat capacity leads to roughly a 3.0% error in $\Delta S$.

**7.37** The entropy of vaporization of water at 85 °C may be carried out through a series of three reversible steps. Namely, reversibly heating the reactants to 100 °C, carrying out the phase change at this temperature, and finally cooling the products back to 85 °C. The sum of the $\Delta S$'s for these three steps will be equivalent to vaporizing water at 85 °C in one irreversible step.

Step 1, heating the reactants to 100 °C:

$$\Delta S_1 = C_{p,m} \ln\left(\frac{T_2}{T_1}\right) = (75.3 \text{ J} \cdot \text{K}^{-1} \cdot \text{mol}^{-1}) \ln\left(\frac{373 \text{ K}}{358 \text{ K}}\right) = 3.09 \text{ J} \cdot \text{K}^{-1} \cdot \text{mol}^{-1}$$

Step 2, the entropy of vaporization of $H_2O$ at 100 °C is

109.0 $\text{J} \cdot \text{K}^{-1} \cdot \text{mol}^{-1}$.

Step 3, cooling the products to 85 °C:

$$\Delta S_3 = C_{p,m} \ln\left(\frac{T_2}{T_1}\right) = \left(33.6 \text{ J} \cdot \text{K}^{-1} \cdot \text{mol}^{-1}\right)\ln\left(\frac{358 \text{ K}}{373 \text{ K}}\right) = -1.38 \text{ J} \cdot \text{K}^{-1} \cdot \text{mol}^{-1}$$

Therefore, the molar entropy of vaporization is $H_2O$ at 85 °C is:

$$\Delta S_{v,m} = \Delta S_1 + \Delta S_2 + \Delta S_3 = 111 \text{ J} \cdot \text{K}^{-1} \cdot \text{mol}^{-1}.$$

**7.39** To calculate the change in entropy for the hot and cold water, the amount of energy which flows from one to the other must first be calculated:

$$q_c = -q_h$$
$$= n_c \cdot C_{p,m}(H_2O) \cdot (T_f - T_{i,c}) = -n_h \cdot C_{p,m}(H_2O) \cdot (T_f - T_{i,h})$$

where $n_c$ and $n_h$ are the moles of cold and hot water, respectively, and $T_{i,c}$ and $T_{i,h}$ are the temperatures of the cold and hot water, respectively.

Dividing both sides by $C_{p,m}(H_2O)$ we obtain:

$$n_c \cdot (T_f - T_{i,c}) = -n_h \cdot (T_f - T_{i,h}).$$

The moles of hot and cold water are:

$$n_c = \frac{50 \text{ g}}{18.015 \text{ g} \cdot \text{mol}^{-1}} = 2.78 \text{ mol and } n_h = \frac{65 \text{ g}}{18.015 \text{ g} \cdot \text{mol}^{-1}} = 3.61 \text{ mol}$$

and $T_f$ is therefore:

$$T_f = \frac{(2.78 \text{ mol} \cdot 293.15 \text{ K}) + (3.61 \text{ mol} \cdot 323.15 \text{ K})}{(2.78 \text{ mol} + 3.61 \text{ mol})} = 310.1 \text{ K}.$$

With $T_f$, we can calculate $\Delta S$ for the hot and cold water, and the total $\Delta S$ for the entire system:

$$\Delta S_c = n_c \cdot C_{p,m} \ln\left(\frac{T_f}{T_{i,c}}\right) = (2.78 \text{ mol})\left(75.3 \text{ J} \cdot \text{K}^{-1} \cdot \text{mol}^{-1}\right)\ln\left(\frac{310.1 \text{ K}}{293 \text{ K}}\right)$$
$$= +11.9 \text{ J} \cdot \text{K}^{-1}$$

$$\Delta S_h = n_h \cdot C_{p,m} \ln\left(\frac{T_f}{T_{i,h}}\right) = (3.61 \text{ mol})\left(75.3 \text{ J} \cdot \text{K}^{-1} \cdot \text{mol}^{-1}\right)\ln\left(\frac{310.1 \text{ K}}{323 \text{ K}}\right)$$
$$= -11.1 \text{ J} \cdot \text{K}^{-1}$$

Therefore, $\Delta S_{tot}$ is :

$$\Delta S_{tot} = \Delta S_c + \Delta S_h = +0.8 \text{ J} \cdot \text{K}^{-1}.$$

**7.41**  (a)  The change in entropy will be given by

(b)  $\Delta S_{surr} = \dfrac{-\Delta H_{system}}{T} = \dfrac{-1.00 \text{ mol} \times 4.60 \times 10^3 \text{ J} \cdot \text{mol}^{-1}}{158.7 \text{ K}} = -29.0 \text{ J} \cdot \text{K}^{-1}$

$\Delta S_{system} = \dfrac{\Delta H_{system}}{T} = \dfrac{1.00 \text{ mol} \times 4.60 \times 10^3 \text{ J} \cdot \text{mol}^{-1}}{158.7 \text{ K}} = +29.0 \text{ J} \cdot \text{K}^{-1}$

(c)  $\Delta S_{surr} = \dfrac{-\Delta H_{system}}{T} = \dfrac{-(1.00 \text{ mol} \times -4.60 \times 10^3 \text{ J} \cdot \text{mol}^{-1})}{158.7 \text{ K}}$

$= +29.0 \text{ J} \cdot \text{K}^{-1}$

$\Delta S_{system} = \dfrac{\Delta H_{system}}{T} = \dfrac{1.00 \text{ mol} \times -4.60 \times 10^3 \text{ J} \cdot \text{mol}^{-1}}{158.7 \text{ K}} = -29.0 \text{ J} \cdot \text{K}^{-1}$

**7.43**  (a)  The total entropy change is given by $\Delta S_{tot} = \Delta S_{surr} + \Delta S. \Delta S$ for an

isothermal, reversible process is calculated from

$\Delta S = \dfrac{q_{rev}}{T} = \dfrac{-w_{rev}}{T} = nR \ln \dfrac{V_2}{V_1}$. To do the calculation we need the value of

$n$, which is obtained by use of the ideal gas law:

$(4.95 \text{ atm})(1.67 \text{ L}) = n(0.082\,06 \text{ L} \cdot \text{atm} \cdot \text{K}^{-1} \cdot \text{mol}^{-1})(323 \text{ K}); n = 0.312 \text{ mol}.$

$\Delta S = (0.312 \text{ mol})(8.314 \text{ J} \cdot \text{K}^{-1} \cdot \text{mol}^{-1}) \ln \dfrac{7.33 \text{ L}}{1.67 \text{ L}} = +3.84 \text{ J} \cdot \text{K}^{-1}$. Because the

process is reversible, $\Delta S_{tot} = 0$, so $\Delta S_{surr} = -\Delta S = -3.84 \text{ J} \cdot \text{K}^{-1}$.

(b)  For the irreversible process, $\Delta S$ is the same, $+3.84 \text{ J} \cdot \text{K}^{-1}$. No work is

done in free expansion (see Section 6.6) so $w = 0$. Because $\Delta U = 0$, it

follows that $q = 0$. Therefore, no heat is transferred into the surroundings,

and their entropy is unchanged: $\Delta S_{surr} = 0$. The total change in entropy is

therefore $\Delta S_{tot} = +3.84 \text{ J} \cdot \text{K}^{-1}$.

**7.45**  Exothermic reactions tend to be spontaneous because the result is an

increase in the entropy of the surroundings. Using the mathematical

relationship $\Delta G_r = \Delta H_r - T\Delta S_r$, it is clear that if $\Delta H_r$ is large and negative compared to $\Delta S_r$, then the reaction will generally be spontaneous.

**7.47**  (a)  $\Delta H°_r = 3(-824.2 \text{ kJ} \cdot \text{mol}^{-1}) - [2(-1118.4 \text{ kJ} \cdot \text{mol}^{-1})]$

$= -235.8 \text{ kJ} \cdot \text{mol}^{-1}$

$\Delta S°_r = 3(87.40 \text{ J} \cdot \text{K}^{-1} \cdot \text{mol}^{-1}) - [2(146.4 \text{ J} \cdot \text{K}^{-1} \cdot \text{mol}^{-1})$

$+ \frac{1}{2}(205.14 \text{ J} \cdot \text{K}^{-1} \cdot \text{mol}^{-1})]$

$= -133.17 \text{ J} \cdot \text{K}^{-1} \cdot \text{mol}^{-1}$

$\Delta G°_r = 3(-742.2 \text{ kJ} \cdot \text{mol}^{-1}) - [2(-1015.4 \text{ kJ} \cdot \text{mol}^{-1})]$

$= -195.8 \text{ kJ} \cdot \text{mol}^{-1}$

$\Delta G_r$ may also be calculated from $\Delta H°_r$ and $\Delta S°_r$ (the numbers calculated differ slightly from the two methods due to rounding differences):

$\Delta G°_r = \Delta H°_r - T\Delta S°_r$

$= -235.8 \text{ kJ} \cdot \text{mol}^{-1} - (298 \text{ K})(-133.17 \text{ J} \cdot \text{K}^{-1} \cdot \text{mol}^{-1})/(100 \text{ J} \cdot \text{kJ}^{-1})$

$= -196.1 \text{ kJ} \cdot \text{mol}^{-1}$

(b)  $\Delta H°_r = -1208.09 \text{ kJ} \cdot \text{mol}^{-1} - [-1219.6 \text{ kJ} \cdot \text{mol}^{-1}]$

$= 11.5 \text{ kJ} \cdot \text{mol}^{-1}$

$\Delta S°_r = -80.8 \text{ J} \cdot \text{K}^{-1} \cdot \text{mol}^{-1} - [68.87 \text{ J} \cdot \text{K}^{-1} \cdot \text{mol}^{-1}]$

$= -149.7 \text{ J} \cdot \text{K}^{-1} \cdot \text{mol}^{-1}$

$\Delta G°_r = -1111.15 \text{ kJ} \cdot \text{mol}^{-1} - [-1167.3 \text{ kJ} \cdot \text{mol}^{-1}]$

$= +56.2 \text{ kJ} \cdot \text{mol}^{-1}$

or

$\Delta G°_r = \Delta H°_r - T\Delta S°_r$

$= 11.5 \text{ kJ} \cdot \text{mol}^{-1} - (298 \text{ K})(-149.7 \text{ J} \cdot \text{K}^{-1} \cdot \text{mol}^{-1})/(1000 \text{ J} \cdot \text{kJ}^{-1})$

$= 56.1 \text{ kJ} \cdot \text{mol}^{-1}$

(c)  $\Delta H°_r = 9.16 \text{ kJ} \cdot \text{mol}^{-1} - [2(33.18 \text{ kJ} \cdot \text{mol}^{-1})] = -57.20 \text{ kJ} \cdot \text{mol}^{-1}$

$\Delta S°_r = 304.29 \text{ J} \cdot \text{K}^{-1} \cdot \text{mol}^{-1} - [2(240.06 \text{ J} \cdot \text{K}^{-1} \cdot \text{mol}^{-1})]$

$= -175.83 \text{ J} \cdot \text{K}^{-1} \cdot \text{mol}^{-1}$

$\Delta G°_r = 97.89 \text{ kJ} \cdot \text{mol}^{-1} - [2(51.31 \text{ kJ} \cdot \text{mol}^{-1})] = -4.73 \text{ kJ} \cdot \text{mol}^{-1}$

or

$$\Delta G^\circ_r = \Delta H^\circ_r - T\Delta S^\circ_r$$
$$= -57.2 \text{ kJ} \cdot \text{mol}^{-1} - (298 \text{ K})(-175.83 \text{ J} \cdot \text{K}^{-1} \cdot \text{mol}^{-1})/(1000 \text{ J} \cdot \text{kJ}^{-1})$$
$$= -4.80 \text{ kJ} \cdot \text{mol}^{-1}$$

**7.49** (a) $\frac{1}{2}N_2(g) + \frac{3}{2}H_2(g) \longrightarrow NH_3(g)$

$$\Delta H^\circ_r = \Delta H^\circ_f(NH_3) = -46.11 \text{ kJ} \cdot \text{mol}^{-1}$$
$$\Delta S^\circ_r = S^\circ_m(NH_3, g) - [\frac{1}{2}S^\circ_m(N_2, g) + \frac{3}{2}S^\circ_m(H_2, g)]$$
$$= 192.45 \text{ J} \cdot \text{K}^{-1} \cdot \text{mol}^{-1} - [\frac{1}{2}(191.61 \text{ J} \cdot \text{K}^{-1} \cdot \text{mol}^{-1})$$
$$+ \frac{3}{2}(130.68 \text{ J} \cdot \text{K}^{-1} \cdot \text{mol}^{-1})]$$
$$= -99.38 \text{ J} \cdot \text{K}^{-1} \cdot \text{mol}^{-1}$$

$$\Delta G^\circ_r = -46.11 \text{ kJ} \cdot \text{mol}^{-1} - (298 \text{ K})(-99.38 \text{ J} \cdot \text{K}^{-1} \cdot \text{mol}^{-1})/(1000 \text{ J} \cdot \text{kJ}^{-1})$$
$$= -16.49 \text{ kJ} \cdot \text{mol}^{-1}$$

$S^\circ_m(NH_3) = 192.45 \text{ J} \cdot \text{K}^{-1} \cdot \text{mol}^{-1}$

$\Delta S^\circ_f(NH_3)$ is negative because several gas molecules combine to form 1

$NH_3$ molecule.

(b) $H_2(g) + \frac{1}{2}O_2(g) \longrightarrow H_2O(g)$

$$\Delta H^\circ_r = \Delta H^\circ_f(H_2O, g) = -241.82 \text{ kJ} \cdot \text{mol}^{-1}$$
$$\Delta S^\circ_r = S^\circ_m(H_2O, g) - [S^\circ_m(H_2, g) + \frac{1}{2}S^\circ_m(O_2, g)]$$
$$= 188.83 \text{ J} \cdot \text{K}^{-1} \cdot \text{mol}^{-1} - [130.68 \text{ J} \cdot \text{K}^{-1} \cdot \text{mol}^{-1} + \frac{1}{2}(205.14 \text{ J} \cdot \text{K}^{-1} \cdot \text{mol}^{-1})]$$
$$= -44.42 \text{ J} \cdot \text{K}^{-1} \cdot \text{mol}^{-1}$$
$$\Delta G^\circ_r = -241.82 \text{ kJ} \cdot \text{mol}^{-1} - (298 \text{ K})(-44.42 \text{ J} \cdot \text{K}^{-1} \cdot \text{mol}^{-1})/(1000 \text{ J} \cdot \text{kJ}^{-1})$$
$$= -228.58 \text{ kJ} \cdot \text{mol}^{-1}$$

$S^\circ_m(H_2O, g) = 188.83 \text{ J} \cdot \text{K}^{-1} \cdot \text{mol}^{-1}$

$\Delta S^\circ_f(H_2O, g)$ is a negative number because there is a reduction in the

number of gas molecules in the reaction when $S^\circ_m$ is positive.

(c) $C(s, \text{graphite}) + \frac{1}{2}O_2(g) \longrightarrow CO(g)$

$$\Delta H°_r = \Delta H°_f(CO, g) = -110.53 \text{ kJ} \cdot \text{mol}^{-1}$$

$$\Delta S°_r = S°_m(CO, g) - [S°_m(C, s) + \tfrac{1}{2}S°_m(O_2, g)]$$

$$= 197.67 \text{ J} \cdot \text{K}^{-1} \cdot \text{mol}^{-1} - [5.740 \text{ J} \cdot \text{K}^{-1} \cdot \text{mol}^{-1} + \tfrac{1}{2}(205.14 \text{ J} \cdot \text{K}^{-1} \cdot \text{mol}^{-1})]$$

$$= +89.36 \text{ J} \cdot \text{K}^{-1} \cdot \text{mol}^{-1}$$

$$\Delta G°_r = -110.53 \text{ kJ} \cdot \text{mol}^{-1} - (298 \text{ K})(89.36 \text{ J} \cdot \text{K}^{-1} \cdot \text{mol}^{-1})/(1000 \text{ J} \cdot \text{kJ}^{-1})$$

$$= -137.2 \text{ kJ} \cdot \text{mol}^{-1}$$

$$S°_m(CO, g) = 197.67 \text{ J} \cdot \text{K}^{-1} \cdot \text{mol}^{-1}$$

The $S°_m(CO, g)$ is larger than $\Delta S°_f(CO, g)$ because in the formation reaction the number of moles of gas is reduced.

(d) $\tfrac{1}{2} N_2(g) + O_2(g) \longrightarrow NO_2(g)$

$$\Delta H°_r = \Delta H°_f(NO_2) = +33.18 \text{ kJ} \cdot \text{mol}^{-1}$$

$$\Delta S°_r = S°_m(NO_2, g) - [\tfrac{1}{2}S°_m(N_2, g) + S°_m(O_2, g)]$$

$$= 240.06 \text{ J} \cdot \text{K}^{-1} \cdot \text{mol}^{-1} - [\tfrac{1}{2}(191.61 \text{ J} \cdot \text{K}^{-1} \cdot \text{mol}^{-1}) + 205.14 \text{ J} \cdot \text{K}^{-1} \cdot \text{mol}^{-1}]$$

$$= -60.89 \text{ J} \cdot \text{K}^{-1} \cdot \text{mol}^{-1}$$

$$\Delta G°_r = 33.18 \text{ kJ} \cdot \text{mol}^{-1} - (298 \text{ K})(-60.89 \text{ J} \cdot \text{K}^{-1} \cdot \text{mol}^{-1})/(1000 \text{ J} \cdot \text{kJ}^{-1})$$

$$= +51.33 \text{ kJ} \cdot \text{mol}^{-1}$$

$$S°_m(NO_2, g) = 240.06 \text{ J} \cdot \text{K}^{-1} \cdot \text{mol}^{-1}$$

The $\Delta S°_f(NO_2, g)$ is somewhat negative due to the reduction in the number of gas molecules during the reaction. For all of these, the important point to gain is that the $S°_m$ value of a compound is not the same as the $\Delta S°_f$ for the formation of that compound. $\Delta S°_f$ is often negative because one is bringing together a number of elements to form that compound.

**7.51** Use the relationship $\Delta G°_r = \sum \Delta G°_f(\text{products}) - \sum \Delta G°_r(\text{reactants})$:

(a) $\Delta G°_r = 2\Delta G°_f(SO_3, g) - [2\Delta G°_f(SO_2, g)]$

$$= 2(-371.06 \text{ kJ} \cdot \text{mol}^{-1}) - [2(-300.19 \text{ kJ} \cdot \text{mol}^{-1})]$$

$$= -141.74 \text{ kJ} \cdot \text{mol}^{-1}$$

The reaction is spontaneous.

(b) $\Delta G^\circ_r = \Delta G^\circ_f(CaO, s) + \Delta G^\circ_f(CO_2, g) - \Delta G^\circ_f(CaCO_3, s)$

$\qquad = (-604.03 \text{ kJ} \cdot \text{mol}^{-1}) + (-394.36 \text{ kJ} \cdot \text{mol}^{-1})$

$\qquad \quad - (-1128.8 \text{ kJ} \cdot \text{mol}^{-1})$

$\qquad = +130.41 \text{ kJ} \cdot \text{mol}^{-1}$

The reaction is not spontaneous.

(c) $\Delta G^\circ_r = 16\Delta G^\circ_f(CO_2, g) + 18\Delta G^\circ_f(H_2O, l) - [2\Delta G^\circ_f(C_8H_{18}, l)]$

$\qquad = 16(-394.36 \text{ kJ} \cdot \text{mol}^{-1}) + 18(-237.13 \text{ kJ} \cdot \text{mol}^{-1})$

$\qquad \quad - [2(6.4 \text{ kJ} \cdot \text{mol}^{-1})]$

$\qquad = -10\ 590.9 \text{ kJ} \cdot \text{mol}^{-1}$

The reaction is spontaneous.

**7.53** The standard free energies of formation of the compounds are: (a) $PCl_5(g)$, $-305.0$ kJ $\cdot$ mol$^{-1}$; (b) HCN(g), $+124.7$ kJ $\cdot$ mol$^{-1}$; (c) NO(g), $+86.55$ kJ $\cdot$ mol$^{-1}$; (d) $SO_2(g)$, $-300.19$ kJ $\cdot$ mol$^{-1}$. Those compounds with a positive free energy of formation are unstable with respect to the elements. Thus (a) and (d) are thermodynamically stable.

**7.55** To understand what happens to $\Delta G^\circ_r$ as temperature is raised, we use the relationship $\Delta G^\circ_r = \Delta H^\circ_r - T\Delta S^\circ_r$. From this it is clear that the free energy of the reaction becomes less favorable (more positive) as temperature increases, only if $\Delta S^\circ_r$ is a negative number. Therefore, we need only to find out whether the standard entropy of formation of the compound is a negative number. This is calculated for each compound as follows:

(a) $P(s) + \frac{5}{2}Cl_2(g) \longrightarrow PCl_5(g)$

$\Delta S^\circ_r = S^\circ_m(PCl_5, g) - [S^\circ_m(P, s) + \frac{5}{2}S^\circ_m(Cl_2, g)]$

$\qquad = 364.6 \text{ J} \cdot \text{K}^{-1} \cdot \text{mol}^{-1} - [41.09 \text{ J} \cdot \text{K}^{-1} \cdot \text{mol}^{-1} + \frac{5}{2}(223.07 \text{ J} \cdot \text{K}^{-1} \cdot \text{mol}^{-1})]$

$\qquad = -234.2 \text{ J} \cdot \text{K}^{-1} \cdot \text{mol}^{-1}$

The compound is less stable at higher temperatures.

(b) $C(s), \text{graphite} + \frac{1}{2}N_2(g) + \frac{1}{2}H_2(g) \longrightarrow HCN(g)$

$$\Delta S^\circ_r = S^\circ_m(\text{HCN, g}) - [S^\circ_m(\text{C, s}) + \tfrac{1}{2}S^\circ_m(\text{N}_2, \text{g}) + \tfrac{1}{2}S^\circ_m(\text{H}_2, \text{g})]$$

$$= 201.78 \text{ J} \cdot \text{K}^{-1} \cdot \text{mol}^{-1} - [5.740 \text{ J} \cdot \text{K}^{-1} \cdot \text{mol}^{-1}$$

$$+ \tfrac{1}{2}(191.61 \text{ J} \cdot \text{K}^{-1} \cdot \text{mol}^{-1}) + \tfrac{1}{2}(130.68 \text{ J} \cdot \text{K}^{-1} \cdot \text{mol}^{-1})]$$

$$= +34.90 \text{ J} \cdot \text{K}^{-1} \cdot \text{mol}^{-1}$$

HCN(g) is more stable at higher $T$.

(c) $\tfrac{1}{2}\text{N}_2(\text{g}) + \tfrac{1}{2}\text{O}_2(\text{g}) \longrightarrow \text{NO(g)}$

$$\Delta S^\circ_r = S^\circ_m(\text{NO, g}) - [\tfrac{1}{2}S^\circ_m(\text{N}_2, \text{g}) + \tfrac{1}{2}S^\circ_m(\text{O}_2, \text{g})]$$

$$= 210.76 \text{ J} \cdot \text{K}^{-1} \cdot \text{mol}^{-1} - [\tfrac{1}{2}(191.61 \text{ J} \cdot \text{K}^{-1} \cdot \text{mol}^{-1})$$

$$+ \tfrac{1}{2}(205.14 \text{ J} \cdot \text{K}^{-1} \cdot \text{mol}^{-1})]$$

$$= +12.38 \text{ J} \cdot \text{K}^{-1} \cdot \text{mol}^{-1}$$

NO(g) is more stable as $T$ increases.

(d) $\text{S(s)} + \text{O}_2(\text{g}) \longrightarrow \text{SO}_2(\text{g})$

$$\Delta S^\circ_r = S^\circ(\text{SO}_2, \text{g}) - [S^\circ(\text{S, s}) + S^\circ(\text{O}_2, \text{g})]$$

$$= 248.22 \text{ J} \cdot \text{K}^{-1} \cdot \text{mol}^{-1} - [31.80 \text{ J} \cdot \text{K}^{-1} \cdot \text{mol}^{-1}$$

$$+ 205.14 \text{ J} \cdot \text{K}^{-1} \cdot \text{mol}^{-1}]$$

$$= +11.28 \text{ J} \cdot \text{K}^{-1} \cdot \text{mol}^{-1}$$

$\text{SO}_2(\text{g})$ is more stable as $T$ increases.

**7.57** (a) $2\text{H}_2\text{O}_2(\text{l}) \longrightarrow 2\text{H}_2\text{O(l)} + \text{O}_2(\text{g})$

$$\Delta S^\circ_r = 2S^\circ_m(\text{H}_2\text{O, l}) + S^\circ_m(\text{O}_2, \text{g}) - 2S^\circ_m(\text{H}_2\text{O}_2, \text{l})$$

$$= 2(69.91 \text{ J} \cdot \text{K}^{-1} \cdot \text{mol}^{-1}) + 205.14 \text{ J} \cdot \text{K}^{-1} \cdot \text{mol}^{-1}$$

$$- 2(109.6 \text{ J} \cdot \text{K}^{-1} \cdot \text{mol}^{-1})$$

$$= +125.8 \text{ J} \cdot \text{K}^{-1} \cdot \text{mol}^{-1}$$

$$\Delta H^\circ_r = 2\Delta H^\circ_f(\text{H}_2\text{O, aq}) - 2\Delta H^\circ_f(\text{H}_2\text{O}_2, \text{l})$$

$$= 2(-285.83 \text{ kJ} \cdot \text{mol}^{-1}) - 2(-187.78 \text{ kJ} \cdot \text{mol}^{-1})$$

$$= -196.10 \text{ kJ} \cdot \text{mol}^{-1}$$

$$\Delta G^\circ_r = 2\Delta G^\circ_f(\text{H}_2\text{O, aq}) - 2\Delta G^\circ_f(\text{H}_2\text{O}_2, \text{l})$$

$$= 2(-237.13 \text{ kJ} \cdot \text{mol}^{-1}) - 2(-120.35 \text{ kJ} \cdot \text{mol}^{-1})$$

$$= -233.56 \text{ kJ} \cdot \text{mol}^{-1}$$

$\Delta G^\circ_r$ can also be calculated from $\Delta S^\circ_r$ and $\Delta H^\circ_r$ using the relationship:

$$\Delta G^\circ_r = \Delta H^\circ_r - T\Delta S^\circ_r$$
$$= -196.1 \text{ kJ} \cdot \text{mol}^{-1} - (298 \text{ K})(+125.8 \text{ J} \cdot \text{K}^{-1} \cdot \text{mol}^{-1})/(1000 \text{ J} \cdot \text{kJ}^{-1})$$
$$= -233.6 \text{ kJ} \cdot \text{mol}^{-1}$$

(b) $2 \text{ F}_2(\text{g}) + 2 \text{ H}_2\text{O}(\text{l}) \longrightarrow 4 \text{ HF}(\text{aq}) + \text{O}_2(\text{g})$

$$\Delta S^\circ_r = 4S^\circ_m(\text{HF, aq}) + S^\circ_m(\text{O}_2, \text{g}) - [2S^\circ_m(\text{F}_2, \text{g}) + 2S^\circ_m(\text{H}_2\text{O}, \text{l})]$$
$$= 4(88.7 \text{ J} \cdot \text{K}^{-1} \cdot \text{mol}^{-1}) + 205.14 \text{ J} \cdot \text{K}^{-1} \cdot \text{mol}^{-1}$$
$$- [2(202.78 \text{ J} \cdot \text{K}^{-1} \cdot \text{mol}^{-1}) + 2(69.91 \text{ J} \cdot \text{K}^{-1} \cdot \text{mol}^{-1})]$$
$$= +14.6 \text{ J} \cdot \text{K}^{-1} \cdot \text{mol}^{-1}$$

$$\Delta H^\circ_r = 4\Delta H^\circ_f(\text{HF, aq}) - 2\Delta H^\circ_f(\text{H}_2\text{O}, \text{l})$$
$$= 4(-330.08 \text{ kJ} \cdot \text{mol}^{-1}) - 2(-285.83 \text{ kJ} \cdot \text{mol}^{-1})$$
$$= -748.66 \text{ kJ} \cdot \text{mol}^{-1}$$

$$\Delta G^\circ_r = 4\Delta G^\circ_f(\text{HF, aq}) - 2\Delta G^\circ_f(\text{H}_2\text{O}, \text{l})$$
$$= 4(-296.82 \text{ kJ} \cdot \text{mol}^{-1}) - 2(-237.13 \text{ kJ} \cdot \text{mol}^{-1})$$
$$= -713.02 \text{ kJ} \cdot \text{mol}^{-1}$$

$\Delta G^\circ_r$ can also be calculated from $\Delta S^\circ_r$ and $\Delta H^\circ_r$ using the relationship:

$$\Delta G^\circ_r = \Delta H^\circ_r - T\Delta S^\circ_r$$
$$= -748.66 \text{ kJ} \cdot \text{mol}^{-1} - (298 \text{ K})(14.6 \text{ J} \cdot \text{K}^{-1} \cdot \text{mol}^{-1})/(1000 \text{ J} \cdot \text{kJ}^{-1})$$
$$= -753.01 \text{ kJ} \cdot \text{mol}^{-1}$$

**7.59** In order to find $\Delta G^\circ_r$ at a temperature other than 298 K, we must first calculate $\Delta H^\circ_r$ and $\Delta S^\circ_r$ and then use the relationship $\Delta G^\circ_r = \Delta H^\circ_r + T\Delta S^\circ_r$ to calculate $\Delta G^\circ_r$.

(a)

$$\Delta H^\circ_r = 2\Delta H^\circ_f(BF_3, g) + 3\Delta H^\circ_f(H_2O, l) - [\Delta H^\circ_f(B_2O_3, s)$$
$$+ 6\Delta H^\circ_f(HF, g)]$$
$$= 2(-1137.0 \text{ kJ} \cdot \text{mol}^{-1}) + 3(-285.83 \text{ kJ} \cdot \text{mol}^{-1})$$
$$- [(-1272.8 \text{ kJ} \cdot \text{mol}^{-1}) + 6(-271.1 \text{ kJ} \cdot \text{mol}^{-1})]$$
$$= -232.1 \text{ kJ} \cdot \text{mol}^{-1}$$

$$\Delta S^\circ_r = 2S^\circ_m(BF_3, g) + 3S^\circ_m(H_2O, l)$$
$$- [S^\circ_m(B_2O_3, S) + 6S^\circ_m(HF, g)]$$
$$= 2(254.12 \text{ J} \cdot \text{K}^{-1} \cdot \text{mol}^{-1}) + 3(69.91 \text{ J} \cdot \text{K}^{-1} \cdot \text{mol}^{-1})$$
$$- [53.97 \text{ J} \cdot \text{K}^{-1} \cdot \text{mol}^{-1} + 6(173.78 \text{ J} \cdot \text{K}^{-1} \cdot \text{mol}^{-1})]$$
$$= -378.68 \text{ J} \cdot \text{K}^{-1} \cdot \text{mol}^{-1}$$

$$\Delta G^\circ_r = -232.1 \text{ J} \cdot \text{K}^{-1} \cdot \text{mol}^{-1} - (353 \text{ K})(-378.68 \text{ J} \cdot \text{K}^{-1} \cdot \text{mol}^{-1})/(1000 \text{ J} \cdot \text{kJ}^{-1})$$
$$= -98.42 \text{ kJ} \cdot \text{mol}^{-1}$$

In order to determine the range over which the reaction will be spontaneous, we consider the relative signs of $\Delta H^\circ_r$ and $\Delta S^\circ_r$ and their effect on $\Delta G^\circ_r$. Because $\Delta H^\circ_r$ is negative and $\Delta S^\circ_r$ is also negative, we expect the reaction to be spontaneous at low temperatures, where the term $T\Delta S^\circ_r$ will be less than $\Delta H^\circ_r$. To find the temperature of the cutoff, we calculate the temperature at which $\Delta G^\circ_r = 0$. For this reaction, that temperature is

$$\Delta G^\circ_r = 0 = -232.1 \text{ kJ} \cdot \text{mol}^{-1} - (T)(-378.68 \text{ J} \cdot \text{K}^{-1} \cdot \text{mol}^{-1})/(1000 \text{ J} \cdot \text{kJ}^{-1})$$
$$T = 612.9 \text{ K}$$

The reaction should be spontaneous below 612.9 K.

(b) 
$$\Delta H°_r = \Delta H°_f(CaCl_2, aq) + \Delta H°_f(C_2H_2, g) - [\Delta H°_f(CaC_2, s)$$
$$+ 2\Delta H°_f(HCl, aq)]$$
$$= (-877.1 \text{ kJ} \cdot \text{mol}^{-1}) + 226.73 \text{ kJ} \cdot \text{mol}^{-1}$$
$$- [(-59.8 \text{ kJ} \cdot \text{mol}^{-1}) + 2(-167.16 \text{ kJ} \cdot \text{mol}^{-1})]$$
$$= -256.3 \text{ kJ} \cdot \text{mol}^{-1}$$
$$\Delta S°_r = S°_m(CaCl_2, aq) + S°_m(C_2H_2, g)$$
$$- [S°_m(CaC_2, s) + 2S°_m(HCl, aq)]$$
$$= 59.8 \text{ J} \cdot \text{K}^{-1} \cdot \text{mol}^{-1} + 200.94 \text{ J} \cdot \text{K}^{-1} \cdot \text{mol}^{-1}$$
$$- [69.96 \text{ J} \cdot \text{K}^{-1} \cdot \text{mol}^{-1} + 2(56.5 \text{ J} \cdot \text{K}^{-1} \cdot \text{mol}^{-1})]$$
$$= +77.8 \text{ J} \cdot \text{K}^{-1} \cdot \text{mol}^{-1}$$
$$\Delta G°_r = -256.2 \text{ kJ} \cdot \text{mol}^{-1} - (353 \text{ K})(+77.8 \text{ J} \cdot \text{K}^{-1} \cdot \text{mol}^{-1})/(1000 \text{ J} \cdot \text{kJ}^{-1})$$
$$= -283.7 \text{ kJ} \cdot \text{mol}^{-1}$$

Because $\Delta H°_r$ is negative and $\Delta S°_r$ is positive, the reaction will be

spontaneous at all temperatures.

(c)

$$\Delta H°_r = \Delta H°_f(C \text{ (s), diamond}) = +1.895 \text{ kJ} \cdot \text{mol}^{-1}$$
$$\Delta S°_r = S°_m(C \text{ (s), diamond}) - S°_m(C \text{ (s), graphite})$$
$$= +2.377 \text{ J} \cdot \text{K}^{-1} \cdot \text{mol}^{-1} - 5.740 \text{ J} \cdot \text{K}^{-1} \cdot \text{mol}^{-1}$$
$$= -3.363 \text{ J} \cdot \text{K}^{-1}$$
$$\Delta G°_r = +1.895 \text{ kJ} \cdot \text{mol}^{-1} - (353 \text{ K})(-3.363 \text{ J} \cdot \text{K}^{-1} \cdot \text{mol}^{-1})/(1000 \text{ J} \cdot \text{kJ}^{-1})$$
$$= +3.082 \text{ kJ} \cdot \text{mol}^{-1}$$

Because $\Delta H°_r$ is positive and $\Delta S°_r$ is negative, the reaction will be

nonspontaneous at all temperatures. Note: This calculation is for

atmospheric pressure. Diamond can be produced from graphite at elevated

pressures and high temperatures.


**7.61** Assuming standard state conditions, $\Delta G_r°$ for $C_2H_4(g) + H_2O(g) \rightarrow$

$CH_3CH_2OH(l)$ is:

$$\Delta G°_r = (-174.8 \text{ kJ} \cdot \text{mol}^{-1}) - (68.15 \text{ kJ} \cdot \text{mol}^{-1}) - (-228.57 \text{ kJ} \cdot \text{mol}^{-1})$$
$$= -14.4 \text{ kJ} \cdot \text{mol}^{-1} \quad \text{Negative } \Delta G°_r \text{ indicates a spontaneous reaction.}$$

$\Delta G_r°$ for $C_2H_6(g) + H_2O(g) \rightarrow CH_3CH_2OH(l) + H_2(g)$ is:

$$\Delta G^{\circ}{}_{r} = \left(-174.8 \text{ kJ} \cdot \text{mol}^{-1}\right) - \left(-32.82 \text{ kJ} \cdot \text{mol}^{-1}\right) - \left(-228.57 \text{ kJ} \cdot \text{mol}^{-1}\right)$$

$= +86.6 \text{ kJ} \cdot \text{mol}^{-1}$  Positive $\Delta G^{\circ}{}_{r}$ indicates a nonspontaneous reaction. Reaction A is spontaneous but reaction B is not spontaneous at any temperature.

**7.63**   (a)  1-propanol $(C_3H_8O)$ and 2-propanone $(C_3H_6O)$ have similar numbers of electrons so that we would expect the molar entropies to be similar. Because 1-propanol exhibits hydrogen bonding, however, we might expect the liquid phase to be more ordered than for 2-propanone. This is observed. The standard molar entropy for 2-propanone is 200 $\text{J} \cdot \text{K}^{-1} \cdot \text{mol}^{-1}$ while that of 1-propanol is 193 $\text{J} \cdot \text{K}^{-1} \cdot \text{mol}^{-1}$.

(b)  In the gas phase, hydrogen bonding will not be important because the molecules are too far apart, so the standard molar entropies should be more similar.

**7.65**   For the cis compound there will be 12 different orientations: For the trans compound there will only be 3 different orientations. Comparing the Boltzmann entropy calculations for the cis and trans forms:

cis:

$$S = k \ln 12^{6.02 \times 10^{23}} = (1.38 \times 10^{-23} \text{ J} \cdot \text{K}^{-1}) \ln 12^{6.02 \times 10^{23}}$$
$$S = 20.6 \text{ J} \cdot \text{K}^{-1}$$
trans:
$$S = k \ln 3^{6.02 \times 10^{23}} = (1.38 \times 10^{-23} \text{ J} \cdot \text{K}^{-1}) \ln 3^{6.02 \times 10^{23}}$$
$$S = 9.13 \text{ J} \cdot \text{K}^{-1}$$

The cis form should have the higher residual entropy.

**7.67**   According to Trouton's rule, the entropy of vaporization of an organic liquid is a constant of approximately 85 $\text{J} \cdot \text{mol}^{-1} \cdot \text{K}^{-1}$. The relationship between entropy of fusion, enthalpy of fusion, and melting point is given

by $\Delta S^{\circ}_{fus} = \dfrac{\Delta H^{\circ}_{fus}}{T_{fus}}$.

For Pb : $\Delta S^{\circ}_{fus} = \dfrac{5100\ J}{600\ K} = 8.50\ J \cdot K^{-1}$

For Hg : $\Delta S^{\circ}_{fus} = \dfrac{2290\ J}{234\ K} = 9.79\ J \cdot K^{-1}$

For Na : $\Delta S^{\circ}_{fus} = \dfrac{2640\ J}{371\ K} = 7.12\ J \cdot K^{-1}$

These numbers are reasonably close but clearly much smaller than the value associated with Trouton's rule.

**7.69**  This is best answered by considering the reaction that interconverts the two compounds

$$4\ Fe_3O_4(s) + O_2(g) \longrightarrow 6\ Fe_2O_3(s)$$

We calculate $\Delta G^{\circ}_r$ using data from Appendix 2A:

$\Delta G^{\circ}_r = 6\Delta G^{\circ}_f(Fe_2O_3,\ s) - [4\Delta G^{\circ}_f(Fe_3O_4,\ s)]$

$\Delta G^{\circ}_r = 6(-742.2\ kJ \cdot mol^{-1}) - [4(-1015.4\ kJ \cdot mol^{-1})]$

$\phantom{\Delta G^{\circ}_r} = -391.6\ kJ \cdot mol^{-1}$

Because $\Delta G^{\circ}_r$ is negative, the process is spontaneous at 25°C.

Therefore, $Fe_2O_3$ is thermodynamically more stable.

**7.71**  We can calculate the free energy changes associated with the conversions:

(a)  $2\ FeS(s) \longrightarrow Fe(s) + FeS_2(s)$

(b)  $FeS_2(s) \longrightarrow S(s) + FeS(s)$

For (a),

$\Delta G^{\circ}_r = -166.9\ kJ \cdot mol^{-1} - 2(-100.4\ kJ \cdot mol^{-1}) = +33.9\ kJ \cdot mol^{-1}$

This process is predicted to be nonspontaneous.

For (b), $\Delta G^{\circ}_r = -100.4\ kJ \cdot mol^{-1} - (-166.9\ kJ \cdot mol^{-1}) = +66.5\ kJ \cdot mol^{-1}$

This process is predicted to be nonspontaneous.

**7.73** (a) Because the enthalpy change for dissolution is positive, the entropy change of the surroundings must be a negative number $\left(\Delta S°_{surr} = -\dfrac{\Delta H°_{system}}{T}\right)$. Because spontaneous processes are accompanied by an increase in entropy, the change in enthalpy does not favor the dissolution process. (b) In order for the process to be spontaneous (because it occurs readily, we know it is spontaneous), the entropy change of the system must be positive. (c) Positional disorder is dominant. (d) Because the surroundings participate in the solution process only as a source of heat, the entropy change of the surroundings is primarily a result of the dispersal of thermal motion. (e) The driving force for the dissolution is the dispersal of matter, resulting in an overall positive $\Delta S$.

**7.75** The values are calculated simply from the Hess's law relationship that the sums of the various energy quantities for the products minus the similar sum for the reactants will give the overall change in the state function desired:

C(s), graphite $\longrightarrow$ C(s), diamond

$\Delta H° = \Delta H°_f\,(\text{C(s), diamond}) = +1.895\ \text{kJ} \cdot \text{mol}^{-1}$
$\Delta S° = \Delta S°_m\,(\text{C(s), diamond}) - \Delta S°_m\,(\text{C(s), graphite})$
$\qquad = 2.377\ \text{J} \cdot \text{K}^{-1} \cdot \text{mol}^{-1} - 5.740\ \text{J} \cdot \text{K}^{-1} \cdot \text{mol}^{-1}$
$\qquad = -3.363\ \text{J} \cdot \text{K}^{-1} \cdot \text{mol}^{-1}$
$\Delta G° = \Delta G°_f\,(\text{C(s), diamond}) = +2.900\ \text{kJ} \cdot \text{mol}^{-1}$

Notice that the values used for the entropy calculation are absolute entropies, not $\Delta S°$ values and that the $S°_m$ value for C(s), graphite, is not 0. Graphite has a delocalized structure similar to that of benzene, whereas in diamond, all the carbon atom are bonded to four other carbon atoms in a very rigid lattice. It is not surprising that the change in entropy upon going from graphite to diamond would decrease, because we would expect graphite to have a higher molar entropy than diamond. Similarly, one can compare the bond formation and breaking that accompanies a change from graphite to diamond. Given the actual numbers, the change is clearly small however. In graphite, the carbon atom is bonded to three other carbon atoms with delocalized bonds (we can approximate this very roughly, using the values of $518 \text{ kJ} \cdot \text{mol}^{-1}$ of delocalized C—C bonds, as given in Table 6.7). In diamond, the carbon atom is bonded to four other carbon atoms by single C—C bonds (approximately by $348 \text{ KJ} \cdot \text{mol}^{-1}$). Even though these approximations overestimate the effect of delocalization on the C—C bond strength in graphite, the trend is expected—the three delocalized bonds in C(s) graphite are actually slightly more exothermic than the four C—C single bonds in diamond, making the standard enthalpy change a positive number. (Notice that if one considered the C—C bonds in graphite to be localized, the opposite prediction would have been made.) The standard free energy change for this reaction is positive as follows, from a positive standard enthalpy change and a negative standard entropy change for the reaction.

**7.77** The entries all correspond to aqueous ions. The fact that they are negative is due to the reference point that has been established. Because ions cannot actually be separated and measured independently, a reference point that defines $S^{\circ}_{m}(\text{H}^{+}, \text{aq}) = 0$ has been established. This definition is then used to calculate the standard entropies for the other ions. The fact that they are negative will arise in part because the solvated ion $M(\text{H}_2\text{O})_{x}^{n+}$ will be more ordered than the isolated ion and solvent molecules $(M^{n+} + x \, \text{H}_2\text{O})$.

**7.79** $\Delta G = \Delta H - T\Delta S$, therefore,

$$\Delta S_r = \frac{\Delta H_r - \Delta G_r}{T}$$

Given the reaction $H_2(g) + \frac{1}{2}O_2(g) \rightarrow H_2O(l)$ and 0.50 g of $H_2(g)$ consumed:

$$\Delta G_r = \left(-237.25 \text{ kJ} \cdot \text{mol}^{-1}\right)\left(\frac{0.50 \text{ g}}{2.016 \text{ g} \cdot \text{mol}^{-1}}\right) = -58.8 \text{ kJ}.$$

Therefore,

$$\Delta S_r = \left(\frac{-70.9 \text{ kJ} - (-58.8 \text{ kJ})}{298 \text{ K}}\right)\left(\frac{10^3 \text{ J}}{1 \text{ kJ}}\right) = -40.6 \text{ J} \cdot \text{K}^{-1}.$$

**7.81** (a) In order to calculate the free energy at different temperatures, we need to know $\Delta H°$ and $\Delta S°$ for the process: $H_2O(l) \longrightarrow H_2O(g)$

$$\Delta H°_r = \Delta H°_f(H_2O, g) - \Delta H°_f(H_2O, l)$$
$$= (-241.82 \text{ kJ} \cdot \text{mol}^{-1}) - [-285.83 \text{ kJ} \cdot \text{mol}^{-1}]$$
$$= 44.01 \text{ kJ} \cdot \text{mol}^{-1}$$
$$\Delta S°_r = S°_m(H_2O, g) - S°_m(H_2O, l)$$
$$= 188.83 \text{ J} \cdot \text{K}^{-1} \cdot \text{mol}^{-1} - [69.91 \text{ J} \cdot \text{K}^{-1} \cdot \text{mol}^{-1}]$$
$$= 118.92 \text{ J} \cdot \text{K}^{-1} \cdot \text{mol}^{-1}$$

$$\Delta G°_r = \Delta H°_r - T\Delta S°_r$$
$$= 44.01 \text{ kJ} \cdot \text{mol}^{-1} - T(118.92 \text{ J} \cdot \text{K}^{-1} \cdot \text{mol}^{-1})/(1000 \text{ J} \cdot \text{kJ}^{-1})$$

$T(K)$  $\Delta G°_r (kJ)$

298    8.57 kJ

373    − 0.35 kJ

423    − 6.29 kJ

The reaction goes from being nonspontaneous near room temperature to being spontaneous above 100°C.

(b) The value at 100°C should be exactly 0, because this is the normal boiling point of water.

(c) The discrepancy arises because the enthalpy and entropy values calculated from the tables are not rigorously constant with temperature. Better values would be obtained using the actual enthalpy and entropy of vaporization measured at the boiling point.

**7.83** The dehydrogenation of cyclohexane to benzene follows the following equation:

$$C_6H_{12}(l) \longrightarrow C_6H_6(l) + 3 H_2(g)$$

We can confirm that this process is nonspontaneous by calculating the $\Delta G^{\circ}_r$ for the process, using data in Appendix 2A:

$$\Delta G^{\circ}_r = \Delta G^{\circ}_f(C_6H_6, l) - \Delta G^{\circ}_f(C_6H_{12}, l)$$
$$= 124.3 \text{ kJ} \cdot \text{mol}^{-1} - 26.7 \text{ kJ} \cdot \text{mol}^{-1}$$
$$= +97.6 \text{ kJ} \cdot \text{mol}^{-1}$$

The reaction of ethane with hydrogen can be examined similarly:

$$C_2H_2(g) + H_2(g) \longrightarrow C_2H_6(g)$$
$$\Delta G^{\circ}_r = \Delta G^{\circ}_f(C_2H_6, g) - \Delta G^{\circ}_f(C_2H_{12}, g)$$
$$= (-32.82 \text{ kJ} \cdot \text{mol}^{-1}) - 68.15 \text{ kJ} \cdot \text{mol}^{-1}$$
$$= -100.97 \text{ kJ} \cdot \text{mol}^{-1}$$

We can now combine these two reactions so that $C_2H_2(g)$ accepts the hydrogen that is formed in the dehydrogenation reaction:

$$C_6H_{12}(l) \longrightarrow C_6H_6(l) + 3 H_2(g)$$
$$+ 3[C_2H_2(g) + H_2(g) \longrightarrow C_2H_6(g)]$$

$$\Delta G^{\circ}_r = +97.6 \text{ kJ} \cdot \text{mol}^{-1}$$
$$\Delta G^{\circ}_r = 3(-100.97 \text{ kJ} \cdot \text{mol}^{-1})$$

$$C_6H_{12}(l) + 3 C_2H_2(g) \longrightarrow C_6H_6(l) + 3 C_2H_6(g)$$
$$\Delta G^{\circ}_r = -205.13 \text{ kJ} \cdot \text{mol}^{-1}$$

We can see that by combining these two reactions, the overall process becomes spontaneous. Essentially, we are using the energy of the favorable reaction to drive the nonfavorable process.

**7.85** (a)

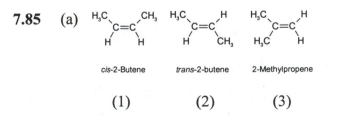

cis-2-Butene     trans-2-butene     2-Methylpropene

(1)        (2)        (3)

(b) For the three reactions, the calculation of $\Delta G°$, $\Delta H°$, and $\Delta S°$ are as follows:

$$\Delta G°_r = \Delta G°_f(2) - \Delta G°_f(1)$$
$$= 62.97 \text{ kJ} \cdot \text{mol}^{-1} - 65.86 \text{ kJ} \cdot \text{mol}^{-1}$$
$$= -2.89 \text{ kJ} \cdot \text{mol}^{-1}$$
$$\Delta H°_r = \Delta H°_f(2) - \Delta H°_f(1)$$
$$= (-11.17 \text{ kJ} \cdot \text{mol}^{-1}) - (-6.99 \text{ kJ} \cdot \text{mol}^{-1})$$
$$= -4.18 \text{ kJ} \cdot \text{mol}^{-1}$$
$$\Delta G°_r = \Delta H°_r - T\Delta S°_r$$
$$-2.89 \text{ kJ} \cdot \text{mol}^{-1} = -4.18 \text{ kJ} \cdot \text{mol}^{-1} - (298 \text{ K})(\Delta S°_r)/(1000 \text{ J} \cdot \text{kJ}^{-1})$$
$$\Delta S°_r = -4.33 \text{ J} \cdot \text{K}^{-1} \cdot \text{mol}^{-1}$$

$$\Delta G°_r = \Delta G°_f(3) - \Delta G°_f(1)$$
$$= 58.07 \text{ kJ} \cdot \text{mol}^{-1} - 65.86 \text{ kJ} \cdot \text{mol}^{-1}$$
$$= -7.79 \text{ kJ} \cdot \text{mol}^{-1}$$
$$\Delta H°_f = \Delta H°_f(3) - \Delta H°_f(1)$$
$$= (-16.90 \text{ kJ} \cdot \text{mol}^{-1}) - (-6.99 \text{ kJ} \cdot \text{mol}^{-1})$$
$$= -9.91 \text{ kJ} \cdot \text{mol}^{-1}$$
$$\Delta G°_r = \Delta H°_r - T\Delta S°_r$$
$$-7.79 \text{ kJ} \cdot \text{mol}^{-1} = -9.91 \text{ kJ} \cdot \text{mol}^{-1} - (298 \text{ K})(\Delta S°_r)/(1000 \text{ J} \cdot \text{kJ}^{-1})$$
$$\Delta S°_r = -7.11 \text{ J} \cdot \text{K}^{-1} \cdot \text{mol}^{-1}$$

$$\Delta G°_r = \Delta G°_f(3) - \Delta G°_f(2)$$
$$= 58.07 \text{ kJ} \cdot \text{mol}^{-1} - 62.97 \text{ kJ} \cdot \text{mol}^{-1}$$
$$= -4.90 \text{ kJ} \cdot \text{mol}^{-1}$$
$$\Delta H°_r = \Delta H°_f(3) - \Delta H°_f(2)$$
$$= (-16.90 \text{ kJ} \cdot \text{mol}^{-1}) - (-11.17 \text{ kJ} \cdot \text{mol}^{-1})$$
$$= -5.73 \text{ kJ} \cdot \text{mol}^{-1}$$
$$\Delta G°_r = \Delta H°_r - T\Delta S°_r$$
$$-4.90 \text{ kJ} \cdot \text{mol}^{-1} = -5.73 \text{ kJ} \cdot \text{mol}^{-1} - (298 \text{ K})(\Delta S°_r)/(1000 \text{ J} \cdot \text{kJ}^{-1})$$
$$\Delta S°_r = -2.78 \text{ J} \cdot \text{K}^{-1} \cdot \text{mol}^{-1}$$

(c) The most stable of the three compounds is 2-methylpropene.

(d) Because $\Delta S°$ is also equal to the difference in the $S°_m$ values for the compounds, we can examine those values to place the three compounds in order of their relative absolute entropies. The ordering is

$$S°_m(1) > S°_m(2) > S°_m(3).$$

**7.87** We need to calculate $\Delta H°_r$ and $\Delta S°_r$ for each process.

(a) $HCOOH(l) \longrightarrow H_2(g) + CO_2(g)$

$$
\begin{aligned}
\Delta H°_r &= \Delta H°_f(CO_2, g) - \Delta H°_f(HCOOH, l) \\
&= (-393.51 \text{ kJ} \cdot \text{mol}^{-1}) - (-424.72 \text{ kJ} \cdot \text{mol}^{-1}) \\
&= +31.21 \text{ kJ} \cdot \text{mol}^{-1}
\end{aligned}
$$

$$
\begin{aligned}
\Delta S°_r &= S°_m(H_2, g) + S°(CO_2, g) - [S°_m(HCOOH, l)] \\
&= 130.68 \text{ J} \cdot \text{K}^{-1} \cdot \text{mol}^{-1} + 213.74 \text{ J} \cdot \text{K}^{-1} \cdot \text{mol}^{-1} \\
&\quad - [128.95 \text{ J} \cdot \text{K}^{-1} \cdot \text{mol}^{-1}] \\
&= +215.47 \text{ J} \cdot \text{K}^{-1} \cdot \text{mol}^{-1}
\end{aligned}
$$

The reaction will become spontaneous above the temperature at which

$\Delta G°_r = 0$:

$$0 = 31.21 \text{ kJ} \cdot \text{mol}^{-1} - T(215.47 \text{ J} \cdot \text{K}^{-1} \cdot \text{mol}^{-1})/(1000 \text{ J} \cdot \text{kJ}^{-1})$$

$$T = 144.8 \text{ K}$$

(b) $CH_3COOH(l) \longrightarrow CH_4(g) + CO_2(g)$

$$
\begin{aligned}
\Delta H°_r &= \Delta H°_f(CH_4, g) + \Delta H°_f(CO_2, g) - \Delta H°_f(CH_3COOH, l) \\
&= (-74.81 \text{ kJ} \cdot \text{mol}^{-1}) + (-393.51 \text{ kJ} \cdot \text{mol}^{-1}) - (-484.5 \text{ kJ} \cdot \text{mol}^{-1}) \\
&= +16.18 \text{ kJ} \cdot \text{mol}^{-1}
\end{aligned}
$$

$$
\begin{aligned}
\Delta S°_r &= S°_m(CH_4, g) + S°(CO_2, g) - [S°_m(CH_3COOH, l)] \\
&= 186.26 \text{ J} \cdot \text{K}^{-1} \cdot \text{mol}^{-1} + 213.74 \text{ J} \cdot \text{K}^{-1} \cdot \text{mol}^{-1} - [159.8 \text{ J} \cdot \text{K}^{-1} \cdot \text{mol}^{-1}] \\
&= +240.2 \text{ J} \cdot \text{K}^{-1} \cdot \text{mol}^{-1}
\end{aligned}
$$

The reaction will become spontaneous above the temperature at which

$\Delta G°_r = 0$:

$$0 = 16.18 \text{ kJ} \cdot \text{mol}^{-1} - T(240.2 \text{ J} \cdot \text{K}^{-1} \cdot \text{mol}^{-1})/(1000 \text{ J} \cdot \text{kJ}^{-1})$$

$$T = 67.36 \text{ K}$$

(c) $C_6H_5COOH(s) \longrightarrow C_6H_6(l) + CO_2(g)$

$$\Delta H^\circ_r = \Delta H^\circ_f(C_6H_6, l) + \Delta H^\circ_f(CO_2, g) - [\Delta H^\circ_f(C_6H_5COOH, s)]$$
$$= +49.0 \text{ kJ} \cdot \text{mol}^{-1} + (-393.51 \text{ kJ} \cdot \text{mol}^{-1}) - [-385.1 \text{ kJ} \cdot \text{mol}^{-1}]$$
$$= +40.6 \text{ kJ} \cdot \text{mol}^{-1}$$
$$\Delta S^\circ_r = S^\circ_m(C_6H_6, l) + S^\circ(CO_2, g) - [S^\circ_m(C_6H_5COOH, s)]$$
$$= +173.3 \text{ J} \cdot \text{K}^{-1} \cdot \text{mol}^{-1} + 213.74 \text{ J} \cdot \text{K}^{-1} \cdot \text{mol}^{-1}$$
$$- (167.6 \text{ J} \cdot \text{K}^{-1} \cdot \text{mol}^{-1})$$
$$= +219.4 \text{ J} \cdot \text{K}^{-1} \cdot \text{mol}^{-1}$$

The reaction will become spontaneous above the temperature at which $\Delta G^\circ_r = 0$:

$$0 = 40.6 \text{ kJ} \cdot \text{mol}^{-1} - T(219.4 \text{ kJ} \cdot \text{mol}^{-1})/(1000 \text{ J} \cdot \text{kJ}^{-1})$$

$$T = 185 \text{ K}$$

It is clear from these calculations that all of these carboxylic acids are thermodynamically unstable with respect to decomposition to produce $CO_2(g)$. A consideration of the parameters shows that this is driven by the entropy increase in the production of the gas, because the enthalpy of the reaction in all cases is endothermic.

# CHAPTER 8
# PHYSICAL EQUILIBRIA

**8.1** In a 1.0 L vessel at 20°C, there will be 17.5 Torr of water vapor. The ideal

gas law can be used to calculate the mass of water present:

$PV = nRT$

let $m$ = mass of water

$$\left(\frac{17.5\ \text{Torr}}{760\ \text{Torr} \cdot \text{atm}^{-1}}\right)(1.0\ \text{L}) = \left(\frac{m}{18.02\ \text{g} \cdot \text{mol}^{-1}}\right)(0.082\ 06\ \text{L} \cdot \text{atm} \cdot \text{K}^{-1} \cdot \text{mol}^{-1})(293\ \text{K})$$

$$m = \frac{(17.5\ \text{Torr})(1.0\ \text{L})(18.02\ \text{g} \cdot \text{mol}^{-1})}{(760\ \text{Torr} \cdot \text{atm}^{-1})(0.082\ 06\ \text{L} \cdot \text{atm} \cdot \text{K}^{-1} \cdot \text{mol}^{-1})(293\ \text{K})}$$

$$m = 0.017\ \text{g}$$

**8.3** (a) 87 °C  (b) 113 °C

**8.5** (a) The quantities $\Delta H°_{\text{vap}}$ and $\Delta S°_{\text{vap}}$ can be calculated using the

relationship $\ln P = -\dfrac{\Delta H°_{\text{vap}}}{R} \cdot \dfrac{1}{T} + \dfrac{\Delta S°_{\text{vap}}}{R}$

Because we have two temperatures with corresponding vapor pressures,

we can set up two equations with two unknowns and solve for $\Delta H°_{\text{vap}}$ and

$\Delta S°_{\text{vap}}$. If the equation is used as is, $P$ must be expressed in atm, which is

the standard reference state. Remember that the value used for $P$ is really

activity, which for pressure is $P$ divided by the reference state of 1 atm, so

that the quantity inside the ln term is dimensionless.

$$8.314\ \text{J} \cdot \text{K}^{-1} \cdot \text{mol}^{-1} \times \ln \frac{58\ \text{Torr}}{760\ \text{Torr}} = -\frac{\Delta H°_{\text{vap}}}{250.4\ \text{K}} + \Delta S°_{\text{vap}}$$

$$8.314\ \text{J} \cdot \text{K}^{-1} \cdot \text{mol}^{-1} \times \ln \frac{512\ \text{Torr}}{760\ \text{Torr}} = -\frac{\Delta H°_{\text{vap}}}{298.2\ \text{K}} + \Delta S°_{\text{vap}}$$

which give, upon combining terms,

$$-21.39 \text{ J} \cdot \text{K}^{-1} \cdot \text{mol}^{-1} = -0.003\ 994 \text{ K}^{-1} \times \Delta H^{\circ}{}_{vap} + \Delta S^{\circ}{}_{vap}$$

$$-3.284 \text{ J} \cdot \text{K}^{-1} \cdot \text{mol}^{-1} = -0.003\ 353 \text{ K}^{-1} \times \Delta H^{\circ}{}_{vap} + \Delta S^{\circ}{}_{vap}$$

Subtracting one equation from the other will eliminate the $\Delta S^{\circ}{}_{vap}$ term and allow us to solve for $\Delta H^{\circ}{}_{vap}$ :

$$-18.11 \text{ J} \cdot \text{K}^{-1} \cdot \text{mol}^{-1} = -0.000\ 641 \times \Delta H^{\circ}{}_{vap}$$

$$\Delta H^{\circ}{}_{vap} = +28.3 \text{ kJ} \cdot \text{mol}^{-1}$$

(b)  We can then use $\Delta H^{\circ}{}_{vap}$ to calculate $\Delta S^{\circ}{}_{vap}$ using either of the two equations:

$$-21.39 \text{ J} \cdot \text{K}^{-1} \cdot \text{mol}^{-1} = -0.003\ 994 \text{ K}^{-1} \times (+28\ 200 \text{ J} \cdot \text{mol}^{-1}) + \Delta S^{\circ}{}_{vap}$$

$$\Delta S^{\circ}{}_{vap} = 91.2 \text{ J} \cdot \text{K}^{-1} \cdot \text{mol}^{-1}$$

$$-3.284 \text{ J} \cdot \text{K}^{-1} \cdot \text{mol}^{-1} = -0.003\ 353 \text{ K}^{-1} \times (+28\ 200 \text{ J} \cdot \text{mol}^{-1}) + \Delta S^{\circ}{}_{vap}$$

$$\Delta S^{\circ}{}_{vap} = 91.3 \text{ J} \cdot \text{K}^{-1} \cdot \text{mol}^{-1}$$

(c)  The $\Delta G^{\circ}{}_{vap}$ is calculated using $\Delta G^{\circ}{}_{r} = \Delta H^{\circ}{}_{r} - T \Delta S^{\circ}{}_{r}$

$$\Delta G^{\circ}{}_{r} = +28.3 \text{ kJ} \cdot \text{mol}^{-1} - (298 \text{ K})(91.2 \text{ J} \cdot \text{K}^{-1} \cdot \text{mol}^{-1})/(1000 \text{ J} \cdot \text{kJ}^{-1})$$

$$\Delta G^{\circ}{}_{r} = +1.1 \text{ kJ} \cdot \text{mol}^{-1}$$

(d)  The boiling point can be calculated using one of several methods. The easiest to use is the one developed in the last chapter:

$$\Delta G^{\circ}{}_{vap} = \Delta H^{\circ}{}_{vap} - T_{B}\ \Delta S^{\circ}{}_{vap} = 0$$

$$\Delta H^{\circ}{}_{vap} = T_{B}\ \Delta S^{\circ}{}_{vap} \text{ or } T_{B} = \frac{\Delta H^{\circ}{}_{vap}}{\Delta S^{\circ}{}_{vap}}$$

$$T_{B} = \frac{28.2 \text{ kJ} \cdot \text{mol}^{-1} \times 1000 \text{ J} \cdot \text{kJ}^{-1}}{91.2 \text{ J} \cdot \text{K}^{-1} \cdot \text{mol}^{-1}} = 309 \text{ K or } 36°\text{C}$$

Alternatively, we could use the relationship $\ln \dfrac{P_2}{P_1} = -\dfrac{\Delta H^{\circ}{}_{vap}}{R}\left[\dfrac{1}{T_2} - \dfrac{1}{T_1}\right]$.

Here we would substitute, in one of the known vapor pressure points, the value of the enthalpy of vaporization and the condition that $P = 1$ atm at the normal boiling point.

**8.7**  (a) The quantities $\Delta H^\circ_{vap}$ and $\Delta S^\circ_{vap}$ can be calculated using the relationship

$$\ln P = -\frac{\Delta H^\circ_{vap}}{R} \cdot \frac{1}{T} + \frac{\Delta S^\circ_{vap}}{R}$$

Because we have two temperatures with corresponding vapor pressures (we know that the vapor pressure = 1 atm at the boiling point), we can set up two equations with two unknowns and solve for $\Delta H^\circ_{vap}$ and $\Delta S^\circ_{vap}$. If the equation is used as is, $P$ must be expressed in atm, which is the standard reference state. Remember that the value used for $P$ is really activity, which for pressure is $P$ divided by the reference state of 1 atm, so that the quantity inside the ln term is dimensionless.

$$8.314 \text{ J} \cdot \text{K}^{-1} \cdot \text{mol}^{-1} \times \ln 1 = -\frac{\Delta H^\circ_{vap}}{292.7 \text{ K}} + \Delta S^\circ_{vap}$$

$$8.314 \text{ J} \cdot \text{K}^{-1} \cdot \text{mol}^{-1} \times \ln \frac{359 \text{ Torr}}{760 \text{ Torr}} = -\frac{\Delta H^\circ_{vap}}{273.2 \text{ K}} + \Delta S^\circ_{vap}$$

which give, upon combining terms,

$$0 \text{ J} \cdot \text{K}^{-1} \cdot \text{mol}^{-1} = -0.003\,416 \text{ K}^{-1} \times \Delta H^\circ_{vap} + \Delta S^\circ_{vap}$$

$$-6.235 \text{ J} \cdot \text{K}^{-1} \cdot \text{mol}^{-1} = -0.003\,660 \text{ K}^{-1} \times \Delta H^\circ_{vap} + \Delta S^\circ_{vap}$$

Subtracting one equation from the other will eliminate the $\Delta S^\circ_{vap}$ term and allow us to solve for $\Delta H^\circ_{vap}$ :

$$+6.235 \text{ J} \cdot \text{K}^{-1} \cdot \text{mol}^{-1} = +0.000\,244 \text{ K}^{-1} \times \Delta H^\circ_{vap}$$

$$\Delta H^\circ_{vap} = +25.6 \text{ kJ} \cdot \text{mol}^{-1}$$

(b)  We can then use $\Delta H^\circ_{vap}$ to calculate $\Delta S^\circ_{vap}$ using either of the two equations:

$$0 = -0.003\,416 \text{ K}^{-1} \times (+25\,600 \text{ J} \cdot \text{mol}^{-1}) + \Delta S^\circ_{vap}$$

$$\Delta S^\circ_{vap} = 87.4 \text{ J} \cdot \text{K}^{-1} \cdot \text{mol}^{-1}$$

$$-6.235 \text{ J} \cdot \text{K}^{-1} \cdot \text{mol}^{-1} = -0.003\,660 \text{ K}^{-1} \times (+25\,600 \text{ J} \cdot \text{mol}^{-1}) + \Delta S^\circ_{vap}$$

$$\Delta S^\circ_{vap} = 87.5 \text{ J} \cdot \text{K}^{-1} \cdot \text{mol}^{-1}$$

(c)  The vapor pressure at another temperature is calculated using

$$\ln \frac{P_2}{P_1} = -\frac{\Delta H^\circ_{vap}}{R}\left[\frac{1}{T_2} - \frac{1}{T_1}\right]$$

We need to insert the calculated value of the enthalpy of vaporization and one of the known vapor pressure points

$(P_{at\ 0.0\ °C} = 359\ \text{Torr} = 0.472\ \text{atml}):$

$$\ln\left(\frac{P_{at\ 8.50\ °C}}{0.472\ \text{atm}}\right) = -\frac{25600\ \text{J} \cdot \text{mol}^{-1}}{8.314\ \text{J} \cdot \text{K}^{-1} \cdot \text{mol}^{-1}}\left[\frac{1}{281.6\ \text{K}} - \frac{1}{273.2\ \text{K}}\right] = 0.66\ \text{atm}$$

**8.9** (a) The quantities $\Delta H^\circ_{vap}$ and $\Delta S^\circ_{vap}$ can be calculated using the relationship

$$\ln P = \frac{\Delta H^\circ_{vap}}{R} \cdot \frac{1}{T} + \frac{\Delta S^\circ_{vap}}{R}$$

Because we have two temperatures with corresponding vapor pressures, we can set up two equations with two unknowns and solve for $\Delta H^\circ_{vap}$ and $\Delta S^\circ_{vap}$. If the equation is used as is, $P$ must be expressed in atm, which is the standard reference state. Remember that the value used for $P$ is really activity which, for pressure, is $P$ divided by the reference state of 1 atm, so that the quantity inside the ln term is dimensionless.

$$8.314\ \text{J} \cdot \text{K}^{-1} \cdot \text{mol}^{-1} \times \ln\frac{35\ \text{Torr}}{760\ \text{Torr}} = -\frac{\Delta H^\circ_{vap}}{161.2\ \text{K}} + \Delta S^\circ_{vap}$$

$$8.314\ \text{J} \cdot \text{K}^{-1} \cdot \text{mol}^{-1} \times \ln\frac{253\ \text{Torr}}{760\ \text{Torr}} = -\frac{\Delta H^\circ_{vap}}{189.6\ \text{K}} + \Delta S^\circ_{vap}$$

which give, upon combining terms,

$$-25.59\ \text{J} \cdot \text{K}^{-1} \cdot \text{mol}^{-1} = -0.006\ 203\ \text{K}^{-1} \times \Delta H^\circ_{vap} + \Delta S^\circ_{vap}$$

$$-9.145\ \text{J} \cdot \text{K}^{-1} \cdot \text{mol}^{-1} = -0.005\ 274\ \text{K}^{-1} \times \Delta H^\circ_{vap} + \Delta S^\circ_{vap}$$

Subtracting one equation from the other will eliminate the $\Delta S^\circ_{vap}$ term and allow us to solve for $\Delta H^\circ_{vap}$ :

$$-16.45\ \text{J} \cdot \text{K}^{-1} \cdot \text{mol}^{-1} = -0.000\ 929 \times \Delta H^\circ_{vap}$$

$$\Delta H^\circ_{vap} = +17.7\ \text{kJ} \cdot \text{mol}^{-1}$$

(b) We can then use $\Delta H°_{vap}$ to calculate $\Delta S°_{vap}$ using either of the two equations:

$-25.59 \text{ J} \cdot \text{K}^{-1} \cdot \text{mol}^{-1} = -0.006\,203 \text{ K}^{-1} \times (+17\,700 \text{ J} \cdot \text{mol}^{-1}) + \Delta S°_{vap}$

$\Delta S°_{vap} = 84 \text{ J} \cdot \text{K}^{-1} \cdot \text{mol}^{-1}$

$-9.145 \text{ J} \cdot \text{K}^{-1} \cdot \text{mol}^{-1} = -0.005\,274 \text{ K}^{-1} \times (+17\,700 \text{ J} \cdot \text{mol}^{-1}) + \Delta S°_{vap}$

$\Delta S°_{vap} = 84 \text{ J} \cdot \text{K}^{-1} \cdot \text{mol}^{-1}$

(c) The $\Delta G°_{vap}$ is calculated using $\Delta G°_r = \Delta H°_r - T\Delta S°_r$

$\Delta G°_r = +17.7 \text{ kJ} \cdot \text{mol}^{-1} - (298 \text{ K})(84 \text{ J} \cdot \text{K}^{-1} \cdot \text{mol}^{-1})/(1000 \text{ J} \cdot \text{kJ}^{-1})$

$\Delta G°_r = -7.3 \text{ kJ} \cdot \text{mol}^{-1}$

Notice that the standard $\Delta G°_r$ is negative, so that the vaporization of arsine is spontaneous, which is as expected; under those conditions arsine is a gas at room temperature.

(d) The boiling point can be calculated from one of several methods. The easiest to use is that developed in the last chapter:

$\Delta G°_{vap} = \Delta H°_{vap} - T_B \Delta S°_{vap} = 0$

$\Delta H°_{vap} = T_B \Delta S°_{vap}$ or $T_B = \dfrac{\Delta H°_{vap}}{\Delta S°_{vap}}$

$T_B = \dfrac{17.7 \text{ kJ} \cdot \text{mol}^{-1} \times 1000 \text{ J} \cdot \text{kJ}^{-1}}{84 \text{ J} \cdot \text{K}^{-1} \cdot \text{mol}^{-1}} = 211 \text{ K}$

Alternatively, we could use the relationship $\ln \dfrac{P_2}{P_1} = -\dfrac{\Delta H°_{vap}}{R}\left[\dfrac{1}{T_2} - \dfrac{1}{T_1}\right]$.

Here, we would substitute, in one of the known vapor pressure points, the value of the enthalpy of vaporization and the condition that $P = 1$ atm at the normal boiling point.

**8.11** (a) The quantities $\Delta H°_{vap}$ and $\Delta S°_{vap}$ can be calculated, using the relationship

$\ln P = -\dfrac{\Delta H°_{vap}}{R} \cdot \dfrac{1}{T} + \dfrac{\Delta S°_{vap}}{R}$

Because we have two temperatures with corresponding vapor pressures (we know that the vapor pressure = 1 atm at the boiling point), we can set up two equations with two unknowns and solve for $\Delta H°_{vap}$ and $\Delta S°_{vap}$. If the equation is used as is, $P$ must be expressed in atm which is the standard reference state. Remember that the value used for $P$ is really activity which, for pressure, is $P$ divided by the reference state of 1 atm, so that the quantity inside the ln term is dimensionless.

$$8.314 \text{ J} \cdot \text{K}^{-1} \cdot \text{mol}^{-1} \times \ln 1 = -\frac{\Delta H°_{vap}}{315.58 \text{ K}} + \Delta S°_{vap}$$

$$8.314 \text{ J} \cdot \text{K}^{-1} \cdot \text{mol}^{-1} \times \ln \frac{140 \text{ Torr}}{760 \text{ Torr}} = -\frac{\Delta H°_{vap}}{273.2 \text{ K}} + \Delta S°_{vap}$$

which give, upon combining terms,

$$0 \text{ J} \cdot \text{K}^{-1} \cdot \text{mol}^{-1} = -0.003\,169 \text{ K}^{-1} \times \Delta H°_{vap} + \Delta S°_{vap}$$

$$-14.06 \text{ J} \cdot \text{K}^{-1} \cdot \text{mol}^{-1} = -0.003\,660 \text{ K}^{-1} \times \Delta H°_{vap} + \Delta S°_{vap}$$

Subtracting one equation from the other will eliminate the $\Delta S°_{vap}$ term and allow us to solve for $\Delta H°_{vap}$.

$$-14.06 \text{ J} \cdot \text{K}^{-1} \cdot \text{mol}^{-1} = -0.000\,491 \text{ K}^{-1} \times \Delta H°_{vap}$$

$$\Delta H°_{vap} = +28.6 \text{ kJ} \cdot \text{mol}^{-1}$$

(b) We can then use $\Delta H°_{vap}$ to calculate $\Delta S°_{vap}$ using either of the two equations:

$$0 = -0.003\,169 \text{ K}^{-1} \times (+28\,600 \text{ J} \cdot \text{mol}^{-1}) + \Delta S°_{vap}$$

$$\Delta S°_{vap} = 90.6 \text{ J} \cdot \text{K}^{-1} \cdot \text{mol}^{-1}$$

$$-14.06 \text{ J} \cdot \text{K}^{-1} \cdot \text{mol}^{-1} = -0.003\,660 \text{ K}^{-1} \times (+28\,600 \text{ J} \cdot \text{mol}^{-1}) + \Delta S°_{vap}$$

$$\Delta S°_{vap} = 90.6 \text{ J} \cdot \text{K}^{-1} \cdot \text{mol}^{-1}$$

(c) The vapor pressure at another temperature is calculated using

$$\ln \frac{P_2}{P_1} = -\frac{\Delta H°_{vap}}{R}\left[\frac{1}{T_2} - \frac{1}{T_1}\right]$$

We need to insert the calculated value of the enthalpy of vaporization and one of the known vapor pressure points:

$$\ln \frac{P_{at\,25.0°C}}{1 \text{ atm}} = -\frac{28\,600 \text{ J} \cdot \text{mol}^{-1}}{8.314 \text{ J} \cdot \text{K}^{-1} \cdot \text{mol}^{-1}}\left[\frac{1}{298.2 \text{ K}} - \frac{1}{315.58 \text{ K}}\right]$$

$$P_{at\,25.0°C} = 0.53 \text{ atm or } 4.0 \times 10^2 \text{ Torr}$$

**8.13** Table 6.2 contains the enthalpy of vaporization and the boiling point of methanol (at which the vapor pressure = 1 atm). Using this data and the equation

$$\ln \frac{P_2}{P_1} = -\frac{\Delta H°_{vap}}{R}\left[\frac{1}{T_2} - \frac{1}{T_1}\right]$$

$$\ln \frac{P_{25.0°C}}{1} = -\frac{35\,300 \text{ J} \cdot \text{mol}^{-1}}{8.314 \text{ J} \cdot \text{K}^{-1} \cdot \text{mol}^{-1}}\left[\frac{1}{298.2 \text{ K}} - \frac{1}{337.2 \text{ K}}\right]$$

$$P_{25.0°C} = 0.19 \text{ atm or } 1.5 \times 10^2 \text{ Torr}$$

**8.15** (a) vapor;  (b) liquid;  (c) vapor

**8.17** (a) 2.4 K;  (b) about 10 atm;  (c) 5.5 K;  (d) no

**8.19** (a) At the lower pressure triple point, liquid helium-I and -II are in equilibrium with helium gas; at the higher pressure triple point, liquid helium-I and -II are in equilibrium with solid helium.  (b) helium-I

**8.21** The pressure increase would bring $CO_2$ into the solid region.

**8.23** $\dfrac{dP}{dT} = \dfrac{\Delta H_{fus}}{T \cdot \Delta V_{fus}}$    $\therefore \Delta H_{fus} = \dfrac{dP}{dT} \cdot T \cdot \Delta V_{fus}$

$dP \approx P_2 - P_1 = 2000 \text{ bar} - 1 \text{ bar} = 1999 \text{ bar}$

$dT \approx T_2 - T_1 = 11.60 °C - 5.52 °C = 6.08 °C = 6.08 \text{ K}$

$$\Delta V_{fus} = V_2 - V_1 = \frac{78 \text{ g} \cdot \text{mol}^{-1}}{0.879 \text{ g} \cdot \text{cm}^{-3}} - \frac{78 \text{ g} \cdot \text{mol}^{-1}}{0.891 \text{ g} \cdot \text{cm}^{-3}} = 1.20 \text{ cm}^3 \cdot \text{mol}^{-1}$$

$$= \left(1.20 \text{ cm}^3 \cdot \text{mol}^{-1}\right)\left(\frac{1 \text{ L}}{1000 \text{ cm}^3}\right) = 0.00120 \text{ L} \cdot \text{mol}^{-1}$$

$$\Delta H_{fus} = \frac{dP}{dT} \cdot T \cdot \Delta V_{fus} = \left( \frac{1999 \text{ bar}}{6.08 \text{ K}} \right)(278.67 \text{ K})(0.00120 \text{ L} \cdot \text{mol}^{-1})$$

$$= 110 \text{ L} \cdot \text{bar} \cdot \text{mol}^{-1}$$

$$= \left(110 \text{ L} \cdot \text{bar} \cdot \text{mol}^{-1}\right)\left(100 \text{ J} \cdot \text{L}^{-1} \cdot \text{bar}^{-1}\right)\left(10^{-3} \text{ kJ} \cdot \text{J}^{-1}\right)$$

$$= 11.0 \text{ kJ} \cdot \text{mol}^{-1}$$

$$\Delta S_{fus} = \frac{\Delta H_{fus}}{T} = \frac{11.0 \text{ kJ} \cdot \text{mol}^{-1}}{278.67 \text{ K}} = 39.3 \text{ kJ} \cdot \text{K}^{-1} \cdot \text{mol}^{-1}$$

**8.25** (a) KCl is an ionic solid, so water would be the best choice;  (b) $CCl_4$ is non-polar, so the best choice is benzene;  (c) $CH_3COOH$ is polar, so water is the better choice.

**8.27** (a) hydrophilic, because $NH_2$ is polar, and has a lone pair and H atoms that can participate in hydrogen bonding to water molecules;

(b) hydrophobic, because the $CH_3$ group is not very polar;

(c) hydrophobic, because the Br group is not very polar;

(d) hydrophilic, because the carboxylic acid group has lone pairs on oxygen and an acidic proton that can participate in hydrogen bonding to water molecules.

**8.29** (a) The solubility of $O_2(g)$ in water is $1.3 \times 10^{-3}$ mol $\cdot$ L$^{-1}$ $\cdot$ atm$^{-1}$.

$$\text{solubility at 50 kPa} = \frac{50 \text{ kPa}}{101.325 \text{ kPa} \cdot \text{atm}^{-1}} \times 1.3 \times 10^{-3} \text{ mol} \cdot \text{L}^{-1} \cdot \text{atm}^{-1}$$

$$= 6.4 \times 10^{-4} \text{ mol} \cdot \text{L}^{-1}$$

(b) The solubility of $CO_2(g)$ in water is $2.3 \times 10^{-2}$ mol $\cdot$ L$^{-1}$ $\cdot$ atm$^{-1}$.

$$\text{solubility at 500 Torr} = \frac{500 \text{ Torr}}{760 \text{ Torr} \cdot \text{atm}^{-1}} \times 2.3 \times 10^{-2} \text{ mol} \cdot \text{L}^{-1} \cdot \text{atm}^{-1}$$

$$= 1.5 \times 10^{-2} \text{ mol} \cdot \text{L}^{-1}$$

(c) solubility $= 0.10 \text{ atm} \times 2.3 \times 10^{-2} \text{ mol} \cdot \text{L}^{-1} \cdot \text{atm}^{-1} = 2.3 \times 10^{-3} \text{ mol} \cdot \text{L}^{-1}$.

**8.31** (a) $4\,mg \cdot L^{-1} \times 1000\,mL \cdot L^{-1} \times 1.00\,g \cdot mL^{-1} \times 1\,kg/1000\,g$

$= 4\,mg \cdot L^{-1}$ or 4 ppm

(b) The solubility of $O_2$ (g) in water is $1.3 \times 10^{-3}\,mol \cdot L^{-1} \cdot atm^{-1}$ which can be converted to parts per million as follows:

In 1.00 L (corresponding to 1 kg) of solution there will be

$1.3 \times 10^{-3}\,mol\,O_2$

$1.3 \times 10^{-3}\,mol \cdot kg^{-1} \cdot atm^{-1}\,O_2 \times 32.00\,g \cdot mol^{-1}\,O_2 \times 10^3\,mg \cdot g^{-1}$

$= 42\,mg \cdot kg^{-1} \cdot atm^{-1}$ or 42 $ppm \cdot atm^{-1}$

$4\,mg \div 41\,mg \cdot kg^{-1} \cdot atm^{-1} = 0.1\,atm$

(c) $P = \dfrac{0.1\,atm}{0.21} = 0.5\,atm$

**8.33** (a) By Henry's law, the concentration of $CO_2$ in solution will double.

(b) No change in the equilibrium will occur; the partial pressure of $CO_2$ is unchanged and the concentration is unchanged.

**8.35** (a) Because it is exothermic, the enthalpy change must be negative;

(b) $Li_2SO_4(s) \rightarrow 2\,Li^+(aq) + SO_4^{2-}(aq) + heat$;   (c) Given that

$\Delta H_L + \Delta H_{hydration} = \Delta H$ of solution, the enthalpy of hydration should be larger. If the lattice energy were greater, the overall process would be endothermic.

**8.37** To answer this question we must first determine the molar enthalpies of solution and multiply this by the number of moles of solid dissolved to get the actual amount of heat released.

(a) $\Delta H^{\circ}_{sol} = 3.9\,kJ \cdot mol^{-1}$

$\Delta H = \dfrac{10.0\,g\,NaCl}{58.44\,g \cdot mol^{-1}} \times (+3.9\,kJ \cdot mol^{-1}) = 0.67\,kJ$ or $6.7 \times 10^2\,J$

(b) $\Delta H^{\circ}_{sol} = -7.5\,kJ \cdot mol^{-1}$

$$\Delta H = \frac{10.0 \text{ g NaI}}{149.89 \text{ g} \cdot \text{mol}^{-1}} \times (-7.5 \text{ kJ} \cdot \text{mol}^{-1}) = -0.50 \text{ kJ} = -5.0 \times 10^2 \text{ J}$$

(c) $\Delta H^\circ_{\text{sol}} = -329 \text{ kJ} \cdot \text{mol}^{-1}$

$$\Delta H = \frac{10.0 \text{ g AlCl}_3}{133.33 \text{ g} \cdot \text{mol}^{-1}} \times (-329 \text{ kJ} \cdot \text{mol}^{-1}) = -24.7 \text{ kJ}$$

(d) $\Delta H^\circ_{\text{sol}} = 6.6 \text{ kJ} \cdot \text{mol}^{-1}$

$$\Delta H = \frac{10.0 \text{ g NH}_4\text{NO}_3}{80.05 \text{ g} \cdot \text{mol}^{-1}} \times (+6.6 \text{ kJ} \cdot \text{mol}^{-1}) = +0.82 \text{ kJ} = +8.2 \times 10^2 \text{ J}$$

**8.39** All the enthalpies of solution are positive. Those of the alkali metal chlorides increase as the cation becomes larger and less strongly hydrated by water. All the alkali metal chlorides are soluble in water, but AgCl is not. AgCl has a relatively large positive enthalpy of solution. When dissolving is highly endothermic, the small increase in disorder due to solution formation may not be enough to compensate for the decrease in disorder for the surroundings, and a solution does not form. This is the case for AgCl.

**8.41** (a) $m_{\text{NaCl}} = \dfrac{\left( \dfrac{25.0 \text{ g NaCl}}{58.44 \text{ g} \cdot \text{mol}^{-1}} \right)}{0.5000 \text{ kg}} = 0.856 \, m$

(b) $m_{\text{NaOH}} = \dfrac{\left( \dfrac{\text{mass NaOH}}{40.00 \text{ g} \cdot \text{mol}^{-1}} \right)}{0.345 \text{ kg}} = 0.18 \, m$

mass NaOH $= 2.5 \text{ g}$

(c) $m_{\text{NaCl}} = \dfrac{\left( \dfrac{0.978 \text{ g urea}}{60.06 \text{ g} \cdot \text{mol}^{-1}} \right)}{(285 \text{ mL})(1 \text{ g} \cdot \text{mL}^{-1})(10^{-3} \text{ kg} \cdot \text{g}^{-1})} = 0.0571 \, m$

**8.43** (a) 1 kg of 5.00% $K_3PO_4$ will contain 50.0 g $K_3PO_4$ and 950.0 g $H_2O$.

$$\frac{\left(\dfrac{50.0 \text{ g } K_3PO_4}{212.27 \text{ g}\cdot\text{mol}^{-1} K_3PO_4}\right)}{0.950 \text{ kg}} = 0.248\, m$$

(b) The mass of 1.00 L of solution will be 1043 g, which will contain

1043 g × 0.0500 = 52.2 g $K_3PO_4$.

$$\frac{\left(\dfrac{52.2 \text{ g } K_3PO_4}{212.27 \text{ g}\cdot\text{mol}^{-1} K_3PO_4}\right)}{1.00 \text{ L}} = 0.246 \text{ M}$$

**8.45** (a) If $x_{MgCl_2}$ is 0.0120, then there are 0.0120 mol $MgCl_2$ for every 0.9880

mol $H_2O$. The mass of water will be

18.02 g · mol$^{-1}$ × 0.9880 mol = 17.80 g or 0.01780 kg.

$$m_{Cl^-} = \frac{\left(\dfrac{2 \text{ mol } Cl^-}{1 \text{ mol } MgCl_2}\right)(0.0120 \text{ mol } MgCl_2)}{0.01780 \text{ kg solvent)}} = 1.35\, m$$

(b) $m_{NaOH} = \dfrac{\left(\dfrac{6.75 \text{ g NaOH}}{40.00 \text{ g}\cdot\text{mol}^{-1}}\right)}{0.325 \text{ kg solvent)}} = 0.519\, m$

(c) 1.000 L of 15.00 M HCl(aq) will contain 15.00 mol with a mass of

15.00 × 36.46 g·mol$^{-1}$ = 546.9 g. The density of the 1.000 L of solution

is 1.0745 g·cm$^{-3}$ so the total mass in the solution is 1074.5 g. This leaves

1074.5 g − 546.9 g = 527.6 g as water.

$$\frac{15.00 \text{ mol HCl}}{0.5276 \text{ kg solvent}} = 28.43\, m$$

**8.47** (a) Molar mass of $CaCl_2 \cdot 6\,H_2O$ = 219.08 g·mol$^{-1}$, which consists of

110.98 g $CaCl_2$ and 108.10 g of water.

$$m_{CaCl_2} = \frac{x \text{ mol } CaCl_2 \cdot 6\,H_2O}{0.500 \text{ kg} + x(6 \times 0.018\,02 \text{ kg } H_2O)}$$

Note: $18.02$ g $H_2O = 0.018\ 02$ kg $H_2O = 1.000$ mol $H_2O$

$x$ = number of moles of $CaCl_2 \cdot 6\ H_2O$ needed to prepare a solution of

molality $m_{CaCl_2}$, in which each mole of $CaCl_2 \cdot 6\ H_2O$ produces $6(0.018$

$02$ kg$)$ of water as solvent (assuming we begin with $0.250$ kg $H_2O$).

For a $0.125\ m$ solution of $CaCl_2 \cdot 6\ H_2O$,

$$0.125\ m = \frac{x}{0.500\ \text{kg} + x(6)(0.01802\ \text{kg}\ H_2O)}$$

$x = 0.0625$ mol $+ x(0.0135$ mol$)$

$x - 0.0135x = 0.0625$ mol

$0.986\ x = 0.0625$ mol

$x = 0.0634$ mol $CaCl_2 \cdot 6\ H_2O$

$$\therefore (0.0634\ \text{mol}\ CaCl_2 \cdot 6\ H_2O) \times \left( \frac{219.08\ \text{g}\ CaCl_2 \cdot 6\ H_2O}{1\ \text{mol}\ CaCl_2 \cdot 6\ H_2O} \right) = 13.9\ \text{g}\ CaCl_2 \cdot 6\ H_2O$$

(b) Molar mass of $NiSO_4 \cdot 6\ H_2O = 262.86\ \text{g} \cdot \text{mol}^{-1}$, which consists of

$154.77$ g $NiSO_4$ and $108.09$ g $H_2O$.

$$m_{NiSO_4} = \frac{x\ \text{mol}\ NiSO_4 \cdot 6\ H_2O}{0.500\ \text{kg} + x\ (6 \times 0.01802\ \text{kg}\ H_2O)}$$

where $x$ = number of moles of $NiSO_4 \cdot 6H_2O$ needed to prepare a solution

of molality $m_{NiSO_4}$, in which each mole of $NiSO_4 \cdot 6\ H_2O$ produces

$6(0.018\ 02$ kg$)$ of water as solvent. Assuming we begin with $0.500$ kg

$H_2O$, for a $0.22\ m$ solution of $NiSO_4 \cdot 6\ H_2O$,

$$0.22\ m = \frac{x}{0.500\ \text{kg} + x(6)(0.01802\ \text{kg}\ H_2))}$$

$x = 0.11$ mol $+ x(0.0238$ mol$)$

$x - 0.0238\ x = 0.11$ mol

$0.976\ x = 0.11$ mol

$x = 0.11$ mol $NiSO_4 \cdot 6\ H_2O$

$$\therefore (0.11\ \text{mol}\ NiSO_4 \cdot 6\ H_2O) \times \left( \frac{262.86\ \text{g}\ NiSO_4 \cdot 6\ H_2O}{1\ \text{mol}\ NiSO_4 \cdot 6\ H_2O} \right) = 29\ \text{g}\ NiSO_4 \cdot 6\ H_2O$$

**8.49** (a) $P = x_{solvent} \times P_{pure\ solvent}$

At 100°C, the normal boiling point of water, the vapor pressure of water is 1.00 atm. If the mole fraction of sucrose is 0.100, then the mole fraction of water is 0.900:

$P = 0.900 \times 1.000\ atm = 0.900\ atm$ or 684 Torr

(b) First, the molality must be converted to mole fraction. If the molality is $0.100\ mol \cdot kg^{-1}$, then there will be 0.100 mol sucrose per 1000 g of water.

$$x_{H_2O} = \frac{n_{H_2O}}{n_{H_2O} + n_{sucrose}} = \frac{\dfrac{1000\ g}{18.02\ g \cdot mol^{-1}}}{\dfrac{1000\ g}{18.02\ g \cdot mol^{-1}} + 0.100\ mol} = 0.998$$

$P = 0.998 \times 1.000\ atm = 0.998\ atm$ or 758 Torr

**8.51** (a) The vapor pressure of water at 0°C is 4.58 Torr. The concentration of the solution must be converted to mole fraction in order to perform the calculation. A solution that is 2.50% ethylene glycol by mass will contain 2.50 g of ethylene glycol per 97.50 g of water.

$$x_{H_2O} = \frac{n_{H_2O}}{n_{H_2O} + n_{ethylene\ glycol}} = \frac{\dfrac{97.50\ g}{18.02\ g \cdot mol^{-1}}}{\dfrac{97.50\ g}{18.02\ g \cdot mol^{-1}} + \dfrac{2.50\ g}{62.07\ g \cdot mol^{-1}}} = 0.993$$

$P = x_{solvent} \times P_{pure\ solvent}$

$P = 0.993 \times 4.58\ Torr = 4.55\ Torr$

(b) The vapor pressure of water is 355.26 Torr at 80°C. The concentration given in $mol \cdot kg^{-1}$ must be converted to mole fraction.

$$x_{H_2O} = \frac{n_{H_2O}}{n_{H_2O} + n_{Na^+} + n_{OH^-}} = \frac{\dfrac{1000\ g}{18.02\ g \cdot mol^{-1}}}{\dfrac{1000\ g}{18.02\ g \cdot mol^{-1}} + 2 \times 0.155\ mol} = 0.9944$$

$P = 0.9972 \times 355.26\ Torr = 354.3\ Torr$

(c) At 10°C, the vapor pressure of water is 9.21 Torr. The concentration must be expressed in terms of mole fraction:

SM-214

$$x_{H_2O} = \frac{n_{H_2O}}{n_{H_2O} + n_{urea}} = \frac{\dfrac{100\text{ g}}{18.02\text{ g}\cdot\text{mol}^{-1}}}{\dfrac{100\text{ g}}{18.02\text{ g}\cdot\text{mol}^{-1}} + \dfrac{5.95\text{ g}}{60.06\text{ g}\cdot\text{mol}^{-1}}} = 0.982$$

$$P = 0.982 \times 9.21\text{ Torr} = 9.04\text{ Torr}$$

The change in the vapor pressure will therefore be $9.21 - 9.04 = 0.17$ Torr

**8.53** (a) From the relationship $P = x_{\text{solvent}} \times P_{\text{pure solvent}}$ we can calculate the mole fraction of the solvent:

$$94.8\text{ Torr} = x_{\text{solvent}} \times 100.0\text{ Torr}$$

$$x_{\text{solvent}} = 0.948$$

The mole fraction of the unknown compound will be $1.000 - 0.948 = 0.052$.

(b) The molar mass can be calculated by using the definition of mole fraction for either the solvent or solute. In this case, the math is slightly easier if the definition of mole fraction of the solvent is used:

$$x_{\text{solvent}} = \frac{n_{\text{solvent}}}{n_{\text{unknown}} + n_{\text{solvent}}}$$

$$0.948 = \frac{\dfrac{100\text{ g}}{78.11\text{ g}\cdot\text{mol}^{-1}}}{\dfrac{100\text{ g}}{78.11\text{ g}\cdot\text{mol}^{-1}} + \dfrac{8.05\text{ g}}{M_{\text{unknown}}}}$$

$$M_{\text{unknown}} = \frac{8.05\text{ g}}{\left[\left(\dfrac{100\text{ g}}{78.11\text{ g}\cdot\text{mol}^{-1}}\right)\Big/0.948\right] - \left(\dfrac{100\text{ g}}{78.11\text{ g}\cdot\text{mol}^{-1}}\right)} = 115\text{ g}\cdot\text{mol}^{-1}$$

**8.55** (a) $\Delta T_b = ik_b m$

Because sucrose is a nonelectrolyte, $i = 1$.

$$\Delta T_b = 0.51\text{ K}\cdot\text{kg}\cdot\text{mol}^{-1} \times 0.10\text{ mol}\cdot\text{kg}^{-1} = 0.051\text{ K or } 0.051°C$$

The boiling point will be $100.000°C + 0.051°C = 100.051°C$.

(b) $\Delta T_b = ik_b m$

For NaCl, $i = 2$.

$$\Delta T_b = 2 \times 0.51\text{ K}\cdot\text{kg}\cdot\text{mol}^{-1} \times 0.22\text{ mol}\cdot\text{kg}^{-1} = 0.22\text{ K or } 0.22°C$$

The boiling point will be $100.00°C + 0.22°C = 100.22°C$.

(c) $\Delta T_b = i k_b m$

$$\Delta T_b = 2 \times 0.51 \, \text{K} \cdot \text{kg} \cdot \text{mol}^{-1} \times \frac{\left( \dfrac{0.230 \, \text{g}}{25.94 \, \text{g} \cdot \text{mol}^{-1}} \right)}{0.100 \, \text{kg}} = 0.090 \, \text{K or } 0.090°C$$

The boiling point will be $100.000°C + 0.090°C$ or $100.090°C$.

**8.57**  (a)  Pure water has a vapor pressure of 760.00 Torr at 100°C. The mole fraction of the solution can be determined from $P = x_{\text{solvent}} \times P_{\text{pure solvent}}$.

$751 \, \text{Torr} = x_{\text{solvent}} \times 760.00 \, \text{Torr}$

$x_{\text{solvent}} = 0.988$

The mole fraction needs to be converted to molality:

$$x_{\text{solvent}} = 0.988 = \frac{n_{\text{H}_2\text{O}}}{n_{\text{H}_2\text{O}} + n_{\text{solute}}}$$

Because the absolute amount of solution is not important, we can assume that the total number of moles = 1.00.

$$0.988 = \frac{n_{\text{H}_2\text{O}}}{1.00 \, \text{mol}}$$

$n_{\text{H}_2\text{O}} = 0.988 \, \text{mol}; \ n_{\text{solute}} = 0.012 \, \text{mol}$

$$\text{molality} = \frac{0.012 \, \text{mol}}{\left( \dfrac{0.988 \, \text{mol} \times 18.02 \, \text{g} \cdot \text{mol}^{-1}}{1000 \, \text{g} \cdot \text{kg}^{-1}} \right)} = 0.67 \, \text{mol} \cdot \text{kg}^{-1}$$

Knowing the mole fraction, one can calculate $\Delta T_b$:

$\Delta T_b = k_b m$

$\Delta T_b = 0.51 \, \text{K} \cdot \text{mol} \cdot \text{kg}^{-1} \times 0.67 \, \text{mol} \cdot \text{kg}^{-1} = 0.34 \, \text{K or } 0.34°C$

Boiling point $= 100.00°C + 0.34°C = 100.34°C$

(b)  The procedure is the same as in (a). Pure benzene has a vapor pressure of 760.00 Torr at 80.1°C. The mole fraction of the solution can be determined from

$P = x_{\text{solvent}} \times P_{\text{pure solvent}}$

$740 \, \text{Torr} = x_{\text{solvent}} \times 760.00 \, \text{Torr}$

$x_{\text{solvent}} = 0.974$

The mole fraction needs to be converted to molality:

$$x_{solvent} = 0.974 = \frac{n_{benzene}}{n_{benzene} + n_{solute}}$$

Because the absolute amount of solution is not important, we can assume that the total number of moles = 1.00:

$$0.974 = \frac{n_{benzene}}{1.00 \text{ mol}}$$

$$n_{benzene} = 0.974 \text{ mol}; \ n_{solute} = 0.026 \text{ mol}$$

$$\text{molality} = \frac{0.026 \text{ mol}}{\left(\dfrac{0.974 \text{ mol} \times 78.11 \text{ g} \cdot \text{mol}^{-1}}{1000 \text{ g} \cdot \text{kg}^{-1}}\right)} = 0.34 \text{ mol} \cdot \text{kg}^{-1}$$

Knowing the mole fraction, one can calculate $\Delta T_b$:

$$\Delta T_b = k_b m$$
$$\Delta T_b = 2.53 \text{ K} \cdot \text{kg} \cdot \text{mol}^{-1} \times 0.34 \text{ mol} \cdot \text{kg}^{-1} = 0.86 \text{ K or } 0.86°C$$
Boiling point $= 80.1°C + 0.86°C = 81.0°C$

**8.59** $\Delta T_b = 61.51°C - 61.20°C = 0.31°C \text{ or } 0.31 \text{ K}$

$$\Delta T_b = k_b \times \text{molality}$$
$$0.31 \text{ K} = 4.95 \text{ K} \cdot \text{kg} \cdot \text{mol}^{-1} \times \text{molality}$$
$$0.31 \text{ K} = 4.95 \text{ K} \cdot \text{kg} \cdot \text{mol}^{-1} \times \frac{\left(\dfrac{1.05 \text{ g}}{M_{unknown}}\right)}{0.100 \text{ kg}}$$
$$\frac{0.100 \text{ kg} \times 0.31 \text{ K}}{4.95 \text{ K} \cdot \text{kg} \cdot \text{mol}^{-1}} = \frac{1.05 \text{ g}}{M_{unknown}}$$
$$M_{unknown} = \frac{1.05 \text{ g} \times 4.95 \text{ K} \cdot \text{kg} \cdot \text{mol}^{-1}}{0.100 \text{ kg} \times 0.31 \text{ K}} = 1.7 \times 10^2 \text{ g} \cdot \text{mol}^{-1}$$

**8.61** (a) $\Delta T_f = i k_f m$

Because sucrose is a nonelectrolyte, $i = 1$.

$$\Delta T_f = 1.86 \text{ K} \cdot \text{kg} \cdot \text{mol}^{-1} \times 0.10 \text{ mol} \cdot \text{kg}^{-1} = 0.19 \text{ K or } 0.19°C$$

The freezing point will be $0.000°C - 0.19°C = -0.19°C$.

(b) $\Delta T_f = i k_f m$

For NaCl, $i = 2$

$\Delta T_f = 2 \times 1.86 \text{ K} \cdot \text{kg} \cdot \text{mol}^{-1} \times 0.22 \text{ mol} \cdot \text{kg}^{-1} = 0.82 \text{ K or } 0.82°C$

The freezing point will be $0.00°C - 0.82°C = -0.82°C$.

(c) $\Delta T_f = i k_f m$

$i = 2$ for LiF

$\Delta T_f = 2 \times 1.86 \text{ K} \cdot \text{kg} \cdot \text{mol}^{-1} \times \dfrac{\left( \dfrac{0.120 \text{ g}}{25.94 \text{ g} \cdot \text{mol}^{-1}} \right)}{0.100 \text{ kg}} = 0.172 \text{ K or } 0.172°C$

The freezing point will be $0.000°C - 0.172°C = -0.172°C$.

**8.63** $\Delta T_f = 179.8°C - 176.9°C = 2.9°C \text{ or } 2.9 \text{ K}$

$\Delta T_f = k_f m$

$2.9 \text{ K} = (39.7 \text{ K} \cdot \text{kg} \cdot \text{mol}^{-1})m$

$2.9 \text{ K} = (39.7 \text{ K} \cdot \text{kg} \cdot \text{mol}^{-1}) \dfrac{\left( \dfrac{1.14 \text{ g}}{M_{unknown}} \right)}{0.100 \text{ kg}}$

$\dfrac{0.100 \text{ kg} \times 2.9 \text{ K}}{39.7 \text{ K} \cdot \text{kg} \cdot \text{mol}^{-1}} = \dfrac{1.14 \text{ g}}{M_{unknown}}$

$M_{unknown} = \dfrac{1.14 \text{ g} \times 39.7 \text{ K} \cdot \text{kg} \cdot \text{mol}^{-1}}{0.100 \text{ kg} \times 2.9 \text{ K}} = 1.6 \times 10^2 \text{ g} \cdot \text{mol}^{-1}$

**8.65** (a) First, calculate the molality using the change in boiling point, and then use that value to calculate the change in freezing point.

$\Delta T_b = k_b m$

$\Delta T_b = 82.0°C - 80.1°C = 1.9°C \text{ or } 1.9 \text{ K}$

$1.9 \text{ K} = 2.53 \text{ K} \cdot \text{Kg} \cdot \text{mol}^{-1} \times \text{molality}$

$\text{molality} = 0.75 \text{ mol} \cdot \text{kg}^{-1}$

$\Delta T_f = k_f m$

$\Delta T_f = 5.12 \text{ K} \cdot \text{kg} \cdot \text{mol}^{-1} \times 0.75 \text{ mol} \cdot \text{kg}^{-1}$

$\Delta T_f = 3.84 \text{ K or } 3.84°C$

The freezing point will be $5.5°C - 3.84°C = 1.7°C$

(b) $\Delta T_f = k_f m$

Because the freezing point of water $= 0.00°C$, the freezing point of the solution equals the freezing point depression.

$3.04 \text{ K} = 1.86 \text{ K} \cdot \text{kg} \cdot \text{mol}^{-1} \times \text{molality}$

$\text{molality} = 1.63 \text{ mol} \cdot \text{kg}^{-1}$

(c) $\Delta T_f = k_f m$

$1.94 \text{ K} = (1.86 \text{ K} \cdot \text{kg} \cdot \text{mol}^{-1}) m$

$1.94 \text{ K} = 1.86 \text{ K} \cdot \text{kg} \cdot \text{mol}^{-1} \times \dfrac{n_{\text{solute}}}{\text{kg (solvent)}}$

$1.94 \text{K} = 1.86 \text{ K} \cdot \text{kg} \cdot \text{mol}^{-1} \times \dfrac{n_{\text{solute}}}{0.200 \text{ kg}}$

$n_{\text{solute}} = 0.209 \text{ mol}$

**8.67** (a) A 1.00% aqueous solution of NaCl will contain 1.00 g of NaCl for 99.0 g of water. To use the freezing point depression equation, we need the molality of the solution:

$$\text{molality} = \dfrac{\left( \dfrac{1.00 \text{ g}}{58.44 \text{ g} \cdot \text{mol}^{-1}} \right)}{0.0990 \text{ kg}} = 0.173 \text{ mol} \cdot \text{kg}^{-1}$$

$\Delta T_f = i k_f m$

$\Delta T_f = i \, (1.86 \text{ K} \cdot \text{kg} \cdot \text{mol}^{-1})(0.173 \text{ mol} \cdot \text{kg}^{-1}) = 0.593 \text{ K}$

$i = 1.84$

(b) molality of all solute species (undissociated NaCl(aq) plus $Na^+$ (aq) $+ Cl^-$ (aq)) $= 1.84 \times 0.173 \text{ mol} \cdot \text{kg}^{-1} = 0.318 \text{ mol} \cdot \text{kg}^{-1}$

(c) If all the NaCl had dissociated, the total molality in solution would have been $0.346 \text{ mol} \cdot \text{kg}^{-1}$, giving an $i$ value equal to 2. If no dissociation had taken place, the molality in solution would have equaled 0.173 $\text{mol} \cdot \text{kg}^{-1}$.

$\text{NaCl(aq)} \qquad \rightarrow \text{Na}^+(\text{aq}) + \text{Cl}^-(\text{aq})$

$0.173 \text{ mol} \cdot \text{kg}^{-1} - x \qquad x \qquad x$

$$0.173 \text{ mol} \cdot \text{kg}^{-1} - x + x + x = 0.318 \text{ mol} \cdot \text{kg}^{-1}$$
$$0.173 \text{ mol} \cdot \text{kg}^{-1} + x = 0.318 \text{ mol} \cdot \text{kg}^{-1}$$
$$x = 0.145 \text{ mol} \cdot \text{kg}^{-1}$$

$$\% \text{ dissociation} = \frac{0.145 \text{ mol} \cdot \text{kg}^{-1}}{0.173 \text{ mol} \cdot \text{kg}^{-1}} \times 100 = 83.8\%$$

**8.69** For an electrolyte that dissociates into two ions, the van't Hoff $i$ factor will be 1 plus the degree of dissociation, in this case 0.075. This can be readily seen for the general case MX. Let A = initial concentration of MX (if none is dissociated) and let Y = the concentration of MX that subsequently dissociates:

$$\text{MX(aq)} \longrightarrow \text{M}^{n+}\text{(aq)} + \text{X}^{n-}\text{(aq)}$$

| A − Y | Y | Y |

The total concentration of solute species is $(A - Y) + Y + Y = A + Y$

The value of $i$ will then be equal to A + Y or 1.075.

The freezing point change is then easy to calculate:

$$\Delta T_f = i k_f m$$
$$= 1.075 \times 1.86 \text{ K} \cdot \text{kg} \cdot \text{mol}^{-1} \times 0.10 \text{ mol} \cdot \text{kg}^{-1}$$
$$= 0.20$$

Freezing point of the solution will be $0.00°C - 0.20°C = -0.20°C$.

**8.71** (a) $\Pi = iRT \times \text{molarity}$
$$= 1 \times 0.082\,06 \text{ L} \cdot \text{atm} \cdot \text{K}^{-1} \cdot \text{mol}^{-1} \times 293 \text{ K} \times 0.010 \text{ mol} \cdot \text{L}^{-1}$$
$$= 0.24 \text{ atm or } 1.8 \times 10^2 \text{ Torr}$$

(b) Because HCl is a strong acid, it should dissociate into two ions, $H^+$ and $Cl^-$, so $i = 2$.

$$\Pi = 2 \times 0.082\,06 \text{ L} \cdot \text{atm} \cdot \text{K}^{-1} \cdot \text{mol}^{-1} \times 293 \text{ K} \times 1.0 \text{ mol} \cdot \text{L}^{-1}$$
$$= 48 \text{ atm}$$

(c) $CaCl_2$ should dissociate into 3 ions in solution, therefore $i = 3$.

$$\Pi = 3 \times 0.082\,06 \text{ L} \cdot \text{atm} \cdot \text{K}^{-1} \cdot \text{mol}^{-1} \times 293 \text{ K} \times 0.010 \text{ mol} \cdot \text{L}^{-1}$$
$$= 0.72 \text{ atm or } 5.5 \times 10^2 \text{ Torr}$$

**8.73** The polypeptide is a nonelectrolye, so $i = 1$.

$\Pi = iRT \times$ molarity

$$\Pi = \frac{3.74 \text{ Torr}}{760 \text{ Torr} \cdot \text{atm}^{-1}} = 1 \times 0.082\,06 \text{ L} \cdot \text{atm} \cdot \text{K}^{-1} \cdot \text{mol}^{-1} \times 300 \text{ K} \times \frac{\left(\dfrac{0.40 \text{ g}}{M_{unknown}}\right)}{1.0 \text{ L}}$$

$$M_{unknown} = \frac{0.082\,06 \text{ L} \cdot \text{atm} \cdot \text{K}^{-1} \cdot \text{mol}^{-1} \times 300 \text{ K} \times 0.40 \text{ g} \times 760 \text{ Torr} \cdot \text{atm}^{-1}}{3.74 \text{ Torr} \times 1.0 \text{ L}}$$

$$= 2.0 \times 10^3 \text{ g} \cdot \text{mol}^{-1}$$

**8.75** We assume the polymer to be a nonelectrolyte, so $i = 1$.

$\Pi = iRT \times$ molarity

$$\Pi = \frac{6.3 \text{ Torr}}{760 \text{ Torr} \cdot \text{atm}^{-1}} = 1 \times 0.08206 \text{ L} \cdot \text{atm} \cdot \text{K}^{-1} \cdot \text{mol}^{-1} \times 293 \text{ K} \times \frac{\left(\dfrac{0.20 \text{ g}}{M_{unknown}}\right)}{0.100 \text{ L}}$$

$$M_{unknown} = \frac{0.08206 \text{ L} \cdot \text{atm} \cdot \text{K}^{-1} \cdot \text{mol}^{-1} \times 293 \text{ K} \times 0.20 \text{ g} \times 760 \text{ Torr} \cdot \text{atm}^{-1}}{6.3 \text{ Torr} \times 0.100 \text{ L}}$$

$$= 5.8 \times 10^3 \text{ g} \cdot \text{mol}^{-1}$$

**8.77** (a) $C_{12}H_{22}O_{11}$ should be a nonelectrolyte, so $i = 1$.

$\Pi = iRT \times$ molarity

$\quad = 1 \times 0.082\,06 \text{ L} \cdot \text{atm} \cdot \text{K}^{-1} \cdot \text{mol}^{-1} \times 293 \text{ K} \times 0.050 \text{ mol} \cdot \text{L}^{-1}$

$\quad = 1.2 \text{ atm}$

(b) NaCl dissociates to give 2 ions in solution, so $i = 2$.

$\Pi = iRT \times$ molarity

$\quad = 2 \times 0.082\,06 \text{ L} \cdot \text{atm} \cdot \text{K}^{-1} \cdot \text{mol}^{-1} \times 293 \text{ K} \times 0.0010 \text{ mol} \cdot \text{L}^{-1}$

$\quad = 0.048 \text{ atm or } 36 \text{ Torr}$

(c) AgCN dissociates in solution to give two ions ($Ag^+$ and $CN^-$), so $i = 2$.

$\Pi = iRT \times$ molarity

We must assume that the AgCN does not significantly affect either the volume or density of the solution, which is reasonable given the very small amount of it that dissolves.

$$\Pi = 2 \times 0.082\ 06\ \text{L} \cdot \text{atm} \cdot \text{K}^{-1} \cdot \text{mol}^{-1} \times 293\ \text{K} \times$$

$$\dfrac{\left(\dfrac{2.3 \times 10^{-5}\ \text{g}}{133.89\ \text{g} \cdot \text{mol}^{-1}}\right)}{\left(\dfrac{100\ \text{g H}_2\text{O}}{1\ \text{g} \cdot \text{cm}^{-3}\ \text{H}_2\text{O} \times 1000\ \text{cm}^3 \cdot \text{L}^{-1}}\right)}$$

$$= 8.3 \times 10^{-5}\ \text{atm}$$

**8.79** $\Pi = n\dfrac{RT}{V} = \dfrac{m}{M}\dfrac{RT}{V}$

where $m$ is the mass of unknown compound.

$$M = \dfrac{mRT}{V\Pi}$$

$$M = \dfrac{(0.166\ \text{g})(0.082\ 06\ \text{L} \cdot \text{atm} \cdot \text{K}^{-1} \cdot \text{mol}^{-1})(293\ \text{K})}{(0.010\ \text{L})\left(\dfrac{1.2\ \text{Torr}}{760\ \text{Torr} \cdot \text{atm}^{-1}}\right)} = 2.5 \times 10^5\ \text{g} \cdot \text{mol}^{-1}$$

**8.81** (a) To determine the vapor pressure of the solution, we need to know the mole fraction of each component.

$$x_{\text{benzene}} = \dfrac{1.50\ \text{mol}}{1.50\ \text{mol} + 0.50\ \text{mol}} = 0.75$$

$$x_{\text{toluene}} = 1 - x_{\text{benzene}} = 0.25$$

$$P_{\text{total}} = (0.75 \times 94.6\ \text{Torr}) + (0.25 \times 29.1\ \text{Torr}) = 78.2\ \text{Torr}$$

The vapor phase composition will be given by

$$x_{\text{benzene in vapor phase}} = \dfrac{P_{\text{benzene}}}{P_{\text{total}}} = \dfrac{0.75 \times 94.6\ \text{Torr}}{78.2\ \text{Torr}} = 0.91$$

$$x_{\text{toluene in vapor phase}} = 1 - 0.91 = 0.09$$

The vapor is richer in the more volatile benzene, as expected.

(b) The procedure is the same as in (a) but the number of moles of each component must be calculated first:

$$n_{\text{benzene}} = \frac{15.0 \text{ g}}{78.11 \text{ g} \cdot \text{mol}^{-1}} = 0.192$$

$$n_{\text{toluene}} = \frac{65.3 \text{ g}}{92.14 \text{ g} \cdot \text{mol}^{-1}} = 0.709$$

$$x_{\text{benzene}} = \frac{0.192 \text{ mol}}{0.192 \text{ mol} + 0.709 \text{ mol}} = 0.213$$

$$x_{\text{toluene}} = 1 - x_{\text{benzene}} = 0.787$$

$$P_{\text{total}} = (0.213 \times 94.6 \text{ Torr}) + (0.787 \times 29.1 \text{ Torr}) = 43.0 \text{ Torr}$$

The vapor phase composition will be given by

$$x_{\text{benzene in vapor phase}} = \frac{P_{\text{benzene}}}{P_{\text{total}}} = \frac{0.213 \times 94.6 \text{ Torr}}{43.0 \text{ Torr}} = 0.469$$

$$x_{\text{toluene in vapor phase}} = 1 - 0.469 = 0.531$$

**8.83**  To calculate this quantity, we must first find the mole fraction of each that will be present in the mixture. This value is obtained from the relationship

$$P_{\text{total}} = [x_{\text{1,1-dichloroethane}} \times P_{\text{pure 1,1-dichloroethane}}]$$
$$+ [x_{\text{1,1-dichlorotetrafluoroethane}} \times P_{\text{pure 1,1-dichlorotetrafluoroethane}}]$$

$$157 \text{ Torr} = [x_{\text{1,1-dichloroethane}} \times 228 \text{ Torr}] + [x_{\text{1,1-dichlorotetrafluoroethane}} \times 79 \text{ Torr}]$$

$$157 \text{ Torr} = [x_{\text{1,1-dichloroethane}} \times 228 \text{ Torr}] + [(1 - x_{\text{1,1-dichloroethane}}) \times 79 \text{ Torr}]$$

$$157 \text{ Torr} = 79 \text{ Torr} + [(x_{\text{1,1-dichloroethane}} \times 228 \text{ Torr}) - (x_{\text{1,1-dichloroethane}} \times 79 \text{ Torr})]$$

$$78 \text{ Torr} = x_{\text{1,1-dichloroethane}} \times 149 \text{ Torr}$$

$$x_{\text{1,1-dichloroethane}} = 0.52$$

$$x_{\text{1,1-dichlorotetrafluoroethane}} = 1 - 0.52 = 0.48$$

To calculate the number of grams of 1,1-dichloroethane, we use the definition of mole fraction. Mathematically, it is simpler to use the

$$x_{\text{1,1-dichlorotetrafluoroethane}} \text{ definition:}$$

$$n_{\text{1,1-dichlorotetrafluoroethane}} = \frac{100.0\ \text{g}}{170.92\ \text{g}\cdot\text{mol}^{-1}} = 0.5851\ \text{mol}$$

$$x_{\text{1,1-dichlorotetrafluoroethane}} = \frac{0.5851\ \text{mol}}{0.5851\ \text{mol} + \dfrac{m}{98.95\ \text{g}\cdot\text{mol}^{-1}}} = 0.48$$

$$0.5851\ \text{mol} = 0.48 \times \left[ 0.5851\ \text{mol} + \frac{m}{98.95\ \text{g}\cdot\text{mol}^{-1}} \right]$$

$$0.5851\ \text{mol} - (0.48 \times 0.5851\ \text{mol}) = 0.48 \times \left[ \frac{m}{98.95\ \text{g}\cdot\text{mol}^{-1}} \right]$$

$$0.3042\ \text{mol} = 0.004\ 851\ \text{mol}\cdot\text{g}^{-1} \times m$$

$$m = 63\ \text{g}$$

**8.85** Raoult's Law applies to the vapor pressure of the mixture, so positive deviation means that the vapor pressure is higher than expected for an ideal solution. Negative deviation means that the vapor pressure is lower than expected for an ideal solution. Negative deviation will occur when the interactions between the different molecules are somewhat stronger than the interactions between molecules of the same kind  (a) For methanol and ethanol, we expect the types of intermolecular attractions in the mixture to be similar to those in the component liquids, so that an ideal solution is predicted.  (b) For HF and $H_2O$, the possibility of intermolecular hydrogen bonding between water and HF would suggest that negative deviation would be observed, which is the case. HF and $H_2O$ form an azeotrope that boils at 111°C, a temperature higher than the boiling point of either HF (19.4°C) or water.  (c) Because hexane is nonpolar and water is polar with hydrogen bonding, we would expect a mixture of these two to exhibit positive deviation (the interactions between the different molecules would be weaker than the intermolecular forces between like molecules). Hexane and water do form an azeotrope that boils at 61.6°C, a temperature below the boiling point of either hexane or water.

**8.87**  $x_{methanol} = 0.65$ and $x_{ethanol} = 0.35$

$P_{methanol} = x_{methanol} \cdot P_{methanol}^* = (0.65)(122.7 \text{ Torr}) = 80 \text{ Torr}$

$P_{ethanol} = x_{ethanol} \cdot P_{ethanol}^* = (0.35)(58.9 \text{ Torr}) = 21 \text{ Torr}$

from the ideal gas law we know that $n \propto P$, therefore:

$$x_{methanol} = \frac{n_{methanol}}{n_{methanol} + n_{ethanol}} = \frac{P_{methanol}}{P_{methanol} + P_{ethanol}} = \frac{80 \text{ Torr}}{80 \text{ Torr} + 21 \text{ Torr}} = 0.79$$

and

$$x_{ethanol} = \frac{n_{ethanol}}{n_{methanol} + n_{ethanol}} = \frac{P_{ethanol}}{P_{methanol} + P_{ethanol}} = \frac{21 \text{ Torr}}{80 \text{ Torr} + 21 \text{ Torr}} = 0.21$$

**8.89**  (a) stronger;  (b) low;  (c) high;  (d) weaker;  (e) weak, low;  (f) low;  (g) strong, high

**8.91**  (a, b) Viscosity and surface tension decrease with increasing temperature; at high temperatures the molecules readily move away from their neighbors because of increased kinetic energy.  (c, d) Evaporation rate and vapor pressure increase with increasing temperature because the kinetic energy of the molecules increase with temperature, and the molecules are more likely to escape into the gas phase.

**8.93**  If the external pressure is lowered, water boils at a lower temperature. The boiling temperature is the temperature at which the vapor pressure equals the external pressure. Vapor pressure increases with temperature. If the external pressure is reduced, the vapor pressure of the water will reach that value at a lower temperature; thus, lowering the external pressure lowers the boiling temperature.

**8.95**  (a) No. Solid helium will melt before converting to a gas at 5 K.  (b) Yes. Raising the temperature of rhombic sulfur at 1 atm will result in a phase change to monoclinic sulfur at 96°C.  (c) Rhombic sulfur is the form observed.  (d) At 100°C and 1 atm, the monoclinic form is more stable.    (e)  No. The phase diagram shows no area where solid diamond and gaseous carbon are adjacent.

**8.97**    (a) $\dfrac{25.0\ \text{Torr}}{31.83\ \text{Torr}} \times 100 = 78.5\%$;    (b) At 25°C the vapor pressure of water is only 23.76 Torr, so some of the water vapor in the air would condense as dew or fog.

**8.99**    (a)  At 30°C, the vapor pressure of pure water = 31.83 Torr. According to Raoult's law, $P = x_{\text{solvent}} \cdot P_{\text{pure}}$. To calculate the $x_{\text{solvent}}$: 0.50 $m$ NaCl gives 0.50 moles $Na^+$ and 0.50 moles $Cl^-$ per kg solvent. The number of moles of solvent is 1000 kg $\div$ 18.02 $g \cdot mol^{-1}$ = 55.49 mol. The total number of moles = 0.50 mol + 0.50 mol + 55.49 mol = 56.49 mol.

$$x_{\text{solvent}} = \frac{55.49\ \text{mol}}{56.49\ \text{mol}} = 0.9823$$

$$P = x_{\text{solvent}} \cdot P_{\text{pure}} = 0.9823 \times 31.83\ \text{Torr} = 31.26\ \text{Torr}$$

(b)

At 100°C, $P_{\text{pure}} = 760\ \text{Torr}$ : $P = x_{\text{solvent}} \cdot P_{\text{pure}} = 0.9823 \times 760\ \text{Torr} = 747\ \text{Torr}$

At 0°C, $P_{\text{pure}} = 4.58\ \text{Torr}$; $P = x_{\text{solvent}} \cdot P_{\text{pure}} = 0.9823 \times 4.58\ \text{Torr} = 4.50\ \text{Torr}$

**8.101**    $\Delta T = 77.19°C - 76.54°C = 0.65°\ 0.65\ \text{K}$

$$\Delta T_{\text{b}} = ik_{\text{b}}m = ik_{\text{b}}\frac{\left(\dfrac{m_{\text{solute}}}{M_{\text{solute}}}\right)}{\text{kg solvent}}$$

$$M_{\text{solute}} = \frac{ik_{\text{b}}m_{\text{solute}}}{(\text{kg solvent})(\Delta T_{\text{b}})} = \frac{(1)(4.95\ \text{K} \cdot \text{kg} \cdot \text{mol}^{-1})(0.30\ \text{g})}{(0.0300\ \text{kg})(0.65\ \text{K})}$$

$$= 76\ \text{g} \cdot \text{mol}^{-1}$$

**8.103**    $\Delta T = k_{\text{f}}m$, $m = \dfrac{n_{\text{solute}}}{\text{mass}_{\text{solvent(kg)}}}$, $n_{\text{solute}} = \dfrac{\text{mass}_{\text{solute}}}{M_{\text{solute}}}$

or

$$\Delta T = k_f \frac{\text{mass}_{\text{solute}}}{M_{\text{solute}} \times \text{mass}_{\text{solvent(kg)}}}$$

Solving for $M_{\text{solute}}$,

$$M_{\text{solute}} = \frac{k_f \times \text{mass}_{\text{solute}}}{\Delta T_f \times \text{mass}_{\text{solvent(kg)}}}$$

(a) If $\text{mass}_{\text{solute}}$ appears greater, $M_{\text{solute}}$ appears greater than actual molar mass, as $\text{mass}_{\text{solute}}$ occurs in the numerator above. Also, the $\Delta T$ measured will be smaller because less solute will actually be dissolved. This has the same effect as increasing the apparent $M_{\text{solute}}$.

(b) Because the true $\text{mass}_{\text{solvent}} = d \times V$, if $d_{\text{solvent}}$ is less than $1.00 \text{ g} \cdot \text{cm}^{-3}$, then the true $\text{mass}_{\text{solvent}}$ will be less than the assumed mass. $M_{\text{solute}}$ is inversely proportional to $\text{mass}_{\text{solvent}}$, so an artificially high $\text{mass}_{\text{solvent}}$ will lead to an artificially low $M_{\text{solute}}$.

(c) If true freezing point is higher than the recorded freezing point, true $\Delta T <$ assumed $\Delta T$, or assumed $\Delta T >$ true $\Delta T$, and $M_{\text{solute}}$ appears less than actual $M_{\text{solute}}$, as $\Delta T$ occurs in the denominator.

(d) If not all solute dissolved, the true $\text{mass}_{\text{solute}} <$ assumed $\text{mass}_{\text{solute}}$ or assumed $\text{mass}_{\text{solute}} >$ true $\text{mass}_{\text{solute}}$, and $M_{\text{solute}}$ appears greater than the actual $M_{\text{solute}}$, as $\text{mass}_{\text{solute}}$ occurs in the numerator.

**8.105** The water Coleridge referred to was seawater. The boards shrank due to osmosis (a net movement of water from the cells of the wood to the saline water). You can't drink seawater: osmosis would cause a net flow of water from the cells of the body to the saline-enriched surround solution and the cells would die.

**8.107** When a drop of aqueous solution containing $Ca(HCO_3)_2$ seeps through a cave ceiling, it encounters a situation where the partial pressure of $CO_2$ is reduced and the reaction $Ca(HCO_3)_2(aq) \rightarrow CaCO_2(s) + CO_2(g) + H_2O(l)$ occurs. The concentration of $CO_2$ decreases as the $CO_2$ escapes as a gas, with $CaCO_3$ precipitating and forming a column that extends downward from the ceiling to form a stalacite. Stalagmite formation is similar, except the drops fall to the floor and the precipitate grows upward.

**8.109** (a) The 5.22 cm or 52.2 mm rise for an aqueous solution must be converted to Torr or mmHg in order to be expressed into consistent units.

$$52.2 \text{ mm} \times \frac{0.998 \text{ g} \cdot \text{cm}^{-3}}{13.6 \text{ g} \cdot \text{cm}^{-3}} = 3.83 \text{ mmHg or } 3.83 \text{ Torr}$$

The molar mass can be calculated using the osmotic pressure equation:

$$\Pi = iRT \times \text{molarity}$$

Assume that the protein is a nonelectrolyte with $i = 1$ and that the amount of protein added does not significantly affect the volume of the solution.

$$\Pi = 1 \times 0.082\,06 \text{ L} \cdot \text{atm} \cdot \text{K}^{-1} \cdot \text{mol}^{-1} \times 293 \text{ K} \times \frac{\left( \dfrac{0.010 \text{ g}}{M_{\text{protein}}} \right)}{0.010 \text{ L}} = \frac{3.83 \text{ Torr}}{760 \text{ Torr} \cdot \text{atm}^{-1}}$$

$$M_{\text{protein}} = \frac{1 \times 0.082\,06 \text{ L} \cdot \text{atm} \cdot \text{K}^{-1} \cdot \text{mol}^{-1} \times 293 \text{ K} \times 0.010 \text{ g} \times 760 \text{ Torr} \cdot \text{atm}^{-1}}{0.010 \text{ L} \times 3.83 \text{ Torr}}$$

$$M_{\text{protein}} = 4.8 \times 10^3 \text{ g} \cdot \text{mol}^{-1}$$

(b) The freezing point can be calculated using the relationship $\Delta T_{\text{f}} = ik_{\text{f}}m$

$$\Delta T_{\text{f}} = ik_{\text{f}}m$$

$$= 1 \times 1.86 \text{ K} \cdot \text{kg} \cdot \text{mol}^{-1} \times \frac{\left( \dfrac{0.010 \text{ g}}{4.8 \times 10^3 \text{ g} \cdot \text{mol}^{-1}} \right)}{\left( \dfrac{10 \text{ mL} \times 1.00 \text{ g} \cdot \text{mL}^{-1}}{1000 \text{ g} \cdot \text{kg}^{-1}} \right)}$$

$$= 3.9 \times 10^{-4} \text{ K or } 3.9 \times 10^{-4} \text{ °C}$$

The freezing point will be $0.00\text{°C} - 3.9 \times 10^{-4}\text{°C} = -3.9 \times 10^{-4}\text{°C}$.

(c) The freezing point change is so small that it cannot be measured accurately, so osmotic pressure would be the preferred method for measuring the molecular weight.

**8.111** (a) The vapor pressure data can be used to obtain the concentration of the solution in terms of mole fraction, which in turn can be converted to molarity and used to calculate the osmotic pressure. Because 80.1°C is the boiling point of benzene, the vapor pressure of pure benzene at that temperature will be 760 Torr.

$$P_{\text{solution}} = x_{\text{solvent}} \times P_{\text{pure solvent}}$$

$$740 \text{ Torr} = x_{\text{solvent}} \times 760 \text{ Torr}$$

$$x_{\text{solvent}} = 0.974$$

The mole fraction of solute is, therefore, $1 - 0.974 = 0.026$. This means that there will be 0.026 moles of solute and 0.974 moles of benzene. From this we can calculate the molarity:

$$\text{molarity} = \frac{0.026 \text{ mol}}{\left( \dfrac{0.974 \text{ mol benzene} \times 78.11 \text{ g} \cdot \text{mol}^{-1} \text{ benzene}}{0.88 \text{ g} \cdot \text{mL}^{-1} \times 1000 \text{ mL} \cdot \text{L}^{-1}} \right)} = 0.30 \text{ mol} \cdot \text{L}^{-1}$$

The osmotic pressure is given by $\Pi = iRT \times \text{molarity}$. Assume $i = 1$.

$$\Pi = 1 \times 0.082\,06 \text{ L} \cdot \text{atm} \cdot \text{K}^{-1} \cdot \text{mol}^{-1} \times 293 \text{ K} \times 0.30 \text{ mol} \cdot \text{L}^{-1} = 7.2 \text{ atm}$$

(b) As in (a), the freezing point will be used to calculate the molality of the solution, which will be converted to molarity for the osmotic pressure calculation.

$$\Delta T_f = 5.5°C - 5.4°C = 0.1°C \text{ or } 0.1 \text{ K}$$

$$\Delta T_f = i k_f m$$

$$0.1 \text{ K} = 1 \times 5.12 \text{ K} \cdot \text{kg} \cdot \text{mol}^{-1} \times \text{molality}$$

$$\text{molality} = 0.02 \text{ mol} \cdot \text{kg}^{-1}$$

A $0.02 \text{ mol} \cdot \text{kg}^{-1}$ solution will contain 0.02 mol of solute and 1 kg of solvent. The volume of the solvent will be

$1000 \text{ g} \div 0.88 \text{ g} \cdot \text{mL}^{-1} = 1.1 \times 10^3 \text{ mL}$ or 1.1 L. The molar concentration

will thus be $\dfrac{0.02 \text{ mol}}{1.1 \text{ L}} = 0.02 \text{ M}$.

The osmotic pressure is given by $\Pi = iRT \times \text{molarity}$. Assume $i = 1$.

$$\Pi = 1 \times 0.082\,06 \text{ L} \cdot \text{atm} \cdot \text{K}^{-1} \cdot \text{mol}^{-1} \times 283 \text{ K} \times 0.02 \text{ mol} \cdot \text{L}^{-1} = 0.5 \text{ atm}$$

(c) The height of the rise of the solution is inversely proportional to the density of that solution. Mercury with a density of $13.6 \text{ g} \cdot \text{cm}^{-3}$ would produce a rise of $0.5 \text{ atm} \times 760 \text{ mmHg} \cdot \text{atm}^{-1}$ or $380 \text{ mmHg}$. For the benzene solution with a density of $0.88 \text{ g} \cdot \text{cm}^{-3}$, we would obtain

$$0.5 \text{ atm} \times \frac{760 \text{ mm}}{1 \text{ atm}} \times \frac{13.6 \text{ g} \cdot \text{cm}^{-3}}{0.88 \text{ g} \cdot \text{cm}^{-3}} = 6 \times 10^3 \text{ mm or 6 m}$$

**8.113** (a) $\Delta T_f = 1.72 \text{ K} = i \cdot k_f \cdot m = \left(1.86 \text{ K} \cdot \text{kg} \cdot \text{mol}^{-1}\right) \dfrac{n_{aa}}{0.95 \text{ kg}}$

where $n_{aa}$ are the moles of acetic acid in solution.
Assuming $i = 1$:

$$n_{aa} = \frac{(1.72 \text{ K})(0.95 \text{ kg})}{\left(1.86 \text{ K} \cdot \text{kg} \cdot \text{mol}^{-1}\right)} = 0.878 \text{ mol}$$

Experimentally, the molar mass of acetic acid is, therefore:

$$\frac{50 \text{ g}}{0.878 \text{ mol}} = 56.9 \text{ g} \cdot \text{mol}^{-1}$$

This experimental molar mass of acetic acid is less than the known molecular mass of the acid ($60.0 \text{ g} \cdot \text{mol}^{-1}$) indicating that the acid is dissociating in solution giving an $i > 1$.

(b) As in part (a) the experimental molar mass is first found:

$$\Delta T_f = 2.32 \text{ K} = i \cdot k_f \cdot m = (5.12 \text{ K} \cdot \text{kg} \cdot \text{mol}^{-1})\left(\frac{n_{aa}}{0.95 \text{ kg}}\right)$$

where $n_{aa}$ are the moles of acetic acid in solution.

Assuming $i = 1$: $n_{aa} = \dfrac{(2.32 \text{ K})(0.95 \text{ kg})}{(5.12 \text{ K} \cdot \text{kg} \cdot \text{mol}^{-1})} = 0.430 \text{ mol}.$

Experimentally, the molar mass of acetic acid is, therefore:

$$\frac{50 \text{ g}}{0.430 \text{ mol}} = 116 \text{ g} \cdot \text{mol}^{-1}.$$

This experimental molar mass of acetic acid is significantly higher than the known molar mass of the acid ($60.0 \text{ g·mol}^{-1}$) indicating that the acid is dissociating in solution giving an $i < 1$. A van't Hoff factor less than 1 indicates that acetic acid is not completely dissolved in the benzene, or acetic acid molecules are aggregating together in solution.

**8.115** (a) The data in Appendix 2A can be used to calculate the change in enthalpy and entropy for the vaporization of methanol:

$$CH_3OH(l) \longleftrightarrow CH_3OH(g)$$

$$\Delta H°_{vap} = \Delta H°_f(CH_3OH, g) - \Delta H°_f(CH_3OH, l)$$
$$= (-200.66 \text{ kJ} \cdot \text{mol}^{-1}) - (-238.86 \text{ kJ} \cdot \text{mol}^{-1})$$
$$= 38.20 \text{ kJ} \cdot \text{mol}^{-1}$$

$$\Delta S°_{vap} = S°_m(CH_3OH(g)) - S°_m(CH_3OH(l))$$
$$= 239.81 \text{ J} \cdot \text{K}^{-1} \cdot \text{mol}^{-1} - 126.8 \text{ J} \cdot \text{K}^{-1} \cdot \text{mol}^{-1}$$
$$= 113.0 \text{ J} \cdot \text{K}^{-1} \cdot \text{mol}^{-1}$$

To derive the general equation, we start with the expression that

$\Delta G°_{vap} = -RT \ln P$, where $P$ is the vapor pressure of the solvent. Because

$\Delta G°_{vap} = \Delta H°_{vap} - T\Delta S°_{vap}$, this is the relationship to use to determine the

temperature dependence of $\ln P$:

$$\Delta H°_{vap} - T\Delta S°_{vap} = -RT \ln P$$

This equation can be rearranged to give

$$\ln P = -\frac{\Delta H°_{vap}}{R} \cdot \frac{1}{T} + \frac{\Delta S°_{vap}}{R}$$

To create an equation specific to methanol, we can plug in the actual

values of $R$, $\Delta H°_{vap}$, and $\Delta S°_{vap}$:

$$\ln P = -\frac{38\,200 \text{ J} \cdot \text{mol}^{-1}}{8.314 \text{ J} \cdot \text{K}^{-1} \cdot \text{mol}^{-1}} \cdot \frac{1}{T} + \frac{113.0 \text{ J} \cdot \text{K}^{-1} \cdot \text{mol}^{-1}}{8.314 \text{ J} \cdot \text{K}^{-1} \cdot \text{mol}^{-1}}$$
$$= -\frac{4595 \text{ K}}{T} + 13.59$$

(b) The relationship to plot is $\ln P$ versus $\frac{1}{T}$. This should result in a

straight line whose slope is $-\dfrac{\Delta H°_{vap}}{R}$ and whose intercept is $\dfrac{\Delta S°_{vap}}{R}$. The pressure must be given in atm for this relationship, because atm is the standard state condition.

(c) Because we have already determined the equation, it is easiest to calculate the vapor by inserting the value of 0.0°C or 273.2 K :

$$\ln P = -\frac{4595\ K}{T} + 13.59 = -\frac{4595\ K}{273.2\ K} + 13.59 = -16.82 + 13.59 = -3.23$$

$P = 0.040$ atm or 30. Torr

(d) As in (c) we can use the equation to find the point where the vapor pressure of methanol = 1 atm.

$$\ln P = \ln 1 = 0 = -\frac{4595\ K}{T} + 13.59$$

$T = 338.1$ K

**8.117** (a) The plot of the data is shown below. On this plot, the slope $= -\dfrac{\Delta H°_{vap}}{R}$

and the intercept $= \dfrac{\Delta S°_{vap}}{R}$.

| Temp. (K) | $T^{-1}$ $(K^{-1})$ | Vapor pressure (Torr) | V.P. (atm) | $\ln P$ |
|-----------|---------------------|------------------------|------------|---------|
| 190 | 0.005 26 | 3.2 | 0.0042 | −5.47 |
| 228 | 0.004 38 | 68 | 0.089 | −2.41 |
| 250 | 0.004 00 | 240 | 0.316 | −1.15 |
| 273 | 0.003 66 | 672 | 0.884 | −0.123 |

From the curve fitting program:

$$y = -3358.714x + 12.247$$

(b) $-\dfrac{\Delta H°_{vap}}{R} = -3359$

$$\Delta H°_{vap} = (3359)(8.314\ J \cdot K^{-1} \cdot mol^{-1}) = 28\ kJ \cdot mol^{-1}$$

(c) $\dfrac{\Delta S°_{vap}}{R} = 12.25$

$$\Delta S°_{vap} = (12.25)(8.314\ J \cdot K^{-1} \cdot mol^{-1}) = 1.0 \times 10^2\ J \cdot K^{-1} \cdot mol^{-1}$$

(d)  The normal boiling point will be the temperature at which the vapor

pressure $= 1$ atm or at which the $\ln 1 = 0$. This will occur when

$T^{-1} = 0.0036$ or $T = 2.7 \times 10^2$ K.

(e) This is most easily done by using an equation derived from $\Delta H°_{vap}$

and $\Delta S°_{vap}$ :

$$\ln P = -\frac{\Delta H°_{vap}}{R} \cdot \frac{1}{T} + \frac{\Delta S°_{vap}}{R}$$

$$\ln \frac{15 \text{ Torr}}{760 \text{ Torr}} = -\frac{28\,000 \text{ J} \cdot \text{mol}^{-1}}{8.314 \text{ J} \cdot \text{K}^{-1} \cdot \text{mol}^{-1}} \cdot \frac{1}{T} + \frac{100 \text{ J} \cdot \text{K}^{-1} \cdot \text{mol}^{-1}}{8.314 \text{ J} \cdot \text{K}^{-1} \cdot \text{mol}^{-1}}$$

$T = 2.1 \times 10^2$ K

**8.119**  The critical temperatures are

Compound $T_C$ (°C)

CH$_4$     −82.1

C$_2$H$_6$     32.2

C$_3$H$_8$     96.8

C$_4$H$_{10}$     152

The critical temperatures increase with increasing mass, showing the

influence of the stronger London forces.

**8.121**  (a)  If sufficient chloroform and acetone are available, the pressures in the

flasks will be the equilibrium vapor pressures at that temperature. We can

calculate these amounts using the ideal gas equation:

$$\left( \frac{195 \text{ Torr}}{760 \text{ Torr} \cdot \text{atm}^{-1}} \right) (1.00 \text{ L}) =$$

$$\left( \frac{m_{chloroform}}{119.37 \text{ g} \cdot \text{mol}^{-1} \text{ chloroform}} \right) (0.08206 \text{ L} \cdot \text{atm} \cdot \text{K}^{-1} \cdot \text{mol}^{-1})(298 \text{ K})$$

$m_{chloroform} = 1.25$ g

$$\left( \frac{225 \text{ Torr}}{760 \text{ Torr} \cdot \text{atm}^{-1}} \right) (1.00 \text{ L}) =$$

$$\left( \frac{m_{acetone}}{58.08 \text{ g} \cdot \text{mol}^{-1} \text{ acetone}} \right) (0.08206 \text{ L} \cdot \text{atm} \cdot \text{K}^{-1} \cdot \text{mol}^{-1})(298 \text{ K})$$

$$m_{acetone} = 0.703 \text{ g}$$

In both cases, sufficient compound is available to achieve the vapor pressure; flask A will have a pressure of 195 Torr and flask B will have a pressure of 222 Torr.

(b) When the stopcock is opened, some chloroform will move into flask B and acetone will move into flask A to restore the equilibrium vapor pressure. Additionally, however, some acetone vapor will dissolve in the liquid chloroform and vice versa. Ultimately the system will reach an equilibrium state in which the compositions of the liquid phases in both flasks are the same and the gas phase composition is uniform. The gas phase and liquid phase compositions will be established by Raoult's law. It is conceptually most convenient for this calculation to start by putting all the material into one liquid phase. Such a solution would have the following composition:

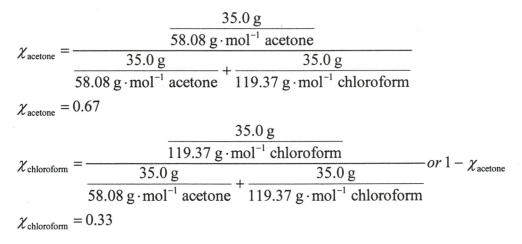

$$\chi_{acetone} = \cfrac{\cfrac{35.0 \text{ g}}{58.08 \text{ g} \cdot \text{mol}^{-1} \text{ acetone}}}{\cfrac{35.0 \text{ g}}{58.08 \text{ g} \cdot \text{mol}^{-1} \text{ acetone}} + \cfrac{35.0 \text{ g}}{119.37 \text{ g} \cdot \text{mol}^{-1} \text{ chloroform}}}$$

$$\chi_{acetone} = 0.67$$

$$\chi_{chloroform} = \cfrac{\cfrac{35.0 \text{ g}}{119.37 \text{ g} \cdot \text{mol}^{-1} \text{ chloroform}}}{\cfrac{35.0 \text{ g}}{58.08 \text{ g} \cdot \text{mol}^{-1} \text{ acetone}} + \cfrac{35.0 \text{ g}}{119.37 \text{ g} \cdot \text{mol}^{-1} \text{ chloroform}}} \quad or \; 1 - \chi_{acetone}$$

$$\chi_{chloroform} = 0.33$$

This gives the composition of the liquid phase. The composition of the gas phase will be determined from the pressures of the gases:

$$P_{\text{acetone}} = \chi_{\text{acetone, liquid}} \cdot P^{\circ}_{\text{acetone}} = (0.67)(222\ \text{Torr}) = 149\ \text{Torr}$$

$$P_{\text{chloroform}} = \chi_{\text{chloroform, liquid}} \cdot P^{\circ}_{\text{chloroform}} = (0.33)(195\ \text{Torr}) = 64\ \text{Torr}$$

$$\chi_{\text{acetone, gas}} = \frac{P_{\text{acetone}}}{P_{\text{acetone}} + P_{\text{chloroform}}}$$

$$\chi_{\text{acetone, gas}} = \frac{149\ \text{Torr}}{149\ \text{Torr} + 64\ \text{Torr}} = 0.70$$

$$\chi_{\text{chloroform, gas}} = 1 - \chi_{\text{acetone, gas}} = 0.30$$

The gas phase composition will, therefore, be slightly richer in acetone than in chloroform. The total pressure in the flask will be 213 Torr.

(c) The solution shows negative deviation from Raoult's law. This means that the molecules of acetone and chloroform attract each other slightly more than molecules of the same kind. Under such circumstances, the vapor pressure is lower than expected from the ideal calculation. This will give rise to a high-boiling azeotrope. The gas phase composition will also be slightly different from that calculated from the ideal state, but whether acetone or chloroform would be richer in the gas phase depends on which side of the azeotrope the composition lies. Because we are not given the composition of the azeotrope, we cannot state which way the values will vary.

**8.123** (a) The vapor pressure above the sucrose solution will be lower than the vapor pressure of the pure solvent. This results in an imbalance in the system that causes pure ethanol to condense into the sucrose solution. As this happens, the sucrose solution becomes more dilute and its vapor pressure would approach that of pure ethanol if there were sufficient pure ethanol. In this case, however, the process will stop once all the pure ethanol has transferred to the solution. This will result in a solution that has 1/2 the original concentration, or 7.5 $m$. (b) The vapor pressure of this solution will be given by

$$P = \chi_{\text{ethanol}} \times P^{\circ}_{\text{ethanol}} = \chi_{\text{ethanol}} \times 60\ \text{Torr}$$

We need to convert the molality of the solution to mole fraction. A 7.5 $m$ solution will contain 7.5 mol sucrose for 1.0 kg solvent. The molar mass of ethanol is $46.07 \text{ g} \cdot \text{mol}^{-1}$, so 1.0 kg represents 22 mol of ethanol. The mole fraction of ethanol will be given by

$$\chi_{\text{ethanol}} = \frac{22 \text{ mol}}{22 \text{ mol} + 7.5 \text{ mol}} = 0.75$$

$$P = 0.75 \times 60 \text{ Torr} = 45 \text{ Torr}$$

**8.125** The vapor pressure is more sensitive if $\Delta H_{\text{vap}}$ is small. The fact that $\Delta H_{\text{vap}}$ is small indicates that it takes little energy to volatilize the sample, which means that the intermolecular forces are weaker. Hence we expect the vapor pressure to be more dramatically affected by small changes in temperature.

**8.127** (a) $\Pi = iRT \times \text{molarity}$

Assuming $i = 1$

$$0.0112 \text{ atm} = \left(0.0821 \text{ L} \cdot \text{atm} \cdot \text{K}^{-1} \cdot \text{mol}^{-1}\right)\left(298 \text{ K}\right)\frac{\left(\frac{3.16 \text{ g}}{X}\right)}{0.500 \text{ L}}$$

$$X = 13\ 800 \text{ g} \cdot \text{mol}^{-1} = \text{molar mass of the polymer}$$

(b) The molecular mass of $CH_3CHCH_2$ is $42.08 \text{ g} \cdot \text{mol}^{-1}$. Dividing the molar mass of the polymer by the molar mass of the monomer, $13800 \text{ g} \cdot \text{mol}^{-1} / 42.08 \text{ g} \cdot \text{mol}^{-1} = 328$, indicates that there are approximately 328 monomers in the average polymer.

(c) $\left(308 \text{ pm} \cdot \text{monomer}^{-1}\right)(328 \text{ monomers}) = 101000 \text{ pm}$ or 101 nm

**8.129** The initial information tells us that the detector is more responsive to A than to B. Under these conditions, the response is

$5.44 \text{ cm}^2 \div 0.52 \text{ mgA} = 11 \text{ cm}^2 \cdot \text{mg}^{-1}$ A whereas the response for B is

$8.72 \text{ cm}^2 \div 2.30 \text{ mg} = 3.79 \text{ cm}^2 \cdot \text{mg}^{-1}$ B. The detector is thus

$11 \text{ cm}^2 \cdot \text{mg}^{-1} \div 3.79 \text{ cm}^2 \cdot \text{mg}^{-1} = 2.9$ times more sensitive for A than B. Because the conditions are different for determining the unknown amount of A present, we cannot use the area ratios directly to determine the quantity of A. Instead, we use the standard B as a reference.

$(X = \text{mg A})$

$$\frac{3.52 \text{ cm}^2}{X} \div \frac{7.58 \text{ cm}^2}{2.00 \text{ mg (B)}} = 2.9$$

$$X = \frac{3.52 \text{ cm}^2}{2.9} \times \frac{2.00 \text{ mg}}{7.58 \text{ cm}^2}$$

$$X = 0.32 \text{ mg A}$$

**8.131** $\Pi = iRT \times \text{molarity}$

Assuming $i = 1$

$$7.7 \text{ atm} = \left(0.0821 \text{ L} \cdot \text{atm} \cdot \text{K}^{-1} \cdot \text{mol}^{-1}\right)(310 \text{ K})(M)$$

$$M = 0.303 \text{ mol} \cdot \text{L}^{-1}$$

If 0.500 L of solution is needed:

$$\left(0.303 \text{ mol} \cdot \text{L}^{-1}\right)(0.500 \text{ L}) = 0.151 \text{ mol of glucose}$$

Molecular mass of glucose is $81.158 \text{ g} \cdot \text{mol}^{-1}$

$$(0.151 \text{ mol})\left(81.158 \text{ g} \cdot \text{mol}^{-1}\right) = 27.3 \text{ g}$$

**8.133** First, calculate the weight % of acetic acid in a solution with $x_{\text{acetic acid}} = 0.15$:

If you have 1 mol total of a solution with $x_{\text{acetic acid}} = 0.15$ and $x_{\text{H}_2\text{O}} = 0.85$, you would have 0.15 mol of acetic acid ($60.053 \text{ g} \cdot \text{mol}^{-1}$) in 0.85 mol of $H_2O$ ($18.015 \text{ g} \cdot \text{mol}^{-1}$). The weight % of acetic acid in such a solution is:

$$(0.15 \text{ mol})(60.053 \text{ g} \cdot \text{mol}^{-1}) = 9.01 \text{ g of acetic acid}$$

$$(0.85 \text{ mol})(18.015 \text{ g} \cdot \text{mol}^{-1}) = 15.3 \text{ g of } H_2O$$

$$\frac{9.01 \text{ g}}{9.01 \text{ g} + 15.3 \text{ g}}(100\%) = 37.0\% \text{ acetic acid by weight}$$

If a solution with $x_{\text{acetic acid}} = 0.15$ is 37.0% acetic acid by weight, 30 g of such a solution will contain:

$(30 \text{ g})(0.370) = 11.1 \text{ g acetic acid}$

$\dfrac{11.1 \text{ g}}{60.053 \text{ g} \cdot \text{mol}^{-1}} = 0.185 \text{ mol of acetic acid}$

given that acetic acid is a monoprotic acid which will react in a 1:1 fashion with NaOH, 0.185 mol of NaOH will completely react with 0.185 mol of acetic acid. The volume of a 0.010 M solution of NaOH needed to completely consume the acetic acid is: $\dfrac{0.185 \text{ mol}}{0.010 \text{ mol} \cdot \text{L}^{-1}} = 18.5 \text{ L}$

**8.135** (a) Assume 100 g of compound:

$(0.590)(100 \text{ g}) = 59 \text{ g C} \rightarrow 4.912 \text{ mol C}$

$(0.262)(100 \text{ g}) = 26.2 \text{ g O} \rightarrow 1.638 \text{ mol O}$

$(0.0710)(100 \text{ g}) = 7.10 \text{ g H} \rightarrow 7.044 \text{ mol H}$

$(0.0760)(100 \text{ g}) = 7.60 \text{ g N} \rightarrow 0.5424 \text{ mol N}$

Dividing by 0.5424 mol and rounding to the nearest whole number we find the C:O:H:N ratio to be 9:3:13:1, giving an empirical formula of $C_9H_{13}O_3N$

(b) Molar mass may be found using the freezing point depression:

$\Delta T_f = i \cdot k_f \cdot m$, assuming $i = 1$:

$$0.50 \text{ K} = (5.12 \text{ K} \cdot \text{kg} \cdot \text{mol}^{-1}) \dfrac{\left( \dfrac{0.64 \text{ g}}{X \text{ g} \cdot \text{mol}^{-1}} \right)}{0.0360 \text{ kg}}$$

$X = 1.8 \times 10^2 \text{ g} \cdot \text{mol}^{-1}$

(c) The molar mass of the compound will be an integral multiple of the molar mass of the empirical formula:

$n[(9 \cdot 12.011 \text{ g} \cdot \text{mol}^{-1}) + (3 \cdot 16.00 \text{ g} \cdot \text{mol}^{-1}) + (13 \cdot 1.0079 \text{ g} \cdot \text{mol}^{-1})$
$+ (14.01 \text{ g} \cdot \text{mol}^{-1})] = 1.8 \times 10^2 \text{ g} \cdot \text{mol}^{-1}$
$\therefore n = 1.$

Therefore, the molecular formula is $C_{9n}H_{13n}O_{3n}N_n = C_9H_{13}O_3N$.

**8.137** The non-polar chains of both the surfactant and pentanol will interact to form a hydrophobic region with the heads of the two molecules pointing away from this region towards the aqueous solution. To prevent the heads of the shorter pentanol molecules from winding up in the hydrophobic region, the layered structure might be comprised of a water region, a surfactant layer (heads pointing toward the water) a pentanol layer (with tails pointing toward the hydrophobic tails of the surfactant) and back to an aqueous region.

# CHAPTER 9

# CHEMICAL EQUILIBRIA

**9.1** (a) False, Equilibrium is dynamic. At equilibrium, the concentrations of reactants and products will not change, but the reaction will continue to proceed in both directions.

(b) False. Equilibrium reactions are affected by the presence of both products and reactants.

(c) False. The value of the equilibrium constant is not affected by the amounts of reactants or products added as long as the temperature is constant.

(d) True.

**9.3**

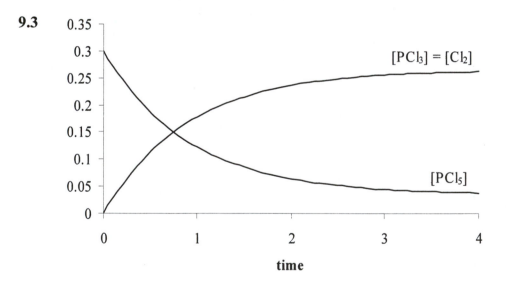

**9.5** (a) $K_C = \dfrac{[COCl][Cl]}{[CO][Cl_2]}$; (b) $K_C = \dfrac{[HBr]^2}{[H_2][Br_2]}$; (c) $K_C = \dfrac{[SO_2]^2 [H_2O]^2}{[H_2S]^2 [O_2]^3}$

**9.7** (a) Because the volume is the same, the number of moles of $O_2$ is larger in the second experiment. (b) Because $K_C$ is a constant and the denominator is larger in the second case, the numerator must also be larger; so the concentration of $O_2$ is larger in the second case. (c) Although $[O_2]^3/[O_3]^2$ is the same, $[O_2]/[O_3]$ will be different, a result seen by solving for $K_C$ in each case. (d) Because $K_C$ is a constant, $[O_2]^3/[O_3]^2$ is the same. (e) Because $[O_2]^3/[O_3]^2$ is the same, its reciprocal must be the same.

**9.9** $K_C = \dfrac{[HI]^2}{[H_2][I_2]}$

for condition 1, $K_C = \dfrac{(0.0137)^2}{(6.47 \times 10^{-3})(0.594 \times 10^{-3})} = 48.8$

for condition 2, $K_C = \dfrac{(0.0169)^2}{(3.84 \times 10^{-3})(1.52 \times 10^{-3})} = 48.9$

for condition 3, $K_C = \dfrac{(0.0100)^2}{(1.43 \times 10^{-3})(1.43 \times 10^{-3})} = 48.9$

**9.11** (a) $\dfrac{1}{P_{O_2}^{\ 3}}$

(b) $P_{H_2O}^{\ 7}$

(c) $\dfrac{P_{NO}P_{NO_2}}{P_{N_2O_3}}$

**9.13** To answer these questions, we will first calculate $\Delta G°$ for each reaction and then use that value in the expression $\Delta G° = -RT \ln K$.

(a) $2\,H_2(g) + O_2(g) \longrightarrow 2\,H_2O(g)$

$$\Delta G_r^{\circ} = 2\left(\Delta G_f^{\circ}(H_2O, g)\right) = 2\left(-228.57 \text{ kJ} \cdot \text{mol}^{-1}\right) = -457.14 \text{ kJ} \cdot \text{mol}^{-1}$$

$$\Delta G_r^{\circ} = -RT \ln K \quad \text{or} \quad \ln K = -\frac{\Delta G_r^{\circ}}{RT}$$

$$\ln K = -\frac{-457140 \text{ J} \cdot \text{mol}^{-1}}{(8.314 \text{ J} \cdot \text{mol}^{-1} \cdot \text{K}^{-1})(298.15 \text{ K})} = +184.42$$

$$K = 1.2 \times 10^{80}$$

(b) $2\,CO(g) + O_2(g) \longrightarrow 2\,CO_2(g)$

$$\Delta G_r^{\circ} = 2\left(\Delta G_f^{\circ}(CO_2, g)\right) - 2\left(\Delta G_f^{\circ}(CO, g)\right)$$
$$= 2\left(-394.36 \text{ kJ} \cdot \text{mol}^{-1}\right) - 2\left(-137.17 \text{ kJ} \cdot \text{mol}^{-1}\right) = -514.38 \text{ kJ} \cdot \text{mol}^{-1}$$

$$\ln K = -\frac{-514380 \text{ J} \cdot \text{mol}^{-1}}{(8.314 \text{ J} \cdot \text{mol}^{-1} \cdot \text{K}^{-1})(298.15 \text{ K})} = +207.51$$

$$K = 1.3 \times 10^{90}$$

(c) $CaCO_3(s) \longrightarrow CaO(s) + CO_2(g)$

$$\Delta G_r^{\circ} = \Delta G_f^{\circ}(CaO, s) + \Delta G_f^{\circ}(CO_2, g) - \Delta G_f^{\circ}(CaCO_3, s)$$
$$= -604.03 \text{ kJ} \cdot \text{mol}^{-1} + -394.36 \text{ kJ} \cdot \text{mol}^{-1} - (-1128.8 \text{ kJ} \cdot \text{mol}^{-1})$$
$$= +130.41 \text{ kJ} \cdot \text{mol}^{-1}$$

$$\ln K = -\frac{-130410 \text{ J} \cdot \text{mol}^{-1}}{(8.314 \text{ J} \cdot \text{mol}^{-1} \cdot \text{K}^{-1})(298.15 \text{ K})} = -52.610$$

$$K = 1.4 \times 10^{-23}$$

**9.15** (a) $\Delta G^{\circ}_r = -RT \ln K$

$$= -(8.314 \text{ J} \cdot \text{K}^{-1} \cdot \text{mol}^{-1})(1200 \text{ K}) \ln 6.8 = -19 \text{ kJ} \cdot \text{mol}^{-1}$$

(b) $\Delta G^{\circ}_r = -RT \ln K = -(8.314 \text{ J} \cdot \text{K}^{-1} \cdot \text{mol}^{-1})(298 \text{ K}) \ln 1.1 \times 10^{-12}$
$$= +68 \text{ kJ} \cdot \text{mol}^{-1}$$

**9.17** First we must calculate $K$ for the reaction, which can be done using data from Appendix 2A:

$$\Delta G^{\circ}_r = 2 \times \Delta G^{\circ}_f(NO, g) = 2 \times 86.55 \text{ kJ} \cdot \text{mol}^{-1} = 173.1 \text{ kJ} \cdot \text{mol}^{-1}$$

$$\Delta G^\circ = -RT \ln K$$

$$\ln K = -\frac{\Delta G^\circ}{RT}$$

$$\ln K = -\frac{+173\,100\ \text{J}\cdot\text{mol}^{-1}}{(8.314\ \text{J}\cdot\text{K}^{-1}\cdot\text{mol}^{-1})(298\ \text{K})} = -69.9$$

$$K = 4 \times 10^{-31}$$

Because $Q > K$, the reaction will tend to proceed to produce reactants.

**9.19** (a) The free energy at a specific set of conditions is given by

$$\Delta G_r = \Delta G^\circ_r + RT \ln Q$$
$$\Delta G_r = -RT \ln K + RT \ln Q$$
$$\Delta G_r = -RT \ln K + RT \ln \frac{[I]^2}{[I_2]}$$
$$= -(8.314\ \text{J}\cdot\text{K}^{-1}\cdot\text{mol}^{-1})(1200\ \text{K}) \ln (6.8)$$
$$+ (8.314\ \text{J}\cdot\text{K}^{-1}\cdot\text{mol}^{-1})(1200\ \text{K}) \ln \frac{(0.98)^2}{(0.13)}$$
$$= 8.3 \times 10^{-1}\ \text{kJ}\cdot\text{mol}^{-1}$$

(b) Because $\Delta G_r$ is positive, the reaction will be spontaneous to produce I$_2$.

**9.21** (a) The free energy at a specific set of conditions is given by

$$\Delta G_r = \Delta G^\circ_r + RT \ln Q$$
$$\Delta G_r = -RT \ln K + RT \ln Q$$
$$\Delta G_r = -RT \ln K + RT \ln \frac{[NH_3]^2}{[N_2][H_2]^3}$$
$$= -(8.314\ \text{J}\cdot\text{K}^{-1}\cdot\text{mol}^{-1})(400\ \text{K}) \ln (41)$$
$$+ (8.314\ \text{J}\cdot\text{K}^{-1}\cdot\text{mol}^{-1})(400\ \text{K}) \ln \frac{(21)^2}{(4.2)(1.8)^3}$$
$$= -27\ \text{kJ}\cdot\text{mol}^{-1}$$

(b) Because $\Delta G_r$ is negative, the reaction will proceed to form products.

**9.23** (a) $K = \dfrac{P_{NO}^{2} P_{Cl_2}}{P_{NOCl}^{2}} = 1.8 \times 10^{-2}$

$$K = \left(\dfrac{T}{12.027 \text{ K}}\right)^{\Delta n} K_C$$

$$K_C = \left(\dfrac{12.027 \text{ K}}{T}\right)^{\Delta n} K = \left(\dfrac{12.027 \text{ K}}{500 \text{ K}}\right)^{(3-2)} 1.8 \times 10^{-2} = 4.3 \times 10^{-4}$$

(b) $K = P_{CO_2} = 167$

$$K_C = \left(\dfrac{12.027 \text{ K}}{T}\right)^{\Delta n} K = \left(\dfrac{12.027 \text{ K}}{1073 \text{ K}}\right)^{(1)} 167 = 1.87$$

**9.25** For the reaction written as $N_2(g) + 3 H_2(g) \rightarrow 2 NH_3(g)$ Eq. 1

$$K = \dfrac{P_{NH_3}^{2}}{P_{N_2} P_{H_2}^{3}} = 41$$

(a) For the reaction written as $2 NH_3(g) \rightarrow N_2(g) + 3 H_2(g)$ Eq. 2

$$K = \dfrac{P_{N_2} P_{H_2}^{3}}{P_{NH_3}^{2}}$$

This is $\dfrac{1}{K_{Eq.\,1}} = \dfrac{1}{41} = 0.024$

(b) For the reaction written as $\frac{1}{2} N_2(g) + \frac{3}{2} H_2(g) \rightarrow NH_3(g)$ Eq. 3

$$K_{Eq.3} = \dfrac{P_{NH_3}}{P_{N_2}^{1/2} P_{H_2}^{3/2}} = \sqrt{K_{Eq.1}} = \sqrt{41} = 6.4$$

Note that Eq. 3 $= \frac{1}{2}$ Eq. 1 and thus $K_{Eq.3} = K_{Eq.1}^{1/2}$.

(c) For the reaction written as $2 N_2(g) + 6 H_2(g) \rightarrow 4 NH_3(g)$ Eq. 4

$$K_{Eq.4} = \dfrac{P_{NH_3}^{4}}{P_{N_2}^{2} P_{H_2}^{6}} = K_{Eq.1}^{2} = 41^{2} = 1.7 \times 10^{3}$$

Note that Eq. 4 = 2 Eq. 1 and thus $K_{Eq.3} = K_{Eq.1}^{2}$.

**9.27** $H_2(g) + I_2(g) \rightarrow 2 HI(g)$ $\qquad K_C = 160$

$$K_C = \frac{[HI]^2}{[H_2][I_2]}$$

$$160 = \frac{(2.21 \times 10^{-3})^2}{[H_2](1.46 \times 10^{-3})}$$

$$[H_2] = \frac{(2.21 \times 10^{-3})^2}{(160)(1.46 \times 10^{-3})}$$

$$[H_2] = 2.1 \times 10^{-5} \text{ mol} \cdot \text{L}^{-1}$$

**9.29**  $$K = \frac{P_{PCl_3} P_{Cl_2}}{P_{PCl_5}}$$

$$25 = \frac{P_{PCl_3}(5.43)}{1.18}$$

$$P_{PCl_3} = \frac{(25)(1.18)}{5.43} = 5.4$$

$$P_{PCl_3} = 5.4 \text{ bar}$$

**9.31**  (a)  $$K = \frac{p_{HI}^2}{p_{H_2} \cdot p_{I_2}} = 160$$

$$Q = \frac{(0.10)^2}{(0.20)(0.10)} = 0.50$$

(b)  $Q \neq K, \therefore$ the system is not at equilibrium

(c)  Because $Q < K$, more products will be formed.

**9.33**  (a)  $$K_C = \frac{[SO_3]^2}{[SO_2]^2[O_2]} = 1.7 \times 10^6$$

$$Q_C = \frac{\left[\dfrac{1.0 \times 10^{-4}}{0.500}\right]^2}{\left[\dfrac{1.20 \times 10^{-3}}{0.500}\right]^2 \left[\dfrac{5.0 \times 10^{-4}}{0.500}\right]} = 6.9$$

(b) Because $Q_C < K_C$, more products will tend to form, which will result in the formation of more $SO_3$.

**9.35** $\dfrac{1.90 \text{ g HI}}{127.91 \text{ g} \cdot \text{mol}^{-1}} = 0.0148 \text{ mol HI}$

$2 \text{ HI(g)} \longrightarrow H_2(g) + I_2$

$0.0172 \text{ mol} - 2x \qquad x \quad x$

$0.0148 \text{ mol} = 0.0172 \text{ mol} - 2x$

$x = 0.0012 \text{ mol}$

$$K_C = \frac{\left[\dfrac{0.0012}{2.00}\right]\left[\dfrac{0.0012}{2.00}\right]}{\left[\dfrac{0.0148}{2.00}\right]^2} = \frac{\left[\dfrac{0.0012}{2.00}\right]^2}{\left[\dfrac{0.0148}{2.00}\right]^2} = 0.0066 \text{ or } 6.6 \times 10^{-3}$$

**9.37** $\dfrac{25.0 \text{ g } NH_4(NH_2CO_2)}{78.07 \text{ g} \cdot \text{mol}^{-1} \, NH_4(NH_2CO_2)} = 0.320 \text{ mol } NH_4(NH_2CO_2)$

$\dfrac{0.0174 \text{ g } CO_2}{44.01 \text{ g} \cdot \text{mol}^{-1} CO_2} = 3.95 \times 10^{-4} \text{ mol } CO_2$

2 mol $NH_3$ are formed per mol of $CO_2$, so mol $NH_3 = 2 \times 3.95 \times 10^{-4}$

$\qquad = 7.90 \times 10^{-4}$

$$K_C = [NH_3]^2[CO_2] = \left(\frac{7.90 \times 10^{-4}}{0.250}\right)^2\left(\frac{3.95 \times 10^{-4}}{0.250}\right) = 1.58 \times 10^{-8}$$

**9.39** (a) The balanced equation is: $Cl_2(g) \rightarrow 2 \, Cl(g)$.

The initial concentration of $Cl_2(g)$ is $\dfrac{0.0020 \text{ mol } Cl_2}{2.0 \text{ L}} = 0.0010 \text{ mol} \cdot \text{L}^{-1}$

| Concentration $(\text{mol} \cdot \text{L}^{-1})$ | $Cl_2(g)$ | $\rightarrow$ | $2 \, Cl(g)$ |
|---|---|---|---|
| initial | 0.0010 | | 0 |
| change | $-x$ | | $+2x$ |
| equilibrium | $0.0010 - x$ | | $+2x$ |

$$K_C = \frac{[Cl]^2}{[Cl_2]} = \frac{(2x)^2}{(0.0010 - x)} = 1.2 \times 10^{-7}$$

$$4x^2 = (1.2 \times 10^{-7})(0.0010 - x)$$

$$4x^2 + (1.2 \times 10^{-7})x - (1.2 \times 10^{-10}) = 0$$

$$x = \frac{-(1.2 \times 10^{-7}) \pm \sqrt{(1.2 \times 10^{-7})^2 - 4(4)(-1.2 \times 10^{-10})}}{2.4}$$

$$x = \frac{-(1.2 \times 10^{-7}) \pm 4.4 \times 10^{-5}}{8}$$

$$x = -5.5 \times 10^{-6} \text{ or } +5.5 \times 10^{-6}$$

The negative answer is not meaningful, so we choose

$x = +5.5 \times 10^{-6}$ mol $\cdot$ L$^{-1}$. The concentration of $Cl_2$ is essentially

unchanged because $0.0010 - 5.5 \times 10^{-6} \cong 0.0010$ mol $\cdot$ L$^{-1}$. The

concentration of Cl atoms is $2 \times (5.5 \times 10^{-6}) = 1.1 \times 10^{-5}$ mol $\cdot$ L$^{-1}$. The

percentage decomposition of $Cl_2$ is given by

$$\frac{5.5 \times 10^{-6}}{0.0010} \times 100 = 0.55\%$$

(b) The balanced equation is: $F_2(g) \rightarrow 2\,F(g)$

The problem is worked in an identical fashion to (a) but the equilibrium

constant is now $1.2 \times 10^{-4}$.

The initial concentration of $F_2(g)$ is $\dfrac{0.0020 \text{ mol } F_2}{2.0 \text{ L}} = 0.0010$ mol $\cdot$ L$^{-1}$

| Concentration (mol $\cdot$ L$^{-1}$) | $F_2(g)$ | $\rightarrow$ | $2\,F(g)$ |
|---|---|---|---|
| initial | 0.0010 | | 0 |
| change | $-x$ | | $+2x$ |
| equilibrium | $0.0010 - x$ | | $+2x$ |

$$K_C = \frac{[F]^2}{[F_2]} = \frac{(2x)^2}{(0.0010 - x)} = 1.2 \times 10^{-4}$$

$$4x^2 = (1.2 \times 10^{-4})(0.0010 - x)$$

$$4x^2 + (1.2 \times 10^{-4})x - (1.2 \times 10^{-7}) = 0$$

$$x = \frac{-(1.2 \times 10^{-4}) \pm \sqrt{(1.2 \times 10^{-4})^2 - 4(4)(-1.2 \times 10^{-7})}}{2.4}$$

$$x = \frac{-(1.2 \times 10^{-4}) \pm 1.4 \times 10^{-3}}{8}$$

$$x = -1.9 \times 10^{-4} \text{ or } +1.6 \times 10^{-4}$$

The negative answer is not meaningful, so we choose

$x = +1.6 \times 10^{-4}$ mol $\cdot$ L$^{-1}$. The concentration of $F_2$ is

$0.0010 - 1.6 \times 10^{-4} = 8 \times 10^{-4}$ mol $\cdot$ L$^{-1}$. The concentration of F atoms is

$2 \times (1.6 \times 10^{-4}) = 3.2 \times 10^{-4}$ mol $\cdot$ L$^{-1}$. The percentage decomposition of $F_2$

is given by

$$\frac{1.6 \times 10^{-4}}{0.0010} \times 100 = 16\%$$

(c) $Cl_2$ is more stable. This can be seen even without the aid of the

calculation from the larger equilibrium constant for the dissociation for $F_2$

compared to $Cl_2$.

**9.41**

| Concentration (bar) | 2 HBr(g) | → | H$_2$(g) | + | Br$_2$(g) |
|---|---|---|---|---|---|
| initial | $1.2 \times 10^{-3}$ | | 0 | | 0 |
| change | $-2x$ | | $+x$ | | $+x$ |
| final | $1.2 \times 10^{-3} - 2x$ | | $+x$ | | $+x$ |

$$K = \frac{p_{H_2} \cdot p_{Br_2}}{p_{HBr}^2}$$

$$7.7 \times 10^{-11} = \frac{(x)(x)}{(1.2 \times 10^{-3} - 2x)^2} = \frac{x^2}{(1.2 \times 10^{-3} - 2x)^2}$$

$$\sqrt{7.7 \times 10^{-11}} = \sqrt{\frac{x^2}{(1.2 \times 10^{-3} - 2x)^2}}$$

$$\frac{x}{(1.2 \times 10^{-3} - 2x)} = 8.8 \times 10^{-6}$$

$$x = (8.8 \times 10^{-6})(1.2 \times 10^{-3} - 2x)$$

$$x + 2(8.8 \times 10^{-6})x = (8.8 \times 10^{-6})(1.2 \times 10^{-3})$$

$$x \cong 1.1 \times 10^{-8}$$

$p_{H_2} = p_{Br_2} = 1.1 \times 10^{-8}$ bar; the pressure of HBr is essentially unaffected by the formation of $Br_2$ and $H_2$.

The percentage decomposition is given by

$$\frac{2\,(1.1 \times 10^{-8}\ \text{bar})}{1.2 \times 10^{-3}\ \text{bar}} \times 100 = 1.8 \times 10^{-3}\%$$

**9.43** (a) Concentration of $PCl_5$ initially

$$= \frac{\left(\dfrac{1.0\ \text{g PCl}_5}{208.22\ \text{g} \cdot \text{mol}^{-1}\ \text{PCl}_5}\right)}{0.250\ \text{L}} = 0.019\ \text{mol} \cdot \text{L}^{-1}$$

| Concentration $(\text{mol} \cdot \text{L}^{-1})$ | $PCl_5(g) \rightarrow$ | $PCl_3(g)$ + | $Cl_2$ |
|---|---|---|---|
| initial | 0.019 | 0 | 0 |
| change | $-x$ | $+x$ | $+x$ |
| final | $0.019 - x$ | $+x$ | $+x$ |

$$K_C = \frac{[PCl_2][Cl_2]}{[PCl_5]} = \frac{(x)(x)}{(0.019 - x)} = \frac{x^2}{(0.019 - x)}$$

$$\frac{x^2}{(0.019 - x)} = 1.1 \times 10^{-2}$$

$$x^2 = (1.1 \times 10^{-2})(0.019 - x)$$

$$x^2 + (1.1 \times 10^{-2})x - 2.1 \times 10^{-4} = 0$$

$$x = \frac{-(1.1 \times 10^{-2}) \pm \sqrt{(1.1 \times 10^{-2})^2 - (4)(1)(-2.1 \times 10^{-4})}}{2.1}$$

$$x = \frac{-(1.1 \times 10^{-2}) \pm 0.031}{2.1} = +0.010 \text{ or } -0.021$$

The negative root is not meaningful, so we choose $x = 0.010 \text{ mol} \cdot \text{L}^{-1}$.

$[PCl_3] = [Cl_2] = 0.010 \text{ mol} \cdot \text{L}^{-1}; [PCl_5] = 0.009 \text{ mol} \cdot \text{L}^{-1}$.

(b) The percentage decomposition is given by

$$\frac{0.010}{0.019} \times 100 = 53\%$$

**9.45** Starting concentration of $NH_3 = \dfrac{0.400 \text{ mol}}{2.00 \text{ L}} = 0.200 \text{ mol} \cdot \text{L}^{-1}$

| Concentration $(\text{mol} \cdot \text{L}^{-1})$ $NH_4HS(s) \rightarrow$ | $NH_3(g)$ | + | $H_2S(g)$ |
|---|---|---|---|
| initial — | 0.200 | | 0 |
| change — | $+x$ | | $+x$ |
| final — | $0.200 + x$ | | $+x$ |

$K_C = [NH_3][H_2S] = (0.200 + x)(x)$

$1.6 \times 10^{-4} = (0.200 + x)(x)$

$x^2 + 0.200x - 1.6 \times 10^{-4} = 0$

$$x = \frac{-(+0.200) \pm \sqrt{(+0.200)^2 - (4)(1)(-1.6 \times 10^{-4})}}{2.1}$$

$$x = \frac{-0.200 \pm 0.2016}{2.1} = +0.0008 \text{ or } -0.2008$$

The negative root is not meaningful, so we choose $x = 8 \times 10^{-4} \text{ mol} \cdot \text{L}^{-1}$

(note that in order to get this number we have had to ignore our normal significant figure conventions).

$[NH_3] = +0.200 \text{ mol} \cdot \text{L}^{-1} + 8 \times 10^{-4} \text{ mol} \cdot \text{L}^{-1} = 0.200 \text{ mol} \cdot \text{L}^{-1}$

$[H_2S] = 8 \times 10^{-4} \text{ mol} \cdot \text{L}^{-1}$

Alternatively, we could have assumed that

$x \ll 0.2$, the $0.200x = 1.6 \times 10^{-4}$, $x = 8.0 \times 10^{-4}$.

**9.47** The initial concentrations of $PCl_5$ and $PCl_3$ are calculated as follows:

$$[PCl_5] = \frac{0.200 \text{ mol}}{4.00 \text{ L}} = 0.0500 \text{ mol} \cdot L^{-1}; \quad [PCl_3] = \frac{0.600 \text{ mol}}{4.00 \text{ L}} = 0.150 \text{ mol} \cdot L^{-1}$$

| Concentrations $(mol \cdot L^{-1})$ | $PCl_5$ | $\rightarrow$ | $PCl_3(g)$ | $+$ | $Cl_2(g)$ |
|---|---|---|---|---|---|
| initial | 0.0500 | | 0.150 | | 0 |
| change | $-x$ | | $+x$ | | $+x$ |
| final | $0.0500 - x$ | | $0.150 + x$ | | $+x$ |

$$K_C = \frac{[PCl_3][Cl_2]}{[PCl_5]} = \frac{(0.150 + x)(x)}{(0.0500 - x)} = 33.3$$

$$x^2 + 0.150x = (33.3)(0.0500 - x)$$

$$x^2 + 33.45x - 1.665 = 0$$

$$x = \frac{-33.45 \pm \sqrt{(33.45)^2 - (4)(1)(-1.665)}}{(2)(1)} = \frac{-33.4 \pm 33.6}{2} = 0.0497$$

The negative root has no physical meaning and so it can be discarded.

$$[PCl_5] = 0.0500 \text{ mol} \cdot L^{-1} - 0.497 \text{ mol} \cdot L^{-1} = 3 \times 10^{-4} \text{ mol} \cdot L^{-1}$$

$$[PCl_3] = 0.150 \text{ mol} \cdot L^{-1} + 0.0497 \text{ mol} \cdot L^{-1} = 0.200 \text{ mol} \cdot L^{-1}$$

$$[Cl_2] = 0.0497 \text{ mol} \cdot L^{-1}$$

**9.49** The initial concentrations of $N_2$ and $O_2$ are equal at $0.114 \text{ mol} \cdot L^{-1}$ because the vessel has a volume of 1.00 L.

| Concentrations $(mol \cdot L^{-1})$ | $N_2(g)$ | $+$ | $O_2(g)$ | $\rightarrow$ | $2 NO(g)$ |
|---|---|---|---|---|---|
| initial | 0.114 | | 0.114 | | 0 |
| change | $-x$ | | $-x$ | | $+2x$ |
| final | $0.114 - x$ | | $0.114 - x$ | | $+2x$ |

$$K_C = \frac{[NO]^2}{[N_2][O_2]} = \frac{(2x)^2}{(0.114 - x)(0.114 - x)} = \frac{(2x)^2}{(0.114 - x)^2}$$

$$1.00 \times 10^{-5} = \frac{(2x)^2}{(0.114 - x)^2}$$

$$\sqrt{1.00 \times 10^{-5}} = \sqrt{\frac{(2x)^2}{(0.114 - x)^2}}$$

$$3.16 \times 10^{-3} = \frac{(2x)}{(0.114 - x)}$$

$$2x = (3.16 \times 10^{-3})(0.114 - x)$$

$$2.00316\,x = 3.60 \times 10^{-4}$$

$$x = 1.8 \times 10^{-4}$$

$[NO] = 2x = 2 \times 1.8 \times 10^{-4} = 3.6 \times 10^{-4}$; the concentrations of $N_2$ and $O_2$

remain essentially unchanged at $0.114$ mol$\cdot$L$^{-1}$.

**9.51** The initial concentrations of $H_2$ and $I_2$ are:

$$[H_2] = \frac{0.400 \text{ mol}}{3.00 \text{ L}} = 0.133 \text{ mol} \cdot L^{-1}; [I_2] = \frac{1.60 \text{ mol}}{3.00 \text{ L}} = 0.533 \text{ mol} \cdot L^{-1}$$

| Concentrations (mol$\cdot$L$^{-1}$) | $H_2$(g) | + | $I_2$(g) | $\rightarrow$ | 2 HI(g) |
|---|---|---|---|---|---|
| initial | 0.133 | | 0.533 | | 0 |
| change | $-x$ | | $-x$ | | $+2x$ |
| final | $0.133 - x$ | | $0.533 - x$ | | $+2x$ |

At equilibrium, 60.0% of the $H_2$ had reacted, so 40.0% of the $H_2$

remains:

$$(0.400)(0.133 \text{ mol} \cdot L^{-1}) = 0.133 \text{ mol} \cdot L^{-1} - x$$

$$x = 0.133 \text{ mol} \cdot L^{-1} - (0.400)(0.133 \text{ mol} \cdot L^{-1})$$

$$x = 0.080 \text{ mol} \cdot L^{-1}$$

at equilibrium: $[H_2] = 0.133 \text{ mol} \cdot L^{-1} - 0.080 \text{ mol} \cdot L^{-1} = 0.053 \text{ mol} \cdot L^{-1}$

$[I_2] = 0.533 \text{ mol} \cdot L^{-1} - 0.080 \text{ mol} \cdot L^{-1} = 0.453 \text{ mol} \cdot L^{-1}$

$[HI] = 2 \times 0.080 \text{ mol} \cdot L^{-1} = 0.16 \text{ mol} \cdot L^{-1}$

$$K_C = \frac{[HI]^2}{[H_2][I_2]} = \frac{0.16^2}{(0.053)(0.453)} = 1.1$$

**9.53** Initial concentrations of CO and $O_2$ are given by

$$[CO] = \frac{\left(\dfrac{0.28 \text{ g CO}}{28.01 \text{ g} \cdot \text{mol}^{-1} \text{CO}}\right)}{2.0 \text{ L}} = 5.0 \times 10^{-3} \text{ mol} \cdot \text{L}^{-1}$$

$$[O_2] = \frac{\left(\dfrac{0.032 \text{ g } O_2}{32.00 \text{ g} \cdot \text{mol}^{-1} \text{ } O_2}\right)}{2.0 \text{ L}} = 5.0 \times 10^{-4} \text{ mol} \cdot \text{L}^{-1}$$

| Concentration (mol·L⁻¹) | 2 CO(g) | + | $O_2$(g) | → | 2 $CO_2$(g) |
|---|---|---|---|---|---|
| initial | $5.0 \times 10^{-3}$ | | $5.0 \times 10^{-4}$ | | 0 |
| change | $-2x$ | | $-x$ | | $+2x$ |
| final | $5.0 \times 10^{-3} - 2x$ | | $5.0 \times 10^{-4} - x$ | | $+2x$ |

$$K_C = \frac{[CO_2]^2}{[CO]^2[O_2]}$$

$$= \frac{(2x)^2}{(5.0 \times 10^{-3} - 2x)^2(5.0 \times 10^{-4} - x)}$$

$$= \frac{4x^2}{(4x^2 - 0.020x + 2.5 \times 10^{-5})(5.0 \times 10^{-4} - x)}$$

$$0.66 = \frac{4x^2}{-4x^3 + 0.022x^2 - 3.5 \times 10^{-5}x + 1.25 \times 10^{-8}}$$

$$4x^2 = (0.66)(-4x^3 + 0.022x^2 - 3.5 \times 10^{-5}x + 1.25 \times 10^{-8})$$

$$6.06x^2 = -4x^3 + 0.022x^2 - 3.5 \times 10^{-5}x + 1.25 \times 10^{-8}$$

$$0 = -4x^3 - 6.04x^2 - 3.5 \times 10^{-5}x + 1.25 \times 10^{-8}$$

$$x = 4.3 \times 10^{-5}$$

$$[CO_2] = 8.6 \times 10^{-5} \text{ mol} \cdot \text{L}^{-1}; [CO] = 4.9 \times 10^{-3} \text{ mol} \cdot \text{L}^{-1}$$

$$[O_2] = 4.6 \times 10^{-4} \text{ mol} \cdot \text{L}^{-1}$$

**9.55** Concentrations

| (mol·L⁻¹) | $CH_3COOH$ | + | $C_2H_5OH$ | → | $CH_3COOC_2H_5$ | + | $H_2O$ |
|---|---|---|---|---|---|---|---|
| initial | 0.32 | | 6.30 | | 0 | | 0 |
| change | $-x$ | | $-x$ | | $+x$ | | $+x$ |
| final | 0.32 - x | | 6.30 - x | | $+x$ | | $+x$ |

$$K_C = \frac{[CH_3COOC_2H_5][H_2O]}{[CH_3COOH][C_2H_5OH]} = \frac{(x)(x)}{(0.32-x)(6.30-x)} = \frac{x^2}{x^2 - 6.62x + 2.02}$$

$$4.0 = \frac{x^2}{x^2 - 6.62x + 2.02}$$

$$4.0x^2 - 26.48x + 8.08 = x^2$$

$$3.0x^2 - 26.48x + 8.08 = 0$$

$$x = \frac{-(-26.48) \pm \sqrt{(-26.48)^2 - (4)(3.0)(8.08)}}{(2)(3.0)} = \frac{+26.48 \pm 24.58}{6.0}$$

$$x = 8.51 \text{ or } 0.317$$

The root 8.51 is meaningless because it is larger than the concentration of acetic acid and ethanol, so the value 0.317 is chosen. The equilibrium concentration of the product ester is, therefore, $0.317 \text{ mol} \cdot L^{-1}$. The numbers can be confirmed by placing them into the equilibrium expression:

$$K_C = \frac{[CH_3COOC_2H_5][H_2O]}{[CH_3COOH][C_2H_5OH]} = \frac{(0.317)(0.317)}{(0.320 - 0.317)(6.300 - 0.317)} = 5.6$$

Note: This number does not appear to agree well with the given value of $K_C = 4.0$.

If 0.316 is used, the agreement is better, giving a quotient of 4.1. If 0.315 is used, the quotient is 3.3. The better answer is thus $0.316 \text{ mol} \cdot L^{-1}$. The discrepancy is caused by rounding errors in places that are really beyond the accuracy of the measurement. Given that $K_C$ is only given to two significant figures, the best report of the concentration of ester would be $0.32 \text{ mol} \cdot L^{-1}$, even though this value will not satisfy the equilibrium expression as well as $0.316 \text{ mol} \cdot L^{-1}$.

**9.57** $\quad K_C = \dfrac{[BrCl]^2}{[Cl_2][Br_2]}$

$$0.031 = \frac{(0.145)^2}{(0.495)[Br_2]}$$

$$[Br_2] = \frac{(0.145)^2}{(0.495)(0.031)} = 1.4 \text{ mol} \cdot L^{-1}$$

**9.59** We can calculate changes according to the reaction stoichiometry:

| Amount (mol) | CO(g) | + | 3 H$_2$(g) | → | CH$_4$(g) | + | H$_2$O(g) |
|---|---|---|---|---|---|---|---|
| initial | 2.00 | | 3.00 | | 0 | | 0 |
| change | $-x$ | | $-3x$ | | $+x$ | | $+x$ |
| final | $2.00 - x$ | | $3.00 - 3x$ | | 0.478 | | $+x$ |

According to the stoichiometry, 0.478 mol $= x$; therefore, at equilibrium, there are 2.00 mol $-$ 0.478 mol $=$ 1.52 mol CO, 3.00 $-$ 3(0.478 mol) $=$ 1.57 mol H$_2$, and 0.478 mol H$_2$O. To employ the equilibrium expression, we need either concentrations or pressures; because $K_C$ is given, we will choose to express these as concentrations. This calculation is easy because $V = 10.0$ L:

$[CO] = 0.152 \text{ mol} \cdot L^{-1}; [H_2] = 0.157 \text{ mol} \cdot L^{-1}; [CH_4] = 0.0478 \text{ mol} \cdot L^{-1};$
$[H_2O] = 0.0478 \text{ mol} \cdot L^{-1}$

$$K_C = \frac{[CH_4][H_2O]}{[CO][H_2]^3} = \frac{(0.0478)(0.0478)}{(0.152)(0.157)^3} = 3.88$$

**9.61** First, we calculate the initial concentrations of each species:

$$[SO_2] = [NO] = \frac{0.100 \text{ mol}}{5.00 \text{ L}} = 0.0200 \text{ mol} \cdot L^{-1};$$

$$[NO_2] = \frac{0.200 \text{ mol}}{5.00 \text{ L}} = 0.0400 \text{ mol} \cdot L^{-1}; [SO_3] = \frac{0.150 \text{ mol}}{5.00 \text{ L}} = 0.0300 \text{ mol} \cdot L^{-1}$$

We can use these values to calculate $Q$ in order to see which direction the reactions will go:

$$Q = \frac{(0.0200)(0.0300)}{(0.0200)(0.0400)} = 0.75.$$ Because $Q < K_C$, the reaction will proceed to produce more products.

Concentration $(\text{mol} \cdot \text{L}^{-1})$

$$\text{SO}_2(g) \quad + \quad \text{NO}_2(g) \quad \rightarrow \quad \text{NO}(g) \quad + \quad \text{SO}_3(g)$$

| | $\text{SO}_2(g)$ | $\text{NO}_2(g)$ | $\text{NO}(g)$ | $\text{SO}_3(g)$ |
|---|---|---|---|---|
| initial | 0.0200 | 0.0400 | 0.0200 | 0.0300 |
| change | $-x$ | $-x$ | $+x$ | $+x$ |
| final | $0.0200 - x$ | $0.0400 - x$ | $0.0200 + x$ | $0.0300 + x$ |

$$K_C = \frac{[\text{NO}][\text{SO}_3]}{[\text{SO}_2][\text{NO}_2]}$$

$$85.0 = \frac{(0.0200 + x)(0.0300 + x)}{(0.0200 - x)(0.0400 - x)}$$

$$= \frac{x^2 + 0.0500\,x + 0.000\,600}{x^2 - 0.0600\,x + 0.000\,800}$$

$$85.0\,(x^2 - 0.0600\,x + 0.000\,800) = x^2 + 0.0500\,x + 0.000\,600$$

$$85.0\,x^2 - 5.10\,x + 0.0680 = x^2 + 0.0500\,x + 0.000\,600$$

$$84.0\,x^2 - 5.15\,x + 0.0674 = 0$$

$$x = \frac{-(-5.15) \pm \sqrt{(-5.15)^2 - (4)(84.0)(0.0674)}}{(2)(84.0)} = \frac{+5.15 \pm 1.97}{168}$$

$$x = +0.0424 \quad \text{or} \quad +0.0189$$

The root 0.0424 is not meaningful because it is larger than the concentration of $\text{NO}_2$. The root of choice is therefore 0.0189.

At equilibrium:

$$[\text{SO}_2] = 0.0200\,\text{mol} \cdot \text{L}^{-1} - 0.0189\,\text{mol} \cdot \text{L}^{-1} = 0.0011\,\text{mol} \cdot \text{L}^{-1}$$

$$[\text{NO}_2] = 0.0400\,\text{mol} \cdot \text{L}^{-1} - 0.0189\,\text{mol} \cdot \text{L}^{-1} = 0.0211\,\text{mol} \cdot \text{L}^{-1}$$

$$[\text{NO}] = 0.0200\,\text{mol} \cdot \text{L}^{-1} + 0.0189\,\text{mol} \cdot \text{L}^{-1} = 0.0389\,\text{mol} \cdot \text{L}^{-1}$$

$$[\text{SO}_3] = 0.0300\,\text{mol} \cdot \text{L}^{-1} + 0.0189\,\text{mol} \cdot \text{L}^{-1} = 0.0489\,\text{mol} \cdot \text{L}^{-1}$$

To check, we can put these numbers back into the equilibrium constant expression:

$$K_C = \frac{[\text{NO}][\text{SO}_3]}{[\text{SO}_2][\text{NO}_2]}$$

$$\frac{(0.0389)(0.0489)}{(0.0011)(0.0211)} = 82.0$$

Compared to $K_C = 85.0$, this is reasonably good agreement given the nature of the calculation. We can check to see, by trial and error, if a better answer could be obtained. Because the $K_C$ value is low for the concentrations we calculated, we can choose to alter $x$ slightly so that this ratio becomes larger. If we let $x = 0.0190$, the concentrations of NO and $SO_3$ are increased to 0.0390 and 0.0490, and the concentrations of $SO_2$ and $NO_2$ are decreased to 0.0010 and 0.0200 (the stoichiometry of the reaction is maintained by calculating the concentrations in this fashion). Then the quotient becomes 91.0, which is further from the value for $K_C$ than the original answer. So, although the agreement is not the best with the numbers we obtained, it is the best possible, given the limitation on the number of significant figures we are allowed to use in the calculation.

**9.63** (a) The initial concentrations are:

$$[PCl_5] = \frac{1.50\ mol}{0.500\ L} = 3.00\ mol \cdot L^{-1}; [PCl_3] = \frac{3.00\ mol}{0.500\ L} = 6.00\ mol \cdot L^{-1};$$

$$[Cl_2] = \frac{0.500\ mol}{0.500\ L} = 1.00\ mol \cdot L^{-1}$$

First calculate $Q$:

$$Q = \frac{[PCl_5]}{[PCl_3][Cl_2]} = \frac{3.00}{(6.00)(1.00)} = 0.500$$

Because $Q \neq K$, the reaction is not at equilibrium.

(b) Because $Q < K_C$, the reaction will proceed to form products.

(c)

| Concentrations (mol·L⁻¹) | $PCl_3(g)$ | + | $Cl_2(g)$ | → | $PCl_5(g)$ |
|---|---|---|---|---|---|
| initial | 6.00 | | 1.00 | | 3.00 |
| change | $-x$ | | $-x$ | | $+x$ |
| final | $6.00 - x$ | | $1.00 - x$ | | $3.00 + x$ |

$$K_C = \frac{[PCl_5]}{[PCl_3][Cl_2]}$$

$$0.56 = \frac{3.00 + x}{(6.00 - x)(1.00 - x)} = \frac{3.00 + x}{x^2 - 7x + 6.00}$$

$$(0.56)(x^2 - 7x + 6.00) = 3.00 + x$$

$$0.56x^2 - 3.92x + 3.36 = 3.00 + x$$

$$0.56x^2 - 4.92x + 0.36 = 0$$

$$x = \frac{-(-4.92) \pm \sqrt{(-4.92)^2 - (4)(0.56)(0.36)}}{(2)(0.56)} = \frac{+4.92 \pm 4.48}{1.12}$$

$x = 9.2$ or $0.07$

Because the root 9.2 is larger than the amount of $PCl_3$ or $Cl_2$ available, it is physically meaningless and can be discarded. Thus, $x = 0.071 \text{ mol} \cdot \text{L}^{-1}$, giving

$$[PCl_5] = 3.00 \text{ mol} \cdot \text{L}^{-1} + 0.07 \text{ mol} \cdot \text{L}^{-1} = 3.07 \text{ mol} \cdot \text{L}^{-1}$$

$$[PCl_3] = 6.00 \text{ mol} \cdot \text{L}^{-1} - 0.07 \text{ mol} \cdot \text{L}^{-1} = 5.93 \text{ mol} \cdot \text{L}^{-1}$$

$$[Cl_2] = 1.00 \text{ mol} \cdot \text{L}^{-1} - 0.07 \text{ mol} \cdot \text{L}^{-1} = 0.93 \text{ mol} \cdot \text{L}^{-1}$$

The number can be checked by substituting them back into the equilibrium constant expression:

$$K_C = \frac{[PCl_5]}{[PCl_3][Cl_2]}$$

$$\frac{(3.07)}{(5.93)(0.93)} \overset{?}{=} 0.56$$

$$0.56 \overset{?}{=} 0.56$$

**9.65**

| Pressures (bar) | 2 HCl(g) | $\rightarrow$ | $H_2(g)$ | + | $Cl_2(g)$ |
|---|---|---|---|---|---|
| initial | 0.22 | | 0 | | 0 |
| change | $-2x$ | | $+x$ | | $+x$ |
| final | $0.22 - 2x$ | | $+x$ | | $+x$ |

$$K = \frac{P_{H_2} P_{Cl_2}}{P_{HCl}^{\ 2}}$$

$$3.2 \times 10^{-34} = \frac{(x)(x)}{(0.22 - 2x)^2}$$

Because the equilibrium constant is small, assume that $x \ll 0.22$

$$3.2 \times 10^{-34} = \frac{x^2}{(0.22)^2}$$

$$x^2 = (3.2 \times 10^{-34})(0.22)^2$$

$$x = \sqrt{(3.2 \times 10^{-34})(0.22)^2}$$

$$x = \pm 3.9 \times 10^{-18}$$

The negative root is not physically meaningful and can be discarded. $x$ is small compared to 0.22, so the initial assumption was valid. The pressures at equilibrium are

$$P_{HCl} = 0.22 \text{ bar}; P_{H_2} = P_{Cl_2} = 3.9 \times 10^{-18} \text{ bar}$$

The values can be checked by substituting them into the equilibrium expression:

$$\frac{(3.9 \times 10^{-18})(3.9 \times 10^{-18})}{(0.22)^2} \stackrel{?}{=} 3.2 \times 10^{-34}$$

$$3.1 \times 10^{-34} \stackrel{\checkmark}{=} 3.2 \times 10^{-34}$$

$$P_{HCl} = 0.22 \text{ bar}; P_{H_2} = P_{Cl_2} = 3.9 \times 10^{-18} \text{ bar}$$

The numbers agree very well for a calculation of this type.

**9.67** (a) To determine on which side of the equilibrium position the conditions lie, we will calculate $Q$:

$$[CO] = \frac{0.342 \text{ mol}}{3.00 \text{ L}} = 0.114 \text{ mol} \cdot \text{L}^{-1}; [H_2] = \frac{0.215 \text{ mol}}{3.00 \text{ L}} = 0.0717 \text{ mol} \cdot \text{L}^{-1};$$

$$[CH_3OH] = \frac{0.125 \text{ mol}}{3.00 \text{ L}} = 0.0417 \text{ mol} \cdot \text{L}^{-1}$$

$$Q = \frac{[CH_3OH]}{[CO][H_2]^2} = \frac{0.0417}{(0.114)(0.0717)^2} = 71.1 \times 10^3$$

Because $Q > K_C$, the reaction will proceed to produce more of the reactants, which means that the concentration of methanol will decrease.

(b)

| Concentrations $(mol \cdot L^{-1})$ | $CO(g) +$ | $2 H_2(g)$ | $\rightarrow$ | $CH_3OH(g)$ |
|---|---|---|---|---|
| initial | 0.114 | 0.0717 | | 0.0417 |
| change | $+x$ | $+2x$ | | $-x$ |
| final | $0.0114 + x$ | $0.0717 + 2x$ | | $0.0417 - x$ |

$$K_C = \frac{0.0417 - x}{(0.0114 + x)(0.0717 + 2x)^2} = 1.1 \times 10^{-2}$$

$$0.0417 - x = (1.1 \times 10^{-2})(0.114 + x)(0.0717 + 2x)^2$$
$$= (1.2 \times 10^{-3} + 1.1 \times 10^{-2}x)(4x^2 + 0.287x + 5.14 \times 10^{-3})$$
$$= 4.4 \times 10^{-2}x^3 + 8.0 \times 10^{-3}x^2 + 4.0 \times 10^{-4}x + 6.2 \times 10^{-6}$$
$$0 = 4.4 \times 10^{-2}x^3 + 8.0 \times 10^{-3}x^2 + 1.00x - 0.0417$$

This equation can be solved approximately, simply by inspection: it is clear that the $x$ term will be very much larger than the $x^3$ and the $x^2$ terms, because their coefficients are very small compared to 1.00. This leads to a prediction that $x = 0.0417 \, mol \cdot L^{-1}$ to within the accuracy of the data. Essentially all of the $CH_3OH$ will react, so that

$$[CO] = 0.114 \, mol \cdot L^{-1} + 0.0417 \, mol \cdot L^{-1} = 0.156 \, mol \cdot L^{-1};$$

$$[H_2] = 0.0717 \, mol \cdot L^{-1} + 2 (0.0417 \, mol \cdot L^{-1}) = 0.155 \, mol \cdot L^{-1}.$$ The mathematical situation is odd in that clearly a $[CH_3OH] = 0$ will not satisfy the equilibrium constant. Knowing that the methanol concentration is very small compared to the CO and $H_2$ concentrations, we can now back-calculate to get a concentration value that will satisfy the equilibrium expression:

$$K_C = \frac{y}{(0.156)(0.155)^2} = 1.1 \times 10^{-2}$$

$$y = (1.1 \times 10^{-2})(0.156)(0.155)^2 = 2.6 \times 10^{-4}$$

Thus, $[CH_3OH] = 2.6 \times 10^{-4}$ mol.$L^{-1}$.

Alternatively, the cubic equation can be solved with the aid of a graphing calculator like the one supplied on the CD accompanying this book.

**9.69** Since reactants are strongly favored it is easier to push the reaction as far to the left as possible then start from new initial conditions.

| Pressures (bar) | $2\,HCl(g)$ | $\rightarrow$ | $H_2(g)$ | $+$ | $Cl_2(g)$ |
|---|---|---|---|---|---|
| original | 2.0 | | 1.0 | | 3.0 |
| new initial | 4.0 | | 0 | | 2.0 |
| change | $-2x$ | | $+x$ | | $+x$ |
| final | $4.0 - 2x$ | | $+x$ | | $2.0 + x$ |

$$K = \frac{P_{H_2}P_{Cl_2}}{P_{HCl}^{\;2}}$$

$$3.2 \times 10^{-34} = \frac{(x)(2.0+x)}{(4.0-2x)^2} \approx \frac{(x)(2.0)}{(4.0)^2} = \frac{x}{16}$$

$$x = 2.6 \times 10^{-33}\,bar$$

$$V = \frac{nRT}{p} = \frac{(1\text{ mole})(8.314 \times 10^{-2}\text{L bar K}^{-1}\text{ mol}^{-1})(298\text{K})}{2.6 \times 10^{-33}\,bar}$$

$$= 9.7 \times 10^{33}\,L$$

**9.71** (a) According to Le Chatelier's principle, an increase in the partial pressure of $CO_2$ will result in creation of reactants, which will decrease the $H_2$ partial pressure.

(b) According to Le Chatelier's principle, if the CO pressure is reduced, the reaction will shift to form more CO, which will decrease the pressure of $CO_2$.

(c) According to Le Chatelier's principle, if the concentration of CO is increased, the reaction will proceed to form more products, which will result in a higher pressure of $H_2$.

(d) The equilibrium constant for the reaction is unchanged, because it is unaffected by any change in concentration.

**9.73** (a) According to Le Chatelier's principle, increasing the concentration of NO will cause the reaction to form reactants in order to reduce the concentration of NO; the amount of water will decrease.

(b) For the same reason as in (a), the amount of $O_2$ will increase.

(c) According to Le Chatelier's principle, removing water will cause the reaction to shift toward products, resulting in the formation of more NO.

(d) According to Le Chatelier's principle, removing a reactant will cause the reaction to shift in the direction to replace the removed substance; the amount of $NH_3$ should increase.

(e) According to Le Chatelier's principle, adding ammonia will shift the reaction to the right, but the equilibrium constant, which is a constant, will not be affected.

(f) According to Le Chatelier's principle, removing NO will cause the formation of more products; the amount of $NH_3$ will decrease.

(g) According to Le Chatelier's principle, adding reactants will promote the formation of products; the amount of oxygen will decrease.

**9.75** As per Le Chatelier's principle, whether increasing the pressure on a reaction will affect the distribution of species within an equilibrium mixture of gases depends largely upon the difference in the number of moles of gases between the reactant and product sides of the equation. If there is a net increase in the amount of gas, then applying pressure will shift the reaction toward reactants in order to remove the stress applied by increasing the pressure. Similarly, if there is a net decrease in the amount of gas, applying pressure will cause the formation of products. If the number of moles of gas is the same on the product and reactant side, then changing the pressure will have little or no effect on the equilibrium distribution of species present. Using this information, we can apply it to the specific reactions given. The answers are: (a) reactants; (b) reactants; (c) reactants; (d) no change (there is the same number of moles of gas on both sides of the equation); (e) reactants

**9.77** (a) If the pressure of NO (a product) is increased, the reaction will shift to form more reactants; the pressure of $NH_3$ should increase.

(b) If the pressure of $NH_3$ (a reactant) is decreased, then the reaction will shift to form more reactants; the pressure of $O_2$ should increase.

**9.79** If a reaction is exothermic, raising the temperature will tend to shift the reaction toward reactants, whereas if the reaction is endothermic, a shift toward products will be observed. For the specific examples given, (a) and (b) are endothermic (the values for (b) can be calculated, but we know that it requires energy to break an X—X bond, so those processes will all be endothermic) and raising the temperature should favor the formation of products; (c) and (d) are exothermic and raising the temperature should favor the formation of reactants.

**9.81** Even though numbers are given, we do not need to do a calculation to answer this qualitative question. Because the equilibrium constant for the formation of ammonia is smaller at the higher temperature, raising the temperature will favor the formation of reactants. Less ammonia will be present at higher temperature, assuming no other changes occur to the system (i.e., the volume does not change, no reactants or products are added or removed from the container, etc.).

**9.83** To answer this question we must calculate $Q$:

$$Q = \frac{[NH_3]^2}{[N_2][H_2]^3} = \frac{(0.500)^2}{(3.00)(2.00)^3} = 0.0104$$

Because $Q \neq K$, the system is not at equilibrium and because $Q < K$, the reaction will proceed to produce more products.

**9.85** Because we want the equilibrium constant at two temperatures, we will need to calculate $\Delta H°_r$ and $\Delta S°_r$ for each reaction:

(a)  $NH_4Cl \rightarrow NH_3(g) + HCl(g)$

$\Delta H°_r = \Delta H°_f(NH_3, g) + \Delta H°_f(HCl, g) - \Delta H°_f(NH_4Cl, s)$

$\Delta H°_r = (-46.11 \text{ kJ} \cdot \text{mol}^{-1}) + (-92.31 \text{ kJ} \cdot \text{mol}^{-1}) - (-314.43 \text{ kJ} \cdot \text{mol}^{-1})$

$\Delta H°_r = 176.01 \text{ kJ} \cdot \text{mol}^{-1}$

$\Delta S°_r = S°(NH_3, g) + S°(HCl, g) - S°(NH_4Cl, s)$

$\Delta S°_r = 192.45 \text{ J} \cdot \text{K}^{-1} \cdot \text{mol}^{-1} + 186.91 \text{ J} \cdot \text{K}^{-1} \cdot \text{mol}^{-1} - 94.6 \text{ J} \cdot \text{K}^{-1} \cdot \text{mol}^{-1}$

$\Delta S°_r = 284.8 \text{ J} \cdot \text{K}^{-1} \cdot \text{mol}^{-1}$

$\Delta G°_r = \Delta H°_r - T\Delta S°_r$

At 298 K:

$\Delta G°_{r(298\,K)} = 176.01 \text{ kJ} - (298 \text{ K})(284.8 \text{ J} \cdot \text{K}^{-1})/(1000 \text{ J} \cdot \text{kJ}^{-1})$

$= 91.14 \text{ kJ} \cdot \text{mol}^{-1}$

$\Delta G°_{r(298\,K)} = -RT \ln K$

$\ln K = -\dfrac{\Delta G°_{r(298\,K)}}{RT}$

$= -\dfrac{91140 \text{ J}}{(8.314 \text{ J} \cdot \text{K}^{-1})(298 \text{ K})} = -36.8$

$K = 1 \times 10^{-16}$

At 423 K:

$\Delta G°_{r(423\,K)} = 176.01 \text{ kJ} - (423 \text{ K})(284.8 \text{ J} \cdot \text{K}^{-1})/(1000 \text{ J} \cdot \text{kJ}^{-1})$

$= 55.54 \text{ kJ} \cdot \text{mol}^{-1}$

$\Delta G°_{r(423\,K)} = -RT \ln K$

$\ln K = -\dfrac{\Delta G°_{r(423\,K)}}{RT}$

$= -\dfrac{55\,540 \text{ J}}{(8.314 \text{ J} \cdot \text{K}^{-1})(423 \text{ K})} = -15.8$

$K = 1 \times 10^{-7}$

(b)  $H_2(g) + D_2O(l) \rightarrow D_2(g) + H_2O(l)$

$$\Delta H^\circ_r = \Delta H^\circ_f(H_2O, l) - [\Delta H^\circ_f(D_2O, l)]$$

$$\Delta H^\circ_r = (-285.83 \text{ kJ} \cdot \text{mol}^{-1}) - [-294.60 \text{ kJ} \cdot \text{mol}^{-1}]$$

$$\Delta H^\circ_r = 8.77 \text{ kJ} \cdot \text{mol}^{-1}$$

$$\Delta S^\circ_r = S^\circ(D_2, g) + S^\circ(H_2O, l) - [S^\circ(H_2, g) + S^\circ(D_2O, l)]$$

$$\Delta S^\circ_r = 144.96 \text{ J} \cdot \text{K}^{-1} \cdot \text{mol}^{-1} + 69.91 \text{ J} \cdot \text{K}^{-1} \cdot \text{mol}^{-1}$$
$$- [130.68 \text{ J} \cdot \text{K}^{-1} \cdot \text{mol}^{-1} + 75.94 \text{ J} \cdot \text{K}^{-1} \cdot \text{mol}^{-1}]$$

$$\Delta S^\circ_r = 8.25 \text{ J} \cdot \text{K}^{-1} \cdot \text{mol}^{-1}$$

At 298 K:

$$\Delta G^\circ_{r(298 \text{ K})} = 8.77 \text{ kJ} \cdot \text{mol}^{-1} - (298 \text{ K})(8.25 \text{ J} \cdot \text{K}^{-1} \cdot \text{mol}^{-1})/(1000 \text{ J} \cdot \text{kJ}^{-1})$$
$$= 6.31 \text{ kJ} \cdot \text{mol}^{-1}$$

$$\Delta G^\circ_{r(298 \text{ K})} = -RT \ln K$$

$$\ln K = -\frac{\Delta G^\circ_{r(298 \text{ K})}}{RT}$$

$$= -\frac{6310 \text{ J}}{(8.314 \text{ J} \cdot \text{K}^{-1})(298 \text{ K})} = -2.55$$

$$K = 7.8 \times 10^{-2}$$

At 423 K:

$$\Delta G^\circ_{r(423 \text{ K})} = 8.77 \text{ kJ} \cdot \text{mol}^{-1} - (423 \text{ K})(8.25 \text{ J} \cdot \text{K}^{-1} \cdot \text{mol}^{-1})/(1000 \text{ J} \cdot \text{kJ}^{-1})$$
$$= 5.28 \text{ kJ} \cdot \text{mol}^{-1}$$

$$\ln K = -\frac{5280 \text{ J}}{(8.314 \text{ J} \cdot \text{K}^{-1})(423 \text{ K})} = -1.50$$

$$K = 0.22$$

**9.87** $\quad K = (RT)^{\Delta n} K_c$

$$\ln \frac{K_2}{K_1} = -\frac{\Delta H^\circ_r}{R}\left(\frac{1}{T_1} - \frac{1}{T_2}\right)$$

$$\ln\left(\frac{(RT_2)^{\Delta n} K_{c2}}{(RT_1)^{\Delta n} K_{c1}}\right) = -\frac{\Delta H^\circ_r}{R}\left(\frac{1}{T_1} - \frac{1}{T_2}\right)$$

$$\Delta n \ln\left(\frac{T_2}{T_1}\right) + \ln\left(\frac{K_{c2}}{K_{c1}}\right) = -\frac{\Delta H_r^\circ}{R}\left(\frac{1}{T_1} - \frac{1}{T_2}\right)$$

$$\ln\left(\frac{K_{c2}}{K_{c1}}\right) = -\frac{\Delta H_r^\circ}{R}\left(\frac{1}{T_1} - \frac{1}{T_2}\right) - \Delta n \ln\left(\frac{T_2}{T_1}\right)$$

**9.89** $\Delta G^\circ = -RT \ln K = \Delta H^\circ - T\Delta S^\circ$

$\Delta G^\circ_{298} = -(8.3145 \text{JK}^{-1}\text{mol}^{-1})(298.15\text{K}) \ln(9.9197 \times 10^{-15})$

$\qquad = 7.9933 \times 10^4 \text{Jmol}^{-1}$

$\Delta G^\circ_{303} = -(8.3145 \text{JK}^{-1}\text{mol}^{-1})(303.15\text{K}) \ln(1.4689 \times 10^{-14})$

$\qquad = 8.0283 \times 10^4 \text{Jmol}^{-1}$

$\Delta G^\circ_{298} - \Delta G^\circ_{303} = \Delta H^\circ - (298.15\text{K})\Delta S^\circ - [\Delta H^\circ - (303.15\text{K})\Delta S^\circ]$

$-3.5097 \times 10^2 \text{Jmol}^{-1} = (5.00\text{K})\Delta S^\circ$

$\Delta S^\circ = -70.194 \text{JK}^{-1}\text{mol}^{-1}$

The sign is negative indicating that the product ions create more order in the system probably because of their interaction with solvating water molecules.

**9.91** (a) According to Le Chatelier's principle, adding a product should cause a shift in the equilibrium toward the reactants side of the equation.

(b) Because there are equal numbers of moles of gas on both sides of the equation, there will be little or no effect upon compressing the system.

(c) If the amount of $CO_2$ is increased, this will cause the reaction to shift toward the formation of products.

(d) Because the reaction is endothermic, raising the temperature will favor the formation of products.

(e) If the amount of $C_6H_{12}O_6$ is removed, this will cause the reaction to shift toward the formation of products.

(f) Because water is a liquid, it is by definition present at unit concentration, so changing the amount of water will not affect the reaction. As long as the glucose solution is dilute, its concentration can be considered unchanged.

(g) Decreasing the concentration of a reactant will favor the production of more reactants.

**9.93** (a) In order to solve this problem, we will manipulate the equations with the known $K$'s so that we can combine them to give the desired overall reaction:

First, reverse equation (1) and multiply it by $\frac{1}{2}$:

$$H_2(g) + \tfrac{1}{2}O_2(g) \rightarrow H_2O(g) \quad K_4 = \frac{1}{(1.6 \times 10^{-11})^{1/2}} \tag{4}$$

Multiply equation (2) by $\frac{1}{2}$ also:

$$CO_2(g) \rightarrow CO(g) + \tfrac{1}{2}O_2(g) \quad K_5 = (1.3 \times 10^{-10})^{1/2} \tag{5}$$

Adding equations (4) and (5) gives the desired reaction. The resultant equilibrium constant will be the product of the $K$'s for (4) and (5):

$$CO_2(g) + H_2(g) \rightarrow H_2O(g) + CO(g) \quad K_5 = \left(\frac{1.3 \times 10^{-10}}{1.6 \times 10^{-11}}\right)^{1/2} = 2.9 \tag{3}$$

(b) To obtain the $K$ value for a net equation from two (or more) others, the $K$'s are multiplied, but $\Delta G^\circ_r$'s are added:

$$2\,H_2O(g) \rightarrow 2\,H_2(g) + O_2(g) \tag{1}$$

$$
\begin{aligned}
\Delta G^\circ_{r(1)} &= -RT \ln K_1 \\
&= -(8.314\,\text{J} \cdot \text{K}^{-1} \cdot \text{mol}^{-1})(1565\,\text{K}) \ln(1.6 \times 10^{-11}) \\
&= +3.2 \times 10^5\,\text{J} \cdot \text{mol}^{-1} = +3.2 \times 10^2\,\text{kJ} \cdot \text{mol}^{-1}
\end{aligned}
$$

$$2\,CO_2(g) \rightarrow 2\,CO(g) + O_2(g) \tag{2}$$

$$
\begin{aligned}
\Delta G^\circ_{r(2)} &= -RT \ln K_2 \\
&= -(8.314\,\text{J} \cdot \text{K}^{-1} \cdot \text{mol}^{-1})(1565\,\text{K}) \ln(1.3 \times 10^{-10}) \\
&= +3.0 \times 10^5\,\text{J} \cdot \text{mol}^{-1} = +3.0 \times 10^2\,\text{kJ} \cdot \text{mol}^{-1}
\end{aligned}
$$

The corresponding values for (4) and (5) are

$$\Delta G^\circ_{r(4)} = -\tfrac{1}{2}(3.2 \times 10^2\,\text{kJ} \cdot \text{mol}^{-1}) = -1.6 \times 10^2\,\text{kJ} \cdot \text{mol}^{-1}$$

$$\Delta G^\circ_{r(5)} = \tfrac{1}{2}(3.0 \times 10^2\,\text{kJ} \cdot \text{mol}^{-1}) = 1.5 \times 10^2\,\text{kJ} \cdot \text{mol}^{-1}$$

Summing these two values will give

$$\Delta G^{\circ}_{r(3)} = -1.6 \times 10^2 \text{ kJ} \cdot \text{mol}^{-1} + 1.5 \times 10^2 \text{ kJ} \cdot \text{mol}^{-1} = -10. \text{ kJ} \cdot \text{mol}^{-1}$$

$$\ln K_3 = -\frac{\Delta G^{\circ}_{r(3)}}{RT} = -\frac{-10000 \text{ J} \cdot \text{mol}^{-1}}{(8.314 \text{ J} \cdot \text{K}^{-1} \cdot \text{mol}^{-1})(1565 \text{ K})} = +0.77$$

$$K_3 = 2.2$$

This value is in reasonable agreement with the one obtained in (a), given the problems in significant figures and due to rounding errors.

**9.95**

| Pressure | $PCl_5(g)$ | $\rightarrow$ | $PCl_3(g)$ | $+$ | $Cl_2(g)$ |
|---|---|---|---|---|---|
| initial | $n$ | | | | |
| change | $-n\alpha$ | | $+n\alpha$ | | $+n\alpha$ |
| final | $n(1-\alpha)$ | | $+n\alpha$ | | $+n\alpha$ |

$$K = \frac{(n\alpha)(n\alpha)}{n(1-\alpha)} = \frac{n^2\alpha^2}{n(1-\alpha)} = \frac{n\alpha^2}{(1-\alpha)}$$

$$P = n(1-\alpha) + n\alpha + n\alpha = n(1+\alpha)$$

$$n = \frac{P}{1+\alpha}$$

$$K = \frac{n\alpha^2}{(1-\alpha)} = \left(\frac{P}{1+\alpha}\right)\left(\frac{\alpha^2}{1-\alpha}\right) = \frac{P\alpha^2}{1-\alpha^2}$$

$$(1-\alpha^2)K = P\alpha^2$$

$$K - K\alpha^2 = P\alpha^2$$

$$K = P\alpha^2 + K\alpha^2$$

$$\alpha^2 = \frac{K}{P+K}$$

$$\alpha = \sqrt{\frac{K}{P+K}}$$

(a) For the specific conditions $K = 4.96$ and $P = 0.50$ bar,

$$\alpha = \sqrt{\frac{4.96}{0.50 + 4.96}} = 0.953$$

(b) For the specific conditions $K = 4.96$ and $P = 1.00$ bar,

$$\alpha = \sqrt{\frac{4.96}{1.00 + 4.96}} = 0.912$$

**9.97** (a) If $K = 1.00$, then $\Delta G°$ must be equal to 0 ($\Delta G° = -RT \ln K$).

(b) This can be calculated by determining the values for $\Delta H°$ and $\Delta S°$ at 25°C.

$$\Delta H° = -393.51 \text{ kJ} \cdot \text{mol}^{-1} - [(-110.53 \text{ kJ} \cdot \text{mol}^{-1}) + (-241.82 \text{ kJ} \cdot \text{mol}^{-1})]$$
$$= -41.16 \text{ kJ} \cdot \text{mol}^{-1}$$

$$\Delta S° = 130.68 \text{ J} \cdot \text{K}^{-1} \cdot \text{mol}^{-1} + 213.74 \text{ J} \cdot \text{K}^{-1} \cdot \text{mol}^{-1}$$
$$- [197.67 \text{ J} \cdot \text{K}^{-1} \cdot \text{mol}^{-1} + 188.83 \text{ J} \cdot \text{K}^{-1} \cdot \text{mol}^{-1}]$$
$$= -42.08 \text{ J} \cdot \text{K}^{-1} \cdot \text{mol}^{-1}$$

$$\Delta G = (-41.16 \text{ kJ} \cdot \text{mol}^{-1})(1000 \text{ J} \cdot \text{kJ}^{-1}) - T(-42.08 \text{ J} \cdot \text{K}^{-1} \cdot \text{mol}^{-1}) = 0$$
$$T = 978 \ K \text{ (or 705°C)}$$

(c)

| | CO(g) + | H$_2$O(g) | → CO$_2$(g) + | H$_2$(g) |
|---|---|---|---|---|
| | 10.00 bar | 10.00 bar | 5.00 bar | 5.00 bar |
| change | $-x$ | $-x$ | $+x$ | $+x$ |
| net | $10.00 - x$ | $10.00 - x$ | $5.00 + x$ | $5.00 + x$ |

$x = 2.50$ bar

All pressures are equal to 7.50 bar.

(d) First, check $Q$ to determine the direction of the reaction:

$$Q = \frac{(10.00)(5.00)}{(6.00)(4.00)} = 2.08$$

Because $Q$ is greater than 1, the reaction will shift to produce reactants.

| | CO(g) + | H$_2$O(g) | → | CO$_2$(g) + | H$_2$(g) |
|---|---|---|---|---|---|
| | 6.00 bar | 4.00 bar | | 10.00 bar | 5.00 bar |
| change | $+x$ | $+x$ | | $-x$ | $-x$ |
| net | $6.00 + x$ | $4.00 + x$ | | $10.00 - x$ | $5.00 - x$ |

$$\frac{(10.00 - x)(5.00 - x)}{(6.00 + x)(4.00 + x)} = 1$$

$$(10.00 - x)(5.00 - x) = (6.00 + x)(4.00 + x)$$

$$x^2 - 15.00x + 50.0 = x^2 + 10.00x + 24.0$$

$$25.00x = 26.0$$

$$x = 1.04 \text{ bar}$$

$P_{CO(g)} = 7.04$ bar; $P_{H_2O(g)} = 5.04$ bar; $P_{CO_2(g)} = 8.96$ bar; $P_{H_2(g)} = 3.96$ bar

**9.99** (a) These values are easily calculated from the relationship $\Delta G° = -RT \ln K$. For the atomic species, the free energy of the reaction will be $\frac{1}{2}$ of this value because the equilibrium reactions are for the formation of two moles of halogen atoms.

The results are

| Halogen | Bond Dissociation Energy (kJ·mol$^{-1}$) | $\Delta G°$ (kJ·mol$^{-1}$) |
|---------|-----------------------------------------|------------------------------|
| fluorine | 146 | 19.2 |
| chlorine | 230 | 47.8 |
| bromine | 181 | 42.8 |
| iodine | 139 | 5.6 |

(b)

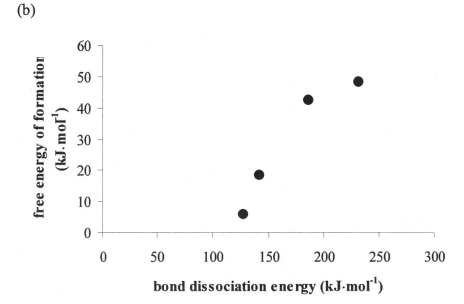

There is a correlation between the bond dissociation energy and the free energy of formation of the atomic species, but the relationship is clearly not linear.

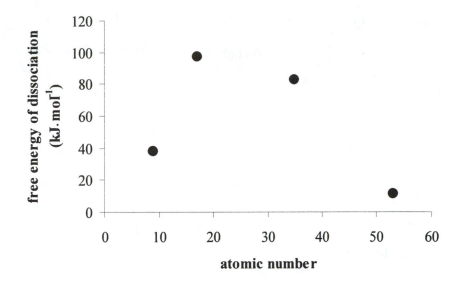

For the heavier three halogens, there is a trend to decreasing free energy of formation of the atoms as the element becomes heavier, but fluorine is anomalous. The F—F bond energy is lower than expected, owing to repulsions of the lone pairs of electrons on the adjacent F atoms because the F—F bond distance is so short.

**9.101** (a) K<1 but will increase at higher temperature because $\Delta S > 0$.

(b) K>1

(c) K>1

(d) K>1

**9.103** (a) Using the thermodynamic data in Appendix 2A:

$Br_2(g) \rightarrow 2\,Br(g)$

$\Delta G° = 2(82.40\,\text{kJ}\cdot\text{mol}^{-1}) - 3.11\,\text{kJ}\cdot\text{mol}^{-1} = 161.69\,\text{kJ}\cdot\text{mol}^{-1}$

$K = e^{-\Delta G°/RT} = 4.5 \times 10^{-29}$

(b) We will use data from Appendix 2A to calculate the vapor pressure of bromine:

$$Br_2(l) \rightarrow Br_2(g)$$

$$\Delta G° = 3.11 \, kJ \cdot mol^{-1}$$

$$K = e^{-\Delta G°/RT} = 0.285$$

The vapor pressure of bromine will, therefore, be 0.285 bar or 0.289 atm. Remember that because the standard state for the thermodynamic quantities is 1 bar, the values in $K$ will be derived in bar as well.

(c) $\quad 4.5 \times 10^{-29} = \dfrac{P_{Br(g)}^2}{P_{Br_2(l)}} = \dfrac{P_{Br(g)}^2}{0.285 \, bar}$

$$P_{Br(g)} = 3.6 \times 10^{15} \, bar \text{ or } 3.6 \times 10^{15} \, atm$$

(d) Use the ideal gas law:

$$PV = nRT$$

$$(0.289 \, atm)V = (0.0100 \, mol)(0.082 \, 06 \, L \cdot atm \cdot K^{-1} \cdot mol^{-1})(298 \, K)$$

$$V = 0.846 \, L \text{ or } 846 \, mL$$

**9.105** First, we calculate the equilibrium constant for the conditions given.

$$K = \dfrac{(23.72)^2}{(3.11)(1.64)^3} = 41.0, \text{ which corresponds to the reaction written as}$$

$$N_2(g) + 3 \, H_2(g) \rightarrow 2 \, NH_3(g)$$

We then set up the table of anticipated changes upon introduction of the nitrogen:

|  | $N_2(g)$ | + | $3 \, H_2(g)$ | $\rightarrow$ | $2 \, NH_3(g)$ |
|---|---|---|---|---|---|
| initial | 4.68 bar | | 1.64 bar | | 23.72 bar |
| change | $-x$ | | $-3x$ | | $+2x$ |
| total | $4.68 - x$ | | $1.64 - 3x$ | | $23.72 + 2x$ |

$$41.0 = \dfrac{(23.72 + 2x)^2}{(4.68 - x)(1.64 - 3x)^3}$$

The equation can be solved using a graphing calculator, other computer software, or by trial and error. The solution is $x = 0.176$. The pressures of gases are

$P_{N_2} = 4.33$ bar or 4.39 atm

$P_{H_2} = 1.11$ bar or 1.12 atm

$P_{NH_3} = 24.07$ bar or 24.39 atm

**9.107**

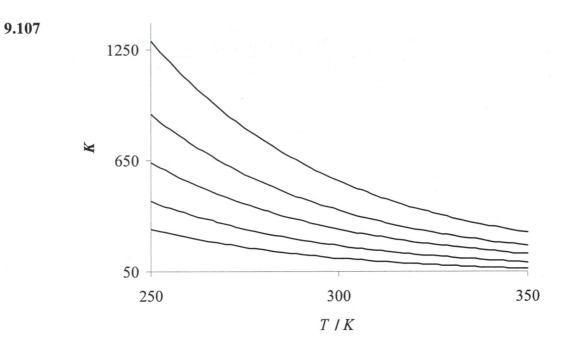

**9.109** (a) $K_C = \dfrac{[NO_2]^2}{[N_2O_4]} = \dfrac{(2.13)^2}{0.405} = 11.2$

(b) If $NO_2$ is added, the equilibrium will shift to produce more $N_2O_4$.

The amount of $NO_2$ will be greater than initially present, but less than the

$3.13 \text{ mol} \cdot L^{-1}$ present immediately upon making the addition. $K_C$ will not

be affected.

(c)

| Concentrations (mol·L$^{-1}$) | $N_2O_4$ | $\rightarrow$ | 2 $NO_2$ |
|---|---|---|---|
| initial | 0.405 | | 3.13 |
| change | +x | | −2x |
| final | 0.405 + x | | 3.13 − 2x |

$$11.2 = \frac{(3.13 - 2x)^2}{0.405 + x}$$

$$(11.2)(0.405 + x) = (3.13 - 2x)^2$$

$$11.2x + 4.536 = 4x^2 - 12.52x + 9.797$$

$$4x^2 - 23.7x + 5.26 = 0$$

$$x = \frac{-(-23.7) \pm \sqrt{(-23.7)^2 - (4)(4)(5.26)}}{(2)(4)} = \frac{23.7 \pm 21.9}{8}$$

$$x = 5.70 \text{ or } 0.23$$

At equilibrium $[N_2O_4] = 0.405 \text{ mol} \cdot L^{-1} + 0.23 \text{ mol} \cdot L^{-1} = 0.64 \text{ mol} \cdot L^{-1}$

$[NO_2] = 3.13 \text{ mol} \cdot L^{-1} - 2(0.23 \text{ mol} \cdot L^{-1}) = 2.67 \text{ mol} \cdot L^{-1}$

These concentrations are consistent with the predictions in (b).

**9.111** To find the vapor pressure, we first calculate $\Delta G°$ for the conversion of the liquid to the gas at 298 K, using the free energies of formation found in the appendix:

$$\Delta G°_{H_2O(l) \rightarrow H_2O(g)} = \Delta G°_{f(H_2O(g))} - \Delta G°_{f(H_2O(l))}$$
$$= (-228.57 \text{ kJ} \cdot \text{mol}^{-1}) - [-237.13 \text{ kJ} \cdot \text{mol}^{-1}]$$
$$= 8.56 \text{ kJ} \cdot \text{mol}^{-1}$$

$$\Delta G°_{D_2O(l) \rightarrow D_2O(g)} = \Delta G°_{f(D_2O(g))} - \Delta G°_{f(D_2O(l))}$$
$$= (-234.54 \text{ kJ} \cdot \text{mol}^{-1}) - [-243.44 \text{ kJ} \cdot \text{mol}^{-1}]$$
$$= 8.90 \text{ kJ} \cdot \text{mol}^{-1}$$

The equilibrium constant for these processes is the vapor pressure of the liquid:

$$K = P_{H_2O} \text{ or } K = P_{D_2O}$$

Using $\Delta G° = -RT \ln K$, we can calculate the desired values.

For $H_2O$:

$$\ln K = -\frac{\Delta G°}{RT} = -\frac{8560 \text{ J} \cdot \text{mol}^{-1}}{(8.314 \text{ J} \cdot \text{K}^{-1} \cdot \text{mol}^{-1})(298 \text{ K})} = -3.45$$

$$K = 0.032 \text{ bar}$$

$$0.032 \text{ bar} \times \frac{1 \text{ atm}}{1.013\,25 \text{ bar}} \times \frac{760 \text{ Torr}}{1 \text{ atm}} = 24 \text{ Torr}$$

For $D_2O$:

$$\ln K = -\frac{\Delta G^\circ}{RT} = -\frac{8900 \text{ J} \cdot \text{mol}^{-1}}{(8.314 \text{ J} \cdot \text{K}^{-1} \cdot \text{mol}^{-1})(298 \text{ K})} = -3.59$$

$$K = 0.028 \text{ bar}$$

$$0.028 \text{ bar} \times \frac{1 \text{ atm}}{1.013\,25 \text{ bar}} \times \frac{760 \text{ Torr}}{1 \text{ atm}} = 21 \text{ Torr}$$

The answer is that D has a lower zero point energy than H. This makes the $D_2O - D_2O$ "hydrogen bond" stronger than the $H_2O - H_2O$ hydrogen bond. Because the hydrogen bond is stronger, the intermolecular forces are stronger, the vapor pressure is lower, and the boiling point is higher.

9.113   First, we derive a general relationship that relates $\Delta G^\circ_f$ values to nonstandard state conditions. We do this by returning to the fundamental definition of $\Delta G$ and $\Delta G^\circ$.

$$\Delta G^\circ_f = \Sigma G^\circ_{m(\text{products})} - \Sigma G^\circ_{m(\text{reactants})}$$

Similarly, for conditions other than standard state, we can write

$$\Delta G_f = \Sigma G_{m(\text{products})} - \Sigma G_{m(\text{reactants})}$$

Because $G_m = G^\circ_m + RT \ln Q$,

$$\Delta G_f = \Sigma (G^\circ_m + RT \ln P_i)_{(\text{products})} - \Sigma (G^\circ_m + RT \ln P_i)_{(\text{reactants})}$$

But because all reactants or products in the same system refer to the same standard state,

$$\Delta G_f = \Sigma G^\circ_{m(\text{products})} - \Sigma G^\circ_{m(\text{reactants})} + \Delta n (RT \ln Q_{(\text{reactants})})$$

$$\Delta G_f = \Delta G^\circ_f + \Delta n (RT \ln Q_{(\text{reactants})})$$

The value $\Delta G_f$, which is the nonstandard value for the conditions of 1 atm,

$1 \text{ mol} \cdot \text{L}^{-1}$, etc. becomes the new standard value.

(a) $\frac{1}{2} H_2(g) + \frac{1}{2} I_2(g) \rightarrow HI(g)$, $\Delta n = 1$ mol $- (\frac{1}{2}$ mol $+ \frac{1}{2}$ mol$) = 0$

1 atm = 1.013 25 bar

$\Delta G_f = 1.70$ kJ $\cdot$ mol$^{-1}$
$\quad + (0)(8.314$ J $\cdot$ K$^{-1} \cdot$ mol$^{-1})(298$ K$)(\ln 1.01325)/(1000$ J $\cdot$ kJ$^{-1})$
$\quad = +1.70.$ kJ $\cdot$ mol$^{-1}$

(b) $C(s) + \frac{1}{2} O_2(g) \rightarrow CO(g)$, $\Delta n = 1$ mol $- \frac{1}{2}$ mol $= \frac{1}{2}$ mol

$\Delta G_f = -137.17$ kJ $\cdot$ mol
$\quad + (\frac{1}{2})(8.314$ J $\cdot$ K$^{-1}.$ mol$^{-1})(298$ K$)(\ln 1.013\ 25)/(1000$ J $\cdot$ kJ$^{-1})$
$\quad = -137.15$ kJ $\cdot$ mol$^{-1}$

(c) $C(s) + \frac{1}{2} N_2(g) + \frac{1}{2} H_2(g) \rightarrow HCN(g)$, $\Delta n = 1$ mol $- (\frac{1}{2}$ mol $+ \frac{1}{2}$ mol$)$
$\quad\quad\quad\quad\quad\quad = 0$

1 Torr $\times \dfrac{1\ \text{atm}}{760\ \text{Torr}} \times \dfrac{1.013\ 25\ \text{bar}}{1\ \text{atm}} = 1.333 \times 10^{-3}$ bar

$\Delta G_f = 124.7$ kJ $\cdot$ mol$^{-1}$
$\quad + (0)(8.314$ J $\cdot$ K$^{-1} \cdot$ mol$^{-1})(298$ K$)(\ln 1.333 \times 10^{-3})/(1000$ J $\cdot$ kJ$^{-1})$
$\quad = +124.7.$ kJ $\cdot$ mol$^{-1}$

(d) $C(s) + 2 H_2(g) \rightarrow CH_4(g)$, $\Delta n = 1$ mol $- 2$ mol $= -1$ mol

1 Pa $= 10^{-5}$ bar

$\Delta G_f = -50.72$ kJ $\cdot$ mol
$\quad + (-1)(8.314$ J $\cdot$ K$^{-1} \cdot$ mol$^{-1})(298$ K$)(\ln 10^{-5})/(1000$ J $\cdot$ kJ$^{-1})$
$\quad = -22.2$ kJ $\cdot$ mol$^{-1}$

**9.115** (a) (i) Increasing the amount of a reactant will push the equilibrium toward the products. More $NO_2$ will form. (ii) Removing a product will pull the equilibrium toward products. More $NO_2$ will form. (iii) Increasing total pressure by adding an inert gas not change the relative partial pressures. There will be no change in the amount of $NO_2$. (b) Since there are two moles of gas on both sides of the reaction, the volume cancels out of the equilibrium constant expression and it is possible to use moles directly for each component.

At equilibrium, $SO_2$ = 2.4 moles -1.2 moles = 1.2 moles, $SO_3$ = 1.2 moles and NO = 1.2 moles since the reaction is 1:1:1:1. The original number of moles of $NO_2$ is set to $x$.

$K = 3.00 = (1.2)(1.2)/(1.2)(x-1.2) = (1.2)/(x-1.2)$

$x=1.2/3.00 + 1.2$

$x=1.6$ moles of $NO_2$

**9.117** (a) Reversing both reactions then adding them together gives the reaction of interest.

$NO + O \rightarrow NO_2$     $K'=1/K=1/6.8 \times 10^{-49}=1.47 \times 10^{48}$

$NO_2 + O_2 \rightarrow O_3 + NO$  $K''=1/K=1/5.8 \times 10^{-34}=1.72 \times 10^{33}$

---------------------------

$O_2 + O \rightarrow O_3$              $K=K'*K''=2.5 \times 10^{81}$

(b) Since the initial mixture is equimolar in the two reactants (i.e. each partial pressure is 2.0 bar) and the equilibrium strongly favors products, we can see that essentially all of both reactants will combine to form an equimolar amount of ozone, or a final equilibrium partial pressure of 2.0 bar. Some very small partial pressure, $x$, of each reactant will be left and must satisfy the equilibrium constant expression:

$K = 2.5 \times 10^{81} = P_{O3}/(P_{O2}*P_O) = 2.0$ bar/$(x^2)$

$x = 2.8 \times 10^{-41}$ bar $= P_{O2} = P_O$ and $P_{O3} = 2.0$ bar at equilibrium

# CHAPTER 10

# ACIDS AND BASES

**10.1** (a) $CH_3NH_3^+$ (b) $NH_2NH_3^+$ (c) $H_2CO_3$ (d) $CO_3^{2-}$

(e) $C_6H_5O^-$ (f) $CH_3CO_2^-$

**10.3** For all parts (a) – (e), $H_2O$ and $H_3O^+$ form a conjugate acid-base pair in which $H_2O$ is the base and $H_3O$ is the acid

(a) $H_2SO_4(aq) + H_2O(l) \leftrightarrow H_3O^+(aq) + HSO_4^-(aq)$

$H_2SO_4$ and $HSO_4^-(aq)$ form a conjugate acid-base pair in which

$H_2SO_4$ is the acid and $HSO_4^-(aq)$ is the base.

(b)

$C_6H_5NH_3^+(aq) + H_2O(l) \leftrightarrow H_3O^+(aq) + C_6H_5NH_2(aq)$
$C_6H_5NH_3^+$ and $C_6H_5NH_2(aq)$ form a conjugate acid-base pair in which

$C_6H_5NH_3^+$ is the acid and $C_6H_5NH_2(aq)$ is the base.

(c) $H_2PO_4^-(aq) + H_2O(l) \leftrightarrow H_3O^+(aq) + HPO_4^{2-}(aq)$

$H_2PO_4^-(aq)$ and $HPO_4^{2-}(aq)$ form a conjugate acid-base pair in which

$H_2PO_4^-(aq)$ is the acid and $HPO_4^{2-}(aq)$ is the base.

(d) $HCOOH(aq) + H_2O(l) \leftrightarrow H_3O^+(aq) + HCO_2^-(aq)$

$HCOOH(aq)$ and $HCO_2^-(aq)$ form a conjugate acid-base pair in which

$HCOOH(aq)$ is the acid and $HCO_2^-(aq)$ is the base.

(e) $NH_2NH_3^+(aq) + H_2O(l) \leftrightarrow H_3O^+(aq) + NH_2NH_2(aq)$

$NH_2NH_3^+(aq)$ and $NH_2NH_2(aq)$ form a conjugate acid-base pair in

which $NH_2NH_3^+(aq)$ is the acid and $NH_2NH_2(aq)$ is the base.

**10.5**   (a)  Brønsted acid: $HNO_3$

Brønsted base: $HPO_4{}^{2-}$

(b)  conjugate base to $HNO_3$: $NO_3{}^-$

conjugate acid to $HPO_4{}^{2-}$: $H_2PO_4{}^-$

**10.7**   (a)  $HCO_3{}^-$, as an acid: $HCO_3^-(aq) + H_2O(l) \leftrightarrow H_3O^+(aq) + CO_3^{2-}(aq)$.

$HCO_3^-$ and $CO_3^{2-}$ form a conjugate acid-base pair in which $HCO_3^-$ is the acid and $CO_3^{2-}$ is the base.

$HCO_3{}^-$, as a base: $H_2O(l) + HCO_3^-(aq) \leftrightarrow H_2CO_3(aq) + OH^-(aq)$.

$HCO_3^-$ and $H_2CO_3$ form a conjugate acid-base pair in which $HCO_3^-$ is the base and $H_2CO_3$ is the acid. $H_2O$ and $OH^-$ form a conjugate acid-base pair in which $H_2O$ is the acid and $OH^-$ is the base.

(b)  $HPO_4{}^{2-}$, as an acid: $HPO_4^{2-}(aq) + H_2O(l) \leftrightarrow H_3O^+(aq) + PO_4^{3-}(aq)$.

$HPO_4{}^{2-}(aq)$ and $PO_4{}^{3-}(aq)$ form a conjugate acid-base pair in which

$HPO_4{}^{2-}(aq)$ is the acid and $PO_4{}^{3-}(aq)$ is the base. $H_2O$ and $H_3O^+$ form a

conjugate acid-base pair in which $H_2O$ is the base and $H_3O^+$ is the acid.

$HPO_4{}^{2-}$, as a base: $HPO_4^{2-}(aq) + H_2O(l) \leftrightarrow H_2PO_4^-(aq) + OH^-(aq)$.

$HPO_4{}^{2-}$ and $H_2PO_4{}^-$ form a conjugate acid-base pair in which $HPO_4{}^{2-}$ is

the base and $H_2PO_4^-$ is the acid. $H_2O$ and $OH^-$ form a conjugate acid-base

pair in which $H_2O$ is the acid and $OH^-$ is the base.

**10.9**   (a) basic;   (b) acidic;   (c) amphoteric;   (d) basic

**10.11**  (a)  $SO_3\,(g) + H_2SO_4\,(l) \leftrightarrow H_2S_2O_7\,(l)$;

(b)
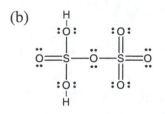

(c) sulfuric acid acts as the Lewis base while SO$_3$ acts as a Lewis acid.

**10.13** In each case, use $K_w = [H_3O^+][OH^-] = 1.0 \times 10^{-14}$, then

$$[OH^-] = \frac{K_w}{[H_3O^+]} = \frac{1.0 \times 10^{-14}}{[H_3O^+]}$$

(a) $[OH^-] = \dfrac{1.0 \times 10^{-14}}{0.02} = 5.0 \times 10^{-13} \ \text{mol} \cdot \text{L}^{-1}$

(b) $[OH^-] = \dfrac{1.0 \times 10^{-14}}{1.0 \times 10^{-5}} = 1.0 \times 10^{-9} \ \text{mol} \cdot \text{L}^{-1}$

(c) $[OH^-] = \dfrac{1.0 \times 10^{-14}}{3.1 \times 10^{-3}} = 3.2 \times 10^{-12} \ \text{mol} \cdot \text{L}^{-1}$

**10.15** (a) $K_w = 2.1 \times 10^{-14} = [H_3O^+][OH^-] = x^2$, where $x = [H_3O^+] = [OH^-]$

$x = \sqrt{2.1 \times 10^{-14}} = 1.4 \times 10^{-7} \ \text{mol} \cdot \text{L}^{-1}$

$pH = -\log[H_3O^+] = 6.80$

(b) $[OH^-] = [H_3O^+] = 1.4 \times 10^7 \ \text{mol} \cdot \text{L}^{-1}$

**10.17** Because Ba(OH)$_2$ is a strong base,

$$\text{Ba(OH)}_2 \, (aq) \longrightarrow \text{Ba}^{2+} (aq) + 2 \, \text{OH}^- (aq), 100\%.$$

Then

$[\text{Ba(OH)}_2]_0 = [\text{Ba}^{2+}]$, $[OH^-] = 2 \times [\text{Ba(OH)}_2]_0$, where $[\text{Ba(OH)}_2]_0 =$ nominal concentration of Ba(OH)$_2$.

$$\text{moles of Ba(OH)}_2 = \frac{0.25 \ \text{g}}{171.36 \ \text{g} \cdot \text{mol}^{-1}} = 1.5 \times 10^{-3} \ \text{mol}$$

$$[\text{Ba(OH)}_2]_0 = \frac{1.5 \times 10^{-3} \ \text{mol}}{0.100 \ \text{L}} = 1.5 \times 10^{-2} \ \text{mol} \cdot \text{L}^{-1} = [\text{Ba}^{2+}]$$

$$[OH^-] = 2 \times [\text{Ba(OH)}_2]_0 = 2 \times 1.5 \times 10^{-2} \ \text{mol} \cdot \text{L}^{-1} = 2.9 \times 10^{-2} \ \text{mol} \cdot \text{L}^{-1}$$

$$[H_3O^+] = \frac{K_w}{[OH^-]} = \frac{1.0 \times 10^{-14}}{2.9 \times 10^{-2}} = 3.4 \times 10^{-13} \ \text{mol} \cdot \text{L}^{-1}$$

**10.19** Because $pH = -\log[H_3O^+]$, $\log[H_3O^+] = -pH$. Taking the antilogs of both

sides gives $[H_3O^+] = 10^{-pH}$ $mol \cdot L^{-1}$

(a) $[H_3O^+] = 10^{-3.3} = 5 \times 10^{-4}$ $mol \cdot L^{-1}$

(b) $[H_3O^+] = 10^{-6.7}$ $mol \cdot L^{-1} = 2 \times 10^{-7}$ $mol \cdot L^{-1}$

(c) $[H_3O^+] = 10^{-4.4}$ $mol \cdot L^{-1} = 4 \times 10^{-5}$ $mol \cdot L^{-1}$

(d) $[H_3O^+] = 10^{-5.3}$ $mol \cdot L^{-1} = 5 \times 10^{-6}$ $mol \cdot L^{-1}$

**10.21** (a) $[HNO_3] = [H_3O^+] = 0.0146$ $mol \cdot L^{-1}$

$pH = -\log(0.0146) = 1.84$, $pOH = 14.00 - (-1.84) = 12.16$

(b) $[HCl] = [H_3O^+] = 0.11$ $mol \cdot L^{-1}$

$pH = -\log(0.11) = 0.96$, $pOH = 14.00 - 0.96 = 13.04$

(c) $[OH^-] = 2 \times [Ba(OH)_2] = 2 \times 0.0092$ M $= 0.018$ $mol \cdot L^{-1}$

$pOH = -\log(0.018) = 1.74$, $pH = 14.00 - 1.74 = 12.26$

(d) $[KOH]_0 = [OH^-]$

$[OH^-] = \left(\dfrac{2.00\ mL}{500\ mL}\right) \times (0.175\ mol \cdot L^{-1}) = 7.0 \times 10^{-4}$ $mol \cdot L^{-1}$

$pOH = -\log(7.0 \times 10^{-4}) = 3.15$, $pH = 14.00 - 3.15 = 10.85$

(e) $[NaOH]_0 = [OH^-]$

number of moles of NaOH $= \dfrac{0.0136\ g}{40.00\ g \cdot mol^{-1}} = 3.40 \times 10^{-4}$ mol

$[NaOH]_0 = \dfrac{3.40 \times 10^{-4}\ mol}{0.350\ L} = 9.71 \times 10^{-4}$ $mol \cdot L^{-1} = [OH^-]$

$pOH = -\log(9.71 \times 10^{-4}) = 3.01$, $pH = 14.00 - 3.01 = 10.99$

(f) $[HBr]_0 = [H_3O^+]$

$[H_3O^+] = \left(\dfrac{75.0\ mL}{500\ mL}\right) \times (3.5 \times 10^{-4}\ mol \cdot L^{-1}) = 5.3 \times 10^{-5}$ $mol \cdot L^{-1}$

$pH = -\log(5.3 \times 10^{-5}) = 4.28$, $pOH = 14.00 - 4.28 = 9.72$

**10.23** $pK_{a1} = -\log K_{a1}$; therefore, after taking antilogs, $K_{a1} = 10^{-pK_{a1}}$

| Acid | $pK_{a1}$ | $K_{a1}$ |
|------|-----------|----------|
| (a) $H_3PO_4$ | 2.12 | $7.6 \times 10^{-3}$ |
| (b) $H_3PO_3$ | 2.00 | $1.0 \times 10^{-2}$ |
| (c) $H_2SeO_3$ | 2.46 | $3.5 \times 10^{-3}$ |
| (d) $HSeO_4$ | 1.92 | $1.2 \times 10^{-2}$ |

(e) The larger $K_{a1}$, the stronger the acid; therefore

$$H_2SeO_3 < H_3PO_4 < H_3PO_3 < HSeO_4^-$$

**10.25** (a) $HClO_2(aq) + H_2O(l) \leftrightarrow H_3O^+(aq) + ClO_2^-(aq) \quad K_a = \dfrac{[H_3O^+][ClO_2^-]}{[HClO_2]}$

$ClO_2^-(aq) + H_2O(l) \leftrightarrow HClO_2(aq) + OH^-(aq) \quad K_b = \dfrac{[HClO_2][OH^-]}{[ClO_2^-]}$

(b) $HCN(aq) + H_2O(l) \leftrightarrow H_3O^+(aq) + CN^-(aq) \quad K_a = \dfrac{[H_3O^+][CN^-]}{[HCN]}$

$CN^-(aq) + H_2O(l) \leftrightarrow HCN(aq) + OH^-(aq) \quad K_b = \dfrac{[HCN][OH^-]}{[CN^-]}$

(c) $C_6H_5OH(aq) + H_2O(l) \leftrightarrow H_3O^+(aq) + C_6H_5O^-(aq)$

$K_a = \dfrac{[H_3O^+][C_6H_5O^-]}{[C_6H_5OH]}$

$C_6H_5O^-(aq) + H_2O(l) \leftrightarrow C_6H_5OH(aq) + OH^-(aq)$

$K_b = \dfrac{[C_6H_5OH][OH^-]}{[C_6H_5O^-]}$

**10.27** Decreasing $pK_a$ will correspond to increasing acid strength because $pK_a = -\log K_a$. The $pK_a$ values (given in parentheses) determine the following ordering:

$(CH_3)_2NH_2^+ \ (14.00 - 3.27 = 10.73) < {}^+NH_3OH \ (14.00 - 7.97 = 6.03)$

$< HNO_2$ (3.37) $< HClO_2$ (2.00).

Remember that the $pK_a$ for the conjugate acid of a weak base will be given by $pK_a + pK_b = 14$.

**10.29** Decreasing $pK_b$ will correspond to increasing base strength because $pK_b = -\log K_b$. The $pK_b$ values (given in parentheses) determine the following ordering:

$$F^- (14.00 - 3.45 = 10.55) < CH_3COO^- (14.00 - 4.75 = 9.25)$$
$$< C_5H_5N (8.75) \ll NH_3 (4.75).$$

Remember that the $pK_b$ for the conjugate base of a weak acid will be given by $pK_a + pK_b = 14$.

**10.31** Any acid whose conjugate base lies above water in Table 10.3 will be a strong acid; that is, the conjugate base of the acid will be a weaker base than water, and so water will accept the $H^+$ preferentially. Based upon this information, we obtain the following analysis: (a) $HClO_3$, strong; (b) $H_2S$, weak; (c) $HSO_4^-$, weak (Note: even though $H_2SO_4$ is a strong acid, $HSO_4^-$ is a weak acid. Its conjugate base is $SO_4^{2-}$); (d) $CH_3NH_3^+$, weak acid; (e) $HCO_3^-$, weak; (f) $HNO_3$, strong; (g) $CH_4$, weak.

**10.33** For oxoacids, the greater the number of highly electronegative O atoms attached to the central atom, the stronger the acid. This effect is related to the increased oxidation number of the central atom as the number of O atoms increases. Therefore, $HIO_3$ is the stronger acid, with the lower $pK_a$.

**10.35** (a) HCl is the stronger acid, because its bond strength is much weaker than the bond in HF, and bond strength is the dominant factor in determining the strength of binary acids.

(b) $HClO_2$ is stronger; there is one more O atom attached to the Cl atom in $HClO_2$ than in $HClO$. The additional O in $HClO_2$ helps to pull the electron of the H atom out of the H—O bond. The oxidation state of Cl is higher in $HClO_2$ than in $HClO$.

(c) $HClO_2$ is stronger; Cl has a greater electronegativity than Br, making the H—O bond $HClO_2$ more polar than in $HBrO_2$.

(d) $HClO_4$ is stronger; Cl has a greater electronegativity than P.

(e) $HNO_3$ is stronger. The explanation is the same as that for part (b). $HNO_3$ has one more O atom.

(f) $H_2CO_3$ is stronger; C has greater electronegativity than Ge. See part (c).

**10.37** (a) The —$CCl_3$ group that is bonded to the carboxyl group, —COOH, in trichloroacetic acid, is more electron withdrawing than the —$CH_3$ group in acetic acid. Thus, trichloroacetic acid is the stronger acid.

(b) The —$CH_3$ group in acetic acid has electron-donating properties, which means that it is less electron withdrawing than the —H attached to the carboxyl group in formic acid, HCOOH. Thus, formic acid is a slightly stronger acid than acetic acid. However, it is not nearly as strong as trichloroacetic acid. The order is $CCl_3COOH \gg HCOOH > CH_3COOH$.

**10.39** (a) Nitrous acid is a stronger acid than acetic acid and, therefore, the acetate ion is a stronger base than the nitrite ion. Since the definition of a strong acid is one that favors products in the deprotonation reaction, the presence of acetate ions will shift the nitrous acid deprotonation reaction toward products by consuming protons making nitrous acid behave like a strong acid. Carbonic acid is a weaker acid and, therefore, will not behave like a strong acid in the presence of acetic acid.

(b) Ammonia will act like a strong base because it's conjugate acid, the ammonium ion, is a weaker acid than acetic acid. The presence of acetic acid will shift the reaction: $NH_3 + H_2O \leftrightarrow NH_4^+ + OH^-$ toward products.

**10.41** The larger the $K_a$, the stronger the corresponding acid. 2,4,6-Trichlorophenol is the stronger acid because the chlorine atoms have a greater electron-withdrawing power than the hydrogen atoms present in the unsubstituted phenol.

**10.43** The larger the $pK_a$ of an acid, the stronger the corresponding conjugate base; hence, the order is aniline < ammonia < methylamine < ethylamine. Although we should not draw conclusions from such a small data set, we might suggest the possibility that

(1) arylamines < ammonia < alkylamines

(2) methyl < ethyl < etc.

(Arylamines are amines in which the nitrogen of the amine is attached to a benzene ring.)

**10.45** (a) Concentration

$(mol \cdot L^{-1})$  $CH_3COOH + H_2O \leftrightarrow H_3O^+ + CH_3CO_2^-$

| | | | | |
|---|---|---|---|---|
| initial | 0.29 | — | 0 | 0 |
| change | $-x$ | — | $+x$ | $+x$ |
| equilibrium | 0.29 $-x$ | — | $x$ | $x$ |

$$K_a = 1.8 \times 10^{-5} = \frac{[H_3O^+][CH_3CO_2^{\,-}]}{[CH_3COOH]} = \frac{x^2}{0.29 - x} \approx \frac{x^2}{0.29}$$

$x = [H_3O^+] = 2.3 \times 10^{-3}\ mol \cdot L^{-1}$

$pH = -\log(1.6 \times 10^{-3}) = 2.64$, $pOH = 14.00 - 2.64 = 11.36$

(b) The equilibrium table for (b) is similar to that for (a).

$$K_a = 3.0 \times 10^{-1} = \frac{[H_3O^+][CCl_3CO_2{}^-]}{[CCl_3COOH]} = \frac{x^2}{0.29 - x}$$

or $x^2 + 3.0 \times 10^{-1}x - 0.087 = 0$

$$x = \frac{-3.0 \times 10^{-1} \pm \sqrt{(3.0 \times 10^{-1})^2 - (4)(-0.087)}}{2} = 0.18, -0.48$$

The negative root is not possible and can be eliminated.

$x = [H_3O^+] = 0.18 \text{ mol} \cdot L^{-1}$

$pH = -\log(0.18) = 0.74, \quad pOH = 14.00 - 0.74 = 13.26$

(c) Concentration $(\text{mol} \cdot L^{-1})$ $HCOOH + H_2O \leftrightarrow H_3O^+ + HCO_2^-$

| | | | | |
|---|---|---|---|---|
| initial | 0.29 | — | 0 | 0 |
| change | $-x$ | — | $+x$ | $+x$ |
| equilibrium | $0.29 - x$ | — | $x$ | $x$ |

$$K_a = \frac{[H_3O^+][HCO_2{}^-]}{[HCOOH]} = \frac{x \cdot x}{0.29 - x} \approx \frac{x^2}{0.29} = 1.8 \times 10^{-4}$$

$x = [H_3O^+] = \sqrt{0.29 \times 1.8 \times 10^{-4}} = 7.2 \times 10^{-3} \text{ mol} \cdot L^{-1}$

$pH = -\log(7.2 \times 10^{-3}) = 2.14, \quad pOH = 14.00 - 2.14 = 11.86$

(d) Acidity increases when the hydrogen atoms in the methyl group of acetic acid are replaced by atoms that have a higher electronegativity, such as chlorine. See also Exercise 10.37.

**10.47** (a) Concentration

| $(\text{mol} \cdot L^{-1})$ | $H_2O$ | $+$ | $NH_3$ | $\leftrightarrow$ | $NH_4^+$ | $+$ | $OH^-$ |
|---|---|---|---|---|---|---|---|
| initial | — | | 0.057 | | 0 | | 0 |
| change | — | | $-x$ | | $+x$ | | $+x$ |
| equilibrium | — | | $0.057 - x$ | | $x$ | | $x$ |

$$K_b = \frac{[NH_4^+][OH^-]}{[NH_3]} = \frac{x \cdot x}{0.057 - x} \approx \frac{x^2}{0.057} = 1.8 \times 10^{-5}$$

$$x = [OH^-] = \sqrt{0.057 \times 1.8 \times 10^{-5}} = 1.0 \times 10^{-3} \text{ mol} \cdot L^{-1}$$

$$pOH = -\log(1.0 \times 10^{-3}) = 3.00, \; pH = 14.00 - 3.00 = 11.00$$

$$\text{percentage protonation} = \frac{1.0 \times 10^{-3}}{0.057} \times 100\% = 1.8\%$$

(b) Concentration

| (mol·L$^{-1}$) | NH$_2$OH | + H$_2$O | $\leftrightarrow$ | $^+$NH$_3$OH | + OH$^-$ |
|---|---|---|---|---|---|
| initial | 0.162 | — | | 0 | 0 |
| change | $-x$ | — | | $+x$ | $+x$ |
| equilibrium | $0.162 - x$ | — | | $x$ | $x$ |

$$K_b = 1.1 \times 10^{-8} = \frac{x^2}{0.162 - x} \approx \frac{x^2}{0.162}$$

$$x = [OH^-] = 4.2 \times 10^{-5} \text{ mol} \cdot L^{-1}$$

$$pOH = -\log(4.2 \times 10^{-5}) = 4.38, \; pH = 14.00 - 4.38 = 9.62$$

$$\text{percentage protonation} = \frac{4.2 \times 10^{-5}}{0.162} \times 100\% = 0.026\%$$

(c) Concentration

| (mol·L$^{-1}$) | (CH$_3$)$_3$N | + H$_2$O | $\leftrightarrow$ | (CH$_3$)$_3$NH$^+$ | + OH$^-$ |
|---|---|---|---|---|---|
| initial | 0.35 | — | | 0 | 0 |
| change | $-x$ | — | | $+x$ | $+x$ |
| equilibrium | $0.35 - x$ | — | | $+x$ | $+x$ |

$$6.5 \times 10^{-5} = \frac{x^2}{0.35 - x}$$

Assume $x \ll 0.35$

Then $x = 4.8 \times 10^{-3} \text{ mol} \cdot L^{-1}$

$[OH^-] = 4.8 \times 10^{-3} \text{ mol} \cdot L^{-1}$

$pOH = -\log(4.8 \times 10^{-3}) = 2.32, \; pH = 14.00 - 2.32 = 11.68$

$$\text{percentage protonation} = \frac{4.8 \times 10^{-3}}{0.35} \times 100\% = 1.4\%$$

(d) $pK_b = 14.00 - pK_a = 14.00 - 8.21 = 5.79$, $K_b = 1.6 \times 10^{-6}$

$$\text{codeine} + H_2O \leftrightarrow \text{codeineH}^+ + OH^-$$

$$K_b = 1.6 \times 10^{-6} = \frac{x^2}{0.073 - x} \approx \frac{x^2}{0.073}$$

$$x = [OH^-] = 1.1 \times 10^{-4} \text{ mol} \cdot L^{-1}$$

$$pOH = -\log(1.1 \times 10^{-4}) = 3.96, \quad pH = 14.00 - 3.96 = 10.04$$

$$\text{percentage protonation} = \frac{1.1 \times 10^{-4}}{0.0073} \times 100\% = 2.5\%$$

**10.49** (a) $HClO_2 + H_2O \leftrightarrow H_3O^+ + ClO_2^-$

$$[H_3O^+] = [ClO_2^-] = 10^{-pH} = 10^{-1.2} = 0.06 \text{ mol} \cdot L^{-1}$$

$$K_a = \frac{[H_3O^+][ClO_2^-]}{[HClO_2]} = \frac{(0.06)^2}{0.10 - 0.06} = 0.09 \text{ (1 sf)}$$

$$pK_a = -\log(0.09) = 1.0$$

(b) $C_3H_7NH_2 + H_2O \leftrightarrow C_3H_7NH_3^+ + OH^-$

$$pOH = 14.00 - 11.86 = 2.14$$

$$[C_3H_7NH_3^+] = [OH^-] = 10^{-2.14} = 7.2 \times 10^{-3} \text{ mol} \cdot L^{-1}$$

$$K_b = \frac{[C_3H_7NH_3^+][OH^-]}{[C_3H_7NH_2]} = \frac{(7.2 \times 10^{-3})^2}{0.10 - 7.2 \times 10^{-3}} = 5.6 \times 10^{-4}$$

$$pK_b = -\log(5.6 \times 10^{-4}) = 3.25$$

**10.51** (a) $pH = 4.60$, $[H_3O^+] = 10^{-pH} = 10^{-4.60} = 2.5 \times 10^{-5} \text{ mol} \cdot L^{-1}$

Let $x$ = nominal concentration of HClO, then

Concentration

| $(\text{mol} \cdot L^{-1})$ | HClO | + | $H_2O$ | $\leftrightarrow$ | $H_3O^+$ | + | $ClO^-$ |
|---|---|---|---|---|---|---|---|
| nominal | $x$ | | — | | 0 | | 0 |
| equilibrium | $x - 2.5 \times 10^{-5}$ | | — | | $2.5 \times 10^{-5}$ | | $2.5 \times 10^{-5}$ |

$$K_a = 3.0 \times 10^{-8} = \frac{(2.5 \times 10^{-5})^2}{x - 2.5 \times 10^{-5}}$$

Solve for $x$; $x = \dfrac{(2.5 \times 10^{-5})^2 + (2.5 \times 10^{-5})(3.0 \times 10^{-8})}{3.0 \times 10^{-8}}$

$$= 2.1 \times 10^{-2} \text{ mol} \cdot \text{L}^{-1} = 0.021 \text{ mol} \cdot \text{L}^{-1}$$

(b)  $\text{pOH} = 14.00 - \text{pH} = 14.00 - 10.20 = 3.80$

$[\text{OH}^-] = 10^{-\text{pOH}} = 10^{-3.80} = 1.6 \times 10^{-4}$

Let $x$ = nominal concentration of $NH_2NH_2$, then

Concentration

| (mol·L$^{-1}$) | $NH_2NH_2$ | + | $H_2O$ | $\leftrightarrow$ | $NH_2NH_3^+$ | + | $OH^-$ |
|---|---|---|---|---|---|---|---|
| nominal | $x$ | | — | | 0 | | 0 |
| equilibrium | $x - 1.6 \times 10^{-4}$ | | — | | $1.6 \times 10^{-4}$ | | $1.6 \times 10^{-4}$ |

$K_b = 1.7 \times 10^{-6} = \dfrac{(1.6 \times 10^{-4})^2}{x - 1.6 \times 10^{-4}}$

Solve for $x$; $x = 1.5 \times 10^{-2} \text{ mol} \cdot \text{L}^{-1}$

**10.53**  Concentration

| (mol·L$^{-1}$) | $C_6H_5COOH$ | + | $H_2O$ | $\leftrightarrow$ | $H_3O^+$ | + | $C_6H_5CO_2^-$ |
|---|---|---|---|---|---|---|---|
| initial | 0.110 | | — | | 0 | | 0 |
| change | $-x$ | | — | | $+x$ | | $+x$ |
| equilibrium | $0.110 - x$ | | — | | $x$ | | $x$ |

$x = 0.024 \times 0.110 \text{ mol} \cdot \text{L}^{-1} = [H_3O^+] = [C_6H_5CO_2^-]$

$K_a = \dfrac{[H_3O^+][C_6H_5COO^-]}{[C_6H_5COOH]} = \dfrac{(0.024 \times 0.110)^2}{(1 - 0.024) \times 0.110} = 6.3 \times 10^{-5}$

$\text{pH} = -\log(2.6 \times 10^{-3}) = 2.58$

**10.55**  The change in the concentration of octylamine is

$x = 0.067 \times 0.10 = 0.0067 \text{ mol} \cdot \text{L}^{-1}$. Thus the equilibrium table is

Concentration

| (mol·L$^{-1}$) | $H_2O$ | + | octylamine | $\leftrightarrow$ | octylamineH$^+$ | + | $OH^-$ |
|---|---|---|---|---|---|---|---|
| initial | — | | 0.100 | | 0 | | 0 |

| change | — | −0.0067 | +0.0067 | +0.0067 |
| equilibrium | — | 0.100 − 0.0067 | 0.0067 | 0.0067 |

The equilibrium concentrations are

$[\text{octylamine}] = 0.100 − 0.067 \times 0.10 = 0.093 \ \text{mol} \cdot \text{L}^{-1}$

$[\text{OH}^-] = [\text{octylamineH}^+] = 0.0067 \ \text{mol} \cdot \text{L}^{-1}$

$\text{pOH} = −\log(0.0067) = 2.17, \ \text{pH} = 14.00 − 2.17 = 11.83$

$K_b = \dfrac{[\text{octylamineH}^+][\text{OH}^-]}{[\text{octylamine}]} = \dfrac{(6.7 \times 10^{-3})^2}{0.093} = 4.8 \times 10^{-4}$

**10.57** $\ \text{CH}_3\text{CH}_2\text{COOH(aq)} + \text{H}_2\text{O(l)} \leftrightarrow \text{CH}_3\text{CH}_2\text{COO}^-\text{(aq)} + \text{H}_3\text{O}^+\text{(aq)}$

Concentration $(\text{mol} \cdot \text{L}^{-1})$

| initial | 0.0147 | — | 0 | 0 |
| change | −x | — | +x | +x |
| equilibrium | 0.0147 − x | — | x | x |

$K_a = \dfrac{[\text{CH}_3\text{CH}_2\text{COO}^-][\text{H}_3\text{O}^+]}{[\text{CH}_3\text{CH}_2\text{COOH}]} = 1.3 \times 10^{-5} = \dfrac{x^2}{0.0147 - x} \approx \dfrac{x^2}{0.0147}$

$x = 4.4 \times 10^{-4}, \ \text{Percent deprotonated} = \dfrac{4.4 \times 10^{-4}}{0.0147} \times 100\% = 3.0\%$

**10.59** (a) less than 7, $\text{NH}_4^+\text{(aq)} + \text{H}_2\text{O(l)} \leftrightarrow \text{H}_3\text{O}^+\text{(aq)} + \text{NH}_3\text{(aq)}$

(b) greater than 7, $\text{H}_2\text{O(l)} + \text{CO}_3^{2-}\text{(aq)} \leftrightarrow \text{HCO}_3^-\text{(aq)} + \text{OH}^-\text{(aq)}$

(c) greater than 7, $\text{H}_2\text{O(l)} + \text{F}^-\text{(aq)} \leftrightarrow \text{HF(aq)} + \text{OH}^-\text{(aq)}$

(d) neutral

(e) less than 7,

$\text{Al(H}_2\text{O)}_6^{3+}\text{(aq)} + \text{H}_2\text{O(l)} \leftrightarrow \text{H}_3\text{O}^+\text{(aq)} + \text{Al(H}_2\text{O)}_5\text{OH}^{2+}\text{(aq)}$

(f) less than 7,

$\text{Cu(H}_2\text{O)}_6^{2+}\text{(aq)} + \text{H}_2\text{O(l)} \leftrightarrow \text{H}_3\text{O}^+\text{(aq)} + \text{Cu(H}_2\text{O)}_5\text{OH}^+\text{(aq)}$

**10.61** (a) $K_b = \dfrac{K_w}{K_a} = \dfrac{1.00 \times 10^{-14}}{1.8 \times 10^{-5}} = 5.6 \times 10^{-10}$

Concentration

$(mol \cdot L^{-1})\ CH_3CO_2^-(aq)\ +\ H_2O(l) \leftrightarrow HCH_3CO_2(aq) + OH^-(aq)$

| | | | | |
|---|---|---|---|---|
| initial | 0.63 | — | 0 | 0 |
| change | $-x$ | — | $+x$ | $+x$ |
| equilibrium | $0.63 - x$ | — | $x$ | $x$ |

$K_b = \dfrac{[HCH_3CO_2][OH^-]}{[CH_3CO_2^-]} = 5.6 \times 10^{-10} = \dfrac{x^2}{0.63 - x} \approx \dfrac{x^2}{0.63}$

$x = 1.9 \times 10^{-5} = [OH^-],\ pOH = -\log(1.9 \times 10^{-5}) = 4.72$

$pH = 14.00 - pOH = 14.00 - 4.72 = 9.28$

(b) $K_a = \dfrac{K_w}{K_b} = \dfrac{1.00 \times 10^{-14}}{1.8 \times 10^{-15}} = 5.6 \times 10^{-10}$

Concentration $(mol \cdot L^{-1})\ NH_4^+(aq)\ +\ H_2O(l) \leftrightarrow H_3O^+(aq) + NH_3(aq)$

| | | | | |
|---|---|---|---|---|
| initial | 0.19 | — | 0 | 0 |
| change | $-x$ | — | $+x$ | $+x$ |
| equilibrium | $0.19 - x$ | — | $x$ | $x$ |

$K_a = \dfrac{[H_3O^+][NH_3]}{[NH_4Cl]} = 5.6 \times 10^{-10} = \dfrac{x^2}{0.19 - x} \approx \dfrac{x^2}{0.19}$

$x = 1.0 \times 10^{-5}\ mol \cdot L^{-1} = [H_3O^+]$

$pH = -\log(1.0 \times 10^{-5}) = 5.00$

(c) Concentration

$(mol \cdot L^{-1})\ Al(H_2O)_6^{3+}(aq) + H_2O(l) \leftrightarrow H_3O^+(aq) + Al(H_2O)_5OH^{2+}(aq)$

| | | | | |
|---|---|---|---|---|
| initial | 0.055 | — | 0 | 0 |
| change | $-x$ | — | $+x$ | $+x$ |
| equilibrium | $0.055 - x$ | — | $x$ | $x$ |

$K_a = \dfrac{[H_3O^+][Al(H_2O)_5OH^{2+}]}{[Al(H_2O)_6^{3+}]} = 1.4 \times 10^{-5} = \dfrac{x^2}{0.055 - x} \approx \dfrac{x^2}{0.055}$

$$x = 8.8 \times 10^{-4} \ mol \cdot L^{-1} = [H_3O^+]$$

$$pH = -\log(8.8 \times 10^{-4}) = 3.06$$

(d) Concentration

| $(mol \cdot L^{-1})$ | $H_2O(l)$ + | $CN^-(aq)$ | $\leftrightarrow$ | $HCN(aq)$ + | $OH^-(aq)$ |
|---|---|---|---|---|---|
| initial | — | 0.65 | | 0 | 0 |
| change | — | $-x$ | | $+x$ | $+x$ |
| equilibrium | — | $0.65 - x$ | | $x$ | $x$ |

$$K_b = \frac{K_w}{K_a} = \frac{1.00 \times 10^{-14}}{4.9 \times 10^{-10}} = 2.0 \times 10^{-5} = \frac{[HCN][OH^-]}{[CN^-]} = \frac{x^2}{0.65 - x} \approx \frac{x^2}{0.65}$$

$$x = [OH^-] = 3.6 \times 10^{-3} \ mol \cdot L^{-1}$$

$$pOH = -\log(3.6 \times 10^{-3}) = 2.44, \ pH = 11.56$$

**10.63** Concentration

| $(mol \cdot L^{-1})$ | $CH_3NH_3^+(aq)$ + | $H_2O(l)$ | $\leftrightarrow$ | $H_3O^+(aq)$ + | $CH_3NH_2(aq)$ |
|---|---|---|---|---|---|
| initial | 0.510 | — | | 0 | 0 |
| change | $-x$ | — | | $+x$ | $+x$ |
| equilibrium | $0.510 - x$ | — | | $x$ | $x$ |

$$K_a = \frac{[H_3O^+][CH_3NH_2]}{[CH_3NH_3^+]} = 2.8 \times 10^{-11} = \frac{x^2}{0.510 - x} \approx \frac{x^2}{0.510}$$

$$x = 3.8 \times 10^{-6} \ mol \cdot L^{-1} = [H_3O^+]$$

$$pH = -\log(3.8 \times 10^{-6}) = 5.42$$

**10.65** (a) 250 mL of solution contains 5.34 g $KC_2H_3O_2$, molar

mass $= 98.14 \ g \cdot mol^{-1}$

$$(5.34 \ g \ KC_2H_3O_2) \left( \frac{1 \ mol \ KC_2H_3O_2}{98.14 \ g \ KC_2H_3O_2} \right) \left( \frac{1}{0.250 \ L} \right) = 0218 \ M \ KC_2H_3O_2$$

Concentration

$$(mol \cdot L^{-1}) \qquad H_2O(l) \ + \ C_2H_3O_2^-(aq) \ \leftrightarrow \ HC_2H_3O_2(aq) \ + \ OH^-(aq)$$

| | | | | |
|---|---|---|---|---|
| initial | — | 0.218 | 0 | 0 |
| change | — | $-x$ | $+x$ | $+x$ |
| equilibrium | — | $0.218 - x$ | $x$ | $x$ |

$$\frac{1.0 \times 10^{-14}}{1.8 \times 10^{-5}} = \frac{x^2}{0.218 - x} \approx \frac{x^2}{0.218}$$

$$[OH^-] = 1.1 \times 10^{-5} \ mol \cdot L^{-1}$$

$$[H_3O^+] = 9.1 \times 10^{-10} \ mol \cdot L^{-1}$$

$$pH = -\log(9.1 \times 10^{-10}) = 9.04$$

(b) 100 mL of solution contains 5.75 g $NH_4Br$, molar mass = 97.95 $g \cdot mol^{-1}$

$$(5.75 \ g \ NH_4Br)\left(\frac{1 \ mol \ NH_4Br}{97.95 \ g \ NH_4Br}\right)\left(\frac{1}{0.100 \ L}\right) = 0.587 \ M \ NH_4Br$$

Concentration

$$(mol \cdot L^{-1}) \qquad NH_4^+(aq) \ + \ H_2O(l) \ \leftrightarrow \ NH_3(aq) \ + \ H_3O^+(aq)$$

| | | | | |
|---|---|---|---|---|
| initial | 0.587 | — | 0 | 0 |
| change | $-x$ | — | $+x$ | $+x$ |
| equilibrium | $0.587 - x$ | — | $x$ | $x$ |

$$\frac{1.0 \times 10^{-14}}{1.8 \times 10^{-5}} = \frac{x^2}{0.587 - x} \approx \frac{x^2}{0.587}$$

$$[H_3O^+] = 1.8 \times 10^{-5} \ mol \cdot L^{-1}$$

$$pH = -\log(1.8 \times 10^{-5}) = 4.74$$

**10.67** (a) $\dfrac{0.020\,\text{mol}\cdot\text{L}^{-1}\,\text{NaCH}_3\text{CO}_2 \times 0.150\,\text{L}}{0.500\,\text{L}} = 0.0060\,\text{mol}\cdot\text{L}^{-1}$

Concentration

$(\text{mol}\cdot\text{L}^{-1})\quad H_2O(l) + CH_3CO_2^-(aq) \leftrightarrow CH_3COOH(aq) + OH^-(aq)$

| | | | |
|---|---|---|---|
| initial | — | 0.0060 | 0 | 0 |
| change | — | $-x$ | $+x$ | $+x$ |
| equilibrium | — | $0.0060 - x$ | $x$ | $x$ |

$K_b = \dfrac{K_w}{K_a} = \dfrac{1.00 \times 10^{-14}}{1.8 \times 10^{-5}} = 5.6 \times 10^{-10} = \dfrac{[CH_3COOH][OH^-]}{[CH_3CO_2{}^-]}$

$5.6 \times 10^{-10} = \dfrac{x^2}{0.0060 - x} \approx \dfrac{x^2}{0.0060}$

$x = 1.8 \times 10^{-6}\,\text{mol}\cdot\text{L}^{-1} = [CH_3COOH]$

(b) $\left(\dfrac{2.16\,\text{g NH}_4\text{Br}}{400\,\text{mL}}\right)\left(\dfrac{1\,\text{mL}}{10^{-3}\,\text{L}}\right)\left(\dfrac{1\,\text{mol NH}_4\text{Br}}{97.95\,\text{g NH}_4\text{Br}}\right)$

$= 0.0551\,(\text{mol NH}_4\text{Br})\cdot\text{L}^{-1}$

Concentration

$(\text{mol}\cdot\text{L}^{-1})\qquad NH_4^+(aq) + H_2O(l) \leftrightarrow H_3O^+(aq) + NH_3(aq)$

| | | | |
|---|---|---|---|
| initial | 0.0551 | — | 0 | 0 |
| change | $-x$ | — | $+x$ | $+x$ |
| equilibrium | $0.0551 - x$ | — | $x$ | $x$ |

$K_a = \dfrac{K_w}{K_b} = \dfrac{1.00 \times 10^{-14}}{1.8 \times 10^{-5}} = 5.6 \times 10^{-10} = \dfrac{[NH_3][H_3O^+]}{[NH_4{}^+]}$

$5.6 \times 10^{-10} = \dfrac{x^2}{0.0551 - x} \approx \dfrac{x^2}{0.0551}$

$x = 5.5 \times 10^{-6}\,\text{mol}\cdot\text{L}^{-1} = [H_3O^+]$ and $pH = -\log(5.5 \times 10^{-6}) = 5.26$

**10.69** (a)

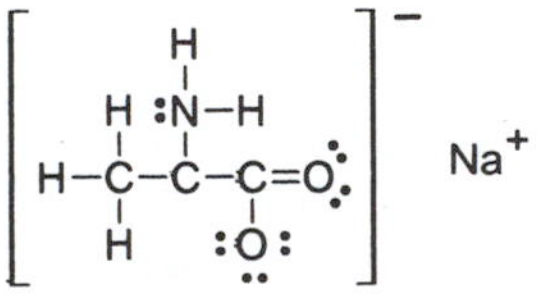

(b) $pH = \frac{1}{2}(pK_{a_1} + pK_{a_2}) = \frac{1}{2}(2.34 + 9.89) = 6.12$

**10.71** (a) $H_2SO_4(aq) + H_2O(l) \leftrightarrow H_3O^+(aq) + HSO_4^-(aq)$

$HSO_4^-(aq) + H_2O(l) \leftrightarrow H_3O^+(aq) + SO_4^{2-}(aq)$

(b) $H_3AsO_4(aq) + H_2O(l) \leftrightarrow H_3O^+(aq) + H_2AsO_4^-(aq)$

$H_2AsO_4^-(aq) + H_2O(l) \leftrightarrow H_3O^+(aq) + HAsO_4^{2-}(aq)$

$HAsO_4^{2-}(aq) + H_2O(l) \leftrightarrow H_3O^+(aq) + AsO_4^{3-}(aq)$

(c)

$C_6H_4(COOH)_2(aq) + H_2O(l) \leftrightarrow H_3O^+(aq) + C_6H_4(COOH)CO_2^-(aq)$

$C_6H_4(COOH)CO_2^-(aq) + H_2O(l) \leftrightarrow H_3O^+(aq) + C_6H_4(CO_2)_2^{2-}(aq)$

**10.73** The initial concentrations of $HSO_4^-$ and $H_3O^+$ are both $0.15 \; mol \cdot L^{-1}$ as a result of the complete ionization of $H_2SO_4$ in the first step. The second ionization is incomplete.

| Concentration (mol · L⁻¹) | $HSO_4^-$ | + | $H_2O$ ↔ | $H_3O^+$ | + | $SO_4^{2-}$ |
|---|---|---|---|---|---|---|
| initial | 0.15 | | — | 0.15 | | 0 |
| change | −x | | — | +x | | +x |
| equilibrium | 0.15 − x | | — | 0.15 + x | | x |

$K_{a2} = 1.2 \times 10^{-2} = \dfrac{[H_3O^+][SO_4^{2-}]}{[HSO_4^-]} = \dfrac{(0.15 + x)(x)}{0.15 - x}$

$x^2 + 0.162x - 1.8 \times 10^{-3} = 0$

$x = \dfrac{-0.162 + \sqrt{(0.162)^2 + (4)(1.8 \times 10^{-3})}}{2} = 0.0104 \; mol \cdot L^{-1}$

$[H_3O^+] = 0.15 + x = (0.15 + 0.0104) \; mol \cdot L^{-1} = 0.16 \; mol \cdot L^{-1}$

$pH = -\log(0.16) = 0.80$

**10.75** (a) Because $K_{a2} \ll K_{a1}$, the second ionization can be ignored.

Concentration

| (mol·L$^{-1}$) | $H_2CO_3$ | + | $H_2O$ | $\leftrightarrow$ | $H_3O^+$ | + | $HCO_3^-$ |
|---|---|---|---|---|---|---|---|
| initial | 0.010 | | — | | 0 | | 0 |
| change | $-x$ | | — | | $+x$ | | $+x$ |
| equilibrium | $0.010 - x$ | | — | | $x$ | | $x$ |

$$K_{a1} = \frac{[H_3O^+][HCO_3^-]}{[H_2CO_3]} = \frac{x^2}{0.010 - x} \approx \frac{x^2}{0.010} = 4.3 \times 10^{-7}$$

$$x = [H_3O^+] = 6.6 \times 10^{-5} \text{ mol} \cdot L^{-1}$$

$$pH = -\log(6.6 \times 10^{-5}) = 4.18$$

(b) Because $K_{a2} \ll K_{a1}$, the second ionization can be ignored.

Concentration

| (mol·L$^{-1}$) | $(COOH)_2$ | + | $H_2O$ | $\leftrightarrow$ | $H_3O^+$ | + | $(COOH)CO_2^-$ |
|---|---|---|---|---|---|---|---|
| initial | 0.10 | | — | | 0 | | 0 |
| change | $-x$ | | — | | $+x$ | | $+x$ |
| equilibrium | $0.10 - x$ | | — | | $x$ | | $x$ |

$$K_{a1} = 5.9 \times 10^{-2} = \frac{[H_3O^+][(COOH)CO_2^-]}{[(COOH)_2]} = \frac{x^2}{0.10 - x}$$

$$x^2 + 5.9 \times 10^{-2} x - 5.9 \times 10^{-3} = 0$$

$$x = \frac{-5.9 \times 10^{-2} + \sqrt{(5.9 \times 10^{-2})^2 + (4)(5.9 \times 10^{-3})}}{2} = 0.053 \text{ mol} \cdot L^{-1}$$

$$pH = -\log(0.053) = 1.28$$

(c) Because $K_{a2} \ll K_{a1}$, the second ionization can be ignored.

| Concentration (mol·L$^{-1}$) | $H_2S$ | + | $H_2O$ | $\leftrightarrow$ | $H_3O^+$ | + | $HS^-$ |
|---|---|---|---|---|---|---|---|
| equilibrium | $0.20 - x$ | | — | | $x$ | | $x$ |

$$K_{a1} = 1.3 \times 10^{-7} = \frac{[H_3O^+][HS^-]}{[H_2S]} = \frac{x^2}{0.20 - x} \approx \frac{x^2}{0.20}$$

$$x = [H_3O^+] = 1.6 \times 10^{-4} \text{ mol} \cdot L^{-1}$$

$$pH = -\log(1.6 \times 10^{-4}) = 3.80$$

**10.77** (a)  The pH is given by $pH = \frac{1}{2}(pK_{a1} + pK_{a2})$. From Table 10.9, we find

$$K_{a1} = 1.5 \times 10^{-2} \quad pK_{a1} = 1.82$$
$$K_{a2} = 1.2 \times 10^{-7} \quad pK_{a2} = 6.92$$
$$pH = \frac{1}{2}(1.82 + 6.92) = 4.37$$

(b)  The pH of a salt solution of a polyprotic acid is independent of the concentration of the salt, therefore $pH = 4.37$.

**10.79** (a)  The pH is given by $pH = \frac{1}{2}(pK_{a1} + pK_{a2})$. For the monosodium salt, the pertinent values are $pK_{a1}$ and $pK_{a2}$ :

$$pH = \frac{1}{2}(3.14 + 5.95) = 4.55$$

(b)  For the disodium salt, the pertinent values are $pK_{a2}$ and $pK_{a3}$ :

$$pH = \frac{1}{2}(5.95 + 6.39) = 6.17$$

**10.81**  The equilibrium reactions of interest are

$$H_2CO_3(aq) + H_2O(l) \leftrightarrow H_3O^+(aq) + HCO_3^-(aq) \quad K_{a1} = 4.3 \times 10^{-7}$$
$$HCO_3^-(aq) + H_2O(l) \leftrightarrow H_3O^+(aq) + CO_3^{2-}(aq) \quad K_{a2} = 5.6 \times 10^{-11}$$

Because the second ionization constant is much smaller than the first, we can assume that the first step dominates:

Concentration

| $(mol \cdot L^{-1})$ | $H_2CO_3(aq)$ | $+ H_2O(l)$ | $\leftrightarrow$ | $H_3O^+(aq)$ | $+ HCO_3^-(aq)$ |
|---|---|---|---|---|---|
| initial | 0.0456 | — | | 0 | 0 |
| change | $-x$ | — | | $+x$ | $+x$ |
| final | $0.0456 - x$ | — | | $+x$ | $+x$ |

$$K_{a1} = \frac{[H_3O^+][HCO_3^-]}{[H_2CO_3]}$$

$$4.3 \times 10^{-7} = \frac{(x)(x)}{0.0456 - x} = \frac{x^2}{0.0456 - x}$$

Assume that $x \ll 0.0456$

Then $x^2 = (4.3 \times 10^{-7})(0.0456)$

$x = 1.4 \times 10^{-4}$

Because $x < 1\%$ of 0.0456, the assumption was valid.

$x = [H_3O^+] = [HCO_3^-] = 1.4 \times 10^{-4}$ mol $\cdot$ L$^{-1}$

This means that the concentration of $H_2CO_3$ is 0.0456 mol $\cdot$ L$^{-1}$ −

0.00014 mol $\cdot$ L$^{-1}$ = 0.0455 mol $\cdot$ L$^{-1}$. We can then use the other equilibria

to determine the remaining concentrations:

$$K_{a2} = \frac{[H_3O^+][CO_3^{2-}]}{[HCO_3^-]}$$

$$5.6 \times 10^{-11} = \frac{(1.4 \times 10^{-4})[CO_3^{2-}]}{(1.4 \times 10^{-4})}$$

$[CO_3^{2-}] = 5.6 \times 10^{-11}$ mol $\cdot$ L$^{-1}$

Because $5.6 \times 10^{-11} \ll 1.4 \times 10^{-4}$, the initial assumption that the first

ionization would dominate is valid.

To calculate [OH$^-$], we use the $K_w$ relationship:

$$K_w = [H_3O^+][OH^-]$$

$$[OH^-] = \frac{K_w}{[H_3O^+]} = \frac{1.00 \times 10^{-14}}{1.4 \times 10^{-4}} = 7.1 \times 10^{-11} \text{ mol} \cdot L^{-1}$$

In summary, $[H_2CO_3] = 0.0455$ mol $\cdot$ L$^{-1}$, $[H_3O^+] = [HCO_3^-]$

$$= 1.4 \times 10^{-4} \text{ mol} \cdot L^{-1},$$

$[CO_3^{2-}] = 5.6 \times 10^{-11}$ mol $\cdot$ L$^{-1}$, $[OH^-] = 7.1 \times 10^{-11}$ mol $\cdot$ L$^{-1}$.

**10.83** The equilibrium reactions of interest are now the base forms of the
carbonic acid equilibria, so $K_b$ values should be calculated for the
following changes:

$$CO_3^{2-}(aq) + H_2O(l) \leftrightarrow HCO_3^-(aq) + OH^-(aq)$$

$$K_{b1} = \frac{K_w}{K_{a2}} = \frac{1.00 \times 10^{-14}}{5.6 \times 10^{-11}} = 1.8 \times 10^{-4}$$

$$HCO_3^-(aq) + H_2O(l) \leftrightarrow H_2CO_3(aq) + OH^-(aq)$$

$$K_{b2} = \frac{K_w}{K_{a1}} = \frac{1.00 \times 10^{-14}}{4.3 \times 10^{-7}} = 2.3 \times 10^{-8}$$

Because the second hydrolysis constant is much smaller than the first, we can assume that the first step dominates:

| Concentration $(mol \cdot L^{-1})$ | $CO_3^{2-}(aq)$ + | $H_2O(l)$ | $\leftrightarrow$ | $HCO_3^-(aq)$ | + | $OH^-(aq)$ |
|---|---|---|---|---|---|---|
| initial | 0.0456 | — | | 0 | | 0 |
| change | $-x$ | — | | $+x$ | | $+x$ |
| final | $0.0456 - x$ | — | | $+x$ | | $+x$ |

$$K_{b1} = \frac{[HCO_3^-][OH^-]}{[CO_3^{2-}]}$$

$$1.8 \times 10^{-4} = \frac{(x)(x)}{0.0456 - x} = \frac{x^2}{[0.0456 - x]}$$

Assume that $x \ll 0.0456$

Then $x^2 = (1.8 \times 10^{-4})(0.0456)$

$x = 2.9 \times 10^{-3}$

Because $x > 5\%$ of 0.0456, the assumption was not valid and the full expression should be solved using the quadratic equation:

$$x^2 + 1.8 \times 10^{-4} x - (1.8 \times 10^{-4})(0.0456) = 0$$

Solving using the quadratic equation gives $x = 0.0028 \ mol \cdot L^{-1}$.

$$x = [HCO_3^-] = [OH^-] = 0.0028 \ mol \cdot L^{-1}$$

Therefore, $[CO_3^{2-}] = 0.0456 \ mol \cdot L^{-1} - 0.0028 \ mol \cdot L^{-1} = 0.0428 \ mol \cdot L^{-1}$

We can then use the other equilibria to determine the remaining concentrations:

$$K_{b2} = \frac{[H_2CO_3][OH^-]}{[HCO_3{}^-]}$$

$$2.3 \times 10^{-8} = \frac{[H_2CO_3](0.0028)}{(0.0028)}$$

$$[H_2CO_3] = 2.3 \times 10^{-8} \text{ mol} \cdot L^{-1}$$

Because $2.3 \times 10^{-8} \ll 0.0028$, the initial assumption that the first hydrolysis would dominate is valid. To calculate $[H_3O^+]$, we use the $K_w$ relationship:

$$K_w = [H_3O^+][OH^-]$$

$$[H_3O^+] = \frac{K_w}{[OH^-]} = \frac{1.00 \times 10^{-14}}{0.0028} = 3.6 \times 10^{-12} \text{ mol} \cdot L^{-1}$$

In summary, $[H_2CO_3] = 2.3 \times 10^{-8} \text{ mol} \cdot L^{-1}$, $[OH^-] = [HCO_3{}^-] =$ 0.0028 mol $\cdot L^{-1}$, $[CO_3{}^{2-}] = 0.0428 \text{ mol} \cdot L^{-1}$, $[H_3O^+] = 3.6 \times 10^{-12} \text{ mol} \cdot L^{-1}$

**10.85** (a) phosphorous acid: The two $pK_a$ values are 2.00 and 6.59. Because pH = 6.30 lies between $pK_{a1}$ and $pK_{a2}$, the dominant form will be the singly deprotonated $HA^-$ ion.

(b) oxalic acid: The two $pK_a$ values are 1.23 and 4.19. Because pH = 6.30 lies above $pK_{a2}$, the species present in largest concentration will be the doubly deprotonated $A^{2-}$ ion.

(c) hydrosulfuric acid: The two $pK_a$ values are 6.89 and 14.15. Because pH = 6.30 lies below both $pK_a$ values, the species present in highest concentrations will be the fully protonated $H_2A$ form.

**10.87** The equilibria present in solution are

$$H_2SO_3(aq) + H_2O(l) \leftrightarrow H_3O^+(aq) + HSO_3^-(aq) \quad K_{a1} = 1.5 \times 10^{-2}$$
$$HSO_3^-(aq) + H_2O(l) \leftrightarrow H_3O^+(aq) + SO_3^{2-}(aq) \quad K_{a2} = 1.2 \times 10^{-7}$$

The calculation of the desired concentrations follows exactly after the method derived in Eq. 25, substituting $H_2SO_3$ for $H_2CO_3$, $HSO_3^-$ for $HCO_3^-$, and $SO_3^{2-}$ for $CO_3^{2-}$. First, calculate the quantity $f$ (at pH = 5.50 $[H_3O^+] = 10^{-5.5} = 3.2 \times 10^{-6}\ mol \cdot L^{-1}$):

$$f = [H_3O^+]^2 + [H_3O^+] K_{a1} + K_{a1} K_{a2}$$
$$= (3.2 \times 10^{-6})^2 + (3.2 \times 10^{-6})(1.5 \times 10^{-2}) + (1.5 \times 10^{-2})(1.2 \times 10^{-7})$$
$$= 5.0 \times 10^{-8}$$

The fractions of the species present are then given by

$$\alpha(H_2SO_3) = \frac{[H_3O^+]}{f} = \frac{(3.2 \times 10^{-6})^2}{5.0 \times 10^{-8}} = 2.1 \times 10^{-4}$$

$$\alpha(HSO_3^-) = \frac{[H_3O^+]K_{a1}}{f} = \frac{(3.2 \times 10^{-6})(1.5 \times 10^{-2})}{5.0 \times 10^{-8}} = 0.96$$

$$\alpha(SO_3^{2-}) = \frac{K_{a1}K_{a2}}{f} = \frac{(1.5 \times 10^{-2})(1.2 \times 10^{-7})}{5.0 \times 10^{-8}} = 0.036$$

Thus, in a $0.150\ mol \cdot L^{-1}$ solution at pH 5.50, the dominant species will be $HSO_3^-$ with a concentration of $(0.150\ mol \cdot L^{-1})(0.96) = 0.14\ mol \cdot L^{-1}$. The concentration of $H_2SO_3$ will be $(2.1 \times 10^{-4})(0.150\ mol \cdot L^{-1}) = 3.2 \times 10^{-5}\ mol \cdot L^{-1}$ and the concentration of $SO_3^{2-}$ will be $(0.036)(0.150\ mol \cdot L^{-1}) = 0.0054\ mol \cdot L^{-1}$.

**10.89** (a) Concentration

| $(mol \cdot L^{-1})$ | $B(OH)_3$ | + | $2 H_2O$ | $\leftrightarrow$ | $H_3O^+$ | + | $B(OH)_4^-$ |
|---|---|---|---|---|---|---|---|
| initial | $1.0 \times 10^{-4}$ | | — | | 0 | | 0 |
| change | $-x$ | | — | | $+x$ | | $+x$ |
| equilibrium | $1.0 \times 10^{-4} - x$ | | — | | x | | x |

$$K_a = 7.2 \times 10^{-10} = \frac{[H_3O^+][B(OH)_4{}^-]}{[B(OH)_3]} = \frac{x^2}{1.0 \times 10^{-4} - x} \approx \frac{x^2}{1.0 \times 10^{-4}}$$

$$x = [H_3O^+] = 2.7 \times 10^{-7} \text{ mol} \cdot L^{-1}$$

$$pH = -\log(2.7 \times 10^{-7}) = 6.57$$

Note: this value of $[H_3O^+]$ is not much different from the value for pure water, $1.0 \times 10^{-7}$ mol $\cdot L^{-1}$; therefore, it is at the lower limit of safely ignoring the contribution to $[H_3O^+]$ from the autoprotolysis of water. The exercise should be solved by simultaneously considering both equilibria.

Concentration

| $(\text{mol} \cdot L^{-1})$ | $B(OH)_3$ | + | $2 H_2O$ | $\leftrightarrow$ | $H_3O^+$ | + | $B(OH)_4^-$ |
|---|---|---|---|---|---|---|---|
| equilibrium | $1.0 \times 10^{-4} - x$ | | — | | $x$ | | $y$ |

Concentration (mol $\cdot L^{-1}$) $2 H_2O \leftrightarrow H_3O^+ + OH^-$

| equilibrium | | — | $x$ | $z$ |
|---|---|---|---|---|

Because there are now two contributions to $[H_3O^+]$, $[H_3O^+]$ is no longer equal to $[B(OH)_4{}^-]$, nor is it equal to $[OH^-]$, as in pure water. To avoid a cubic equation, $x$ will again be ignored relative to $1.0 \times 10^{-4}$ mol $\cdot L^{-1}$. This approximation is justified by the approximate calculation above, and because $K_a$ is very small relative to $1.0 \times 10^{-4}$. Let $a = $ initial concentration of $B(OH)_3$, then

$$K_a = 7.2 \times 10^{-10} = \frac{xy}{a - x} \approx \frac{xy}{a} \text{ or } y = \frac{aK_a}{x}$$

$$K_w = 1.0 \times 10^{-14} = xz$$

Electroneutrality requires

$x = y + z$ or $z = x - y$; hence, $K_w = xz = x(x - y)$.

Substituting for $y$ from above:

$$x \times \left( x - \frac{aK_a}{x} \right) = K_w$$

$$x^2 - aK_a = K_w$$

$$x^2 = K_w + aK_a$$

$$x = \sqrt{K_w + aK_a} = \sqrt{1.0 \times 10^{-14} + 1.0 \times 10^{-4} \times 7.2 \times 10^{-10}}$$

$$x = 2.9 \times 10^{-7} \text{ mol} \cdot \text{L}^{-1} = [H_3O^+]$$

$$pH = -\log(2.9 \times 10^{-7}) = 6.54$$

This value is slightly, but measurably, different from the value 6.57

obtained by ignoring the contribution to $[H_3O^+]$ from water.

(b) In this case, the second ionization can safely be ignored; $K_{a2} \ll K_{a1}$.

Concentration

| (mol·L$^{-1}$) | H$_3$PO$_4$ | + | H$_2$O | ↔ | H$_3$O$^+$ | + | H$_2$PO$_4^-$ |
|---|---|---|---|---|---|---|---|
| initial | 0.015 | | — | | 0 | | 0 |
| change | −x | | — | | +x | | +x |
| equilibrium | 0.015 − x | | — | | x | | x |

$$K_{a1} = 7.6 \times 10^{-3} = \frac{x^2}{0.015 - x}$$

$$x^2 + 7.6 \times 10^{-3} x - 1.14 \times 10^{-4} = 0$$

$$x = [H_3O^+] = \frac{-7.6 \times 10^{-3} + \sqrt{(7.6 \times 10^{-3})^2 + 4.56 \times 10^{-4}}}{2}$$

$$= 7.5 \times 10^{-3} \text{ mol} \cdot \text{L}^{-1}$$

$$pH = -\log(7.5 \times 10^{-3}) = 2.12$$

(c) In this case, the second ionization can safely be ignored; $K_{a2} \ll K_{a1}$.

Concentration

| (mol·L$^{-1}$) | H$_2$SO$_3$ | + | H$_2$O | ↔ | H$_3$O$^+$ | + | HSO$_3^-$ |
|---|---|---|---|---|---|---|---|
| initial | 0.1 | | — | | 0 | | 0 |
| change | −x | | — | | +x | | +x |
| equilibrium | 0.1 − x | | — | | x | | x |

$$K_{a1} = 1.5 \times 10^{-2} = \frac{x^2}{0.10 - x}$$

$$x^2 + 1.5 \times 10^{-2} x - 1.5 \times 10^{-3} = 0$$

$$x = [H_3O^+] = \frac{-1.5 \times 10^{-2} + \sqrt{(1.5 \times 10^{-2})^2 + 6.0 \times 10^{-3}}}{2} = 0.032 \text{ mol} \cdot L^{-1}$$

$$pH = -\log(0.032) = 1.49$$

**10.91** The three equilibria involved are:

$$H_3PO_4(aq) \leftrightarrow H_2PO_4^-(aq) + H_3O^+(aq), \quad K_{a_1} = 7.6 \times 10^{-3} = \frac{[H_3O^+][H_2PO_4^-]}{[H_3PO_4]}$$

$$H_2PO_4^-(aq) \leftrightarrow HPO_4^{2-}(aq) + H_3O^+(aq), \quad K_{a_2} = 6.2 \times 10^{-8} = \frac{[H_3O^+][HPO_4^{2-}]}{[H_2PO_4^-]}$$

$$HPO_4^{2-}(aq) \leftrightarrow PO_4^{3-}(aq) + H_3O^+(aq), \quad K_{a_3} = 2.1 \times 10^{-13} = \frac{[H_3O^+][PO_4^{3-}]}{[HPO_4^{2-}]}$$

We also know that the combined concentration of all the phosphate species is:

$$[H_3PO_4] + [H_2PO_4^-] + [HPO_4^{2-}] + [PO_4^{3-}] = 1.5 \times 10^{-2} \text{ mol} \cdot L^{-1}$$

and the hydronium ion concentration is:

$$[H_3O^+] = 10^{-pH} = 10^{-2.25} = 5.6 \times 10^{-2} \text{ mol} \cdot L^{-1}$$

At this point it is a matter of solving this set of simultaneous equations to obtain the concentrations of the phosphate containing species. We start by dividing both sides of the equilibrium constant expressions above by the given hydronium ion concentration to obtain three ratios:

$$1.35 = \frac{[H_2PO_4^-]}{[H_3PO_4]}, \quad 1.10 \times 10^{-5} = \frac{[HPO_4^{2-}]}{[H_2PO_4^-]}, \quad \text{and} \quad 3.74 \times 10^{-11} = \frac{[PO_4^{3-}]}{[HPO_4^{2-}]}$$

Through rearrangement and substitution of these three ratios, we can obtain the following expressions:

$[H_2PO_4^-] = 1.35 \cdot [H_3PO_4],$

$[HPO_4^{2-}] = 1.10 \times 10^{-5} \cdot [H_2PO_4^-] = 1.10 \times 10^{-5} \cdot 1.35 \cdot [H_3PO_4]$

$\qquad = 1.48 \times 10^{-5} \cdot [H_3PO_4],$ and

$[PO_4^{3-}] = 3.74 \times 10^{-11} \cdot [HPO_4^{2-}] = 3.74 \times 10^{-11} \cdot 1.48 \times 10^{-5} \cdot [H_3PO_4]$

$\qquad = 5.54 \times 10^{-16} \cdot [H_3PO_4].$

Substituting these expressions back into the sum:

$[H_3PO_4] + [H_2PO_4^-] + [HPO_4^{2-}] + [PO_4^{3-}]$

$= [H_3PO_4] + \left(1.35 \cdot [H_3PO_4]\right) + \left(1.48 \times 10^{-5} \cdot [H_3PO_4]\right) + \left(5.54 \times 10^{-16} \cdot [H_3PO_4]\right)$

$= 1.5 \times 10^{-2} \; mol \cdot L^{-1},$

we find :

$[H_3PO_4] = 6.4 \times 10^{-3} \; mol \cdot L^{-1},$

$[H_2PO_4^-] = 1.35 \cdot [H_3PO_4] = 8.6 \times 10^{-3} \; mol \cdot L^{-1},$

$[HPO_4^{2-}] = 1.48 \times 10^{-5} \cdot [H_3PO_4] = 9.4 \times 10^{-8} \; mol \cdot L^{-1},$ and

$[PO_4^{3-}] = 5.54 \times 10^{-16} \cdot [H_3PO_4] = 3.5 \times 10^{-18} \; mol \cdot L^{-1}.$

**10.93** We can use the relationship derived in the text:

$[H_3O^+]^2 - [HA]_{initial}[H_3O^+] - K_w = 0$, in which HA is any strong acid.

$[H_3O^+]^2 - (6.55 \times 10^{-7})[H_3O^+] - (1.00 \times 10^{-14}) = 0$

Solving using the quadratic equation gives

$[H_3O^+] = 6.70 \times 10^{-7}$, pH = 6.174.

This value is slightly lower than the value calculated, based on the acid concentration alone (pH $= -\log(6.55 \times 10^{-7}) = 6.184$).

**10.95** We can use the relationship derived in the text:

$[H_3O^+]^2 + [B]_{initial}[H_3O^+] - K_w = 0$, in which B is any strong base.

$[H_3O^+]^2 + (9.78 \times 10^{-8})[H_3O^+] - (1.00 \times 10^{-14}) = 0$

Solving using the quadratic equation gives

$[H_3O^+] = 6.24 \times 10^{-8}$, pH = 7.205.

This value is higher than the value calculated, based on the base concentration alone ($pOH = -\log(9.78 \times 10^{-8}) = 7.009$).

**10.97** (a) In the absence of a significant effect due to the autoprotolysis of water, the pH values of the $1.00 \times 10^{-4}$ M and $1.00 \times 10^{-6}$ M HBrO solutions can be calculated as described earlier.

For $1.00 \times 10^{-4}$ mol·L$^{-1}$ :

Concentration

| (mol·L$^{-1}$) HBrO(aq) | + | H$_2$O(l) | $\leftrightarrow$ | H$_3$O$^+$(aq) | + | BrO$^-$(aq) |
|---|---|---|---|---|---|---|
| initial | $1.00 \times 10^{-4}$ | — | | 0 | | 0 |
| change | $-x$ | — | | $+x$ | | $+x$ |
| final | $1.00 \times 10^{-4} - x$ | — | | $+x$ | | $+x$ |

$$K_a = \frac{[H_3O^+][BrO^-]}{[HBrO]}$$

$$2.0 \times 10^{-9} = \frac{(x)(x)}{1.00 \times 10^{-4} - x} = \frac{x^2}{[1.00 \times 10^{-4} - x]}$$

Assume $x \ll 1.00 \times 10^{-4}$

$$x^2 = (2.0 \times 10^{-9})(1.00 \times 10^{-4})$$

$$x = 4.5 \times 10^{-7}$$

Because $x < 1\%$ of $1.00 \times 10^{-4}$, the assumption was valid. Given this value, the pH is then calculated to be $-\log(4.5 \times 10^{-7}) = 6.35$.

For $1.00 \times 10^{-6}$ mol·L$^{-1}$ :

Concentration

| (mol·L$^{-1}$) | HBrO(aq) | + | H$_2$O(l) | $\leftrightarrow$ | H$_3$O$^+$(aq) | + | BrO$^-$(aq) |
|---|---|---|---|---|---|---|---|
| initial | $1.00 \times 10^{-6}$ | | — | | 0 | | 0 |
| change | $-x$ | | — | | $+x$ | | $+x$ |
| final | $1.00 \times 10^{-6} - x$ | | — | | $+x$ | | $+x$ |

$$K_a = \frac{[H_3O^+][BrO^-]}{[HBrO]}$$

$$2.0 \times 10^{-9} = \frac{(x)(x)}{1.00 \times 10^{-6} - x} = \frac{x^2}{[1.00 \times 10^{-6} - x]}$$

Assume $x \ll 1.00 \times 10^{-6}$

$$x^2 = (2.0 \times 10^{-9})(1.00 \times 10^{-6})$$
$$x = 4.5 \times 10^{-8}$$

$x$ is 4.5% of $1.00 \times 10^{-6}$, so the assumption is less acceptable. The pH is calculated to be $-\log(4.5 \times 10^{-8}) = 7.35$. Because this predicts a basic solution, it is not reasonable.

(b) To calculate the value taking into account the autoprotolysis of water, we can use equation (22):

$$x^3 + K_a x^2 - (K_w + K_a \cdot [HA]_{initial})x - K_w \cdot K_a = 0, \text{ where } x = [H_3O^+].$$

To solve the expression, you substitute the values of $K_w = 1.00 \times 10^{-14}$, the initial concentration of acid, and $K_a = 2.0 \times 10^{-9}$ into this equation and then solve the expression either by trial and error or, preferably, using a graphing calculator such as the one found on the CD accompanying this text.

Alternatively, you can use a computer program designed to solve simultaneous equations. Because the unknowns include $[H_3O^+], [OH^-], [HBrO],$ and $[BrO^-],$ you will need four equations. As seen in the text, pertinent equations are

$$K_a = \frac{[H_3O^+][BrO^-]}{[HBrO]}$$

$$K_w = [H_3O^+][OH^-]$$
$$[H_3O^+] = [OH^-] + [BrO^-]$$
$$[HBrO]_{initial} = [HBrO] + [BrO^-]$$

Both methods should produce the same result.

The values obtained are

$[H_3O^+] = 4.6 \times 10^{-7}$ mol $\cdot$ L$^{-1}$, pH = 6.34 (compare to 6.35 obtained in (a))

$[BrO^-] = 4.4 \times 10^{-7}$ mol $\cdot$ L$^{-1}$

$[HBrO] \cong 1.0 \times 10^{-5}$ mol $\cdot$ L$^{-1}$

$[OH^-] = 2.2 \times 10^{-8}$ mol $\cdot$ L$^{-1}$

Similarly, for $[HBrO]_{initial} = 1.00 \times 10^{-6}$ :

$[H_3O^+] = 1.1 \times 10^{-7}$ mol $\cdot$ L$^{-1}$, pH = 6.96 (compare to 7.32 obtained in (a))

$[BrO^-] = 1.8 \times 10^{-8}$ mol $\cdot$ L$^{-1}$

$[HBrO] \cong 9.8 \times 10^{-7}$ mol $\cdot$ L$^{-1}$

$[OH^-] = 9.1 \times 10^{-8}$ mol $\cdot$ L$^{-1}$

Note that for the more concentrated solution, the effect of the autoprotolysis of water is very small. Notice also that the less concentrated solution is more acidic, due to the autoprotolysis of water, than would be predicted if this effect were not operating.

**10.99** (a) In the absence of a significant effect due to the autoprotolysis of water, the pH values of the $8.50 \times 10^{-5}$ M and $7.37 \times 10^{-6}$ M HCN solutions can be calculated as described earlier.

For $8.50 \times 10^{-5}$ mol $\cdot$ L$^{-1}$ :

Concentration

| (mol $\cdot$ L$^{-1}$) | HCN(aq) | + | H$_2$O(l) | $\leftrightarrow$ | H$_3$O$^+$(aq) | + | CN$^-$(aq) |
|---|---|---|---|---|---|---|---|
| initial | $8.50 \times 10^{-5}$ | | — | | 0 | | 0 |
| change | $-x$ | | — | | $+x$ | | $+x$ |
| final | $8.50 \times 10^{-5} - x$ | | — | | $+x$ | | $+x$ |

$$K_a = \frac{[H_3O^+][CN^-]}{[HCN]}$$

$$4.9 \times 10^{-10} = \frac{(x)(x)}{8.50 \times 10^{-5} - x} = \frac{x^2}{8.50 \times 10^{-5} - x}$$

Assume $x \ll 8.50 \times 10^{-5}$.

$$x^2 = (4.9 \times 10^{-10})(8.50 \times 10^{-5})$$

$$x = 2.0 \times 10^{-7}$$

Because $x < 1\%$ of $8.50 \times 10^{-5}$, the assumption was valid. Given this value, the pH is then calculated to be $-\log(2.0 \times 10^{-7}) = 6.70$.

For $7.37 \times 10^{-6}\ \text{mol} \cdot \text{L}^{-1}$:

Concentration

| $(\text{mol} \cdot \text{L}^{-1})$ | $HCN(aq)$ | $+$ | $H_2O(l)$ | $\leftrightarrow$ | $H_3O^+(aq)$ | $+$ | $CN^-(aq)$ |
|---|---|---|---|---|---|---|---|
| initial | $7.37 \times 10^{-6}$ | | — | | $0$ | | $0$ |
| change | $-x$ | | — | | $+x$ | | $+x$ |
| final | $7.37 \times 10^{-6} - x$ | | — | | $+x$ | | $+x$ |

$$K_a = \frac{[H_3O^+][CN^-]}{[HCN]}$$

$$4.9 \times 10^{-10} = \frac{(x)(x)}{7.37 \times 10^{-6} - x} = \frac{x^2}{[7.37 \times 10^{-6} - x]}$$

Assume $x \ll 7.37 \times 10^{-6}$

$$x^2 = (4.9 \times 10^{-10})(7.37 \times 10^{-6})$$

$$x = 6.0 \times 10^{-8}$$

$x$ is $< 1\%$ of $7.37 \times 10^{-6}$, so the assumption is still reasonable. The pH is then calculated to be $-\log(6.0 \times 10^{-8}) = 7.22$. This answer is not reasonable because we know HCN is an acid.

(b) To calculate the value, taking into account the autoprotolysis of water, we can use equation (20):

$$x^3 + K_a x^2 - (K_w + K_a \cdot [HA]_{initial})x - K_w \cdot K_a = 0, \text{ where } x = [H_3O^+].$$

To solve the expression, you substitute the values of $K_w = 1.00 \times 10^{-14}$, the initial concentration of acid, and $K_a = 4.9 \times 10^{-10}$ into this equation and then solve the expression either by trial and error or, preferably, using a graphing calculator such as the one found on the CD accompanying this text.

Alternatively, you can use a computer program designed to solve simultaneous equations. Because the unknowns include

$[H_3O^+]$, $[OH^-]$, $[HBrO]$, and $[BrO^-]$, you will need four equations. As seen in the text, the pertinent equations are

$$K_a = \frac{[H_3O^+][CN^-]}{[HCN]}$$

$$K_w = [H_3O^+][OH^-]$$

$$[H_3O^+] = [OH^-] + [CN^-]$$

$$[HCN]_{initial} = [HCN] + [CN^-]$$

Both methods should produce the same result.

For $[HCN] = 8.50 \times 10^{-5}$ mol·L$^{-1}$, the values obtained are

$[H_3O^+] = 2.3 \times 10^{-7}$ mol·L$^{-1}$, pH=6.64 (compare to 6.70 obtained in (a))

$[CN^-] = 1.8 \times 10^{-7}$ mol·L$^{-1}$

$[HCN] \cong 8.5 \times 10^{-5}$ mol·L$^{-1}$

$[OH^-] = 4.4 \times 10^{-8}$ mol·L$^{-1}$

Similarly, for $[HCN]_{initial} = 7.37 \times 10^{-6}$:

$[H_3O^+] = 1.2 \times 10^{-7}$ mol·L$^{-1}$, pH = 6.92 (compare to 7.22 obtained in (a))

$[CN^-] = 3.1 \times 10^{-8}$ mol·L$^{-1}$

$[HCN] \cong 7.3 \times 10^{-6}$ mol·L$^{-1}$

$[OH^-] = 8.6 \times 10^{-8}$ mol·L$^{-1}$

Note that for the more concentrated solution, the effect of the autoprotolysis of water is smaller. Notice also that the less concentrated solution is more acidic, due to the autoprotolysis of water, than would be predicted if this effect were not operating.

**10.101** (a) Assuming all sulfur is converted to SO$_2$, the amount of SO$_2$ produced is:

$$\frac{(1.00 \times 10^3 \text{ kg})(0.025)(1000 \text{ g·kg}^{-1})}{32.07 \text{ g·mol}^{-1}} = 780 \text{ mol of S}$$

Therefore, $(780 \text{ mol})(64.07 \text{ g·mol}^{-1}) = 50,000$ g or 50 kg of SO$_2$ is produced

(b) To determine the pH we first must calculate the volume of water in which this 50 kg of SO$_2$ is dissolved:

$$V = (2 \text{ cm})(2.6 \text{ Km})\left(\frac{1000 \text{ m}}{1 \text{ Km}}\right)^2\left(\frac{100 \text{ cm}}{1 \text{ m}}\right)^2\left(\frac{1 \text{ L}}{1000 \text{ cm}^3}\right) = 5.2 \times 10^7 \text{ L}$$

The concentration of $SO_2(aq)$ is then:

$(780 \text{ mol})/(5.2 \times 10^7 \text{ L}) = 1.5 \times 10^{-5}$ M. We can assume that upon

solution the $SO_2(aq)$ is converted to sulfurous acid and the pH is

calculated as described earlier:

Concentration

| $(\text{mol} \cdot \text{L}^{-1})$ | $H_2SO_3(aq)$ | + | $H_2O(l)$ | $\leftrightarrow$ | $H_3O^+(aq)$ | + | $HSO_3^-(aq)$ |
|---|---|---|---|---|---|---|---|
| initial | $1.50 \times 10^{-5}$ | | — | | 0 | | 0 |
| change | $-x$ | | — | | $+x$ | | $+x$ |
| final | $1.50 \times 10^{-5} - x$ | | — | | $+x$ | | $+x$ |

$$K_a = \frac{[H_3O^+][HSO_3^-]}{[H_2SO_3]}$$

$$1.55 \times 10^{-2} = \frac{(x)(x)}{1.5 \times 10^{-5} - x}$$

Employing the quadratic formula, we find $x = [H_3O^+] = 1.5 \times 10^{-5}$, giving

a pH of: $pH = -\log(1.5 \times 10^{-5}) = 4.82$.


(c) When dissolved in water, $SO_3$ will form sulfuric acid:

$$SO_3(aq) + 2 H_2O(l) \rightarrow H_3O^+(aq) + HSO_4^-(aq)$$

The fist deprotonation of sulfuric acid is complete, the concentration of

hydronium ion due to the second deprotonation of sulfuric acid may be

found as described earlier:

Concentration

| $(\text{mol} \cdot \text{L}^{-1})$ | $HSO_3^-(aq)$ | + | $H_2O(l)$ | $\leftrightarrow$ | $H_3O^+(aq)$ | + | $SO_3^{2-}(aq)$ |
|---|---|---|---|---|---|---|---|
| initial | $1.50 \times 10^{-5}$ | | — | | $1.50 \times 10^{-5}$ | | 0 |
| change | $-x$ | | — | | $+x$ | | $+x$ |
| final | $1.50 \times 10^{-5} - x$ | | — | | $1.50 \times 10^{-5} - x$ | | $+x$ |

$$K_a = \frac{[H_3O^+][HSO_3^-]}{[H_2SO_3]}$$

$$1.2 \times 10^{-2} = \frac{(1.5 \times 10^{-5} + x)(x)}{1.5 \times 10^{-5} - x}$$

Employing the quadratic formula, we find $x = [H_3O^+] = 1.495 \times 10^{-5}$, and

the total hydronium ion concentration to be

$$1.495 \times 10^{-5} + 1.5 \times 10^{-5} = 3.0 \times 10^{-5}.$$

Therefore, $pH = -\log(3.0 \times 10^{-5}) = 4.52$.

**10.103** (a) $K_w = [H_3O^+][OH^-] = x^2 = 3.8 \times 10^7$

$[H_3O^+] = x = 1.9 \times 10^7 \text{ mol} \cdot L^{-1}$

$pH = -\log[H_3O^+] = 6.72$

(b) There are three data points available:

25°C ($K_w = 1.0 \times 10^{-14}$), 40°C ($K_w =$

$3.8 \times 10^{-14}$), and 37°C ($K_w = 2.1 \times 10^{-14}$)

| $T$ | $1/T$ (K$^{-1}$) | $\ln K_w$ |
|---|---|---|
| 25°C | 0.003356 | −32.2362 |
| 37°C | 0.003226 | −31.4943 |
| 40°C | 0.003195 | −31.1376 |

The slope is equal to $-\Delta H°/R$ and the intercept equals $\Delta S°/R$.

$\Delta H° = -(-6502 \text{ K})(8.314 \text{ J} \cdot K^{-1} \cdot \text{mol}^{-1}) = 54 \text{ kJ} \cdot \text{mol}^{-1}$

$\Delta S° = -(-10.43)(8.314 \text{ J} \cdot K^{-1} \cdot \text{mol}^{-1}) = 87 \text{ J} \cdot K^{-1} \cdot \text{mol}^{-1}$

(c) The equation determined from the graph is for $\ln K_w$. In order to

write an equation for the pH dependence of pure water, we must rearrange

the equation.

First we note the relationship between $K_w$ and the pH of pure water.

$$[H_3O^+] = K_w^{1/2}$$

$$pH = -\log[H_3O^+] = -\frac{1}{2}\log K_w$$

$$\ln K_w = -\frac{6501}{T} - 10.43$$

$$2.303 \log K_w = -\frac{6501}{T} - 10.43$$

$$\log K_w = -\frac{2823}{T} - 4.529$$

$$pH = -\frac{1}{2}\log K_w$$

$$pH = \frac{1411}{T} + 2.264$$

**10.105** (a) We begin by finding the empirical formula of the compound:

C: $\dfrac{0.942 \text{ g } CO_2}{44.011 \text{ g} \cdot \text{mol}^{-1}} = 0.0214 \text{ mol } CO_2$ ∴ 0.214 mol C

$(0.214 \text{ mol C})(12.011 \text{ g} \cdot \text{mol}^{-1}) = 0.257 \text{ g C}$

H: $\dfrac{0.0964 \text{ g } H_2O}{18.0158 \text{ g} \cdot \text{mol}^{-1}} = 0.00535 \text{ mol } H_2O$ ∴ 0.0107 mol H

$(0.0107 \text{ mol H})(1.008 \text{ g} \cdot \text{mol}^{-1}) = 0.0108 \text{ g H}$

Na: $\dfrac{0.246 \text{ g } H_2O}{22.99 \text{ g} \cdot \text{mol}^{-1}} = 0.0107 \text{ mol Na}$

$(0.0107 \text{ mol Na})(22.99 \text{ g} \cdot \text{mol}^{-1}) = 0.0246 \text{ g Na}$

O: mass of O $= 1.200 \text{ g} - 0.257 \text{ g} - 0.0108 \text{ g} - 0.0246 \text{ g} = 0.686 \text{ g of O}$

$\dfrac{0.686 \text{ g O}}{16.00 \text{ g} \cdot \text{mol}^{-1}} = 0.0429 \text{ mol O}$

Dividing through by 0.0107 moles, we find the empirical formula to be:
$NaC_2HO_4$.

A molar mass of 112.02 g·mol$^{-1}$ indicates that this is also the molecular formula.

(b)

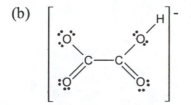

(c) The dissolved substance is sodium oxalate, it is capable of gaining or losing a proton and, therefore, amphiprotic. $pH = \frac{1}{2}(pK_{a_1} + pK_{a_2}) = 2.71$.

**10.107** We wish to calculate $K_a$ for the reaction

$$HF(aq) + H_2O(l) \leftrightarrow H_3O^+(aq) + F^-(aq)$$

This equation is equivalent to

$$HF(aq) \leftrightarrow H^+(aq) + F^-(aq)$$

This latter writing of the expression is simpler for the purpose of the thermodynamic calculations.

The $\Delta G°$ value for this reaction is easily calculated from the free energies given in the appendix:

$$= (-278.79 \text{ kJ} \cdot \text{mol}^{-1}) - (-296.82 \text{ kJ} \cdot \text{mol}^{-1}) = 18.03 \text{ kJ} \cdot \text{mol}^{-1}$$
$$= -RT \ln K_a$$
$$K_a = e^{-\Delta G°/RT}$$
$$K_a = e^{-(18030 \text{ J} \cdot \text{mol}^{-1})/[(8.314 \text{ J} \cdot \text{K}^{-1} \cdot \text{mol}^{-1})(298 \text{ K})]} = 6.9 \times 10^{-4}.$$

**10.109** (a) $\quad D_2O + D_2O \leftrightarrow D_3O^+ + OD^-$

(b) $\quad K_{D_2O} = [D_3O^+][OD^-] = 1.35 \times 10^{-15}$, $pK_{D_2O} = -\log K_{D_2O} = 14.870$

(c) $\quad [D_3O^+] = [OD^-] = \sqrt{1.35 \times 10^{-15}} = 3.67 \times 10^{-8} \text{ mol} \cdot L^{-1}$

(d) $\quad pD = -\log(3.67 \times 10^{-8}) = 7.435 = pOD$

(e) $\quad pD + pOD = pK_{D_2O}(D_2O) = 14.870$

**10.111**

$$CH_3CH(OH)COOH(aq) + H_2O(l) \leftrightarrow H_3O^+(aq) + CH_3CH(OH)COO^-(aq)$$
Concentration $(\text{mol} \cdot L^{-1})$

| | | | | |
|---|---|---|---|---|
| initial | 1.00 | — | 0 | 0 |
| change | $-x$ | — | $+x$ | $+x$ |
| equilibrium | $1.00 - x$ | — | $x$ | $x$ |

$$K_a = \frac{[H_3O^+][CH_3CH(OH)COO^-]}{[CH_3CH(OH)COOH]} = \frac{x^2}{1.00-x} = 8.4 \times 10^{-4}$$

$$x = 0.029$$

$$\text{Percent deprotonation} = \frac{0.029}{1.00}(100\%) = 2.9\% \text{ deprotonated}$$

To calculate $T_f$ for this solution we need the concentration in moles of solute per kilogram of solvent. Assuming a density of 1 g·cm-1, 1.00 L of solution will weigh 1000 g. One liter of solution will contain 1 mole (90.08 g) of lactic acid. Therefore, one liter of solution will contain 1000 g − 90.08 g = 910 g or 0.910 kg of solvent.

The molarity of the solution is: $\frac{1.029 \text{ mol solute}}{0.910 \text{ kg solvent}} = 1.13 \, m$.

Using the freezing-point constant for water listed in Table 8.8, the freezing point decreases by: $\Delta T_f = (1.86 \text{ K} \cdot \text{kg} \cdot \text{mol}^{-1})(1.13 \text{ mol} \cdot \text{kg}^{-1}) = 2.11 \text{ K}$.

The temperature at which the solution will freeze is 271.04 K or −2.11 °C.

**10.113** $T' = 30 \, °C = 303 \text{ K}, T' = 20 \, °C = 293 \text{ K}$.

$$\ln\left(\frac{K'_a}{K_a}\right) = \frac{\Delta H°}{R}\left(\frac{1}{T} - \frac{1}{T'}\right) = \frac{\Delta H°}{R}\left(\frac{T'-T}{TT'}\right).$$

This is the van't Hoff equation.

$$\ln\left(\frac{1.768 \times 10^{-4}}{1.765 \times 10^{-4}}\right) = \frac{\Delta H°}{8.314 \text{ J} \cdot \text{K}^{-1} \cdot \text{mol}^{-1}}\left(\frac{303 \text{ K} - 293 \text{ K}}{(293 \text{ K})(303 \text{ K})}\right) = 0.001698$$

$$\Delta H° = (8.314 \text{ J} \cdot \text{K}^{-1} \cdot \text{mol}^{-1})\left(\frac{(293 \text{ K})(303 \text{ K})}{303 \text{ K} - 293 \text{ K}}\right)(0.001698)$$

$$= 1.2 \times 10^2 \text{ J} \cdot \text{mol}^{-1}$$

**10.115** (a) The equilibrium constant for the autoprotolysis of pure deuterium oxide is given by:

$$K = e^{-\Delta G/_{R \cdot T}} = e^{-84800 \text{ J·mol}^{-1}/(8.3145 \text{J·K}^{-1}\text{·mol}^{-1})(298 \text{ K})} = 1.37 \times 10^{-15}$$

The concentration of $D_3O^+(aq)$ at equilibrium is found in the familiar way:

Concentration

| (mol · L$^{-1}$) | 2 D$_2$O | ↔ | D$_3$O$^+$ | + | OD$^-$ |
|---|---|---|---|---|---|
| initial | — | | 0 | | 0 |
| change | — | | +x | | +x |
| equilibrium | — | | +x | | +x |

$K = [D_3O^+][OD^-] = 1.37 \times 10^{-15}$

$[D_3O^+] = 3.7 \times 10^{-8}$

$pD = -\log(3.7 \times 10^{-8}) = 7.43$

(b) Given the expression for $K$ in part (b) above, it is apparent that as $T$ increases $K$ will increase resulting in an increase in $[D_3O^+(aq)]$ and a decrease in pD.

**10.117** (a) and (b) Buffer regions are marked A, B, and C.

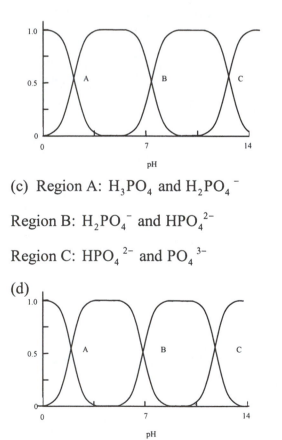

(c) Region A: $H_3PO_4$ and $H_2PO_4^-$

Region B: $H_2PO_4^-$ and $HPO_4^{2-}$

Region C: $HPO_4^{2-}$ and $PO_4^{3-}$

(d)

(e) The major species present are similar for both $H_3PO_4$ and $H_3AsO_4$: $H_2EO_4^-$ and $HEO_4^{2-}$ where $E = P$ or As. For As, there is more $HAsO_4^{2-}$ than $H_2AsO_4^-$, with a ratio of approximately 0.63 to 0.37, or 1.7 : 1. For P, the situation is reversed, with more $H_2PO_4^-$ than $HPO_4^{2-}$ in a ratio of about 0.61 to 0.39, or 2.2 : 1.

**10.119** Given that $CO_2$ will react with water to form carbonic acid, $H_2CO_3$, it only remains to determine the concentration of $H_3O^+$ due to the deprotonation of $H_2CO_3$.

$$[CO_2] = (2.3 \times 10^{-2}\ mol \cdot L^{-1} \cdot atm^{-1})(3.04 \times 10^{-4}\ atm) = 7.0 \times 10^{-6}\ mol \cdot L^{-1}$$

$$H_2CO_3(aq) \quad + \quad H_2O(l) \quad \leftrightarrow \quad H_3O^+(aq) \quad + \quad HCO_3^-(aq)$$

|  | $H_2CO_3$ | $H_2O$ | $H_3O^+$ | $HCO_3^-$ |
|---|---|---|---|---|
| initial | $7.0 \times 10^{-6}$ | — | 0 | 0 |
| change | $-x$ | — | $+x$ | $+x$ |
| equilibrium | $7.0 \times 10^{-6} - x$ | — | $x$ | $x$ |

$$K_a = 10^{-pK_a} = 10^{-6.37} = 4.26 \times 10^{-7} = \left( \frac{x^2}{7.0 \times 10^{-6} - x} \right)$$

Solving the quadratic equation for $x$, we can calculate the pH:

$$x = 1.5 \times 10^{-6} = [H_3O^+]$$
$$pH = -\log(1.5 \times 10^{-6}) = 5.82.$$

(Note: The mantissa of the pH has two significant figures because the mantissa of the $pK_a$ has two significant figures.)

# CHAPTER 11

# AQUEOUS EQUILIBRIA

**11.1** (a) When solid sodium acetate is added to an acetic acid solution, the concentration of $H_3O^+$ decreases because the equilibrium

$$HC_2H_3O_2(aq) + H_2O(l) \leftrightarrow H_3O^+(aq) + C_2H_3O_2^-(aq)$$

shifts to the left to relieve the stress imposed by the increase of $[C_2H_3O_2^-]$ (Le Chatelier's principle).

(b) When HCl is added to a benzoic acid solution, the percentage of benzoic acid that is deprotonated decreases because the equilibrium

$$C_6H_5COOH(aq) + H_2O(l) \leftrightarrow H_3O^+(aq) + C_6H_5CO_2^-(aq)$$

shifts to the left to relieve the stress imposed by the increased $[H_3O^+]$ (Le Chatelier's principle).

(c) When solid $NH_4Cl$ is added to an ammonia solution, the concentration of $OH^-$ decreases because the equilibrium

$$NH_3(aq) + H_2O(l) \leftrightarrow NH_4^+(aq) + OH^-(aq)$$ shifts to the left to relieve the stress imposed by the increased $[NH_4^+]$ (Le Chatelier's principle). Because $[OH^-]$ decreases, $[H_3O^+]$ increases and pH decreases.

**11.3** (a) $K_a = \dfrac{[H_3O^+][A^-]}{[HA]}$; $pK_a = pH - \log \dfrac{[A^-]}{[HA]}$. If $[A^-] = [HA]$, then

$pK_a = pH$.

$pH = pK_a = 3.08$, $K_a = 8.3 \times 10^{-4}$

(b) Let $x = [\text{lactate ion}] = [L^-]$ and $y = [H_3O^+]$

Concentration

| (mol·L$^{-1}$) | HL(aq) | + | H$_2$O(l) | $\leftrightarrow$ | H$_3$O$^+$(aq) | + | L$^-$(aq) |
|---|---|---|---|---|---|---|---|
| initial | $2x$ | | — | | — | | $x$ |
| change | $-y$ | | — | | $+y$ | | $+y$ |
| equilibrium | $2x-y$ | | — | | $y$ | | $y+x$ |

$$K_a = \frac{[H_3O^+][L^-]}{[HL]} = \frac{(y)(y+x)}{(2x-y)} \cong \frac{(y)(x)}{(2x)} = 8.3 \times 10^{-4}$$

$$y = 2(8.3 \times 10^{-4}) \cong 1.7 \times 10^{-3} \ \text{mol} \cdot L^{-1} \cong [H_3O^+]$$

$$\text{pH} \approx 2.77$$

**11.5** In each case, the equilibrium involved is

$$HSO_4^-(aq) + H_2O(l) \leftrightarrow H_3O^+(aq) + SO_4^{2-}(aq)$$

HSO$_4^-$(aq) and SO$_4^{2-}$(aq) are conjugate acid and base; therefore, the pH calculation is most easily performed with the Henderson-Hasselbalch equation:

$$\text{pH} = pK_a + \log\left(\frac{[\text{base}]}{[\text{acid}]}\right) = pK_a + \log\left(\frac{[SO_4^{2-}]}{[HSO_4^-]}\right)$$

(a)  $\text{pH} = 1.92 + \log\left(\dfrac{0.25 \ \text{mol} \cdot L^{-1}}{0.5 \ \text{mol} \cdot L^{-1}}\right) = 1.62, \quad \text{pOH} = 14.00 - 1.62 = 12.38$

(b)  $\text{pH} = 1.92 + \log\left(\dfrac{0.10 \ \text{mol} \cdot L^{-1}}{0.50 \ \text{mol} \cdot L^{-1}}\right) = 1.22, \quad \text{pOH} = 12.78$

(c)  $\text{pH} = pK_a = 1.92, \quad \text{pOH} = 12.08$

See solution to Exercise 11.3.

**11.7** $\left(\dfrac{0.356 \ \text{g NaF}}{0.050 \ \text{L}}\right)\left(\dfrac{1 \ \text{mol NaF}}{41.99 \ \text{g NaF}}\right) = 0.17 \ \text{mol} \cdot L^{-1}$

Concentration

| (mol·L$^{-1}$) | HF(aq) | + | H$_2$O(l) | $\leftrightarrow$ | H$_3$O$^+$(aq) | + | F$^-$(aq) |
|---|---|---|---|---|---|---|---|

| initial | 0.40 | — | 0 | 0.17 |
| change | $-x$ | — | $+x$ | $+x$ |
| equilibrium | $0.40 - x$ | — | $x$ | $0.17 + x$ |

$$K_a = \frac{[H_3O^+][F^-]}{[HF]} = \frac{(x)(0.17 + x)}{(0.40 - x)} \approx \frac{(x)(0.17)}{(0.40)} = 3.5 \times 10^{-4}$$

$$x \cong 8.2 \times 10^{-4} \ mol \cdot L^{-1} \cong [H_3O^+]$$

$$pH = -\log[H_3O^+] = -\log(8.2 \times 10^{-4}) = 3.09$$

$$change \ in \ pH = 3.09 - 1.93 = 1.16$$

**11.9** (a) $HCN(aq) + H_2O(l) \leftrightarrow H_3O^+(aq) + CN^-(aq)$

total volume $= 100 \ mL = 0.100 \ L$

moles of $HCN = 0.0300 \ L \times 0.050 \ mol \cdot L^{-1} = 1.5 \times 10^{-3} \ mol \ HCN$

moles of $NaCN = 0.0700 \ L \times 0.030 \ mol \cdot L^{-1} = 2.1 \times 10^{-3} \ mol \ NaCN$

$$initial \ [HCN]_0 = \frac{1.5 \times 10^{-3} \ mol}{0.100 \ L} = 1.5 \times 10^{-2} \ mol \cdot L^{-1}$$

$$initial \ [CN^-]_0 = \frac{2.1 \times 10^{-3} \ mol}{0.100 \ L} = 2.1 \times 10^{-2} \ mol \cdot L^{-1}$$

Concentration

| $(mol \cdot L^{-1})$ | $HCN(aq)$ | $+$ | $H_2O(l)$ | $\leftrightarrow$ | $H_3O^+(aq)$ | $+$ | $CN^-(aq)$ |
| initial | $1.5 \times 10^{-2}$ | | — | | 0 | | $2.1 \times 10^{-2}$ |
| change | $-x$ | | — | | $+x$ | | $+x$ |
| equilibrium | $1.5 \times 10^{-2} - x$ | | — | | $x$ | | $2.1 \times 10^{-2} + x$ |

$$K_a = \frac{[H_3O^+][CN^-]}{[HCN]} = \frac{(x)(2.1 \times 10^{-2} + x)}{(1.5 \times 10^{-2} - x)} \approx \frac{(x)(2.1 \times 10^{-2})}{(1.5 \times 10^{-2})} = 4.9 \times 10^{-10}$$

$$x \approx 3.5 \times 10^{-10} \ mol \cdot L^{-1} \approx [H_3O^+]$$

$$pH = -\log[H_3O^+] = -\log(3.5 \times 10^{-10}) = 9.46$$

(b) The solution here is the same as for part (a), except for the initial concentrations:

$$[HCN]_0 = \frac{0.0400 \text{ L} \times 0.030 \text{ mol} \cdot \text{L}^{-1}}{0.100 \text{ L}} = 1.2 \times 10^{-2} \text{ mol} \cdot \text{L}^{-1}$$

$$[CN^-]_0 = \frac{0.0600 \text{ L} \times 0.050 \text{ mol} \cdot \text{L}^{-1}}{0.100 \text{ L}} = 3.0 \times 10^{-2} \text{ mol} \cdot \text{L}^{-1}$$

$$K_a = 4.9 \times 10^{-10} = \frac{(x)(3.0 \times 10^{-2})}{(1.2 \times 10^{-2})}$$

$$x = [H_3O^+] = 2.0 \times 10^{-10} \text{ mol} \cdot \text{L}^{-1}$$

$$pH = -\log(2.0 \times 10^{-10}) = 9.71$$

(c) $[HCN]_0 = [NaCN]_0$ after mixing; therefore,

$$K_a = 4.9 \times 10^{-10} = \frac{(x)[NaCN]_0}{[HCN]_0} = x = [H_3O^+]$$

$$pH = pK_a = -\log(4.9 \times 10^{-10}) = 9.31$$

**11.11**  In a solution containing $HClO(aq)$ and $ClO^-(aq)$, the following equilibrium occurs:

$$HClO(aq) + H_2O(l) \leftrightarrow H_3O^+(aq) + ClO^-(aq)$$

The ratio $[ClO^-]/[HClO]$ is related to pH, as given by the Henderson-Hasselbalch equation: $pH = pK_a + \log\left(\frac{[ClO^-]}{[HClO]}\right)$, or

$$\log\left(\frac{[ClO^-]}{[HClO]}\right) = pH - pK_a = 6.50 - 7.53 = -1.03$$

$$\frac{[ClO^-]}{[HClO]} = 9.3 \times 10^{-2}$$

**11.13**  The rule of thumb we use is that the effective range of a buffer is roughly within plus or minus one pH unit of the $pK_a$ of the acid. Therefore,

(a)  $pK_a = 3.08$; pH range, 2–4

(b)  $pK_a = 4.19$; pH range, 3–5

(c) $pK_{a3} = 12.68$; pH range, 11.5–13.5

(d) $pK_{a2} = 7.21$; pH range, 6–8

(e) $pK_b = 7.97$, $pK_a = 6.03$; pH range, 5–7

**11.15** Choose a buffer system in which the conjugate acid has a $pK_a$ close to the desired pH. Therefore,

(a) $HClO_2$ and $NaClO_2$, $pK_a = 2.00$

(b) $NaH_2PO_4$ and $Na_2HPO_4$, $pK_{a2} = 7.21$

(c) $CH_2ClCOOH$ and $NaCH_2ClCO_2$, $pK_a = 2.85$

(d) $Na_2HPO_4$ and $Na_3PO_4$, $pK_a = 12.68$

**11.17** (a) $HCO_3^-(aq) + H_2O(l) \leftrightarrow CO_3^{2-}(aq) + H_3O^+(aq)$

$$K_{a2} = \frac{[H_3O^+][CO_3^{2-}]}{[HCO_3^-]}, \quad pK_{a2} = 10.25$$

$$pH = pK_{a2} + \log\left(\frac{[CO_3^{2-}]}{[HCO_3^-]}\right)$$

$$\log\left(\frac{[CO_3^{2-}]}{[HCO_3^-]}\right) = pH - pK_{a2} = 11.0 - 10.25 = 0.75$$

$$\frac{[CO_3^{2-}]}{[HCO_3^-]} = 5.6$$

(b) $[CO_3^{2-}] = 5.6 \times [HCO_3^-] = 5.6 \times 0.100 \text{ mol} \cdot L^{-1} = 0.56 \text{ mol} \cdot L^{-1}$

moles of $CO_3^{2-}$ = moles of $K_2CO_3$ = 0.56 mol $\cdot L^{-1} \times 1$ L = 0.56 mol

mass of $K_2CO_3$ = 0.56 mol $\times \left(\dfrac{138.21 \text{ g } K_2CO_3}{1 \text{ mol } K_2CO_3}\right) = 77 \text{ g } K_2CO_3$

(c) $[HCO_3^-] = \dfrac{[CO_3^{2-}]}{5.6} = \dfrac{0.100 \text{ mol} \cdot L^{-1}}{5.6} = 1.8 \times 10^{-2} \text{ mol} \cdot L^{-1}$

moles of $HCO_3^-$ = moles of $KHCO_3$ = $1.8 \times 10^{-2}$ mol $\cdot L^{-1} \times 1$ L

$$= 1.8 \times 10^{-2} \text{ mol}$$

mass $KHCO_3 = 1.8 \times 10^{-2}$ mol $\times 100.12$ g·mol$^{-1} = 1.8$ g $KHCO_3$

(d) $[CO_3{}^{2-}] = 5.6 \times [HCO_3{}^-]$

moles of

$HCO_3{}^- =$ moles $KHCO_3 = 0.100$ mol·L$^{-1} \times 0.100$ L $= 1.00 \times 10^{-2}$ mol.

Because the final total volume is the same for both

$KHCO_3$ and $K_2CO_3$, the number of moles of $K_2CO_3$ required is

$5.6 \times 1.00 \times 10^{-2}$ mol $= 5.6 \times 10^{-2}$ mol.

Thus,

volume of $K_2CO_3$ solution $= \dfrac{5.6 \times 10^{-2} \text{ mol}}{0.200 \text{ mol·L}^{-1}} = 0.28$ L $= 2.8 \times 10^2$ mL

**11.19** (a) $pH = pK_a + \log\left(\dfrac{[CH_3CO_2{}^-]}{[CH_3COOH]}\right)$ (see Exercise 11.23)

$pH = pK_a + \log\left(\dfrac{0.100}{0.100}\right) = 4.75$ (initial pH)

final pH: $(0.0100 \text{ L})(0.950 \text{ mol·L}^{-1}) = 9.50 \times 10^{-3}$ mol NaOH (strong

base) produces $9.50 \times 10^{-3}$ mol $CH_3CO_2{}^-$ from $CH_3COOH$

$0.100$ mol·L$^{-1} \times 0.100$ L $= 1.00 \times 10^{-2}$ mol $CH_3COOH$ initially
$0.100$ mol·L$^{-1} \times 0.100$ L $= 1.00 \times 10^{-2}$ mol $CH_3CO_2{}^-$ initially

After adding NaOH:

$[CH_3COOH] = \dfrac{(1.00 \times 10^{-2} - 9.50 \times 10^{-3}) \text{ mol}}{0.110 \text{ L}} = 4.54 \times 10^{-3}$ mol·L$^{-1}$

$[CH_3CO_2{}^-] = \dfrac{(1.00 \times 10^{-2} + 9.50 \times 10^{-3}) \text{ mol}}{0.110 \text{ L}} = 1.77 \times 10^{-1}$ mol·L$^{-1}$

$pH = 4.75 + \log\left(\dfrac{1.77 \times 10^{-1}}{4.54 \times 10^{-3}}\right) = 6.34,$

where $\Delta pH = \log\left(\dfrac{1.77 \times 10^{-1}}{4.54 \times 10^{-3}}\right) = \log(3.90 \times 10^{+1}) = 1.591.$

(b) $(0.0200 \text{ L})(0.100 \text{ mol} \cdot \text{L}^{-1}) = 2.00 \times 10^{-3}$ mol $HNO_3$ (strong acid)

produces

$2.00 \times 10^{-3}$ mol $CH_3COOH$ from $CH_3CO_2^-$.

After adding $HNO_3$ [see part (a) of this exercise]:

$$[CH_3COOH] = \frac{(1.00 \times 10^{-2} + 2.00 \times 10^{-3}) \text{ mol}}{0.120 \text{ L}} = 1.00 \times 10^{-1} \text{ mol} \cdot \text{L}^{-1}$$

$$[CH_3CO_2^-] = \frac{(1.00 \times 10^{-2} - 2.00 \times 10^{-3}) \text{ mol}}{0.120 \text{ L}} = 6.7 \times 10^{-2} \text{ mol} \cdot \text{L}^{-1}$$

$$pH = 4.75 + \log\left(\frac{6.7 \times 10^{-2} \text{ mol} \cdot \text{L}^{-1}}{1.00 \times 10^{-1} \text{ mol} \cdot \text{L}^{-1}}\right) = 4.75 - 0.17 = 4.58$$

$\Delta pH = -0.17$

**11.21**

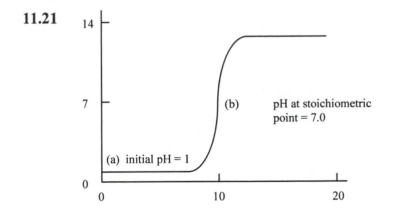

**11.23** $HCl(aq) + NaOH(aq) \longrightarrow H_2O(l) + Na^+(aq) + Cl^-(aq)$

(a) $V_{HCl} = (\frac{1}{2})(25.0 \text{ mL})\left(\frac{10^{-3} \text{ L}}{1 \text{ mL}}\right)\left(\frac{0.110 \text{ mol NaOH}}{1 \text{ L}}\right)$

$\left(\frac{1 \text{ mol HCl}}{1 \text{ mol NaOH}}\right)\left(\frac{1 \text{ L HCl}}{0.150 \text{ mol HCl}}\right)$

$= 9.17 \times 10^{-3}$ L HCl(aq)

(b) $2 \times 9.17 \times 10^{-3}$ L $= 0.0183$ L

(c)  volume $= (0.0250 + 0.0183)$ L $= 0.0433$ L

$$[Na^+] = (0.0250 \text{ L})\left(\frac{0.110 \text{ mol NaOH}}{1 \text{ L}}\right)\left(\frac{1 \text{ mol Na}^+}{1 \text{ mol NaOH}}\right)\left(\frac{1}{0.0433 \text{ L}}\right)$$

$$= 0.0635 \text{ mol} \cdot \text{L}^{-1}$$

(d)  number of moles of $H_3O^+$ (from acid) $= (0.0200 \text{ L})\left(\frac{0.150 \text{ mol}}{1 \text{ L}}\right)$

$$= 3.00 \times 10^{-3} \text{ mol } H_3O^+$$

number of moles of $OH^-$ (from base) $= (0.0250 \text{ L})\left(\frac{0.110 \text{ mol Na}^+}{1 \text{ L}}\right)$

$$= 2.75 \times 10^{-3} \text{ mol } OH^-$$

excess $H_3O^+ = (3.00 - 2.75) \times 10^{-3} \text{ mol} = 2.5 \times 10^{-4} \text{ mol } H_3O^+$

$$[H_3O^+] = \frac{2.5 \times 10^{-4} \text{ mol}}{0.0450 \text{ L}} = 5.6 \times 10^{-3} \text{ mol} \cdot \text{L}^{-1}$$

$$pH = -\log(5.6 \times 10^{-3}) = 2.25$$

**11.25**  (a) The moles of $OH^-$ added are equivalent to the number of moles of HA present:

$$\left(0.0350 \text{ mol} \cdot \text{L}^{-1}\right)(0.050 \text{ L}) = 0.0182 \text{ mol } OH^-,$$

$\therefore$  0.0182 mol of HA were present in solution.

$$\text{molar mass} = \frac{4.25 \text{ g}}{0.0182 \text{ mol}} = 234 \text{ g} \cdot \text{mol}^{-1}$$

(b) Since 52.0 mL of strong base is required to reach the stoichiometric point, adding 26.0 mL puts the titration halfway to the stoichiometric point.  At this halfway point $[HA] = [A^-]$, and since

$$pK_a = pH + \log\frac{[A^-]}{[HA]}, \text{ then } pK_a = pH = 3.82.$$

**11.27**  mass of pure NaOH in original sample

$$= (0.0342 \text{ L HCl})\left(\frac{0.0695 \text{ mol HCl}}{1 \text{ L HCl}}\right)$$

$$\left(\dfrac{1 \text{ mol NaOH}}{1 \text{ mol HCl}}\right)\left(\dfrac{40.00 \text{ g NaOH}}{1 \text{ mol NaOH}}\right)\left(\dfrac{300 \text{ mL}}{25.0 \text{ mL}}\right)$$

$$= 1.14 \text{ g}$$

$$\text{percent purity} = \dfrac{1.14 \text{ g}}{1.436 \text{ g}} \times 100\% = 79.4\%$$

**11.29** (a) $pOH = -\log(0.110) = 0.959$, $pH = 14.00 - 0.959 = 13.04$

(b) initial moles of $OH^-$ (from base) $= (0.0250 \text{ L})\left(\dfrac{0.110 \text{ mol}}{1 \text{ L}}\right)$

$$= 2.75 \times 10^{-3} \text{ mol OH}^-$$

moles of $H_3O^+$ added $= (0.0050 \text{ L})\left(\dfrac{0.150 \text{ mol}}{1 \text{ L}}\right) = 7.5 \times 10^{-4} \text{ mol H}_3\text{O}^+$

excess $OH^- = (2.75 - 0.75) \times 10^{-3} \text{ mol} = 2.00 \times 10^{-3} \text{ mol OH}^-$

$$[OH^-] = \dfrac{2.00 \times 10^{-3} \text{ mol}}{0.030 \text{ L}} = 0.067 \text{ mol} \cdot \text{L}^{-1}$$

$pOH = -\log(0.067) = 1.17$, $pH = 14.00 - 1.17 = 12.83$

(c) moles of $H_3O^+$ added $= 2 \times 7.5 \times 10^{-4} \text{ mol} = 1.50 \times 10^{-3} \text{ mol H}_3\text{O}^+$

excess $OH^- = (2.75 - 1.50) \times 10^{-3} \text{ mol} = 1.25 \times 10^{-3} \text{ mol OH}^-$

$$[OH^-] = \dfrac{1.25 \times 10^{-3} \text{ mol}}{0.035 \text{ L}} = 0.036 \text{ mol} \cdot \text{L}^{-1}$$

$pOH = -\log(0.036) = 1.44$, $pH = 14.00 - 1.44 = 12.56$

(d) $pH = 7.00$

$$V_{HCl} = (2.75 \times 10^{-3} \text{ mol NaOH})\left(\dfrac{1 \text{ mol HCl}}{1 \text{ mol NaOH}}\right)\left(\dfrac{1 \text{ L HCl}}{0.150 \text{ mol HCl}}\right)$$

$$= 0.0183 \text{ L}$$

(e) $[H_3O^+] = (0.0050 \text{ L})\left(\dfrac{0.150 \text{ mol}}{1 \text{ L}}\right)\left(\dfrac{1}{(0.0250 + 0.0183 + 0.0050) \text{ L}}\right)$

$$= 0.016 \text{ mol} \cdot \text{L}^{-1}$$

$pH = -\log(0.016) = 1.80$

(f) $[H_3O^+] = \left(\dfrac{0.010 \text{ L}}{0.0533 \text{ L}}\right)\left(\dfrac{0.150 \text{ mol}}{1 \text{ L}}\right) = 0.028 \text{ mol} \cdot \text{L}^{-1}$

$$pH = -\log(0.028) = 1.55$$

**11.31** (a) Concentration

| $(\text{mol} \cdot \text{L}^{-1})$ | $CH_3COOH(aq)$ | $+ H_2O(l)$ | $\leftrightarrow$ | $H_3O^+(aq)$ | $+ CH_3CO_2^-(aq)$ |
|---|---|---|---|---|---|
| initial | 0.10 | — | | 0 | 0 |
| change | $-x$ | — | | $+x$ | $+x$ |
| equilibrium | $0.10 - x$ | — | | $x$ | $x$ |

$$K_a = \frac{[H_3O^+][CH_3CO_2^-]}{[CH_3COOH]} = 1.8 \times 10^{-5} = \frac{x^2}{0.10 - x} \approx \frac{x^2}{0.10}$$

$$x^2 = 1.8 \times 10^{-6}$$

$$x = 1.3 \times 10^{-3} \text{ mol} \cdot \text{L}^{-1} = [H_3O^+]$$

initial pH $= -\log(1.3 \times 10^{-3}) = 2.89$

(b)   moles of $CH_3COOH = (0.0250 \text{ L})(0.10 \text{ M})$
$$= 2.50 \times 10^{-3} \text{ mol } CH_3COOH$$

moles of NaOH $= (0.0100 \text{ L})(0.10 \text{ M}) = 1.0 \times 10^{-3}$ mol $OH^-$

After neutralization,

$$\frac{1.50 \times 10^{-3} \text{ mol } CH_3COOH}{0.0350 \text{ L}} = 4.29 \times 10^{-2} \text{ mol} \cdot \text{L}^{-1} \, CH_3COOH$$

$$\frac{1.0 \times 10^{-3} \text{ mol } CH_3CO_2^-}{0.0350 \text{ L}} = 2.86 \times 10^{-2} \text{ mol} \cdot \text{L}^{-1} \, CH_3CO_2^-$$

Then consider equilibrium, $K_a = \dfrac{[H_3O^+][CH_3CO_2^-]}{[CH_3COOH]}$

Concentration

| $(\text{mol} \cdot \text{L}^{-1})$ | $CH_3COOH(aq)$ | $+ H_2O(l)$ | $\leftrightarrow$ | $H_3O^+(aq)$ | $+ CH_3CO_2^-(aq)$ |
|---|---|---|---|---|---|
| initial | $4.29 \times 10^{-2}$ | — | | 0 | $2.86 \times 10^{-2}$ |
| change | $-x$ | — | | $+x$ | $+x$ |
| equilibrium | $4.29 \times 10^{-2} - x$ | — | | $x$ | $2.86 \times 10^{-2} + x$ |

$$1.8 \times 10^{-5} = \frac{(x)(x + 2.86 \times 10^{-2})}{(4.29 \times 10^{-2} - x)}; \text{ assume } +x \text{ and } -x \text{ negligible.}$$

$[H_3O^+] = x = 2.7 \times 10^{-5} \text{ mol} \cdot L^{-1}$ and $pH = -\log(2.7 \times 10^{-5}) = 4.56$

(c)  Because acid and base concentrations are equal, their volumes are equal at the stoichiometric point. Therefore, 25.0 mL NaOH is required to reach the stoichiometric point and 12.5 mL NaOH is required to reach the half stoichiometric point.

(d)  At the half stoichiometric point, $pH = pK_a = 4.75$

(e)  25.0 mL; see part (c)

(f)  The pH is that of 0.050 M $NaCH_3CO_2$.

Concentration

$(\text{mol} \cdot L^{-1})$ $H_2O(l) + CH_3CO_2^-(aq) \leftrightarrow CH_3COOH(aq) + OH^-(aq)$

| | | | |
|---|---|---|---|
| initial | — | 0.050 | 0 | 0 |
| change | — | $-x$ | $+x$ | $+x$ |
| equilibrium | — | $0.050 - x$ | $x$ | $x$ |

$$K_b = \frac{K_w}{K_a} = \frac{1.00 \times 10^{-14}}{1.8 \times 10^{-5}} = 5.6 \times 10^{-10} = \frac{x^2}{0.050 - x} \approx \frac{x^2}{0.050}$$

$x^2 = 2.8 \times 10^{-11}$

$x = 5.3 \times 10^{-6} \text{ mol} \cdot L^{-1} = [OH^-]$

$pOH = 5.28, pH = 14.00 - 5.28 = 8.72$

**11.33** (a)  $K_b = \dfrac{[NH_4^+][OH^-]}{[NH_3]} = 1.8 \times 10^{-5}$

Concentration

$(\text{mol} \cdot L^{-1})$ $H_2O(l) + NH_3(aq) \leftrightarrow NH_4^+(aq) + OH^-(aq)$

| | | | |
|---|---|---|---|
| initial | — | 0.15 | 0 | 0 |
| change | — | $-x$ | $+x$ | $+x$ |
| equilibrium | — | $0.15 - x$ | $x$ | $x$ |

$$1.8 \times 10^{-5} = \frac{x^2}{0.15 - x} \approx \frac{x^2}{0.15}$$

$[OH^-] = x = 1.6 \times 10^{-3} \text{ mol} \cdot L^{-1}$

$pOH = 2.80$, initial $pH = 14.00 - 2.80 = 11.20$

(b) initial moles of

$NH_3 = (0.0150 \text{ L})(0.15 \text{ mol} \cdot L^{-1}) = 2.3 \times 10^{-3} \text{ mol NH}_3$

moles of $HCl = (0.0150 \text{ L})(0.10 \text{ mol} \cdot L^{-1}) = 1.5 \times 10^{-3} \text{ mol HCl}$

$$\frac{(2.3 \times 10^{-3} - 1.5 \times 10^{-3}) \text{ mol NH}_3}{0.0300 \text{ L}} = 2.7 \times 10^{-2} \text{ mol} \cdot L^{-1} \text{ NH}_3$$

$$\frac{1.5 \times 10^{-3} \text{ mol HCl}}{0.0300 \text{ L}} = 5.0 \times 10^{-2} \text{ mol} \cdot L^{-1} \text{ HCl} \approx 5.0 \times 10^{-2} \text{ mol} \cdot L^{-1} \text{ NH}_4^+$$

Then consider the equilibrium:

Concentration

| $(\text{mol} \cdot L^{-1})$ | $H_2O(l)$ | $+$ | $NH_3(aq)$ | $\leftrightarrow$ | $NH_4^+(aq)$ | $+$ | $OH^-(aq)$ |
|---|---|---|---|---|---|---|---|
| initial | — | | $2.7 \times 10^{-2}$ | | $5.0 \times 10^{-2}$ | | $0$ |
| change | — | | $-x$ | | $+x$ | | $+x$ |
| equilibrium | — | | $2.7 \times 10^{-2} - x$ | | $5.0 \times 10^{-2} + x$ | | $x$ |

$$K_b = \frac{[NH_4^+][OH^-]}{[NH_3]} = 1.8 \times 10^{-5}$$

$$= \frac{(x)(5.0 \times 10^{-2} + x)}{(2.7 \times 10^{-2} - x)}; \text{ assume that } +x \text{ and } -x \text{ are negligible}$$

$[OH^-] = x = 9.7 \times 10^{-6} \text{ mol} \cdot L^{-1}$ and $pOH = 5.01$

Therefore, $pH = 14.00 - 5.01 = 8.99$

(c) At the stoichiometric point, moles of $NH_3$ = moles of HCl

$$\text{volume HCl added} = \frac{(0.15 \text{ mol} \cdot L^{-1} \text{ NH}_3)(0.0150 \text{ L})}{0.10 \text{ mol} \cdot L^{-1} \text{ HCl}} = 0.0225 \text{ L HCl}$$

Therefore, halfway to the stoichiometric point, volume HCl added

$= 22.5/2 = 11.25 \text{ mL}$

(d) At half stoichiometric point, $pOH = pK_b$ and $pOH = 4.75$

Therefore, $pH = 14.00 - 4.75 = 9.25$

(e)  22.5 mL;  see part (c)

(f)  $NH_4^+(aq) + H_2O(l) \leftrightarrow H_3O^+(aq) + NH_3(aq)$

The initial moles of $NH_3$ have now been converted to moles of $NH_4^+$ in a

(15 + 22.5 = 37.5) mL volume:

$$[NH_4^+] = \frac{2.25 \times 10^{-3} \text{ mol}}{0.0375 \text{ L}} = 0.060 \text{ mol} \cdot L^{-1}$$

$$K_a = \frac{K_w}{K_b} = \frac{1.00 \times 10^{-14}}{1.8 \times 10^{-5}} = 5.6 \times 10^{-10}$$

Concentration

| (mol·L$^{-1}$) | $NH_4^+(aq)$ | + | $H_2O(l)$ | $\leftrightarrow$ | $H_3O^+(aq)$ | + | $NH_3(aq)$ |
|---|---|---|---|---|---|---|---|
| initial | 0.060 | | — | | 0 | | 0 |
| change | $-x$ | | — | | $+x$ | | $+x$ |
| equilibrium | $0.060 - x$ | | — | | $x$ | | $x$ |

$$K_a = 5.6 \times 10^{-10} = \frac{x^2}{0.060 - x} \approx \frac{x^2}{0.060}$$

$$x = [H_3O^+] = 5.8 \times 10^{-6} \text{ mol} \cdot L^{-1}$$

$$pH = -\log(5.8 \times 10^{-6}) = 5.24$$

**11.35**  At the stoichiometric point, the volume of solution will have doubled;

therefore, the concentration of $CH_3CO_2^-$ will be 0.10 M. The equilibrium

is

Concentration

| (mol·L$^{-1}$) | $CH_3CO_2^-(aq)$ | + | $H_2O(l)$ | $\leftrightarrow$ | $CH_3COOH(aq)$ | + | $OH^-(aq)$ |
|---|---|---|---|---|---|---|---|
| initial | 0.10 | | — | | 0 | | 0 |
| change | $-x$ | | — | | $+x$ | | $+x$ |
| equilibrium | $0.10 - x$ | | — | | $x$ | | $x$ |

$$K_b = \frac{K_w}{K_a} = \frac{1.00 \times 10^{-14}}{1.8 \times 10^{-5}} = 5.6 \times 10^{-10}$$

$$K_b = \frac{[HCH_3CO_2][OH^-]}{[CH_3CO_2{}^-]} = \frac{x^2}{0.10-x} \approx \frac{x^2}{0.10} = 5.6 \times 10^{-10}$$

$$x = 7.5 \times 10^{-6}\ mol \cdot L^{-1} = [OH^-]$$

$$pOH = -\log(7.5 \times 10^{-6}) = 5.12,\ pH = 14.00 - 5.12 = 8.88$$

From Table 11.2, we see that this pH value lies within the range for thymol blue and phenolphthalein; so these indicators are suitable, while the others are not.

**11.37** Exercise 11.31: thymol blue or phenolphthalein; Exercise 11.33: methyl red or bromocresol green.

**11.39** (a) To reach the first stoichiometric point, we must add enough solution to neutralize one $H^+$ on the $H_3AsO_4$. To do this, we will require

$0.0750\ L \times 0.137\ mol \cdot L^{-1} = 0.0103\ mol\ OH^-$. The volume of base required will be given by the number of moles of base required, divided by the concentration of base solution:

$$\frac{0.0750\ L \times 0.137\ mol \cdot L^{-1}}{0.275\ mol \cdot L^{-1}} = 0.0374\ L\ or\ 37.4\ mL$$

(b) and (c) To reach the second stoichiometric point will require double the amount calculated in (a), or 74.8 mL, and the third stoichiometric point will be reached with three times the amount added in (a), or 112 mL.

**11.41** (a) The base $HPO_3{}^{2-}$ is the fully deprotonated form of phosphorous acid $H_3PO_3$ (the remaining H attached to P is not acidic). It will require an equal number of moles of $HNO_3$ to react with $HPO_3{}^{2-}$, in order to reach the first stoichiometric point (formation of $H_2PO_3^-$). The value will be

given by $\dfrac{0.0355 \text{ L} \times 0.158 \text{ mol} \cdot \text{L}^{-1}}{0.255 \text{ mol} \cdot \text{L}^{-1}} = 0.0220$ L or 22.0 mL

(b) To reach the second stoichiometric point would require double the amount of solution calculated in (a), or 44.0 mL.

**11.43** (a) This value is calculated as described in Example 10.12. First we calculate the molarity of the starting phosphorous acid solution:

$\dfrac{0.122 \text{ g}}{81.99 \text{ g} \cdot \text{mol}^{-1}} \bigg/ 0.0500 \text{ L} = 0.0298 \text{ mol} \cdot \text{L}^{-1}$. We then use the first acid

dissociation of phosphorous acid as the dominant equilibrium. The $K_{a1}$ is $1.0 \times 10^{-2}$. Let $H_2P$ represent the fully-protonated phosphorus acid.

Concentration

| (mol $\cdot$ L$^{-1}$) | $H_2P$(aq) + | $H_2O$(l) $\leftrightarrow$ | $HP^-$(aq) + | $H_3O^+$(aq) |
|---|---|---|---|---|
| initial | 0.0298 | — | 0 | 0 |
| change | $-x$ | — | $+x$ | $+x$ |
| final | $0.0298 - x$ | — | $+x$ | $+x$ |

$K_a = \dfrac{[H_3O^+][HP^-]}{[H_2P]} = 1.0 \times 10^{-2}$

$1.0 \times 10^{-2} = \dfrac{x \cdot x}{0.0298 - x} = \dfrac{x^2}{0.0298 - x}$

If we assume $x \ll 0.0298$, then the equation becomes

$x^2 = (1.0 \times 10^{-2})(0.0298) = 2.98 \times 10^{-4}$

$x = 1.73 \times 10^{-2}$

Because this value is more than 10% of 0.0400, the full quadratic solution should be undertaken. The equation is

$x^2 = (1.0 \times 10^{-2})(0.0298 - x)$ or

$x^2 + (1.0 \times 10^{-2}x) - (2.98 \times 10^{-4}) = 0$

Using the quadratic formula, we obtain $x = 0.013$.

$pH = 1.89$

(b) First, carry out the reaction between phosphorous acid and the strong base to completion:

$$H_2P(aq) + OH^-(aq) \longrightarrow HP^-(aq) + H_2O(l)$$

moles of $H_2P = (0.0298 \text{ mol} \cdot L^{-1})(0.0500 \text{ L}) = 1.49 \times 10^{-3}$ mol

moles of $OH^- = (0.00500 \text{ L})(0.175 \text{ mol} \cdot L^{-1}) = 8.76 \times 10^{-4}$ mol

$8.75 \times 10^{-3}$ mol $OH^-$ will react completely with $1.49 \times 10^{-3}$ mol $H_2P$ to give $8.75 \times 10^{-3}$ mol $HP^-$ with $6.1 \times 10^{-4}$ moles of $H_2P$ remaining.

$$[H_2P] = \frac{6.1 \times 10^{-4} \text{ mol}}{0.0565 \text{ L}} = 0.0109 \text{ mol} \cdot L^{-1}$$

$$[HP^-] = \frac{8.75 \times 10^{-3} \text{ mol}}{0.0565 \text{ L}} = 0.0155 \text{ mol} \cdot L^{-1}$$

Concentration

| $(\text{mol} \cdot L^{-1})$ | $H_2P(aq)$ | $+ \ H_2O(l)$ | $\leftrightarrow$ | $HP^-(aq)$ | $+$ | $H_3O^+(aq)$ |
|---|---|---|---|---|---|---|
| initial | 0.0109 | — | | 0.0155 | | 0 |
| change | $-x$ | — | | $+x$ | | $+x$ |
| final | $0.0109 - x$ | — | | $0.0155 + x$ | | $+x$ |

The calculation is performed as in part (a):

$$1.0 \times 10^{-2} = \frac{(0.0155 + x)x}{0.0109 - x}$$

$$x = 1.08 \times 10^{-2}$$

$$pH = 1.96$$

(c) moles of $H_2P = 1.49 \times 10^{-3}$

moles of $OH^- = 8.75 \times 10^{-4} + 8.75 \times 10^{-4} = 1.75 \times 10^{-3}$

Following the reaction between $H_2P$ and $OH^-$, 0 mol of $H_2P$ remain and $2.62 \times 10^{-4}$ mol $OH^-$ remain.

$[OH^-] = 4.37 \times 10^{-3}$ M, $\therefore$ $[H_3O^+] = 2.29 \times 10^{-12}$ M.

$1.49 \times 10^{-3}$ mol of $HP^-$ remain, $[HP^-] = \dfrac{1.49 \times 10^{-3} \text{ mol}}{0.060 \text{ L}} = 2.48 \times 10^{-2}$ M

Concentration

| (mol·L$^{-1}$) | HP$^-$ | + | H$_3$O$^+$ | $\leftrightarrow$ | H$_2$P | + | H$_2$O |
|---|---|---|---|---|---|---|---|
| initial | $2.48 \times 10^{-2}$ | | $2.29 \times 10^{-12}$ | | 0 | | — |
| change | $-x$ | | $-x$ | | $+x$ | | — |
| final | $2.48 \times 10^{-2} - x$ | | $2.29 \times 10^{-12} - x$ | | $x$ | | — |

The calculation is performed as in part (a):

$$100 = \frac{x}{\left(2.48 \times 10^{-2} - x\right)\left(2.29 \times 10^{-12} - x\right)}$$

Using the quadratic equation, we find $x = 5.6 \times 10^{-16}$,

$[H_3O] = 2.29 \times 10^{-12}$, and pH $= -\log(2.29 \times 10^{-12}) = 11.6$

**11.45** (a) The reaction of the base $Na_2HPO_4$ with the strong acid will be taken

to completion first:

$HPO_4^{2-}(aq) + H_3O^+(aq) \longrightarrow H_2PO_4^- + H_2O(l)$

Initially, moles of

$HPO_4^{2-} =$ moles of $H_3O^+ = 0.0500$ L $\times 0.275$ mol·L$^{-1} = 0.0138$ mol

Because this reaction proceeds with no excess base or acid, we are dealing

with a solution that can be viewed as being composed of $H_2PO_4^-$. The

problem then becomes one of estimating the pH of this solution, which

can be done from the relationship

pH $= \frac{1}{2}(pK_{a1} + pK_{a2})$

pH $= \frac{1}{2}(2.12 + 7.21) = 4.66$

(b) This reaction proceeds as in (a), but there is more strong acid

available, so the excess acid will react with $H_2PO_4^-$ to produce $H_3PO_4$.

Addition of the first 50.0 mL of acid solution will convert all the $HPO_4^{2-}$

into $H_2PO_4^-$. The additional 25.0 mL of the strong acid will react with

$H_2PO_4^-$:

$H_2PO_4^-(aq) + H_3O^+(aq) \longrightarrow H_3PO_4(aq) + H_2O(l)$

0.0138 mol $H_2PO_4^-$ will react with 0.006 88 mol $H_3O^+$ to give 0.0069

mol $H_3PO_4$ with 0.069 mol $H_2PO_4^-$ in excess. The concentrations will be

$[H_3PO_4] = [H_2PO_4^-] = \dfrac{0.0069 \text{ mol}}{0.125 \text{ L}} = 0.055.$ The appropriate relationship

to use is then

Concentration

$(mol \cdot L^{-1})$  $H_3PO_4(aq)$  +  $H_2O(l)$  $\leftrightarrow$  $H_2PO_4^-(aq)$  +  $H_3O^+(aq)$

| | | | | |
|---|---|---|---|---|
| initial | 0.055 | — | 0.055 | 0 |
| change | $-x$ | — | $+x$ | $+x$ |
| final | $0.055 - x$ | — | $0.055 + x$ | $+x$ |

$K_{a1} = \dfrac{[H_2PO_4^-][H_3O^+]}{[H_3PO_4]}$

Because the equilibrium constant is not small compared to 0.055, the full

quadratic solution must be calculated:

$x^2 + 0.055x = 7.6 \times 10^{-3}\,(0.055 - x)$

$x^2 + 0.063x - 4.2 \times 10^{-4} = 0$

$x = 1.6 \times 10^{-3}$

$pH = -\log(1.6 \times 10^{-3}) = 2.80$

(c) The reaction of $Na_2HPO_4$ with strong acid goes only halfway to

completion. 0.275 mol of $HPO_4^{2-}$ will react with

$(0.025 \text{ L} \times 0.275 \text{ mol} \cdot L^{-1}) = 6.9 \times 10^{-3}$ mol HCl to produce $6.9 \times 10^{-3}$ mol

$H_2PO_4^-$ and leave $6.9 \times 10^{-3}$ $HPO_4^{2-}$ unreacted.

$6.9 \times 10^{-3}$ mol $\div$ 0.075 L $= 0.092$ mol $\cdot L^{-1}$

Concentration

$(mol \cdot L^{-1})$  $H_2PO_4^-(aq)$  +  $H_2O(l)$ $\leftrightarrow$ $HPO_4^{2-}(aq)$  +  $H_3O^+(aq)$

| | | | | |
|---|---|---|---|---|
| initial | 0.092 | — | 0.092 | 0 |
| change | $-x$ | — | $+x$ | $+x$ |

| final | $0.092 - x$ | — | $0.092 + x$ | $+x$ |

$$K_{a2} = 6.2 \times 10^{-8} = \frac{[\text{HPO}_4{}^{2-}][\text{H}_3\text{O}^+]}{[\text{H}_2\text{PO}_4{}^-]}$$

$$\frac{[0.092 + x][\text{H}_3\text{O}^+]}{[0.092 - x]} = 6.2 \times 10^{-8}$$

assuming $x \ll$ than 0.092

$$x = [\text{H}_3\text{O}^+] = 6.2 \times 10^{-8}$$

$$\text{pH} = -\log(6.2 \times 10^{-8}) = 7.21$$

**11.47** (a) The solubility equilibrium is $\text{AgBr(s)} \leftrightarrow \text{Ag}^+(\text{aq}) + \text{Br}^-(\text{aq})$.

$$[\text{Ag}^+] = [\text{Br}^-] = 8.8 \times 10^{-7} \text{ mol} \cdot \text{L}^{-1} = S = \text{solubility}$$

$$K_{sp} = [\text{Ag}^+][\text{Br}^-] = (8.8 \times 10^{-7})(8.8 \times 10^{-7}) = 7.7 \times 10^{-13}$$

(b) The solubility equilibrium is $\text{PbCrO}_4(\text{s}) \leftrightarrow \text{Pb}^{2+}(\text{aq}) + \text{CrO}_4^{2-}(\text{aq})$.

$$[\text{Pb}^{2+}] = 1.3 \times 10^{-7} \text{ mol} \cdot \text{L}^{-1} = S, \quad [\text{CrO}_4{}^{2-}] = 1.3 \times 10^{-7} \text{ mol} \cdot \text{L}^{-1} = S$$

$$K_{sp} = [\text{Pb}^{2+}][\text{CrO}_4{}^{2-}] = (1.3 \times 10^{-7})(1.3 \times 10^{-7}) = 1.7 \times 10^{-14}$$

(c) The solubility equilibrium is $\text{Ba(OH)}_2(\text{s}) \leftrightarrow \text{Ba}^{2+}(\text{aq}) + 2\,\text{OH}^-(\text{aq})$.

$$[\text{Ba}^{2+}] = 0.11 \text{ mol} \cdot \text{L}^{-1} = S, \quad [\text{OH}^-] = 0.22 \text{ mol} \cdot \text{L}^{-1} = 2S$$

$$K_{sp} = [\text{Ba}^{2+}][\text{OH}^-]^2 = (0.11)(0.22)^2 = 5.3 \times 10^{-3}$$

(d) The solubility equilibrium is $\text{MgF}_2(\text{s}) \leftrightarrow \text{Mg}^{2+}(\text{aq}) + 2\,\text{F}^-(\text{aq})$.

$$[\text{Mg}^{2+}] = 1.2 \times 10^{-3} \text{ mol} \cdot \text{L}^{-1} = S, \quad [\text{F}^-] = 2.4 \times 10^{-3} \text{ mol} \cdot \text{L}^{-1} = 2S$$

$$K_{sp} = [\text{Mg}^{2+}][\text{F}^-]^2 = (1.2 \times 10^{-3})(2.4 \times 10^{-3})^2 = 6.9 \times 10^{-9}$$

**11.49** (a) Equilibrium equation: $\text{Ag}_2\text{S(s)} \leftrightarrow 2\,\text{Ag}^+(\text{aq}) + \text{S}^{2-}(\text{aq})$.

$$K_{sp} = [\text{Ag}^+]^2[\text{S}^{2-}] = (2S)^2(S) = 4S^3 = 6.3 \times 10^{-51}$$

$$S = 1.2 \times 10^{-17} \text{ mol} \cdot \text{L}^{-1}$$

(b) Equilibrium equation: $\text{CuS(s)} \leftrightarrow \text{Cu}^{2+}(\text{aq}) + \text{S}^{2-}(\text{aq})$.

$$K_{sp} = [Cu^{2+}][S^{2-}] = S \times S = S^2 = 1.3 \times 10^{-36}$$
$$S = 1.1 \times 10^{-18} \text{ mol} \cdot L^{-1}$$

(c) Equilibrium equation: $CaCO_3(s) \leftrightarrow Ca^{2+}(aq) + CO_3^{2-}(aq)$.

$$K_{sp} = [Ca^{2+}][CO_3{}^{2-}] = S \times S = S^2 = 8.7 \times 10^{-9}$$
$$S = 9.3 \times 10^{-5} \text{ mol} \cdot L^{-1}$$

**11.51**  $TlCrO_4(s) \leftrightarrow 2\,Tl^+(aq) + CrO_4^{2-}(aq)$.

$$[CrO_4{}^{2-}] = S = 6.3 \times 10^{-5} \text{ mol} \cdot L^{-1}$$
$$[Tl^+] = 2S = 2(6.3 \times 10^{-5}) \text{ mol} \cdot L^{-1}$$
$$K_{sp} = [Tl^+]^2[CrO_4{}^{2-}] = (2S)^2 \times (S)$$
$$K_{sp} = [2(6.3 \times 10^{-5})]^2 \times (6.3 \times 10^{-5}) = 1.0 \times 10^{-12}$$

**11.53**  (a)

| Concentration $(\text{mol} \cdot L^{-1})$ | $AgCl(s) \leftrightarrow$ | $Ag^+(aq)$ | $+$ | $Cl^-(aq)$ |
|---|---|---|---|---|
| initial | — | 0 | | 0.20 |
| change | — | $+S$ | | $+S$ |
| equilibrium | — | $S$ | | $S + 0.20$ |

$$K_{sp} = [Ag^+][Cl^-] = (S) \times (S + 0.20) = 1.6 \times 10^{-10}$$

Assume $S$ in $(S + 0.20)$ is negligible, so $0.20\,S = 1.6 \times 10^{-10}$

$S = 8.0 \times 10^{-10} \text{ mol} \cdot L^{-1} = [Ag^+] =$ molar solubility of AgCl in

0.15 M NaCl

(b)

| Concentration $(\text{mol} \cdot L^{-1})$ | $Hg_2Cl_2(s) \leftrightarrow$ | $Hg_2^{2+}(aq)$ | $+$ | $2\,Cl^-(aq)$ |
|---|---|---|---|---|
| initial | — | 0 | | 0.150 |
| change | — | $+S$ | | $+2S$ |
| equilibrium | — | $S$ | | $0.150 + 2S$ |

$$K_{sp} = [Hg_2^{2+}][Cl^-]^2 = (S) \times (2S + 0.150)^2 = 2.6 \times 10^{-18}$$

Assume $2S$ in $(2S + 0.150)$ is negligible, so $(0.150)^2 S = 2.6 \times 10^{-18}$

$S = 1.2 \times 10^{-16} \text{ mol} \cdot \text{L}^{-1} = [\text{Hg}_2^{2+}] = \text{molar solubility of } \text{Hg}_2\text{Cl}_2$

in $0.225$ M NaCl

(c) Concentration $(\text{mol} \cdot \text{L}^{-1})$ $\text{PbCl}_2(\text{s}) \leftrightarrow \text{Pb}^{2+}(\text{aq}) + 2\,\text{Cl}^-(\text{aq})$

| | | $\text{Pb}^{2+}$ | $\text{Cl}^-$ |
|---|---|---|---|
| initial | — | 0 | $2 \times 0.025 = 0.05$ |
| change | — | $+S$ | $+S$ |
| equilibrium | — | $S$ | $S + 0.05$ |

$K_{sp} = [\text{Pb}^{2+}][\text{Cl}^-]^2 = S \times (2S + 0.05)^2 = 1.6 \times 10^{-5}$

$S$ may not be negligible relative to 0.05, so the full cubic form may be required. We do it both ways:

For $S^3 + 0.20\,S^2 + (0.0025 \times 10^{-2}\,S) - (1.6 \times 10^{-5}) = 0$, the solution by standard methods is $S = 4.6 \times 10^{-3} \text{ mol} \cdot \text{L}^{-1}$.

If $S$ had been neglected, the answer would have been the same, $4.6 \times 10^{-3}$, to within two significant figures.

(d) Concentration $(\text{mol} \cdot \text{L}^{-1})$ $\text{Fe(OH)}_2(\text{s}) \leftrightarrow \text{Fe}^{2+}(\text{aq}) + 2\,\text{OH}^-(\text{aq})$

| | | $\text{Fe}^{2+}$ | $\text{OH}^-$ |
|---|---|---|---|
| initial | — | $2.5 \times 10^{-3}$ | 0 |
| change | — | $+S$ | $+2S$ |
| equilibrium | — | $2.5 \times 10^{-3} + S$ | $2S$ |

$K_{sp} = [\text{Fe}^{2+}][\text{OH}^-]^2 = (S + 2.5 \times 10^{-3}) \times (2S)^2 = 1.6 \times 10^{-14}$

Assume $S$ in $(S + 2.5 \times 10^{-3})$ is negligible, so

$4S^2 \times (2.5 \times 10^{-3}) = 1.6 \times 10^{-1}$.

$S^2 = 1.6 \times 10^{-12}$

$S = 1.3 \times 10^{-6} \text{ mol} \cdot \text{L}^{-1} = \text{molar solubility of Fe(OH)}_2 \text{ in } 2.5 \times 10^{-3}$ M $\text{FeCl}_2$

**11.55** (a) $\text{Ag}^+(\text{aq}) + \text{Cl}^-(\text{aq}) \leftrightarrow \text{AgCl(s)}$

| Concentration $(\text{mol} \cdot \text{L}^{-1})$ | $\text{Ag}^+$ | $\text{Cl}^-$ |
|---|---|---|
| initial | 0 | $1.0 \times 10^{-5}$ |
| change | $+x$ | 0 |

equilibrium $\qquad x \qquad 1.0 \times 10^{-5}$

$$K_{sp} = [Ag^+][Cl^-] = 1.6 \times 10^{-10} = (x)(1.0 \times 10^{-5})$$

$$x = [Ag^+] = 1.6 \times 10^{-5} \text{ mol} \cdot L^{-1}$$

(b) mass $AgNO_3$

$$= \left( \frac{1.6 \times 10^{-5} \text{ mol } AgNO_3}{1 \text{ L}} \right)(0.100 \text{ L}) \left( \frac{169.88 \text{ g } AgNO_3}{1 \text{ mol } AgNO_3} \right) \left( \frac{1 \mu g}{10^{-6} \text{ g}} \right)$$

$$= 2.7 \times 10^2 \ \mu g \ AgNO_3$$

**11.57** (a) $Ni^{2+}(aq) + 2 OH^-(aq) \leftrightarrow Ni(OH)_2(s)$

| Concentration $(\text{mol} \cdot L^{-1})$ | $Ni^{2+}$ | $OH^-$ |
|---|---|---|
| initial | 0.060 | 0 |
| change | 0 | $+x$ |
| equilibrium | 0.060 | $x$ |

$$K_{sp} = [Ni^{2+}][OH^-]^2 = 6.5 \times 10^{-18} = (0.060)(x)^2$$

$$[OH^-] = x = 1.0 \times 10^{-8} \text{ mol} \cdot L^{-1}$$

$$pOH = -\log(1.0 \times 10^{-8}) = 8.00, \ pH = 14.00 - 8.00 = 6.00$$

(b) A similar set up for $[Ni^{2+}] = 0.030 \text{ M}$

gives $x = 1.5 \times 10^{-8}$

$$pOH = -\log(1.5 \times 10^{-8}) = 7.82$$

$$pH = 14.00 - 7.82 = 6.18$$

**11.59** $\left( \dfrac{1 \text{ mL}}{20 \text{ drops}} \right) \times 1 \text{ drop} = 0.05 \text{ mL} = 0.05 \times 10^{-3} \text{ L} = 5 \times 10^{-5} \text{ L}$

and $(5 \times 10^{-5} \text{ L})(0.010 \text{ mol} \cdot L^{-1}) = 5 \times 10^{-7} \text{ mol NaCl} = 5 \times 10^{-7} \text{ mol Cl}^-$

(a) $Ag^+(aq) + Cl^-(aq) \leftrightarrow AgCl(s), \quad [Ag^+][Cl^-] = K_{sp}$

$$Q_{sp} = \left[ \frac{(0.010 \text{ L})(0.0040 \text{ mol} \cdot L^{-1})}{0.010 \text{ L}} \right] \left[ \frac{5 \times 10^{-7} \text{ mol}}{0.010 \text{ L}} \right] = 2 \times 10^{-7}$$

Will precipitate, because $Q_{sp} (2 \times 10^{-7}) > K_{sp} (1.6 \times 10^{-10})$

(b)  $Pb^{2+}(aq) + 2\,Cl^-(aq) \leftrightarrow PbCl_2(s)$,  $[Pb^{2+}][Cl^-]^2 = K_{sp}$

$$Q_{sp} = \left[\frac{(0.0100\ L)(0.0040\ mol\cdot L^{-1})}{0.010\ L}\right]\left[\frac{5\times10^{-7}\ mol}{0.010\ L}\right]^2 = 1\times10^{-11}$$

Will not precipitate, because $Q_{sp}\ (1\times10^{-11}) < K_{sp}\ (1.6\times10^{-5})$

**11.61** (a)  $K_{sp}[Ni(OH)_2] < K_{sp}[Mg(OH)_2] < K_{sp}[Ca(OH)_2]$

This is the order for the solubility products of these hydroxides. Thus, the order of precipitation is (first to last):  $Ni(OH)_2,\ Mg(OH)_2,\ Ca(OH)_2$.

(b)  $K_{sp}[Ni(OH)_2] = 6.5\times10^{-18} = [Ni^{2+}][OH^-]^2$

$$[OH^-]^2 = \frac{6.5\times10^{-18}}{0.0010} = 6.5\times10^{-15}$$
$$[OH^-] = 8.1\times10^{-8}$$
$pOH = -\log[OH^-] = 7.09\ \ pH \approx 7$
$K_{sp}[Mg(OH)_2] = 1.1\times10^{-11} = [Mg^{2+}][OH^-]^2$

$$[OH^-] = \sqrt{\frac{1.1\times10^{-11}}{0.0010}} = 1.0\times10^{-4}$$
$pOH = -\log(1.0\times10^{-4}) = 4.00\ \ pH = 14.00 - 4.00 = 10.00,\ pH \approx 10$
$K_{sp}[Ca(OH)^2] = 5.5\times10^{-6} = [Ca^{2+}][OH^-]^2$

$$[OH^-] = \sqrt{\frac{5.5\times10^{-6}}{0.0010}} = 7.4\times10^{-2}$$
$pOH = -\log(7.4\times10^{-2}) = 1.13\ \ pH = 14.00 - 1.13 = 12.87,\ pH \approx 13$

**11.63** The $K_{sp}$ values are  $MgF_2$      $6.4\times10^{-9}$

$BaF_2$      $1.7\times10^{-6}$

$MgCO_3$      $1.0\times10^{-5}$

$BaCO_3$      $8.1\times10^{-9}$

The difference in these numbers suggests that there is a greater solubility difference between the carbonates, and thus this anion should give a better separation. Because different numbers of ions are involved, it is instructive to convert the $K_{sp}$ values into molar solubility. For the fluorides the reaction is

$$MF_2(s) \leftrightarrow M^{2+}(aq) + 2\,F^-(aq)$$

Change $\qquad\qquad +x \qquad +2x$

$$K_{sp} = x(2x)^2$$

Solving this for $MgF_2$ gives 0.0012 M and for $BaF_2$ gives 0.0075 M.

For the carbonates:

$$MCO_3(s) \leftrightarrow M^{2+}(aq) + CO_3^{2-}(aq)$$

$$\qquad\qquad\qquad +x \qquad\quad +x$$

$$K_{sp} = x^2$$

Solving this for $MgCO_3$ gives 0.0032 M and for $BaCO_3$ gives $9.0 \times 10^{-5}$ M. Clearly, the solubility difference is greatest between the two carbonates, and $CO_3^{2-}$ is the better choice of anion.

**11.65** $Cu(IO_3)_2$ $(K_{sp} = 1.4 \times 10^{-7})$ is more soluble than $Pb(IO_3)_2$ $(K_{sp} = 2.6 \times 10^{-13})$ so $Cu^{2+}$ will remain in solution until essentially all the $Pb(IO_3)_2$ has precipitated. Thus, we expect very little $Pb^{2+}$ to be the left in solution by the time we reach the point at which $Cu(IO_3)_2$ begins to precipitate.

The concentration of $IO_3^-$ at which $Cu^{2+}$ begins to precipitate will be given by

$$K_{sp} = [Cu^{2+}][IO_3^-]^2 = 1.4 \times 10^{-7} = [0.0010][IO_3^-]^2$$
$$[IO_3^-] = 0.012 \text{ mol} \cdot L^{-1}$$

The concentration of Pb in solution when the $[IO_3^-] = 0.012 \text{ mol} \cdot L^{-1}$ is given by

$$K_{sp} = [Pb^{2+}][IO_3^-]^2 = 2.6 \times 10^{-13} = [Pb^{2+}][0.012]^2$$
$$[Pb^{2+}] = 1.8 \times 10^{-9} \text{ mol} \cdot L^{-1}$$

**11.67** (a) $pH = 7.0; [OH^-] = 1.0 \times 10^{-7} \text{ mol} \cdot L^{-1}$

$$Al^{3+}(aq) + 3\,OH^-(aq) \leftrightarrow Al(OH)_3(s)$$

$$[Al^{3+}][OH^-]^3 = K_{sp} = 1.0 \times 10^{-33}$$

$$S \times (10^{-7})^3 = 1.0 \times 10^{-33}$$

$$S = \frac{1.0 \times 10^{-33}}{1 \times 10^{-21}} = 1.0 \times 10^{-12} \text{ mol} \cdot L^{-1} = [Al^{3+}]$$

= molar solubility of $Al(OH)_3$ at pH = 7.0.

(b) $pH = 4.5; pOH = 9.5; [OH^-] = 3.2 \times 10^{-10} \text{ mol} \cdot L^{-1}$

$$[Al^{3+}][OH^-]^3 = K_{sp} = 1.0 \times 10^{-33}$$

$$S \times (3.2 \times 10^{-10})^3 = 1.0 \times 10^{-33}$$

$$S = \frac{1.0 \times 10^{-33}}{3.3 \times 10^{-29}} = 3.1 \times 10^{-5} \text{ mol} \cdot L^{-1} = [Al^{3+}]$$

= molar solubility of $Al(OH)_3$ at pH = 4.5

(c) $pH = 7.0; [OH^-] = 1.0 \times 10^{-7} \text{ mol} \cdot L^{-1}$

$$Zn^{2+}(aq) + 2\,OH^-(aq) \leftrightarrow Zn(OH)_2(s)$$

$$[Zn^{2+}][OH^-]^2 = K_{sp} = 2.0 \times 10^{-17}$$

$$S \times (1.0 \times 10^{-7})^2 = 2.0 \times 10^{-17}$$

$$S = \frac{2.0 \times 10^{-17}}{1.0 \times 10^{-14}} = 2.0 \times 10^{-3} \text{ mol} \cdot L^{-1} = [Zn^{2+}]$$

= molar solubility of $Zn(OH)_2$ at pH = 7.0

(d) $pH = 6.0; pOH = 8.0; [OH^-] = 1.0 \times 10^{-8} \text{ mol} \cdot L^{-1}$

$$[Zn^{2+}][OH^-]^2 = 2.0 \times 10^{-17} = K_{sp}$$

$$S \times (1.0 \times 10^{-8})^2 = 2.0 \times 10^{-17}$$

$$S = \frac{2.0 \times 10^{-17}}{1.0 \times 10^{-16}} = 2.0 \times 10^{-1} = 0.20 \text{ mol} \cdot L^{-1} = [Zn^{2+}]$$

= molar solubility of $Zn(OH)_2$ at pH = 6.0

**11.69** $\quad CaF_2(s) \leftrightarrow Ca^{2+}(aq) + 2\,F^-(aq) \qquad\qquad K_{sp} = 4.0 \times 10^{-11}$

$\quad F^-(aq) + H_2O(l) \leftrightarrow HF(aq) + OH^-(aq) \qquad K_b(F^-) = 2.9 \times 10^{-11}$

(a) Multiply the second equilibrium equation by 2 and add to the first equilibrium:

$$CaF_2(s) + 2 H_2O(l) \leftrightarrow Ca^{2+}(aq) + 2 HF(aq) + 2 OH^-(aq)$$

$$K = K_w \cdot K_b^2 = (4.0 \times 10^{-11})(2.9 \times 10^{-11})^2 = 3.4 \times 10^{-32}$$

(b) The calculation of $K_{sp}$ is complicated by the fact that the anion of the salt is part of a weak base-acid pair. If we wish to solve the equation algebraically, then we need to consider which equilibrium is the dominant one at pH = 7.0, for which $[H_3O^+] = 1 \times 10^{-7}$.

To determine whether $F^-$ or HF is the dominant species at this pH (if either), consider the base hydrolysis reaction:

$$F^-(aq) + H_2O(l) \leftrightarrow HF(aq) + OH^-(aq) \qquad K_b = 2.9 \times 10^{-11}$$

$$K_b = \frac{[HF][OH^-]}{[F^-]}$$

$$2.9 \times 10^{-11} = \frac{[HF][1 \times 10^{-7}]}{[F^-]}$$

$$\frac{[HF]}{[F^-]} = \frac{2.9 \times 10^{-11}}{1 \times 10^{-7}} = 3 \times 10^{-4}$$

Given that the ratio of HF to $F^-$ is on the order of $10^{-4}$ to 1, the $F^-$ species is still dominant. The appropriate equation to use is thus the original one for the $K_{sp}$ of $CaF_2(s)$:

$$CaF_2(s) \leftrightarrow Ca^{2+}(aq) + 2 HF(aq) \quad K_{sp} = 4.0 \times 10^{-11}$$

$$K_{sp} = [Ca^{2+}][F^-]^2$$

$$4.0 \times 10^{-11} = x(2x)^2 = 4x^3$$

$$x = 2.2 \times 10^{-4}$$

molar solubility = $2.2 \times 10^{-4}$ mol·L$^{-1}$

(c) At

pH = 3.0, $[H_3O^+] = 1 \times 10^{-3}$ mol·L$^{-1}$ and $[OH^-] = 1 \times 10^{-11}$ mol·L$^{-1}$.

Under these conditions

$$K_b = \frac{[HF][OH^-]}{[F^-]}$$

$$2.9 \times 10^{-11} = \frac{[HF][1 \times 10^{-11}]}{[F^-]}$$

$$\frac{[HF]}{[F^-]} = \frac{2.9 \times 10^{-11}}{1 \times 10^{-11}} = 3$$

$$[HF] = 3\,[F^-]$$

As can be seen, at pH 3.0 the amounts of $F^-$ and HF are comparable, so the protonation of $F^-$ to form HF cannot be ignored. The relation $2\,[Ca^{2+}] = [F^-] + [HF]$ is required by the mass balance as imposed by the stoichiometry of the dissolution equilibrium.

$$2\,[Ca^{2+}] = [F^-] + 3\,[F^-]$$
$$2\,[Ca^{2+}] = 4\,[F^-]$$
$$[Ca^{2+}] = 2\,[F^-]$$

Using this with $K_{sp}$ relationship:

$$(2\,[F^-])\,[F^-]^2 = 4.0 \times 10^{-11}$$
$$2\,[F^-]^3 = 4.0 \times 10^{-11}$$
$$[F^-] = 2 \times 10^{-4}$$
$$[Ca^{2+}] = (2)(2 \times 10^{-4}) = 4 \times 10^{-4}\,mol \cdot L^{-1}$$

The solubility is about double that at pH = 7.0.

**11.71**

$$AgBr(s) \leftrightarrow Ag^+(aq) + Br^-(aq) \qquad\qquad K_{sp} = 7.7 \times 10^{-13}$$
$$Ag^+(aq) + 2\,CN^-(aq) \leftrightarrow Ag(CN)_2^-(aq) \qquad\qquad K_f = 5.6 \times 10^8$$
$$AgBr(s) + 2\,CN^-(aq) \leftrightarrow Ag(CN)_2^-(aq) + Br^-(aq) \qquad K = 4.3 \times 10^{-4}$$

Hence, $K = \dfrac{[Ag(CN)_2{}^-][Br^-]}{[CN^-]^2} = 4.3 \times 10^{-4}$

Concentration

| $(mol \cdot L^{-1})$ | AgBr(s) | + 2 $CN^-$(aq) | $\leftrightarrow$ | $Ag(CN)_2^-$(aq) | + | $Br^-$(aq) |
|---|---|---|---|---|---|---|
| initial | — | 0.10 | | 0 | | 0 |

| change | — | $-2S$ | $+S$ | $+S$ |
| equilibrium | — | $0.10 - 2S$ | $S$ | $S$ |

$$\frac{[Ag(CN)_2^-][Br^-]}{[CN^-]^2} = \frac{S^2}{(0.10 - 2S)^2} = 4.3 \times 10^{-4}$$

$$\frac{S}{0.10 - 2S} = \sqrt{4.3 \times 10^{-4}} = 2.1 \times 10^{-2}$$

$$S = (2.1 \times 10^{-3}) - (4.2 \times 10^{-2} S)$$

$$1.042 S = 2.1 \times 10^{-3}$$

$$S = 2.0 \times 10^{-3} \text{ mol} \cdot \text{L}^{-1} = \text{molar solubility of AgBr}$$

**11.73** The two salts can be distinguished by their solubility in $NH_3$. The equilibria that are pertinent are

$$AgCl(s) + 2\,NH_3(aq) \leftrightarrow Ag(NH_3)_2^+(aq) + Cl^-(aq)$$
$$K = K_{sp} \cdot K_f = 2.6 \times 10^{-3}$$

$$AgI(s) + 2\,NH_3(aq) \leftrightarrow Ag(NH_3)_2^+(aq) + I^-(aq)$$

$$K = K_{sp} \cdot K_f = 2.4 \times 10^{-9}$$

For example, let's consider the solubility of these two salts in 1.00 M $NH_3$ solution:

For AgCl $\quad K = \dfrac{[Ag(NH_3)_2^+][Cl^-]}{[NH_3]^2} = 2.6 \times 10^{-3}$

Concentration

| (mol·L$^{-1}$) | AgCl(s) | + | $2\,NH_3$(aq) | $\leftrightarrow$ | $Ag(NH_3)_2^+$(aq) | + | $Cl^-$(aq) |
|---|---|---|---|---|---|---|---|
| initial | — | | 1.00 | | 0 | | 0 |
| change | — | | $-2x$ | | $+x$ | | $+x$ |
| final | — | | $1.00 - 2x$ | | $+x$ | | $+x$ |

$$K = \frac{[Ag(NH_3)_2{}^+][Cl^-]}{[NH_3]^2} = 2.6 \times 10^{-3}$$

$$2.6 \times 10^{-3} = \frac{[x][x]}{[1.00 - 2x]^2} = \frac{x^2}{[1.00 - 2x]^2}$$

$$0.051 = \frac{x}{1.00 - 2x}$$

$$x = 0.046$$

0.046 mol AgCl will dissolve in 1.00 L of aqueous solution. The molar mass of AgCl is

143.32 $g \cdot mol^{-1}$; this corresponds to 0.046

$mol \cdot L^{-1} \times 142.32 \ g \cdot mol^{-1} = 6.5 \ g \cdot L^{-1}$.

For AgI, the same calculation gives $x = 4.9 \times 10^{-5}$ $mol \cdot L^{-1}$. The molar mass of AgI is 234.77 $g \cdot mol^{-1}$, giving a solubility of

$4.9 \times 10^{-5}$ $mol \cdot L^{-1} \times 234.77 \ g \cdot mol^{-1} = 0.023 \ g \cdot L^{-1}$.

Thus, we could treat a 0.10 g sample of the compound with 20.0 mL of 1.00 M $NH_3$. The AgCl would all dissolve, whereas practically none of the AgI would.

Note: AgI is also slightly yellow in color, whereas AgCl is white, so an initial distinction could be made based upon the color of the sample.

**11.75** In order to use qualitative analyses, the sample must first be dissolved. This can be accomplished by digesting the sample with concentrated $HNO_3$ and then diluting the resulting solution. HCl or $H_2SO_4$ could not be used, because some of the metal compounds formed would be insoluble, whereas all of the nitrates would dissolve. Once the sample is dissolved and diluted, an aqueous solution containing chloride ions can be introduced. This should precipitate the $Ag^+$ as AgCl but would leave the bismuth and nickel in solution, as long as the solution was acidic. The remaining solution can then be treated with $H_2S$. In acidic solution, $Bi_2S_3$ will precipitate but NiS will not. Once the $Bi_2S_3$ has been precipitated, the pH of the solution can be raised by addition of base. Once this is done, NiS should precipitate.

**11.77** The suggested reaction is:

$$C_6H_5CH_2CH(CH_3)NH_3^+ + H_2O \leftrightarrow C_6H_5CH_2CH(CH_3)NH_2 + H_3O^+,$$

$$pK_a = pK_w - pK_b = 14.00 - 3.11 = 10.89$$

$$K_a = 10^{-10.89} = 1.3 \times 10^{-11} = \frac{[C_6H_5CH_2CH(CH_3)NH_2][H_3O^+]}{[C_6H_5CH_2CH(CH_3)NH_3^+]}$$

Substituting the given pH of 1.7, $[H_3O^+] = 10^{-1.7} = 2 \times 10^{-2}$ into the equation above, we obtain the ratio:

$$\frac{[C_6H_5CH_2CH(CH_3)NH_2]}{[C_6H_5CH_2CH(CH_3)NH_3^+]} = \frac{1.3 \times 10^{-11}}{2 \times 10^{-2}} = 6 \times 10^{-10}.$$

This small ratio reveals that the drug will remain as $C_6H_5CH_2CH(CH_3)NH_3^+$ at this pH of 1.7.

**11.79** The relation to use is $pH = pK_a + \log\dfrac{[\text{Base form}]}{[\text{Acid form}]}$. The $K_a$ value for acetic acid is $1.8 \times 10^{-5}$ and $pK_a = 4.74$. Because we are adding acid, the pH will fall upon the addition and we want the final pH to be no more than 0.20 pH units different from the initial pH, or 4.54.

$$4.54 = 4.74 + \log\frac{[\text{Base form}]}{[\text{Acid form}]}$$

$$-0.20 = \log\frac{[\text{Base form}]}{[\text{Acid form}]}$$

$$\frac{[\text{Base form}]}{[\text{Acid form}]} = 0.63$$

We want the concentration of the base form to be 0.63 times that of the acid form. We do not know the initial number of moles of base or acid forms, but we know that the two amounts were equal. Let C = initial number of moles of acetic acid and the initial number of moles of sodium acetate. The number of moles of $H_3O^+$ to be added (in the form of HCl(aq)) is $0.001\ 00\ L \times 6.00\ mol \cdot L^{-1} = 0.006\ 00\ mol$. The total final volume will be $0.1010\ L$.

$$\frac{\dfrac{C - 0.006\ 00}{0.1010\ L}}{\dfrac{C + 0.006\ 00}{0.1010\ L}} = 0.63$$

$$\frac{C - 0.006\ 00}{C + 0.006\ 00} = 0.63$$

$$C - 0.006\ 00 = 0.63(C + 0.006\ 00)$$

$$C - 0.006\ 00 = 0.63\ C + 0.003\ 78$$

$$0.37\ C = 0.009\ 78$$

$$C = 0.026$$

The initial buffer solution must contain at least 0.026 mol acetic acid and 0.026 mol sodium acetate. The concentration of the initial solution will then be $0.026\ mol \div 0.100\ L = 0.260\ M$ in both acetic acid and sodium acetate.

**11.81** It stands to reason that if the solution desired has a higher pH than the one available, we should add conjugate base (sodium benzoate). The solution available on the shelf has a $[A^-]/[HA]$ ratio of:

$$pH = 3.95 = pKa + \log\frac{[A^-]}{[HA]} = 4.19 + \log\frac{[A^-]}{[HA]}$$

$$\frac{[A^-]}{[HA]} = 10^{3.95-4.19} = 0.575$$

If $[A^-] = 0.200\ M$ as advertised on the label, then $[HA] = \dfrac{0.200\ M}{0.575} = 0.348\ M$

The desired solution will have a $[A^-]/[HA]$ ratio of:

$$pH = 4.35 = pKa + \log\frac{[A^-]}{[HA]} = 4.19 + \log\frac{[A^-]}{[HA]}$$

$$\frac{[A^-]}{[HA]} = 10^{4.35-4.19} = 1.45$$

To achieve this ratio, we must increase $[A^-]$. Fixing $[HA]$ at 0.348 M, the concentration of $[A^-]$ needed is: $[A^-] = 1.45 \times 0.348$ M $= 0.502$ M.

Therefore, $[A^-]$ needs to increase by 0.502 M – 0.200 M = 0.302 M.

To determine the number of grams of conjugate base needed:

(0.302 M)(0.100 L) = 0.0302 mol

$(0.0302 \text{ mol})(144.04 \text{ g} \cdot \text{mol}^{-1}) = 4.35$ g

**11.83** (a) $M_a = 0.0567$, $M_b = 0.0296$, $V_a = 15.0$, $V_b = 0.0$ to $50.0$

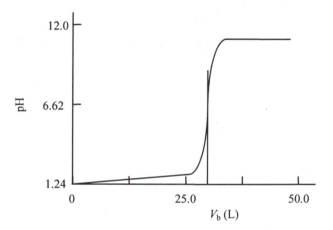

(b) 28.6 mL

(c) 1.24

(d) Because this is a titration of a strong acid with a strong base, the pH at the equivalence point will be 7.00.

**11.85** The strong acid, HCl, will protonate the $HCO_2^-$ ion.

moles of HCl $= 0.0040$ L $\times 0.070$ mol $\cdot$ L$^{-1} = 2.8 \times 10^{-4}$ mol HCl ($H^+$)

moles of $HCO_2^- = 0.0600$ L $\times 0.10$ mol $\cdot$ L$^{-1} = 6.0 \times 10^{-3}$ mol $HCO_2^-$

After protonation, there are

$(6.0 - 0.28) \times 10^{-3}$ mol $HCO_2^- = 5.7 \times 10^{-3}$ mol $HCO_2^-$

and $2.8 \times 10^{-4}$ mol HCOOH.

$$[HCO_2^-] = \frac{5.7 \times 10^{-3} \text{ mol}}{0.0640 \text{ L}} = 8.9 \times 10^{-2} \text{ mol} \cdot L^{-1}$$

$$[HCOOH] = \frac{2.8 \times 10^{-4} \text{ mol}}{0.0640 \text{ L}} = 4.4 \times 10^{-3} \text{ mol} \cdot L^{-1}$$

Concentration

| (mol·L$^{-1}$) | HCOOH(aq) | + | H$_2$O(l) | $\leftrightarrow$ | H$_3$O$^+$(aq) | + | HCO$_2^-$(aq) |
|---|---|---|---|---|---|---|---|
| initial | $4.4 \times 10^{-3}$ | | — | | 0 | | $8.9 \times 10^{-2}$ |
| change | $-x$ | | — | | $+x$ | | $+x$ |
| equilibrium | $4.4 \times 10^{-3} - x$ | | — | | $x$ | | $8.9 \times 10^{-2} + x$ |

$$K_a = 1.8 \times 10^{-4} = \frac{[H_3O^+][HCO_2^-]}{[HCOOH]} = \frac{(x)(8.9 \times 10^{-2} + x)}{(4.4 \times 10^{-3} - x)}$$

Assume that $+x$ and $-x$ are negligible; so $1.8 \times 10^{-4} = \dfrac{(8.9 \times 10^{-2}) \times (x)}{4.4 \times 10^{-3}}$

$$= 20.2x$$

$$[H_3O^+] = x = 8.9 \times 10^{-6} \text{ mol} \cdot L^{-1}$$

$$[HCOOH] = (4.4 \times 10^{-3}) - (8.9 \times 10^{-6}) \approx 4.4 \times 10^{-3} \text{ mol} \cdot L^{-1}$$

$$pH = -\log(8.9 \times 10^{-6}) = 5.05$$

**11.87** (a) The acidity constant may be found using:

$$K_a = e^{-\Delta G_r / R \cdot T}$$

$$\Delta G_r = -369.31 \text{ kJ} \cdot mol^{-1} - (-396.46 \text{ kJ} \cdot mol^{-1}) = 27.15 \text{ kJ} \cdot mol^{-1}$$

$$= 27150 \text{ J} \cdot mol^{-1}$$

$$K_a = e^{-27150 \text{ J·mol}^{-1} / (8.3145 \text{ J·K}^{-1} \cdot mol^{-1})(298 \text{ K})} = 1.74 \times 10^{-5}$$

(b) Given the acidity constant above, the mass of sodium acetate needed is:

$$pK_a = -\log(K_a) = 4.76$$

$$pH = pK_a + \log\frac{[A^-]}{[HA]} = 4.8 = 4.76 + \log\frac{[A^-]}{[HA]}$$

$$\frac{[A^-]}{[HA]} = 10^{4.8-4.76} = 1.10.$$

Given $[HA] = 1.0$ M, $[A^-] = 1.10 \times 1.0$ M $= 1.1$ M

and $(1.1 \text{ M})(2.5 \text{ L})(82.03 \text{ g} \cdot \text{mol}^{-1}) = 2.2 \times 10^2$ g.

**11.89** Let novocaine $= N$; $N(aq) + H_2O(l) \leftrightarrow HN^+(aq) + OH^-(aq)$

$$K_b = \frac{[HN^+][OH^-]}{[N]}$$

$$pK_a = pK_w - pK_b = 14.00 - 5.05 = 8.95$$

$$pH = pK_a + \log\left(\frac{[N]}{[HN^+]}\right)$$

$$\log\left(\frac{[N]}{[HN^+]}\right) = pH - pK_a = 7.4 - 8.95 = -1.55$$

Therefore, the ratio of the concentrations of novocaine and its conjugate

acid is $[N]/[HN^+] = 10^{-1.55} = 2.8 \times 10^{-2}$.

**11.91** $CH_3COOH(aq) + H_2O(l) \leftrightarrow H_3O^+(aq) + CH_3CO_2^-(aq)$

$$K_a = \frac{[H_3O^+][CH_3CO_2^-]}{[CH_3COOH]}$$

$$pH = pK_a + \log\frac{[CH_3CO_2^-]}{[CH_3COOH]} = pK_a + \log\left(\frac{0.50}{0.150}\right)$$

$$= 4.75 + 0.52 = 5.27 \text{ (initial pH)}$$

(a) $(0.0100 \text{ L})(1.2 \text{ mol} \cdot \text{L}^{-1}) = 1.2 \times 10^{-2}$ mol HCl added (a strong acid)

Produces $1.2 \times 10^{-2}$ mol $CH_3COOH$ from $CH_3CO_2^-$ after adding HCl:

$$[CH_3COOH] = \frac{(0.100 \text{ L})(0.150 \text{ mol} \cdot \text{L}^{-1})}{0.110 \text{ L}} + \frac{1.2 \times 10^{-2} \text{ mol}}{0.110 \text{ L}}$$

$$= 0.245 \text{ mol} \cdot \text{L}^{-1}$$

$$[CH_3CO_2^-] = \frac{(0.100 \text{ L})(0.50 \text{ mol} \cdot \text{L}^{-1})}{0.110 \text{ L}} - \frac{1.2 \times 10^{-2} \text{ mol}}{0.110 \text{ L}}$$

$$= 0.345 \text{ mol} \cdot \text{L}^{-1}$$

$$pH = 4.75 + \log\left(\frac{0.345}{0.245}\right) = 4.75 + 0.15 = 4.90 \text{ (after adding HCl)}$$

(b) $(0.0500 \text{ L})(0.095 \text{ mol} \cdot \text{L}^{-1}) = 4.7 \times 10^{-3}$ mol NaOH added (a strong base)

Produces $4.7 \times 10^{-3}$ mol $CH_3CO_2^-$ from $CH_3COOH$ after adding NaOH:

$$[CH_3COOH] = \frac{(0.100 \text{ L})(0.150 \text{ mol} \cdot \text{L}^{-1})}{0.1500 \text{ L}} - \frac{4.7 \times 10^{-3} \text{ mol}}{0.150 \text{ L}}$$

$$= 6.9 \times 10^{-2} \text{ mol} \cdot \text{L}^{-1}$$

$$[CH_3CO_2^-] = \frac{(0.100 \text{ L})(0.50 \text{ mol} \cdot \text{L}^{-1})}{0.150 \text{ L}} + \frac{4.7 \times 10^{-3} \text{ mol}}{0.150 \text{ L}}$$

$$= 3.6 \times 10^{-1} \text{ mol} \cdot \text{L}^{-1}$$

$$pH = 4.75 + \log\left(\frac{3.6 \times 10^{-1}}{6.9 \times 10^{-2}}\right) = 4.75 + 0.72 = 5.47 \text{ (after adding base)}$$

**11.93** $CaF_2(s) \leftrightarrow Ca^{2+}(aq) + 2 F^-(aq)$

$[Ca^{2+}][F^-]^2 = 4.0 \times 10^{-11}$

$[Ca^{2+}](5 \times 10^{-5})^2 = 4.0 \times 10^{-11}$  $[Ca^{2+}] = 0.016 \text{ mol} \cdot \text{L}^{-1}$

The maximum concentration of $[Ca^{2+}]$ allowed will be $0.016 \text{ mol} \cdot \text{L}^{-1}$.

**11.95** moles of $Ag^+ = (0.100 \text{ L})(1.0 \times 10^{-4} \text{ mol} \cdot \text{L}^{-1}) = 1.0 \times 10^{-5} \text{ mol } Ag^+$

moles of $CO_3^{2-} = (0.100 \text{ L})(1.0 \times 10^{-4} \text{ mol} \cdot \text{L}^{-1}) = 1.0 \times 10^{-5} \text{ mol } CO_3^{2-}$

$Ag_2CO_3(s) \leftrightarrow 2 Ag^+(aq) + CO_3^{2-}(aq)$

$[Ag^+]^2[CO_3^{2-}] = Q_{sp}$

$$\left(\frac{1.0 \times 10^{-5}}{0.200}\right)^2\left(\frac{1.0 \times 10^{-5}}{0.200}\right) = 1.2 \times 10^{-13} = Q_{sp}$$

Because $Q_{sp}(1.2 \times 10^{-13})$ is less than $K_{sp}(6.2 \times 10^{-12})$, no precipitate will form.

**11.97** In addition to the reaction corresponding to the dissolution of $PbF_2(s)$:

$$PbF_2(s) \leftrightarrow Pb^{2+}(aq) + 2\,F^-(aq) \qquad K = 3.7 \times 10^{-8}$$

The buffer will provide a source of $H_3O^+(aq)$ ions which will allow the reaction:

$$H_3O^+(aq) + F^- \leftrightarrow HF(aq) + H_2O(l) \qquad K = 1/(K_a(HF)) = 1/(3.5 \times 10^{-4}) = 2.86 \times 10^3$$

These two coupled reactions give two equilibrium expressions which must be simultaneously satisfied:

$$[F^-]^2[Pb^{2+}] = 1.7 \times 10^{-6} \qquad \text{and} \qquad \frac{[HF]}{[H_3O^+][F^-]} = 2.86 \times 10^{-3}$$

Given that all fluoride ions come from $PbF_2(s)$ and wind up as either $F^-(aq)$ or $HF(aq)$, and that for every one mole of $Pb^{2+}(aq)$ generated two moles of $F^-(aq)$ are also produced, we can write a third equation which relates the concentration of the fluoride containing species to the concentration of dissolved barium:

$$[Pb^{2+}] = \frac{1}{2}([F^-] + [HF]) .$$

In the end, the concentration of $Pb^{2+}(aq)$ will be equal to the solubility of $PbF_2(s)$. To determine the equilibrium concentration of $Pb^{2+}(aq)$, we first determine $[H_3O^+]$, which is fixed by the buffer system, and then use the three simultaneous equations above to solve for $[Pb^{2+}]_{eq}$.

The buffer determines the equilibrium concentration of $H_3O^+(aq)$. The initial concentration of $H_3O^+(aq)$ and $NaCH_3CO_2(aq)$ are:

$$[H_3O^+]_i = \frac{(0.055\ \text{L})(0.15\ \text{mol} \cdot \text{L}^{-1})}{0.10\ \text{L}} = 0.0825\ \text{M} \quad \text{and}$$

$$[CH_3CO_2^-]_i = \frac{(0.045\ \text{L})(0.65\ \text{mol} \cdot \text{L}^{-1})}{0.10\ \text{L}} = 0.293\ \text{M}.$$

To determine their equilibrium concentrations we solve using the familiar method:

Concentration

$$\text{(mol·L}^{-1}\text{)} \quad CH_3COOH(aq) \;+\; H_2O(l) \;\leftrightarrow\; H_3O^+(aq) \;+\; CH_3CO_3^-(aq)$$

| | | | |
|---|---|---|---|
| initial | 0 | — | 0.0825 | 0.293 |
| change | $+x$ | — | $-x$ | $-x$ |
| equilibrium | $x$ | — | $0.0825-x$ | $0.292-x$ |

$$K_a = 1.8\times10^{-5} = \frac{[H_3O^+][CH_3CO_2^-]}{[CH_3COOH]} = \frac{(0.0825-x)(0.292-x)}{(x)}$$

Rearranging this expression we obtain:

$$0.024072 - 0.3745\,x + x^2 = 0.$$

Using the quadratic formula we find $x = 0.082493$, and $[H_3O^+] = 7.1\times10^{-6}$ M

With this equilibrium concentration of $H_3O^+$(aq) we revisit the three simultaneous equations from above, namely

$$[F^-]^2[Pb] = 3.7\times10^{-8};$$

$$[Pb^{2+}] = \tfrac{1}{2}\big([F^-]+[HF]\big); \text{ and}$$

$$\frac{[HF]}{[H_3O^+][F^-]} = 2.86\times10^{-3}.$$

Due to the presence of the buffer, $[H_3O^+] = 7.1\times10^{-6}$ and this last equation simplifies to

$$\frac{[HF]}{[F^-]} = 2.03\times10^{-2}.$$

Rearranging these three simultaneous equations we find:

$$[Pb^{2+}] = \frac{3.7\times10^{-8}}{[F^-]^2}, \quad [HF] = [F^-]\times\big(2.03\times10^{-2}\big), \quad \text{and}$$

$$[Pb^{2+}] = \tfrac{1}{2}\big([F^-]+[HF]\big)$$

$$\frac{3.7\times10^{-8}}{[F^-]^2} = \tfrac{1}{2}\Big[[F^-]+\big([F^-]\times(2.03\times10^{-2})\big)\Big]. \quad \text{Solving this expression for } [F^-]:$$

$$[F^-] = \sqrt[3]{\frac{3.7\times10^{-8}}{0.5203}} = 4.14\times10^{-3}.$$

The equilibrium concentration of $Pb^{2+}$ (aq) is then:

$$[Pb^{2+}] = \frac{3.7 \times 10^{-8}}{[F^-]^2} = \frac{3.7 \times 10^{-8}}{(4.14 \times 10^{-3})^2} = 2.16 \times 10^{-3} \text{ M}$$

Therefore, the solubility of $PbF_2$ (s) is $2.2 \times 10^{-3}$ M

**11.99** The $K_{sp}$ values from Tables 11.5 and 11.7 are

$Cu^{2+}, 1.3 \times 10^{-36}; Co^{2+}, 5 \times 10^{-22}; Cd^{2+}, 4 \times 10^{-29}$. All the salts have the same expression for $K_{sp}$, $K_{sp} = [M^{2+}][S^{2-}]$, so the compound with the smallest $K_{sp}$ will precipitate first, in this case CuS.

**11.101** (1) first stoichiometric point:

$$OH^- (aq) + H_2SO_4 (aq) \longrightarrow H_2O(l) + HSO_4^- (aq)$$

Then, because the volume of the solution has doubled, $[HSO_4^-] = 0.10$ M,

Concentration

| (mol·L$^{-1}$) | $HSO_4^-$ (aq) + $H_2O$(l) | ↔ | $SO_4^{2-}$ (aq) | + $H_3O^+$ (aq) |
|---|---|---|---|---|
| initial | 0.10 | — | 0 | 0 |
| change | $-x$ | — | $+x$ | $+x$ |
| equilibrium | $0.10 - x$ | — | $x$ | $x$ |

$$K_{a2} = 0.012 = \frac{x^2}{0.10 - x}$$

$$0.0012 - 0.012x = x^2$$

$$x^2 + 0.012x - 0.0012 = 0$$

$$x = \frac{-0.012 + \sqrt{(0.012)^2 + (4)(0.0012)}}{2}$$

$$x = [H_3O^+] = 0.029 \text{ mol·L}^{-1}$$

$$pH = -\log(0.029) = 1.54$$

(2) second stoichiometric point:

$HSO_4^- (aq) + OH^- (aq) \longrightarrow SO_4^{2-} (aq) + H_2O(l)$. Because the volume of the solution has increased by an equal amount,

Concentration

| $(mol \cdot L^{-1})$ | $SO_4^{2-}(aq)$ | $+ H_2O(l)$ | $\leftrightarrow$ | $HSO_4^-(aq)$ | $+ OH^-(aq)$ |
|---|---|---|---|---|---|
| initial | 0.067 | — | | 0 | $1 \times 10^{-7}$ |
| change | $-x$ | — | | $+x$ | $+x$ |
| equilibrium | $0.067 - x$ | — | | $x$ | $1 \times 10^{-7} + x$ |

$$K_b = 8.3 \times 10^{-3} = \frac{(x)(1 \times 10^{-7} + x)}{0.067 - x}$$

$$(5.6 \times 10^{-14}) - (8.3 \times 10^{-13} x) = 1 \times 10^{-7} x + x^2$$

$$x^2 + (1 \times 10^{-7} x) - (5.6 \times 10^{-14}) = 0$$

$$x = \frac{-1 \times 10^{-7} + \sqrt{(1 \times 10^{-7})^2 + (4)(5.6 \times 10^{-14})}}{2}$$

$$x = [HSO_4^-] = 1.9 \times 10^{-7} \, mol \cdot L^{-1}$$

$$[OH^-] = (1.0 \times 10^{-7}) + (1.9 \times 10^{-7}) = 2.9 \times 10^{-7} \, mol \cdot L^{-1}$$

$$pOH = -\log(2.9 \times 10^{-7}) = 6.54, \quad pH = 14.00 - 6.54 = 7.46$$

**11.103** (a)   $PbCl_2(s) \leftrightarrow Pb^{2+}(aq) + 2\,Cl^-(aq)$

$$[Pb^{2+}][Cl^-]^2 = K_{sp} = 1.6 \times 10^{-5}$$

$$[0.010][Cl^-]^2 = 1.6 \times 10^{-5}$$

$$[Cl^-]^2 = 1.6 \times 10^{-3}$$

$[Cl^-] = 4.0 \times 10^{-2} \, mol \cdot L^{-1}$; will precipitate lead (II) ion

$Ag^+(aq) + Cl^-(aq) \rightleftharpoons AgCl(s)$

$$[Ag^+][Cl^-] = K_{sp} = 1.6 \times 10^{-10}$$

$$[0.010][Cl^-] = 1.6 \times 10^{-10}$$

$[Cl^-] = 1.6 \times 10^{-8} \, mol \cdot L^{-1}$; will precipitate $Ag^+$

(b)  From part (a), $Ag^+$ will precipitate first.

(c)  $[Ag^+](4.0 \times 10^{-2}) = 1.6 \times 10^{-10}$

$$[Ag^+] = \frac{1.6 \times 10^{-10}}{4.0 \times 10^{-2}} = 4.0 \times 10^{-9} \, mol \cdot L^{-1}$$

(d) percentage $Ag^+$ remaining $= \dfrac{4.0 \times 10^{-9}}{0.010} \times 10^2 = 4.0 \times 10^{-5}$ %

unprecipitated; virtually 100% of the first cation ($Ag^+$) is precipitated

**11.105** The $K_{sp}$ value for $PbF_2$ obtained from Table 11.5 is $3.7 \times 10^{-8}$. Using this value, the $\Delta G°$ of the dissolution reaction can be obtained from $\Delta G° = -RT \ln K$.

$\Delta G° = -(8.314 \text{ J} \cdot \text{K}^{-1} \cdot \text{mol}^{-1})(298.2 \text{ K}) \ln (3.7 \times 10^{-8})$

$\Delta G° = +42.43 \text{ kJ} \cdot \text{mol}^{-1}$

From the Appendices we find that $\Delta G°_f (F^-, aq) = -278.79 \text{ kJ} \cdot \text{mol}^{-1}$ and $\Delta G°_f (Pb^{2+}, aq) = -24.43 \text{ kJ} \cdot \text{mol}^{-1}$.

$\Delta G° = +42.43 \text{ kJ} \cdot \text{mol}^{-1} = \Delta G°_f (Pb^{2+}, aq) + \Delta G°_f (F^-, aq) - \Delta G°_f (PbF_2, s)$

$+42.43 \text{ kJ} \cdot \text{mol}^{-1} = (-24.43 \text{ kJ} \cdot \text{mol}^{-1}) + (-278.79 \text{ kJ} \cdot \text{mol}^{-1})$

$- \Delta G°_f (PbF_2, s)$

$\Delta G°_f (PbF_2, s) = -345.65 \text{ kJ} \cdot \text{mol}^{-1}$

**11.107** (a) The dissolution reaction for calcium oxalate is:

$CaC_2O_4(s) \leftrightarrow Ca^{2+}(aq) + C_2O_4^{2-}(aq)$

$K_{sp} = 10^{-pK_{sp}} = 10^{-8.59} = 2.57 \times 10^{-9}$

The concentration of each ion at equilibrium is equal to the molar solubility, $s$. Therefore,

$K_{sp} = [Ca^{2+}][C_2O_4^{2-}] = s \times s = 2.57 \times 10^{-9}$, and $s = 5.07 \times 10^{-5} \text{ mol} \cdot \text{L}^{-1}$.

(b) Given $[Mg^{2+}] = 0.020$ M and $[C_2O_4^{2-}] = 0.035$ M:

$Q_{sp} = [Mg^{2+}][C_2O_4^{2-}] = (0.020)(0.035) = 7 \times 10^{-4}$

$Q_{sp}$ is significantly less than $K_{sp}$, which is given as $10^{-4.07} = 8.51 \times 10^{-5}$, indicating that a precipitate will form.

**11.109** (a) The amount of $CO_2$ present at equilibrium may be found using the equilibrium expression for the reaction of interest:

$$H_3O^+(aq) + HCO_3^-(aq) \leftrightarrow 2\,H_2O(l) + CO_2(aq)$$

$$K = 7.9 \times 10^{-7} = \frac{[CO_2]}{[H_3O^+][HCO_3^-]}$$

Solving for $[CO_2]$:

$$[CO_2] = (7.9 \times 10^{-7})[H_3O^+][HCO_3^-]$$

Given: $[H_3O^+] = 10^{-6.1} = 7.9 \times 10^{-7}$ M and $[HCO_3^-] = 5.5\ \mu mol \cdot L^{-1} = 5.5 \times 10^{-6}$ M

$$[CO_2] = (7.9 \times 10^{-7})(7.9 \times 10^{-7})(5.5 \times 10^{-6}) = 3.5 \times 10^{-18}\ mol \cdot L^{-1}$$

In 1.0 L of solution there will be $3.5 \times 10^{-18}$ mol of $CO_2$ (aq).

(b) Adding $0.65 \times 10^{-6}$ mol of $H_3O^+$(aq) to the equilibrium system in (a) will give an initial $[H_3O^+]$ of $1.44 \times 10^{-6}$ M. To determine the equilibrium concentration of $H_3O^+$(aq) we set up the familiar problem:

Concentration

| (mol $\cdot$ L$^{-1}$) | $H_3O^+$(aq) | + | $HCO_3^-$(aq) | $\leftrightarrow$ | $2\,H_2O$(l) | + | $CO_2$(aq) |
|---|---|---|---|---|---|---|---|
| initial | $1.44 \times 10^{-6}$ | | $5.5 \times 10^{-6}$ | | — | | $3.45 \times 10^{-18}$ |
| change | $-x$ | | $-x$ | | — | | $+x$ |
| equilibrium | $1.44 \times 10^{-6} - x$ | | $5.5 \times 10^{-6} - x$ | | — | | $3.45 \times 10^{-18} + x$ |

$$K = 7.9 \times 10^{-7} = \frac{\left(3.45 \times 10^{-18} + x\right)}{\left(1.44 \times 10^{-6} - x\right)\left(5.5 \times 10^{-6} - x\right)}$$

Rearranging to obtain a polynomial in $x$:

$$2.81 \times 10^{-18} - x + 7.9 \times 10^{-7} x^2 = 0$$

Using the quadratic formula one finds:

$$x = 2.81 \times 10^{-18}$$

Giving: $[H_3O^+] = 1.4 \times 10^{-6}$ M and pH = 5.8. The pH decreases by 0.3.

# CHAPTER 12
# ELECTROCHEMISTRY

**12.1** (a) Cr reduced from 6+ to 3+; C oxidized from 2- to 1-;

(b) $C_2H_5OH(aq) \rightarrow C_2H_4O(aq) + 2\,H^+(aq) + 2\,e^-$;

(c) $Cr_2O_7{}^{2-}(aq) + 14\,H^+(aq) + 6\,e^- \rightarrow 2\,Cr^{3+}(aq) + 7\,H_2O(l)$;

(d) $8\,H^+(aq) + Cr_2O_7{}^{2-}(aq) + 3\,C_2H_5OH(aq) \rightarrow$
$\qquad 2\,Cr^{3+}(aq) + 3\,C_2H_4O(aq) + 7\,H_2O(l)$

**12.3** In each case, first obtain the balanced half-reactions. Multiply the oxidation and reduction half-reactions by appropriate factors that will result in the same number of electrons being present in both half-reactions. Then add the half-reactions, canceling electrons in the process, to obtain the balanced equation for the whole reaction. Check to see that the final equation is balanced.

(a) $4[Cl_2(g) + 2\,e^- \rightarrow 2\,Cl^-(aq)]$

$1[S_2O_3{}^{2-}(aq) + 5\,H_2O(l) \rightarrow 2\,SO_4{}^{2-}(aq) + 10\,H^+(aq) + 8\,e^-]$
$4\,Cl_2(g) + S_2O_3{}^{2-}(aq) + 5\,H_2O(l) + 8\,e^- \rightarrow$
$\quad 8\,Cl^-(aq) + 2\,SO_4{}^{2-}(aq) + 10\,H^+(aq) + 8\,e^-$
$4\,Cl_2(g) + S_2O_3{}^{2-}(aq) + 5\,H_2O(l) \rightarrow 8\,Cl^-(aq) + 2\,SO_4{}^{2-}(aq) + 10\,H^+(aq)$
$Cl_2$ is the oxidizing agent and $S_2O_3{}^{2-}$ is the reducing agent.

(b) $2[MnO_4{}^-(aq) + 8\,H^+(aq) + 5\,e^- \rightarrow Mn^{2+}(aq) + 4\,H_2O(l)]$

$$5[H_2SO_3(aq) + H_2O(l) \rightarrow HSO_4^-(aq) + 3\,H^+(aq) + 2\,e^-]$$

$$2\,MnO_4^-(aq) + 16\,H^+(aq) + 5\,H_2SO_3(aq) + 5\,H_2O(l) + 10\,e^- \rightarrow$$
$$2\,Mn^{2+}(aq) + 8\,H_2O(l) + 5\,HSO_4^-(aq) + 15\,H^+(aq) + 10\,e^-$$

$$2\,MnO_4^-(aq) + H^+(aq) + 5\,H_2SO_3(aq) \rightarrow$$
$$2\,Mn^{2+}(aq) + 3\,H_2O(l) + 5\,HSO_4^-(aq)$$

$MnO_4^-$ is the oxidizing agent and $H_2SO_3$ is the reducing agent.

(c)  $Cl_2(g) + 2\,e^- \rightarrow 2\,Cl^-(aq)$

$$H_2S(aq) \rightarrow S(s) + 2\,H^+(aq) + 2\,e^-$$
$$Cl_2(g) + H_2S(aq) + 2\,e^- \rightarrow 2\,Cl^-(aq) + S(s) + 2\,H^+(aq) + 2\,e^-$$
$$Cl_2(g) + H_2S(aq) \rightarrow 2\,Cl^-(aq) + S(s) + 2\,H^+(aq)$$

$Cl_2$ is the oxidizing agent and $H_2S$ is the reducing agent.

(d)  $Cl_2(g) + 2\,e^- \rightarrow 2\,Cl^-(aq)$

$$2\,H_2O(l) + Cl_2(g) \rightarrow 2\,HOCl(aq) + 2\,H^+(aq) + 2\,e^-$$
$$2\,H_2O(l) + 2\,Cl_2(g) + 2\,e^- \rightarrow 2\,HOCl(aq) + 2\,H^+(aq) + 2\,Cl^-(aq) + 2\,e^-$$

or  $H_2O(l) + Cl_2(g) \rightarrow HOCl(aq) + H^+(aq) + Cl^-(aq)$

$Cl_2$ is both the oxidizing and the reducing agent.

12.5  (a)  $O_3(g) \rightarrow O_2(g)$

$O_3(g) \rightarrow O_2(g) + H_2O(l)$     (balances O's)

$2\,H_2O(l) + O_3(g) \rightarrow O_2(g) + H_2O(l) + 2\,OH^-(aq)$     (balances H's)

$H_2O(l) + O_3(g) \rightarrow O_2(g) + 2\,OH^-(aq)$     (cancels $H_2O$)

$H_2O(l) + O_3(g) + 2\,e^- \rightarrow O_2(g) + 2\,OH^-(aq)$     (balances charge);

$Br^-(aq) \rightarrow BrO_3^-(aq)$

$3\,H_2O(l) + Br^-(aq) \rightarrow BrO_3^-(aq)$     (balances O's)

$6\,OH^-(aq) + 3\,H_2O(l) + Br^-(aq) \rightarrow BrO_3^-(aq) + 6\,H_2O(l)$   (balances H's)

$6 OH^-(aq) + 3 H_2O(l) + Br^-(aq) \rightarrow BrO_3^-(aq) + 6 H_2O(l) + 6 e^-$

(balances charge)

Combining half-reactions yields

$3[H_2O(l) + O_3(g) + 2 e^- \rightarrow O_2(g) + 2 OH^-(aq)]$

$6 OH^-(aq) + 3 H_2O(l) + Br^-(aq) \rightarrow BrO_3^-(aq) + 6 H_2O(l) + 6 e^-$

$6 H_2O(l) + 3 O_3(g) + 6 OH^-(aq) + Br^-(aq) + 6 e^- \rightarrow$
$$3 O_2(g) + 6 OH^-(aq) + BrO_3^-(aq) + 6 H_2O(l) + 6 e^-$$

and $3 O_3(g) + Br^-(aq) \rightarrow 3 O_2(g) + BrO_3^-(aq)$

$O_3$ is the oxidizing agent and $Br^-$ is the reducing agent.

(b) $Br_2(l) + 2 e^- \rightarrow 2 Br^-(aq)$      (balanced reduction half-reaction)

$Br_2(l) + 6 H_2O(l) \rightarrow 2 BrO_3^-(aq)$      (O's balanced); then

$Br_2(l) + 6 H_2O(l) + 12 OH^-(aq) \rightarrow 2 BrO_3^-(aq) + 12 H_2O(l)$  (H's balanced);

and $Br_2(l) + 12 OH^-(aq) \rightarrow 2 BrO_3^-(aq) + 6 H_2O(l) + 10 e^-$    (electrons balanced)

Combining half-reactions yields

$5[Br_2(l) + 2 e^- \rightarrow 2 Br^-(aq)]$

$1[Br_2(l) + 12 OH^-(aq) \rightarrow 2 BrO_3^-(aq) + 6 H_2O(l) + 10 e^-]$

$6 Br_2(l) + 12 OH^-(aq) + 10 e^- \rightarrow$
   $10 Br^-(aq) + 2 BrO_3^-(aq) + 6 H_2O(l) + 10 e^-$

$6 Br_2(l) + 12 OH^-(aq) \rightarrow 10 Br^-(aq) + 2 BrO_3^-(aq) + 6 H_2O(l)$

Dividing by 2 gives

$3 Br_2(l) + 6 OH^-(aq) \rightarrow 5 Br^-(aq) + BrO_3^-(aq) + 3 H_2O(l)$

$Br_2$ is both the oxidizing agent and the reducing agent.

(c) $Cr^{3+}(aq) + 4 H_2O(l) \rightarrow CrO_4^{2-}(aq)$      (O's balanced); then

$Cr^{3+}(aq) + 4 H_2O(l) + 8 OH^-(aq) \rightarrow CrO_4^{2-}(aq) + 8 H_2O(l)$   (H's balanced); and

$Cr^{3+}(aq) + 8 OH^-(aq) \rightarrow CrO_4^{2-}(aq) + 4 H_2O(l) + 3 e^-$      (charge balanced)

$MnO_2(s) \rightarrow Mn^{2+}(aq) + 2 H_2O(l)$;  then

$MnO_2(s) + 4 H_2O(l) \rightarrow Mn^{2+}(aq) + 2 H_2O(l) + 4 OH^-(aq)$  (H's balanced); and

$MnO_2(s) + 2 H_2O(l) + 2 e^- \rightarrow Mn^{2+}(aq) + 4 OH^-(aq)$  (charge balanced)

Combining half-reactions yields

$2[Cr^{3+}(aq) + 8 OH^-(aq) \rightarrow CrO_4^{2-}(aq) + 4 H_2O(l) + 3 e^-]$

$3[MnO_2(s) + 2 H_2O(l) + 2 e^- \rightarrow Mn^{2+}(aq) + 4 OH^-(aq)]$

$2 Cr^{3+}(aq) + 16 OH^-(aq) + 3 MnO_2(s) + 6 H_2O(l) + 6 e^- \rightarrow$
$\qquad 2 CrO_4^{2-}(aq) + 8 H_2O(l) + 3 Mn^{2+}(aq) + 12 OH^-(aq) + 6 e^-$

$2 Cr^{3+}(aq) + 4 OH^-(aq) + 3 MnO_2(s) \rightarrow$
$\qquad 2 CrO_4^{2-}(aq) + 2 H_2O(l) + 3 Mn^{2+}(aq)$

$Cr^{3+}$ is the reducing agent and $MnO_2$ is the oxidizing agent.

(d)  $3[P_4(s) + 8 OH^-(aq) \rightarrow 4 H_2PO_2^-(aq) + 4 e^-]$

$P_4(s) + 12 H_2O(l) + 12 e^- \rightarrow 4 PH_3(g) + 12 OH^-(aq)$

$4 P_4(s) + 12 H_2O(l) + 24 OH^-(aq) + 12 e^- \rightarrow$
$\qquad 12 H_2PO_2^-(aq) + 4 PH_3(g) + 12 OH^-(aq) + 12 e^-$

$4 P_4(s) + 12 H_2O(l) + 12 OH^-(aq) \rightarrow 12 H_2PO_2^-(aq) + 4 PH_3(g)$

or $P_4(s) + 3 H_2O(l) + 3 OH^-(aq) \rightarrow 3 H_2PO_2^-(aq) + PH_3(g)$

$P_4(s)$ is both the oxidizing and the reducing agent.


**12.7**  $P_4S_3(aq) \rightarrow H_3PO_4(aq) + SO_4^{2-}(aq)$

For the oxidation of $P_4S_3$, both the P and S atoms are oxidized. The assignment of oxidation states to the P and S atoms is complicated by the presence of P—P bonds in the molecule, which leads to non-integral values. As long as we are consistent in our assignments, the end result should be the same. We will assume that S in $P_4S_3$ is 2– and, therefore, loses 8 electrons on going to $S^{6+}$ in the sulfate ion. Because $P_4S_3$ is a neutral molecule and, if S has an oxidation number of −2, then each phosphorus atom will have an oxidation number of +1.5. Phosphorus in phosphoric acid has an oxidation number of +5. so each P atom of $P_4S_3$ must lose 3.5 electrons. The total number of electrons lost is $(4 \times 3.5) + (3 \times 8) = 38$.

$$P_4S_3(aq) \rightarrow 4\,H_3PO_4(aq) + 3\,SO_4{}^{2-}(aq) + 38\,e^-$$

We balance the charge by adding $H^+$ in an acidic solution:

$$P_4S_3(aq) \rightarrow 4\,H_3PO_4(aq) + 3\,SO_4{}^{2-}(aq) + 44\,H^+(aq) + 38\,e^-$$

The final balance is achieved by adding water to provide the oxygen and hydrogen atoms:

$$P_4S_3(aq) + 28\,H_2O(l) \rightarrow 4\,H_3PO_4(aq) + 3\,SO_4{}^{2-}(aq) + 44\,H^+(aq) + 38\,e^-$$

The other half-reaction is simpler.

$$NO_3{}^-(aq) \rightarrow NO(g)$$

N has an oxidation number of +5 in the nitrate ion and +2 in nitric oxide. Each nitrogen atom gains three electrons in the course of the reaction.

$$NO_3{}^-(aq) + 3\,e^- \rightarrow NO(g)$$

Charge balance is again achieved by adding $H^+$:

$$NO_3{}^-(aq) + 4\,H^+(aq) + 3\,e^- \rightarrow NO(g)$$

The number of hydrogen and oxygen atoms is completed by the addition of water:

$$NO_3{}^-(aq) + 4\,H^+(aq) + 3\,e^- \rightarrow NO(aq) + 2\,H_2O(l)$$

Combining the two half-reactions gives

$$38\,[NO_3{}^-(aq) + 4\,H^+(aq) + 3\,e^- \rightarrow NO(g) + 2\,H_2O(l)]$$
$$+3\,[P_4S_3(aq) + 28\,H_2O(l) \rightarrow$$
$$4\,H_3PO_4(aq) + 3\,SO_4{}^{2-}(aq) + 44\,H^+(aq) + 38\,e^-\,]$$
$$3\,P_4S_3(aq) + 38\,NO_3{}^-(aq) + 20\,H^+(aq) + 8\,H_2O(l) \rightarrow$$
$$12\,H_3PO_4(aq) + 9\,SO_4{}^{2-}(aq) + 38\,NO(g)$$

**12.9**  (a)  $Ag^+(aq) + e^- \rightarrow Ag(s)$    $E°(\text{cathode}) = +0.80$ V

$Ni^{2+}(aq) + 2 e^- \rightarrow Ni(s)$    $E°(\text{anode}) = -0.23$ V

Reversing the anode half-reaction yields

$Ni(s) \rightarrow Ni^{2+}(aq) + 2 e^-$

and the cell reaction is, upon addition of the half-reactions,

$2 Ag^+(aq) + Ni(s) \rightarrow 2 Ag(s) + Ni^{2+}(aq)$    $E°_{\text{cell}} = +0.80$ V $- (-0.23)$ V
$$= +1.03 \text{ V}$$

(b)  $2 H^+(aq) + 2 e^- \rightarrow H_2(g)$    $E°(\text{anode}) = 0.00$ V

$Cl_2(g) + 2 e^- \rightarrow 2 Cl^-(aq)$    $E°(\text{cathode}) = +1.36$ V

Therefore, at the anode, after reversal,

$H_2(g) \rightarrow 2 H^+(aq) + 2 e^-$

and, the cell reaction is, upon addition of the half-reactions,

$Cl_2(g) + H_2(g) \rightarrow 2 H^+(aq) + 2 Cl^-(aq)$    $E°_{\text{cell}} = +1.36$ V $- 0.00$ V
$$= +1.36 \text{ V}$$

(c)  $Cu^{2+}(aq) + 2 e^- \rightarrow Cu(s)$    $E°(\text{anode}) = +0.34$ V

$Ce^{4+}(aq) + e^- \rightarrow Ce^{3+}(aq)$    $E°(\text{cathode}) = +1.61$ V

Therefore, at the anode, after reversal,

$Cu(s) \rightarrow Cu^{2+}(aq) + 2 e^-$

and, the cell reaction is, upon addition of the half-reactions,

$2 Ce^{4+}(aq) + Cu(s) \rightarrow Cu^{2+}(aq) + 2 Ce^{3+}(aq)$    $E°_{\text{cell}} = 1.61$ V $- (0.34$ V$)$
$$= +1.27 \text{ V}$$

(d)  $O_2(g) + 2 H_2O(l) + 4 e^- \rightarrow 4 OH^-(aq)$    $E°(\text{cathode}) = 0.40$ V

$O_2(g) + 4 H^+(aq) + 4 e^- \rightarrow 2 H_2O(l)$    $E°(\text{anode}) = 1.23$ V

Reversing the anode half-reaction yields

$2 H_2O(l) \rightarrow O_2(g) + 4 H^+(aq) + 4 e^-$

and the cell reaction is, upon addition of the half-reactions,

$4 H_2O(l) \rightarrow 4 H^+(aq) + 4 OH^-(aq)$    $E°_{\text{cell}} = 0.40$ V $- 1.23$ V $= -0.83$ V

or, $H_2O(l) \rightarrow H^+(aq) + OH^-(aq)$

Note: This balanced equation corresponds to the cell notation given. The spontaneous process is the reverse of this reaction.

(e) $Sn^{4+}(aq) + 2 e^- \rightarrow Sn^{2+}(aq)$ $\quad\quad\quad E°(\text{anode}) = +0.15$ V

$Hg_2Cl_2(s) + 2 e^- \rightarrow 2 Hg(l) + 2 Cl^-(aq)$ $\quad E°(\text{cathode}) = +0.27$ V

Therefore, at the anode, after reversal,

$Sn^{2+}(aq) \rightarrow Sn^{4+}(aq) + 2 e^-$

and the cell reaction is, upon addition of the half-reactions,

$Sn^{2+}(aq) + Hg_2Cl_2(s) \rightarrow 2 Hg(l) + 2 Cl^-(aq) + Sn^{4+}(aq)$

$E°_{\text{cell}} = 0.27$ V $- 0.15$ V $= 0.12$ V


**12.11** (a) $Ni^{2+}(aq) + 2 e^- \rightarrow Ni(s)$ $\quad\quad E°(\text{cathode}) = -0.23$ V

$Zn^{2+}(aq) + 2 e^- \rightarrow Zn(s)$ $\quad\quad E°(\text{anode}) = -0.76$ V

Reversing the anode reaction yields

$Zn(s) \rightarrow Zn^{2+}(aq) + 2 e^-$ $\quad$ (at anode); then, upon addition,

$Ni^{2+}(aq) + Zn(s) \rightarrow Ni(s) + Zn^{2+}(aq)$ $\quad$ (overall cell)

$E°_{\text{cell}} = -0.23$ V $- (-0.76$ V$) = +0.53$ V

and $Zn(s) \,|\, Zn^{2+}(aq) \,\|\, Ni^{2+}(aq) \,|\, Ni(s)$

(b) $2[Ce^{4+}(aq) + e^- \rightarrow Ce^{3+}(aq)]$ $\quad E°(\text{cathode}) = +1.61$ V

$I_2(s) + 2 e^- \rightarrow 2 I^-(aq)$ $\quad\quad\quad\quad E°(\text{anode}) = +0.54$ V

Reversing the anode reaction yields

$2 I^-(aq) \rightarrow 2 e^- + I_2(s)$ $\quad$ (at anode); then, upon addition,

$2 I^-(aq) + 2 Ce^{4+}(aq) \rightarrow 2 Ce^{3+}(aq) + I_2(s)$ $\quad$ (overall cell)

$E°_{\text{cell}} = +1.61$ V $- 0.54$ V $= +1.07$ V

and $Pt(s) \,|\, I^-(aq) \,|\, I_2(s) \,\|\, Ce^{4+}(aq), Ce^{3+}(aq) \,|\, Pt(s)$

An inert electrode such as Pt is necessary when both oxidized and reduced species are in the same solution.

(c) $Cl_2(g) + 2 e^- \rightarrow 2 Cl^-(aq)$ $\quad E°(\text{cathode}) = +1.36$ V

$2 H^+(aq) + 2 e^- \rightarrow H_2(g)$ $E°$(anode) = 0.00 V

Reversing the anode reaction yields

$H_2(g) \rightarrow 2 H^+(aq) + 2 e^-$ (at anode); then, upon addition,

$H_2(g) + Cl_2(g) \rightarrow 2 HCl(aq)$ (overall cell) $E°_{cell}$ = +1.36 V − 0.00 V

$= +1.36$ V

and $Pt(s) \,|\, H_2(g) \,|\, H^+(aq) \,\|\, Cl^-(aq) \,|\, Cl_2(g) \,|\, Pt(s)$

An inert electrode such as Pt is necessary for gas/ion electrode reactions.

(d)  $3[Au^+(aq) + e^- \rightarrow Au(s)]$ $E°$(cathode) = +1.69 V

$Au^{3+}(aq) + 3 e^- \rightarrow Au(s)$ $E°$(anode) = +1.40 V

Reversing the anode reaction yields

$Au(s) \rightarrow Au^{3+}(aq) + 3 e^-$ then, upon addition (anode),

$3 Au^+(aq) \rightarrow$
$\quad 2 Au(s) + Au^{3+}(aq)$ (overall cell) $E°_{cell}$ = +1.69 V − 1.40 V

$= +0.29$ V

and $Au(s) \,|\, Au^{3+}(aq) \,\|\, Au^+(aq) \,|\, Au(s)$

**12.13** (a)  $Ag^+(aq) + e^- \rightarrow Ag(s)$ $E°$(cathode) = +0.80 V

$AgBr(s) + e^- \rightarrow Ag(s) + Br^-(aq)$ $E°$(anode) = +0.07 V

Reversing the anode reaction yields

$Ag(s) + Br^-(aq) \rightarrow AgBr(s) + e^-$ then, upon addition,

$Ag^+(aq) + Br^-(aq) \rightarrow AgBr(s)$ (overall cell) $E°_{cell}$ = +0.80 V − 0.07 V

$= +0.73$ V

This is the direction of the spontaneous standard cell reaction that could be used to study the reverse of the given solubility equilibrium. A cell diagram for this favorable process is

$Ag(s) \,|AgBr(s) \,|\, Br^-(aq) \,\|\, Ag^+(aq) \,|\, Ag(s)$

(b) To conform to the notation of this chapter, the neutralization is rewritten as

$H^+(aq) + OH^- \rightarrow H_2O(l)$

$O_2(g) + 4 H^+(aq) + 4 e^- \rightarrow 2 H_2O(l)$  $E°(\text{cathode}) = +1.23$ V

$O_2(g) + 2 H_2O(l) + 4 e^- \rightarrow 4 OH^-(aq)$  $E°(\text{anode}) = +0.40$ V

Reversing the anode reaction yields

$4 OH^-(aq) \rightarrow O_2(g) + 2 H_2O(l) + 4 e^-$; then, upon addition,

$4 H^+(aq) + 4 OH^-(aq) \rightarrow 4 H_2O(l)$

or $H^+(aq) + OH^-(aq) \rightarrow H_2O(l)$  (overall cell)  $E° = +1.23$ V $- 0.40$ V

$\qquad\qquad\qquad\qquad = +0.83$ V

and $Pt(s) |O_2(g) |OH^-(aq) \| H^+(aq) |O_2(g) |Pt(s)$

(c)  $Cd(OH)_2(s) + 2 e^- \rightarrow Cd(s) + 2 OH^-(aq)$  $E°(\text{anode}) = -0.81$ V

$Ni(OH)_3(s) + e^- \rightarrow Ni(OH)_2(s) + OH^-(aq)$  $E°(\text{cathode}) = +0.49$ V

Reversing the anode reaction and multiplying the cathode reaction by 2 yields

$Cd(s) + 2 OH^-(aq) \rightarrow Cd(OH)_2(s) + 2 e^-$

$2 Ni(OH)_3 + 2 e^- \rightarrow 2 Ni(OH)_2(s) + 2 OH^-(aq)$  then, upon addition,

$2 Ni(OH)_2(s) + Cd(s) \rightarrow Cd(OH)_2(s) + 2 Ni(OH)_2(s)$

$\qquad\qquad$ overall cell $E° = +1.30$ V

and $Cd(s) | Cd(OH)_2(s) | KOH(aq) \| Ni(OH)_3(s) | Ni(OH)_2(s) | Ni(s)$

**12.15**  (a)  $MnO_4^-(aq) + 8 H^+(aq) + 5 e^- \rightarrow Mn^{2+}(aq) + 4 H_2O(l)$ (cathode half-reaction)

$5[Fe^{2+}(aq) \rightarrow Fe^{3+}(aq) + e^-]$ (anode half-reaction)

(b)  Reversing the anode reaction and adding the two equations yields

$MnO_4^-(aq) + 5 Fe^{2+}(aq) + 8 H^+(aq) \rightarrow Mn^{2+}(aq) + 5 Fe^{3+}(aq) + 4 H_2O(l)$

The cell diagram is

$Pt(s) | Fe^{3+}(aq), Fe^{2+}(aq) \| H^+(aq), MnO_4^-(aq), Mn^{2+}(aq) | Pt(s)$

**12.17** A galvanic cell has a positive potential difference; therefore, identify as cathode and anode the electrodes that make $E°$ (cell) positive upon calculating

$E°(cell) = E°(cathode) - E°(anode)$

There are only two possibilities: If your first guess gives a negative $E°$ (cell), switch your identification.

(a) $Cu^{2+}(aq) + 2 e^- \rightarrow Cu(s)$  $E°(cathode) = +0.34$ V

$Cr^{3+}(aq) + e^- \rightarrow Cr^{2+}(aq)$  $E°(anode) = -0.41$ V
$E°(cell) = +0.34$ V $- (-0.41$ V$) = +0.75$ V

(b) $AgCl(s) + e^- \rightarrow Ag(s) + Cl^-(aq)$  $E°(cathode) = +0.22$ V

$AgI(s) + e^- \rightarrow Ag(s) + I^-(aq)$  $E°(anode) = -0.15$ V
$E°(cell) = +0.22$ V $- (-0.15$ V$) = +0.37$ V

(c) $Hg_2^{2+}(aq) + 2 e^- \rightarrow 2 Hg(l)$  $E°(cathode) = +0.79$ V

$Hg_2Cl_2(s) + 2 e^- \rightarrow 2 Hg(l) + 2 Cl^-(aq)$  $E°(anode) = +0.27$ V
$E°(cell) = +0.79$ V $- (+0.27$ V$) = +0.52$ V

(d) $Pb^{4+}(aq) + 2 e^- \rightarrow Pb^{2+}(aq)$  $E°(cathode) = +1.67$ V

$Sn^{4+}(aq) + 2 e^- \rightarrow Sn^{2+}(aq)$  $E°(anode) = +0.15$ V
$E°(cell) = +1.67$ V $- (+0.15$ V$) = +1.52$ V

**12.19** See Exercise 12.17 solutions for $E°$ (cell) values. In each case,

$\Delta G°_r = -nFE°$.

$1$ V $= 1$ J $\cdot$ C$^{-1}$. $n$ is determined by balancing the equation for the cell reaction constructed from the half-reactions given in Exercise 12.17.

(a) $Cu^{2+}(aq) + 2 Cr^{2+}(aq) \rightarrow Cu(s) + 2 Cr^{3+}(aq)$, $n = 2$

$E°_{cell} = +0.75$ V and $\Delta G°_r = -nFE°$
$\qquad = -(2)(9.6485 \times 10^4$ C $\cdot$ mol$^{-1})(0.75$ J $\cdot$ C$^{-1})$
$\qquad = -1.4 \times 10^5$ J $\cdot$ mol$^{-1} = -1.4 \times 10^2$ kJ $\cdot$ mol$^{-1}$

(b) $AgCl(s) + I^-(aq) \rightarrow AgI(s) + Cl^-(aq)$, $n = 1$

$E°_{cell} = +0.37$ V and $\Delta G°_r = -nFE°$

$$= -1 \times 9.6485 \times 10^4 \text{ C} \cdot \text{mol}^{-1} \times 0.37 \text{ J} \cdot \text{C}^{-1} = -36 \text{ kJ} \cdot \text{mol}^{-1}$$

(c) $Hg_2^{2+}(aq) + 2 Cl^-(aq) \rightarrow Hg_2Cl_2(s)$, $n = 2$

$E°_{cell} = +0.52$ V and $\Delta G°_r = -nFE°$

$$= -(2)(9.6485 \times 10^4 \text{ C} \cdot \text{mol}^{-1})(0.52 \text{ J} \cdot \text{C}^{-1})$$

$$= -1.0 \times 10^5 \text{ J} \cdot \text{mol}^{-1} = -1.0 \times 10^2 \text{ kJ} \cdot \text{mol}^{-1}$$

(d) $Pb^{4+}(aq) + Sn^{2+}(aq) \rightarrow Pb^{2+}(aq) + Sn^{4+}(aq)$, $n = 2$

$E°_{cell} = +1.52$ V and $\Delta G°_r = -nFE°$

$$= -(2)(9.6485 \times 10^4 \text{ C} \cdot \text{mol}^{-1})(1.52 \text{ J} \cdot \text{C}^{-1})$$

$$= -293 \text{ kJ} \cdot \text{mol}^{-1}$$

**12.21** The cell, as written $Cu(s)\,|\,Cu^{2+}(aq)\,\|\,M^{2+}(aq)\,|\,M(s)$, makes the $Cu/Cu^{2+}$ electrode the anode, because this is where oxidation is occurring; the $M^{2+}/M$ electrode is the cathode. The calculation is

$E° = E°(\text{cathode}) - E°(\text{anode})$

$-0.689$ V $= E°(\text{cathode}) - (+0.34$ V$)$

$E°(\text{cathode}) = -0.349$ V

**12.23** Refer to Appendix 2B. The more negative (less positive) the standard reduction potential, the stronger is the metal as a reducing agent.

(a) Cu < Fe < Zn < Cr

(b) Mg < Na < K < Li

(c) V < Ti < Al < U

(d) Au < Ag < Sn < Ni

**12.25** In each case, identify the couple with the more positive reduction potential. This will be the couple at which reduction occurs, and therefore which contains the oxidizing agent. The other couple contains the reducing agent.

(a) $Co^{2+}/Co$   $E° = -0.28$ V, $Co^{2+}$ is the oxidizing agent (cathode)

$Ti^{3+}/Ti^{2+}$   $E° = -0.37$ V, $Ti^{2+}$ is the reducing agent (anode)

$Pt(s) | Ti^{2+}(aq), Ti^{3+}(aq) \| Co^{2+}(aq) | Co(s)$

$E°_{cell} = E°(cathode) - E°(anode) = -0.28$ V $- (-0.37$ V$) = +0.09$ V

(b) $U^{3+}/U$   $E° = -1.79$ V, $U^{3+}$ is the oxidizing agent (cathode)

$La^{3+}/La$   $E° = -2.52$ V, La is the reducing agent (anode)

$La(s) | La^{3+}(aq) \| U^{3+}(aq) | U(s)$

$E°_{cell} = -1.79$ V $- (-2.52$ V$) = +0.73$ V

(c) $Fe^{3+}/Fe^{2+}$   $E° = +0.77$ V, $Fe^{3+}$ is the oxdizing agent (cathode)

$H^+/H_2$   $E° = 0.00$ V, $H_2$ is the reducing agent (anode)

$Pt(s) | H_2(g) | H^+(aq) \| Fe^{2+}(aq), Fe^{3+}(aq) | Pt(s)$

$E°_{cell} = +0.77$ V $- 0.00$ V $= +0.77$ V

(d) $O_3/O_2,OH^-$   $E° = +1.24$ V, $O_3$ is the oxidizing agent (cathode)

$Ag^+/Ag$   $E° = +0.80$ V, Ag is the reducing agent (anode)

$Ag(s) | Ag^+(aq) \| OH^-(aq) | O_3(g), O_2(g) | Pt(s)$

$E°_{cell} = +1.24$ V $- 0.80$ V $= +0.44$ V

**12.27** (a) $E°(Cl_2, Cl^-) = +1.36$ V (cathode)

$E°(Br_2, Br^-) = +1.09$ V (anode)

Because $E°(Cl_2, Cl^-) > E°(Br_2, Br^-)$ the reaction favors products.

$E°_{cell} = +1.36$ V $- 1.09$ V $= +0.27$ V

$Cl_2(g)$ is the oxidizing agent.

(b) $E°(Ce^{4+}/Ce^{3+}) = +1.61$ V          (anode)

$E°(MnO_4^-/Mn^{2+}) = +1.51$ V          (cathode)

Because $E°(Ce^{4+}/Ce^{3+}) > E°(MnO_4^-/Mn^{2+})$, the reaction does not favor products.

(c) $E°(Pb^{4+}/Pb^{2+}) = +1.67$ V          (anode)

$E°(Pb^{2+}/Pb) = -0.13$ V          (cathode)

Because $E°(Pb^{4+}/Pb^{2+}) > E°(Pb^{2+}/Pb)$, the reaction does not favor products.

(d) $E°(NO_3^-/NO_2/H^+) = +0.80$ V   (cathode)

$E°(Zn^{2+}/Zn) = -0.76$ V                    (anode)

Because $E°(NO_3^-/NO_2/H^+) > E°(Zn^{2+}/Zn)$, the reaction favors products.

$E°_{cell} = +0.80$ V $- (-0.76$ V$) = +1.56$ V

$NO_3^-$ is the oxidizing agent.

**12.29** (a) $3\,Au^+(aq) \rightarrow 2\,Au\,(s) + Au^{3+}(aq)$

(b) $Au^+(aq) + e^- \rightarrow Au(s)$  $E° = +1.69$ V

$Au^{3+}(aq) + 3\,e^- \rightarrow Au(s)$   $E° = +1.40$ V

Multiplying the first equation by three and subtracting the second equation gives the net equation desired. The potential is given simply be subtracting the second from the first:

$E° = 1.69$ V $- 1.40$ V $= +0.29$ V

Because $E°$ is positive, the process should be spontaneous for standard state conditions.

**12.31** The appropriate half-reactions are:

$U^{4+} + e^- \rightarrow U^{3+}$          $E° = -0.61$          (A)

$U^{3+} + 3\,e^- \rightarrow U$          $E° = -1.79$          (B)

(A) and (B) add to give the desired half-reaction (C):

$U^{4+} + 4\,e^- \rightarrow U$              $E° = ?$          (C)

In order to calculate the potential of a *half-reaction*, we need to convert the $E°$ values into $\Delta G°$ values:

$$\Delta G°(A) = -nFE°(A) = -1F(-0.61 \text{ V})$$
$$\Delta G°(B) = -nFE°(B) = -3F(-1.79 \text{ V})$$
$$\Delta G°(C) = -nFE°(C) = -4FE°(C)$$
$$\Delta G°(C) = \Delta G°(A) + \Delta G°(B)$$
$$-4FE°(C) = -1F(-0.61 \text{ V}) + [-3F(-1.79 \text{ V})]$$

The constant $F$ will cancel from both sides, leaving:

$$-4E°(C) = -1(-0.61 \text{ V}) - 3(-1.79 \text{ V})$$
$$E°(C) = -[0.61 \text{ V} + 5.37 \text{ V}]/4 = -1.50 \text{ V}$$

**12.33** (a)  $Ti^{2+}(aq) + 2 e^- \rightarrow Ti(s)$   $E°(\text{cathode}) = -1.63 \text{ V}$

$Mn^{2+}(aq) + 2 e^- \rightarrow Mn(s)$ $\qquad\qquad$ $E°(\text{anode}) = -1.18 \text{ V}$

Note: These equations represent the cathode and anode half-reactions for the overall reaction as written. The spontaneous direction of this reaction under standard conditions is the opposite of that given.

$E°_{cell} = E°(\text{cathode}) - E°(\text{anode}) = -1.63 \text{ V} - (-1.18 \text{ V}) = -0.45 \text{ V}$, and

$$\ln K = \frac{nFE°}{RT}. \qquad \text{At } 25°C \quad \ln K = \frac{nE°}{0.02569 \text{ V}}.$$

$$\therefore \quad \ln K = \frac{(2)(-0.45 \text{ V})}{0.02569 \text{ V}} = -35 \quad \text{and} \quad K = 6 \times 10^{-16}.$$

(b)  $In^{3+}(aq) + 2 e^- \rightarrow In^{2+}(aq)$ $\qquad$ $E°(\text{cathode}) = -0.49 \text{ V}$

$U^{4+}(aq) + e^- \rightarrow U^{3+}(aq)$ $\qquad\qquad$ $E°(\text{anode}) = -0.61 \text{ V}$

$E°_{cell} = E°(\text{cathode}) - E°(\text{anode}) = -0.49 \text{ V} - (-0.61 \text{ V}) = +0.12 \text{ V}$, and

at 25°C $\quad \ln K = \dfrac{(2)(+0.12 \text{ V})}{0.02569 \text{ V}} = +9.3.$

$$\therefore \quad K = 1 \times 10^4.$$

**12.35** (a)  $Pb^{4+}(aq) + 2 e^- \rightarrow Pb^{2+}(aq)$   $E°(\text{cathode}) = +1.67 \text{ V}$

$Sn^{2+}(aq) \rightarrow Sn^{4+}(aq) + 2e^-$   $E°(\text{anode}) = +0.15 \text{ V}$

$Pb^{4+}(aq) + Sn^{2+}(aq) \rightarrow Pb^{2+}(aq) + Sn^{4+}(aq)$   $E°_{cell}$

$\qquad = 1.67 \text{ V} - (0.15 \text{ V}) = +1.52 \text{ V}$

Then, $E = E^\circ - \left(\dfrac{0.025\,693\text{ V}}{n}\right)\ln Q$; $1.33\text{ V} = 1.52\text{ V} - \left(\dfrac{0.025\,693\text{ V}}{2}\right)\ln Q$

$$\ln Q = \dfrac{1.52\text{ V} - 1.33\text{ V}}{0.0129\text{ V}} = \dfrac{0.19\text{ V}}{0.0129\text{ V}} = 15 \quad Q = 10^6$$

(b) $\begin{aligned}2[Cr_2O_7^{\,2-}(aq) + 14\,H^+(aq) + 6\,e^- &\rightarrow \\ 2\,Cr^{3+}(aq) + 7\,H_2O(l)]\ E^\circ(\text{cathode}) &= 1.33\text{ V}\end{aligned}$

$3[2\,H_2O(l) \rightarrow O_2(g) + 4\,H^+(aq) + 4\,e^-]\ E^\circ(\text{anode}) = +1.23\text{ V}$

$2\,Cr_2O_7^{\,2-}(aq) + 16\,H^+(aq) \rightarrow 4\,Cr^{3+}(aq) + 8\,H_2O(l) + 3\,O_2(g)\ E^\circ_{cell}$
$\quad = 0.10\text{ V}$

Then, $E = E^\circ - \left(\dfrac{0.0257\text{ V}}{n}\right)\ln Q$; $0.10\text{ V} = +0.10\text{ V} - \left(\dfrac{0.0257\text{ V}}{12}\right)\ln Q$

$\ln Q = 0.00 \quad Q = 1.0$

**12.37** (a) $Cu^{2+}(aq,\,0.010\,M) + 2\,e^- \rightarrow Cu(s)$ (cathode)

$Cu^{2+}(aq,\,0.0010\,M) + 2\,e^- \rightarrow Cu(s)$ (anode)

$Cu^{2+}(aq,\,0.010\,M) \rightarrow Cu^{2+}(aq,\,0.0010\,M),\ n = 2$

$E^\circ_{cell} = E^\circ(\text{cathode}) - E^\circ(\text{anode}) = 0\text{ V}$

$E_{cell} = E^\circ_{cell} - \left(\dfrac{RT}{nF}\right)\ln Q = -\left(\dfrac{0.025\,693\text{ V}}{2}\right)\ln Q$ at 25°C

$E_{cell} = -\left(\dfrac{0.025\,693\text{ V}}{2}\right)\ln\left(\dfrac{0.0010\,M}{0.010\,M}\right) = +0.030\text{ V}$

(b) at $pH = 3.0,\ [H^+] = 1 \times 10^{-3}\,M$

at $pH = 4.0,\ [H^+] = 1 \times 10^{-4}\,M$

Cell reaction is $H^+(aq,\,1 \times 10^{-3}\,M) \rightarrow H^+(aq,\,1 \times 10^{-4}\,M),\ n = 1$

$E^\circ_{cell} = 0\text{ V} \quad E_{cell} = E^\circ_{cell} - \left(\dfrac{RT}{nF}\right)\ln Q = -\left(\dfrac{0.025\,693\text{ V}}{1}\right)\ln\left(\dfrac{1 \times 10^{-4}}{1 \times 10^{-3}}\right)$
$\quad = +6 \times 10^{-2}\text{ V}$

**12.39** In each case, $E^\circ_{cell} = E^\circ(\text{cathode}) - E^\circ(\text{anode})$. Recall that the values for
$E^\circ$ at the electrodes refer to the electrode potential for the half-reaction

written as a reduction reaction. In balancing the cell reaction, the half-reaction at the anode is reversed. However, this does not reverse the sign of electrode potential used at the anode, because the value always refers to the reduction potential.

(a) $2 H^+(aq, 1.0 \text{ M}) + 2 e^- \rightarrow H_2(g, 1 \text{ atm})$   $E°(\text{cathode}) = 0.00 \text{ V}$

$H_2(g, 1 \text{ atm}) \rightarrow 2 H^+(aq, 0.075 \text{ M}) + 2 e^-$   $E°(\text{anode}) = 0.00 \text{ V}$

$2 H^+(aq, 1.0 \text{ M}) + H_2(g, 1 \text{ atm}) \rightarrow 2 H^+(aq, 0.075 \text{ M}) + H_2(g, 1 \text{ atm})$

$E°_{cell} = 0.00 \text{ V}$

Then, $E = E° - \left( \dfrac{0.025\ 693 \text{ V}}{n} \right) \ln \left( \dfrac{[H^+, 0.075 \text{ M}]^2 P_{H_2}}{[H^+, 1.0 \text{ M}]^2 P_{H_2}} \right)$

$E = 0.00 \text{ V} - \left( \dfrac{0.025\ 693 \text{ V}}{2} \right) \ln \left( \dfrac{(0.075 \text{ M})^2 \times 1 \text{ atm}}{(1.0 \text{ M})^2 \times 1 \text{ atm}} \right)$

$E = -0.0129 \text{ V} \ln (0.075)^2 = +0.067 \text{ V}$

(b) $Ni^{2+}(aq) + 2 e^- \rightarrow Ni(s)$   $E°(\text{cathode}) = -0.23 \text{ V}$

$Zn(s) \rightarrow Zn^{2+}(aq) + 2 e^-$   $E°(\text{anode}) = -0.76 \text{ V}$

$Ni^{2+}(aq) + Zn(s) \rightarrow Ni(s) + Zn^{2+}(aq)$   $E°_{cell} = +0.53 \text{ V}$

Then, $E = E° - \left( \dfrac{0.025\ 693 \text{ V}}{n} \right) \ln \left( \dfrac{[Zn^{2+}]}{[Ni^{2+}]} \right)$

$E = 0.53 \text{ V} - \left( \dfrac{0.025\ 693 \text{ V}}{2} \right) \ln \left( \dfrac{0.37}{0.059} \right) = 0.53 \text{ V} - 0.02 \text{ V} = 0.51 \text{ V}$

(c) $2 H^+(aq) + 2 e^- \rightarrow H_2(g)$   $E°(\text{cathode}) = 0.00 \text{ V}$

$2 Cl^-(aq) \rightarrow Cl_2(g) + 2 e^-$   $E°(\text{anode}) = +1.36 \text{ V}$

$2 H^+(aq) + 2 Cl^-(aq) \rightarrow H_2(g) + Cl_2(g)$   $E°_{cell} = -1.36 \text{ V}$

Then,

$$E = E° - \left(\frac{0.025\ 693\ V}{n}\right) \ln \left(\frac{P_{H_2} P_{Cl_2}}{[H^+]^2 [Cl^-]^2}\right)$$

$$E = -1.36\ V - \left(\frac{0.025\ 693\ V}{2}\right) \ln \left(\frac{\left(\frac{125}{760}\right)\left(\frac{250}{760}\right)}{(0.85)^2\ (1.0)^2}\right) (1.01325)^2$$

$$E = -1.36\ V + 0.03\ V$$
$$= -1.33\ V$$

(d)  $Sn^{4+}(aq,\ 0.867\ M) + 2\ e^- \rightarrow Sn^{2+}(aq,\ 0.55\ M)$

$\qquad E°(\text{cathode}) = +0.15\ V$

$Sn(s) \rightarrow Sn^{2+}(aq,\ 0.277\ M) + 2\ e^-\quad E°(\text{anode}) = -0.14\ V$

$Sn^{4+}(aq,\ 0.867\ M) + Sn(s) \rightarrow Sn^{2+}(aq,\ 0.55) + Sn^{2+}(aq,\ 0.277\ M)$

$E°_{cell} = 0.29\ V$

$$E = E° - \left(\frac{0.025\ 693\ V}{2}\right) \ln \left(\frac{(0.55)(0.277)}{(0.867)}\right)$$

$$E = 0.29\ V + 0.02\ V = 0.31\ V$$

**12.41**  In each case, obtain the balanced equation for the cell reaction from the half-cell reactions at the electrodes, by reversing the reduction equation for the half-reaction at the anode, multiplying the half-reaction equations by an appropriate factor to balance the number of electrons, and then adding the half-reactions. Calculate $E°_{cell} = E°(\text{cathode}) - E°(\text{anode})$. Then write the Nernst equation for the cell reaction and solve for the unknown.

(a)  $Hg_2Cl_2(s) + 2\ e^- \rightarrow 2\ Hg(l) + 2\ Cl^-(aq)\quad E°(\text{cathode}) = +0.27\ V$

$H_2(g) \rightarrow 2\ H^+(aq) + 2\ e^-\quad E°(\text{anode}) = 0.00\ V$

$H_2(g) + Hg_2Cl_2(s) \rightarrow 2\ H^+(aq) + 2\ Hg(l) + 2\ Cl^-(aq)\quad E°_{cell} = +0.27\ V$

$$E = E° - \left( \frac{0.025\ 693\ V}{n} \right) \ln \left( \frac{[H^+]^2 [Cl^-]^2}{[H_2]} \right)$$

$$0.33\ V = 0.27\ V - \left( \frac{0.025\ 693\ V}{2} \right) \ln \left( \frac{[H^+]^2 (1)^2}{(1)} \right)$$

$$= 0.27\ V - (0.0129\ V) \ln [H^+]^2$$

$$0.06\ V = -0.0257\ V \ln [H^+] = -0.0257\ V \times (2.303 \log [H^+])$$

$$pH = \frac{0.06\ V}{(2.303)\ (0.025\ 693\ V)} = 1.0$$

(b)
$$2[MnO_4^-(aq) + 8\ H^+(aq) + 5\ e^- \rightarrow$$
$$Mn^{2+}(aq) + 4\ H_2O(l)] \quad E°(cathode) = +1.51\ V$$

$$5[2\ Cl^-(aq) \rightarrow Cl_2(g) + 2\ e^-] \quad E°(anode) = +1.36\ V$$

$$2\ MnO_4^-(aq) + 16\ H^+(aq) + 10\ Cl^-(aq) \rightarrow$$
$$5\ Cl_2(g) + 2\ Mn^{2+}(aq) + 8\ H_2O(l) \quad E°_{cell} = +0.15\ V$$

$$E = E° - \left( \frac{0.0257\ V}{n} \right) \ln \left( \frac{[Cl_2]^5 [Mn^{2+}]^2}{[MnO_4]^2 [H^+]^{16} [Cl^-]^{10}} \right)$$

$$-0.30\ V = +0.15\ V - \left( \frac{0.0257\ V}{10} \right) \ln \left( \frac{(1)^5 (0.10)^2}{(0.010)^2\ (1 \times 10^{-4})^{16} (Cl^-)^{10}} \right)$$

$$-0.45\ V = -(0.002\ 5693\ V) \log \left( \frac{1 \times 10^{-2}}{(1 \times 10^{-4})\ (1 \times 10^{-64})\ [Cl^-]^{10}} \right)$$

$$= -0.002\ 5693\ V \left[ \ln (1 \times 10^{66}) + \ln \left( \frac{1}{[Cl^-]^{10}} \right) \right]$$

$$= -0.390\ V + (0.0025\ 693\ V) \ln [Cl^-]^{10}$$

$$-0.0594\ V = 0.002\ 5693\ V \ln[Cl^-]^{10}$$

$$= (0.025\ 693\ V) \ln[Cl^-]$$

$$\ln[Cl^-] = \frac{-0.06\ V}{0.025\ 693\ V} = -2$$

$$[Cl^-] = 10^{-1}\ mol \cdot L^{-1}$$

**12.43** Since the reduction potential of tin(II) is negative relative to the S.H.E., we will assume the tin electrode to be the anode such that the standard cell potential would be positive. Then we can use the Nernst equation to solve for the hydrogen ion activity in order to calculate the pH. The cell reaction is $Sn(s) + 2 H^+ \rightarrow Sn^{2+} + H_2(g)$.

$$E = E° - \left(\frac{0.025\ 693\ V}{n}\right) \ln\left(\frac{[Sn^{2+}]p_{H_2}}{[H^+]^2}\right)$$

$$0.061\ V = 0.14\ V - \left(\frac{0.025\ 693\ V}{2}\right) \ln\left(\frac{[0.015][1]}{[H^+]^2}\right)$$

$$= 0.14\ V - (0.01284\ V)(\ln(0.015) - \ln[H^+]^2)$$

$$\frac{0.079\ V}{0.01284\ V} = \ln(0.015) - 2\ln[H^+]$$

$$-10.352 = 2\ln[H^+]$$

$$\ln[H^+] = -5.176, \quad [H^+] = 5.650 \times 10^{-3}$$

$$pH = -\log(5.650 \times 10^{-3}) = 2.25$$

**12.45**  To calculate this value, we need to determine the $E°$ value for the solubility reaction:

$Hg_2Cl_2(s) \rightarrow Hg_2^{2+}(aq) + 2\ Cl^-(aq) \quad E° = ?$

The relationship $\Delta G° = -nRT \ln K = -nFE°$ can be used to calculate the value of $K_{sp}$.

The equations that will add to give the net equation we want are

$Hg_2Cl_2(s) + 2\ e^- \rightarrow 2\ Hg(l) + 2\ Cl^-(aq) \qquad E° = +0.27\ V$

$2\ Hg(l) \rightarrow Hg_2^{2+}(aq) + 2\ e^- \qquad\qquad\qquad E° = 0.79\ V$

Notice that the second equation is reversed from the reduction reaction given in the Appendix, and consequently the $E°$ value is changed in sign. Adding these two equations together gives the desired net reaction, and summing the $E°$ values will give the $E°$ value for that process:

$E° = (+0.29\ V) + (-0.79\ V) = -0.50\ V$

$$\ln K_{sp} = \frac{nFE°}{RT} = \frac{(2)(9.65 \times 10^4\ C \cdot mol^{-1})(-0.50\ V)}{(8.314\ J \cdot K^{-1} \cdot mol^{-1})(298\ K)} = -38.95$$

$K_{sp} = 1.2 \times 10^{-17}$

(b)  This value is a factor of 10 greater than the value in Table 11.6 $(1.3 \times 10^{-18})$.

**12.47**  This cell uses two silver electrodes, so $E° = 0$ and $E$ is determined by the ratio of $[Ag^+]_{anode}$ to $[Ag^+]_{cathode}$. Since $[Ag^+]_{anode} < [Ag^+]_{cathode}$, the ratio is less than 1 and $E > 0$, so the cell can do work because

$$\Delta G = w_{max} = -nFE.$$

$$E = -\left(\frac{0.025\ 693\ V}{n}\right) \ln\left(\frac{[Ag^+]_{anode}}{[Ag^+]_{cathode}}\right) = -\left(\frac{0.025\ 693\ V}{1}\right) \ln\left(\frac{5.0 \times 10^{-3}}{0.15}\right)$$

$$= 0.0874\ V$$

$$\Delta G = w_{max} = -nFE = -(1\ mol)(96\ 485\ J\ V^{-1}mol^{-1})(0.0874\ V)$$

$$= -8.4\ kJ$$

Therefore, the maximum work that the cell can perform is 8.4 kJ per mole of Ag.

**12.49**  For the standard calomel electrode, $E° = +0.27\ V$. If this were set equal to 0, all other potentials would also be decreased by 0.27 V.   (a) Therefore, the standard hydrogen electrode's standard reduction potential would be 0.00 V − 0.27 V or −0.27 V.   (b) The standard reduction potential for $Cu^{2+}/Cu$ would be 0.34 V − 0.27 V or +0.07 V.

**12.51**  The strategy is to consider the possible competing cathode and anode reactions. At the cathode, choose the reduction reaction with the most positive (least negative) standard reduction potential ($E°$ value). At the anode, choose the oxidation reaction with the least positive (most negative) standard reduction potential ($E°$ value, as given in the table). Then calculate $E°_{cell} = E°(\text{cathode}) - E°(\text{anode})$. The negative of this value is the minimum potential that must be supplied.

(a)  cathode: $Ni^{2+}$ (aq) $+ 2\ e^- \rightarrow Ni(s)$　　　　　　$E° = -0.23\ V$

(rather than $2\ H_2O(l) + 2\ e^- \rightarrow H_2(g) + 2\ OH^-(aq)$ $E° = -0.83\ V$)

(b)  anode: $2\ H_2O(l) \rightarrow O_2(g) + 4\ H^+(aq) + 4\ e^-$　$E° = +1.23\ V$

(the $SO_4^{2-}$ ion will not oxidize)

(c) $E^\circ_{cell} = E^\circ(\text{cathode}) - E^\circ(\text{anode}) = -0.23 \text{ V} - (+1.23 \text{ V}) = -1.46 \text{ V}$

Therefore E (supplied) must be $> +1.46 \text{ V}$ (1.46 V is the minimum).

**12.53** In each case, compare the reduction potential of the ion to the reduction potential of water ($E^\circ = -0.42 \text{ V}$) and choose the process with the least negative $E^\circ$ value.

(a) $\text{Mn}^{2+}(\text{aq}) + 2 \text{ e}^- \rightarrow \text{Mn(s)}$     $E^\circ = -1.18 \text{ V}$

(b) $\text{Al}^{3+}(\text{aq}) + 3 \text{ e}^- \rightarrow \text{Al(s)}$     $E^\circ = -1.66 \text{ V}$

The reactions in (a) and (b) evolve hydrogen rather than yield a metallic deposit because water is reduced, according to

$2 \text{ H}_2\text{O(l)} + 2 \text{ e}^- \rightarrow \text{H}_2(\text{g}) + 2 \text{ OH}^-(\text{aq})$ ($E^\circ = -0.42 \text{ V}$, at pH = 7)

(c) $\text{Ni}^{2+}(\text{aq}) + 2 \text{ e}^- \rightarrow \text{Ni(s)}$     $E^\circ = -0.23 \text{ V}$

(d) $\text{Au}^{3+}(\text{aq}) + 3 \text{ e}^- \rightarrow \text{Au(s)}$     $E^\circ = +1.69 \text{ V}$

In (c) and (d) the metal ion will be reduced.

**12.55** $4500 \text{ C} \div 9.65 \times 10^4 \text{ C} \cdot \text{F}^{-1} = 0.047 \text{ F} = 0.047 \text{ mol e}^-$

(a) $(0.047 \div 3) \text{ mol Bi}^{3+} + 0.047 \text{ mol e}^- \rightarrow$
$(0.047 \div 3) \text{ mol Bi, or } 0.016 \text{ mol Bi} = 3.3 \text{ g}$

(b) $0.047 \text{ mol H}^+ + 0.047 \text{ mol e}^- \rightarrow (0.047 \div 2) \text{ mol H}_2$

$0.024 \text{ mol H}_2 \times 24.45 \text{ L} \cdot \text{mol}^{-1}$ (at 298 K) = 0.59 L

(c) $(0.047 \div 3) \text{ mol Co}^{3+} + 0.047 \text{ mol e}^- \rightarrow$
$(0.047 \div 3) \text{ mol Co} = 0.016 \text{ mol Co or } 0.94 \text{ g}$

**12.57** (a) $\text{Ag}^+(\text{aq}) + \text{e}^- \rightarrow \text{Ag(s)}$

$\text{time} = (1.50 \text{ g Ag})\left(\dfrac{1 \text{ mol Ag}}{107.98 \text{ g Ag}}\right)\left(\dfrac{1 \text{ mol e}^-}{1 \text{ mol Ag}}\right)$

$\left(\dfrac{9.65 \times 10^4 \text{ C}}{1 \text{ mol e}^-}\right)\left(\dfrac{1 \text{ A} \cdot \text{s}}{1 \text{ C}}\right)\left(\dfrac{1}{0.0136}\right) = 9.9 \times 10^4 \text{ s or 27 h}$

(b) $Cu^{2+}(aq) + 2e^- \rightarrow Cu(s)$

$$\text{mass Cu} = (9.9 \times 10^4 \text{ s})(0.0136 \text{ A})\left(\frac{1 \text{ C}}{1 \text{ A} \cdot \text{s}}\right)\left(\frac{1 \text{ mol } e^-}{9.65 \times 10^4 \text{ C}}\right)$$

$$\left(\frac{0.50 \text{ mol Cu}}{1 \text{ mol } e^-}\right)\left(\frac{63.5 \text{ g Cu}}{1 \text{ mol Cu}}\right) = 0.44 \text{ g Cu}$$

**12.59** (a) $Cr(VI) + 6e^- \rightarrow Cr(s)$

$$\text{current} = \frac{\text{charge}}{\text{time}}$$

$$= \frac{2.5 \text{ g Cr}\left(\dfrac{1 \text{ mol Cr}}{52.00 \text{ g Cr}}\right)\left(\dfrac{6 \text{ mol } e^-}{1 \text{ mol Cr}}\right)\left(\dfrac{9.65 \times 10^4 \text{ C}}{1 \text{ mol } e^-}\right)}{12 \text{ h} \times 3600 \text{ s} \cdot \text{h}^{-1}}$$

$$= 0.64 \text{ C} \cdot \text{s}^{-1} = 0.64 \text{ A}$$

(b) $Na^+ + e^- \rightarrow Na(s)$

$$\text{current} = \frac{2.5 \text{ g Na}\left(\dfrac{1 \text{ mol Na}}{22.99 \text{ g Na}}\right)\left(\dfrac{1 \text{ mol } e^-}{1 \text{ mol Na}}\right)\left(\dfrac{9.65 \times 10^4 \text{ C}}{1 \text{ mol } e^-}\right)}{12 \text{ h} \times 3600 \text{ s} \cdot \text{h}^{-1}}$$

$$= 0.24 \text{ C} \cdot \text{s}^{-1} = 0.24 \text{ A}$$

**12.61** $Ru^{n+}(aq) + n e^- \rightarrow Ru(s)$; solve for $n$

$$\text{moles of Ru} = (0.0310 \text{ g Ru})\left(\frac{1 \text{ mol}}{101.07 \text{ g Ru}}\right) = 3.07 \times 10^{-4} \text{ mol}$$

$$\text{total charge} = (500 \text{ s}) (120 \text{ mA})\left(\frac{10^{-3} \text{ A}}{1 \text{ mA}}\right)\left(\frac{1 \text{ C} \cdot \text{s}^{-1}}{1 \text{ A}}\right) = 60 \text{ C}$$

$$\text{moles of } e^- = (60 \text{ C})\left(\frac{1 \text{ mol } e^-}{96\,500 \text{ C}}\right) = 6.2 \times 10^{-4} \text{ mol } e^-$$

$$n = \frac{6.2 \times 10^{-4} \text{ mol } e^-}{3.07 \times 10^{-4} \text{ mol}} = \frac{2 \text{ mol charge}}{1 \text{ mol}}$$

Therefore, oxidation number of $Ru^{2+}$ is +2.

**12.63**  $Hf^{n+} + n\,e^- \rightarrow Hf(s)$; solve for $n$.

charge consumed $= 15.0\ C \cdot s^{-1} \times 2.00\ h \times 3600\ s \cdot h^{-1} = 1.08 \times 10^5\ C$

moles of charge consumed $= (1.08 \times 10^5\ C)\left(\dfrac{1\ mol\ e^-}{9.65 \times 10^4\ C}\right) = 1.12\ mol\ e^-$

moles of Hf plated $= (50.0\ g\ Hf)\left(\dfrac{1\ mol\ Hf}{178.49\ g\ Hf}\right) = 0.280\ mol\ Hf$

Then, $n = \dfrac{1.12\ mol\ e^-}{0.280\ mol\ Hf} = 4.00\ mol\ e^- /mol\ Hf$

Therefore, the oxidation number is 4, that is, $Hf^{4+}$.

**12.65**  $MCl_3 \rightarrow M^{3+} + 3\ Cl^- \qquad M^{3+} + 3\ e^- \rightarrow M(s)$

First, determine the number of moles of electrons consumed; the number of moles of $M^{3+}$ reduced is one-third of this number.

charge used $= (6.63\ h)\left(\dfrac{3600\ s}{1\ h}\right)\left(\dfrac{0.700\ C}{1\ s}\right) = 1.67 \times 10^4\ C$

number of moles of $e^- = (1.67 \times 10^4\ C)\left(\dfrac{1\ mol\ e^-}{9.65 \times 10^4\ C}\right) = 0.173$

number of moles of $M^{3+}$ (and M) $= 0.173\ mol\ e^- \times \dfrac{1\ mol\ M^{3+}}{3\ mol\ e^-}$

$$= 0.0577$$

molar mass $M = \dfrac{3.00\ g}{0.0577\ mol} = 52.0\ g \cdot mol^{-1}$ (Cr)

**12.67**  Assuming all the energy comes from reduction of oxygen focuses attention on this half reaction:

$O_2(g) + 4\ H^+(aq) + 4\ e^- \rightarrow 2\ H_2O(l) \qquad E° = +1.23\ V$

Body conditions are far from standard state values, so the actual value of $E$ would be reduced by about 0.5 V if we take pH, $p_{O_2}$ and T into account.  However, we are only estimating an average current to one significant digit, so $E = 1.23 \pm 0.5$ V $\approx 1$ V is adequate. With these approximations in mind, we can calculate the current.

$$It = nF = \frac{\Delta G}{-E} \quad \text{or}$$

$$I = \frac{\Delta G}{-Et} = \frac{(-10 \times 10^6 \text{ J})}{-(1 \text{ V})(24 \text{ h})(3600 \text{ s} \cdot \text{h}^{-1})} \cdot \frac{1 \text{ V} \cdot \text{C}}{1 \text{ J}} = 115 \text{ A} \approx 100 \text{ A}$$

**12.69** (a) The electrolyte is KOH(aq)/HgO(s), which will have the consistency of a moist paste.

(b) The oxidizing agent is HgO(s).

(c) $HgO(s) + Zn(s) \rightarrow Hg(l) + ZnO(s)$

**12.71** See Table 12.1.

The anode reaction is $Zn(s) \rightarrow Zn^{2+}(aq) + 2 \text{ e}^-$; this reaction supplies the electrons to the external circuit. The cathode reaction is

$MnO_2(s) + H_2O(l) + e^- \rightarrow MnO(OH)_2(s) + OH^-(aq)$. The $OH^-(aq)$

produced reacts with $NH_4^+(aq)$ from the $NH_4Cl(aq)$ present:

$NH_4^+(aq) + OH^-(aq) \rightarrow H_2O(l) + NH_3(g)$. The $NH_3(g)$ produced

complexes with the $Zn^{2+}(aq)$ produced in the anode reaction

$Zn^{2+}(aq) + 4 NH_3(g) \rightarrow [Zn(NH_3)_4]^{2+}(aq)$.

The overall reaction is complicated.

**12.73** See Table 12.1 (a) KOH(aq) (b) In the charging process, the cell reaction is the reverse of what occurs in discharge. Therefore, at the anode, $2 Ni(OH)_2(s) + 2 OH^-(aq) \rightarrow 2 Ni(OH)_3 + 2 \text{ e}^-$.

**12.75** $Fe^{3+}(aq) + 3 \text{ e}^- \rightarrow Fe(s) \qquad E° = -0.04$ V

$Cr^{3+}(aq) + 3 \text{ e}^- \rightarrow Cr(s) \qquad E° = -0.74$ V

$$Fe^{2+}(aq) + 2\,e^- \rightarrow Fe(s) \qquad E° = -0.44\ V$$

$$Cr^{2+}(aq) + 2\,e^- \rightarrow Cr(s) \qquad E° = -0.91\ V$$

Comparison of the reduction potentials shows that Cr is more easily oxidized than Fe, so the presence of Cr retards the rusting of Fe. At the position of the scratch, the gap is filled with oxidation products of Cr, thereby preventing contact of air and water with the iron.

**12.77** (a) $n_{e^-} = n_{Ag^+} = \dfrac{It}{F} = \dfrac{(3.5\ A)(395.0\ s)}{(96\ 485\ C \cdot mol^{-1})} = 1.43 \times 10^{-2}$ mol Ag

$1.43 \times 10^{-2}$ mol Ag $\left(\dfrac{107.87\ g\ Ag}{mol\ Ag}\right) = 1.55$ g Ag

$\dfrac{1.55\ g}{2.69\ g} \times 100 = 57.4\%$ Ag

(b) 2.69 g − 1.55 g = 1.14 g X

Since the salt is 1:1 Ag:X, the molar mass of X is

$\dfrac{1.14\ g}{1.43 \times 10^{-2}\ mol} = 79.7\ g \cdot mol^{-1}$

This molar mass is closest to bromine, so the formula is AgBr.

**12.79** (a) $Fe_2O_3 \cdot H_2O$ (b) $H_2O$ and $O_2$ jointly oxidize iron. (c) Water is more highly conducting if it contains dissolved ions, so the rate of rusting is increased.

**12.81** (a) aluminum or magnesium; both are below titanium in the electrochemical series.

(b) cost, availability, and toxicity of products in the environment

(c) $Cu^{2+} + 2\,e^- \rightarrow Cu(s) \qquad E° = +0.34\ V$

$Cu^+ + e^- \rightarrow Cu(s) \qquad E° = +0.52\ V$

$Fe^{3+} + 3\,e^- \rightarrow Fe(s) \qquad E° = -0.04\ V$

$Fe^{2+} + 2\,e^- \rightarrow Fe(s) \qquad E° = -0.44\ V$

Fe could act as the anode of an electrochemical cell if $Cu^{2+}$ or $Cu^+$ were present; therefore, it could be oxidized at the point of contact. Water with dissolved ions would act as the electrolyte.

**12.83**  $2[Zn^{2+}(aq) + 2e^- \rightarrow Zn(s)]$ $\qquad E°(\text{cathode}) = -0.76$ V

$M(s) \rightarrow M^{4+}(aq) + 4e^-$ $\qquad\qquad E°(\text{anode}) = x$

$M(s) + 2Zn^{2+}(aq) \rightarrow 2Zn(s) + M^{4+}(aq)$  $E°_{\text{cell}} = 0.16$ V

$E°_{\text{cell}} = E°(\text{cathode}) - E°(\text{anode})$

$+0.16$ V $= -0.76$ V $- (x)$

$x = -0.92$ V $= E°(M^{4+}/M)$

**12.85**  The strategy is to find the $E°$ value for the solubility reaction and then find appropriate half-reactions that add to give that solubility reaction. One of these half-reactions is our unknown, the other is obtained from Appendix 2B:

$Cu(IO_3)_2(s) + 2e^- \rightarrow Cu(s) + 2IO_3^-(aq)$ $\qquad E° = ?$ (A)

$Cu(s) \qquad \rightarrow Cu^{2+}(aq) + 2e^-$ $\qquad\qquad E° = -0.34$ V $\qquad$ (B)

$Cu(IO_3)_2(s) + \qquad \rightarrow Cu^{2+}(aq) + 2IO_3^-(aq)$ $\qquad E° = \dfrac{RT \ln K_{sp}}{nF}$ $\quad$ (C)

$E° = \dfrac{RT \ln K_{sp}}{nF}$

$\quad = \dfrac{(8.314 \text{ J} \cdot \text{K}^{-1} \cdot \text{mol}^{-1})(298.2 \text{ K}) \ln (1.4 \times 10^{-7})}{2(9.65 \times 10^4 \text{ C} \cdot \text{mol}^{-1})}$

$\quad = -0.20$ V

$-0.20$ V $= E°(A) + (-0.34$ V$)$

$E°(A) = +0.14$ V

**12.87**  (a)  In acidic solution, the relevant reactions are

$O_2 + 4H^+ + 4e^- \rightarrow 2H_2O$ $\qquad\qquad E° = +1.23$ V

$$Ag \rightarrow Ag^+ + e^- \qquad\qquad\qquad\qquad E° = -0.80 \text{ V}$$

Overall reaction:

$$O_2(g) + 4\,H^+(aq) + 4\,Ag(s) \rightarrow 4\,Ag^+(aq) + 2\,H_2O(l) \qquad E° = +0.43 \text{ V}$$

Because the potential is positive, the reaction should be spontaneous and would be expected to occur. We should also consider the conditions; because air is only 20.95% $O_2$, the potential may be different from that calculated for standard conditions. If air is the source of oxygen, then it will be present at $0.2095 \times 1.013\,25$ bar $= 0.2123$ bar.

$$E = E° - \frac{0.0592}{4} \log \frac{[Ag^+]^4}{P_{O_2}[H^+]^4}$$

$$= +0.43 \text{ V} - \frac{0.0592}{4} \log \frac{[1.0]^4}{(0.2123)[1.0]^4}$$

$$= +0.43 \text{ V} - \frac{0.0592}{4} \log \frac{1}{0.2123}$$

$$= +0.43 \text{ V} - 0.010 \text{ V}$$

$$= +0.42 \text{ V}$$

The potential is still positive and the reaction is expected to be spontaneous.

(b)  In basic solution, the relevant reactions are

$$O_2 + 2\,H_2O + 4\,e^- \rightarrow 4\,OH^- \qquad\qquad E° = +0.40 \text{ V}$$

$$Ag \rightarrow Ag^+ + e^- \qquad\qquad\qquad\qquad E° = -0.80 \text{ V}$$

Overall reaction:

$$O_2(g) + 2\,H_2O(l) + 4\,Ag(s) \rightarrow 4\,Ag^+(aq) + 4\,OH^-(aq)$$
$$E° = -0.40 \text{ V}$$

This process as written is nonspontaneous and is not predicted to occur. However, AgOH forms an insoluble precipitate, changing the nature of the reaction. The $K_{sp}$ value for AgOH is $1.5 \times 10^{-8}$. We use the Nernst equation to calculate the potential under these conditions:

$$E = E° - \frac{0.0592}{4} \log \frac{[Ag^+]^4 [OH^-]^4}{P_{O_2}}$$

$$= -0.40 \text{ V} - \frac{0.0592}{4} \log \frac{K_{sp}^{\ 4}}{P_{O_2}}$$

$$= -0.40 \text{ V} - \frac{0.0592}{4} \log \frac{(1.5 \times 10^{-8})^4}{0.2132}$$

$$= -0.40 \text{ V} + 0.45 \text{ V}$$

$$= +0.05 \text{ V}$$

Under these conditions, the potential is slightly positive and the oxidation should be spontaneous.

**12.89** In each case, determine the cathode and anode half-reactions corresponding to the reactions *as written*. Look up the standard reduction potentials for these half-reactions and then calculate $E°_{cell} = E°(\text{cathode}) - E°(\text{anode})$. If $E°_{cell}$ is positive, the reaction is spontaneous under standard conditions.

(a) $E°_{cell} = E°(\text{cathode}) - E°(\text{anode}) = +0.96 \text{ V} - (+0.79 \text{ V}) = +0.17 \text{ V}$.

Therefore, spontaneous galvanic cell:

$$Hg(l) | Hg_2^{2+}(aq) \| NO_3^-(aq), H^+(aq) | NO(g) | Pt(s)$$
$$\Delta G°_r = -nFE° = -(6)(9.65 \times 10^4 \text{ C} \cdot \text{mol}^{-1})(+0.17 \text{ J} \cdot \text{C}^{-1}) = -98 \text{ kJ} \cdot \text{mol}^{-1}$$

(b) $E°_{cell} = E°(\text{cathode}) - E°(\text{anode}) = +0.92 \text{ V} - (+1.09 \text{ V}) = -0.17 \text{ V}$

Therefore, not spontaneous.

(c) $E°_{cell} = E°(\text{cathode}) - E°(\text{anode}) = +1.33 \text{ V} - (+0.97 \text{ V}) = +0.36 \text{ V}$

Therefore, spontaneous galvanic cell.

$$Pt(s) | Pu^{3+}(aq), Pu^{4+}(aq) \| Cr_2O_7^{2-}(aq), Cr^{3+}(aq), H^+(aq) | Pt(s)$$
$$\Delta G°_r = -nFE° = -(6)(9.65 \times 10^4 \text{ C} \cdot \text{mol}^{-1})(0.36 \text{ J} \cdot \text{C}^{-1}) = -208 \text{ kJ} \cdot \text{mol}^{-1}$$

**12.91** (a) $M_{Ag^+} V_{Ag^+} = M_{I^-} V_{I^-}$

$$M_{Ag^+} = \frac{M_{I^-} V_{I^-}}{V_{Ag^+}} = \frac{(0.015 \text{ M})(16.7 \text{ mL})}{(25.0 \text{ mL})}$$

$$= 1.0 \times 10^{-2} \text{ M}$$

(b) We can find $[Ag^+]$ by using the Nernst equation appropriately.

$$E = E° - \left(\frac{0.025\ 693 \text{ V}}{n}\right) \ln\left(\frac{1}{[Ag^+]}\right)$$

Since the standard reduction potential of silver(I) is +0.80 V, it will be the reduction half reaction versus the S.H.E., so $[Ag^+]$ appears in the denominator of $Q$. In addition, $n = 1$ and $E° = 0.080$ V .

$$0.325 \text{ V} = 0.80 \text{ V} - \left(\frac{0.025\ 693 \text{ V}}{1}\right) \ln\left(\frac{1}{[Ag^+]}\right)$$

$$-0.475 \text{ V} = (-2.567 \times 10^{-2} \text{ V})(\ln 1 - \ln[Ag^+])$$

$$-18.50 = \ln[Ag^+]$$

$$[Ag^+] = 9.23 \times 10^{-9} \text{ M}$$

Recalling that $K_{sp} = [Ag^+][I^-]$, and assuming $[Ag^+]=[I^-]$ at the stoichiometric point of the titration,

$$K_{sp} = [Ag^+][I^-] = (9.23 \times 10^{-9})^2 = 8.5 \times 10^{-17}$$

**12.93** $F_2(g) + 2 e^- \rightarrow 2 F^-(aq)$  $E°(\text{cathode}) = +2.87$ V

$2 HF(aq) \rightarrow F_2(g) + 2 H^+(aq) + 2 e^-$  $E°(\text{anode}) = +3.03$ V

$2 HF(aq) \rightarrow 2 H^+(aq) + 2 F^-(aq)$

$E°_{cell} = E°(\text{cathode})-E°(\text{anode}) = +2.87 \text{ V} - (+3.03 \text{ V}) = -0.16$ V

For the above reaction, $K = \dfrac{[H^+]^2[F^-]^2}{[HF]^2}$ and $\ln K = \dfrac{nFE°}{RT}$

at $25°C = \dfrac{nE°}{0.025\ 69 \text{ V}} = \dfrac{(2)(-0.16 \text{ V})}{0.025\ 69 \text{ V}} = -12$

$K = 10^{-5}$

$K_a = \sqrt{K} = \sqrt{10^{-5}} = 10^{-3}$

**12.95** The wording of this exercise suggests that $K^+$ ions participate in an electrolyte concentration cell reaction. Therefore, $E°_{cell} = 0.00$ V, because the two half cells would be identical under standard conditions. Then,

$$E = E° - \left(\frac{0.0257 \text{ V}}{n}\right) \ln\left(\frac{[K_{out}^+]}{[K_{in}^+]}\right) = 0.00 \text{ V} - \left(\frac{0.0257 \text{ V}}{1}\right) \ln\left(\frac{1}{30}\right)$$

$$= +0.09 \text{ V}$$

and $E = 0.00 \text{ V} - \left(\frac{0.0257 \text{ V}}{1}\right) \ln\left(\frac{1}{20}\right) = +0.08 \text{ V}$

The range of potentials is 0.08 V to 0.09 $V$.

**12.97** $Ag^+(aq) + e^- \rightarrow Ag$ $E°(\text{cathode}) = +0.80$ V

$Fe^{2+}(aq) \rightarrow Fe^{3+}(aq) + e^-$ $E°(\text{anode}) = +0.77$ V

$Ag^+(aq) + Fe^{2+}(aq) \rightarrow Fe^{3+}(aq) + Ag(s)$ $E°_{cell} = +0.03$ V

$$E_{cell} = E°_{cell} - \left(\frac{0.0257 \text{ V}}{n}\right) \ln\left(\frac{[Fe^{3+}]}{[Ag^+][Fe^{2+}]}\right)$$

$$= 0.03 \text{ V} - (0.0257 \text{ V}) \ln\left(\frac{0.20}{(0.020)(0.0010)}\right) = 0.03 \text{ V} - 0.30 \text{ V}$$

$$= -0.21 \text{ V}$$

*Comment:* The cell changes from spontaneous to nonspontaneous as a function of concentration.

**12.99** Since the number of electrons transferred in each half-reaction is different, the Gibbs free energy relationship must be used rather than just the reduction potentials themselves (see Example 12.6).

$$\Delta G_3° = \Delta G_1° + \Delta G_2°$$

$$(-n_3 FE_3°) = (-n_1 FE_1°) + (-n_2 FE_2°)$$

$$n_3 E_3° = n_1 E_1° + n_2 E_2°$$

$$E_3° = \frac{(1)(-0.256 \text{ V}) + (2)(-1.175 \text{ V})}{3}$$

$$= -0.869 \text{ V}$$

**12.101** buffer system $= HA \rightarrow H^+ + A^-$

$$Q = \frac{(H^+)(A^-)}{(HA)}$$

Note: $(H^+)$, as opposed to $[H^+]$, indicates a nonequilibrium molarity.

Because in a buffer system $(A) \approx (HA)$, we can write

$$Q = (H^+)$$

$$E_{cell} = E^\circ_{cell} - \frac{RT}{nF} \ln (H^+)$$

$$0.060 \text{ V} = E^\circ_{cell} - \left( \frac{0.025\ 693}{1} \right) (2.303) (\log(H^+))$$

Because $\log (H^+) = - [-\log(H^+)] = -pH$, we have

$$0.060 \text{ V} = E^\circ_{cell} - 0.0592 \times (-pH)$$

$$0.060 \text{ V} = E^\circ_{cell} + 0.0592 \times pH$$

$$0.060 \text{ V} = E^\circ_{cell} + 0.0592 \times 9.40$$

$$0.060 \text{ V} = E^\circ_{cell} + 0.556 \text{ V}$$

$$E^\circ = 0.060 \text{ V} - 0.556 \text{ V} = -0.496 \text{ V}$$

Similarly, $0.22 \text{ V} = -0.496 \text{ V} + 0.0592 \text{ V} \times pH$

$$pH = \frac{0.22 \text{ V} + 0.496 \text{ V}}{0.0592 \text{ V}} = 12$$

**12.103** (1) $ClO_4^- + 2 H^+ + 2 e^- \rightarrow ClO_3^- + H_2O \quad E^\circ = +1.23 \text{ V}$

(2) $ClO_4^- + H_2O + 2 e^- \rightarrow ClO_3^- + 2 OH^- \quad E^\circ = +0.36 \text{ V}$

(a) The Nernst equation can be used to derive the potential as a function of pH:

$$E' = E^\circ - \frac{RT}{nF} \ln Q$$

For (1), $E'(1) = 1.23 \text{ V} - \dfrac{0.05916 \text{ V}}{2} \log \dfrac{[ClO_3^-]}{[ClO_4^-][H^+]^2}$

We are only interested in varying $[H^+]$, so the $[ClO_3^-]$ and $[ClO_4^-]$ will be left at the standard values of 1 M.

$$E'(1) = 1.23\ V - \frac{0.05916\ V}{2}\log\frac{1}{[H^+]^2}$$

$$= 1.23\ V - \frac{0.05916\ V}{2}\times 2\log\frac{1}{[H^+]}$$

$$= 1.23\ V - (0.05916\ V)(-\log[H^+])$$

$$= 1.23\ V - (0.05916\ V)\ pH$$

Similarly, for (2):

$$E'(2) = 0.36\ V - \frac{0.05916\ V}{2}\log\frac{[ClO_3^-][OH^-]^2}{[ClO_4^-]}$$

As above, we are only interested in varying

$[OH^-]$, so $[ClO_3^-]$ and $[ClO_4^-]$ will be left at the standard value of 1 M.

$$E'(2) = 0.36\ V - \frac{0.05916\ V}{2}\log[OH^-]^2$$

$$= 0.36\ V - \frac{0.05916\ V}{2}\times 2\log[OH^-]$$

$$= 0.36\ V - (0.05916\ V)\times\log[OH^-]$$

$$= 0.36\ V + (0.05916\ V)\ pOH$$

Because $pOH + pH = pK_w = 14.00$, we can write:

$pOH = 14.00 - pH$

$$E'(2) = 0.36\ V + (0.05916\ V)(14.00 - pH)$$

$$= 0.36\ V + 0.83\ V - (0.05916\ V)\ pH$$

$$= 1.19\ V - (0.05916\ V)\ pH$$

If we compare this to $E'(1)$, we find that the equations are essentially the same. They should be identical, the difference being due to the limitation of the number of significant figures available for the calculations.

(b) From the discussion above, we can see that the potential in neutral solution should be the same, regardless of which half-reaction we use to calculate the value.

Using $E'(1) = +1.23\ V - (0.05916\ V)\ pH$,

$E'(1) = +1.23 \text{ V} - (0.05916 \text{ V})(7.00) = +0.82 \text{ V}.$

Using $E'(2) = +0.36 \text{ V} + (0.05916 \text{ V}) \text{pOH},$

$E'(2) = +0.36 \text{ V} + (0.05916 \text{ V})(7.00) = +0.77 \text{ V}.$

Although these numbers differ slightly, they should be identical; again the difference lies in the limitation of the number of significant figures.

**12.105 (a)** $\quad Fe^{2+} + 2\,e^- \rightarrow Fe \qquad\qquad E° = -0.44 \text{ V}$

$\qquad\qquad Mn^{2+} + 2\,e^- \rightarrow Mn \qquad\quad E° = -1.18$

Because these are reduction reactions, we need a corresponding oxidation. The nitrate ion contains N in its highest oxidation state so it cannot be oxidized further. The logical choice is the oxidation of water. The appropriate reduction potential is

$O_2 + 4\,H^+ + 4\,e^- \rightarrow 2\,H_2O \quad E° = +1.23 \text{ V}$

The two overall reactions will be:

$2\,Fe^{2+} + 2\,H_2O \rightarrow 2\,Fe + O_2 + 4\,H^+ \quad E° = -0.44 \text{ V} - 1.23 \text{ V}$
$$= -1.67 \text{ V}$$

$2\,Mn^{2+} + 2\,H_2O \rightarrow 2\,Mn + O_2 + 4\,H^+ \qquad E° = -1.18 \text{ V} - 1.23 \text{ V}$
$$= -2.41 \text{ V}$$

**(b)** The actual potentials, however, will differ from these standard potentials because the concentrations of the metal ions and hydrogen ions are not 1 M, and the pressure of $O_2$ is not 1 bar. To calculate the actual values, the Nernst equation is used.

For the Fe reaction: $E = -1.67 \text{ V} - \dfrac{0.05916}{4} \log \dfrac{P_{O_2}[H^+]^4}{[Fe^{2+}]^2}$

In an open beaker, with the metal ions dissolved in water with pH = 5.00, the pressure of $O_2$ will be $0.2095 \times 1.00 \text{ atm} \times 0.987 \text{ bar} \cdot \text{atm}^{-1} = 0.207$ bar. Substituting the specific values will give

$E = -1.67 \text{ V} - \dfrac{0.05916}{4} \log \dfrac{(0.207)(1.00 \times 10^{-5})^4}{(0.0950)^2}$

$\qquad = -1.67 \text{ V} + 0.276 \text{ V} = -1.39 \text{ V}$

For the Mn reaction:

$$E = -2.41 \text{ V} - \frac{0.059\,16}{4} \log \frac{(0.207)(1.00 \times 10^{-5})^4}{(0.115)^2}$$

$$= -2.41 \text{ V} + 0.278 \text{ V} = -2.13 \text{ V}$$

In order to plate out iron from this mixture, 1.39 V must be applied, and 2.13 V must be applied to cause the reduction of $Mn^{2+}$.

(c) Because the potential for reducing iron(II) is more positive than the potential for reducing manganese(II), the iron will plate out first.

(d) The answer to this question is obtained from the Nernst equation by determining the concentration of $Fe^{2+}$ when the applied potential reaches 2.13 V:

$$-2.13 \text{ V} = -1.67 \text{ V} - \frac{0.059\,16}{4} \log \frac{(0.207)[H^+]^4}{[Fe^{2+}]^2}$$

$$-0.46 \text{ V} = -\frac{0.059\,16}{4} \log \frac{(0.207)[H^+]}{[Fe^{2+}]^2}$$

$$31.10 = \log 0.207 + \log \frac{[H^+]^4}{[Fe^{2+}]^2}$$

$$31.78 = \log \frac{[H^+]^4}{[Fe^{2+}]^2}$$

$$\frac{[H^+]^4}{[Fe^{2+}]^2} = 6.0 \times 10^{31}$$

$$\frac{[H^+]^2}{[Fe^{2+}]} = 7.7 \times 10^{15}$$

For the last ratio to be $7.7 \times 10^{15}$, essentially all of the $Fe^{2+}$ must be converted to Fe(s). This means that $[H^+]$ will essentially be 0.190 mol·$L^{-1}$. Substituting this number gives $[Fe^{2+}] = 5 \times 10^{-18}$ mol·$L^{-1}$. We can say that the iron is quantitatively precipitated by this point.

We might note, however, that the potential of 2.13 V is now no longer the potential at which $Mn^{2+}$ will begin to be reduced. Because the reduction of $Fe^{2+}$ has produced a considerable amount of acid, the original reduction of $Mn^{2+}$ should be recalculated:

$$E = -2.41 \text{ V} - \frac{0.059\,16}{4} \log \frac{(0.207)(0.190)^4}{(0.115)^2} = -2.39 \text{ V}$$

Thus, even less iron will remain in solution.

**12.107** (a) Addition of an electron to any molecule should have the electron enter the molecule's lowest unoccupied molecular orbital (LUMO) first. (b) For $CH_3X$, one would predict that the LUMO would be antibonding between C and one of the atoms attached to it. Because the C—H bond strength ($412 \text{ kJ} \cdot \text{mol}^{-1}$) is greater than all of the C—X bond strengths given (C—Cl, $338 \text{ kJ} \cdot \text{mol}^{-1}$; C—Br, $276 \text{ kJ} \cdot \text{mol}^{-1}$; C—I, $238$ $\text{kJ} \cdot \text{mol}^{-1}$) we would expect the LUMO to be the antibonding orbital for the C—X bond. Adding an electron to this orbital should then result in a weakening of the C—X bond. The result is the elimination of $X^-$ and the formation of a $CH_3$ radical:

$$CH_3X + e^- \rightarrow CH_3 + X^-$$

(c) We would expect this reduction process to follow the C—X bond strengths so that the formation of $X^-$ and generation of $CH_3$ radicals would be easiest for X = I, followed by Br, and then Cl.

**12.109** In order to determine the current applied, we need to find the number of moles of electrons transferred. The electrolysis of water to produce gaseous oxygen and hydrogen,

$2 H_2O(l) \rightarrow O_2(g) + 2 H_2(g)$, transfers 4 moles of electrons for each mole of oxygen gas produced: $4 OH^-(aq) \rightarrow O_2(g) + 2 H_2O(l) + 4 e^-$. We can determine the number of moles of oxygen from its volume, partial pressure, and temperature.

$$n_{O_2} = \frac{p_{O_2}V}{RT} = \frac{(p_{tot} - p_{H_2O})V}{RT}$$

$$= \frac{(722 \text{ Torr} - 19.83 \text{ Torr})(25.0 \text{ mL})}{(0.08206 \text{ L} \cdot \text{atm} \cdot \text{K}^{-1} \cdot \text{mol}^{-1})(295 \text{ K})} \cdot \frac{1 \text{ L}}{1000 \text{ mL}} \cdot \frac{1 \text{ atm}}{760 \text{ Torr}}$$

$$= 9.542 \times 10^{-4} \text{ mol O}_2 \text{ produced}$$

$$n_{e^-} = n_{O_2} \times \frac{4 \text{ mol e}^-}{\text{mol O}_2} = 9.542 \times 10^{-4} \text{ mol O}_2 \times \frac{4 \text{ mol e}^-}{\text{mol O}_2}$$

$$= 3.817 \times 10^{-3} \text{ mol e}^-$$

$$It = nF$$

$$I = \frac{nF}{t} = \frac{(3.817 \times 10^{-3} \text{ mol e}^-)}{(30.0 \text{ min})(60 \text{ s} \cdot \text{min}^{-1})} \cdot \left( 96\,485 \frac{\text{C}}{\text{mol e}^-} \right)$$

$$= 0.205 \text{ A}$$

**12.111** The strategy for working this problem is to create a set of equations that will add to the desired equilibrium reaction:

$$HBrO(aq) + H_2O(l) \rightarrow H_3O^+(aq) + BrO^-(aq)$$

From Appendix 2B, we find

$$2 \text{ HBrO} + 2 \text{ H}^+ + 2 \text{ e}^- \rightarrow Br_2 + 2 \text{ H}_2O \qquad E^\circ = +1.60 \text{ V}$$

$$BrO^- + H_2O + 2 \text{ e}^- \rightarrow Br^- + 2 \text{ OH}^- \qquad E^\circ = +0.76$$

On examination of these equations, it is clear that we will also need a half-reaction that, when combined with the two above, will eliminate $Br_2$ and $Br^-$. The obvious choice is

$$Br_2 + 2 \text{ e}^- \rightarrow 2 \text{ Br}^- \qquad E^\circ = +1.09$$

We combine these by adding twice the reverse reaction to the other two:

$$2 \text{ HBrO} + 2 \text{ H}^+ + 2 \text{ e}^- \rightarrow Br_2 + 2 \text{ H}_2O$$
$$2(Br^- + 2 \text{ OH}^- \rightarrow BrO^- + H_2O + 2 \text{ e}^-)$$
$$Br_2 + 2 \text{ e}^- \rightarrow 2 \text{ Br}^-$$

$$2 \text{ HBrO} + 2 \text{ H}^+ + 4 \text{ OH}^- \rightarrow 2 \text{ BrO}^- + 4 \text{ H}_2O$$

*Caution:* We must be careful here in adding the $E°$ values—we have created essentially a new half-reaction by summing these reactions, which requires that we convert to $\Delta G$ values. Whenever one sums more than two half-reactions, it is necessary to convert to the $\Delta G$ values using $\Delta G° = nFE°$, in order to work the problem:

$$2\ HBrO + 2\ H^+ + 2\ e^- \rightarrow Br_2 + 2\ H_2O$$
$$\Delta G° = -2(9.65 \times 10^4\ C \cdot mol^{-1})(+1.60\ V) = -309\ kJ \cdot mol^{-1}$$

$$BrO^- + H_2O + 2\ e^- \rightarrow Br^- + 2\ OH^-$$
$$\Delta G° = -2(9.65 \times 10^4\ C \cdot mol^{-1})(+0.76\ V) = -147\ kJ \cdot mol^{-1}$$

$$Br_2^- + 2\ e^- \rightarrow 2\ Br^-$$
$$\Delta G° = -2(9.65 \times 10^4\ C \cdot mol^{-1})(+1.09\ V) = -210\ kJ \cdot mol^{-1}$$

For $2\ HBrO + 2\ H^+ + 4\ OH^- \rightarrow 2\ BrO^- + 4\ H_2O$

$\Delta G° = -309\ kJ + 2(+147\ kJ) - 210\ kJ = -225\ kJ \cdot mol^{-1}$

We now see that we will need to eliminate $OH^-$ from the left side of the equation. This can be done in one of two ways: we can use the $K_w$ value for the autoprotolysis of water or, equivalently, we can use appropriate half-reactions that sum to the autoprotolysis of water. The appropriate half-reactions are

$$2\ H_2O \rightarrow O_2 + 4\ H^+ + 4\ e^- \qquad E° = -1.23\ V$$

$$O_2 + 2\ H_2O + 4\ e^- \rightarrow 4\ OH^- \qquad E° = +0.40\ V$$

These sum to give

$$4\ H_2O \rightarrow 4\ H^+ + 4\ OH^- \qquad E° = -0.83\ V$$

This is a $4\ e^-$ reaction. Alternatively, one can write the $1\ e^-$ process that will have the same $E°$ value.

$$H_2O \rightarrow H^+ + OH^- \qquad E° = -0.83\ V$$

$$\Delta G° = -(1)(9.65 \times 10^4\ C \cdot mol^{-1})(-0.83\ V) = +80\ kJ \cdot mol^{-1}$$

$$2\ HBrO + 2\ H^+ + 4\ OH^- \rightarrow 2\ BrO^- + 4\ H_2O \qquad \Delta G° = -225\ kJ \cdot mol^{-1}$$

$$4(H_2O \rightarrow H^+ + OH^-) \qquad 4(\Delta G° = +80\ kJ \cdot mol^{-1})$$

$$2 \text{ HBrO} \rightarrow 2 \text{ H}^+ + 2 \text{ BrO}^- \qquad \Delta G^\circ = -225 \text{ kJ} \cdot \text{mol}^{-1} + 4(+80 \text{ kJ} \cdot \text{mol}^{-1})$$
$$= +95 \text{ kJ} \cdot \text{mol}^{-1}$$

The desired reaction is half of this, for which $\Delta G^\circ = +48 \text{ kJ} \cdot \text{mol}^{-1}$.

Using $\Delta G^\circ = -RT \ln K$, we obtain $K = 4 \times 10^{-9}$, which is in reasonable agreement for this type of calculation with the value of $2 \times 10^{-9}$ given in Table 10.1.

# CHAPTER 13
# CHEMICAL KINETICS

**13.1** (a) $\text{rate } (N_2) = \text{rate}(H_2) \times \left( \dfrac{1 \text{ mol } N_2}{3 \text{ mol } H_2} \right) = \dfrac{1}{3} \times \text{rate}(H_2)$

    (b) $\text{rate } (NH_3) = \text{rate}(H_2) \times \left( \dfrac{2 \text{ mol } NH_3}{3 \text{ mol } H_2} \right) = \dfrac{2}{3} \times \text{rate}(H_2)$

    (c) $\text{rate } (NH_3) = \text{rate}(N_2) \times \left( \dfrac{2 \text{ mol } NH_3}{1 \text{ mol } N_2} \right) = 2 \times \text{rate}(N_2)$

**13.3** (a) The rate of formation of dichromate ions =

$$\left( \dfrac{0.14 \text{ mol } Cr_2O_7^{2-}}{L \cdot s} \right) \left( \dfrac{2 \text{ mol } CrO_4^{2-}}{1 \text{ mol } Cr_2O_7^{2-}} \right) = 0.28 \text{ mol} \cdot L \cdot s^{-1}$$

    (b) $0.14 \text{ mol} \cdot L^{-1} \cdot s^{-1} \div 1 = 0.14 \text{ mol} \cdot L^{-1} \cdot s^{-1}$

**13.5** (a) $\text{rate of formation of } O_2 = \left( 6.5 \times 10^{-3} \dfrac{\text{mol } NO_2}{L \cdot s} \right) \times \left( \dfrac{1 \text{ mol } O_2}{2 \text{ mol } NO_2} \right)$

$$= 3.3 \times 10^{-3} \text{ (mol } O_2) \cdot L^{-1} \cdot s^{-1}$$

    (b) $6.5 \times 10^{-3} \text{ mol} \cdot L^{-1} \cdot s^{-1} \div 2 = 3.3 \times 10^{-3} \text{ mol} \cdot L^{-1} \cdot s^{-1}$

**13.7**  (a) and (c)

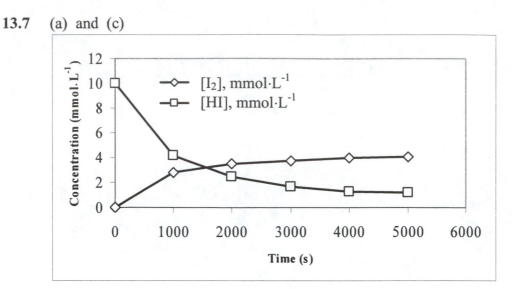

Note that the curves for the $[I_2]$ and $[H_2]$ are identical and only the $[I_2]$

curve is shown.

(b)  The rates at individual points are given by the slopes of the lines

tangent to the points in question. If these are determined graphically, there

may be some variation from the numbers given below.

time, s  rate, $mmol \cdot L^{-1} \cdot s^{-1}$

0       20.0060

1000    20.003

2000    20.000 98

3000    20.000 61

4000    20.000 40

5000    20.000 31

**13.9**  For A $\longrightarrow$ products, rate $= (mol\ A) \cdot L^{-1} \cdot s^{-1}$

(a) rate

$[(mol\ A) \cdot L^{-1} \cdot s^{-1}] = k_0[A]^0 = k_0$, so units of $k_0$ are $(mol\ A) \cdot L^{-1} \cdot s^{-1}$

(same as the units for the rate, in this case)

(b) rate

$[(mol\ A) \cdot L^{-1} \cdot s^{-1}] = k_1[A]$, so units of $k_1$ are $\dfrac{(mol\ A) \cdot L^{-1} \cdot s^{-1}}{(mol\ A) \cdot L^{-1}} = s^{-1}$

(c) rate

$[(mol\ A) \cdot L^{-1} \cdot s^{-1}] = k_1[A]^2$, so units of $k_1$ are $\dfrac{(mol\ A) \cdot L^{-1} \cdot s^{-1}}{\left[(mol\ A) \cdot L^{-1}\right]^2}$

$= L \cdot (mol\ A)^{-1} \cdot s^{-1}$

**13.11** From the units of the rate constant, $k$, it follows that the reaction is first

order, thus rate $= k[N_2O_5]$.

$[N_2O_5] = \left(\dfrac{3.45\ g\ N_2O_5}{0.750\ L}\right)\left(\dfrac{1\ mol\ N_2O_5}{108.02\ g\ N_2O_5}\right) = 0.0426\ mol \cdot L^{-1}$

rate $= 5.2 \times 10^{-3}\ s^{-1} \times 0.0426\ mol \cdot L^{-1} = 2.2 \times 10^{-4}\ (mol\ N_2O_5) \cdot L^{-1} \cdot s^{-1}$

**13.13** (a) From the units of the rate constant, it follows that the reaction is

second order; therefore,

rate $= k[H_2][I_2]$

$= (0.063\ L \cdot mol^{-1} \cdot s^{-1})\left(\dfrac{0.52\ g\ H_2}{0.750\ L}\right)\left(\dfrac{1\ mol\ H_2}{2.016\ g\ H_2}\right)\left(\dfrac{0.19\ g\ I_2}{0.750\ L}\right)\left(\dfrac{1\ mol\ I_2}{253.8\ g\ I_2}\right)$

$= 2.2 \times 10^{-5}\ mol \cdot L^{-1} \cdot s^{-1}$

(b) rate(new) $= k \times 2 \times [H_2]_{initial}[I_2] = 2 \times$ rate(initial), so, by a factor of 2

**13.15** Because the rate increased in direct proportion to the concentrations of

both reactants, the rate is first order in both reactants.

rate $= k[CH_3Br][OH^-]$

**13.17** When the concentration of ICl is doubled, the rate doubles (experiments 1

and 2). Therefore, the reaction is first order in ICl. When the concentration

of $H_2$ is tripled, the rate triples (experiments 2 and 3); thus, the reaction is

first order in $H_2$.

(a) rate $= k[ICl][H_2]$

(b) $k = \left( \dfrac{22 \times 10^{-7}\ \text{mol}}{\text{L} \cdot \text{s}} \right) \left( \dfrac{\text{L}}{3.0 \times 10^{-3}\ \text{mol}} \right) \left( \dfrac{\text{L}}{4.5 \times 10^{-3}\ \text{mol}} \right)$

$= 0.16\ \text{L} \cdot \text{mol}^{-1} \cdot \text{s}^{-1}$

(c) $\text{rate} = \left( \dfrac{0.16\ \text{L}}{\text{mol} \cdot \text{s}} \right) \left( \dfrac{4.7 \times 10^{-3}\ \text{mol}}{\text{L}} \right) \left( \dfrac{2.7 \times 10^{-3}\ \text{mol}}{\text{L}} \right)$

$= 2.0 \times 10^{-6}\ \text{mol} \cdot \text{L}^{-1} \cdot \text{s}^{-1}$

**13.19** (a) Doubling the concentration of A (experiments 1 and 2) doubles the rate; therefore, the reaction is first order in A. Increasing the concentration of B by the ratio 3.02/1.25 (experiments 2 and 3) increases the rate by $(3.02/1.25)^2$; hence, the reaction is second order in B. Tripling the concentration of C (experiments 3 and 4) increases the rate by $3^2 = 9$; thus, the reaction is second order in C. Therefore, $\text{rate} = k[\text{A}][\text{B}]^2[\text{C}]^2$.

(b) overall order $= 5$

(c) $k = \dfrac{\text{rate}}{[\text{A}][\text{B}]^2[\text{C}]^2}$

Using the data from experiment 4, we get

$k = \left( \dfrac{0.457\ \text{mol}}{\text{L} \cdot \text{s}} \right) \left( \dfrac{\text{L}}{1.25 \times 10^{-3}\ \text{mol}} \right) \left( \dfrac{\text{L}}{3.02 \times 10^{-3}\ \text{mol}} \right)^2 \left( \dfrac{\text{L}}{3.75 \times 10^{-3}\ \text{mol}} \right)^2$

$= 2.85 \times 10^{12}\ \text{L}^4 \cdot \text{mol}^{-4} \cdot \text{s}^{-1}$

From experiment 3, we get

$k = \left( \dfrac{5.08 \times 10^{-2}\ \text{mol}}{\text{L} \cdot \text{s}} \right) \left( \dfrac{\text{L}}{1.25 \times 10^{-3}\ \text{mol}} \right) \left( \dfrac{\text{L}}{3.02 \times 10^{-3}\ \text{mol}} \right)^2$

$\times \left( \dfrac{\text{L}}{1.25 \times 10^{-3}\ \text{mol}} \right)^2$

$= 2.85 \times 10^{12}\ \text{L}^4 \cdot \text{mol}^{-4} \cdot \text{s}^{-1}$ (Checks!)

(d) $\text{rate} = \left( \dfrac{2.85 \times 10^{12}\ L^4}{mol^4 \cdot s} \right)\left( \dfrac{3.01 \times 10^{-3}\ mol}{L} \right)\left( \dfrac{1.00 \times 10^{-3}\ mol}{L} \right)^2$

$\times \left( \dfrac{1.15 \times 10^{-3}\ mol}{L} \right)^2$

$= 1.13 \times 10^{-2}\ mol \cdot L^{-1} \cdot s^{-1}$

**13.21** (a) $k = \dfrac{0.693}{t_{1/2}} = \dfrac{0.693}{1000\ s} = 6.93 \times 10^{-4}\ s^{-1}$

(b) We use $\ln \left( \dfrac{[A]_0}{[A]_t} \right) = kt$ and solve for $k$.

$k = \dfrac{\ln\,([A]_0/[A]_t)}{t} = \dfrac{\ln \left( \dfrac{0.67\ mol \cdot L^{-1}}{0.53\ mol \cdot L^{-1}} \right)}{25\ s} = 9.4 \times 10^{-3}\ s^{-1}$

(c) $[A]_t = \left( \dfrac{0.153\ mol\ A}{L} \right) - \left[ \left( \dfrac{2\ mol\ A}{1\ mol\ B} \right)\left( \dfrac{0.034\ mol\ B}{L} \right) \right]$

$= 0.085\ (mol\ A) \cdot L^{-1}$

$k = \dfrac{\ln \left( \dfrac{0.153\ mol \cdot L^{-1}}{0.085\ mol \cdot L^{-1}} \right)}{115\ s} = 5.1 \times 10^{-3}\ s^{-1}$

**13.23** (a) $t_{1/2} = \dfrac{0.693}{k} = \left( \dfrac{0.693\ s}{3.7 \times 10^{-5}} \right)\left( \dfrac{1\ min}{60\ s} \right)\left( \dfrac{1\ h}{60\ min} \right) = 5.2\ h$

(b) $[A]_t = [A]_0\ e^{-kt}$

$t = 3.5\ h \times 3600\ s \cdot h^{-1} = 1.3 \times 10^4\ s$

$[N_2O_5] = 0.0567\ mol \cdot L^{-1} \times e^{-(3.7 \times 10^{-5}\ s^{-1})(1.3 \times 10^4\ s)} = 3.5 \times 10^{-2}\ mol \cdot L^{-1}$

(c) Solve for $t$ from $\ln \left( \dfrac{[A]_0}{[A]_t} \right) = kt$, which gives

$$t = \frac{\ln\left(\dfrac{[A]_0}{[A]_t}\right)}{k} = \frac{\ln\left(\dfrac{[N_2O_5]_0}{[N_2O_5]_t}\right)}{k} = \frac{\ln\left(\dfrac{0.0567}{0.0135}\right)}{3.7 \times 10^{-5} \text{ s}^{-1}} = 3.9 \times 10^4 \text{ s}$$

$$= (3.9 \times 10^4 \text{ s})\left(\frac{1 \text{ min}}{60 \text{ s}}\right) = 6.5 \times 10^2 \text{ min}$$

**13.25** (a) $\dfrac{[A]}{[A]_0} = \dfrac{1}{4} = \left(\dfrac{1}{2}\right)^2$ ; so the time elapsed is 2 half-lives.

$t = 2 \times 355 \text{ s} = 710 \text{ s}$

(b) Because 15% is not a multiple of $\frac{1}{2}$, we cannot work directly from the

half-life. But $k = 0.693/t_{1/2}$

so $k = \dfrac{0.693}{355 \text{ s}} = 1.95 \times 10^{-3} \text{ s}^{-1}$

Then [see the solution to Exercise 13.23(c)],

$$t = \frac{\ln\left(\dfrac{[A]_0}{[A]_t}\right)}{k} = \frac{\ln\left(\dfrac{1}{0.15}\right)}{1.95 \times 10^{-3} \text{ s}^{-1}} = 9.7 \times 10^2 \text{ s}$$

(c) $t = \dfrac{\ln\dfrac{[A]_0}{\frac{1}{9}[A]_0}}{k} = \dfrac{\ln 9}{1.95 \times 10^{-3} \text{ s}^{-1}} = 1.1 \times 10^3 \text{ s}$

**13.27** (a) $t_{1/2} = \dfrac{0.693}{k} = \dfrac{0.693}{2.81 \times 10^{-3} \text{ min}^{-1}} = 247 \text{ min}$

(b) See the solutions to Exercises 13.31(c) and 13.33(c).

$$t = \frac{\ln\left(\dfrac{[SO_2Cl_2]_0}{[SO_2Cl_2]_t}\right)}{k} = \frac{\ln 10}{2.81 \times 10^{-3} \text{ min}^{-1}} = 819 \text{ min}$$

(c) $[A]_t = [A]_0 \, e^{-kt}$

Because the vessel is sealed, masses and concentrations are proportional,

and we write

$$(\text{mass left})_t = (\text{mass})_0 \; e^{-kt}$$

$$= 14.0 \text{ g} \times e^{-(2.81 \times 10^{-3} \text{ min}^{-1} \times 60 \text{ min} \cdot \text{h}^{-1} \times 1.5 \text{ h})}$$

$$= 10.9 \text{ g}$$

Note: Knowledge of the volume of the vessel is not required. However, we could have converted mass to concentration, solved for the new concentration at 1.5 h. and finally converted back to the new (remaining) mass. But this is not necessary

**13.29** (a) We first calculate the concentration of A at 3.0 min.

$$[A]_t = [A]_0 - \left( \frac{1 \text{ mol A}}{3 \text{ mol B}} \right) \times [B]_t$$

$$= 0.015 \text{ mol} \cdot \text{L}^{-1} - \left( \frac{1 \text{ mol A}}{3 \text{ mol B}} \right) \times 0.018 \text{ (mol B)} \cdot \text{L}^{-1}$$

$$= 0.009 \text{ mol} \cdot \text{L}^{-1}$$

The rate constant is then determined from the first-order integrated rate law.

$$k = \frac{\ln \left( \dfrac{[A]_0}{[A]_t} \right)}{t} = \frac{\ln \left( \dfrac{0.015}{0.009} \right)}{3.0 \text{ min}} = 0.17 \text{ min}^{-1}$$

(b) $[A]_t = 0.015 \text{ mol} \cdot \text{L}^{-1} - \left( \dfrac{1 \text{ mol A}}{3 \text{ mol B}} \right) \times 0.030 \text{ (mol B)} \cdot \text{L}^{-1}$

$$= 0.005 \text{ mol} \cdot \text{L}^{-1}$$

$$t = \frac{\ln \left( \dfrac{[A]_0}{[A]_t} \right)}{k} = \frac{\ln \left( \dfrac{0.015}{0.005} \right)}{0.17 \text{ min}^{-1}} = 6.5 \text{ min}$$

additional time $= 6.5 \text{ min} - 3.0 \text{ min} = 3.5 \text{ min}$

**13.31** (a)

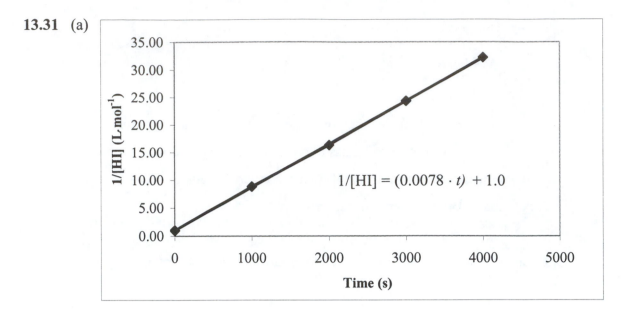

1/[HI] = (0.0078 · t) + 1.0

Equation 17.b in the text can be rearranged as

$$\frac{1}{[A]_t} = \frac{1 + [A]_0\,kt}{[A]_0} = \frac{1}{[A]_0} + kt$$

Thus, if the reaction is second order, a plot of 1/[HI] against time should give a straight line of slope $k$.

As can be seen from the graph, the data fit the equation for a second-order reaction quite well. The slope is determined by a least squares fit of the data by the graphing program.

(b)  i. The rate constant for the rate law for the loss of HI is simply the slope of the best fit line, $7.8 \times 10^{-3}$ L · mol$^{-1}$ · s$^{-1}$. ii. Since two moles of HI are consumed per mole of reaction, the rate constant for the unique rate law is half the slope or $3.9 \times 10^{-3}$ L · mol$^{-1}$ · s$^{-1}$.

**13.33**  It is convenient to obtain an expression for the half-life of a second-order reaction. We work with Eq. 17.b.

$$[A]_t = \frac{[A]_0}{1 + [A]_0 kt} \quad (17.b)$$

$$\frac{[A]_{t_{1/2}}}{[A]_0} = \frac{1}{2} = \frac{1}{1 + [A]_0 kt_{1/2}}$$

Therefore, $1 + [A]_0 kt_{1/2} = 2$, or $[A]_0 kt_{1/2} = 1$, or

$$t_{1/2} = \frac{1}{k[A]_0} \text{ and } k = \frac{1}{t_{1/2}[A]_0}$$

It is also convenient to rewrite Eq. 17.b to solve for $t$. We take reciprocals:

$$\frac{1}{[A]_t} = \frac{1}{[A]_0} + kt$$

giving

$$t = \frac{\dfrac{1}{[A]_t} - \dfrac{1}{[A]_0}}{k}$$

(a) $k = \dfrac{1}{t_{1/2}[A]_0} = \dfrac{1}{(50.5 \text{ s})(0.84 \text{ mol} \cdot \text{L}^{-1})} = 0.024 \text{ L} \cdot \text{mol}^{-1} \cdot \text{s}^{-1}$

$$t = \frac{\dfrac{1}{[A]} - \dfrac{1}{[A]_0}}{k} = \frac{\dfrac{16}{[A]_0} - \dfrac{1}{[A]_0}}{k} = \frac{15}{k[A]_0}$$

$$= \frac{15}{(0.024 \text{ L} \cdot \text{mol}^{-1} \cdot \text{s}^{-1})(0.84 \text{ mol} \cdot \text{L}^{-1})} = 7.4 \times 10^2 \text{ s}$$

(b) $t = \dfrac{\dfrac{4}{[A]_0} - \dfrac{1}{[A]_0}}{k} = \dfrac{3}{k[A]_0}$

$$= \frac{3}{(0.024 \text{ L} \cdot \text{mol}^{-1} \cdot \text{s}^{-1})(0.84 \text{ mol} \cdot \text{L}^{-1})} = 1.5 \times 10^2 \text{ s}$$

(c) $t = \dfrac{\dfrac{5}{[A]_0} - \dfrac{1}{[A]_0}}{k} = \dfrac{4}{k[A]_0}$

$$= \frac{4}{(0.024 \text{ L} \cdot \text{mol}^{-1} \cdot \text{s}^{-1})(0.84 \text{ mol} \cdot \text{L}^{-1})} = 2.0 \times 10^2 \text{ s}$$

**13.35** See the solution to Exercise 13.33 for the derivation of the formulas needed here.

(a) $t = \dfrac{\dfrac{1}{[A]} - \dfrac{1}{[A]_0}}{k} = \dfrac{\dfrac{1\,L}{0.080\,mol} - \dfrac{1\,L}{0.10\,mol}}{0.015\,L\cdot mol^{-1}\cdot min^{-1}} = 1.7 \times 10^2\,min$

(b) $[A] = \dfrac{0.15\,mol\,A}{L} - \left[\left(\dfrac{0.19\,mol\,B}{L}\right)\left(\dfrac{1\,mol\,A}{2\,mol\,B}\right)\right]$

$= 0.055(mol\,A)\cdot L^{-1} = 0.37[A]_0$

$t = \dfrac{\dfrac{1}{[A]_t} - \dfrac{1}{[A]_0}}{k}$

$= \dfrac{\dfrac{1}{0.055\,mol\cdot L^{-1}} - \dfrac{1}{0.15\,mol\cdot L^{-1}}}{0.0035\,L\cdot mol^{-1}\cdot min^{-1}}$

$= 3.3 \times 10^3\,min$

**13.37** (a) $rate = -\dfrac{1}{a}\dfrac{d[A]}{dt} = k[A]$

$\dfrac{d[A]}{[A]} = -ak\,dt$

integrate from $[A]_0$ at $t = 0$ to $[A]_t$ at $t$ :

$\displaystyle\int_{[A]_0}^{[A]_t} \dfrac{d[A]}{[A]} = -ak\int_0^t dt$

$\ln\dfrac{[A]_t}{[A]_0} = -akt$, and $[A]_t = [A]_0\exp(-akt)$

(b) at $t_{1/2}$, $[A]_t = \frac{1}{2}[A]_0$. Therefore:

$\ln\dfrac{[A]_0}{[A]_t} = \ln 2 = akt_{1/2}$, and

$t_{1/2} = \dfrac{\ln 2}{ak}$

**13.39**   Given: $\dfrac{d[A]}{dt} = -k[A]^3$, we can derive an expression for the amount

of time needed for the inital concentration of A, $[A]_0$, to decrease
by 1/2.  Begin by obtaining the integrated rate law for a third-order
reaction by separation of variables:

$$\int_{[A]_0}^{[A]_t} [A]^{-3} d[A] = \int_0^t -k\, dt = -\frac{1}{2}\left[[A]_t^{-2} - [A]_0^{-2}\right] = -kt$$

To obtain an expression for the half-life, let $[A]_t = \tfrac{1}{2}[A]_0$ and $t = t_{1/2}$ :

$$-\frac{1}{2}\left[(\tfrac{1}{2}[A]_0)^{-2} - [A]_0^{-2}\right] = -kt_{1/2}$$

solving for the half-life:

$$t_{1/2} = \frac{3}{2k[A]_0^2}$$

**13.41**   The overall reaction is $CH_2=CHCOOH + HCl \rightarrow ClCH_2CH_2COOH$.  The
intermediates include chloride ion, $CH_2=CHC(OH)_2^+$ and
$ClCH_2CHC(OH)_2$.

**13.43**   The first elementary reaction is the rate-controlling step, because it is the
slow step. The second elementary reaction is fast and does not affect the
overall reaction order, which is second order as a result of the fact that the
rate-controlling step is bimolecular.

rate $= k[NO][Br_2]$

**13.45**   If mechanism (a) were correct, the rate law would be rate $= k[NO_2][CO]$.
But this expression does not agree with the experimental result and can be
eliminated as a possibility. Mechanism (b) has rate $= k[NO_2]^2$ from the
slow step. Step 2 does not influence the overall rate, but it is necessary to
achieve the correct overall reaction; thus this mechanism agrees with the
experimental data. Mechanism (c)  is not correct, which can be seen from
the rate expression for the slow step, rate $= k[NO_3][CO]$.  [CO] cannot be
eliminated from this expression to yield the experimental result, which
does not contain [CO].

**13.47** (a) True;   (b) False. At equilibrium, the *rates* of the forward and reverse reactions are equal, *not the rate constants*.   (c) False. Increasing the concentration of a reactant causes the rate to increase by providing more reacting molecules. It does not affect the rate constant of the reaction.

**13.49** The overall rate of formation of A is $\text{rate} = -k[A] + k'[B]$. The first term accounts for the forward reaction and is negative as this reaction reduces [A]. The second term, which is positive, accounts for the back reaction which increases [A]. Given the 1:1 stoichiometry of the reaction, if no B was present at the beginning of the reaction, [A] and [B] at any time are related by the equation: $[A] + [B] = [A]_o$ where $[A]_o$ is the initial concentration of A. Therefore, the rate law may be written:

$$\frac{d[A]}{dt} = -k[A] + k'([A]_o - [A]) = -(k + k')[A] + k'[A]_o$$

The solution of this first-order differential equation is:

$$[A] = \frac{k' + ke^{-(k'+k)t}}{k' + k}[A]_o$$

As $t \to \infty$ the exponential term in the numerator goes to zero and the concentrations reach their equilibrium values given by:

$$[A]_{eq} = \frac{k'[A]_o}{k' + k} \quad \text{and} \quad [B]_{eq} = [A]_o - [A]_\infty = \frac{k[A]_o}{k + k'}$$

taking the ratio of products over reactants we see that:

$$\frac{[B]_{eq}}{[A]_{eq}} = \frac{k}{k'} = K \quad \text{where } K \text{ is the equilibrium constant for the reaction.}$$

**13.51**

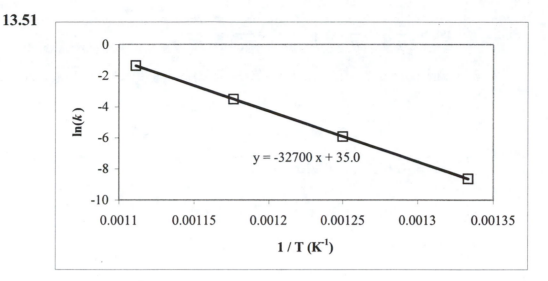

(a) Given the Arrhenius equation, $\ln k = \ln A - E_a/RT$, we see that the slope of the best fit line to the data $(-3.27 \times 10^4 \text{ K})$ is $E_a/R$ and the y-intercept (35.0) is $\ln A$. Therefore,

$$E_a = (3.27 \times 10^4 \text{ K})(8.31 \times 10^{-3} \text{ kJ} \cdot \text{mol}^{-1} \cdot \text{K}^{-1}) = 2.71 \times 10^2 \text{ kJ} \cdot \text{mol}^{-1}.$$

(b) At 600 °C (or 873 K), the rate constant is:

$$\ln(k) = \left(-3.27 \times 10^4 \text{ K}\right)\frac{1}{873 \text{ K}} + 35.0 = -2.46$$

$$k = 0.088$$

**13.53** We use $\ln\left(\dfrac{k'}{k}\right) = \dfrac{E_a}{R}\left(\dfrac{1}{T} - \dfrac{1}{T'}\right) = \dfrac{E_a}{R}\left(\dfrac{T'-T}{T'T}\right)$

$$\ln\left(\frac{k'}{k}\right) = \ln\left(\frac{0.87 \text{ s}^{-1}}{0.76 \text{ s}^{-1}}\right)$$

$$= \left(\frac{E_a}{8.31 \times 10^{-3} \text{ kJ} \cdot \text{K}^{-1} \cdot \text{mol}^{-1}}\right)\left(\frac{1030 \text{ K} - 1000 \text{ K}}{1030 \text{ K} \times 1000 \text{ K}}\right)$$

$$E_a = \frac{(8.31 \times 10^{-3} \text{ kJ} \cdot \text{K}^{-1} \cdot \text{mol}^{-1})(1000 \text{ K})(1030 \text{ K})}{(1030 \text{ K} - 1000 \text{ K})}\ln\left(\frac{0.87 \text{ s}^{-1}}{0.76 \text{ s}^{-1}}\right)$$

$$= 39 \text{ kJ} \cdot \text{mol}^{-1}$$

**13.55** We use $\ln\left(\dfrac{k'}{k}\right) = \dfrac{E_a}{R} = \left(\dfrac{1}{T} - \dfrac{1}{T'}\right) = \dfrac{E_a}{R}\left(\dfrac{T'-T}{TT'}\right)$

$k'$ = rate constant at 700°C, $T' = (700 + 273)\ \text{K} = 973\ \text{K}$

$$\ln\left(\frac{k'}{k}\right) = \left(\frac{315\ \text{kJ}\cdot\text{mol}^{-1}}{8.314\times10^{-3}\ \text{kJ}\cdot\text{mol}^{-1}}\right)\left(\frac{973\ \text{K} - 1073\ \text{K}}{973\ \text{K} \times 1073\ \text{K}}\right)$$

$$= -3.63;\quad \frac{k'}{k} = 0.026$$

$$k' = 0.026 \times 9.7 \times 10^{10}\ \text{L}\cdot\text{mol}^{-1}\cdot\text{s}^{-1} = 2.5 \times 10^{9}\ \text{L}\cdot\text{mol}^{-1}\cdot\text{s}^{-1}$$


**13.57** $\ln\left(\dfrac{k'}{k}\right) = \dfrac{E_a}{R}\left(\dfrac{1}{T} - \dfrac{1}{T'}\right) = \dfrac{E_a}{R}\left(\dfrac{T'-T}{TT'}\right)$

$$= \left(\frac{103\ \text{kJ}\cdot\text{mol}^{-1}}{8.314\times10^{-3}\ \text{kJ}\cdot\text{K}^{-1}\cdot\text{mol}^{-1}}\right)\left(\frac{323\ \text{K} - 318\ \text{K}}{318\ \text{K} \times 323\ \text{K}}\right) = 0.60$$

$$\frac{k'}{k} = 1.8$$

$$k' = 1.8 \times 5.1 \times 10^{-4}\ \text{s}^{-1} = 9.2 \times 10^{-4}\ \text{s}^{-1}$$


**13.59** (a) The equilibrium constant will be given by the ratio of the rate constant of the forward reaction to the rate constant of the reverse reaction:

$$K = \frac{k}{k'} = \frac{265\ \text{L}\cdot\text{mol}^{-1}\cdot\text{min}^{-1}}{392\ \text{L}\cdot\text{mol}^{-1}\cdot\text{min}^{-1}} = 0.676$$

(b) The reaction profile corresponds to a plot similar to that shown in Fig. 13.31a. The reaction is endothermic—the reverse reaction has a lower activation barrier than the forward reaction.

(c) Raising the temperature will increase the rate constant of the reaction with the higher activation barrier more than it will the rate constant of the reaction with the lower energy barrier. We expect the rate of the forward reaction to go up substantially more than for the reverse reaction in this case. $k$ will increase more than $k'$ and consequently the equilibrium constant $K$ will increase. This is consistent with Le Chatelier's principle.

**13.61** (a) cat = catalyzed, uncat = uncatalyzed $E_{a,cat} = \frac{75}{125}E_{a,uncat} = 0.60\,E_{a,uncat}$

$$\frac{\text{rate(cat)}}{\text{rate(uncat)}} = \frac{k_{cat}}{k_{uncat}} = \frac{Ae^{-E_{a,cat}/RT}}{Ae^{-E_{a,uncat}/RT}} = \frac{e^{-(0.60)E_{a,uncat}/RT}}{e^{-E_{a,uncat}/RT}}$$

$$= e^{(-0.60+1.00)E_{a,uncat}/RT} = e^{(0.40)E_{a,uncat}/RT}$$

$$= e^{[(0.40)(125\,\text{kJ·mol}^{-1})/(8.314\times10^{-3}\,\text{kJ·K}^{-1}\text{·mol}^{-1}\times 298\,\text{K})]} = 5.8\times10^{8}$$

(b) The last step of the calculation in (a) is repeated with $T = 350$ K.

$$e^{[(0.40)(125\,\text{kJ·mol}^{-1})/(8.314\times10^{-3}\,\text{kJ·K}^{-1}\text{·mol}^{-1}\times 350\,\text{K})]} = 2.9\times10^{7}$$

The rate enhancement is lower at higher temperatures.


**13.63** cat = catalyzed

$$\frac{\text{rate(cat)}}{\text{rate(uncat)}} = \frac{k_{cat}}{k_{uncat}} = 1000 = \frac{Ae^{-E_{a,cat}/RT}}{Ae^{-E_a/RT}} = \frac{e^{-E_{a,cat}/RT}}{e^{-E_a/RT}}$$

ln 1000

$$E_{a,cat} = E_a - RT\ln 1000$$

$$= 98\,\text{kJ·mol}^{-1} - (8.31\times10^{-3}\,\text{kJ·K}^{-1}\text{·mol}^{-1})(298\,\text{K})(\ln 1000)$$

$$= 81\,\text{kJ·mol}^{-1}$$


**13.65** The overall reaction is $RCN + H_2O \longrightarrow RC(RO)NH_2$. The intermediates include

The hydroxide ion serves as a catalyst for the reaction.


**13.67** (a) False. A catalyst increases the rate of both the forward and reverse reactions by providing a completely different pathway. (b) True, although a catalyst may be poisoned and lose activity.
(c) False. There is a completely different pathway provided for the reaction in the presence of a catalyst.
(d) False. The position of the equilibrium is unaffected by the presence of a catalyst.

**13.69** (a) To obtain the Michaelis-Menten rate equation, we will begin by employing the steady-state approximation, setting the rate of change in the concentration of the ES intermediate equal to zero:

$$\frac{d[ES]}{dt} = k_1[E][S] - k'_1[ES] - k_2[ES] = 0.$$

Rearranging gives: $\quad [E][S] = \left(\frac{k_2 + k'_1}{k_1}\right)[ES] = K_M[ES].$

The total bound and unbound enzyme concentration, $[E]_0$, is given by:

$[E]_0 = [E] + [ES]$, and, therefore, $[E] = [ES] = [E]_0$.

Substituting this expression for $[E]$ into the preceding equation, we obtain:

$([ES] - [E]_0)[S] = K_M[ES].$

Rearranging to obtain $[ES]$ gives: $\quad [ES] = \dfrac{[E]_0[S]}{K_M + [S]}.$

From the mechanism, the rate of appearance of the product is given by *rate* = $k_2[ES]$. Substituting the preceding equation for $[ES]$, we obtain:

$$\text{Rate} = \frac{k_2[E]_0[S]}{K_M + [S]},$$

the Michaelis-Menten rate equation, which can be rearranged to obtain:

$$\frac{1}{\text{rate}} = \frac{K_M}{k_2[E]_0[S]} + \frac{1}{k_2[E]_0}.$$

If one plots $\dfrac{1}{\text{rate}}$ versus $\dfrac{1}{[S]}$, the slope will be $\dfrac{K_M}{k_2[E]_0}$ and the $y$-intercept

will be $\dfrac{1}{k_2[E]_0}.$

(b)

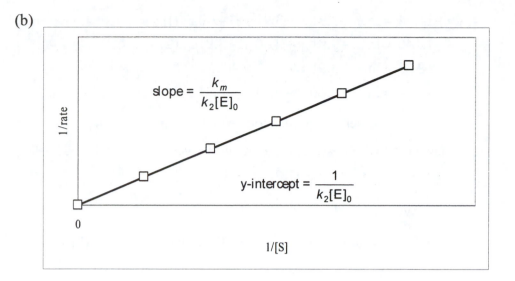

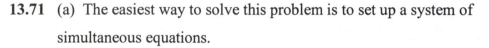

**13.71** (a) The easiest way to solve this problem is to set up a system of simultaneous equations.

| $[H_2SeO_3]$ | $[I^-]$ | $[H^+]$ | Rate, $mol \cdot L^{-1} \cdot s^{-1}$ |
|---|---|---|---|
| 0.020 | 0.020 | 0.010 | $8.0 \times 10^{-6}$ |
| 0.020 | 0.010 | 0.020 | $4.0 \times 10^{-6}$ |
| 0.020 | 0.030 | 0.030 | $2.4 \times 10^{-4}$ |
| 0.010 | 0.020 | 0.020 | $1.6 \times 10^{-5}$ |

We have the following general relationship:

rate $= k[H_2SeO_3]^x[I^-]^y[H^+]^z$, which can be rewritten for ease of computation as

$\ln (\text{rate}) = \ln k + x \ln [H_2SeO_3] + y \ln [I^-] + z \ln [H^+]$

Using the data above, we can create four equations, which should be enough to solve the system of four unknown variables:

$\ln(8.0 \times 10^{-6}) = \ln k + x \ln 0.020 + y \ln 0.020 + z \ln 0.010$

$\ln(4.0 \times 10^{-6}) = \ln k + x \ln 0.020 + y \ln 0.010 + z \ln 0.020$

$\ln(2.4 \times 10^{-4}) = \ln k + x \ln 0.020 + y \ln 0.030 + z \ln 0.030$

$\ln(1.6 \times 10^{-5}) = \ln k + x \ln 0.010 + y \ln 0.020 + z \ln 0.020$

which give, upon calculating the numerical logarithms:

$-11.74 = \ln k - 3.91\,x - 3.91\,y - 4.60\,z$ (1)

$-12.40 = \ln k - 3.91\,x - 4.60\,y - 3.91\,z$ (2)

$-8.33 = \ln k - 3.91\,x - 3.51\,y - 3.51\,z$ (3)

$-11.04 = \ln k - 4.60\,x - 3.91\,y - 3.91\,z$ (4)

Solving this set of simultaneous equations and rounding the $x$, $y$, and $z$ answers to the nearest whole number gives $x = 1$, $y = 3$, and $z = 2$, with $k = 5.0 \times 10^5 \ \mathrm{L^5 \cdot mol^{-5} \cdot s^{-1}}$.

(b) With

$[\mathrm{H_2SeO_3}] = 0.030 \ \mathrm{mol \cdot L^{-1}}$, $[\mathrm{I^-}] = 0.025 \ \mathrm{mol \cdot L^{-1}}$ and $[\mathrm{H^+}] = 0.015 \ \mathrm{mol \cdot L^{-1}}$,

$\text{rate} = (5.0 \times 10^5 \ \mathrm{L^5 \cdot mol^{-5} \cdot s^{-1}})(0.030 \ \mathrm{mol \cdot L^{-1}})^1 (0.025 \ \mathrm{mol \cdot L^{-1}})^3$
$\times (0.015 \ \mathrm{mol \cdot L^{-1}})^2$

$\text{rate} = 5.3 \times 10^{-5} \ \mathrm{mol \cdot L^{-1} \cdot s^{-1}}$

**13.73** (a) $k = \dfrac{\dfrac{1}{[A]_t} - \dfrac{1}{[A]_0}}{t} = \dfrac{\dfrac{1}{0.0050 \ \mathrm{mol \cdot L^{-1}}} - \dfrac{1}{0.040 \ \mathrm{mol \cdot L^{-1}}}}{12 \ \mathrm{h}}$

$= 15 \ \mathrm{L \cdot mol^{-1} \cdot h^{-1}}$

(b) $[\mathrm{EX_2}] = 0.040\,\dfrac{\mathrm{mol\ EX_2}}{\mathrm{L}} - \left(0.070\,\dfrac{\mathrm{mol\ X}}{\mathrm{L}}\right)\left(\dfrac{1\ \mathrm{mol\ EX_2}}{2\ \mathrm{mol\ X}}\right)$

$= 0.005 \ \mathrm{mol \cdot L^{-1}}$

$k = \dfrac{\dfrac{1}{[\mathrm{EX_2}]_t} - \dfrac{1}{[\mathrm{EX_2}]_0}}{t} = \dfrac{\dfrac{1}{0.005 \ \mathrm{mol \cdot L^{-1}}} - \dfrac{1}{0.040 \ \mathrm{mol \cdot L^{-1}}}}{15 \ \mathrm{h}}$

$= 10 \ \mathrm{L \cdot mol^{-1} \cdot h^{-1}}$ (1 sf)

**13.75**

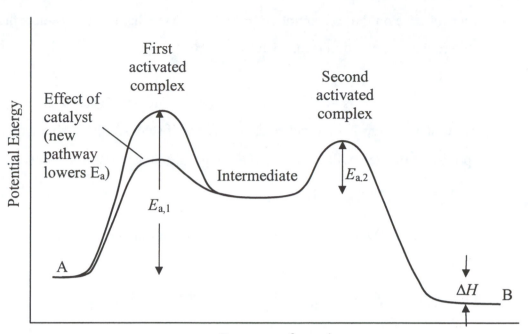

**13.77** $x$ = amount of original sample = 25.0 mg

$n$ = number of half-lives

$$\left(\frac{1}{2}\right)^n \times x = \text{amount remaining}$$

$$\frac{10.9}{12.3} = 0.886 \text{ half-lives}$$

$$\left(\frac{1}{2}\right)^{0.886} \times 25.0 \text{ mg} = 13.5 \text{ mg}$$

**13.79** The anticipated rate for mechanism (i) is: rate = $k[C_{12}H_{22}O_{11}]$, while the expected rate for mechanism (ii) is: rate = $k[C_{12}H_{22}O_{11}][H_2O]$. The rate for mechanism (ii) will be pseudo-first-order in dilute solutions of sucrose because the concentration of water will not change. Therefore, in dilute solutions kinetic data can not be used to distinguish between the two mechanisms. However, in a highly concentrated solution of sucrose, the concentration of water will change during the course of the reaction. As a result, if mechanism (ii) is correct the kinetics will display a first-order dependence on the concentration of $H_2O$ while mechanism (i) predicts that the rate of the reaction is independent of $[H_2O]$.

**13.81** (a) The objective is to reproduce the observed rate law. If step 2 is the slow step, if step 1 is a rapid equilibrium, and if step 3 is fast also, then our proposed rate law will be rate $= k_2[N_2O_2][H_2]$. Consider the equilibrium of Step 1: $k_1[NO]^2 = k_1'[N_2O_2]$

$[N_2O_2] = \dfrac{k_1}{k_1'}[NO]^2$ Substituting in our proposed rate law, we have

$$\text{rate} = k_2\left(\frac{k_1}{k_1'}\right)[NO]^2[H_2] = k[NO]^2[H_2] \text{ where } k = k_2\left(\frac{k_1}{k_1'}\right)$$

The assumptions made above reproduce the observed rate law; therefore, step 2 is the slow step.

(b)

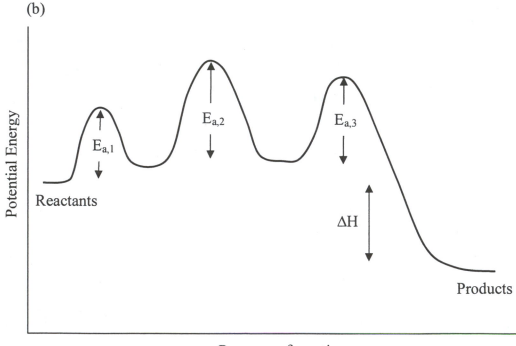

Note: The dips that represent the formation of the intermediate $N_2O_2$ and $N_2O$ will not be at the same energy, but we have no information to determine which should be lower.

**13.83** $\dfrac{\text{rate at } 28°C}{\text{rate at } 5°C} = \dfrac{k'}{k} = \dfrac{t}{t'} = \dfrac{48\ \text{h}}{4\ \text{h}}$

We use $\ln\left(\dfrac{k'}{k}\right) = \dfrac{E_a}{R}\left(\dfrac{1}{T} - \dfrac{1}{T'}\right)$ and solve for $E_a$.

$$E_a = \frac{R\ln\left(\dfrac{k'}{k}\right)}{\left(\dfrac{1}{T} - \dfrac{1}{T'}\right)} = \frac{(8.314\times10^{-3}\ \text{kJ}\cdot\text{K}^{-1}\cdot\text{mol}^{-1})\ln\left(\dfrac{48}{4}\right)}{\left(\dfrac{1}{278\ \text{K}} - \dfrac{1}{301\ \text{K}}\right)} = 75\ \text{kJ}\cdot\text{mol}^{-1}$$

**13.85** (a) To obtain the Michaelis-Menten rate equation assuming a pre-equilibrium between the bound and unbound states of the substrate we begin with the expression for the equilibrium constant of the fast equilibrium between the bound and unbound substrate:

$K = \dfrac{[\text{ES}]}{[\text{E}][\text{S}]}$,    solving for [ES] we obtain:    $[\text{ES}] = K[\text{E}][\text{S}]$

as before in problem 13.69, the total bound and unbound enzyme concentration, $[\text{E}]_0$, is given by:

$[\text{E}]_0 = [\text{E}] + [\text{ES}]$, and, therefore,

$[\text{E}] = [\text{E}]_0 - [\text{ES}]$

Substituting this expression for [E] into the equation above we obtain:

$K([\text{E}]_0 - [\text{ES}])[\text{S}] = [\text{ES}]$.

Rearrainging to obtain [ES]:

$[\text{ES}] = \dfrac{K[\text{E}]_0[\text{S}]}{1 + K[\text{S}]} = \dfrac{[\text{E}]_0[\text{S}]}{K^{-1} + [\text{S}]}$

From the mechanism, the rate of appearance of the product is given by:

rate $= k_2[\text{ES}]$. Substituting the equation above for [ES] one obtains:

rate $= \dfrac{k_2[\text{E}]_0[\text{S}]}{K^{-1} + [\text{S}]}$, the Michaelis-Menten rate equation.

**13.87** (a) ClO is the reaction intermediate; Cl is the catalyst.

(b) Cl, ClO, O, $O_2$

(c) Step 1 and step 2 are propagating.

(d) $Cl + Cl \longrightarrow Cl_2$

**13.89** For a third-order reaction,

$$t_{1/2} \propto \frac{1}{[A_0]^2} \text{ or } t_{1/2} = \frac{\text{constant}}{[A_0]^2}$$

(a) The time necessary for the concentration to fall to one-half of the initial concentration is one half-life:

$$\text{first half-life} = t_1 = t_{1/2} = \frac{\text{constant}}{[A_0]^2}$$

(b) This time, $t_{1/4}$, is two half-lives, but because of different starting concentrations, the half-lives are not the same:

$$\text{second half-life} = t_2 = \frac{\text{constant}}{(\frac{1}{2}[A_0])^2} = \frac{4(\text{constant})}{[A_0]^2} = 4t_1$$

$$\text{total time} = t_1 + t_2 = t_1 + 4t_1 = 5t_1 = t_{1/4}$$

(c) This time, $t_{1/16}$, is four half-lives; again, the half-lives are not the same:

$$\text{third half-life} = t_3 = \frac{\text{constant}}{(\frac{1}{4}[A_0])^2} = \frac{16(\text{constant})}{[A_0]^2} = 16t_1$$

$$\text{fourth half-life} = t_4 = \frac{\text{constant}}{(\frac{1}{8}[A_0])^2} = \frac{64(\text{constant})}{[A_0]^2} = 64t_1$$

$$\text{total time} = t_1 + t_2 + t_3 + t_4 = t_1 + 4t_1 + 16t_1 + 64t_1 = 85t_1 = t_{1/16}$$

If $t_1$ is known, the times $t_{1/4}$ and $t_{1/16}$ can be calculated easily.

**13.91** By analogy with the reaction in Exercise 13.82, the overall reaction here is

$$CH_4(g) + Cl_2(g) \longrightarrow CH_3Cl + HCl$$

(a) Initiation: $Cl_2 \longrightarrow 2\,Cl$

Propagation: $Cl + CH_4 \longrightarrow CH_3Cl + H$

$H + Cl_2 \longrightarrow HCl + Cl$

Termination: $Cl + Cl \longrightarrow Cl_2$

$H + H \longrightarrow H_2$

$H + Cl \longrightarrow HCl$

(b) $CH_3Cl$ and HCl

**13.93** The strategy for working this problem is to obtain the equilibrium constants for the reaction at two or more temperatures and then use those values to calculate $\Delta H°_r$ and $\Delta S°_r$. From Table 13.1 we can obtain $K$ values at 4 temperatures:

$$K = \frac{k}{k'}$$

$$K_{500} = \frac{6.4 \times 10^{-9} \text{ L} \cdot \text{mol}^{-1} \cdot \text{s}^{-1}}{4.3 \times 10^{-7} \text{ L} \cdot \text{mol}^{-1} \cdot \text{s}^{-1}} = 0.015$$

$$K_{600} = \frac{9.7 \times 10^{-6} \text{ L} \cdot \text{mol}^{-1} \cdot \text{s}^{-1}}{4.4 \times 10^{-4} \text{ L} \cdot \text{mol}^{-1} \cdot \text{s}^{-1}} = 0.022$$

$$K_{700} = \frac{1.8 \times 10^{-3} \text{ L} \cdot \text{mol}^{-1} \cdot \text{s}^{-1}}{6.3 \times 10^{-2} \text{ L} \cdot \text{mol}^{-1} \cdot \text{s}^{-1}} = 0.028$$

$$K_{700} = \frac{9.7 \times 10^{-2} \text{ L} \cdot \text{mol}^{-1} \cdot \text{s}^{-1}}{2.6 \text{ L} \cdot \text{mol}^{-1} \cdot \text{s}^{-1}} = 0.037$$

We can choose to calculate the desired quantities from any two of these points, or we can plot the data and determine the values from the slope and intercept of the graph:

$$\ln K = -\frac{\Delta H°_r}{R}\left(\frac{1}{T_1}\right) + \frac{\Delta S°_r}{R}$$

The plot should be ln $K$ versus $\frac{1}{T}$. The slope will be $-\frac{\Delta H°_r}{R}$ and the intercept will be $\frac{\Delta S°_r}{R}$.

| $T(K)$ | $\frac{1}{T}(K^{-1})$ | $K$ | $\ln K$ |
|--------|-----------|-------|---------|
| 500 | 0.0200 | 0.015 | −4.20 |
| 600 | 0.001 67 | 0.022 | −3.82 |
| 700 | 0.001 43 | 0.028 | −3.58 |
| 800 | 0.001 25 | 0.037 | −3.30 |

$-\dfrac{\Delta H°_r}{R} = -1177$, $\Delta H°_r = 9.8 \text{ kJ} \cdot \text{mol}^{-1}$

$\dfrac{\Delta S°_r}{R} = -1.86$, $\Delta S°_r = -15 \text{ J} \cdot \text{K}^{-1} \cdot \text{mol}^{-1}$

**13.95** In order for the reaction to be catalyzed heterogeneously, the reacting species must attach themselves to the surface of the catalyst. The concentration of the reactants is usually much greater than the number of active sites available on the catalyst so that the rate is determined by the surface area of the catalyst and not by the concentrations or pressures of reactants.

**13.97** (a) $OCl^- + H_2O \underset{k_1'}{\overset{k_1}{\rightleftharpoons}} HOCl + OH^-$    fast equilibrium

$HOCl + I^- \xrightarrow{k_2} HOI + Cl^-$    very slow

$HOI + OH^- \underset{k_3'}{\overset{k_3}{\rightleftharpoons}} OI^- + H_2O$    fast equilibrium

The overall reaction is $OCl^- + I^- \longrightarrow OI^- + Cl^-$

(b) The rate law will be based upon the slow step of the reaction:

rate $= k_2[\text{HOCl}][\text{I}^-]$

Even though HOCl is a stable species because it is an intermediate in the reaction as written, technically we should not leave the rate law in this form. The concentration of HOCl can be expressed in terms of the reactants and products using the fast equilibrium approach:

$$K = \frac{k_1}{k_19} = \frac{[\text{HOCl}][\text{OH}^-]}{[\text{OCl}^-]}$$

$$[\text{HOCl}] = \frac{k_1}{k_19}\frac{[\text{OCl}^-]}{[\text{OH}^-]}$$

$$\text{rate} = \frac{k_2 k_1}{k_19}\frac{[\text{OCl}^-][\text{I}^-]}{[\text{OH}^-]}$$

(c) An examination of the rate law shows that the rate is dependent upon the concentration of $\text{OH}^-$, which means that the rate will be dependent upon the pH of the solution.

(d) If the reaction is carried out in an organic solvent, then $\text{H}_2\text{O}$ is no longer the solvent and its concentration must be included in calculating the equilibrium concentration of HOCl:

$$K = \frac{k_1}{k_19} = \frac{[\text{HOCl}][\text{OH}^-]}{[\text{OCl}^-][\text{H}_2\text{O}]}$$

$$[\text{HOCl}] = \frac{k_1}{k_19}\frac{[\text{OCl}^-][\text{H}_2\text{O}]}{[\text{OH}^-]}$$

$$\text{rate} = \frac{k_2 k_1}{k_19}\frac{[\text{OCl}^-][\text{I}^-][\text{H}_2\text{O}]}{[\text{OH}^-]}$$

The rate of reaction will then show a dependence upon the concentration of water, which will be obscured when the reaction is carried out with water as the solvent.

**13.99** Concentration (mol $\cdot$ L$^{-1}$) $\quad 2\,N_2O_5 \quad \rightleftharpoons \quad 4\,NO_2 \quad + \quad O_2$

| | | | |
|---|---|---|---|
| initial | $P_0$ | 0 | 0 |
| change | $-x$ | $+2x$ | $+0.5x$ |
| at time $t$ | $P_0 - x$ | $2x$ | $0.5x$ |

Therefore, $P_{total}$ at time $t = P_0 + 1.5x$. This allows calculation of $x$ at each time, which in turn allows calculation of $P_{N_2O_5} (= P_0 - x)$ at these times. Converting the units to atmospheres by dividing by 101.325 kPa $\cdot$ atm$^{-1}$ and to $[N_2O_5]$ by dividing by $RT$ allows us to make the following table:

| $t$, min | $x$, kPa | $P_{N_2O_5}$, kPa | $P_{N_2O_5}$, atm | $[N_2O_5]$, mol $\cdot$ L$^{-1}$ | ln $[N_2O_5]$ |
|---|---|---|---|---|---|
| 0 | 0 | 27.3 | 0.269 | 0.0100 | $-4.605$ |
| 5 | 10.9 | 16.4 | 0.162 | $6.01 \times 10^{-3}$ | $-5.114$ |
| 10 | 17.5 | 9.85 | 0.0972 | $3.61 \times 10^{-3}$ | $-5.624$ |
| 15 | 21.4 | 5.9 | 0.058 | $2.2 \times 10^{-3}$ | $-6.12$ |
| 20 | 23.8 | 3.5 | 0.035 | $1.3 \times 10^{-3}$ | $-6.65$ |
| 30 | 26.0 | 1.3 | 0.013 | $4.8 \times 10^{-4}$ | $-7.64$ |

The data fit closely to a straight line; therefore, this is a first-order reaction. The rate constant can be obtained from the slope, which is

$$\frac{-4.605 - (-7.64)}{30 \text{ min}} = 0.101 \text{ min}^{-1} = k$$

rate $= k[N_2O_5] = 0.101 \text{ min}^{-1}[N_2O_5]$, which gives the results in the table below.

| $t$, min | rate, mol $\cdot$ L$^{-1} \cdot$ min$^{-1}$ |
|---|---|
| 0 | $1.01 \times 10^{-3}$ |
| 5 | $6.07 \times 10^{-4}$ |
| 10 | $3.65 \times 10^{-4}$ |
| 15 | $2.2 \times 10^{-4}$ |

| | |
|---|---|
| 20 | $1.3 \times 10^{-4}$ |
| 30 | $4.8 \times 10^{-5}$ |

# CHAPTER 14

# THE ELEMENTS: THE FIRST FOUR MAIN GROUPS

**14.1** (a) carbon  (b) lithium  (c) indium  (d) iodine

**14.3** (a) sulfur  (b) selenium  (c) sodium  (d) oxygen

**14.5** iodine < bromine < chlorine

**14.7** chlorine

**14.9** antimony

**14.11** (a) KCl because the ionic radius of $K^+$ is larger than that of $Na^+$

(b) Na—O. The higher charge on $Mg^{2+}$ makes its ionic radius much smaller than that of $Na^+$. (c) Thallium(I) chloride because the Tl(I) ion is larger.

**14.13** $2 K(s) + H_2(g) \rightarrow 2 KH(s)$

**14.15** (a) saline  (b) molecular  (c) molecular  (d) metallic

**14.17** (a) acidic  (b) amphoteric  (c) acidic  (d) basic

**14.19** (a) $CO_2$  (b) $B_2O_3$

**14.21** (a) $C_2H_2(g) + H_2(g) \rightarrow H_2C{=}CH_2(g)$

Oxidation number of C in $C_2H_2 = -1$; of C in $H_2C{=}CH_2 = -2$; carbon has been reduced.

(b) $CO(g) + H_2O(g) \rightarrow CO_2(g) + H_2(g)$

(c) $BaH_2(s) + 2\,H_2O(l) \rightarrow Ba(OH)_2(aq) + 2\,H_2(g)$

**14.23** (a) $CH_4(g) + H_2O(g) \rightarrow CO(g) + 3\,H_2(g)$

$$\Delta H°_r = \Delta H°_f(CO,\, g) - [\Delta H°_f(CH_4,\, g) + \Delta H°_f(H_2O,\, g)]$$
$$= (-110.53\text{ kJ} \cdot \text{mol}^{-1}) - [(-74.81\text{ kJ} \cdot \text{mol}^{-1}) + (-241.82\text{ kJ} \cdot \text{mol}^{-1})]$$
$$= +206.10\text{ kJ} \cdot \text{mol}^{-1}$$

(b)
$$\Delta S°_r = S°(CO,\, g) + 3S°(H_2,\, g) - [S°(CH_4,\, g) - S°(H_2O,\, g)]$$
$$= 197.67\text{ J} \cdot \text{K}^{-1} \cdot \text{mol}^{-1} + 3(130.68\text{ J} \cdot \text{K}^{-1} \cdot \text{mol}^{-1})$$
$$- [186.26\text{ J} \cdot \text{K}^{-1} \cdot \text{mol}^{-1} + 188.83\text{ J} \cdot \text{K}^{-1} \cdot \text{mol}^{-1}]$$
$$= +214.62\text{ J} \cdot \text{K}^{-1} \cdot \text{mol}^{-1}$$

(c)
$$\Delta G°_r = \Delta G°_f(CO,\, g) - [\Delta G°_f(CH_4,\, g) + \Delta G°_f(H_2O,\, g)]$$
$$= (-137.17\text{ kJ} \cdot \text{mol}^{-1})$$
$$- [(-50.72\text{ kJ} \cdot \text{mol}^{-1}) + (-228.57\text{ kJ} \cdot \text{mol}^{-1})]$$
$$= +142.12\text{ kJ} \cdot \text{mol}^{-1}$$

$\Delta G°_r$ can also be calculated from $\Delta H°_r$ and $\Delta S°_r$:

$$\Delta G°_r = \Delta H°_r - T\Delta S°_r$$
$$= +206.10\text{ kJ} \cdot \text{mol}^{-1} - (298\text{ K})(+214.62\text{ J} \cdot \text{K}^{-1} \cdot \text{mol}^{-1})/(1000\text{ J} \cdot \text{kJ}^{-1})$$
$$= +142.14\text{ kJ} \cdot \text{mol}^{-1}$$

**14.25** (a) $H_2(g) + Cl_2(g) \xrightarrow{\text{light}} 2\,HCl(g)$

(b) $H_2(g) + 2\,Na(l) \xrightarrow{\Delta} 2\,NaH(s)$

(c) $P_4(s) + 6\,H_2(g) \rightarrow 4\,PH_3(g)$

(d) $2\,Cu(s) + H_2(g) \rightarrow 2\,CuH(s)$

**14.27** (a)  $H_2(g) \rightarrow 2\,H^+(aq) + 2\,e^-$    $E° = 0.00$ V

$O_2(g) + 4\,H^+(aq) + 4\,e^- \rightarrow 2\,H_2O(l)$    $E° = +1.23$ V

$2\,H_2(g) + O_2(g) \rightarrow 2\,H_2O(l)$    $E° = +1.23$ V

The maximum potential possible is 1.23 V.

(b) The difficulty is isolating the two half cells but still maintaining electrical contact. Ions need to flow through the system to maintain charge balance in the reaction. In this case, a material that allows hydrogen ions but not hydrogen gas or oxygen gas to pass through would be necessary.

**14.29** Lithium is the only Group 1 element that reacts directly with nitrogen to form lithium nitride:

$6\,Li(s) + N_2(g) \xrightarrow{\Delta} 2\,Li_3N(s)$

Lithium reacts with oxygen to form mainly the oxide:

$4\,Li(s) + O_2(g) \rightarrow 2\,Li_2O(s)$

The other members of the group form mainly the peroxide or superoxide. Lithium exhibits the diagonal relationship that is common to many first members of a group. Li is similar in many of its compounds to the compounds of Mg. This behavior is related to the small ionic radius of $Li^+$, 58 pm, which is closer to the ionic radius of $Mg^{2+}$, 72 pm, but substantially less than that of $Na^+$, 102 pm.

**14.31** (a)  $4\,Li(s) + O_2(g) \rightarrow 2\,Li_2O(s)$

(b)  $6\,Li(s) + N_2(g) \xrightarrow{\Delta} 2\,Li_3N(s)$

(c)  $2\,Na(s) + 2\,H_2O(l) \rightarrow 2\,NaOH(aq) + H_2(g)$

(d)  $4\,KO_2(s) + 2\,H_2O(g) \rightarrow 4\,KOH(s) + 3\,O_2(g)$

**14.33** 1 mol $Na_2CO_3 \cdot 10\,H_2O$ yields 1 mol $Na_2CO_3$ in water.

$$\text{mass of Na}_2\text{CO}_3 \cdot 10\,\text{H}_2\text{O} = 0.500\,\text{L} \times 0.135\,\text{mol} \cdot \text{L}^{-1}$$
$$\times\, 286.15\,\text{g Na}_2\text{CO}_3 \cdot 10\,\text{H}_2\text{O} \cdot \text{mol}^{-1}$$
$$= 19.3\,\text{g Na}_2\text{CO}_3 \cdot 10\,\text{H}_2\text{O}$$

**14.35** $Mg(s) + 2\,H_2O(l) \rightarrow Mg(OH)_2 + H_2(g)$

**14.37** (a) $CaO(s) + H_2O(l) \rightarrow Ca(OH)_2(s)$

(b)
$$\Delta G^\circ{}_r = \Delta G^\circ{}_f(Ca(OH)_2,\,s) - [\Delta G^\circ{}_f(CaO,\,s) + \Delta G^\circ{}_f(H_2O,\,l)]$$
$$= -898.49\,\text{kJ} \cdot \text{mol}^{-1} - [(-604.03\,\text{kJ} \cdot \text{mol}^{-1}) + (-237.13\,\text{kJ} \cdot \text{mol}^{-1})]$$
$$= -57.33\,\text{kJ} \cdot \text{mol}^{-1}$$

**14.39** Be is the weakest reducing agent; Mg is stronger, but weaker than the remaining members of the group, all of which have approximately the same reducing strength. This effect is related to the very small radius of the $Be^{2+}$ ion, 27 pm; its strong polarizing power introduces much covalent character into its compounds. Thus, Be attracts electrons more strongly and does not release them as readily as other members of the group. $Mg^{2+}$ is also a small ion, 58 pm, so the same reasoning applies to it as well, but to a lesser extent. The remaining ions of the group are considerably larger, release electrons more readily, and are better reducing agents.

**14.41** $2\,Al(s) + 2\,OH^-(aq) + 6\,H_2O(l) \rightarrow 2[Al(OH)_4]^-(aq) + 3\,H_2(g)$

$Be(s) + 2\,OH^-(aq) + 2\,H_2O(l) \rightarrow [Be(OH)_4]^{2-}(aq) + H_2(g)$

Be and Al are diagonal neighbors in the periodic table and exhibit similar chemical behavior.

**14.43** (a) $Mg(OH)_2(s) + 2\,HCl(aq) \rightarrow MgCl_2(aq) + 2\,H_2O(l)$

(b) $Ca(s) + 2\,H_2O(l) \rightarrow Ca(OH)_2(aq) + H_2(g)$

(c) $BaCO_3(s) \xrightarrow{\Delta} BaO(s) + CO_2(g)$

14.45  (a)  $:\!\overset{\cdot\cdot}{\underset{\cdot\cdot}{Cl}}\!-\!Be\!-\!\overset{\cdot\cdot}{\underset{\cdot\cdot}{Cl}}\!:$   $Mg^{2+}$  $2\left[:\overset{\cdot\cdot}{\underset{\cdot\cdot}{Cl}}:\right]^{-}$

$MgCl_2$ is ionic; $BeCl_2$ is a molecular compound. (b) 180° (c) *sp*

14.47  $CaC_2(s) + 2\,H_2O(l) \rightarrow C_2H_2(g) + Ca(OH)_2(aq)$

$$25.0\ g\ CaC_2 \times \frac{1\ mole}{64.10\ g} = 0.3900\ mol\ CaC_2$$

$$25.0\ mL\ H_2O \times \frac{1.00\ g}{1\ mL} \times \frac{1\ mole}{18.02\ g} = 13.87\ mol\ H_2O$$

Since the reaction is 1:2, only 0.780 mol of water would be needed to completely react with all of the calcium carbide. Calcium carbide is the limiting reagent while water is present in excess.

mass ethyne =

$$0.3900\ mol\ CaC_2\left(\frac{1\ mol\ C_2H_2}{1\ mol\ CaC_2}\right)\left(\frac{26.04\ g\ C_2H_2}{1\ mol\ C_2H_2}\right) = 10.2\ g\ C_2H_2$$

14.49  The overall equation for the electrolytic reduction in the Hall process is

$$4\,Al^{3+}(melt) + 6\,O^{2-}(melt) + 3\,C(s,\ gr) \rightarrow 4\,Al(s)\ + 3\,CO_2(g)$$

14.51  (a)  $B_2O_3(s) + 3\,Mg(l) \xrightarrow{\Delta} 2\,B(s) + 3\,MgO(s)$

(b)  $2\,Al(s) + 3\,Cl_2(g) \rightarrow 2\,AlCl_3(s)$

(c)  $4\,Al(s) + 3\,O_2(g) \rightarrow 2\,Al_2O_3(s)$

14.53  (a) The hydrate of $AlCl_3$, that is, $AlCl_3 \cdot 6H_2O$, functions as a deodorant and antiperspirant. (b) $\alpha$- Alumina is corundum. It is used as an abrasive in sandpaper. (c) $B(OH)_3$ is an antiseptic and insecticide.

**14.55** Since there are only 22 valence electrons, or 11 electron pairs, it is not possible to draw a good conventional Lewis structure for tetraborane, $B_4H_{10}$, that includes four B-H-B bridges. For the suggested structure given below, each bridging H and each four-coordinate B would have a formal charge of 1- while each of the six terminal H atoms and each three-coordinate B would have a formal charge of 0. The total formal charge adds up to 6- in this case even though the molecule is neutral. However, if the bridges are viewed as 3-center 2-electron bonds, then every atom can be assigned a formal charge of 0.

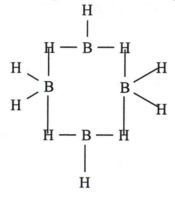

This drawing shows the structure of tetraborane

See:

www.chem.leeds.ac.uk/boronweb/articles/incredible_boron/incredible.htm

**14.57** The cathode reaction is $Al^{3+}(melt) + 3\,e^- \rightarrow Al(l)$

$$\text{charge consumed} = (12.0\ \text{h})\left(\frac{3600\ \text{s}}{1\ \text{h}}\right)(3.0\times10^6\ \text{C}\cdot\text{s}^{-1}) = 1.3\times10^{11}\ \text{C}$$

mass of Al produced

$$= (1.3 \times 10^{11} \text{ C}) \left( \frac{1 \text{ mol e}^-}{9.65 \times 10^4 \text{ C}} \right) \left( \frac{1 \text{ mol Al}}{3 \text{ mol e}^-} \right) \left( \frac{26.98 \text{ g Al}}{1 \text{ mol Al}} \right)$$

$$= 1.2 \times 10^7 \text{ g Al}$$

**14.59** (a) We want $E°$ for $\text{Tl}^{3+}(\text{aq}) + 3 \text{ e}^- \rightarrow \text{Tl}(\text{s})$, $n = 3$.

This reaction is the reverse of the formation reaction:

$$\text{Tl}(\text{s}) \rightarrow \text{Tl}^{3+}(\text{aq}) + 3 \text{ e}^-$$

Therefore, for the $\text{Tl}^{3+}/\text{Tl}$ couple, $\Delta G°_r = -215 \text{ kJ} \cdot \text{mol}^{-1}$

$$\Delta G°_r = -nFE° = -215 \text{ kJ} \cdot \text{mol}^{-1}$$

$$E° = \frac{\Delta G°_r}{-nF} = \frac{-2.15 \times 10^5 \text{ J} \cdot \text{mol}^{-1}}{-3 \times 9.65 \times 10^4 \text{ C} \cdot \text{mol}^{-1}} = +0.743 \text{ V}$$

(b) Using the potential from part (a) and the potential from Appendix 2B for the reduction of $\text{Tl}^+$, we can decide whether or not $\text{Tl}^+$ will disproportionate in solution. The equation of interest is

$$3 \text{ Tl}^+(\text{aq}) \rightarrow 2 \text{ Tl}(\text{s}) + \text{Tl}^{3+}(\text{aq})$$

The half-reactions to combine are

$$\text{Tl}^+ \rightarrow \text{Tl}^{3+} + 2 \text{ e}^- \qquad (1)$$

$$\text{Tl}^+ + \text{e}^- \rightarrow \text{Tl} \qquad (2)$$

The potential for reaction (1) must be obtained using the $\Delta G°$ values of the two known half-reactions:

$$\text{Tl} \rightarrow \text{Tl}^{3+} + 3 \text{ e}^- \qquad \Delta G° = +215 \text{ kJ} \cdot \text{mol}^{-1}$$

$$\text{Tl}^+ + \text{e}^- \rightarrow \text{Tl}$$

$$\Delta G° = -nFE°$$
$$\qquad = -1(9.65 \times 10^4 \text{ J} \cdot \text{V}^{-1} \cdot \text{mol}^{-1})(-0.34 \text{ V})/(1000 \text{ J} \cdot \text{kJ}^{-1})$$
$$\qquad = +33 \text{ kJ} \cdot \text{mol}^{-1}$$

$\Delta G°$ for the combined half-reaction $\text{Tl}^+ \rightarrow \text{Tl}^{3+} + 2 \text{ e}^-$ is the sum of these two numbers:

$$+215 \text{ kJ} \cdot \text{mol}^{-1} + 33 \text{ kJ} \cdot \text{mol}^{-1} = +248 \text{ kJ} \cdot \text{mol}^{-1}$$

We can now combine this with the reduction process for $Tl^+$ to get the desired equation:

$$Tl^+ \rightarrow Tl^{3+} + 2\,e^- \qquad \Delta G° = +248\ kJ \cdot mol^{-1}$$

$$+2(Tl^+ + e^- \rightarrow Tl) \qquad \Delta G° = 2(+33\ kJ \cdot mol^{-1})$$

$$3\,Tl^+ \rightarrow 2\,Tl + Tl^{3+} \qquad \Delta G° = +248\ kJ \cdot mol^{-1} + 2(+33\ kJ \cdot mol^{-1})$$

$$= +314\ kJ \cdot mol^{-1}$$

Since $\Delta G°$ is large and positive, K<<1. This disproportionation reaction does not favor products.

**14.61** Silicon occurs widely in the Earth's crust in the form of silicates in rocks and as silicon dioxide in sand. It is obtained from quartzite, a form of quartz $(SiO_2)$, by the following processes:

(1) reduction in an electric arc furnace

$$SiO_2(s) + 2\,C(s) \rightarrow Si(s,\ crude) + 2\,CO(g)$$

(2) purification of the crude product in two steps

$$Si(s,\ crude) + 2\,Cl_2(g) \rightarrow SiCl_4(l)$$

followed by reduction with hydrogen to the pure element

$$SiCl_4(l) + 2\,H_2(g) \rightarrow Si(s,\ pure) + 4\,HCl(g)$$

**14.63** In diamond, carbon is $sp^3$ hybridized and forms a tetrahedral, three-dimensional network structure, which is extremely rigid. Graphite carbon is $sp^2$ hybridized and planar. Its application as a lubricant results from the fact that the two-dimensional sheets can "slide" across one another, thereby reducing friction. In graphite, the unhybridized $p$-electrons are free to move from one carbon atom to another, which results in its high electrical conductivity. In diamond, all electrons are localized in $sp^3$ hybridized C—C $\sigma$-bonds, so diamond is a poor conductor of electricity.

**14.65** (a) $SiCl_4(l) + 2 H_2(g) \rightarrow Si(s) + 4 HCl(g)$

(b) $SiO_2(s) + 3 C(s) \xrightarrow{2000°C} SiC(s) + 2 CO(g)$

(c) $Ge(s) + 2 F_2(g) \rightarrow GeF_4(s)$

(d) $CaC_2(s) + 2 H_2O(l) \rightarrow Ca(OH)_2(s) + C_2H_2(g)$

**14.67**

formal charges: Si = 0, O = -1; oxidation numbers: Si = +4, O = -2; This is an $AX_4$ VSEPR structure; therefore, the shape is tetrahedral.

**14.69** $SiO_2(s) + 2 C(s) \rightarrow Si(s) + 2 CO(g)$

$$\Delta H°_r = \Delta H°_f(\text{products}) - \Delta H°_f(\text{reactants})$$
$$= [(2)(-110.53 \text{ kJ} \cdot \text{mol}^{-1})] - [-910.94 \text{ kJ} \cdot \text{mol}^{-1}]$$
$$= +689.88 \text{ kJ} \cdot \text{mol}^{-1}$$

$$\Delta S°_r = S°(\text{products}) - S°(\text{reactants})$$
$$= [18.83 \text{ J} \cdot \text{K}^{-1} \cdot \text{mol}^{-1} + (2)(197.67 \text{ J} \cdot \text{K}^{-1} \cdot \text{mol}^{-1})]$$
$$- [41.84 \text{ J} \cdot \text{K}^{-1} \cdot \text{mol}^{-1} + (2)(5.740 \text{ J} \cdot \text{K}^{-1} \cdot \text{mol}^{-1})]$$
$$= +360.85 \text{ J} \cdot \text{K}^{-1} \cdot \text{mol}^{-1}$$

$$\Delta G°_r = \Delta H°_r - T\Delta S°_r$$
$$= 689.88 \text{ kJ} \cdot \text{mol}^{-1} - (298.15 \text{ K})(360.85 \text{ J} \cdot \text{K}^{-1} \cdot \text{mol}^{-1})/1000 \text{ J} \cdot \text{kJ}^{-1}$$
$$= +582.29 \text{ kJ} \cdot \text{mol}^{-1}$$

The temperature at which the equilibrium constant becomes greater than 1 is the temperature at which $\Delta G°_r = -RT \ln K = 0$, because $\ln 1 = 0$.

Above this temperature, the equilibrium constant is greater than 1. $\Delta G°_r = 0$ when $T\Delta S°_r = \Delta H°_r$, or

$$T = \frac{\Delta H°_r}{\Delta S°_r} = \frac{+689.88 \times 10^3 \text{ J} \cdot \text{mol}^{-1}}{+360.85 \text{ J} \cdot \text{K}^{-1} \cdot \text{mol}^{-1}} = 1912 \text{ K}$$

**14.71**   mass of HF $= (3.00 \times 10^{-3} \text{ g}) \left( \dfrac{1 \text{ mol } SiO_2}{60.09 \text{ g } SiO_2} \right) \left( \dfrac{6 \text{ mol HF}}{1 \text{ mol } SiO_2} \right) \left( \dfrac{20.01 \text{ g HF}}{1 \text{ mol HF}} \right)$

   $= 5.99 \times 10^{-3} \text{ g HF} = 5.99 \text{ mg HF}$

**14.73**   (a) The $Si_2O_7^{6-}$ ion is built from two $SiO_4^{4-}$ tetrahedral ions in which the silicate tetrahedra share one O atom. This is the only case in which one O is shared.   (b) The pyroxenes, for example, jade, $NaAl(SiO_3)_2$, consist of chains of $SiO_4$ units in which two O atoms are shared by neighboring units. The repeating unit has the formula $SiO_3^{2-}$. See Fig. 14.39.

**14.75**   $SiF_6^{2-}$

**14.77**   Ionic fluorides react with water to liberate HF, which then reacts with the glass. Glass bottles used to store metal fluorides become brittle and may disintegrate upon standing on the shelf.

**14.79**   The iron ions impart a deep red color to the clay, which is not desirable for the manufacture of fine china; a white base is aesthetically more pleasing.

**14.81**   In the majority of its reactions, hydrogen acts as a reducing agent. Examples are $2 H_2(g) + O_2(g) \rightarrow 2 H_2O(l)$ and various ore reduction processes, such as $NiO(s) + H_2(g) \xrightarrow{\Delta} Ni(s) + H_2O(g)$. With highly electropositive elements, such as the alkali metals, $H_2(g)$ acts as an oxidizing agent and forms metal hydrides, for example, $2 K(s) + H_2(g) \rightarrow 2 KH(s)$.

**14.83** (a)

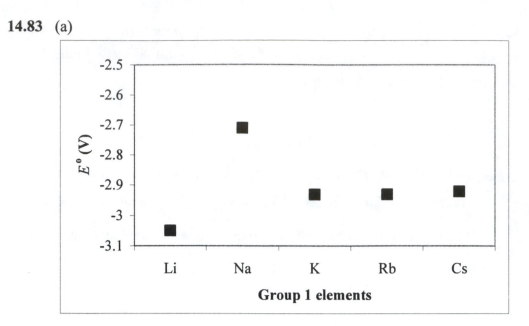

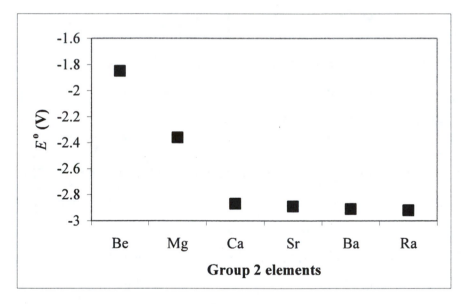

(b) For both groups, the trend in standard potentials with increasing atomic number is overall downward (they become more negative), but lithium is anomalous. This overall downward trend makes sense, because we expect that it is easier to remove electrons that are farther away from the nuclei. However, because there are several factors that influence ease of removal, the trend is not smooth. The potentials are a net composite of the free energies of sublimation of solids, dissociation of gaseous molecules, ionization enthalpies, and enthalpies of hydration of gaseous ions. The origin of the anomalously strong reducing power of Li is the strongly exothermic energy of hydration of the very small $Li^+$ ion, which favors the ionization of the element in aqueous solution.

**14.85** (a) The oxide ion, $O^{2-}$, in CaO acts as a Lewis base and reacts with the Lewis acid, $SiO_2$, in a Lewis acid-base reaction:

$$CaO(s) + SiO_2(s) \rightarrow CaSiO_3(l)$$

$SiO_2$, which is an impurity in iron ore, is removed by this reaction. The calcium oxide in this reaction can be obtained from limestone $[CaCO_3(s) \xrightarrow{\Delta} CaO(s) + CO_2(g)]$. This is the reason that limestone is important in the iron industry.

(b) $CaO(s) + CO_2(g) \rightarrow CaCO_3(s)$ (not an efficient preparation of $CaCO_3$, because of the weak Lewis acidity of $CO_2$).


**14.87** (a) $H_3BO_3$, acid; $B(OH)_4^-$, conjugate base

(b) $B(OH)_3(aq) + 2 H_2O(l) \rightarrow H_3O^+(aq) + B(OH)_4^-(aq)$


**14.89** (a) The ionization energies decrease and atomic radii increase down Groups 13/III and 14/IV.

|  | Element | Ionization energy, kJ·mol$^{-1}$ | Atomic radius, pm |
|---|---|---|---|
| Group 13 | B | 799 | 88 |
|  | Al | 577 | 143 |
|  | Ga | 577 | 153 |
|  | In | 556 | 167 |
|  | Tl | 590 | 171 |
| Group 14 | C | 1090 | 77 |
|  | Si | 786 | 118 |
|  | Ge | 784 | 122 |
|  | Sn | 707 | 158 |
|  | Pb | 716 | 175 |

(b) The ionization energies generally decrease down a group. As the atomic number of an element increases, atomic shells and subshells that are farther from the nucleus are filled. The outermost valence electrons are consequently easier to remove. The radii increase down a group for the same reason. The radii are primarily determined by the outer shell electrons, which are farther from the nucleus in the heavy elements.

(c) The trends correlate well with elemental properties; for example, the greater ease of outermost electron removal correlates with increased metallic character, that is, the ability to form positive ions by losing one or more electrons.

**14.91** In the majority of its reactions, hydrogen acts as a reducing agent, that is, $H_2(g) \rightarrow 2 H^+(aq) + 2 e^-$, $E° = 0$ V. In these reactions, hydrogen resembles Group 1 elements, such as Na and K. However, as described in the text and in the answer to Exercise 14.81, it may also act as an oxidizing agent; that is, $H_2(g) + 2 e^- \rightarrow 2 H^-(aq)$, $E° = -2.25$ V. In these reactions, hydrogen resembles Group 17 elements, such as Cl and Br. Consequently, $H_2$ will oxidize elements with standard reduction potentials more negative than $-2.25$ V, such as the alkali and alkaline earth metals (except Be). The compounds formed are hydrides and contain the $H^-$ ion; the singly charged negative ion is reminiscent of the halide ions. Hydrogen also forms diatomic molecules and covalent bonds like the halogens. The atomic radius of H is 78 pm, which compares rather well to that of F (64 pm) but not as well to that of Li (157 pm). The ionization energy of H is 1310 $kJ \cdot mol^{-1}$, which is similar to that of F (1680 $kJ \cdot mol^{-1}$) but not similar to that of Li (519 $kJ \cdot mol^{-1}$). The electron affinity of H is $+73$ $kJ \cdot mol^{-1}$, that of F is $+328$ $kJ \cdot mol^{-1}$, and that of Li is 60 $kJ \cdot mol^{-1}$. So in its atomic radius and ionization energy, H more closely resembles the Period 2 halogen, fluorine, in Group 17, than the Period 2 alkali metal, lithium, in Group 1; whereas in electron affinity, it more closely resembles lithium, Group 1. In electronegativity, H does not resemble elements in either Group 1 or Group 17, although its electronegativity is somewhat closer to those of Group 1. Consequently, hydrogen could be placed in either Group 1 or Group 17. But it is best to think of hydrogen as a unique element that has properties in common with both metals and nonmetals; therefore, it should probably be centered in the periodic table, as it is shown in the table in the text.

**14.93** (a)

| Ion | Radius, pm | Polarizing ability ($\times 1000$) | Ion | Radius, pm | Polarizing ability ($\times 1000$) |
|-----|-----|-----|-----|-----|-----|
| $Li^+$ | 58 | 17 | $Be^{2+}$ | 27 | 74 |
| $Na^+$ | 102 | 9.80 | $Mg^{2+}$ | 72 | 28 |
| $K^+$ | 138 | 7.25 | $Ca^{2+}$ | 100 | 20.0 |
| $Rb^+$ | 149 | 6.71 | $Sr^{2+}$ | 116 | 17.2 |
| $Cs^+$ | 170 | 5.88 | $Ba^{2+}$ | 136 | 14.7 |

(b) These data roughly support the diagonal relationship. $Li^+$ is more like $Mg^{2+}$ than $Be^{2+}$, and $Na^+$ is more like $Ca^{2+}$ than $Mg^{2+}$; but further down the group, the correlation fails. Charge divided by $r^3$ would be a better measure of polarizing ability.

**14.95** $H_2(g) + Br_2(l) \rightarrow 2\ HBr(g)$

$$\text{number of moles of HBr} = (0.135\text{ L H}_2)\left(\frac{1\text{ mol H}_2}{22.4\text{ L H}_2}\right)\left(\frac{2\text{ mol HBr}}{1\text{ mol H}_2}\right)$$
$$= 0.0121\text{ mol}$$

$$\text{molar concentration of HBr} = \frac{0.0121\text{ mol}}{0.225\text{ L}} = 0.0538\text{ mol}\cdot L^{-1}$$

**14.97** The smaller the cation, the greater is the ability of the cation to polarize and weaken the carbonate ion, $CO_3^{2-}$. On that basis, we would predict that within a group the carbonates of the first members of the group are less stable than those of the later members. Thus, $Li_2CO_3 <$

$Na_2CO_3 < K_2CO_3 < Rb_2CO_3 < Cs_2CO_3$ and

$BeCO_3 < MgCO_3 < CaCO_3 < SrCO_3 < BaCO_3$. Between groups, we

would expect the stability of the carbonates in one period to decrease from

Group 1 to Group 13 because of the smaller size of Group 13 ions. Thus,

$Al_2(CO_3)_2 < MgCO_3 < Na_2CO_3$. Carbonates of Group 13 $M^{3+}$ ions are,

in fact, so unstable that they do not exist. $Tl_2CO_3$ where Tl has an

oxidation number of $+1$ is known, however.

**14.99** (a) The unit cell described will contain a total of 4 B atoms and 4 N

atoms. The volume of the cell is

$(361.5 \text{ pm})^3 = (3.615 \times 10^{-8} \text{ cm})^3 = $ or $4.724 \times 10^{-23} \text{ cm}^{-3}$. The mass in the

unit cell will be

$(4 \times 10.81 \text{ g} \cdot \text{mol}^{-1} + 4 \times 14.01 \text{ g} \cdot \text{mol}^{-1}) \div 6.022 \times 10^{23} \text{ mol}^{-1}$
$= 1.649 \times 10^{-22} \text{ g}.$

$$d = \frac{1.649 \times 10^{-22} \text{ g}}{4.724 \times 10^{-23} \text{ cm}^3} = 3.491 \text{ g} \cdot \text{cm}^{-3}$$

(b) Because the density of cubic boron nitride is greater than that of

hexagonal BN, we would expect the cubic form to be favored at high

pressures, exactly as found for the cubic (diamond) and hexagonal

(graphite) forms of carbon.

**14.101** The metal hydride compounds have a molecular orbital structure that is

very asymmetric. Because the hydrogen atom is much more

electronegative than the metal atom, its orbital lies much lower in energy.

Consequently, when a bond is formed, it is a strongly ionic bond with the

electrons heavily localized on the H atom.

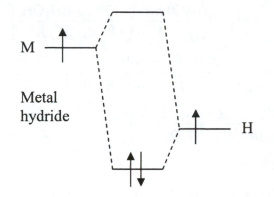

M

Metal
hydride

H

However, in the case of the lighter $p$-block elements, the electronegativity difference is not so great and the bonding is much more covalent. If anything, the $p$-block element is more electronegative than hydrogen; the bond polarity would lie in the other direction but would be much less pronounced than in the saline hydrides.

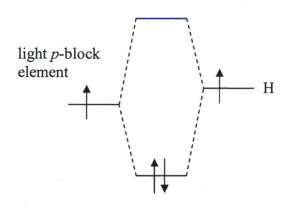

light $p$-block
element

H

**14.103** Species (a), (b), (c) and (d) can all function as greenhouse gases while (e) cannot. Any molecule other than a homonuclear diatomic can exhibit a changing dipole moment as it vibrates with certain vibrational modes. Since argon is monoatomic it has no covalent bonds, no vibrational modes, and no dipole moment.

**14.105** $CH_3OH(l) + \frac{3}{2} O_2(g) \rightarrow CO_2(g) + 2 H_2O(l)$

$$\text{number of kg CO}_2 = 1.00 \text{ L CH}_3\text{OH}\left(\frac{1000 \text{ mL}}{1 \text{ mL}}\right)\left(\frac{0.791 \text{ g CH}_3\text{OH}}{1 \text{ mL CH}_3\text{OH}}\right)$$

$$\times\left(\frac{1 \text{ mole CH}_3\text{OH}}{32.04 \text{ g CH}_3\text{OH}}\right)\left(\frac{1 \text{ mole CO}_2}{1 \text{ mole CH}_3\text{OH}}\right)$$

$$\times\left(\frac{44.02 \text{ g CO}_2}{1 \text{ mole CO}_2}\right)\left(\frac{1 \text{ kg}}{1000 \text{ g}}\right)$$

$$= 1.09 \text{ kg CO}_2$$

This mass of carbon dioxide is about half the amount generated by combusting an equivalent volume of octane (2.16 kg per liter). However, we also need to consider how much energy is produced per liter of fuel and how the mass of carbon dioxide produced compares for a given amount of energy produced. Standard enthalpies of combustion given in Appendix 2 are $\Delta H_c^\circ = -5471 \text{ kJ} \cdot \text{mol}^{-1}$ for octane and $\Delta H_c^\circ = -726 \text{ kJ} \cdot \text{mol}^{-1}$ for methanol.

$$\text{energy per L methanol} = 1.00 \text{ L CH}_3\text{OH}\left(\frac{1000 \text{ mL}}{1 \text{ L}}\right)\left(\frac{0.791 \text{ g CH}_3\text{OH}}{1 \text{ mL CH}_3\text{OH}}\right)$$

$$\times\left(\frac{1 \text{ mole CH}_3\text{OH}}{32.04 \text{ g CH}_3\text{OH}}\right)\left(\frac{726 \text{ kJ}}{1 \text{ mole CH}_3\text{OH}}\right)$$

$$= 1.79 \times 10^4 \text{ kJ}$$

$$\text{energy per L octane} = 1.00 \text{ L C}_8\text{H}_{18}\left(\frac{1000 \text{ mL}}{1 \text{ L}}\right)\left(\frac{0.703 \text{ g C}_8\text{H}_{18}}{1 \text{ mL C}_8\text{H}_{18}}\right)$$

$$\times\left(\frac{1 \text{ mole C}_8\text{H}_{18}}{114.22 \text{ g C}_8\text{H}_{18}}\right)\left(\frac{5471 \text{ kJ}}{1 \text{ mole C}_8\text{H}_{18}}\right)$$

$$= 3.37 \times 10^4 \text{ kJ}$$

So the combustion of octane produces almost twice as much energy per liter as methanol. (octane/methanol=1.88)

For an equivalent amount of combustion energy, methanol produces $1.88 \text{ L} \times 1.09 \text{ kg CO}_2 \cdot \text{L}^{-1} = 2.05 \text{ kg CO}_2$, which is still slightly less than octane. (However, it requires that the vehicle carry about 90% more fuel by volume, 1.9 vs. 1 L, and more than twice as much fuel by mass, 1.5 kg vs. 0.7 kg.)

**14.107** (a) Diborane, $B_2H_6$, and $Al_2Cl_6$ (g) have the same basic structure in the way in which the atoms are arranged in space. (b) The bonding between the boron atoms and the bridging hydrogen atoms is electron deficient. There are three atoms and only two electrons to hold them together in a 3-center-2-electron bond. The bonding in $Al_2Cl_6$ is conventional in that all the bonds involve two atoms and two electrons. Here, the lone pair of a Cl atom is donated to an adjacent Al. (c) The hybridization is $sp^3$ at the B and Al atoms. (d) The molecules are not planar. The Group 13 element and the terminal atoms to which it is bound lie in a plane that is perpendicular to the plane that contains the main group element and the bridging atoms.

**14.109** (a) By viewing the unit cell from different directions, it is clear that it belongs to a hexagonal crystal system. (b) There are eight carbonate ions on edges, giving $\frac{1}{4} \times 8 = 2$ carbonate ions, plus there are four carbonate ions completely within the unit cell. The total number of carbonate ions in the unit cell is six. Calcium ions lie at the corners of the unit cell ($\frac{1}{8} \times 8 = 1$) as well as on the edges ($4 \times \frac{1}{4}$) and within the cell (4). The total number of calcium ions in the cell is six, agreeing with the overall stoichiometry of calcite, $CaCO_3$.

**14.111** There are two $Ca^{2+}$ ions located completely within the cell plus four located on faces ($4 \times \frac{1}{2}$) for a total of four $Ca^{2+}$ ions in the unit cell. Similarly, there are two sulfate ions located completely within the unit cell and four on faces, also giving a total of four sulfate ions in the unit cell.

There are eight water molecules completely inside the unit cell. The overall formula is $Ca_4(SO_4)_4(H_2O)_8$ or $CaSO_4 \cdot 2H_2O$.

**14.113** (a) $C(s) + 2\,H_2O(l) \rightarrow CO_2(g) + 4\,H^+(aq) + 4\,e^-$    oxidation

$PbO(s) + 2\,H^+(aq) + 2\,e^- \rightarrow Pb(s) + 2\,H_2O(l)$     reduction

$C(s) + 2\,PbO(s) \rightarrow CO_2(g) + 2\,Pb(s)$        overall

(b)  $3\,PbO(s) + \tfrac{1}{2}O_2(g) \rightarrow Pb_3O_4(s)$    Lead is oxidized from $2+$ up to $2\tfrac{2}{3}+$.

$$\text{electrons transferred} = 5.0\ \text{g PbO}\left(\frac{1\ \text{mole PbO}}{223.2\ \text{g PbO}}\right)\left(\frac{1\ \text{mole Pb}_3O_4}{3\ \text{mole PbO}}\right)$$

$$\times \left(\frac{2\ \text{mole e}^-}{1\ \text{mole Pb}_3O_4}\right)$$

$$= 1.5 \times 10^{-2}\ \text{mole electrons}$$

**14.115** The spherical structures require the formation of five-membered rings (see structure **10**, C$_{60}$).  Boron nitride cannot form these rings because they would require high-energy boron-boron or nitrogen-nitrogen bonds.

# CHAPTER 15

# THE ELEMENTS:

# THE LAST FOUR MAIN GROUPS

**15.1**  23      $NH_3$, $Li_3N$, $LiNH_2$, $NH_2^-$

      22      $H_2NNH_2$

      21      $N_2H_2$, $NH_2OH$

       0      $N_2$

     +1      $N_2O$, $N_2F_2$

     +2      $NO$

     +3      $NF_3$, $NO_2^-$, $NO^+$

     +4      $NO_2$, $N_2O_4$

     +5      $HNO_3$, $NO_3^-$, $NO_2F$

**15.3**  $CO(NH_2)_2(aq) + 2\,H_2O(l) \rightarrow (NH_4)_2CO_3(aq)$

$$\text{mass of } (NH_4)_2CO_3 = (4.0 \text{ kg urea})\left(\frac{10^3 \text{ g}}{1 \text{ kg}}\right)\left(\frac{1 \text{ mol}}{60.06}\right)$$

$$\left(\frac{1 \text{ mol } (NH_4)_2CO_3}{1 \text{ mol urea}}\right)\left(\frac{96.09 \text{ g } (NH_4)_2CO_3}{1 \text{ mol } (NH_4)_2CO_3}\right)$$

$$= 6.4 \times 10^3 \text{ g (or 6.4 kg) } (NH_4)_2CO_3$$

**15.5**  (a)  1 mole of $N_2(g)$ occupies 22.4 L at STP. For the reaction

$Pb(N_3)_2 \longrightarrow Pb + 3\,N_2$, the volume of $N_2(g)$ produced is

$$(1.5 \text{ g Pb}(N_3)_2)\left(\frac{1 \text{ mol Pb}(N_3)_2}{291.25 \text{ g Pb}(N_3)_2}\right)\left(\frac{3 \text{ mol } N_2}{1 \text{ mol Pb}(N_3)_2}\right)\left(\frac{22.4 \text{ L } N_2}{1 \text{ mol } N_2}\right)$$

$$= 0.35 \text{ L } N_2(g)$$

(b) $Hg(N_3)_2$ would produce a larger volume, because its molar mass is less. Note that molar mass occurs in the denominator in this calculation.

(c) Metal azides are good explosives because the azide ion is thermodynamically unstable with respect to the production of $N_2(g)$. This is because the N≡N triple bond is so strong and also because the production of a gas is favored entropically.

**15.7** $N_2O: H_2N_2O_2; \quad N_2O(g) + H_2O(l) \longrightarrow H_2N_2O_2(aq)$

$N_2O_3: HNO_2; \quad N_2O_3(g) + H_2O(l) \longrightarrow 2 \, HNO_2(aq)$

$N_2O_5: HNO_3; \quad N_2O_5(g) + H_2O(l) \longrightarrow 2 \, HNO_3(aq)$

**15.9** (a)

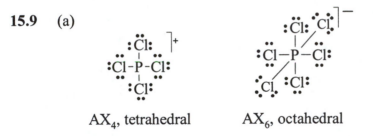

AX$_4$, tetrahedral    AX$_6$, octahedral

(b) and (c) The $PX_6^-$ ions (where X = halogen) are octahedral. The calculation of the X—X separation can easily be done using the Pythagorean theorem.

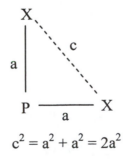

$$c^2 = a^2 + a^2 = 2a^2$$

For Br, a = 220 pm, giving c = 311 pm. For Cl, a = 204 pm, giving c = 288 pm. Both these distances are much shorter than twice the van der Waals radius (Cl, 362 pm; Br, 390 pm) but are substantially longer than twice the atomic radius (Cl, 198; Br, 228 pm). It is not clear from these numbers that the steric interaction is the main consideration. Weaker P—Br bond energies may also play a role in the nonobservation of the $PBr_6^-$ ion.

**15.11** (a) The reaction is $3\ NO_2(g) + H_2O(l) \longrightarrow 2\ HNO_3(aq) + NO(g)$

$$\Delta H°_r = 2(-207.36\ kJ \cdot mol^{-1}) + 90.25\ kJ \cdot mol^{-1}$$
$$-[3(33.18\ kJ \cdot mol^{-1}) + (285.83\ kJ \cdot mol^{-1})]$$
$$= -138.18\ kJ \cdot mol^{-1}$$

(b) $HNO_3(l) \longrightarrow HNO_3(aq)$

$$\Delta H°_r = -207.36\ kJ \cdot mol^{-1} - (-174.10\ kJ \cdot mol^{-1})$$
$$= -33.26\ kJ.mol^{-1}$$

**15.13** (a) $4\ Li(s) + O_2(g) \xrightarrow{\Delta} 2\ Li_2O(s)$

(b) $2\ Na(s) + 2\ H_2O(l) \longrightarrow 2\ NaOH(aq) + H_2(g)$

(c) $2\ F_2(g) + 2\ H_2O(l) \longrightarrow 4\ HF(aq) + O_2(g)$

(d) $2\ H_2O(l) \longrightarrow O_2(g) + 4\ H^+(aq) + 4\ e^-$

**15.15** (a) $2\ H_2S(g) + 3\ O_2(g) \xrightarrow{\Delta} 2\ SO_2(g) + 2\ H_2O(g)$

(b) $CaO(s) + H_2O(l) \longrightarrow Ca(OH)_2(aq)$

(c) $2\ H_2S(g) + SO_2(g) \xrightarrow{300°C, Al_2O_3} 3\ S(s) + 2\ H_2O(l)$

**15.17** (a)

$$
\begin{array}{c}
\text{H} \\
| \\
\ddot{\text{O}}-\ddot{\text{O}}: \\
| \\
\text{H}
\end{array}
$$

Each O in $H_2O_2$ is an $AX_2E_2$ structure; therefore, the bond angle is predicted to be $<109.5°$. In actuality, it is $97°$.

(b)–(e), The reduction potential of $H_2O_2$ is $+1.78$ V in acidic solution. It should, therefore, be able to oxidize any ion that has a reduction potential that is less than $+1.78$ V. For the ions listed, $Cu^+$ and $Mn^{2+}$ will be oxidized. It would require an input of 1.98 V to oxidize $Ag^+$ to $Ag^{2+}$ and 2.87 V to oxidize $F^-$.

**15.19** $O_2^{2-} + H_2O \leftrightarrow HO_2^- + OH^-$  essentially complete

$$HO_2^- + H_2O \leftrightarrow H_2O_2 + OH^- \qquad K_b = \frac{K_w}{K_{a1}}$$

$$K_{a1} = 1.8 \times 10^{-12} \quad K_b = \frac{1.00 \times 10^{-14}}{1.8 \times 10^{-12}} = 5.6 \times 10^{-3}$$

Because this $K_b$ is relatively small, we can assume that essentially all the $OH^-$ is formed in the first ionization; therefore,

$$[OH^-] = \left(\frac{2.00 \text{ g Na}_2O_2}{0.200 \text{ L}}\right)\left(\frac{1 \text{ mol Na}_2O_2}{77.98 \text{ g Na}_2O_2}\right)\left(\frac{1 \text{ mol OH}^-}{1 \text{ mol Na}_2O_2}\right)$$

$$= 0.128 \text{ mol} \cdot L^{-1}$$

$$pOH = -\log(0.128) = 0.893 \quad pH = 14.00 - 0.893 = 13.11$$

If we do not ignore the second ionization, then the additional contribution to $[OH^-]$ can be approximately calculated as follows:

$$K_b = \frac{[H_2O_2][OH^-]}{[HO_2^-]} = \frac{x(0.128 + x)}{(0.128 - x)} = 5.6 \times 10^{-3}$$

To a first approximation, $x = 5.6 \times 10^{-3} \text{ mol} \cdot L^{-1}$

To a second approximation, $x = \dfrac{K_b(0.128 - 0.0056)}{(0.128 + 0.0056)} = 0.005$

Then $[OH^-] = 0.128 + 0.005 = 0.133$; $pOH = -\log(0.133) = 0.876$; and pH $= 13.12$. The difference between calculations is slight.

**15.21** The weaker the H—X bond, the stronger the acid. $H_2Te$ has the weakest bond; $H_2O$, the strongest. Therefore, the acid strengths are

$$H_2Te > H_2Se > H_2S > H_2O$$

**15.23** (a) The reaction is $H_2SO_4(l) \longrightarrow H_2SO_4(aq)$

where $H_2SO_4(aq)$ is $H^+(aq) + HSO_4^-(aq)$ because $H_2SO_4$ is a strong acid. (Note: $\Delta H°_f(H^+)$ is defined as 0).

$\Delta H°_r = -887.34 \text{ kJ} \cdot \text{mol}^{-1} - (-813.99 \text{ kJ} \cdot \text{mol}^{-1}) = -73.35 \text{ kJ} \cdot \text{mol}^{-1}$

(b) The number of moles of $H_2SO_4$ is

$10.00 \div 98.07 \text{ g} \cdot \text{mol}^{-1} = 0.1020 \text{ mol}$

The amount of heat generated should be

$0.1020 \text{ mol} \times -73.35 \text{ kJ} \cdot \text{mol}^{-1} = -7.482 \text{ kJ}$.

The heat capacity of water is $4.18 \text{ J} \cdot (°C)^{-1} \cdot g^{-1}$. Adding 7.482 kJ of heat to the water should raise the temperature by

$$\Delta t = \frac{7.482 \text{ kJ} \times 1000 \text{ J} \cdot \text{kJ}^{-1}}{4.18 \text{ J} \cdot (°C)^{-1} \cdot g^{-1} \times 500.0 \text{ g}} = 3.56°$$

The final temperature should be $25.0°C + 3.56°C = 28.6°C$.

**15.25** Fluorine comes from the minerals fluorspar, $CaF_2$; cryolite, $Na_3AlF_6$; and the fluorapatites, $Ca_5F(PO_4)_3$. The free element is prepared from HF and the KF by electrolysis, but the HF and KF needed for the electrolysis are prepared in the laboratory. Chlorine primarily comes from the mineral rock salt, NaCl. The pure element is obtained by electrolysis of liquid NaCl.

Bromine is found in seawater and brine wells as the $Br^-$ ion; it is also found as a component of saline deposits; the pure element is obtained by oxidation of $Br^-(aq)$ by $Cl_2(g)$.

Iodine is found in seawater, seaweed, and brine wells as the $I^-$ ion; the pure element is obtained by oxidation of $I^-(aq)$ by $Cl_2(g)$.

**15.27** (a) $HIO(aq)$ $H = +1, O = -2$; therefore, $I = +1$

(b) $ClO_2$ $O = -2$; therefore, $Cl = +4$

(c) $Cl_2O_7$ $O = -2$; therefore, $Cl = +14/2 = +7$

(d) $NaIO_3$ $Na = +1, O = -2$; therefore, $I = +5$

**15.29** (a) $4 KClO_3(l) \xrightarrow{\Delta} 3 KClO_4(s) + KCl(s)$

(b) $Br_2(l) + H_2O(l) \longrightarrow HBrO(aq) + HBr(aq)$

(c) $NaCl(s) + H_2SO_4(aq) \longrightarrow NaHSO_4(aq) + HCl(g)$

(d) (a) and (b) are redox reactions. In (a), Cl is both oxidized and reduced. In (b), Br is both oxidized and reduced. (c) is a Brønsted acid-base reaction; $H_2SO_4$ is the acid, and $Cl^-$ the base.

**15.31** (a) $HClO < HClO_2 < HClO_3 < HClO_4$ ($HClO_4$ is strongest; HClO, weakest)

(b) The oxidation number of Cl increases from HClO to $HClO_4$. In $HClO_4$, chlorine has its highest oxidation number of +7, so $HClO_4$ will be the strongest oxidizing agent.

**15.33** $: \overset{..}{Cl} - \overset{..}{O} - \overset{..}{Cl} :$, $AX_2E_2$, angular, about $109°$. The actual bond angle is $110.9°$.

**15.35**

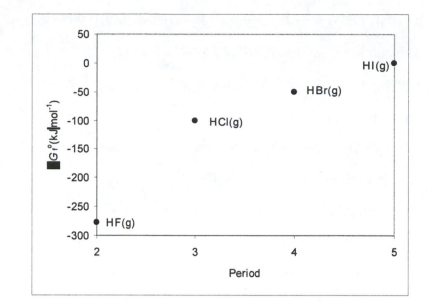

The thermodynamic stability of the hydrogen halides decreases down the group. The $\Delta G°_f$ values of HCl, HBr, and HI fit nicely on a straight line; whereas HF is anomalous. In other properties, HF is also the anomalous member of the group, in particular, its acidity. Also see Exercise 15.42.

**15.37** $Cl_2(g) + 2 e^- \longrightarrow 2 Cl^-(aq) \quad E° = +1.36 \text{ V}$

$MnO_4^-(aq) + 8 H^+(aq) + 5 e^- \longrightarrow Mn^{2+}(aq) + 4 H_2O(l) \quad E° = +1.51 \text{ V}$

$E°_{cell} = (1.36 - 1.51) \text{ V} = -0.15 \text{ V}$

Because $E°_{cell}$ is negative, $Cl_2(g)$ will not oxidize $Mn^{2+}$ to form the permanganate ion in an acidic solution.

**15.39** $F^-(aq) + PbCl_2(s) \longrightarrow Cl^-(aq) + PbClF(s)$

$\text{molarity of F}^- \text{ ions} = \left( \dfrac{0.765 \text{ g PbClF}}{0.0250 \text{ L}} \right) \left( \dfrac{1 \text{ mol PbClF}}{261.64 \text{ g PbClF}} \right) \left( \dfrac{1 \text{ mol F}^-}{1 \text{ mol PbClF}} \right)$

$= 0.117 \text{ mol} \cdot \text{L}^{-1}$

**15.41**  $3 \, NH_4ClO_4(s) + 3 \, Al(s) \longrightarrow Al_2O_3(s) + AlCl_3(s) + 6 \, H_2O(g) + 3 \, NO(g)$

The number of moles of $NH_4ClO_4$ is

$1.00 \, kg \times 1000 \, g \cdot kg^{-1} \div 117.49 \, g \cdot mol^{-1} = 8.51 \, mol$

The number of moles of Al is

$1.00 \, kg \times 1000 \, g \cdot kg^{-1} \div 26.98 \, g \cdot mol^{-1} = 37.06 \, mol$

The limiting reagent is the $NH_4ClO_4$.

The standard enthalpy for the reaction is given by

$$\Delta H°_r = (-1675.7 \, kJ \cdot mol^{-1}) + (-704.2 \, kJ \cdot mol^{-1}) + 6(-241.82 \, kJ \cdot mol^{-1})$$
$$+ 3(90.25 \, kJ \cdot mol^{-1}) - [3(-295.31 \, kJ \cdot mol^{-1})]$$
$$= -2674.1 \, kJ \cdot mol^{-1}$$

This value is the amount of heat released for 3 mol $NH_4ClO_4$. The amount released for 8.51 mol will be

$$8.51 \, mol \, NH_4ClO_4 \times \frac{-2674.1 \, kJ \cdot mol^{-1}}{3 \, mol \, NH_4ClO_4} = -7.59 \times 10^3 \, kJ$$

There will be $7.59 \times 10^3 \, kJ$ of heat released.

**15.43**  Helium occurs as a component of natural gases found under rock formations in certain locations, especially some in Texas. Argon is obtained by distillation of liquid air.

**15.45**  (a)  $KrF_2 : F = -1$; therefore, $Kr = +2$

(b)  $XeF_6 : F = -1$; therefore, $Xe = +6$

(c)  $KrF_4 : F = -1$; therefore, $Kr = +4$

(d)  $XeO_4^{2-} : O = -2$, $N_{ox}(Xe) - 8 = -2$; therefore, $N_{ox}(Xe) = +6$

**15.47**  $XeF_4(aq) + 4 \, H^+(aq) + 4 \, e^- \rightarrow Xe(g) + 4 \, HF(aq)$

**15.49**  Because $H_4XeO_6$ has more highly electronegative O atoms bonded to Xe, we predict that $H_4XeO_6$ is more acidic than $H_2XeO_4$.

**15.51**  A sol is colloid comprised of solid particles suspended in a liquid. Muddy water is a type of sol. A foam is a suspension of a gas in a solid or liquid. Styrofoam, foam rubber, soapsuds, and aerogels are all types of foams.

**15.53**  (a) both a sol and an emulsion   (b) a foam   (c) a sol

**15.55**  In fluorescence, light absorbed by molecules is immediately emitted, whereas, in phosphorescence, molecules remain in an excited state for a period of time before emitting the absorbed light.  In both phenomena, the energy of the emitted photon is lower than that of the absorbed photon (emitted photons have a longer wavelength).

**15.57**  The mercury atoms are an energy transfer agent. They absorb the energy from a high-voltage discharge and emit ultraviolet light in the region of 254 and 185 nm. This emitted light then excites a fluorescent material that emits radiation in the visible region of the spectrum, which is the light observed when the lamp is turned on.

**15.59**  (a) and (b). The formal charges are given under each atom.

$$\overset{\displaystyle ..\quad ..\quad ..}{\underset{\displaystyle ..\quad ..\quad ..}{N=N=N}}\Big]^{-}$$

$$\;\;-1\quad +1\quad -1$$

(c) The value of $-\frac{1}{3}$ is an average oxidation number based solely upon the number of nitrogen atoms and the overall charge. From the Lewis structure, we can see that the molecule is asymmetric and it is possible that the different nitrogen atoms may have different oxidation numbers. It would be extreme to state that the terminal nitrogens have oxidation numbers of –1 and the central nitrogen atom has an oxidation number of

+1, but assigning the same oxidation number of $-\frac{1}{3}$ to all three atoms is also not strictly accurate.

(d) This situation most often arises when an element has bonds to other atoms of the same type.

**15.61** The larger the value of $E°$ for the reduction $X_2 + 2\,e^- \longrightarrow 2\,X^-$, the greater the oxidizing strength of the halogen $X_2$. From Appendix 2B,

$$F_2 + 2\,e^- \longrightarrow 2\,F^- \qquad E° = +2.87 \text{ V}$$
$$Cl_2 + 2\,e^- \longrightarrow 2\,Cl^- \qquad E° = +1.36 \text{ V}$$
$$Br_2 + 2\,e^- \longrightarrow 2\,Br^- \qquad E° = +1.09 \text{ V}$$
$$I_2 + 2\,e^- \longrightarrow 2\,I^- \qquad E° = +0.54 \text{ V}$$

Thus, $I_2 < Br_2 < Cl_2 < F_2$.

**15.63** oxidation:

$$As_2S_3(s) + 8\,H_2O(l) \longrightarrow 2\,H_2AsO_4(aq) + 3\,H_2S(aq) + 4\,H^+(aq) + 4\,e^-$$

reduction: $H_2O_2(aq) + 2\,H^+(aq) + 2\,e^- \longrightarrow 2\,H_2O(l)$

Multiply the reduction reaction by 2, cancel electrons, and add.

overall:

$$As_2S_3(s) + 2\,H_2O_2(aq) + 4\,H_2O(l) \longrightarrow 2\,H_3AsO_4(aq) + 3\,H_2S(aq)$$

**15.65** This ratio, $\Delta H_{vap}/T_b$, is the entropy of vaporization. Hydrogen bonding is much stronger in $H_2O(l)$ than in $H_2S(l)$. Thus $H_2O(l)$ has a more ordered arrangement than $H_2S(l)$. Consequently, the change in entropy upon transformation to the gaseous state is greater for $H_2O$ than for $H_2S$.

**15.67** The two molecules are shown below:

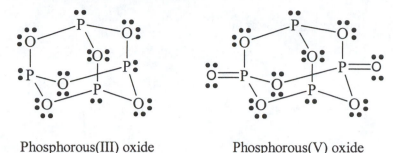

Phosphorous(III) oxide          Phosphorous(V) oxide

The basic structure of the two molecules is the same. The phosphorus atoms lie in a tetrahedral arrangement in which there are bridging oxygen atoms to the other phosphorus atoms. In phosphorus(V) oxide, there is an additional terminal oxygen atom bonded to each phosphorus atom. In phosphorus(III) oxide, each oxygen atom has a formal charge of 0 as does each phosphorus atom. In phosphorus(V) oxide, this is also true. According to the Lewis structures, all P—O bonds in phosphorus(III) oxide have a bond order of 1, while in phosphorus(V) oxide, the terminal oxygen atoms have a bond order of 2 between them and the phosphorus atoms to which they are attached. If one examines the molecular parameters, one sees that all of the P—O$_{bridging}$ distances for phosphorus(III) oxide are slightly longer than those of phosphorus(V) oxide (163.8 pm versus 160.4 pm). This is expected because the radius of phosphorus(V) should be smaller than that of phosphorus(III). The terminal P==O distances for phosphorus(V) oxide are considerably shorter (142.9 pm); this agrees with the higher bond order between phosphorus and these atoms.

**15.69**  (a) $SO_2(g) + H_2O(l) \rightarrow H_2SO_3(l)$  This is a Lewis acid-base reaction. $SO_2$ is the acid and $H_2O$ is the base.

(b)  $2 F_2(g) + 2 NaOH(aq) \longrightarrow OF_2(g) + 2 NaF(aq) + H_2O(l)$  This is a redox reaction illustrating the oxidizing ability of $F_2$ in basic solution and is used for the preparation of $OF_2(g)$.

O is oxidized and F is reduced.

(c) $S_2O_3^{2-}(aq) + 4\,Cl_2(g) + 13\,H_2O(l) \longrightarrow$
$$2\,HSO_4^-(aq) + 8\,H_3O^+(aq) + 8\,Cl^-(aq)$$

This is a redox reaction illustrating the oxidizing power of $Cl_2(g)$ in acidic solution. S is oxidized and Cl is reduced.

(d) $2\,XeF_6(s) + 16\,OH^-(aq) \longrightarrow$
$$XeO_6^{4-}(aq) + Xe(g) + 12\,F^-(aq) + 8\,H_2O(l) + O_2(g)$$

This is a redox reaction that is also a disproportionation reaction in that Xe goes from oxidation number $+6$ to $+8$ and to 0. Xe is both oxidized and reduced.

**15.71** (a) $I_2(s) + 3\,F_2(g) \longrightarrow 2\,IF_3(s); I_2(s) + 5\,F_2(g) \longrightarrow 2\,IF_5(s)$, etc.

(b) $I_2(aq) + I^-(aq) \longrightarrow I_3^-(aq)$

(c) $Cl_2(g) + H_2O(l) \longrightarrow HCl(aq) + HOCl(aq)$

But there are competing reactions, such as

$Cl_2(g) + H_2O(l) \longrightarrow 2\,HCl(aq) + \frac{1}{2}O_2(g)$. The predominant reaction is determined by the temperature and pH.

(d) $2\,F_2(g) + 2\,H_2O(l) \longrightarrow 4\,HF(aq) + O_2(g)$

**15.73** The reaction of interest is: $XeF_4(s) + 2\,SF_4(s) \longrightarrow 2\,SF_6(s) + Xe(g)$

To determine how much product is formed, we must first identify the limiting reagent:

$$\text{moles of } XeF_4(s) = \frac{330.0\text{ g}}{207.29\text{ g} \cdot \text{mol}^{-1}} = 1.592\text{ mol}$$

$$\text{moles of } SF_4(s) = \frac{250.0\text{ g}}{108.07\text{ g} \cdot \text{mol}^{-1}} = 2.313\text{ mol}$$

Given the 1:2 stoichiometry of the reaction, $SF_4(s)$ is the limiting reagent and 2.313 mol of $SF_6(s)$ will be produced.

$(2.313 \text{ mol})(146.07 \text{ g} \cdot \text{mol}^{-1}) = 337.9 \text{ g of } SF_6(s)$ produced.

**15.75** Orpiment is $As_2S_3$ and realgar is $As_4S_4$. Orpiment is yellow and realgar is orange-red. They are both used as pigments.

**15.77** (a)

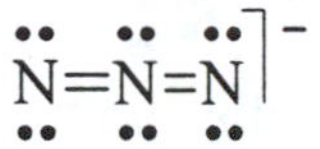

$$\overset{\cdot\cdot}{N}=\overset{\cdot\cdot}{N}=\overset{\cdot\cdot}{\underset{\cdot\cdot}{N}} \Big]^{-}$$

$AX_2$, linear 180°

(b) $F^-$, 133 pm; $N_3^-$, 148 pm; $Cl^-$, 181 pm; therefore, between fluorine and chlorine.

(c) HCl, HBr, and HI are all strong acids. For HF, $K_a = 3.5 \times 10^{-4}$, so HF is slightly more acidic than $HN_3$. The small size of the azide ion suggests that the H—N bond in $HN_3$ is similar in strength to that of the H—F bond, so it is expected to be a weak acid.

(d) ionic: $NaN_3$, $Pb(N_3)_2$, $AgN_3$, etc.

covalent: $HN_3$, $B(N_3)_3$, $FN_3$, etc.

**15.79** $[ClO^-] = (0.028\,34 \text{ L}) \left( \dfrac{0.110 \text{ mol } S_2O_3^{2-}}{1 \text{ L } Na_2S_2O_3} \right)$

$\left( \dfrac{1 \text{ mol } I_2}{2 \text{ mol } S_2O_3^{2-}} \right) \left( \dfrac{1 \text{ mol } ClO^-}{1 \text{ mol } I_2} \right) \left( \dfrac{1}{0.010\,00 \text{ L } ClO^-} \right)$

$= 0.156 \text{ mol} \cdot L^{-1}$

**15.81** The solubility of the ionic halides is determined by a variety of factors, especially the lattice enthalpy and enthalpy of hydration. There is a delicate balance between the two factors, with the lattice enthalpy usually being the determining one. Lattice enthalpies decrease from chloride to iodide, so water molecules can more readily separate the ions in the latter. Less ionic halides, such as the silver halides, generally have a much lower solubility, and the trend in solubility is the reverse of the more ionic halides. For the less ionic halides, the covalent character of the bond allows the ion pairs to persist in water. The ions are not easily hydrated, making them less soluble. The polarizability of the halide ions, and thus, the covalency of their bonding, increases down the group.

**15.83** To answer this question, we can compare equilibrium "vapor pressures" of water over each of these reagents. The one with the lowest equilibrium vapor pressure will be the better drying agent. The two reactions of interest are

$$CaO(s) + H_2O(g) \longrightarrow Ca(OH)_2(s)$$
$$P_4O_{10}(s) + 6\,H_2O(g) \longrightarrow 4\,H_3PO_4(s)$$

First, the free energies of the reactions are calculated and from these the equilibrium pressure of water can be obtained.

For CaO:

$$\Delta G°_r = -898.49\ \text{kJ}\cdot\text{mol}^{-1} - [(-604.03\ \text{kJ}\cdot\text{mol}^{-1}) + (-228.57\ \text{kJ}\cdot\text{mol}^{-1})]$$
$$= -65.89\ \text{kJ}\cdot\text{mol}^{-1}$$

$$\Delta G°_r = -RT\ln K$$

$$K = e^{-\frac{\Delta G°}{RT}} = e^{-\frac{-65\,890\ \text{J}\cdot\text{mol}^{-1}}{(8.314\ \text{J}\cdot\text{K}^{-1}\cdot\text{mol}^{-1})(298\ \text{K})}} = 3.5\times10^{11}$$

$$K = \frac{1}{P_{H_2O}} = 3.5\times10^{11}$$

$$P_{H_2O} = 2.8\times10^{-12}\ \text{bar}$$

For $P_4O_{10}$:

$$\Delta G°_r = 4(-1119.2\ \text{kJ}\cdot\text{mol}^{-1})$$
$$\qquad\qquad - [(-2697.0\ \text{kJ}\cdot\text{mol}^{-1}) + 6(-228.57\ \text{kJ}\cdot\text{mol}^{-1})]$$
$$= -408.4\ \text{kJ}\cdot\text{mol}^{-1}$$

$$K = e^{\frac{\Delta G^\circ}{RT}} = e^{-\frac{-408\,400 \text{ J·mol}^{-1}}{(8.314 \text{ J·K}^{-1}\text{·mol}^{-1})(298 \text{ K})}} = 3.9 \times 10^{71}$$

$$K = \frac{1}{(P_{H_2O})^6} = 3.9 \times 10^{71}$$

$$P_{H_2O} = 1.2 \times 10^{-12} \text{ bar}$$

Because the pressure of water possible above CaO is greater than that above $P_4O_{10}$, CaO will be a poorer drying agent.

**15.85** (a) $Ag^+ + e^- \longrightarrow Ag$ $\qquad\qquad$ $E^\circ = +0.80$ V

$I_2 + 2\,e^- \longrightarrow 2\,I^-$ $\qquad\qquad\qquad$ $E^\circ = +0.54$ V

For $2\,Ag^+ + 2\,I^- \longrightarrow Ag + I_2$, $E^\circ = +0.80$ V $- 0.54$ V $= +0.26$ V

The process should be spontaneous.

(b) The formation of AgI precipitate means that the concentration of $Ag^+$ ions is never high enough to achieve the conditions necessary for the redox reaction to take place. The solubility product $K_{sp}$ limits the concentrations in solution, so that the actual redox potential is not the value calculated, which represents the values when $[Ag^+] = 1$ M and $[I^-] = 1$ M. If we use the concentrations established by the solubility equilibrium and the Nernst equation, we can calculate the actual redox potential:

$$E = E^\circ - \frac{RT}{nF} \ln Q$$

where, in this case, $Q = \dfrac{1}{K_{sp}^{\,2}}$ for the reaction as written

$K_{sp} = 1.5 \times 10^{-16}$ for AgI

$$E = +0.26 \text{ V} - \frac{(8.314 \text{ J} \cdot \text{K}^{-1} \cdot \text{mol}^{-1})(298 \text{ K})}{(2)(96\,485 \text{ J} \cdot \text{V}^{-1} \cdot \text{mol}^{-1})} \ln \frac{1}{(1.5 \times 10^{-16})^2}$$

$$= +0.26 \text{ V} - \frac{(8.314 \text{ J} \cdot \text{K}^{-1} \cdot \text{mol}^{-1})(298 \text{ K})}{(96\,485 \text{ J} \cdot \text{V}^{-1} \cdot \text{mol}^{-1})} \ln \frac{1}{(1.5 \times 10^{-16})}$$

$$= +0.26 \text{ V} - 0.94 \text{ V}$$

$$= -0.68 \text{ V}$$

The fact that the concentrations of $Ag^+$ and $I^-$ are limited in solution means that the redox potential for a spontaneous reaction is never achieved.

15.87 (a) The molecular orbital diagram for $NO^+$ should have the oxygen orbitals slightly lower in energy than the nitrogen orbitals, because oxygen is more electronegative. This will cause the bonding to be more ionic than in either $N_2$ or $O_2$. There is an ambiguity, however, in that the MO diagram could be similar to either that of $N_2$ or that of $O_2$. Refer to Figures 3.34 and 3.35 where you will see that the $\sigma_{2p}$ and the $\pi_{2p}$ have different relative energies. There are consequently two possibilities for the orbital energy diagram:

(b) The two orbital diagrams predict the same bond order (3) and same magnetic properties (diamagnetic), and so these properties cannot be used to determine which diagram is the correct one. That must be determined by more complex spectroscopic measurements.

15.89 (a) Pyrite adopts a face-centered cubic unit cell. (b) The iron atoms lie at the corners and at the face centers of the unit cell. Eight sulfur atoms lie completely within the unit cell. (c) The coordination number of iron is six (octahedral). (d) Each sulfur atom is bonded to one other sulfur atom and three iron atoms. (e) The locations of the sulfur atoms can be considered in one of two ways. An examination of the structure shows that these are best thought of as $S_2^{2-}$ ions. The locations of the midpoints of the S-S bonds are at the centers of each edge of the unit cell $(12\ S_2^{2-} \times \frac{1}{4})$ plus 1 $S_2^{2-}$ ion in the center of the unit cell (only half the ions have sulfur atoms within a given cell). This gives a total of 4 $S_2^{2-}$ ions in the unit cell. Alternatively, identify eight S atoms within the unit cell.

**15.91** (a) State A: 94.72 kJ · mol⁻¹, State B: 157.85 kJ · mol⁻¹. State B is the higher energy state because it requires more energy to pair electrons in the same orbital than it does to force the spins of two electrons in different orbitals to be antiparallel

(b)

$$\lambda = \frac{h \cdot c}{E} = \frac{\left(6.62608 \times 10^{-34}\ \text{J} \cdot \text{s}\right)\left(2.9979 \times 10^8\ \text{m} \cdot \text{s}^{-1}\right)\left(6.02214 \times 10^{23}\ \text{mol}^{-1}\right)}{\left(94720\ \text{J} \cdot \text{mol}^{-1}\right)}$$

$$= 1.263 \times 10^{-6}\ \text{m}\quad \text{or}\quad 1.263\ \mu\text{m}$$

**15.93**

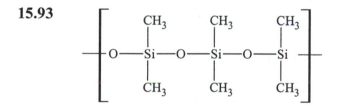

# CHAPTER 16

# THE ELEMENTS: THE *d* BLOCK

**16.1**   Elements at the left of the *d* block tend to have strongly negative standard potentials.

**16.3**   (a) Sc   (b) Au   (c) Nb

(d) One might expect osmium to be larger than ruthenium because it is a third row transition metal and ruthenium is in row two; however, because of the lanthanide contraction, they are about the same size (Os, 135 pm; Ru, 134 pm).

**16.5**   (a) Ti   (b) Cu   (c) Zn   (d) Fe   (e) Os

**16.7**   Hg is much more dense than Cd, because the shrinkage in atomic radius that occurs between $Z = 58$ and $Z = 71$ (the lanthanide contraction) causes the atoms following the rare earths to be smaller than might have been expected for their atomic masses and atomic numbers. Zn and Cd have densities that are not too dissimilar, because the radius of Cd is subject only to a smaller *d*-block contraction.

**16.9**   (a) Proceeding down a group in the *d* block (for example, from Cr to Mo to W), there is an increasing probability of finding the elements in higher oxidation states. That is, higher oxidation states become more stable on going down a group.

(b) The trend for the *p*-block elements is reversed. Because of the inert pair effect, the higher oxidation states tend to be less stable as one descends a group.

**16.11** In $MO_3$, M has an oxidation number of +6. Of these three elements, the +6 oxidation state is most stable for Cr. See Fig. 16.6.

**16.13** (a) Ti(s), $MgCl_2$(s)

$$TiCl_4(g) + 2\ Mg(l) \xrightarrow{\Delta} Ti(s) + 2\ MgCl_2(s)$$

(b) $Co^{2+}$(aq), $HCO_3^-$(aq), $NO_3^-$(aq)

$$CoCO_3(s) + HNO_3\ (aq) \rightarrow Co^{2+}(aq) + HCO_3^-\ (aq) + NO_3^-\ (aq)$$

(c) V(s), CaO(s)

$$V_2O_5(s) + 5\ Ca(l) \xrightarrow{\Delta} 2\ V(s) + 5\ CaO(s)$$

**16.15** (a) titanium(IV) oxide, $TiO_2$

(b) iron(III) oxide, $Fe_2O_3$

(c) manganese(IV) oxide, $MnO_2$

**16.17** (a) $V^{2+} + 2\ e^- \rightarrow V(s)$            $E° = -1.19\ V$

$V^{3+} + e^- \rightarrow V^{2+}$            $E° = -0.26\ V$

$2\ H^+ + 2\ e^- \rightarrow H_2(g)$            $E° = 0.00\ V$

Therefore, V(s) will be oxidized to $V^{3+}$. The products are $V^{3+}$, $H_2$, and $Cl^-$.

(b) $Hg_2^{2+} + 2\ e^- \rightarrow 2\ Hg$    $E° = +0.79\ V$

$Hg^{2+} + 2\ e^- \rightarrow Hg$            $E° = +1.62\ V$

$2\ H^+ + 2\ e^- \rightarrow H_2(g)$          $E° = +0.00\ V$

Therefore, no reaction.

(c) $Co^{2+} + 2 e^- \rightarrow Co(s)$  $E° = -0.28$ V

$Co^{3+} + e^- \rightarrow Co^{2+}$  $E° = +1.81$ V

$2 H^+ + 2 e^- \rightarrow H_2(g)$  $E° = =0.00$ V

Therefore, Co(s) will be oxidized to $Co^{2+}$. The products are

$Co^{2+}$, $H_2$, and $Cl^-$.

The further oxidation to $Co^{3+}$ is not favorable by reaction of $H^+$ with

$Co^{2+}$.

However, $Co^{2+}$ is oxidized in air to $Co^{3+}$.


**16.19** (a) $V_2O_5(s) + 2 H_3O^+(aq) \rightarrow 2 VO_2^+(aq) + 3 H_2O(l)$

(b) $V_2O_5(s) + 6 OH^-(aq) \rightarrow 2 VO_4^{3-}(aq) + 3 H_2O(l)$


**16.21** Even though all three Group 1B/11 metal atoms have the valence shell electron configuration $(n-1)d^{10}ns^1$, Cu is more reactive than Ag or Au. Metals ordinarily lose one or more electrons to form cations when they react with some other species. As the value of $n$ increases, $d$ and $f$ electrons become less effective at shielding the outermost, highest energy electron(s) from the attractive charge of the nucleus. This higher effective nuclear charge makes it more difficult to oxidize the metal atom or ion. So, for example, $Cu^{2+}$ exists in many common compounds (and can be formed by $Cu^+$ disproportionation in water) while $Ag^{2+}$ does not. Furthermore, the valence electron orbital energies for most common Lewis bases match the orbitals of Cu and its cations more closely and would interact with them more favorably to form products.


**16.23** (a) $Cr^{3+}$ ions in water form the complex $[Cr(H_2O)_6]^{3+}(aq)$, which behaves as a Brønsted acid:

$[Cr(H_2O)_6]^{3+}(aq) + H_2O(l) \rightarrow [Cr(H_2O)_5OH]^{2+}(aq) + H_3O^+(aq)$

(b) The gelatinous precipitate is the hydroxide $Cr(OH)_3$. The precipitate dissolves as the $Cr(OH)_4^-$ complex ion is formed:

$$Cr^{3+}(aq) + 3\,OH^-(aq) \rightarrow Cr(OH)_3(s)$$
$$Cr(OH)_3(s) + OH^-(aq) \rightarrow Cr(OH)_4^-(aq)$$

16.25 (a) hexacyanoferrate(II) ion

Let $x$ = the oxidation number to be determined

$$x(Fe) + 6 \times (-1) = -4$$
$$x(Fe) = -4 - (-6) = +2$$

(b) hexaamminecobalt(III)ion

$$x(Co) + 6 \times (0) = +3$$
$$x(Co) = +3$$

(c) aquapentacyanocobaltate(III) ion

$$x(Co) + 5 \times (-1) + 1 \times (0) = -2$$
$$x(Co) = -2 - (-5) = +3$$

(d) pentaamminesulfatocobalt(III) ion

$$x(Co) + 1 \times (-2) + 5 \times (0) = +1$$
$$x(Co) = +1 - (-2) = +3$$

16.27 (a) $K_3[Cr(CN)_6]$

(b) $[Co(NH_3)_5(SO_4)]Cl$

(c) $[Co(NH_3)_4(H_2O)_2]Br_3$

(d) $Na[Fe(H_2O)_2(C_2O_4)_2]$

16.29 (a) The molecule $HN(CH_2CH_2NH_2)_2$ has three nitrogen atoms, each with a lone pair of electrons that may be used for bonding to a metal center. The molecule can thus function as a tridentate ligand.

(b) The $CO_3^{2-}$ ion can bind to a metal ion through either one or two oxygen atoms. It may, therefore, serve as a mono- or bidentate ligand.

(c) $H_2O$ is always a monodenate ligand.

(d) The oxalate ion can bind through two oxygen atoms and is usually a bidentate ligand.

**16.31** As shown below, only the molecule (b) can function as a chelating ligand. The two amine groups in (a) and (c) are arranged so that they would not be able to coordinate simultaneously to the same metal center. It is possible for each of the amine groups in (a) and (c) to coordinate to two different metal centers, however. This is not classified as chelating. When a single ligand binds to two different metal centers, it is known as a *bridging* ligand.

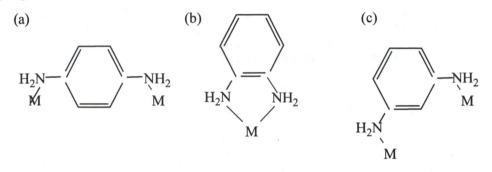

**16.33** (a) 4  (b) 2  (c) 6 (en is bidentate)  (d) 6 (EDTA is hexadentate)

**16.35** (a) structural isomers, linkage isomers

(b) structural isomers, ionization isomers

(c) structural isomers, linkage isomers

(d) structural isomers, ionization isomers

**16.37** (a) yes

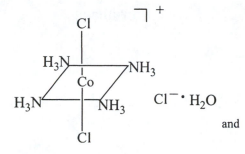

and

*trans*-tetraamminedichlorcobalt(III)
chloride monohydrate

*cis*-tetraamminedichlorcobalt(III)
chloride monohydrate

(b) no

(c) yes

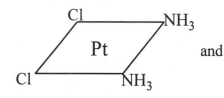 and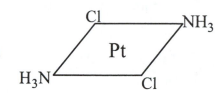

*cis*-diamminedichloroplatinum(II)

*trans*-diamminedichloroplatinum(II)

**16.39** (a)

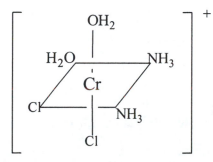

(b) Yes, the species is optically active as it has a nonsuperimposable mirror image.

**16.41** (a)

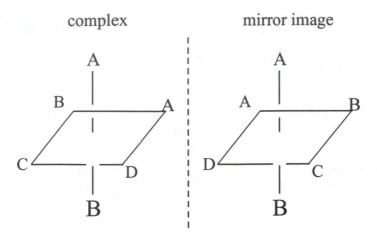

No rotation will make the complex and its mirror image match; therefore, it is chiral.

(b)

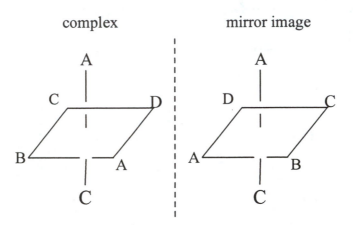

A double rotation shows that the complex and its mirror image are superimposable; therefore, it is not chiral.

The two complexes are not enantiomers; they are not even isomers.

**16.43** (a) 1; (b) 6; (c) 5; (d) 3; (e) 6; (f) 6

**16.45** (a) 2; (b) 5; (c) 8; (d) 10; (e) 0 (or 8); (f) 10

**16.47** (a) octahedral; strong-field ligand, 6 e$^-$

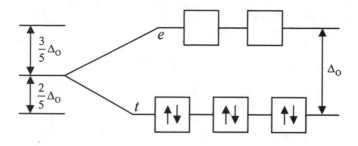

diamagnetic, no unpaired electrons

(b) tetrahedral: weak-field ligand, 8 e$^-$

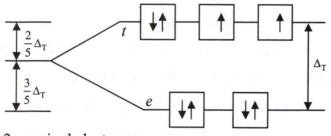

2 unpaired electrons

(c) octahedral: weak-field ligand, 5 e$^-$

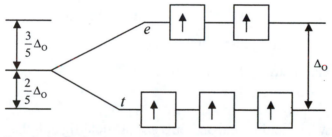

5 unpaired electrons

(d) octahedral: strong-field ligand, 5 e$^-$

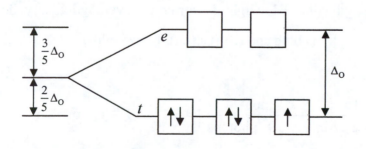

one unpaired electron

**16.49** (a) $[Co(en)_3]^{3+}$     6 e⁻

(upper level: two empty orbitals) ___ ___

(lower level) ↑↓ ↑↓ ↑↓

$[Co(en)_3]^{3+}$ has no unpaired electrons.

(b) $[Mn(CN)_6]^{3-}$     4 e⁻

(upper level: two empty orbitals) ___ ___

(lower level) ↑↓ ↑ ↑

$[Mn(CN)_6]^{3-}$ has two unpaired electrons.

**16.51** Weak-field ligands do not interact strongly with the $d$-electrons in the metal ion, so they produce only a small crystal field splitting of the $d$-electron energy states. The opposite is true of strong-field ligands. With weak-field ligands, unpaired electrons remain unpaired if there are unfilled orbitals; hence, a weak-field ligand is likely to lead to a high-spin complex. Strong-field ligands cause electrons to pair up with electrons in lower energy orbitals. A strong-field ligand is likely to lead to a low-spin complex. Ligands arranged in the spectrochemical series help to distinguish strong-field and weak-field ligands. Measurement of magnetic susceptibility (paramagnetism) can be used to determine the number of unpaired electrons, which, in turn, establishes whether the associated ligand is weak-field or strong-field in nature.

**16.53** Because F⁻ is a weak-field ligand and en a strong-field ligand, the splitting between levels is less in (a) than in (b). Therefore, (a) will absorb light of longer wavelength than will (b) and consequently will display a shorter wavelength color. Blue light is shorter in wavelength than yellow light, so (a) $[CoF_6]^{3-}$ is blue and (b) $[Co(en)_3]^{3+}$ is yellow.

**16.55** $E_{photon} = \left( \dfrac{209 \text{ kJ}}{\text{mol}} \right) \left( \dfrac{1 \text{ mol}}{6.02 \times 10^{23} \text{ photons}} \right) = 3.47 \times 10^{-22} \text{ kJ} \cdot \text{photon}^{-1}$

$$v = \frac{E}{h} = \frac{3.47 \times 10^{-19}\,\text{J}}{6.626 \times 10^{-34}\,\text{J} \cdot \text{s}} = 5.24 \times 10^{14}\,\text{s}^{-1}$$

$$\lambda = \frac{c}{v} = \left( \frac{3.00 \times 10^{8}\,\text{m} \cdot \text{s}^{-1}}{5.24 \times 10^{14}\,\text{s}^{-1}} \right)\left( \frac{10^{9}\,\text{nm}}{1\,\text{m}} \right) = 573\,\text{nm}$$

This wavelength is in the yellow region of the visible spectrum. Since the complex absorbs yellow light, it transmits the complement, or purple (a.k.a. violet).

**16.57** In $Zn^{2+}$, the $3d$-orbitals are filled ($d^{10}$). Therefore, there can be no electronic transitions between the $t$ and $e$ levels; hence, no visible light is absorbed and the aqueous ion is colorless. The $d^{10}$ configuration has no unpaired electrons, so Zn compounds would be diamagnetic, not paramagnetic.

**16.59** (a) $\Delta_O = \dfrac{hc}{\lambda} = \dfrac{(6.63 \times 10^{-34}\,\text{J} \cdot \text{s}^{-1})(3.00 \times 10^{8}\,\text{m} \cdot \text{s}^{-1})}{740 \times 10^{-9}\,\text{m}} = 2.69 \times 10^{-19}\,\text{J}$

(b) $\Delta_O = \dfrac{hc}{\lambda} = \dfrac{(6.63 \times 10^{-34}\,\text{J} \cdot \text{s}^{-1})(3.00 \times 10^{8}\,\text{m} \cdot \text{s}^{-1})}{460 \times 10^{-9}\,\text{m}} = 4.32 \times 10^{-19}\,\text{J}$

(c) $\Delta_O = \dfrac{hc}{\lambda} = \dfrac{(6.63 \times 10^{-34}\,\text{J} \cdot \text{s}^{-1})(3.00 \times 10^{8}\,\text{m} \cdot \text{s}^{-1})}{575 \times 10^{-9}\,\text{m}} = 3.46 \times 10^{-19}\,\text{J}$

These numbers can be multiplied by $6.02 \times 10^{23}$ to obtain $\text{kJ} \cdot \text{mol}^{-1}$.

(a) $2.69 \times 10^{-19}\,\text{J} \times 6.02 \times 10^{23}\,\text{mol}^{-1} = 162\,\text{kJ} \cdot \text{mol}^{-1}$

(b) $4.32 \times 10^{-19}\,\text{J} \times 6.02 \times 10^{23}\,\text{mol}^{-1} = 260\,\text{kJ} \cdot \text{mol}^{-1}$

(c) $3.46 \times 10^{-19}\,\text{J} \times 6.02 \times 10^{23}\,\text{mol}^{-1} = 208\,\text{kJ} \cdot \text{mol}^{-1}$

(d) $Cl^- < H_2O < NH_3$     (spectrochemical series)

**16.61** The $e_g$ set, which comprises the $d_{x^2-y^2}$ and $d_{z^2}$ orbitals.

**16.63** (a) The CN⁻ ion is a $\pi$-acid ligand accepting electrons into the empty $\pi^*$ orbital created by the C—N multiple bond. (b) The Cl⁻ ion has extra lone pairs in addition to the one that is used to form the $\sigma$-bond to the metal, and so it can act as a $\pi$-base, donating electrons in a $p$-orbital to an empty $d$-orbital on the metal. (c) $H_2O$, like Cl⁻, also has an "extra" lone pair of electrons that can be donated to a metal center, making it a weak $\pi$-base; (d) en is neither a $\pi$-acid nor a $\pi$-base, because it does not have any empty $\pi$-type antibonding orbitals nor does it have any extra lone pairs of electrons to donate. (e) Cl⁻ < $H_2O$ < en < CN⁻. Note that the spectrochemical series orders the ligands as $\pi$-bases < $\sigma$-bond only ligands < $\pi$-acceptors.

**16.65** Nonbonding or slightly antibonding. In a complex that forms only $\sigma$-bonds, the $t_{2g}$ set of orbitals is nonbonding. If the ligands can function as weak $\pi$-donors (those close to the middle of the spectrochemical series, such as $H_2O$), the $t_{2g}$ set becomes slightly antibonding by interacting with the filled $p$-orbitals on the ligands.

**16.67** Antibonding. The $e_g$ set of orbitals on an octahedral metal ion are always antibonding because of interactions with ligand orbitals that form the $\sigma$-bonds. This is true regardless of whether the ligands are $\pi$-acceptors, $\pi$-donors, or neither.

**16.69** Water has two lone pairs of electrons. Once one of these is used to form the $\sigma$-bond to the metal ion, the second may be used to form a $\pi$-bond. This causes the $t_{2g}$ set of orbitals to move up in energy, making $\Delta_O$ smaller; therefore, water is a weak-field ligand. Ammonia does not have this extra lone pair of electrons and consequently cannot function as a $\pi$-donor ligand.

**16.71** (a) CO

(b) In Zones D & C,

$$3 Fe_2O_3(s) + CO(g) \rightarrow 2 Fe_3O_4(s) + CO_2(g)$$
$$Fe_3O_4(s) + CO(g) \rightarrow 3 FeO(s) + CO_2(g)$$

These reactions combine to give

$$Fe_2O_3(s) + CO(g) \rightarrow 2 FeO(s) + CO_2(g)$$

In Zone B,

$$Fe_2O_3(s) + 3 CO(g) \rightarrow 2 Fe(s) + 3 CO_2(g)$$

$$FeO(s) + CO(g) \rightarrow Fe(s) + CO_2(g)$$

(c) carbon

**16.73** The major impurity is carbon; it is removed by oxidation of the carbon to $CO_2$, followed by capture of the $CO_2$ by base to form a slag.

**16.75** Copper and zinc

**16.77** Alloys are usually (1) harder and more brittle, and (2) poorer conductors of electricity than the metals from which they are made.

**16.79** The compound is ferromagnetic below $T_C$ because the magnetization is higher. Above the Curie temperature, the compound is a simple paramagnet with randomly oriented spins, but below that temperature, the spins align and the magnetization increases.

**16.81** (a) More than one kind of reduction occurs.

In Zone C,

$$Fe_2O_3(s) + 3 CO(g) \rightarrow 2 Fe(s) + 3 CO_2(g)$$

In Zone D,

$$3 Fe_2O_3(s) + CO(g) \rightarrow 2 Fe_3O_4(s) + CO_2(g)$$
$$Fe_3O_4(s) + CO(g) \rightarrow 3 FeO(s) + CO_2(g)$$

These reactions combine to give

$$Fe_2O_3(s) + CO(g) \rightarrow 2\ FeO(s) + CO_2(g)$$

In Zone C,

$$FeO(s) + CO(g) \rightarrow Fe(s) + CO_2(g)$$

(b) $TiCl_4(g) + 2\ Mg(l) \xrightarrow{\Delta} Ti(s) + 2\ MgCl_2(s)$

(c) $CaO(s) + SiO_2(s) \xrightarrow{\Delta} CaSiO_3(l)$

**16.83** $Cr(OH)_3(s) + 3\ e^- \rightarrow Cr(s) + 3\ OH^- \qquad E° = -1.34\ V$

$Cr(s) \rightarrow Cr^{3+} + 3\ e^- \qquad\qquad\qquad E° = +0.74\ V$

$Cr(OH)_3(s) \rightarrow Cr^{3+} + 3\ OH^- \qquad\quad E° = -0.60\ V$

$\Delta G° = -nFE° = -RT \ln K$

$\ln K = \dfrac{nFE°}{RT} = \dfrac{(3)(96\ 485\ J \cdot V^{-1} \cdot mol^{-1})(-0.60\ V)}{(8.314\ J \cdot K^{-1} \cdot mol^{-1})(298\ K)} = -70.1$

$K_{sp} = e^{-70.1} = 3.6 \times 10^{-31}$

**16.85** (a) $[PtBrCl(NH_3)_2]$

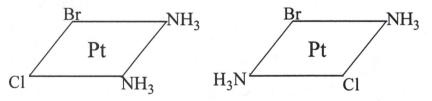

*cis*-Diamminebromochloroplatinum(II)      *trans*-Diamminebromochloroplatinum(II)

(b) If the compound were tetrahedral, there would be only one compound, not two.

**16.87** (a) The first, $[Ni(SO_4)(en)_2]Cl_2$, will give a precipitate of AgCl when

AgNO$_3$ is added; the second will not. (b) The second, $[NiCl_2(en)_2]I_2$,

will show free $I_2$ when mildly oxidized with, for example, $Br_2$, but the

first will not.

**16.89** (a)

[MnCl$_6$]$^{4-}$

5 e$^-$, Cl$^-$ is a weak-field ligand

[Mn(CN)$_6$]$^{4-}$

5 e$^-$, CN$^-$ is a strong-field ligand

(b) [MnCl$_6$]$^{4-}$: five; [Mn(CN)$_6$]$^{4-}$: one.

(c) Complexes with weak-field ligands absorb longer wavelength light, therefore, [MnCl$_6$]$^{4-}$ absorbs longer wavelengths.

**16.91** High spin Mn$^{2+}$ ions have a $d^5$ configuration with 5 unpaired electrons, as shown below.

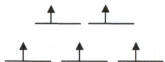

In order for light to be absorbed in the visible region of the spectrum, an electron from the $t_{2g}$ set has to be moved into the $e_g$ set of orbitals.

Because each orbital is already singly occupied by an electron and all five electrons have parallel spins, the electron making the transition must change spin in order to spin pair in the upper orbital. This sort of transition is called "spin-forbidden" because it has a very low quantum mechanical probability of occurring, and so the complexes usually are only faintly colored.

**16.93**

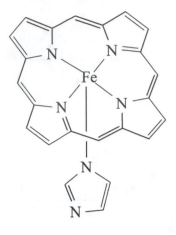

**16.95**  The correct structure for $[Co(NH_3)_6]Cl_3$ consists of four ions,

$Co(NH_3)_6^{3+}$, and 3 $Cl^-$ in aqueous solution. The chloride ions can be

easily precipitated as AgCl. This would not be possible if they were

bonded to the other $(NH_3)$ ligands. If the structure were

$Co(NH_3-NH_3-Cl)_3$, VSEPR theory would predict that the $Co^{3+}$ ion

would have a trigonal planar ligand arrangement. The splitting of the $d$-

orbital energies would not be the same as the octahedral arrangement and

would lead to different spectroscopic and magnetic properties inconsistent

with the experimental evidence. In addition, neither optical nor

geometrical isomers would be observed.

**16.97**  The spectrochemical series given in Figure 16.32 shows the relative ligand

field strengths of the halide ions to lie in the order $I^- < Br^- < Cl^- < F^-$.

Since their electronegativities follow the same trend, there is a positive

correlation between ligand field strength and electronegativity for the

halide ions. The value of $\Delta_O$ correlates with the ability of the ligand's

extra lone pairs of electrons to interact with the $t_{2g}$ set of the octahedral

metal ion. This means that the less electronegative the ligand is, the easier

it is for the ligand to donate electrons to the metal ion, the more the energy of the $t_{2g}$ in the complex is raised, and the smaller $\Delta_O$ becomes (see Figure 16.40(a)).

**16.99** In order to determine these relationships, we need to consider the types of interactions that the ligands on the ends of the spectrochemical series will have with the metal ions. Those that are weak-field (form high spin complexes, $\pi$-bases) have extra lone pairs of electrons that can be donated to a metal ion in a $\pi$ fashion. The strong-field ligands (form low spin complexes, $\pi$-acids) accept electrons from the metals. The complexes that will be more stable will be produced in general by the match between ligand and metal. Thus, the early transition metals in high oxidation states will have few or no electrons in the $d$-orbitals. These metal ions will become stabilized by ligands that can donate more electrons to the metal—the $d$-orbitals that are empty can readily accept electrons. The more stable complexes will be formed with the weak-field ligands. The opposite is true for metals with many electrons, which are the ones at the right side of the periodic table, in low oxidation states. These metals generally have most of the $d$-orbitals filled, so they would, in fact, be destabilized by $\pi$ donation. They form instead more stable complexes with the $\pi$-acceptor ligands (strong-field, $\pi$-acids) that can remove some of the electron density from the metal ions.

**16.101** (a) $ZrO_2$ : there are 4 zirconium atoms in the unit cell. ($1/8 \times 8$ corner atoms + $1/2 \times 6$ face-centered atoms) and 8 oxygen atoms (All O atoms lie completely inside the unit cell.); (b) face-centered cubic; (c) eight; (d) coordination number = 4, tetrahedral;

(e) An obvious difference is that zirconia is composed of two different types of atoms, whereas diamond is made up solely of carbon; however, both structures are based upon a face-centered cubic unit cell. In zirconia, all of the tetrahedral holes created in the cubic close-packed arrangement of zirconium atoms are occupied by oxide ligands. In the case of diamond, only half of the tetrahedral holes are filled. The consequence is that the zirconium atoms are connected to twice as many oxygen atoms as carbon is connected to other carbon atoms.

**16.103** (a)  $PtCl_2(NH_3)_2$, *cis*-diamminedichloroplatinum(II);  (b)  The only other isomer is the trans form. Neither the cis nor the trans form is optically active.

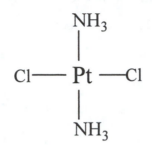

(c)  square planar

**16.105**  $\text{number of moles of complex} = \dfrac{2.11 \text{ g}}{211.42 \text{ g} \cdot \text{mol}^{-1}}$

$= 9.98 \times 10^{-3} \text{ mol complex}$

$\text{number of moles of } Cl^- = 2.87 \text{ g AgCl}\left(\dfrac{1 \text{ mole AgCl}}{143.32 \text{ g} \cdot \text{mol}^{-1}}\right)\left(\dfrac{1 \text{ mole } Cl^-}{1 \text{ mole AgCl}}\right)$

$= 2.00 \times 10^{-2} \text{ mole } Cl^-$

Therefore the compound contains 2 moles of $Cl^-$ counterion for every mole of complex (assuming that the incorrect formula gives the correct molar mass).  Then the correct formula of the compound would be $[CrNH_3Cl(H_2O)_2]Cl_2$.  Since the $Cr^{3+}$ is $d^3$ and coordinated by four ligands, the complex cation is likely to be tetrahedral.  There are no linkage isomers possible and no enantiomers for this complex ion.

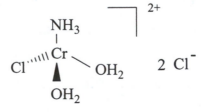

However, the compound as a whole has several isomers, including one that corresponds to the incorrect formula, $[CrNH_3Cl_3] \cdot 2 H_2O$.

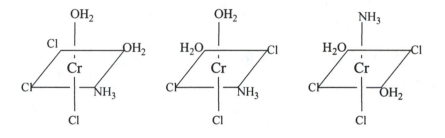

The other reasonable isomers include octahedral species that result if the chloride ions as well as both water molecules are all attached directly to the metal ion.

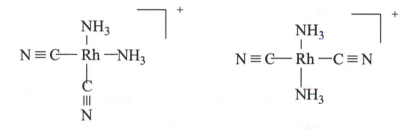

**16.107**

Since the rhodium ion has a 3+ charge, its electron configuration is $d^6$.

# CHAPTER 17
# NUCLEAR CHEMISTRY

**17.1** $\lambda = \dfrac{c}{v}$, $E = N_A hv$, $1\,\text{Hz} = 1\,\text{s}^{-1}$

(a) $\lambda = \dfrac{3.00 \times 10^8 \text{ m} \cdot \text{s}^{-1}}{5.3 \times 10^{20} \text{ s}^{-1}} = 5.7 \times 10^{-13} \text{ m}$

$E = 6.02 \times 10^{23} \text{ mol}^{-1} \times 6.63 \times 10^{-34} \text{ J} \cdot \text{s} \times 5.3 \times 10^{20} \text{ s}^{-1}$
$= 2.1 \times 10^{11} \text{ J} \cdot \text{mol}^{-1}$

(b) $\lambda = \dfrac{3.00 \times 10^8 \text{ m} \cdot \text{s}^{-1}}{4.7 \times 10^{22} \text{ s}^{-1}} = 6.4 \times 10^{-15} \text{ m}$

$E = 6.02 \times 10^{23} \text{ mol}^{-1} \times 6.63 \times 10^{-34} \text{ J} \cdot \text{s} \times 4.7 \times 10^{22} \text{ s}^{-1}$
$= 1.9 \times 10^{13} \text{ J} \cdot \text{mol}^{-1}$

(c) $\lambda = \dfrac{3.00 \times 10^8 \text{ m} \cdot \text{s}^{-1}}{2.8 \times 10^{21} \text{ s}^{-1}} = 1.1 \times 10^{-13} \text{ m}$

$E = 6.02 \times 10^{23} \text{ mol}^{-1} \times 6.63 \times 10^{-34} \text{ J} \cdot \text{s} \times 2.8 \times 10^{21} \text{ s}^{-1}$
$= 1.1 \times 10^{12} \text{ J} \cdot \text{mol}^{-1}$

(d) $\lambda = \dfrac{3.00 \times 10^8 \text{ m} \cdot \text{s}^{-1}}{6.5 \times 10^{19} \text{ s}^{-1}} = 4..6 \times 10^{-12} \text{ m}$

$E = 6.02 \times 10^{23} \text{ mol}^{-1} \times 6.63 \times 10^{-34} \text{ J} \cdot \text{s} \times 6.5 \times 10^{19} \text{ s}^{-1}$
$= 2.6 \times 10^{10} \text{ J} \cdot \text{mol}^{-1}$

**17.3** We assume that all the change in energy goes into the energy of the $\gamma$ ray emitted. Then, in each case,

$$v = \frac{\Delta E}{h}, \qquad \lambda = \frac{c}{v}$$

$$\text{energy of 1 MeV} = \left(\frac{10^6 \text{ eV}}{1 \text{ MeV}}\right)\left(\frac{1.602 \times 10^{-19} \text{ J}}{1 \text{ eV}}\right)$$

$$= 1.602 \times 10^{-13} \text{ J} \cdot \text{MeV}^{-1}$$

(a) $\Delta E = (1.33 \text{ MeV})\left(\dfrac{1.602 \times 10^{-13} \text{ J}}{1 \text{ MeV}}\right) = 2.13 \times 10^{-13} \text{ J}$

$$\nu = \frac{\Delta E}{h} = \frac{2.13 \times 10^{-13} \text{ J}}{6.63 \times 10^{-34} \text{ J} \cdot \text{s}} = 3.21 \times 10^{20} \text{ s}^{-1} = 3.21 \times 10^{20} \text{ Hz}$$

$$\lambda = \frac{c}{\nu} = \frac{3.00 \times 10^8 \text{ m} \cdot \text{s}^{-1}}{3.21 \times 10^{20} \text{ s}^{-1}} = 9.35 \times 10^{-13} \text{ m}$$

(b) $\Delta E = (1.64 \text{ MeV})\left(\dfrac{1.602 \times 10^{-13} \text{ J}}{1 \text{ MeV}}\right) = 2.63 \times 10^{-13} \text{ J}$

$$\nu = \frac{\Delta E}{h} = \frac{2.63 \times 10^{-13} \text{ J}}{6.63 \times 10^{-34} \text{ J} \cdot \text{s}} = 3.97 \times 10^{20} \text{ s}^{-1} = 3.97 \times 10^{20} \text{ Hz}$$

$$\lambda = \frac{3.00 \times 10^8 \text{ m} \cdot \text{s}^{-1}}{3.95 \times 10^{20} \text{ s}^{-1}} = 7.59 \times 10^{-13} \text{ m}$$

(c) $\Delta E = (1.10 \text{ MeV})\left(\dfrac{1.602 \times 10^{-13} \text{ J}}{1 \text{ MeV}}\right) = 1.76 \times 10^{-13} \text{ J}$

$$\nu = \frac{\Delta E}{h} = \frac{1.76 \times 10^{-13} \text{ J}}{6.63 \times 10^{-34} \text{ J} \cdot \text{s}} = 2.65 \times 10^{20} \text{ s}^{-1} = 2.65 \times 10^{20} \text{ Hz}$$

$$\lambda = \frac{c}{\nu} = \frac{3.00 \times 10^8 \text{ m} \cdot \text{s}^{-1}}{2.65 \times 10^{20} \text{ s}^{-1}} = 1.13 \times 10^{-12} \text{ m}$$

**17.5** (a) $^{3}_{1}\text{T} \rightarrow {}^{0}_{-1}\text{e} + {}^{3}_{2}\text{He}$

(b) $^{83}_{39}\text{Y} \rightarrow {}^{0}_{1}\text{e} + {}^{83}_{38}\text{Sr}$

(c) $^{87}_{36}\text{Kr} \rightarrow {}^{0}_{-1}\text{e} + {}^{87}_{37}\text{Rb}$

(d) $^{225}_{91}\text{Pa} \rightarrow {}^{4}_{2}\alpha + {}^{221}_{89}\text{Ac}$

**17.7** (a) $^{8}_{5}\text{B} \rightarrow {}^{0}_{1}\text{e} + {}^{A}_{Z}\text{E}$ $\qquad A = 8 - 0 = 8, Z = 5 - 1 = 4, \text{E} = \text{Be}$

so, $^{8}_{5}\text{B} \rightarrow {}^{0}_{1}\text{e} + {}^{8}_{4}\text{Be}$

(b) $^{63}_{28}\text{Ni} \rightarrow \, ^{0}_{-1}\text{e} + \, ^{A}_{Z}\text{E}$   $A= 63 - 0 = 63$, $Z = 28 - (-1) = 29$, E = Cu

so, $^{63}_{28}\text{Ni} \rightarrow \, ^{0}_{-1}\text{e} + \, ^{63}_{29}\text{Cu}$

(c) $^{185}_{79}\text{Au} \rightarrow \, ^{4}_{2}\alpha + \, ^{A}_{Z}\text{E}$   $A= 185 - 4 = 181$, $Z= 79 - 2 = 77$, E = Ir

so, $^{185}_{79}\text{Au} \rightarrow \, ^{4}_{2}\alpha + \, ^{181}_{77}\text{Ir}$

(d) $^{7}_{4}\text{Be} + \, ^{0}_{-1}\text{e} \rightarrow \, ^{A}_{Z}\text{E}$   $A= 7 + 0 = 7$, $Z = 4 - 1 = 3$, E = Li

so, $^{7}_{4}\text{Be} + \, ^{0}_{-1}\text{e} \rightarrow \, ^{7}_{3}\text{Li}$


**17.9** (a) $^{24}_{11}\text{Na} \rightarrow \, ^{24}_{12}\text{Mg} + \, ^{0}_{-1}\text{e}$; a β particle is emitted.

(b) $^{128}_{50}\text{Sn} \rightarrow \, ^{128}_{51}\text{Sb} + \, ^{0}_{-1}\text{e}$; a β particle is emitted.

(c) $^{140}_{57}\text{La} \rightarrow \, ^{140}_{56}\text{Ba} + \, ^{0}_{1}\text{e}$; a positron ($\beta^+$) is emitted.

(d) $^{228}_{90}\text{Th} \rightarrow \, ^{224}_{88}\text{Ra} + \, ^{4}_{2}\alpha$; an α particle is emitted.


**17.11** (a) $^{11}_{5}\text{B} + \, ^{4}_{2}\alpha \rightarrow 2 \, ^{1}_{0}\text{n} + \, ^{13}_{7}\text{N}$

(b) $^{35}_{17}\text{Cl} + \, ^{2}_{1}\text{D} \rightarrow \, ^{1}_{0}\text{n} + \, ^{36}_{18}\text{Ar}$

(c) $^{96}_{42}\text{Mo} + \, ^{2}_{1}\text{D} \rightarrow \, ^{1}_{0}\text{n} + \, ^{97}_{43}\text{Tc}$

(d) $^{45}_{21}\text{Sc} + \, ^{1}_{0}\text{n} \rightarrow \, ^{4}_{2}\alpha + \, ^{42}_{19}\text{K}$


**17.13** (a) $A/Z = 68/29 = 2.34 > (A/Z)_{\text{based}}$; hence, $^{68}_{29}\text{Cu}$

is neutron rich, and β decay is most likely. $^{68}_{29}\text{Cu} \rightarrow \, ^{0}_{-1}\text{e} + \, ^{68}_{30}\text{Zn}$

(b) $A/Z = 103/48 = 2.15 < (A/Z)_{\text{based}}$; therefore,

$^{103}_{48}\text{Cd}$ is proton rich, and $\beta^+$ decay is most likely. $^{103}_{48}\text{Cd} \rightarrow \, ^{0}_{1}\text{e} + \, ^{103}_{47}\text{Ag}$

(c) $^{243}_{97}\text{Bk}$ has $Z > 83$ and is proton rich; therefore, α decay is most likely.

$^{243}_{97}\text{Bk} \rightarrow \, ^{4}_{2}\alpha + \, ^{239}_{95}\text{Am}$

(d) $^{260}_{105}\text{Db}$ has $Z > 83$; therefore, α decay is most likely.

$^{260}_{105}\text{Db} \rightarrow \, ^{4}_{2}\alpha + \, ^{256}_{103}\text{Lr}$

**17.15** α $\quad {}^{235}_{92}\text{U} \rightarrow {}^{4}_{2}\alpha + {}^{231}_{90}\text{Th}$

β $\quad {}^{231}_{90}\text{Th} \rightarrow {}^{0}_{-1}\text{e} + {}^{231}_{91}\text{Pa}$

α $\quad {}^{231}_{91}\text{Pa} \rightarrow {}^{4}_{2}\alpha + {}^{227}_{89}\text{Ac}$

β $\quad {}^{227}_{89}\text{Ac} \rightarrow {}^{0}_{-1}\text{e} + {}^{227}_{90}\text{Th}$

α $\quad {}^{227}_{90}\text{Th} \rightarrow {}^{4}_{2}\alpha + {}^{223}_{88}\text{Ra}$

α $\quad {}^{223}_{88}\text{Ra} \rightarrow {}^{4}_{2}\alpha + {}^{219}_{86}\text{Rn}$

α $\quad {}^{219}_{86}\text{Rn} \rightarrow {}^{4}_{2}\alpha + {}^{215}_{84}\text{Po}$

β $\quad {}^{215}_{84}\text{Po} \rightarrow {}^{0}_{-1}\text{e} + {}^{215}_{85}\text{At}$

α $\quad {}^{215}_{85}\text{At} \rightarrow {}^{4}_{2}\alpha + {}^{211}_{83}\text{Bi}$

β $\quad {}^{211}_{83}\text{Bi} \rightarrow {}^{0}_{-1}\text{e} + {}^{211}_{84}\text{Po}$

α $\quad {}^{211}_{84}\text{Po} \rightarrow {}^{4}_{2}\alpha + {}^{207}_{82}\text{Pb}$

**17.17** To determine the charge and mass of the unknown particle, it helps to write ${}^{1}_{1}\text{p}$ and ${}^{1}_{0}\text{n}$ for the proton and neutron, respectively; and ${}^{0}_{-1}\text{e}$ and ${}^{0}_{1}\text{e}$ for the β particle and positron, respectively.

(a) ${}^{14}_{7}\text{N} + {}^{4}_{2}\alpha \rightarrow {}^{17}_{8}\text{O} + {}^{1}_{1}\text{p}$

(b) ${}^{248}_{96}\text{Cm} + {}^{1}_{0}\text{n} \rightarrow {}^{249}_{97}\text{Bk} + {}^{0}_{-1}\text{e}$

(c) ${}^{243}_{95}\text{Am} + {}^{1}_{0}\text{n} \rightarrow {}^{244}_{96}\text{Cm} + {}^{0}_{-1}\text{e} + \gamma$

(d) ${}^{13}_{6}\text{C} + {}^{1}_{0}\text{n} \rightarrow {}^{14}_{6}\text{C} + \gamma$

**17.19** (a) ${}^{20}_{10}\text{Ne} + {}^{4}_{2}\alpha \rightarrow {}^{8}_{4}\text{Be} + {}^{16}_{8}\text{O}$

(b) ${}^{20}_{10}\text{Ne} + {}^{20}_{10}\text{Ne} \rightarrow {}^{16}_{8}\text{O} + {}^{24}_{12}\text{Mg}$

(c) ${}^{44}_{20}\text{Ca} + {}^{4}_{2}\alpha \rightarrow \gamma + {}^{48}_{22}\text{Ti}$

(d) ${}^{27}_{13}\text{Al} + {}^{2}_{1}\text{H} \rightarrow {}^{1}_{1}\text{p} + {}^{28}_{13}\text{Al}$

**17.21** In each case, identify the unknown particle by performing a mass and charge balance as you did in the solutions to Exercises 17.5 and 17.7. Then write the complete nuclear equation.

(a) $^{14}_{7}N + ^{4}_{2}\alpha \rightarrow ^{17}_{8}O + ^{1}_{1}p$

(b) $^{239}_{94}Pu + ^{1}_{0}n \rightarrow ^{240}_{95}Am + ^{0}_{-1}e$

**17.23** (a) untriquadium, Utq  (b) unquadpentium, Uqp  (c) binilunium, Bnu

**17.25** $\text{activity} = (4.7 \times 10^{5} \text{ Bq})\left(\dfrac{1 \text{ Ci}}{3.7 \times 10^{10} \text{ Bq}}\right) = 1.3 \times 10^{-5} \text{ Ci}$

**17.27** $1 \text{ Bq} = 1$ disintegration per second (dps)

(a) $(2.5 \text{ } \mu\text{Ci})\left(\dfrac{10^{-6} \text{ Ci}}{1 \text{ } \mu\text{Ci}}\right)\left(\dfrac{3.7 \times 10^{10} \text{ dps}}{1 \text{ Ci}}\right) = 9.2 \times 10^{4} \text{ dps}$

$= 9.2 \times 10^{4} \text{ Bq}$

(b) $142 \text{ Ci} = (142)(3.7 \times 10^{10} \text{ dps}) = 5.3 \times 10^{12} \text{ Bq}$

(c) $(7.2 \text{ mCi})\left(\dfrac{10^{-3} \text{ Ci}}{1 \text{ mCi}}\right)\left(\dfrac{3.7 \times 10^{10} \text{ dps}}{1 \text{ Ci}}\right) = 2.7 \times 10^{8} \text{ dps}$

$= 2.7 \times 10^{8} \text{ Bq}$

**17.29** $\text{dose in rads} = 1.0 \text{ J} \cdot \text{kg}^{-1} \times \left(\dfrac{1 \text{ rad}}{10^{-2} \text{ J} \cdot \text{kg}^{-1}}\right) = 1.0 \times 10^{2} \text{ rad}$

$\text{dose equivalent in rems} = Q \times \text{dose in rads}$

$= \left(\dfrac{1 \text{ rem}}{1 \text{ rad}}\right)(1.0 \times 10^{2} \text{ rad}) = 1.0 \times 10^{2} \text{ rem}$

$1.0 \times 10^{2} \text{ rem} \div 100 \text{ rem/Sv} = 1.0 \text{ Sv}$

**17.31** $1.0 \text{ rad} \cdot \text{day}^{-1} = (1.0 \text{ rad} \cdot \text{day}^{-1})\left(\dfrac{1 \text{ rem}}{1 \text{ rad}}\right) = 1 \text{ rem} \cdot \text{day}^{-1}$

$$100 \text{ rem} = 1 \text{ rem} \cdot \text{day}^{-1} \times \text{time}$$

$$\text{time} = 100 \text{ day}$$

**17.33** $k = \dfrac{0.693}{t_{1/2}}$

    (a) $k = \dfrac{0.693}{12.3 \text{ a}} = 5.63 \times 10^{-2} \text{ a}^{-1}$

    (b) $k = \dfrac{0.693}{0.84 \text{ s}} = 0.83 \text{ s}^{-1}$

    (c) $k = \dfrac{0.693}{10.0 \text{ min}} = 0.0693 \text{ min}^{-1}$

**17.35** We know that initial activity $\propto N_0$, and final activity $\propto N$. Therefore,

$$\frac{\text{final activity}}{\text{initial activity}} = \frac{N}{N_0} = e^{-kt}$$

$$k = \frac{0.693}{t_{1/2}} = \frac{0.693}{5.26 \text{ a}} = 0.132 \text{ a}^{-1}$$

$$\begin{aligned}
\text{final activity} &= \text{initial activity} \times e^{-kt} \\
&= 4.4 \text{ Ci} \times e^{-(0.132 \text{ a}^{-1} \times 50 \text{ a})} \\
&= 6.0 \times 10^{-3} \text{ Ci}
\end{aligned}$$

**17.37** In each case, $k = \dfrac{0.693}{t_{1/2}}$, $N = N_0 e^{-kt}$, $\dfrac{N}{N_0} = e^{-kt}$, and the percentage

remaining

$$= 100\% \times (N/N_0)$$

    (a) $k = \dfrac{0.693}{5.73 \times 10^3 \text{ a}} = 1.21 \times 10^{-4} \text{ a}^{-1}$

        percentage remaining $= 100\% \times e^{-(1.21 \times 10^{-4} \text{ a}^{-1} \times 2000 \text{ a})} = 78.5\%$

    (b) $k = \dfrac{0.693}{12.3 \text{ a}} = 0.0563 \text{ a}^{-1}$

    percentage remaining $= 100\% \times e^{-(0.0563 \text{ a}^{-1} \times 11.0 \text{ a})} = 53.8\%$

**17.39** (a) From Table 17.5, $t_{1/2} = 4.5 \times 10^9$ a

$$k = \frac{0.693}{t_{1/2}} = \frac{0.693}{4.5 \times 10^9 \text{ a}} = 1.54 \times 10^{-10} \text{ a}^{-1}$$

$$\text{fraction remaining} = \frac{N}{N_0} = e^{-kt}$$

$$= e^{-(1.54 \times 10^{-10} \text{ a}^{-1} \times 4.5 \times 10^9 \text{ a})}$$

$$= e^{-1.4} = 0.50$$

After 1 half-life, 50% remains.

(b) fraction remaining $= \dfrac{N}{N_0} = \dfrac{3}{5}$;

$$t_{1/2} = 1.26 \times 10^9 \text{ a}, \ k = \frac{0.693}{1.26 \times 10^9 \text{ a}} = 5.50 \times 10^{-10} \text{ a}^{-1}$$

$$\frac{3}{5} = e^{-kt}$$

$$\frac{3}{5} = e^{-(5.50 \times 10^{-10} \text{ a}^{-1} \times x)}$$

$$x = 9.3 \times 10^8 \text{ a}$$

**17.41** Let dis = disintegrations

$$\text{activity from "old" sample} = \frac{1500 \text{ dis}/0.250 \text{ g}}{10.0 \text{ h}} = 600 \text{ dis} \cdot \text{g}^{-1} \cdot \text{h}^{-1}$$

activity from current sample $= 921 \text{ dis} \cdot \text{g}^{-1} \cdot \text{h}^{-1}$

$$k = \frac{0.693}{t_{1/2}} = \frac{0.693}{5.73 \times 10^3 \text{ a}} = 1.21 \times 10^{-4} \text{ a}^{-1}$$

"old" activity $\propto N$, current activity $\propto N_0$

$$\frac{\text{"old" activity}}{\text{current activity}} = \frac{N}{N_0} = e^{-kt}, \ \frac{N_0}{N} = e^{kt}, \ \ln\left(\frac{N_0}{N}\right) = kt$$

Solve for $t \ (= \text{age}):$

$$t = \frac{\ln\left(\dfrac{N_0}{N}\right)}{k} = \frac{\ln\left(\dfrac{921}{600}\right)}{1.21 \times 10^{-4} \text{ a}^{-1}} = 3.54 \times 10^3 \text{ a}$$

**17.43** In each case, $k = \dfrac{0.693}{t_{1/2} \text{ (in s)}}$, activity in $Bq = k \times N$

$$\text{activity in Ci} = \frac{\text{activity in } Bq}{3.7 \times 10^{10} \; Bq \cdot Ci^{-1}}$$

Note: $Bq$ (= disintegrating nuclei per second) has the units of $nuclei \cdot s^{-1}$

(a) $k = \left(\dfrac{0.693}{1.60 \times 10^3 \; a}\right)\left(\dfrac{1\,a}{3.17 \times 10^7 \; s}\right) = 1.37 \times 10^{-11} \; s^{-1}$

$N = (1.0 \times 10^{-3} \; g)\left(\dfrac{1 \; mol}{226 \; g}\right)\left(\dfrac{6.02 \times 10^{23} \; nuclei}{1 \; mol}\right) = 2.7 \times 10^{18} \; nuclei$

$\text{activity} = 1.37 \times 10^{-11} \; s^{-1} \times 2.7 \times 10^{18} \; nuclei \times \left(\dfrac{1\,Ci}{3.7 \times 10^{10} \; Bq}\right)$

$\qquad = 1.0 \times 10^{-3} \; Ci$

(b) $k = \left(\dfrac{0.693}{28.1 \; a}\right)\left(\dfrac{1y}{3.17 \times 10^7 \; s}\right) = 7.80 \times 10^{-10} \; s^{-1}$

$N = (2.0 \times 10^{-6} \; g)\left(\dfrac{1 \; mol}{90 \; g}\right)\left(\dfrac{6.02 \times 10^{23} \; nuclei}{1 \; mol}\right) = 1.3 \times 10^{16} \; nuclei$

$\text{activity} = (7.80 \times 10^{-10} \; s^{-1})(1.3 \times 10^{16} \; nuclei)\left(\dfrac{1 \; Ci}{3.7 \times 10^{10} \; Bq}\right)$

$\qquad = 2.7 \times 10^{-4} \; Ci$

(c) $k = \left(\dfrac{0.693}{2.6 \; a}\right)\left(\dfrac{1\,y}{3.17 \times 10^7 \; s}\right) = 8.4 \times 10^{-9} \; s^{-1}$

$N = (0.43 \times 10^{-3} \; g)\left(\dfrac{1 \; mol}{147 \; g}\right)\left(\dfrac{6.02 \times 10^{23} \; nuclei}{1 \; mol}\right) = 1.8 \times 10^{18} \; nuclei$

$\text{activity} = (8.4 \times 10^{-9} \; s^{-1})(1.8 \times 10^{18} \; nuclei)\left(\dfrac{1 \; Ci}{3.7 \times 10^{10} \; Bq}\right) = 0.41 \; Ci$

**17.45** $k = \dfrac{0.693}{t_{1/2}} = \dfrac{0.693}{8.05 \; d} = 0.0861 \; d^{-1}$

$$N = N_0 e^{-kt} \text{ and } \frac{N}{N_0} = e^{-kt}$$

Taking natural log of both sides gives

$$\ln\left(\frac{N}{N_0}\right) = -kt$$

Because activity is proportional to $N$ (Eq. 2), we can write

$$\ln\left(\frac{\text{final activity}}{\text{initial activity}}\right) = -kt$$

Solving for $t$ gives

$$t = -\left(\frac{1}{k}\right)\ln\left(\frac{\text{final activity}}{\text{initial activity}}\right) = -\left(\frac{1}{0.0861\ \text{d}^{-1}}\right)\ln\left(\frac{10}{500}\right) = 45\ \text{d}$$

**17.47** (a) activity $\propto N$; and, because $\ln\left(\dfrac{N}{N_0}\right) = -kt$

$$\ln\left(\frac{\text{final activity}}{\text{initial activity}}\right) = -kt$$

$$\ln\left(\frac{32}{58}\right) = -k \times 12.3\ \text{d}$$

$$k = 0.048\ \text{d}^{-1}$$

$$t_{1/2} = \frac{0.693}{k} = \frac{0.693}{0.048\ \text{d}^{-1}} = 14\ \text{d}$$

(b) $\ln\left(\dfrac{N}{N_0}\right) = -0.048\ \text{d}^{-1} \times 30\ \text{d} = -1.4$

$$\frac{N}{N_0} = \text{fraction remaining} = 0.25$$

**17.49** $k = \dfrac{0.693}{t_{1/2}} = \dfrac{0.693}{5.27\ \text{a}} = 0.131\ \text{a}$

$$\frac{N}{N_0} = e^{-kt} = e^{-(0.131\ a^{-1})(2.50\ a)} = \frac{0.266\ g}{N_0}$$

$$N_0 = 0.370\ g$$

$$\frac{0.370\ g}{1.40\ g} \times 100 = 26.4\%$$

**17.51** Since radioactive decay follows first-order kinetics, the rate of loss of X is

$$\frac{d[X]}{dt} = -k_1[X]$$

(1)  Y, which is an intermediate, is lost in the first reaction but formed in the second one, so its rate equation can be expressed as

$$\frac{d[Y]}{dt} = k_1[X] - k_2[Y]$$

(2)  Z is the final product of the two consecutive reactions so its rate law is

$$\frac{d[Z]}{dt} = k_2[Y]$$

(3)  As discussed in Chapter 13, the integrated form of equation (1) is

$$[X] = [X]_0 e^{-k_1 t}$$

(4)  Substituting this expression into the rate law for Y and rearranging gives

$$\frac{d[Y]}{dt} + k_2[Y] = k_1[X]_0 e^{-k_1 t}$$

(5)  This linear first-order differential equation has the solution

$$[Y] = \frac{k_1}{k_2 - k_1}(e^{-k_1 t} - e^{-k_2 t})[X]_0 \qquad \text{when } [Y]_0 = 0.$$

(6)  Since $[X]+[Y]+[Z]=[X]_0$ at all times, $[Z]= [X]_0 -([X]+[Y])$, or

$$[Z] = [X]_0 - \left([X]_0 e^{-k_1 t} + \frac{k_1}{k_2 - k_1}(e^{-k_1 t} - e^{-k_2 t})[X]_0\right) = [X]_0\left(1 + \frac{k_1 e^{-k_2 t} - k_2 e^{-k_1 t}}{k_2 - k_1}\right)$$

(7)  The values of the rate constants can be found from the half-lives:

$$k_1 = \frac{0.693}{27.4\ d} = 0.0253\ d^{-1} \qquad k_2 = \frac{0.693}{18.7\ d} = 0.0371\ d^{-1}$$

Using these constants and assuming $[X]_0 = 2.00$ g, equations (4), (6) and (7) are graphed below.

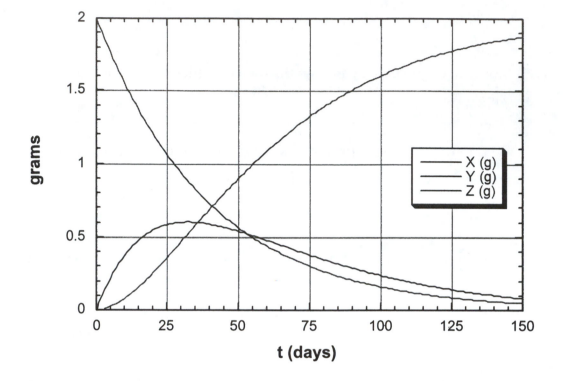

**17.53** If isotopically enriched water, such as $H_2{}^{18}O$, is used in the reaction, the label can be followed. Once the products are separated, a suitable technique, such as vibrational spectroscopy or mass spectrometry, can be used to determine whether the product has incorporated the $^{18}O$. For example, if the methanol ends up with the O atom from the water molecules, then its molar mass would be $34$ g $\cdot$ mol$^{-1}$, rather than $32$ g $\cdot$ mol$^{-1}$ found for methanol with elements present at their natural isotopic abundance.

**17.55** The vibrational frequency is proportional to the reduced mass of the two atoms that form the bond according to the equation:

$$v = \frac{1}{2\pi}\sqrt{\frac{k}{\mu}}$$

where $\mu = \dfrac{m_A m_B}{m_A + m_B}$

Because we are not given $v$, it is easiest to make a relative comparison by taking the ratio of $v$ for the C—D molecule versus $v$ for the C—H molecule:

$$\frac{v_{C-D}}{v_{C-H}} = \frac{\frac{1}{2\pi}\sqrt{\frac{k}{\mu_{C-D}}}}{\frac{1}{2\pi}\sqrt{\frac{k}{\mu_{C-H}}}} = \sqrt{\frac{\mu_{C-H}}{\mu_{C-D}}} = \sqrt{\frac{\dfrac{m_C m_H}{m_C + m_H}}{\dfrac{m_C m_D}{m_C + m_D}}} = \sqrt{\frac{\dfrac{(12.011)(1.0078)}{12.011 + 1.0078}}{\dfrac{(12.011)(2.0140)}{12.011 + 2.0140}}}$$

$$= \sqrt{\frac{\left(\dfrac{12.105}{13.019}\right)}{\left(\dfrac{24.190}{14.025}\right)}} = 0.73422$$

We would thus expect the vibrational frequency for the C—D bond to be approximately 0.73 times the value for the C—H bond (lower in energy).

**17.57** Remember to convert g to kg.

(a) $E = mc^2 = (1.0 \times 10^{-3}\ \text{kg})(3.00 \times 10^8\ \text{m} \cdot \text{s}^{-1})^2$
$\qquad = 9.0 \times 10^{13}\ \text{kg} \cdot \text{m}^2 \cdot \text{s}^{-2} = 9.0 \times 10^{13}\ \text{J}$

(b) $E = mc^2 = (9.109 \times 10^{-31}\ \text{kg})(3.00 \times 10^8\ \text{m} \cdot \text{s}^{-1})^2$
$\qquad = 8.20 \times 10^{-14}\ \text{kg} \cdot \text{m}^2 \cdot \text{s}^{-2} = 8.20 \times 10^{-14}\ \text{J}$

(c) $E = mc^2 = (1.0 \times 10^{-15}\ \text{kg})(3.00 \times 10^8\ \text{m} \cdot \text{s}^{-1})^2 = 90\ \text{kg} \cdot \text{m}^2 \cdot \text{s}^{-2} = 90\ \text{J}$

(d) $E = mc^2$

$E = (1.673 \times 10^{-27}\ \text{kg})(3.00 \times 10^8\ \text{m} \cdot \text{s}^{-1})^2 = 1.51 \times 10^{-10}\ \text{J}$

**17.59** $\Delta m = \dfrac{\Delta E}{c^2} = \dfrac{-3.9 \times 10^{26}\ \text{J} \cdot \text{s}^{-1}}{(3.00 \times 10^8\ \text{m} \cdot \text{s}^{-1})^2} = -4.3 \times 10^9\ \text{kg} \cdot \text{s}^{-1}$

**17.61**  $1\,\text{u} = 1.6605 \times 10^{-27}\,\text{kg}$

In each case, calculate the difference in mass between the nucleus and the free particles from which it may be considered to have been formed. Then obtain the binding energy from the relation $E_{\text{bind}} = \Delta m c^2$.

(a)  $^{62}_{28}\text{Ni}$:  $28\,^{1}\text{H} + 34\,\text{n} \rightarrow\,^{62}_{28}\text{Ni}$

$$\Delta m = 61.928346\,\text{u} - (28 \times 1.0078\,\text{u} + 34 \times 1.0087\,\text{u}) = -0.5858\,\text{u}$$

$$\Delta m = (-0.5858\,\text{u})\left(\frac{1.6605 \times 10^{-27}\,\text{kg}}{1\,\text{u}}\right) = -9.728 \times 10^{-28}\,\text{kg}$$

$$E_{\text{bind}} = \left|-9.728 \times 10^{-28}\,\text{kg}\right| \times (2.99792 \times 10^{8}\,\text{m}\cdot\text{s}^{-1})^2$$

$$= 8.743 \times 10^{-11}\,\text{kg}\cdot\text{m}^2\cdot\text{s}^{-2} = 8.743 \times 10^{-11}\,\text{J}$$

$$E_{\text{bind}}\,/\,\text{nucleon} = \frac{8.743 \times 10^{-11}\,\text{J}}{62\,\text{nucleons}} = 1.410 \times 10^{-12}\,\text{J}\cdot\text{nucleon}^{-1}$$

(b)  $^{239}_{94}\text{Pu}$:  $94\,^{1}\text{H} + 145\,\text{n} \rightarrow\,^{239}_{94}\text{Pu}$

$$\Delta m = 239.0522\,\text{u} - (94 \times 1.0078\,\text{u} + 145 \times 1.0087\,\text{u}) = -1.9425\,\text{u}$$

$$\Delta m = (-1.9425\,\text{u})\left(\frac{1.6605 \times 10^{-27}\,\text{kg}}{1\,\text{u}}\right) = -3.2255 \times 10^{-27}\,\text{kg}$$

$$E_{\text{bind}} = \left|-3.2255 \times 10^{-27}\,\text{kg}\right| \times (2.99792 \times 10^{8}\,\text{m}\cdot\text{s}^{-1})^2$$

$$= 2.8989 \times 10^{-10}\,\text{kg}\cdot\text{m}^2\cdot\text{s}^{-2} = 2.8989 \times 10^{-10}\,\text{J}$$

$$E_{\text{bind}}\,/\,\text{nucleon} = \frac{2.8989 \times 10^{-10}\,\text{J}}{239\,\text{nucleons}} = 1.2129 \times 10^{-12}\,\text{J}\cdot\text{nucleon}^{-1}$$

(c)  $^{2}_{1}\text{H}$:  $^{1}\text{H} + \text{n} \rightarrow\,^{2}_{1}\text{H}$

$$\Delta m = 2.0141\,\text{u} - (1.0078\,\text{u} + 1.0087\,\text{u}) = -0.0024\,\text{u}$$

$$\Delta m = (-0.0024\,\text{u})\left(\frac{1.6605 \times 10^{-27}\,\text{kg}}{1\,\text{u}}\right) = -4.0 \times 10^{-30}\,\text{kg}$$

$$E_{\text{bind}} = \left|-4.0 \times 10^{-30}\,\text{kg}\right| \times (2.99792 \times 10^{8}\,\text{m}\cdot\text{s}^{-1})^2$$

$$= 3.6 \times 10^{-13}\,\text{kg}\cdot\text{m}^2\cdot\text{s}^{-2} = 3.6 \times 10^{-13}\,\text{J}$$

$$E_{\text{bind}}\,/\,\text{nucleon} = \frac{3.6 \times 10^{-13}\,\text{J}}{2\,\text{nucleons}} = 1.8 \times 10^{-13}\,\text{J}\cdot\text{nucleon}^{-1}$$

(d)  $^{3}_{1}\text{H}$:  $^{1}\text{H} + 2\,\text{n} \rightarrow\,^{3}_{1}\text{H}$

$$\Delta m = 3.01605 \text{ u} - (1.0078 \text{ u} + 2 \times 1.0087 \text{ u}) = -0.0092 \text{ u}$$

$$\Delta m = (-0.0092 \text{ u})\left(\frac{1.6605 \times 10^{-27} \text{ kg}}{1 \text{ u}}\right) = -1.5 \times 10^{-29} \text{ kg}$$

$$E_{\text{bind}} = \left|-1.5 \times 10^{-29} \text{ kg}\right| \times (2.99792 \times 10^8 \text{ m} \cdot \text{s}^{-1})^2$$
$$= 1.3 \times 10^{-12} \text{ kg} \cdot \text{m}^2 \cdot \text{s}^{-2} = 1.3 \times 10^{-12} \text{ J}$$

$$E_{\text{bind}} / \text{nucleon} = \frac{1.3 \times 10^{-12} \text{ J}}{3 \text{ nucleons}} = 4.3 \times 10^{-13} \text{ J} \cdot \text{nucleon}^{-1}$$

(e) $^{62}\text{Ni}$ is the most stable, because it has the largest binding energy per nucleon.

**17.63** In each case, we first determine the change in mass, $\Delta m = $ (mass of products) $-$ (mass of reactants). We then calculate the energy released from $\Delta E = (\Delta m)c^2$.

(a) $D + D \rightarrow {}^3\text{He} + n$

$2.0141 \text{ u} + 2.0141 \text{ u} \rightarrow 3.0160 \text{ u} + 1.0087 \text{ u}$

$4.0282 \text{ u} \rightarrow 4.0247 \text{ u}$

$$\Delta m = (-0.0035 \text{ u})\left(\frac{1.661 \times 10^{-27} \text{ kg}}{1 \text{ u}}\right) = -5.8 \times 10^{-30} \text{ kg}$$

$$\Delta E = \Delta mc^2 = (-5.8 \times 10^{-30} \text{ kg})(3.00 \times 10^8 \text{ m} \cdot \text{s}^{-1})^2 = -5.2 \times 10^{-13} \text{ J}$$

$$\left(\frac{-5.2 \times 10^{-13} \text{ J}}{4.0282 \text{ u}}\right)\left(\frac{1 \text{ u}}{1.661 \times 10^{-24} \text{ g}}\right) = -7.8 \times 10^{10} \text{ J} \cdot \text{g}^{-1}$$

(b) $^3\text{He} + D \rightarrow {}^4\text{He} + {}^1_1\text{H}$

$3.0160 \text{ u} + 2.0141 \text{ u} \rightarrow 4.0026 \text{ u} + 1.0078 \text{ u}$

$5.0301 \text{ u} \rightarrow 5.0104 \text{ u}$

$$\Delta m = (-0.0197 \text{ u})\left(\frac{1.661 \times 10^{-27} \text{ kg}}{1 \text{ u}}\right) = -3.27 \times 10^{-29} \text{ kg}$$

$$\Delta E = \Delta mc^2 = -(3.27 \times 10^{-29} \text{ kg})(3.00 \times 10^8 \text{ m} \cdot \text{s}^{-1})^2 = -2.94 \times 10^{-12} \text{ J}$$

$$\left(\frac{-2.94 \times 10^{-12} \text{ J}}{5.0301 \text{ u}}\right)\left(\frac{1 \text{ u}}{1.661 \times 10^{-24} \text{ g}}\right) = -3.52 \times 10^{11} \text{ J} \cdot \text{g}^{-1}$$

(c) $^7\text{Li} + {}^1_1\text{H} \rightarrow 2\,{}^4\text{He}$

$7.0160 \text{ u} + 1.0078 \text{ u} \rightarrow 2(4.0026 \text{ u})$

$8.0238 \text{ u} \rightarrow 8.0052 \text{ u}$

$$\Delta m = (-0.0186 \text{ u})\left(\frac{1.661 \times 10^{-27} \text{ kg}}{1 \text{ u}}\right) = -3.09 \times 10^{-29} \text{ kg}$$

$$\Delta E = \Delta mc^2 = (-3.09 \times 10^{-29} \text{ kg})(3.00 \times 10^8 \text{ m} \cdot \text{s}^{-1})^2 = -2.78 \times 10^{-12} \text{ J}$$

$$\left(\frac{-2.78 \times 10^{-12} \text{ J}}{8.0238 \text{ u}}\right)\left(\frac{1 \text{ u}}{1.661 \times 10^{-24} \text{ g}}\right) = -2.09 \times 10^{11} \text{ J} \cdot \text{g}^{-1}$$

(d)  $D + T \rightarrow {}^4\text{He} + \text{n}$

$2.0141 \text{ u} + 3.0160 \text{ u} \rightarrow 4.0026 \text{ u} + 1.0087 \text{ u}$

$5.0301 \text{ u} \rightarrow 5.0113 \text{ u}$

$$\Delta m = (-0.0188 \text{ u})\left(\frac{1.661 \times 10^{-27} \text{ kg}}{1 \text{ u}}\right) = -3.12 \times 10^{-29} \text{ kg}$$

$$\Delta E = \Delta mc^2 = (-3.12 \times 10^{-29} \text{ kg})(3.00 \times 10^8 \text{ m} \cdot \text{s}^{-1})^2 = -2.81 \times 10^{-12} \text{ J}$$

$$\left(\frac{-2.81 \times 10^{-12} \text{ J}}{5.0301 \text{ u}}\right)\left(\frac{1 \text{ u}}{1.661 \times 10^{-24} \text{ g}}\right) = -3.36 \times 10^{11} \text{ J} \cdot \text{g}^{-1}$$

**17.65**  (a)  ${}^{24}_{11}\text{Na} \rightarrow {}^{24}_{12}\text{Mg} + {}^{0}_{-1}\text{e}$

mass $({}^{24}_{11}\text{Na}) = 23.990\,96 \text{ u}$

mass $({}^{24}_{12}\text{Mg}) = 23.985\,04 \text{ u}$

The mass of the electron does not need to be explicitly included in the calculation because it is already included in the mass of Mg.

$$\Delta m = \text{mass}\,({}^{24}_{12}\text{Mg}) - \text{mass}\,({}^{24}_{11}\text{Na}) = 23.985\,04 \text{ u} - 23.990\,96 \text{ u}$$
$$= -5.92 \times 10^{-3} \text{ u}$$

$\Delta m$ (in kg) $= -5.92 \times 10^{-3} \text{ u} \times 1.661 \times 10^{-27} \text{ kg u}^{-1} = -9.83 \times 10^{-30} \text{ kg}$

(b)  $\Delta E = \Delta mc^2 = -(9.83 \times 10^{-30} \text{ kg})(3.00 \times 10^8 \text{ m} \cdot \text{s}^{-1})^2 = -8.85 \times 10^{-13} \text{ J}$

(c)  $\Delta E$ (per nucleon) $= \dfrac{-8.85 \times 10^{-13} \text{ J}}{24 \text{ nucleons}} = -3.69 \times 10^{-14} \text{ J} \cdot \text{nucleon}^{-1}$

This simple calculation works because the number of nucleons is the same on both sides of the equation.

**17.67** (a) $^{244}_{95}\text{Am} \rightarrow \, ^{134}_{53}\text{I} + \, ^{107}_{42}\text{Mo} + 3\, ^{1}_{0}\text{n}$

(b) $^{235}_{92}\text{U} + \, ^{1}_{0}\text{n} \rightarrow \, ^{96}_{40}\text{Zr} + \, ^{138}_{52}\text{Te} + 2\, ^{1}_{0}\text{n}$

(c) $^{235}_{92}\text{U} + \, ^{1}_{0}\text{n} \rightarrow \, ^{101}_{42}\text{Mo} + \, ^{132}_{50}\text{Sn} + 3\, ^{1}_{0}\text{n}$

**17.69** (a) $1\,\text{Ci} = 3.7 \times 10^{10}$ decays per second (dps)

decays per minute (dpm) for

$$4\,\text{pCi} = 4 \times 10^{-12}\,\text{Ci} \times 3.7 \times 10^{10}\,\text{dps} \times \left(\frac{60\,\text{s}}{1\,\text{min}}\right)$$

$$= 9\,\text{dpm}$$

(b) $\text{volume(L)} = (2.0 \times 3.0 \times 2.5)\,\text{m}^3 \times \left(\frac{10^3\,\text{L}}{1\,\text{m}^3}\right) = 1.5 \times 10^4\,\text{L}$

$$\text{number of decays} = (1.5 \times 10^4\,\text{L})\left(\frac{4\,\text{pCi}}{1\,\text{L}}\right)\left(\frac{9\,\text{decays} \cdot \text{min}^{-1}}{4\,\text{pCi}}\right)(5.0\,\text{min})$$

$$= 7 \times 10^5\,\text{decays}$$

**17.71** $N_0 = $ number of $^{222}\text{Rn}$ atoms $= 2.0 \times 10^{-5}\,\text{mol} \times 6.0 \times 10^{23}\,\text{atoms} \cdot \text{mol}^{-1}$

$$= 1.2 \times 10^{19}\,\text{atoms}$$

$$k = \frac{\ln 2}{t_{1/2}} = \frac{0.693}{3.82\,\text{d}} = 0.181\,\text{d}^{-1}$$

(a) rate of decay $= k \times N = \left(\frac{0.181}{\text{d}}\right)\left(\frac{1\,\text{d}}{8.64 \times 10^4\,\text{s}}\right)(1.2 \times 10^{19}\,\text{atoms})$

$$= 2.52 \times 10^{13}\,\text{atoms} \cdot \text{s}^{-1}\,(\text{dps or Bq})$$

initial activity $= (2.52 \times 10^{13}\,\text{Bq})\left(\frac{1\,\text{Ci}}{3.7 \times 10^{10}\,\text{Bq}}\right)\left(\frac{1\,\text{pCi}}{10^{-12}\,\text{Ci}}\right)$

$$\times \left(\frac{1}{2000\,\text{m}^3}\right)\left(\frac{1\,\text{m}^3}{10^3\,\text{L}}\right)$$

$$= 3.4 \times 10^8\,\text{pCi} \cdot \text{L}^{-1}$$

(b) $N = N_0 e^{-kt} = 1.2 \times 10^{19}\,\text{atoms} \times e^{-0.181\,\text{d}^{-1} \times 1\,\text{d}} = 1.0 \times 10^{19}\,\text{atoms}$

(c) $\ln\left(\dfrac{\text{activity}}{\text{initial acitivity}}\right) = -kt$

$$t = -\left(\dfrac{1}{k}\right)\ln\left(\dfrac{\text{activity}}{\text{initial activity}}\right) = -\left(\dfrac{1}{0.181\ \text{d}^{-1}}\right)\ln\left(\dfrac{4}{3.4 \times 10^{8}}\right)$$

$$= 1 \times 10^{2}\ \text{days}$$

**17.73** (a) At first thought, it might seem that a fusion bomb would be more suitable for excavation work, because the fusion process itself does not generate harmful radioactive waste products. However, in practice, fusion cannot be initiated in a bomb in the absence of the high temperatures that can only be generated by a fission bomb. So, there is no environmental advantage to the use of a fusion bomb. The fission bomb has the advantage that its destructive power can be more carefully controlled. It is possible to make small fission bombs whose destructive effect can be contained within a small area.

(b) The principal argument for the use of bombs in excavation is speed, and therefore cost-effectiveness, of the process. The principal argument against their use is environmental damage.

**17.75** $k = \dfrac{0.693}{4.5 \times 10^{9}\ \text{a}} = 1.5 \times 10^{-10}\ \text{a}^{-1}$

$$t(= \text{age}) = -\left(\dfrac{1}{k}\right)\ln\left(\dfrac{N}{N_0}\right)$$

$$\dfrac{N}{N_0} = \dfrac{\text{mass of } ^{238}\text{U}}{\text{initial mass of } ^{238}\text{U}} = \dfrac{1}{1 + \dfrac{\text{mass of } ^{206}\text{Pb}}{\text{mass of } ^{238}\text{U}}}$$

(a) $\dfrac{N}{N_0} = \dfrac{1}{1 + 1.00} = \dfrac{1}{2.00}$, therefore age $= t_{1/2} = 4.5 \times 10^{9}\ \text{a}$

(b) $\dfrac{N}{N_0} = \dfrac{1}{1 + \dfrac{1}{1.25}} = 0.556$

$$t(= \text{age}) = -\left(\frac{1}{1.5 \times 10^{-10} \text{ a}^{-1}}\right) \ln(0.556) = 3.9 \times 10^9 \text{ a}$$

**17.77** (a) activity $= (17.3 \text{ Ci})\left(\dfrac{3.7 \times 10^{10} \text{ Bq}}{1 \text{ Ci}}\right)$

$$= 6.4 \times 10^{11} \text{ Bq} = 6.4 \times 10^{11} \text{ nuclei} \cdot \text{s}^{-1}$$

$$N = (2.0 \times 10^{-6} \text{ g})\left(\frac{1 \text{ u}}{1.661 \times 10^{-24} \text{ g}}\right)\left(\frac{1 \text{ nucleus}}{24 \text{ u}}\right) = 5.0 \times 10^{16} \text{ nuclei}$$

$$k = \frac{\text{activity}}{N} = \frac{6.4 \times 10^{11} \text{ nuclei} \cdot \text{s}^{-1}}{5.0 \times 10^{16} \text{ nuclei}} = 1.3 \times 10^{-5} \text{ s}^{-1} = 1.1 \text{ d}^{-1}$$

$$t_{1/2} = \frac{0.693}{k} = \frac{0.693}{1.3 \times 10^{-5} \text{ s}^{-1}} = 5.3 \times 10^4 \text{ s} = 15 \text{ h} = 0.63 \text{ d}$$

(b) $m = m_0 e^{-kt} = 2.0 \text{ mg} \times e^{-1.11 \text{ d}^{-1} \times 2.0 \text{ d}} = 0.22 \text{ mg}$

**17.79** (a) Radioactive substances which emit $\gamma$ radiation are most effective for diagnosis because they are the least destructive of the types of radiation listed. Additionally, $\gamma$ rays pass easily through body tissues and can be counted, whereas $\alpha$ and $\beta$ particles are stopped by the body tissues.

(b) $\alpha$ particles tend to be best for this application because they cause the most destruction.   (c) and (d) $^{131}$I, 8d (used to image the thyroid); $^{67}$Ga, 78 h (used most often as the citrate complex); $^{99m}$Tc, 6 h (used for various body tissues by varying the ligands attached to the Tc atom).

**17.81** (a)

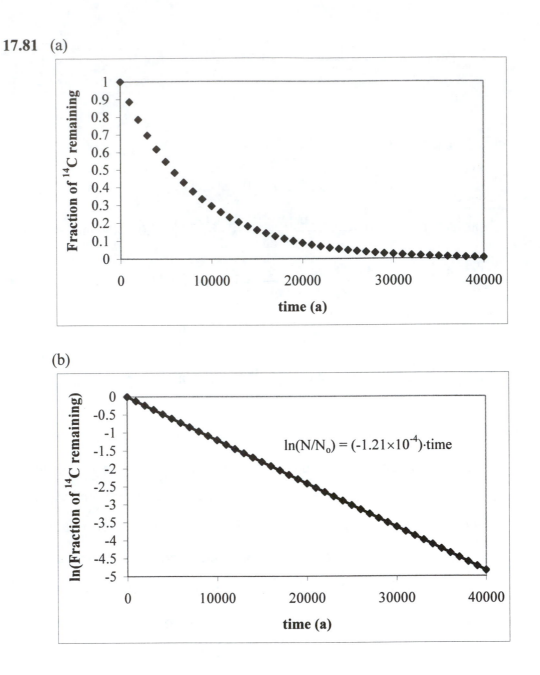

(b)

(c) The information can be obtained from the graphs or from the equation

$$-\ln \frac{N}{N_0} = kt$$

If we want to have less than 1% of the original amount of $^{14}C$ present, then we will want the value for which $N/N_0$ is 0.01 or less.

$$-\ln 0.01 = (1.21 \times 10^{-4} \ a^{-1})(t)$$
$$t = 3.8 \times 10^4 \ a$$

**17.83** Radon-222 decays to polonium-218 by alpha emission with a half-life of 3.824 days.

$$^{222}_{86}\text{Rn} \rightarrow ^{218}_{84}\text{Po} + ^{4}_{2}\alpha \qquad\qquad t_{1/2} = 3.824 \text{ d}$$

An alpha particle is the nucleus of $^4\text{He}$. Assuming that (1) the alpha particles are captured inside the container, (2) they behave as an ideal gas and (3) the temperature is constant at 298 K, we can find the volume of the container by calculating the number of moles of $^4_2\alpha$ formed in 15 days and then applying the ideal gas law.

$$n_{He} = n_{Po} = n_{initial\ Rn} - n_{final\ Rn} \qquad n_{final\ Rn} \propto N_{15\ days}$$

$$k = \frac{0.693}{t_{1/2}} = \frac{0.693}{3.824 \text{ d}} = 0.181 \text{ d}^{-1}$$

$$N_{15\ days} = N_0 e^{-kt} = (2.5 \text{ g})e^{-(0.181\ \text{d}^{-1})(15\ \text{d})} = 0.165 \text{ g}$$

$$n_{final\ Rn} = (0.165 \text{ g } ^{222}\text{Rn})\left(\frac{1 \text{ mol } ^{222}\text{Rn}}{222.0175 \text{ g } ^{222}\text{Rn}}\right)$$

$$= 7.43 \times 10^{-4} \text{ mol } ^{222}\text{Rn}$$

$$n_{initial\ Rn} = (2.5 \text{ g } ^{222}\text{Rn})\left(\frac{1 \text{ mol } ^{222}\text{Rn}}{222.0175 \text{ g } ^{222}\text{Rn}}\right)$$

$$= 1.13 \times 10^{-2} \text{ mol } ^{222}\text{Rn}$$

$$n_{He} = n_{initial\ Rn} - n_{final\ Rn} = 1.13 \times 10^{-2} \text{ mol} - 7.43 \times 10^{-4} \text{ mol}$$

$$= 1.06 \times 10^{-2} \text{ mol}$$

$$PV = nRT$$

$$V = \frac{nRT}{P} = \frac{(1.06 \times 10^{-2} \text{ mol})(0.08206 \text{ L} \cdot \text{atm} \cdot \text{K}^{-1} \cdot \text{mol}^{-1})(298 \text{ K})}{1.00 \text{ atm}}$$

$$= 0.26 \text{ L}$$

**17.85** (a) The rate will decrease because the heavier D atom makes the zero-point energy of the X-D bond lower than that of the X-H bond. The X-D bond energy is deeper in the potential energy well and requires slightly more energy for dissociation.

(b) The activation energy would correspond to the bond dissociation energy. As long as the force constant for the transition state is low, the bond dissociation energies for C-H and C-D will be the same except for the small difference in their zero-point energies.

$$E_{a(C-D)} = E_{a(C-H)} + \tfrac{1}{2}h(\nu_{\text{C-H}} - \nu_{\text{C-D}})$$

Recall from problem 17.55:

$$\nu = \frac{1}{2\pi}\sqrt{\frac{k}{\mu}}$$

where $\mu = \dfrac{m_A m_B}{m_A + m_B}$

$$\frac{\nu_{\text{C-D}}}{\nu_{\text{C-H}}} = \sqrt{\frac{\mu_{\text{C-H}}}{\mu_{\text{C-D}}}} = 0.73422$$

The ratio of rates will be the same as the ratio of the rate constants.

$$\frac{k_{\text{C-H}}}{k_{\text{C-D}}} = \frac{Ae^{-\frac{E_{a(C-H)}}{RT}}}{Ae^{-\frac{E_{a(C-D)}}{RT}}} = e^{\frac{E_{a(C-D)}}{RT}} \cdot e^{-\frac{E_{a(C-H)}}{RT}} = e^{\frac{E_{a(C-D)} - E_{a(C-H)}}{RT}}$$

$$E_{a(C-D)} - E_{a(C-H)} = \tfrac{1}{2}h(\nu_{\text{C-H}} - \nu_{\text{C-D}}) = \tfrac{1}{2}h(\nu_{\text{C-H}} - 0.73422\nu_{\text{C-H}})$$
$$= \tfrac{1}{2}h(0.26578\nu_{\text{C-H}})$$

Noting that a typical C-H stretching frequency is about $3000\ \text{cm}^{-1}$ or $9 \times 10^{13}\,\text{Hz}$ (see, for example, Exercise 2.101)

$$\frac{k_{\text{C-H}}}{k_{\text{C-D}}} = e^{\frac{\frac{1}{2}(6.626\times10^{-34}\,\text{J·s})(0.26578)(9\times10^{13}\,\text{s}^{-1})(6.02\times10^{23}\,\text{mol}^{-1})}{(8.314\,\text{J·K}^{-1}\text{·mol}^{-1})(298\text{K})}} = e^{1.925} = 6.86 \approx 7$$

(c) If the carbon atom is infinitely heavy, then it acts like a stationary wall and does not vibrate. In that case, the H or D atomic mass can be used instead of the reduced mass since only that atom is moving.

$$\frac{\nu_{\text{D}}}{\nu_{\text{H}}} = \sqrt{\frac{m_{\text{H}}}{m_{\text{D}}}} = 0.70739$$

$$E_{a(D)} - E_{a(H)} = \tfrac{1}{2}h(\nu_{\text{H}} - \nu_{\text{D}}) = \tfrac{1}{2}h(\nu_{\text{H}} - 0.70739\nu_{\text{H}}) = \tfrac{1}{2}h(0.29261\nu_{\text{H}})$$

$$\frac{k_{\text{C-H}}}{k_{\text{C-D}}} = e^{\frac{\frac{1}{2}(6.626\times10^{-34}\,\text{J·s})(0.29261)(9\times10^{13}\,\text{s}^{-1})(6.02\times10^{23}\,\text{mol}^{-1})}{(8.314\,\text{J·K}^{-1}\text{·mol}^{-1})(298\text{K})}} = e^{2.119} = 8.32 \approx 8$$

This result suggests that the kinetic isotope effect can become even more pronounced for heavier molecules.

**17.87**  $m_{tot} = m_{e^-} + m_{e^+}$

$$= 2m_{e^-} = 2(9.109\ 39 \times 10^{-31}\ \text{kg})$$

$$= 1.821\ 88 \times 10^{-30}\ \text{kg}$$

$E = mc^2$

$$= (1.821\ 88 \times 10^{-30}\text{kg})(2.997\ 92 \times 10^8\text{m} \cdot \text{s}^{-1})^2$$

$$= 1.637\ 42 \times 10^{-13}\text{J}$$

**17.89**  (a) $^{8}_{2}\text{He} \rightarrow\ ^{8}_{3}\text{Li} +\ ^{0}_{-1}\text{e}$  (*Note*: This product is only the first daughter product.  Because it is not itself stable, we can expect further steps, such as $^{8}_{3}\text{Li} \rightarrow\ ^{8}_{4}\text{Be} +\ ^{0}_{-1}\text{e}$, followed by $^{8}_{4}\text{Be} \rightarrow 2\ ^{4}_{2}\text{He}$.)

(b) $^{14}_{8}\text{O} +\ ^{0}_{-1}\text{e} \rightarrow\ ^{14}_{7}\text{N}$

# CHAPTER 18
# ORGANIC CHEMISTRY I:
# THE HYDROCARBONS

**18.1**

(a)

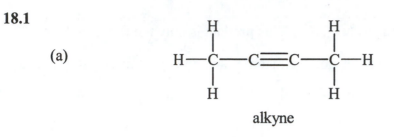

alkyne

(b)

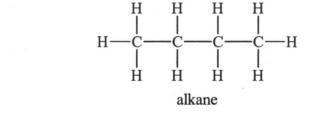

alkane

(c)

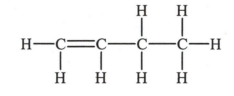

alkene

(d)

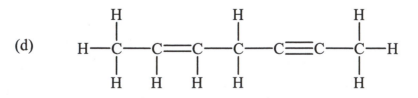

alkene and alkyne

(e)

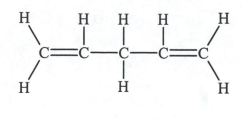

alkene

**18.3** (a) $(CH_3)_3CH$ or $C_4H_{10}$, alkane; (b) $C_6H_7CH_3$ or $C_7H_{10}$ alkene; (c) $C_6H_{12}$, alkane; (d) $C_6H_{12}$ alkane

**18.5** (a) $C_{12}H_{26}$, alkane; (b) $C_{13}H_{20}$, alkene; (c) $C_7H_{14}$, alkane; (d) $C_{14}H_8$ aromatic hydrocarbon

**18.7** (a) propane; (b) butane; (c) heptane; (d) decane

**18.9** (a) methyl; (b) pentyl; (c) propyl; (d) hexyl

**18.11** (a) propane; (b) ethane; (c) pentane; (d) 2,3-dimethylbutane

**18.13** (a) 4-methyl-2-pentene; (b) 2,3-dimethyl-2-phenylpentane

**18.15** (a) $CH_2=CHCH(CH_3)CH_2CH_3$;
(b) $CH_3CH_2C(CH_3)_2CH(CH_2CH_3)(CH_2)_2CH_3$;
(c) $HC{\equiv}C(CH_2)_2C(CH_3)_3$; (d) $CH_3CH(CH_3)CH(CH_2CH_3)CH(CH_3)_2$

**18.17**

(a)

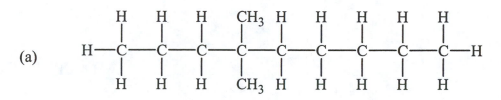

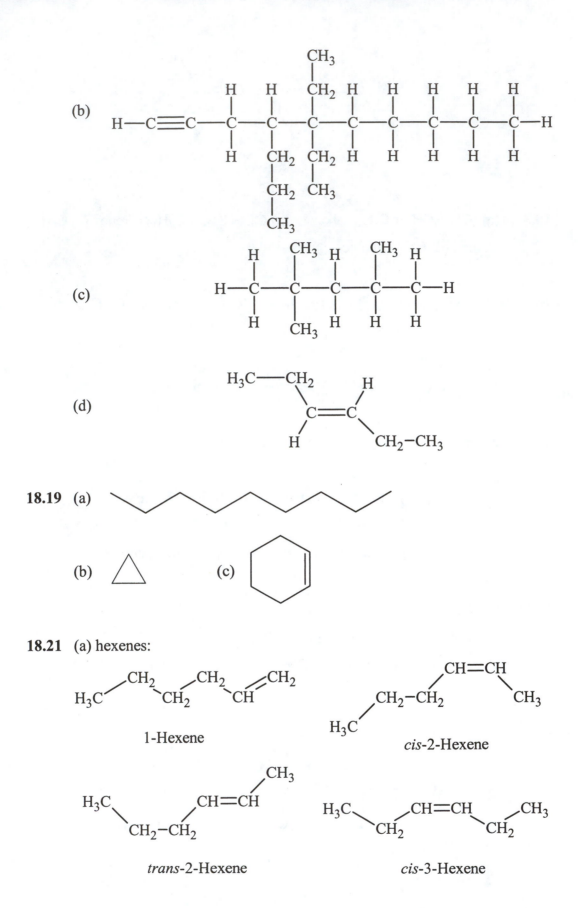

(b)

(c)

(d)

**18.19** (a)

(b)     (c)

**18.21** (a) hexenes:

1-Hexene

*cis*-2-Hexene

*trans*-2-Hexene

*cis*-3-Hexene

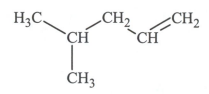

*trans*-3-Hexene

Pentenes:

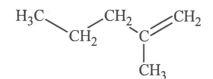

4-Methyl-1-pentene

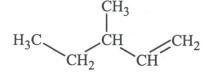

3-Methyl-1-pentene

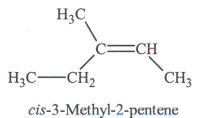

2-Methyl-1-pentene

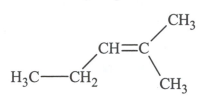

2-Methyl-2-pentene

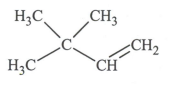

*cis*-3-Methyl-2-pentene

(+ trans isomer)

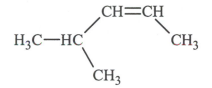

*cis*-4-Methyl-2-pentene

(+ trans isomer)

butanes:

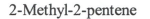

3,3-Dimethyl-1-butene

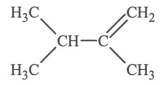

2,3-Dimethyl-1-butene

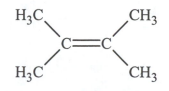

2,3-Dimethyl-2-butene

(b) cyclic molecules:

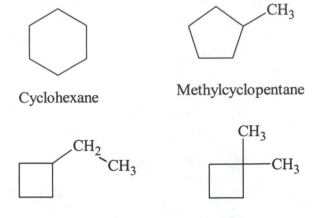

Cyclohexane          Methylcyclopentane

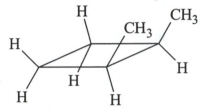

ethylcyclobutane     1,1-dimethylcyclobutane

The following structures are drawn to emphasize the stereochemistry

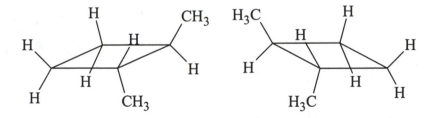

*cis*-1,2-Dimethylcyclobutane

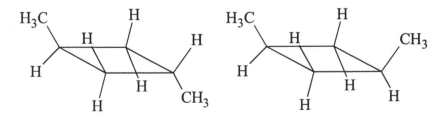

*trans*-1,2-Dimethylcyclobutane

(nonsuperimposable mirror images

*trans*-1,3-Dimethylcyclobutane     *cis*-1,3-Dimethylcyclobutane

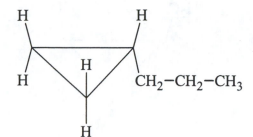

Propylcyclopropane

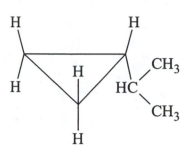

Isoropylcyclopropane

or 2-cyclopropylpropane

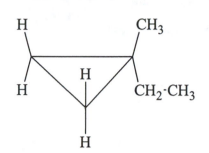

1-Ethyl-1-methylcyclopropane

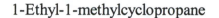
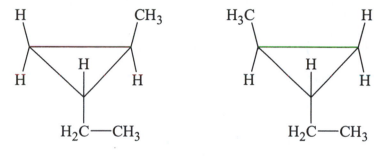

*trans*-1-Ethyl-2-methylcyclopropane

(nonsuperimposable mirror images)

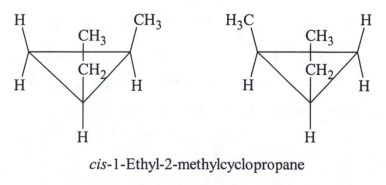

*cis*-1-Ethyl-2-methylcyclopropane

(nonsuperimposable mirror images)

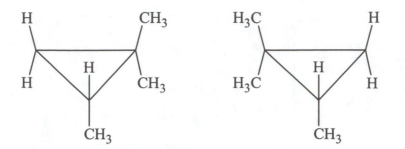

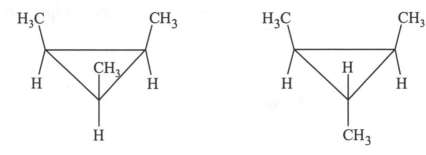

1,1,2-Trimethylcyclopropane

(nonsuperimposable mirror images)

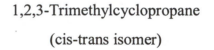

1,2,3-Trimethylcyclopropane

(all cis isomer)

1,2,3-Trimethylcyclopropane

(cis-trans isomer)

**18.23** (a) Butane is $C_4H_{10}$, cyclobutane is $C_4H_8$. Because they have different formulas, they are not isomers.

(b) Same formula, but different structures; therefore, they are structural isomers.

(c) Same formula ($C_5H_{10}$), same structure (bonding arrangement is the same), but different geometry; therefore, they are geometrical isomers.

(d) Not isomers, because only their positions in space are different and these positions can be interchanged. Same molecule.

**18.25** (a)

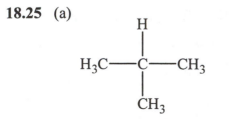

(b) If only two isomeric products are formed and they are both branched, then the only possibilities are

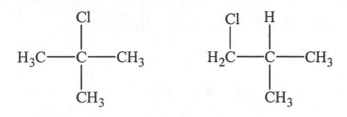

**18.27** An * designates a chiral carbon.

(a) optically active,

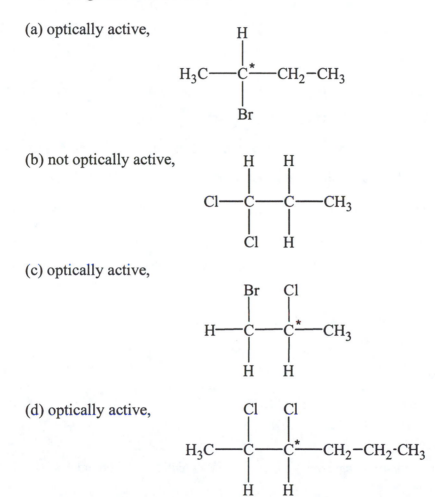

(b) not optically active,

(c) optically active,

(d) optically active,

**18.29** The difference can be traced to the weaker London forces that exist in branched molecules. Atoms in neighboring branched molecules cannot lie as close together as they can in unbranched isomers.

**18.31** The balanced equations are

$$C_3H_8(g) + 5\,O_2(g) \longrightarrow 3\,CO_2(g) + 4\,H_2O(l)$$

$$C_4H_{10}(g) + 13/2\,O_2(g) \longrightarrow 4\,CO_2(g) + 5\,H_2O(l)$$

$$C_5H_{12}(g) + 8\,O_2(g) \longrightarrow 5\,CO_2(g) + 6\,H_2O(l)$$

The enthalpies of combustion that correspond to these reactions are listed in the Appendix:

| Compound | (a) Enthalpy of combustion $kJ \cdot mol^{-1}$ | (b) Heat released per g $kJ \cdot g^{-1}$ |
|---|---|---|
| Propane | 22220 | 50.3 |
| Butane | 22878 | 49.5 |
| Pentane | 23537 | 49.0 |

The molar enthalpy of combustion increases with molar mass as might be expected, because the number of moles of $CO_2$ and $H_2O$ formed will increase as the number of carbon and hydrogen atoms in the compounds increases. The heat released per gram of these hydrocarbons is essentially the same because the H to C ratio is similar in the three hydrocarbons.

**18.33** There are nine possible products:

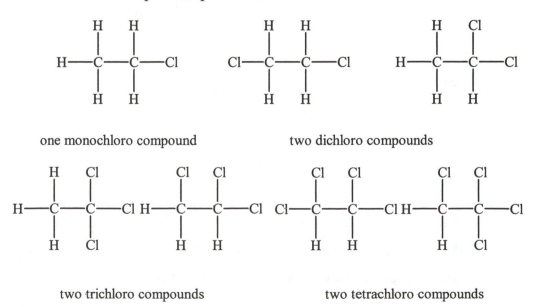

one monochloro compound     two dichloro compounds

two trichloro compounds              two tetrachloro compounds

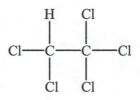

one pentachloro compound

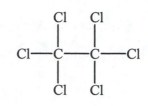

one hexachloro compound

None of these form optical isomers.

**18.35**

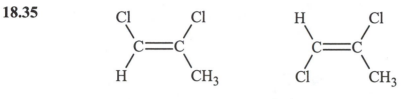

*cis*-1,2-Dichloropropene     *trans*-1,2-Dichloropropene

*cis*-1,2-Dichloropropene is polar, although *trans*-1,2-Dichloropropene is slightly polar also.

**18.37** (a)

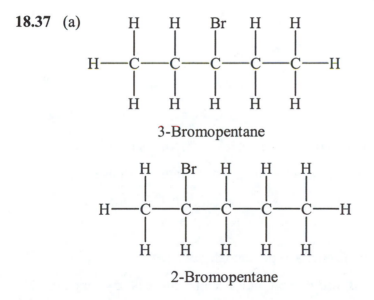

3-Bromopentane

2-Bromopentane

(b) addition reaction

**18.39** (a) $C_6H_{11}Br + NaOCH_2CH_3 \rightarrow C_6H_{10} + NaBr + HOCH_2CH_3$

(b)

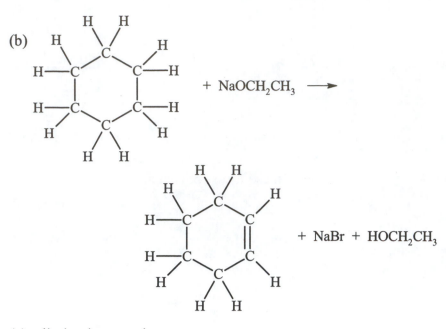

+ NaOCH$_2$CH$_3$ $\longrightarrow$

+ NaBr + HOCH$_2$CH$_3$

(c) elimination reaction

**18.41** C$_2$H$_4$ + X$_2$ $\longrightarrow$ C$_2$H$_4$X$_2$

We will break one X—X bond and form two C—X bonds.

Using bond enthalpies:

| Halogen | Cl | Br | I |
|---|---|---|---|
| X—X bond breakage (kJ·mol$^{-1}$) | +242 | +193 | +151 |
| C—X bond formation (kJ·mol$^{-1}$) | 22(338) | 22(276) | 22(238) |
| Total (kJ·mol$^{-1}$) | 2434 | 2359 | 2325 |

The reaction is less exothermic as the halogen becomes heavier. In general, the reactivity, and also the danger associated with use of the halogens in reactions, decreases as one descends the periodic table.

**18.43** (a) 1-ethyl-3-methylbenzene;  (b) pentamethylbenzene (1,2,3,4,5-pentamethylbenzene is also correct, but, because there is only one possible pentamethylbenzene, the use of the numbers is not necessary)

**18.45** (a)

(b)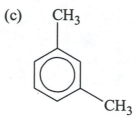

(c)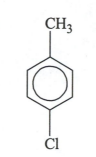

(d)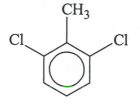

**18.47**

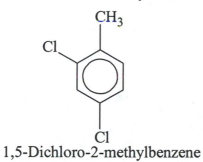

1,3-Dichloro-2-methylbenzene     1,4-Dichloro-2-methylbenzene

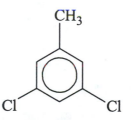

1,5-Dichloro-2-methylbenzene     1,3-Dichloro-5-methylbenzene

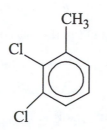 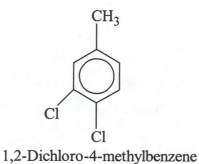

1,2-Dichloro-3-methylbenzene

1,2-Dichloro-4-methylbenzene

(b) All of these molecules are at least slightly polar.

**18.49**

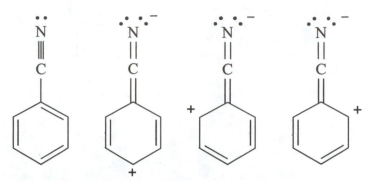

Electrophiles tend to avoid the ortho and para positions that develop slight + charges in the resonance forms.

**18.51** Two compounds can be produced. Resonance makes positions 1, 4, 6, and 9 equivalent. It also makes positions 2, 3, 7, and 8 equivalent. Positions 5 and 10 are equivalent but have no H atom.

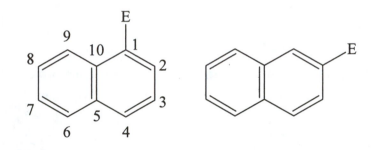

**18.53** These hydrocarbons are too volatile (they are all gases at room temperature) and would not remain in the liquid state.

**18.55** Cracking is the process of breaking down hydrocarbons with many carbon atoms into smaller units, whereas alkylation is the process of combining smaller hydrocarbons into larger units. Both processes are carried out catalytically, and both are used to convert hydrocarbons into units having from 5 to 11 carbon atoms suitable for use in gasoline.

**18.57** (a) 4 σ-type single bonds

(b) 2 σ-type single bonds and 1 double bond with a σ- and a π-bond

(c) 1 σ-type single bond and 1 triple bond with a σ-bond and 2 π-bonds

**18.59** (a) substitution, $CH_4 + Cl_2 \rightarrow CH_3Cl + HCl$

(b) addition $CH_2=CH_2 + Br_2 \rightarrow CH_2Br\text{-}CH_2Br$

**18.61** Water is not used as the nonpolar reactants will not readily dissolve in a highly polar solvent like water. Also, the ethoxide ion reacts with water.

**18.63** The double bond in alkenes makes them more rigid than alkanes. Some of the atoms of alkene molecules are locked into a planar arrangement by the π-bond; hence, they cannot roll up into a ball as compactly as alkanes can. Because they do not pack together as compactly as alkanes do, they have lower boiling and melting points.

**18.65** (a) 2-methyl-1-propene, no geometrical isomers; (b) *cis*-3-methyl-2-pentene, *trans*-3-methyl-2-pentene; (c) 1-hexyne, no geometrical isomers; (d) 3-hexyne, no geometrical isomers; (e) 2-hexyne, no geometrical isomers.

**18.67** (a) $C_{10}H_{18}$; (b) naphthalene, ⬡⬡ , $C_{10}H_8$; (c) Yes. Cis and trans forms (relative to the C-C bond common to the two six-membered rings) are possible.

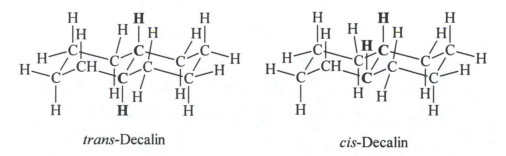

*trans*-Decalin                    *cis*-Decalin

**18.69**

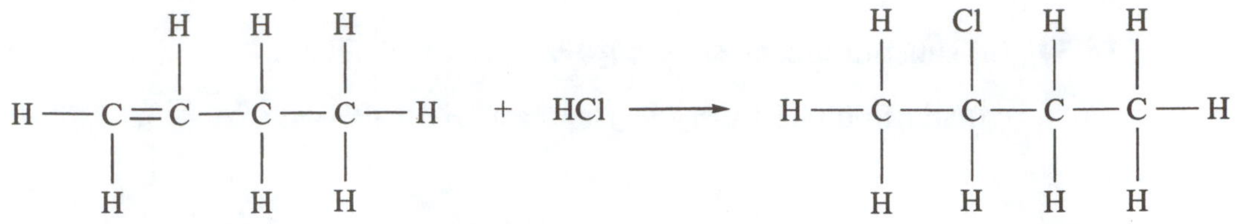

$$H-C=C-C-C-H \quad + \quad HCl \longrightarrow \quad H-C-C-C-C-H$$

(structural formulas with H substituents; left molecule has C=C double bond, right molecule has Cl substituent)

Bonds Broken: H-Cl, C=C

Bonds Formed: C-Cl, C-H, C-C

The reaction enthalpy will be equivalent to the enthalpy gained by the system in breaking bonds minus the enthalpy lost during bond formation:

$\Delta H_r = \Delta H_B(H–Cl) + \Delta H_B(C=C) – \Delta H_B(C–Cl) – \Delta H_B(C–H) – \Delta H_B(C–C)$

From tables 6.7 and 6.8,

$\Delta H_r = 431$ kJ·mol$^{-1}$ + 612 kJ·mol$^{-1}$ – 338 kJ·mol$^{-1}$

$\quad\quad - 412$ kJ·mol$^{-1}$ – 348 kJ·mol$^{-1}$

$\quad\quad = - 55$ kJ·mol$^{-1}$

**18.71** number of moles of H $= \left( \dfrac{4.48 \text{ g } H_2O}{18.02 \text{ g } H_2O/\text{mol } H_2O} \right) \left( \dfrac{2 \text{ mol H}}{1 \text{ mol } H_2O} \right)$

$\quad\quad = 0.497$ mol H

number of moles of C $= \left( \dfrac{9.72 \text{ g } CO_2}{44.01 \text{ g } CO_2/\text{mol } CO_2} \right) \left( \dfrac{1 \text{ mol C}}{1 \text{ mol } CO_2} \right)$

$\quad\quad = 0.221$ mol C

$\dfrac{0.497 \text{ mol H}}{0.221 \text{ mol C}} = \dfrac{9 \text{ mol H}}{4 \text{ mol C}}$

The empirical formula is $C_4H_9$; the molecular formula might be $C_8H_{18}$, which matches the formula for alkanes ($C_nH_{2n+2}$). It is not likely an alkene or alkyne, because there is no reasonable Lewis structure for a compound having the empirical formula $C_4H_9$ and multiple bonds.

**18.73** (a) 4-methyl-3-propylheptane

The longest chain has eight carbon atoms in it. The systematic name of the compound is 4-ethyl-5-methyloctane.

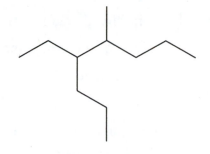

4-ethyl-5-methyloctane;

(b) 4,6-dimethyloctane

The compound name is almost correct, but the numbering scheme with the lowest numbers would be 3,5-dimethyloctane.

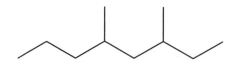

3,5-dimethyloctane;

(c) 2,2-dimethyl-4-propylhexane

The longest carbon chain in the molecule is seven carbon atoms long. The systematic name is 2,2-dimethyl-4-propylheptane.

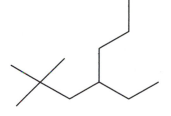

2,2-dimethyl-4-ethylheptane;

(d) 2,2-dimethyl-3-ethylhexane.

The name is essentially correct except that ethyl should be listed first The systematic name is 3-ethyl-2,2-dimethylhexane

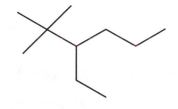

3-ethyl-2,2-dimethylhexane

**18.75** Bromine is an electrophile which will undergo an addition reaction with alkenes in the dark. The lack of a reaction in the dark with $Br_2$ indicates that the molecule is not an alkene. In the presence of light, bromine will undergo a substitution reaction with alkanes. Therefore, the molecule is most likely an alkane and the only alkane with the molecular formula $C_3H_6$ is cyclopropane.

**18.77** The $NO_2$ group is a meta-directing group and the Br atom is an ortho, para-directing group. Because the position para to Br is already substituted with the $NO_2$ group, further bromination will not occur there. The resonance forms show that the bromine atom will activate the position ortho to it as expected. The $NO_2$ group will deactivate the group ortho to itself, thus in essence enhancing the reactivity of the position meta to the $NO_2$ group. This position is ortho to the Br atom, so the effects of the Br and $NO_2$ groups reinforce each other. Bromination is thus expected to occur as shown:

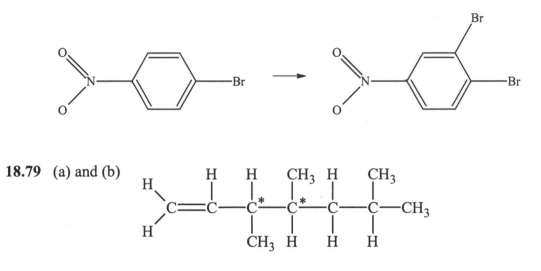

**18.79** (a) and (b)

$$H_2C=CH-C^*H(CH_3)-C^*H(CH_3)-CH_2-CH(CH_3)-CH_3$$

(c) No, there are no cis/trans isomers for this molecule.

**18.81** If the molecule contains two carbon centers that have four different substituents attached but are arranged such that they are mirror images of one another within the molecule, the molecule will not be optically active. Such an example is shown below in general for a 1,2-X-1,2-Y-1,2-Z substituted ethane. Many other examples are possible, the only criterion being that the carbon atoms that have four substituents must have a mirror image carbon center within the molecule.

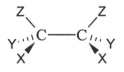

**18.83** For a molecule such as 1,2-dichloro-4-diethylbenzene, $C_6H_3Cl_2(CH_2CH_3)$, 175.04 u, it is relatively easy to lose heavy atoms such as chlorine and groups of atoms such as methyl and ethyl fragments. Molecules can also lose hydrogen atoms. In mass spectrometry, $P$ is used to represent the *parent ion*, which is the ion formed from the molecule without fragmentation. Fragments are then represented as $P - x$, where $x$ is the particular fragment lost from the parent ion to give the observed mass. Because the mass spectrum will measure the masses of individual molecules, the mass of carbon used will be 12.00 u (by definition) because the large majority of the molecules will have all $^{12}C$. The mass of H is 1.0078 u. Some representative peaks that may be present are listed below.

| Fragment formula | Relation to parent ion | Mass, u |
|---|---|---|
| $C_6H_3{}^{35}Cl_2(CH_2CH_3)$ | $P$ | 174.00 |
| $C_6H_3{}^{35}Cl{}^{37}Cl(CH_2CH_3)$ | $P$ | 176.00 |
| $C_6H_3{}^{37}Cl_2(CH_2CH_3)$ | $P$ | 177.99 |
| $C_6H_3{}^{35}Cl(CH_2CH_3)$ | $P$-Cl | 139.03 |
| $C_6H_3{}^{37}Cl(CH_2CH_3)$ | $P$-Cl | 141.03 |
| $C_6H_3{}^{35}Cl_2(CH_2)$ | $P$-CH$_3$ | 158.98 |
| $C_6H_3{}^{35}Cl{}^{37}Cl(CH_2)$ | $P$-CH$_3$ | 160.97 |
| $C_6H_3{}^{37}Cl_2(CH_2)$ | $P$-CH$_3$ | 162.97 |
| $C_6H_3{}^{35}Cl_2$ | $P$-CH$_2$CH$_3$ | 144.96 |
| $C_6H_3{}^{35}Cl{}^{37}Cl$ | $P$-CH$_2$CH$_3$ | 146.96 |
| $C_6H_3{}^{37}Cl_2$ | $P$-CH$_2$CH$_3$ | 148.96 |
| $C_6H_3{}^{35}Cl$ | $P$-CH$_2$CH$_3$-Cl | 109.99 |
| $C_6H_3{}^{37}Cl$ | $P$-CH$_2$CH$_3$-Cl | 111.99 |
| Etc. | | |

**18.85** The presence of one bromine atom will produce in the ions that contain Br companion peaks that are separated by 2 u. Any fragment that contains Br will show this "doublet" in which the peaks are nearly but not exactly equal in intensity. Thus, seeing a mass spectrum of a compound that is known to have Br or that was involved in a reaction in which Br could have been added or substituted with such doublets, is almost a sure sign that Br is present in the compound. It is also fairly easy to detect Br atoms in the mass spectrum at 79 and 81 u, confirming their presence. If more than one Br atom is present, then a more complicated pattern is observed for the presence of the two isotopes. The possible combinations for a molecule of unknown formula with two Br atoms is $^{79}Br^{79}Br$, $^{79}Br^{81}Br$, $^{81}Br^{79}Br$, and $^{81}Br^{81}Br$. Thus, a set of three peaks (the two possibilities $^{79}Br^{81}Br$ and $^{81}Br^{79}Br$ have identical masses) will be generated that differ in mass by two units. The center peak, which is produced by the $^{79}Br^{81}Br$ and $^{81}Br^{79}Br$ combinations, will have twice the intensity of the outer two peaks, because statistically there are twice as many combinations that produce this mass. All modern mass spectrometers have spectral simulation programs that can readily calculate and print our the relative isotopic distribution pattern expected for any compound formulation, so that it is possible to easily match the expected pattern for a particular ion with the experimental result.

**18.87** $C_8H_{10}$ will have an absorption maximum at a longer wavelength. Molecular orbital theory predicts that in conjugated hydrocarbons (molecules which contain a chain of carbon atoms with alternating single and double bonds) electrons become delocalized and are free to move up and down the chain of carbon atoms. Such electrons may be described using the one dimensional "particle in a box" model introduced in Chapter 1. According to this model, as the box to which electrons are confined lengthens, the quantized energy states available to the electrons get closer together. As a result, the energy needed to excite an electron from the ground state to the next higher state is lower for electrons confined to longer boxes. Therefore, lower energy photons, i.e. photons with longer wavelengths, will be absorbed by the $C_8H_{10}$ molecule because it provides a longer "box" than $C_6H_8$.

# CHAPTER 19
# ORGANIC CHEMISTRY II: FUNCTIONAL GROUPS

**19.1** (a) $RNH_2, R_2NH, R_3N$ (b) $ROH$ (c) $RCOOH$ (d) $RCHO$

**19.3** (a) ether; (b) ketone; (c) amine; (d) ester

**19.5** (a) 2-iodo-2-butene; (b) 2,4-dichloro-4-methylhexane; (c) 1,1,1,-triiodoethane; (d) dichloromethane

**19.7** (a) $ClC_6H_4Cl(OH)$, phenol;

(b) $CH_3CH(CH_3)CH(OH)CH_2CH_3$, secondary alcohol; (c)

$CH_3CH_2CH(CH_3)CH_2CH(CH_3)CH_2OH$, primary alcohol;

(d) $CH_3C(CH_3)(OH)CH_2CH_3$, tertiary alcohol

**19.9** (a) $CH_3CH_2OCH_3$; (b) $CH_3CH_2OCH_2CH_2CH_3$; (c) $CH_3OCH_3$

**19.11** (a) butyl propyl ether; (b) methyl phenyl ether; (c) pentyl propyl ether

**19.13** (a) aldehyde, ethanal; (b) ketone, propanone; (c) ketone, 3-pentanone

**19.15** (a)

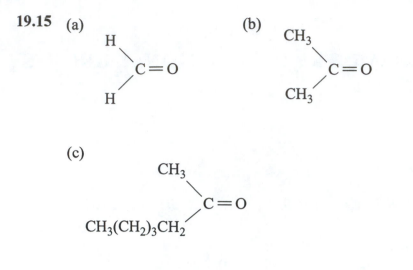

(c)

**19.17** (a) ethanoic acid;   (b) butanoic acid;   (c) 2-aminoethanoic acid

**19.19** (a)

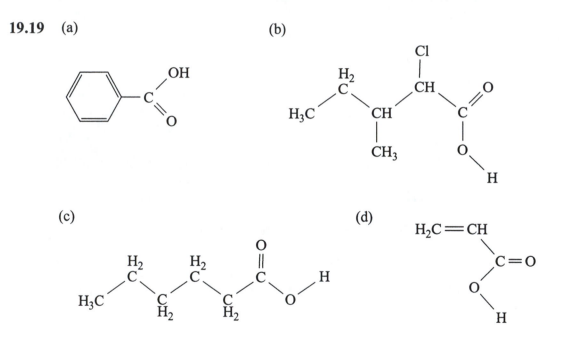

**19.21** (a) methylamine;   (b) diethylamine;   (c) *o*-methylaniline, 2-methylaniline, *o*-methylphenylamine, or 1-amino-2-methylbenzene

**19.23**

(a)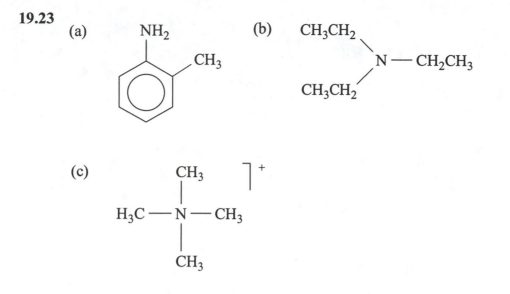

(b)

(c)

**19.25** Only (a) and (c) may function as nucleophiles, because they have lone pairs of electrons that will be attracted to a positively charged carbon center. $CO_2$ and $SiH_4$ have no lone pairs and cannot function as nucleophiles.

**19.27** (a) ethanol;  (b) 2-octanol;  (c) 5-methyl-1-octanol. These reactions can be accomplished with an oxidizing agent such as acidified sodium dichromate, $Na_2Cr_2O_7$.

**19.29**

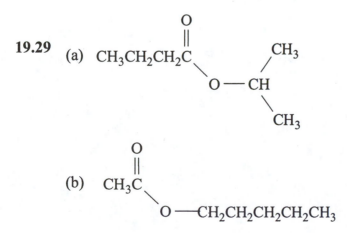

(a)

(b)

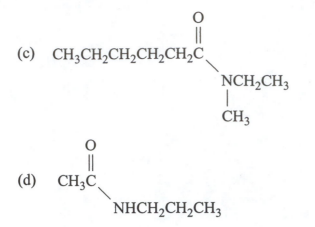

(c) $CH_3CH_2CH_2CH_2CH_2C$
<br>
$\overset{O}{\overset{\|}{}}$ , $NCH_2CH_3$ , $CH_3$

(d) $CH_3C$
<br>
$\overset{O}{\overset{\|}{}}$ , $NHCH_2CH_2CH_3$

**19.31** The following procedures can be used:

(1) Dissolve the compounds in water and use an acid-base indicator to look for a color change.

(2) $CH_3CH_2CHO \xrightarrow{\text{Tollens reagent}} CH_3CH_2COOH + Ag(s)$

(3) $CH_3COCH_3 \xrightarrow{\text{Tollens reagent}}$ no reaction

Procedure (1) will distinguish ethanoic acid; (2) and (3) will distinguish propanal from 2-propanone.

**19.33** $CH_3CH_2COOH < CH_3COOH < ClCH_2COOH < Cl_3CCOOH$

The greater the electronegativities of the groups attached to the carboxyl group, the stronger the acid (see Chapter 10).

**19.35** (a) $-CH_2-C(CH_3)_2-CH_2-C(CH_3)_2-CH_2-C(CH_3)_2-$

(b) $-CH-CH_2-CH-CH_2-CH-CH_2-$
<br>
$\quad\; | \qquad\qquad | \qquad\qquad |$
<br>
$\quad CN \qquad\quad CN \qquad\quad CN$

(c)

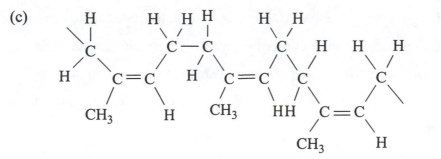

cis version

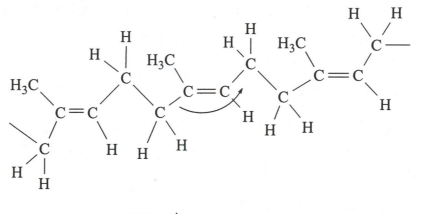

trans version

**19.37** (a) $CHCl=CH_2$; (b) $CFCl=CF_2$

**19.39** (a) $-OCCONH(CH_2)_4NHCOCONH(CH_2)_4NH-$ ;

(b) $-OC-CH(CH_3)-NH-OC-CH(CH_3)-NH-$

**19.41** An isotactic polymer is a polymer in which the substituents are all on the same side of the chain. A syndiotactic polymer is a polymer in which the substituent groups alternate, from one side of the chain to the other. An atactic polymer is a polymer in which the groups are randomly attached, one side or the other, along the chain.

**19.43** block copolymer

**19.45** Larger average molar mass corresponds to longer average chain length. Longer chain length allows for greater intertwining of the chains, making them more difficult to pull apart. This twining results in (a) higher softening points, (b) greater viscosity, and (c) greater mechanical strength.

**19.47** Highly linear, unbranched chains allow for maximum interaction between chains. The greater the intermolecular contact between chains, the stronger the forces between them, and the greater the strength of the material.

**19.49** (a)

$$-\overset{\overset{\displaystyle O}{\|}}{C}\diagdown_{\displaystyle NH-}$$

(b) amide; (c) condensation

**19.51** Side groups that contain hydroxyl, carbonyl, amino, and sulfide groups are all potentially capable of participating in hydrogen bonding that could contribute to the tertiary structure of the protein. Thus, serine, threonine, tyrosine, aspartic acid, glutamic acid, lysine, arginine, histidine, asparagine, and glutamine satisfy the criteria. Proline and tryptophan generally do not contribute through hydrogen bonding, because they are typically found in hydrophobic regions of proteins.

**19.53**

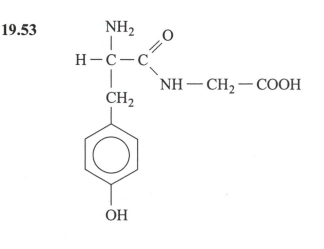

**19.55**  (a)  The functional groups are alcohols and aldehydes.

(b)  The chiral carbon atoms are marked with asterisks (*).

$$OHC - \overset{\displaystyle H}{\underset{\displaystyle OH}{\overset{|}{\underset{|}{C}}}} * - \overset{\displaystyle H}{\underset{\displaystyle OH}{\overset{|}{\underset{|}{C}}}} * - \overset{\displaystyle OH}{\underset{\displaystyle H}{\overset{|}{\underset{|}{C}}}} * - \overset{\displaystyle OH}{\underset{\displaystyle H}{\overset{|}{\underset{|}{C}}}} * - CH_2OH$$

**19.57**  (a)  GTACTCAAT;  (b)  ACTTAACGT

**19.59**  (a)  $C_5H_5N_5O$   (b)  $C_6H_{12}O_6$   (c)  $C_3H_7NO_2$

**19.61**  (a)  alcohol, ether, aldehyde;  (b)  ketone, alkene;  (c)  amine, amide

**19.63**  (a)  carboxylic acid, ester;  (b)  ether, ketone, phenol, alkene;  (c)  aromatic amine, tertiary amine;  (d)  ketone, alcohol, alkene

**19.65**

(a)  (b)

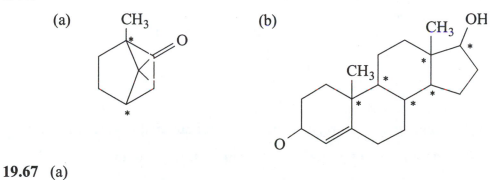

**19.67**  (a)

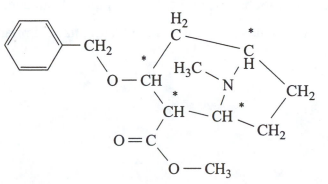

---

* An asterisk (*) denotes a chiral carbon atom.

(b)

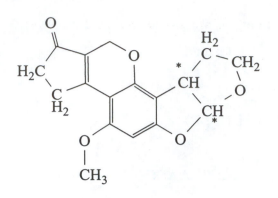

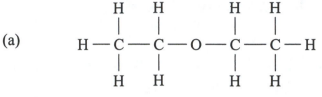

**19.69** (a)

Diethyl ether

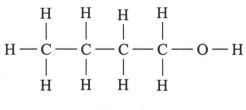

1-butanol

(b) 1-Butanol can hydrogen bond with itself but diethyl ether cannot, so 1-butanol molecules are held together more strongly in the liquid and therefore 1-butanol has the higher boiling point. Both compounds can form hydrogen bonds with water and therefore have similar solubilities.

**19.71** (a)

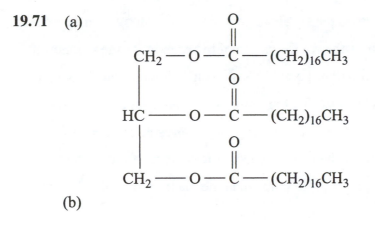

(b)

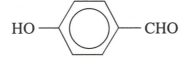

**19.73** (a) addition; (b) condensation; (c) addition; (d) addition; (e) condensation

**19.75** (a) $HOCH_2CH_2OH + 2\,CH_3(CH_2)_{16}COOH \rightarrow$

$$CH_3(CH_2)_{16}COOCH_2CH_2OOC(CH_2)_{16}CH_3 + 2\,H_2O$$

(b)

$2\,CH_3CH_2OH + HOOCCOOH \rightarrow CH_3CH_2OOCCOOCH_2CH_3 + 2\,H_2O$

(c) $CH_3CH_2CH_2CH_2OH + CH_3CH_2COOH \xrightarrow{\Delta}$

$$CH_3CH_2COOCH_2CH_2CH_2CH_3 + H_2O$$

**19.77** Polyalkenes < polyesters < polyamides, due to the increasing strength of intermolecular forces between the chains. The three types of polymer have about the same London forces if their chains are about the same length. However, polyesters also have dipole forces contributing to the strength of intermolecular forces, and polyamides form very strong hydrogen bonds between their chains.

**19.79** (a) Primary structure is the sequence of amino acids along a protein chain. Secondary structure is the conformation of the protein, or the manner in which the chain is coiled or layered, as a result of interactions between amide and carboxyl groups. Tertiary structure is the shape into which sections of the proteins twist and intertwine, as a result of interactions between side groups of the amino acids in the protein. If the protein consists of several polypeptide units, then the manner in which the units stick together is the quaternary structure.

(b) The primary structure is held together by covalent bonds. Intermolecular forces provide the major stabilizing force of the secondary structure. The tertiary structure is maintained by a combination of London forces, hydrogen bonding, and sometimes ion-ion interactions. The same forces are responsible for the quaternary structure.

**19.81** (a) $^+H_3NCH_2COOH(aq) + H_2O(l) \rightarrow {}^+H_3NCH_2COO^-(aq) + H_3O^+(aq)$

$^+H_3NCH_2COO^-(aq) + H_2O(l) \rightarrow H_2NCH_2COO^-(aq) + H_3O^+(aq)$

(b) $pK_{a1} = 2.35$ $\qquad\qquad$ $pK_{a2} = 9.78$

$pH = 2,\ {}^+H_3NCH_2COOH$

$pH = 5,\ {}^+H_3NCH_2COO^-$

$pH = 12,\ H_2NCH_2COO^-$

**19.83** Condensation polymerization involves the loss of a small molecule, often water or HCl, when monomers are combined. Dacron is more linear than the polymer obtained from benzene-1,2-dicarboxylic acid and ethylene glycol, so Dacron can be more readily spun into yarn.

**19.85**

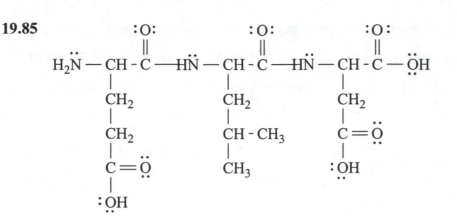

**19.87** (a)

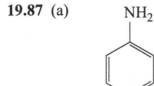

(b) $sp^3$; (c) $sp^2$; (d) each N atom carries one lone pair of electrons; (e) Yes, the N atoms help to carry the current because the unhybridized $p$ orbital on each N atom is part of the extended $\pi$ conjugation (delocalized $\pi$-bonds) that allows electrons to move freely along the polymer.

**19.89** (a) reduced; (b) oxidized; $O = -1$ in the peroxide anion, $O_2^{2-}$, but $O = -2$ in water; $C = +\frac{2}{3}$ in ascorbic acid, but $C = +1$ in dehydroxyascorbic acid.

**19.91** Two peaks are observed with relative overall intensities 3 : 1. The larger peak is due to the three methyl protons and is split into two lines with equal intensities. The smaller peak is due to the proton on the carbonyl carbon atom and is split into four lines with relative intensities 1 : 3 : 3 : 1.

**19.93** The peaks in the spectrum can be assigned on the basis of the intensities and the coupling to other peaks. The hydrogen atoms of the $CH_3$ unit of the ethyl group will have an intensity of 3 and will be split into a triplet by the two protons on the $CH_2$ unit. This peak is found at $\delta \approx 1.2$. The $CH_2$ unit will have an intensity of 2 and will be split into a quartet by the three protons on the methyl group. This peak is found at $\delta = 4.1$.. The $CH_2$ group that is part of the butyl function will have an intensity of two but will appear as a singlet because there are no protons on adjacent carbon atoms. This is the signal found at $\delta = 2.1$. The remaining $CH_3$ groups are equivalent and also will not show coupling. They can be attributed to the signal at $\delta = 1.0$. Notice that the peak that is most downfield is the one for the $CH_2$ group attached directly to the electronegative oxygen atom, and that the second most downfield peak is the one attached to the carbonyl group.

**19.95** If one considers the reaction, the products should be those arising from substitution of hydrogen atoms on the propane by chlorine atoms. We would expect then to form a chloropropane or, perhaps a dichloropropane. Remember that in the halogenation of alkanes, substitution becomes more difficult as more halogen atoms are introduced. If we then consider the NMR spectrum, we see that there is one large peak that sees a single proton, because it is split into a doublet. There is also a weaker feature, corresponding most likely to one proton, which is split into a septet. This indicates that the proton sees six equivalent protons. The structure that is consistent with this spectrum is 2-chloropropane.

**19.97** (a) $^{13}C$; (b) 1.11%; (c) No. The reason is that the probability of finding two $^{13}C$ nuclei next to each other is very low. A $^{12}C$ nucleus next to a $^{13}C$ nucleus will not interact with the $^{13}C$ nucleus because the $^{12}C$ nucleus has no spin. Because the natural abundance of $^{13}C$ is 1.11%, the probability of finding two $^{13}C$ nuclei next to each other in an organic molecule will be $0.0111 \times 0.0111$ or $1.23 \times 10^{-4}$. Although such coupling is possible, it is generally not observed because the signal is so much weaker than the signal due to the molecules with a single $^{13}C$ nucleus.

(d) Maybe. Because most of the carbon to which the protons are attached is $^{12}C$, the bulk of the signal will not be split. The protons that are attached to the $^{13}C$ atoms will be split, but this amounts to only 1.11% of the sample, so the peaks are very small. Peaks that result from coupling to a small percentage of a magnetically-active isotope are referred to as satellites and may be observed if one has a very good spectrum.

(e) Yes. Although the splitting of protons by $^{13}C$ may not be observed because the amount of $^{13}C$ present is low, the opposite situation is not true. If a $^{13}C$ is attached to H atoms, the large majority of those H atoms will have a spin and so the $^{13}C$ will show fine structure due to splitting by the H atoms.

**19.99** (a) 1; (b) 1; (c) 1; (d) 2; (e) Free rotation around the C-C and C-Cl bonds averages out the environments so that both $CH_2$ hydrogens are equivalent and all three $CH_3$ hydrogens are equivalent, resulting in just two different hydrogen signals.